4th International Conference on Green Energy and Sustainable Development (GESD 2021)

E3S Web of Conferences Volume 329 (2021)

Shenyang, China
13 - 14 November 2021

Editors:

Zhihui Tan
Lin Liu

ISBN: 978-1-7138-4174-6

Printed from e-media with permission by:

Curran Associates, Inc.
57 Morehouse Lane
Red Hook, NY 12571

Some format issues inherent in the e-media version may also appear in this print version.

For additional information, please contact EDP Sciences – Web of Conferences at the address below.

EDP Sciences – Web of Conferences
17, Avenue du Hoggar
Parc d'Activité de Courtabœuf
BP 112
F-91944 Les Ulis Cedex A
France

Phone: +33 (0) 1 69 18 75 75
Fax: +33 (0) 1 69 28 84 91

contact-edps@webofconferences.org

Additional copies of this publication are available from:

Curran Associates, Inc.
57 Morehouse Lane
Red Hook, NY 12571 USA
Phone: 845-758-0400
Fax: 845-758-2633
Email: curran@proceedings.com
Web: www.proceedings.com

TABLE OF CONTENTS

https://doi.org/10.1051/e3sconf/202132900001

Preface: 4th International Conference on Green Energy and Sustainable Development (GESD 2021)

This proceeding includes original and peer-reviewed research papers from the 4th International Conference on Green Energy and Sustainable Development (GESD 2021), which held on November 13-14, 2021 in Shenyang City, China.

GESD 2021 is organized by Research Center of Engineering and Science, co-organized by Chongqing University, Northeast Petroleum University, Huazhong University of Science and Technology, Asian Union of Information Technology, sponsored by Wanxiang Group Technology Center, CCCC Second Harbour Engineering Company Ltd., Chishun Chemical.

During the COVID-19 outbreak, many academic exchange activities were affected, such as visas, possible 14-day quarantine across regions, epidemic prevention measures, etc. In the end, the number of overseas participants in the conference was far less than expected.

The aim of GESD 2021 was to provide a platform for researchers, engineers, academicians, as well as industrial professionals, to present their research results and development activities in Green Energy and Sustainable Development. It provides opportunities for the delegates to exchange new ideas and application experiences, to establish business or research relations and to find global partners for future collaboration. This two-day conference was being organized in invited keynote talks and three Parallel Sessions - Green Energy & Sustainable Energy and Power; Resource Extraction, Utilization & Sustainable Development; Environmental Sustainability and Green Technologies.

All full paper submissions to this GESD would be sent to at least two reviewers and evaluated based on originality, technical or research content, correctness, relevance to conference, contributions, and readability. The full paper submissions were chosen based on technical merit, interest, applicability, and how well they fit a coherent and balanced technical program. GESD 2021 received 171 manuscripts. And 90 submissions had been accepted by our reviewers. The Committees of GESD 2021 invested great efforts in reviewing the papers submitted to the conference and organizing the sessions to enable the participants to gain maximum benefit.

Zhihui Tan
Lin Liu

Conference Organizing Committee

edp sciences

Web of Conferences

Statement of Peer review

In submitting conference proceedings to *Web of Conferences*, the editors of the proceedings certify to the Publisher that

1. They adhere to its **Policy on Publishing Integrity** in order to safeguard good scientific practice in publishing.
2. All articles have been subjected to peer review administered by the proceedings editors.
3. Reviews have been conducted by expert referees, who have been requested to provide unbiased and constructive comments aimed, whenever possible, at improving the work.
4. Proceedings editors have taken all reasonable steps to ensure the quality of the materials they publish and their decision to accept or reject a paper for publication has been based only on the merits of the work and the relevance to the journal.

Title, date and place of the conference

GESD 2021
November 13-14, 2021
Shenyang City, China

Proceedings editor(s):

Zhihui Tan
Lin Liu

Date and editor's signature November 12, 2021

Research on small-scale sand fracturing and puff technology

Qian Wang*, Zuohao Wu, Jiapeng Zheng, Junkai Lu, and Menghong Yu

Drilling &Production Technology Research Institute of Jidong Oilfield, Tangshan, China

Abstract. Affected by complex fault blocks, sedimentary environment and reservoir physical properties, Jidong Oilfield generally develops small-scale sand bodies, accounting for 24.2% of the produced reserves of medium and low permeability reservoirs. Such sand bodies rely on elastic energy to develop and are difficult to use effectively. In this paper, through the three-dimensional hydraulic physical model experiment and the analysis of the oil layer-dry layer-mudstone mixed sand body, the reasons that affect the model parameters are obtained. The results show that: the length of the high-permeability reservoir area, the length of the intermediate-permeability dry layer area, the rate of change of permeability parameters, the viscosity of the construction fluid, and the construction displacement have an impact on the effect of fracturing enhanced injection energy storage.

1 Introduction

Small-scale sand bodies refer to sand bodies whose scale is small and the existing well pattern is difficult to establish an effective displacement relationship. Affected by complex fault blocks, sedimentary environment and reservoir physical properties, Jidong Oilfield generally develops small-scale sand bodies, accounting for 24.2% of the produced reserves of medium and low permeability reservoirs. Such sand bodies rely on elastic energy to develop and are difficult to use effectively. It is necessary to strengthen technical research on development methods to achieve beneficial development. According to the characteristics of small sand bodies in G12 block of Jidong Oilfield, a simulation study of crack propagation was carried out. On this basis, a deterministic modeling method was used to establish a fine geological model of the block's work area, and on the basis of cutting and coarsening and reservoir numerical simulation, guided by reservoir numerical simulation results, establish fracturing model, obtain fracturing construction parameters, and guide fracturing design and construction.

2 Three-dimensional hydraulic model experiment

This experiment uses a large-scale full three-dimensional hydraulic fracturing physical simulation experimental system. The size of the experimental rock sample is 762 mm × 762 mm × 914 mm, which can effectively reduce the boundary effect of fracture propagation. The system consists of four parts: a rock sample loading frame, a confining pressure loading system, an injection system, and a sound wave monitoring system. The rock sample is placed in the loading frame. The vertical stress is loaded by jack hydraulics, up to 35MPa, and the horizontal ground stress passes through The pressure plate placed in the gap between the rock sample and the frame is loaded. Water is injected into the pressure plate. The increase in water pressure causes the pressure plate to expand. Then the water pressure is transmitted to the surface of the rock sample to achieve the purpose of simulating horizontal ground stress, up to 69MPa.

The injection system is controlled by a hydraulic servo system. The maximum injection displacement in the experiment can reach 200cm3/s, and the maximum injection pressure at the wellhead is 69MPa. The experimental fracturing fluid adopts two systems of slippery water and cross-linked jelly, and it is convenient to observe the fracture morphology afterwards. A certain amount of phosphor particles are mixed into the fracturing fluid. The fluid injection rate of the shale fracturing experiment was controlled at 1-170 cm3/s. The experimental computer control system can record the wellhead pressure, pumping displacement, and three-way confining pressure data in real time, and can adjust the pumping rate in real time according to the experimental conditions.

The macrophysical model experiment of the rock with natural fractures shows that a number of fractures that are perpendicular to the horizontal bedding and extend along the north-south direction are formed in the open-hole section, and stop at a large natural bedding to the north, and a few cracks extend out of the rock to the south. Like the boundary, part of the fracturing fluid flows along the southern horizontal bedding. Therefore, in the vertical direction, the vertical expansion of hydraulic fractures is restricted by the horizontal bedding, but while communicating with the horizontal bedding, it will also further communicate with the vertical fractures in other parts, which is conducive to the complication of fracture morphology and increases the volume of reservoir reconstruction.

* Corresponding author: ls_wq@petrochina.com.cn

The large-scale model experiment of the rock without natural fractures shows that the fractures form simple fractures with double wings, there are no existing complex fractures, and the reservoir reconstruction volume is limited,see Figure 1.

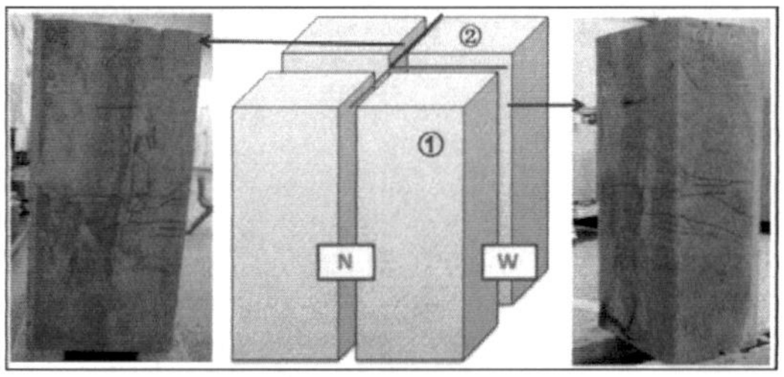

Figure 1. Fracture morphology with natural fractures

At the same time, it can be seen through digital simulation software that: reservoirs without natural fractures form a single fracture; reservoirs with natural fractures can form complex fractures. Numerical simulation shows that natural cracks are the most critical factor in the formation of complex cracks,see Figure 2,Figure 3.

Figure 2. Fracture fracture morphology without natural fractures

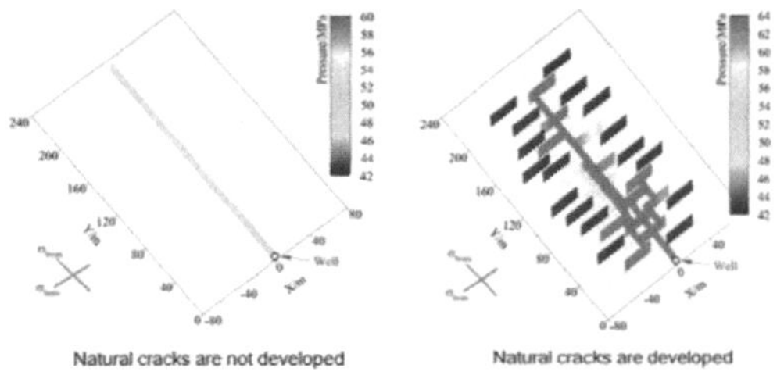

Figure 3. Fracture shape simulated by digital simulation

3. Oil layer-dry layer-mudstone mixed sand body

Through triaxial large-scale physical model experiments, it has been pointed out that small-scale sand bodies have the corresponding characteristics, except for the mechanical behavior of artificial fractures when they pass through weak structural surfaces such as natural fractures. When the reservoir physical properties are discontinuously distributed, different process parameters will also lead to different fracturing fluid injection pressurization effects. In order to study the effect of fracturing injection enhancement under different reservoir physical conditions, the fracture extension effects and injection enhancement effects of the reservoir physical properties along the fracture extension direction of oil layer, dry layer, mudstone layer, and oil layer directly contacting the mudstone layer were simulated.

3.1 Oil layer-dry layer-mudstone transition type

The difference in reservoir physical parameters is mainly reflected in the difference in permeability, porosity and lithology of the formation. During the simulation, the simulation results of oil layer-dry layer-mudstone layer were compared with the simulation results of oil layer-dry layer-dry layer, so as to analyze the blocking effect of mudstone on fracture propagation. It can be seen from the two figures that when ignoring the existence of mudstone intervals, the extension of fractures is almost unimpeded, the fractures extend longer, and the formation pressure changes caused by injection increase only appear corresponding to the permeability of different intervals. Characteristics. When considering the existence of the mudstone interval, the interface between the dry layer and the mudstone layer has an inhibitory effect on the fracture extension, so that the fracture no longer extends forward, and the injected fluid mainly increases the pressure in the width direction of the fracture in the reservoir. Makes the distribution of the pressurized area more even and a larger area.

From the perspective of fracture geometry, when the existence of mudstone layer is not considered, the fracture continues to extend before, and the length and width of the fracture are both increased. When considering the existence of mudstone layers, the extension of fractures is blocked, and the length and width of the fractures are small. The fluid is mainly used for filtration, thereby increasing the pressure of the reservoir and enhancing the elastic energy of the reservoir,see Figure 4.

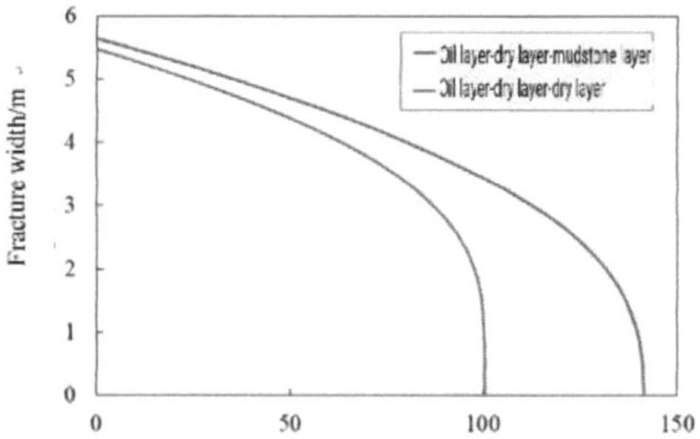

Figure 4. Fracture geometry of oil layer-dry layer-mudstone transition sandbody

The simulation results of the oil layer-dry layer-mudstone transition type show that when there is a variation in permeability in the oil layer, the pressure presents an obvious step-like shape. When the cracks extend and encounter mudstone, due to the poor porosity and

https://doi.org/10.1051/e3sconf/202132901001

permeability of the mudstone, it is difficult for the pressure to diffuse effectively in the mudstone, and the pressure diffusion occurs at the sand-shale interface and mainly formed in the dry layer, which is useful for improving the elasticity of the formation. Energy is unfavorable.

3.2 Oil layer-mudstone mutation

Further study the effect of oil layer-mudstone breakthrough type fracturing injection. The existence of the interface between oil layer and mudstone will affect the effect of fracture extension. The performance is that when the mudstone exists, the extension of the fracture is hindered, while the extension of the fracture is not considered when the mudstone is not considered. interference. When the fracture extension is blocked, the injected fluid along the sand-shale interface enhances the pressurization effect at the far end of the well, and the elastic energy inside the oil layer increases better; when the fracture extends normally, the injected fluid has a poor pressurization effect on the formation at the far end of the well. And it is mainly used to enlarge the vicinity of the crack tip.

It can be seen from the fracture geometry of the two that the existence of the mudstone layer reflects the influence on the length of the fracture, and the difference in the width of the fracture between the two is proportional to the distance from the wellbore within the fracture. This is because the abrupt change in permeability inside the sandstone caused by the dry layer is not considered at this time, so that the difference between the two is mainly concentrated at the front of the fracture,see Figure 5.

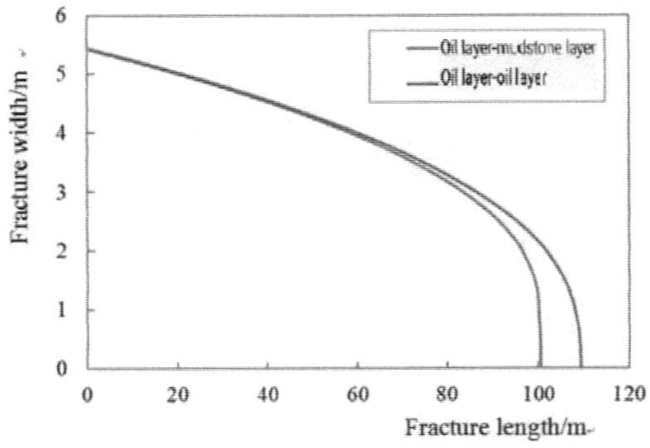

Figure 5. Fracture geometry of oil layer-mudstone catastrophic sandbody

The same study on the oil layer-mudstone abrupt layer shows that after the fracture extends into the mudstone, the pressure spreads along the abrupt interface accelerates. It can be inferred that the conductivity of the far end of the oil well will be improved to a certain extent, which is conducive to water injection and injection. Gas vomiting operation.

4. Model parameter influence analysis

In order to optimize the design of the construction of fracturing enhanced energy storage, the influence of different parameters on the effect of fracturing enhanced energy storage was studied. These parameters include the length of the high-permeability oil zone, the length of the intermediate-permeability dry zone, and the permeability Parameter change speed, construction fluid viscosity and construction displacement, etc.

4.1 Length of high permeability zone

When the length of the high-permeability oil zone is 50m, 35m, and 20m respectively, as the length of the high-permeability zone becomes longer, the energy-enhancing effect of the corresponding high-permeability zone becomes more obvious. On the premise of not changing the length of the dry zone with moderate permeability, and when the fracture extends to the mudstone interface, the longer the length of the high permeability zone, the better the supplementary effect of fracturing injection on the elastic energy of the formation, and the increase in the position of the fracture front The better the energy effect.

4.2 Length of low permeability zone

When the length of the low- and medium-permeability dry zone is set to 50m, 35m, and 20m, the length of the low-permeability zone has an impact on the injection enhancement effect, but the effect is not as obvious as that of the change in the length of the high-permeability zone.

4.3 Permeability change rate

The influence of the change rate of formation permeability in the reservoir on the pressurization effect of fracturing energy storage is studied. The abrupt change in permeability is considered to be that the permeability changes more drastically, while the gradual change means that the permeability changes more slowly. When the permeability changes greatly, the formation pressure distribution during fracturing pressurization has an obvious step-like shape. When the permeability change is small, the change area of the formation pressure distribution cloud map is smooth, and the slower the change speed, the smoother the pressure distribution. At the same time, when the rate of change of permeability becomes slower, it can be seen that the distribution area of formation pressure near the fracture tip is smaller, indicating that the effect of fracturing energy storage and pressurization at this time has become worse.

4.4 Viscosity

The fracturing is the influence of fluid viscosity on the construction effect. The higher the fluid viscosity, the greater the pressure raised in the formation, and it is mainly concentrated at the end of the fracture close to the wellbore. When the fluid viscosity is low, the increase in pressure in the formation becomes smaller, but the corresponding distribution area is increased, so that the sweeping range is more uniform, which is more beneficial for recovering the elastic energy of the reservoir.

4.5 Displacement

Under the premise that the total liquid volume of the construction remains unchanged, the influence of the construction displacement on the fracturing energization effect is studied. When using high displacement liquid injection, the maximum pressure increase in the formation is greater, and the pressure The distribution of the lifting distribution area along the cracks is also more uniform, which shows that increasing the construction displacement can increase the pressure diffusion effect at the far end of the cracks, and the effect of supplementing energy becomes better.

5 Conclusion

The large-scale model experiment of the rock without natural fractures shows that the fractures form simple fractures with double wings, there are no existing complex fractures, and the reservoir reconstruction volume is limited.

At the same time, it can be seen through digital simulation software that: reservoirs without natural fractures form a single fracture; reservoirs with natural fractures can form complex fractures. Numerical simulation shows that natural cracks are the most critical factor in the formation of complex cracks.

Through triaxial large-scale physical model experiments, it has been pointed out that small-scale sand bodies have the corresponding characteristics, except for the mechanical behavior of artificial fractures when they pass through weak structural surfaces such as natural fractures. When the reservoir physical properties are discontinuously distributed, different process parameters will also lead to different fracturing fluid injection pressurization effects.

In order to optimize the design of the construction of fracturing enhanced energy storage, different parameters have an impact on the effect of fracturing enhanced energy storage. These parameters include the length of the high-permeability reservoir area, the length of the intermediate-permeability dry layer area, and the permeability parameters. Change speed, construction fluid viscosity and construction displacement, etc.

References

1. Zhang Min.Fine description of the single sand body and the division of reservoir units in the third member of the east 3rd member of the Cha 71 fault block in Chaheji Oilfield [D]. Northwest University 2019
2. Li Yichao.Single sand body description and the corresponding relationship between injection and production in Baima Central District [D]. China University of Petroleum (Beijing) 2018
3. Guo Xulin.Reservoir characteristics of Lianhua oil layer in block Gao 3-6-18 of Liaohe Oilfield[D]. Yangtze University 2019
4. Liu Wei.Single sand body characterization of Baijiantan Formation in the Seven Middle East Area of Junggar Basin[D]. Yangtze University 2019
5. Xiong Zhaochuan.Single sand body fine description and three-dimensional geological modeling research of Chang 61 reservoir in S392 oil field [D]. Northwest University 2018
6. Li Hang.Recognition of thin-bed sand single sand body and analysis of water-flooded characteristics in the outer front of the delta of the Sa II and III oil layer groups in the eastern area of Xingshi District, Daqing Oilfield [D]. Northeast Petroleum University 2018
7. Xiang Yao.Research on the internal architecture model of a meandering channel single sand body based on dense coring wells [D]. Northeast Petroleum University 2018
8. Han Lingling.Research on fine division of sedimentary units and potential tapping of remaining oil in P108 block [D]. Northeast Petroleum University 2018
9. Wang Wen.Reservoir geological characteristics and reserves evaluation of Kumkol South oil field in Kazakhstan [D]. Yangtze University 2012
10. Wang Beibei.Study on the architecture of Chang 6 sandbody in the southern area of Wanggoumen, Wuqi Oilfield [D]. Xi'an Shiyou University 2018

Frequency analysis of representative flood control water level stations in Puyang River

Yang Wang, Shuhui Zhang*, Ziyi Zhang, Caichao Su, Peng Ding, Guangtian Cao

Zhejiang Tongji Vocational College of Science and Technology, Hangzhou 311231, China

Abstract. Puyang river basin is located in the north central part of Zhejiang Province, which is one of the most important river basins in Zhejiang Province. The lower reaches of Puyang River are easily influenced by tide in Hangzhou Bay and flood in Qiantang River. When Puyang river floods, it often meets Fuchun River and floods occur at the same time. The flood discharge of Puyang river is blocked and the water level rises and rises, which makes the Puyang River vulnerable to disaster. Water level frequency analysis is the basis of Puyang river planning and flood control plan. The representative flood control water level stations of Puyang River include Zhuji station and wenjiayan station. The frequency analysis of these representative stations is helpful to determine the water level of these key nodes under different frequencies, and to provide basic data for accurate flood control of Puyang River and ensure the safety of flood control.

1 Introduction

Puyang river is located in the north central part of Zhejiang Province and the west part of xiaoshaoning area. It is adjacent to Longmen Mountain and its remaining veins in the West and south, Kuaiji mountain in the East, and Hangzhou Bay of Taihu Lake in the north, with a drainage area of 3431km2. The Puyang river basin is slightly elliptical, surrounded by mountains in the East, South and West. The hilly area accounts for 70% of the total area, and the plain accounts for 30%. The highest mountain in the basin is Dongbai mountain, with an altitude of 1193 M. The lower reaches of Puyang River are easily influenced by tide in Hangzhou Bay and flood in Qiantang River. When the Puyang river floods, it often meets the Fuchun River and floods at the same time. At this time, the Puyang River flood (the reach below Zhuji Chengguan) will be blocked, and the water level will rise and rise. Therefore, if the Puyang river floods, if the downstream spring tide or flood butts, the flood discharge is not smooth, which is very easy to cause flood stagnation in the vicinity of Zhuji Chengguan Meichi area, seriously endangering the safety of farmland villages and towns on both sides of the Strait and the Zhejiang trunk railway, and easily lead to serious disasters.

Water level frequency analysis is the basis of Puyang river planning and flood control plan. The representative flood control water level stations of Puyang River include Zhuji station and wenjiayan station. The frequency analysis of these representative stations is helpful to determine the water level of these key nodes under different frequencies, and to provide basic data for accurate flood control of Puyang River and ensure the safety of flood control.

2.Overview of main flood control projects of Puyang River

Since the founding of new China, Puyang river has carried out flood control four times. At present, the embankment along the main stream of Puyang River on one side of the railway basically meets the design standard of 50-year return period, other embankments along the main stream basically meet the design standard of 20-year return period, and the construction of Fengqiao river reaches the design standard of 10-year return period.

Since the founding of new China, a large number of reservoir projects have been built in the upper reaches of the basin, including 2 large (II) reservoirs, 9 medium-sized reservoirs, 37 small (I) reservoirs and 199 small (II) reservoirs. Only large and medium-sized reservoirs have a total storage capacity of 536.4 million m3, of which the flood control storage capacity is 186 million m3 (including high lake), accounting for 37.3% of the total storage capacity. In addition, a high Lake flood detention reservoir will be built at Chengguan, Zhuji in the middle reaches. In the past floods, these reservoirs have played a great role in retaining floods, reducing flood peaks and reducing the flood control pressure downstream.

* Corresponding author: 1411658545@qq.com

3.Disastrous rain pattern and flood

Puyang river basin is a subtropical monsoon climate area with mild climate and significant monsoon. Frontal rain often stays over the basin in spring and summer to form strong Meiyu rainstorm. Short duration and high-intensity typhoon rainstorm will be formed in case of tropical storm and typhoon in summer and autumn. Puyang River Basin belongs to the concurrent area of Meiyu and Taiyu. The middle and upper reaches are mainly controlled by Taiyu and the lower reaches are mainly controlled by Meiyu.
Puyang river is a typical mountain stream River, with a large mountainous area, the shape of the basin is quite fan-shaped, and mountain torrents are easy to gather; The middle and lower reaches have flat and low-lying terrain, slow water flow, narrow river channel and poor drainage; Affected by the tide of Hangzhou Bay and the flood jacking of Qiantang River, the water level in the lower reaches is easy to rise.
Puyang river basin is known as the "Little Yellow River" because of its frequent floods. According to historical records, floods mostly occur in spring and summer. In the 915 years from 1034 to 1949, there were 106 major floods, of which 13 were flooded in Zhuji Chengguan. The flood in 1922 was the most serious flood in the modern history of Puyang river. Since the founding of new China, there have been 13 major floods, with an average of once every four years. Among them, the "6.19" flood in 1956, the "9.6" flood in 1962, the "6.17" flood in 1977, the "6.17" flood in 1994 and the "7.9" flood in 1997 are the most serious.

Table 1. Causes of major floods in Puyang River after the founding of the people's Republic of China

time	Average rainfall in the basin		Water level of Zhuji Taiping Bridge	origin
	One day	Three days	(m)	
1950.6.24			12.55 -14.37	mould rains
1952.6.1			12.7 -14.52	mould rains
1953.7.1	83.8	108.1	12.63 -14.45	mould rains
1954.5.26	76.6	118.5	12.4 -14.22	mould rains
1955.6.19	96	106.8	12.49 -14.31	mould rains
1956.8.2	181.2	203.7	13.05 -14.87	typhoon
1961.10.5	90	143.8	12.18 -14	typhoon
1962.9.6	196.6	257.5	13.01 -14.83	typhoon
1970.6.26	71.8	132.8	12.71 -14.53	mould rains
1973.5.17	67.8	133.3	12.76 -14.58	soft spring rain
1977.6.17	117	152.9	13 -14.82	mould rains
1994.6.17	115	206	12.4 -14.22	mould rains
1997.7.9	137.8	244	12.7 -14.52	mould rains

4 Pearson type III curve analysis method

Pearson type III frequency curve is recommended to be used in hydrological statistics [1-3]. Pearson type III curve [4-5] is an unsymmetrical unimodal and positive skew curve with a finite end and an infinite end,The probability density function is as follows:

$$f(x)=\frac{\beta^{\alpha}}{\Gamma(\alpha)}(x-\alpha_0)^{\alpha-1}e^{-\beta(x-a_0)} \quad (1)$$

Where: $\Gamma(\alpha)$ - gamma function of α ;

α β and α_0 are the shape scale and position unknown parameters of Pearson type III distribution, $\alpha>0$, $\beta>0$, respectively. Obviously, the density function can be determined after the three parameters are determined. It can be inferred that these three parameters have the following relationship with the overall three parameters $\bar{x}$, c_v and c_s:

$$\alpha=\frac{4}{C_s^2}$$

$$\beta=\frac{2}{\bar{x}c_v c_s} \quad (2)$$

$$\alpha_0=\bar{x}(1-\frac{2c_v}{c_s})$$

In hydrological calculation, it is generally necessary to calculate the value of random variable xp corresponding to the specified frequency P, that is, by integrating the density curve, that is:

$$P=P(x\geq x_p)=\frac{\beta^{\alpha}}{\Gamma(\alpha)}\int_{x_p}^{\infty}(x-\alpha_0)^{\alpha-1}e^{-\beta(x-a_0)}dx \quad (3)$$

Find the cumulative frequency p value equal to or greater than xp. It is very troublesome to calculate the value of P directly from the above formula. In practice, it is transformed into the following integral form through variable transformation:

$$P(\Phi \geq \Phi_p) = \int_{\Phi_p}^{\infty} f(\Phi \cdot C_s) d\Phi \quad (4)$$

In the formula, the integrand contains only one undetermined parameter c_s, the other two parameters $\bar{x}$ and c_v are included in Φ. x is a standardized variable called the mean deviation coefficient. $\Phi = \frac{x - \bar{x}}{\bar{x} C_V}$ The mean value of Φ is 0 and the standard deviation is 1. Therefore, we only need to assume a c_v value, then we can get the relationship between and by integration. For some given c_s values,The relationship between and Φ can be obtained by integrating.For several given c_s values, Table of corresponding values for Φ_p and p have been made by foster of the United States and leibkin of the former Soviet Union successively. Then the x value of the corresponding frequency can be obtained:

$$x = \bar{x}(1 + C_V \Phi) \quad (5)$$

5 Frequency analysis of representative flood control water level stations in Puyang River

Zhuji station and wenjiayan station are representative flood control stations of Puyang river. Pearson type III curve is used to analyze their frequency. The analysis results are shown in Figure 1.

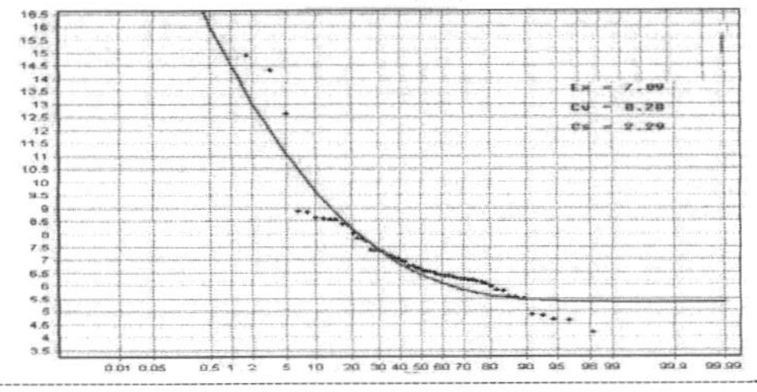

Figure 1. Frequency analysis results of Zhuji water level station

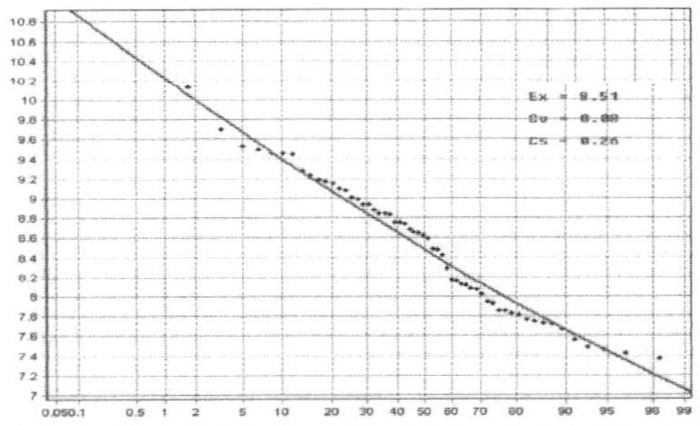

Figure 2. Frequency analysis results of Wenjiayan water level station

Table 2. Water level values of each frequency at Zhuji station

Frequency (%)	0.1	0.2	0.3333	0.5	1
Design value	19.56	18.04	16.92	16.04	14.53
Frequency (%)	2	3.3333	5	10	20
Design value	13.04	11.94	11.08	9.62	8.2

Table 3. Water level values of each frequency at wenjiayan station

frequency(%)	0.1	0.2	0.3333	0.5	1
Design value	10.87	10.69	10.54	10.43	10.22
frequency(%)	2	3.3333	5	10	20
Design value	10	9.83	9.68	9.4	9.07

6 Conclusion

This paper analyzes the frequency of Puyang River flood control representative station by Pearson type III curve analysis method. Through the analysis, the water level values of Zhuji station and wenjiayan station under different frequencies are obtained. These water level values are the basis of flood control planning and flood control plan of Puyang river. The data can provide basic support for flood control decision-making and flood control advance deployment of Puyang river.

Acknowledgments

This work was financially supported by Water conservancy science and technology plan project of Zhejiang Province (Project No.RC2008) fund.

https://doi.org/10.1051/e3sconf/202132901002

References

1. Xie chongbao, Yuan Hongyuan, Guo Yuanyu. Parameter estimation of P - Ⅲ theoretical frequency curve -- fuzzy extreme value method [J]. Hydrology, 1997 (03): 2-9
2. Li Shicai, Peng Yueying, Wei Wenzhan. A new numerical algorithm for Pearson type III curve and its application [J]. Guangxi water resources and hydropower, 2001 (01): 18-23
3. Che Guowen. Excel drawing method of Pearson type III curve [J]. journal of liaoning teachers college (natural science edtitton), 2018,20 (04): 9-11 + 108.
4. John L. Carlson. A Study of the Distribution of Means Estimated From Small Samples by the Method of Maximum Likelihood for Pearson's Type II Curve. 1932, 3(2):86-107.
5. Small-Sample Asymptotic Distributions of M-Estimators of Location. 1982, 69(1):29-46.

Multi scale characteristics of Antarctic temperature change in January from 1982 to 2020

Guo Huifang, Yin Fangfang*, Guangtian Cao

Zhejiang Tongji Vocational College of science and technology, Hangzhou, Zhejiang 311231, China

Abstract. Antarctica is a sensitive area of global temperature change. Analyzing the temperature change in Antarctica can provide a basis for the analysis of global climate change The shared data provided by the automatic weather station plan are analyzed, and the relatively complete January temperature of Antarctic 8905 station from 1981 to 2020 is selected for analysis. Using the multi-scale analysis function of wavelet analysis, Morlet wavelet function is selected to carry out continuous wavelet transform on the daily temperature data, and the real part, modulus square and variance of continuous wavelet change are analyzed. It is found that the daily temperature of Antarctica is There are 14, 27, 72, 268, 380 and 665 day main periods for temperature changes. The trend analysis of wavelet transform coefficients under each period shows that the Antarctic temperature is in a high temperature period of 14, 27, 72, 268, 380 and 665 day periods in the years after 2021. The maximum temperature in January in Antarctica has a 20-year main period, and the minimum temperature has a 14-year and 23-year main period, The average value has a main cycle of 6 years, 14 years and 24 years.

1 Introduction

The blue book on world climate change points out that from 1880 to 2012, the global average surface temperature increased by about 0.85 °. In the next 100 years, the global average surface temperature will increase by 1.4 to 5.8 degrees Celsius, becoming the fastest warming period in recent 10000 years. Global warming has led to the rise of sea level and the frequent occurrence of various climate disasters in land areas. As an important part of the earth system, the polar climate system is the most sensitive area to climate warming. Analyzing the law of temperature change in the polar region is helpful to provide a reference basis for global research on the characteristics of climate change. At present, many scholars at home and abroad have carried out research and Analysis on temperature changes in Antarctica. Nicolas et al. [1,2] studied extreme high temperature events in Antarctica through era interim reanalysis data. Li Jingfeng [3] and others analyzed the characteristics of temperature in the north of the Antarctic Peninsula and its relationship with sea ice. Liu Yu [4] analyzed the impact of tropical intraseasonal oscillation on Antarctic temperature change. Wang Yicheng et al. [5] used polar WRF to analyze extreme low temperature events in Antarctica. Ma Qiang et al. [6] analyzed the variation characteristics of temperature, wind and precipitation at the Great Wall Station in Antarctica. Using the shared data provided by the automatic weather station program of the University of Wisconsin (Madison), the temperature data of 8905 station is selected for analysis by analyzing the data of all observation stations. The temperature observation data of the station is relatively sound, and the observation data are from 1982 to 2020. However, some monthly data are missing. This time, relatively continuous January temperature data are selected for analysis to issue the internal laws of Antarctic temperature change and find out its internal scale and change trend at each scale. Wavelet analysis is a good multi-scale analysis method. It can find the internal laws of data through Fourier change. It is also known as the "magnifying glass" of data analysis.

2 Analysis of Antarctic temperature change based on CWT

If the Antarctic temperature data sequence is $f(t)$, its continuous wavelet transform CWT is [7,8]:

$$W_f(a,b) = |a|^{-1/2} \int_{t=-\infty}^{\infty} f(t)\Psi(\frac{t-b}{a})dt \qquad (1)$$

Among them: $W_f(a,b)$ is the wavelet transform coefficient, $\Psi_{a,b}(t)$ is the continuous wavelet function, which is also called the basic wavelet. It has the characteristics of small energy, small shape, oscillation in time domain, good vanishing moment and good self phasing. Namely:

$$\int_{-\infty}^{\infty} |\Psi_{a,b}(t)|^2 dt = \int_{-\infty}^{\infty} |\Psi(t)|^2 dt = 1 \qquad (2)$$

*Corresponding author: 61681034@qq.com

 https://doi.org/10.1051/e3sconf/202132901003

$$\int_{-\infty}^{\infty} \Psi_{a,b}(t)dt = \int_{-\infty}^{\infty} \Psi(t)dt = 0 \quad (3)$$

$$\int_{-\infty}^{\infty} t^p \Psi(t)dt = 0 \quad (4)$$

The Antarctic temperature signal can be changed and reduced by adjusting the size of a and B. If you need to analyze the signal details, you can adjust a to reduce, $\Psi_{a,b}(t)$' s value and shrink in the direction of the time axis, so as to obtain high-frequency information. If it is necessary to analyze the general picture of wavelet, a can be increased, which $\Psi_{a,b}(t)$ extends in the direction of time axis, so as to obtain the low-frequency information contained in the signal.

In practical application, the continuous wavelet function must meet the admissibility condition. At present, the commonly used Morlet function is the wavelet function that can meet this condition. Morlet wavelet is a single frequency complex sinusoidal function. The expression is:

$$\Psi(t) = e^{iw_0 t} e^{-t^2/2} \quad (5)$$

Morlet wavelet function is used to continue the multi-scale analysis of the Antarctic temperature change.

3 Multiscale characteristic analysis of Antarctic temperature

The analysis data is the temperature data of Antarctic 8905 station from January 1981 to 2020. The observation data is measured once every 3 hours, but some of them are missing. The average data of the first three hours and the last three hours are used to analyze the missing data. The data of three years are missing, and the data of these three years are not involved in the analysis. The data analyzed in this time are daily temperature data series, annual maximum temperature data series, annual minimum temperature series and annual average temperature series. The above Morlet wavelet function is selected to carry out continuous wavelet change on the Antarctic temperature change, and the existing law of Antarctic temperature change is found by calculating the main eigenvalues of wavelet transform coefficients.

3.1 Multiscale analysis of daily temperature in Antarctica

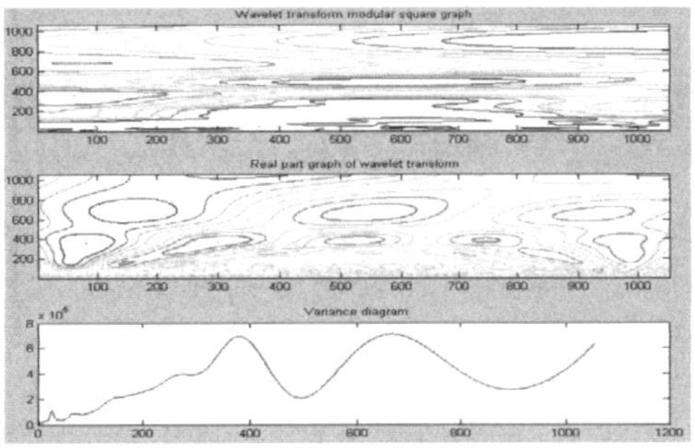

Fig. 1 Multiscale analysis of daily temperature in Antarctica

By analyzing the modulus square, real part and variance of the wavelet transform of daily temperature in Antarctica, and enlarging the turning point, it is found that there are 14, 27, 72, 268, 380 and 665 day main cycles in the daily temperature change series in Antarctica in January. By analyzing the real part change process of wavelet transform at each main scale, the change process of temperature series at various scales can be obtained.

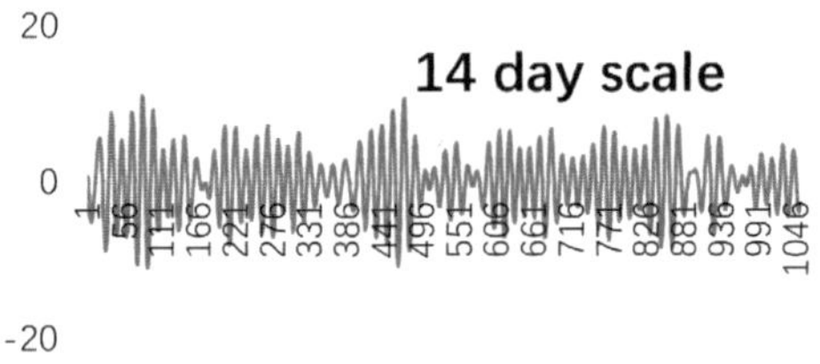

Fig. 2 Analysis of Antarctic temperature change on 14 day scale

In the 14 day scale, the periodic change of temperature is more intense. According to the analysis above, the temperature of Antarctica in January 2021 is in a high temperature period on the 14 day scale.

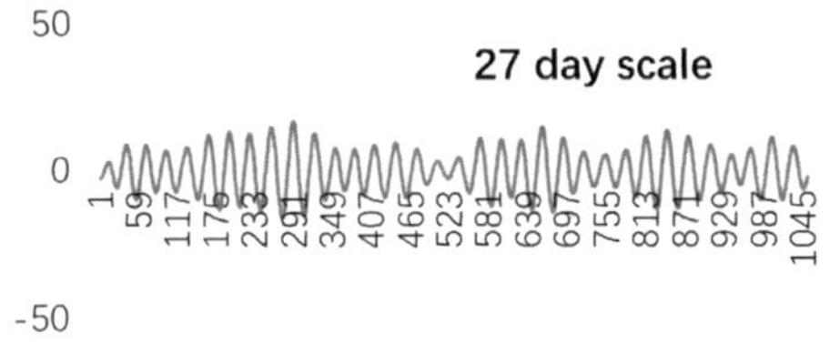

Fig. 3 Analysis of Antarctic temperature change on 27 day scale

In the 27 day scale, the periodic change of temperature is more intense. According to the analysis of the above figure, the temperature in January 2021 in Antarctica is in a high temperature period on the scale of 27 days.

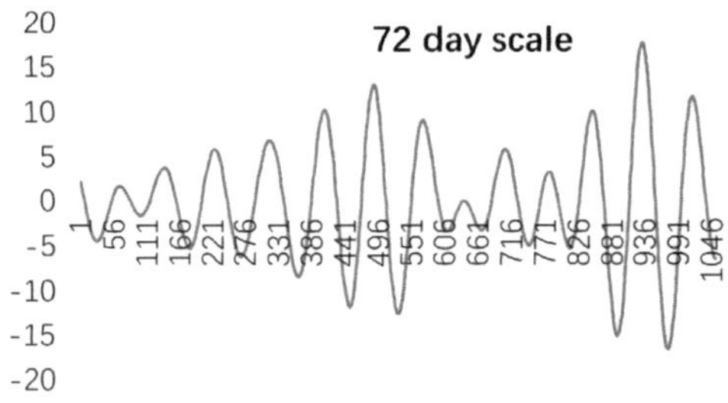

Fig. 4 Analysis of Antarctic temperature change on 72 day scale

It can be seen from the above figure. In January 2021, the temperature in Antarctica is in the low temperature period of 72 day scale. In January 2022, the temperature in Antarctica is in the high temperature period of 72 day scale.

 https://doi.org/10.1051/e3sconf/202132901003

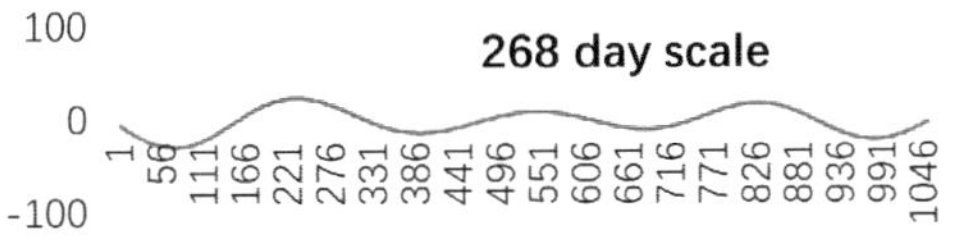

Fig. 5 Analysis of Antarctic temperature change on 268 day scale

It can be seen from the above figure that the temperature of Antarctica in January in 2021, 2022, 2023 and 2024 is in the period of high temperature on the scale of 268 days. After 2006, the turning years of temperature from low to high are 2020 and 2009. The turning years of temperature from high to low are 2015 and 2006.

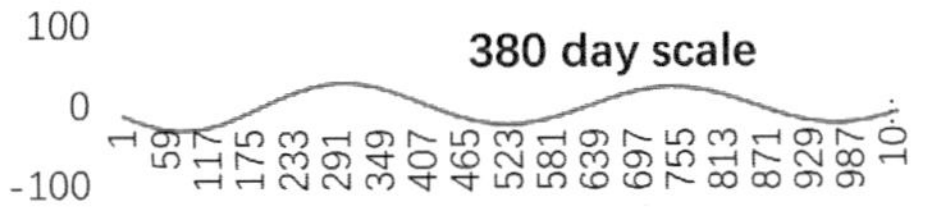

Fig. 6 Analysis of Antarctic temperature change on 380 day scale

It can be seen from the above figure that the temperature in January of Antarctica in 2021, 2022, 2023, 2024, 2025 and 2026 is in a high temperature period under the 380 day scale. The turning years of temperature from low to high after 2006 are 2020 and 2006. The turning year of temperature from high to low is 2013.

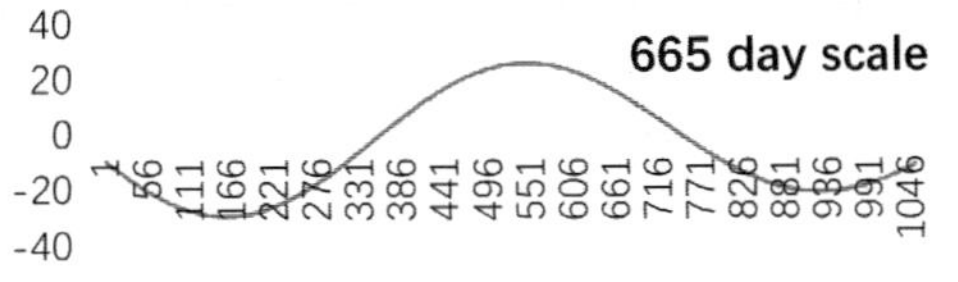

Fig. 7 Analysis of Antarctic temperature change on 665 day scale

It can be seen from the above figure that the temperature of Antarctica in January from 2022 to 2029 is in a high temperature period on the 665 day scale. After 2006, the turning year of temperature from low to high is 2021; The turning year of temperature from high to low is 2010.

3.2 Multi scale analysis of temperature variation in Antarctica in January

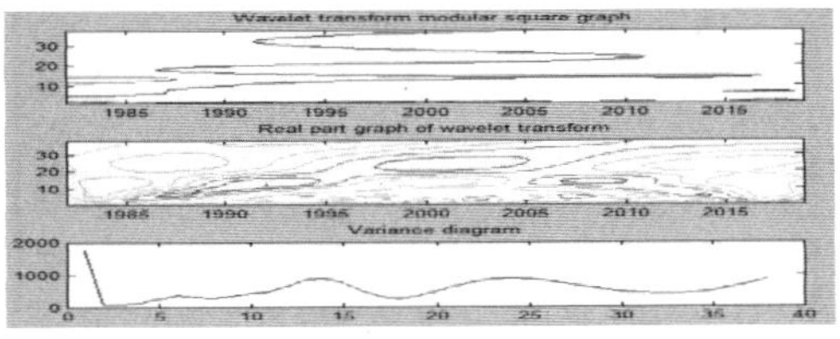

Fig. 8 Multi scale characteristic analysis of temperature change in Antarctica in January

The modulus square, real part and variance of the wavelet transform of the average temperature in January in Antarctica are analyzed, and the turning point is enlarged. It is found that it has a main cycle of 6 years, 14 years and 24 years.

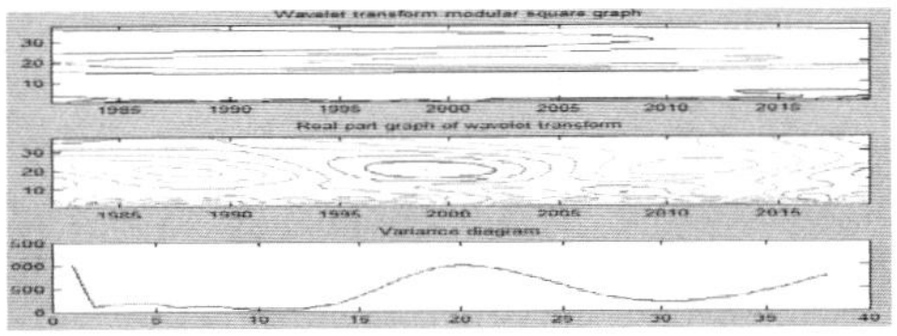

Fig. 9 Multi scale characteristic analysis of temperature change in Antarctica in January

The modulus square, real part and variance of the wavelet transform of the maximum temperature in January in Antarctica are analyzed, and the turning point is enlarged. It is found that there is a main period of 20 years.

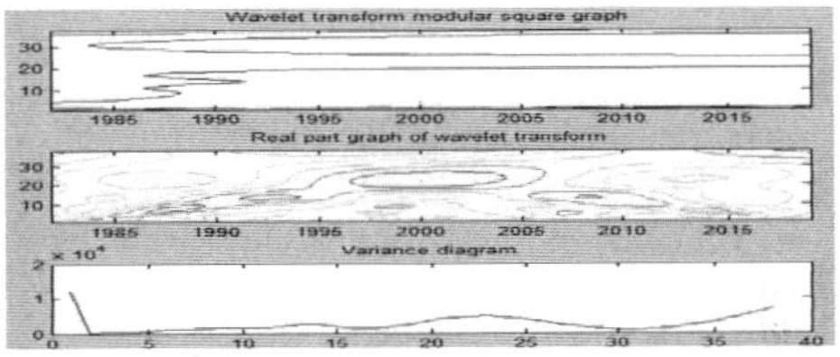

Fig. 10 Multi scale characteristic analysis of temperature change in Antarctica in January

The modulus square, real part and variance of the wavelet transform of the maximum temperature in January in Antarctica are analyzed, and the turning point is enlarged. It is found that there are 14-year and 23-year main cycles.

4. Conclusion

Antarctica is a sensitive area of global temperature change. Analyzing the temperature change in Antarctica can provide a basis for the analysis of global climate change. Using the shared data provided by the automatic weather station plan of the University of Wisconsin (Madison), the temperature data of 8905 station in January are selected for analysis. Using the multi-scale analysis method of continuous wavelet transform, the main periods of daily temperature change in Antarctica are analyzed, which are 14, 27, 72, 268, 380 and 665 days. The trend of wavelet transform coefficients in each period is analyzed. It is found that the Antarctic temperature is in the high temperature period of 14, 27, 72, 268, 380 and 665 days in the years after 2021. The maximum temperature in January in Antarctica has a main cycle of 20 years, the minimum has a main cycle of 14 and 23 years, and the average has a main cycle of 6, 14 and 24 years. In the next few years, the global temperature will be in a period of high temperature. All climate change related units should take response measures such as energy conservation and emission reduction and green environmental protection to reduce carbon dioxide emissions, so as to reduce the impact of human activities on temperature and slow down the rising trend of global temperature.

Acknowledgments

This work was financially supported by Water conservancy science and technology plan project of Zhejiang Province (Project No.RC2008) fund.

References

1. Nicolas J P, Vogelmann A M, Scott R C, et al. January 2016 extensive summer melt in West Antarctica favoured by strong El Niño[J]. Nature Communications, 2017, 8: 15799.
2. de Los Milagros Skansi M, King J, Lazzara M A, et al. Evalu-ating the highest temperature extremes in the antarctic[J]. Eos, 2017, 98(5): 18 - 23.
3. Li Jingfeng, Bian Lingen, Duan Juqi, Ding Zhuoming, Wei Qinwu, Lu Changgui. Temperature characteristics in the northern Antarctic Peninsula and their relationship with sea ice [J]. Meteorological science and technology, 2016,44 (05): 763-768
4. Liu Yu. Influence of tropical intraseasonal oscillation on Antarctic temperature change [D]. Nanjing University of information engineering, 2019
5. Wang Yicheng, Xie Aihong, Ding Minghu, Xiao cunde. Analysis of extreme low temperature event in Antarctic Dome A based on polar WRF [J]. Glacier and frozen soil, 2020,42 (03): 727-733
6. Ma Qiang, Wang Yating, Zhang Cui, he Yun, Guo Guangyuan. Analysis on variation characteristics of temperature, wind and precipitation at the Great Wall Station in Antarctica [J]. Shaanxi meteorology, 2016 (01): 25-28
7. Cui Jintai. Introduction to wavelet analysis [M]. Xi'an: Xi'an Jiaotong University Press, 1995:1 ~ 20
8. Ma Xixia, Hu Tiecheng. Annual runoff prediction of reservoir based on ANFIS [J]. Journal of hydropower, 2008 (4): 33-37

E3S Web of Conferences **329**, 01004 (2021) https://doi.org/10.1051/e3sconf/202132901004
GESD 2021

Load Forecasting of Distributed Energy in Industrial Park and Its Investment Linkage Analysis

Shuo Yin [1], Yao Lu [1], Shiqian Wang [1], Zhe Chai [1], Man Jin [1], Xing Chen [1,*], Jie Hu [2]

[1] State Grid Henan Economic Research Institute, Zhengzhou, China
[2] Ohio State University, USA

Abstract. Power grid enterprises are changing their function orientation and profit model under the background of new power system reform. Vector autoregressive model used for industrial park to forecast load of distributed energy is created in this paper, which provides the data to the analysis of the investment environment of grid enterprises. We excavate the core influencing factors of grid investment in incremental distribution parks based on the linkage of grid investment influencing factors in system dynamics.

Keywords: Park; Incremental distribution; Investment impact; Linkage.

1 Preface

The "No. 9 Document" clearly puts forward the requirements of steadily promoting the reform of the electricity sales side, promoting the construction and development of distribution networks, exploring effective ways for social capital to invest in the electricity distribution business, gradually liberalizing the investment business of incremental electricity distribution, and improving the efficiency of electricity distribution operations, and encouraging the development of electricity distribution and sales business by means of mixed ownership. Load forecasting of energy and its linkage investment are key ways to provide scientific basis for optimum resources matching and raising productivity. In the short term, the objectives of this electricity reform are clear, aiming to build on the achievements of the previous round of electricity reform, consolidate and strengthen the effectiveness of the reform, further promote the opening up of the electricity sales side, open up the field of investment in incremental electricity distribution, attract social capital injection, strengthen competition in the electricity sales market, reasonably optimise resources on the distribution side, improve efficiency, while accounting for transmission and distribution prices, change the profitability model of grid enterprises, strengthen government regulation in those segments with a strong natural monopoly in transmission and distribution The Government should also allow the power generation side to participate directly in the competition on the electricity sales side, conduct direct power purchase with large users, establish a unified trading centre, improve the electricity trading system, and realise the marketisation of electricity mainly in pilot areas, so as to provide experience and lessons for the future marketisation reform of the distribution and sales side and lead the way for future reform.

2 Load forecasting of distributed energy in industrial park based on vector autoregressive model

Based on the historical data of distributed electricity power $\{\ l_e(t)\ \}$, gas energy$\{\ l_g(t)\ \}$, cold energy $\{l_c(t)\}$and heat energy $\{l_h(t)\}$ in industrial park, vector autoregressive model (VAR) can be used to describe and simulate their load, of which the autoregressive model order should be decided in accordance with minimum information criterion (AIC). Load forecasting vector autoregressive model has the advantage of reflecting dynamic persistence characteristics, which means it can reveal the relationship among different forms of energy from the historical data and accurately forecast distributed energy load in the future.

(1) Vector autoregressive model of electricity power

In the vector autoregressive model of hybrid forecasting electricity power, of which the autoregressive model order is assumed to be 2, electricity power and cold energy are assumed to be endogenous variables, while gas and heat energy are assumed to be exogenous variables. Electricity power can be expressed as:

$$l_e(t) = (A_1 \quad A_2)\begin{pmatrix} L_e \\ L_c \end{pmatrix} + (B_1 \quad B_2)\begin{pmatrix} l_g(t) \\ l_h(t) \end{pmatrix} + \varepsilon(t)\ , \qquad (1)$$

where L_e can be expressed as $[l_e(t-1), l_e(t-2), l_e(t-3)]^T$ and L_c has the expression of $[l_c(t-1), l_c(t-2), l_c(t-3)]^T$.

(2) Vector autoregressive model of gas energy

In the vector autoregressive model of hybrid forecasting gas energy, of which during the autoregressive model

* Corresponding author: chenxing@ha.sgcc.com.cn

order from 1 to threshold value, gas and heat energy are assumed to be endogenous variables, while electricity power and cold energy are assumed to be exogenous variables; during the autoregressive model order above threshold value, gas and heat energy are assumed to be endogenous variables, while only electricity power is assumed to be exogenous variables. Gas energy can be expressed as:

$$l_g(t)=\begin{cases}(A_1 \quad A_2)\begin{pmatrix}L_g\\L_h\end{pmatrix}+(B_1 \quad B_2)\begin{pmatrix}l_e(t)\\l_c(t)\end{pmatrix}+\varepsilon(t) & ,1\le t\le 21, t=30\\(A_1^{'} \quad A_2^{'})\begin{pmatrix}L_g^{'}\\L_h^{'}\end{pmatrix}+B_1 l_e^{'}(t)+\varepsilon(t) & ,22\le t\le 29\end{cases}, \tag{2}$$

(3) Vector autoregressive model of cold energy

In the vector autoregressive model of hybrid forecasting cold energy, of which during the autoregressive model order from 1 to threshold value, gas and heat energy are assumed to be endogenous variables, while electricity power and cold energy are assumed to be exogenous variables; during the autoregressive model order above threshold value, gas and heat energy are assumed to be endogenous variables. Cold energy can be expressed as:

$$l_c(t)=\begin{cases}(A_1 \quad A_2)\begin{pmatrix}L_e\\L_c\end{pmatrix}+(B_1 \quad B_2)\begin{pmatrix}l_g(t)\\l_h(t)\end{pmatrix}+\varepsilon(t) & ,1\le t\le 21, t=30\\(A_1 \quad A_2)\begin{pmatrix}L_e^{'}\\L_c^{'}\end{pmatrix} & ,22\le t\le 29\end{cases}, \tag{3}$$

(4) Vector autoregressive model of heat energy

In the vector autoregressive model of hybrid forecasting heat energy, of which during the autoregressive model order from 1 to threshold value, gas and heat energy are assumed to be endogenous variables, while electricity power and cold energy are assumed to be exogenous variables; during the autoregressive model order above threshold value, gas and heat energy are assumed to be endogenous variables, while only electricity power is assumed to be exogenous variables. Heat energy can be expressed as:

$$l_h(t)=\begin{cases}(A_1 \quad A_2)\begin{pmatrix}L_g\\L_h\end{pmatrix}+(B_1 \quad B_2)\begin{pmatrix}l_e(t)\\l_c(t)\end{pmatrix}+\varepsilon(t) & ,1\le t\le 20, t=30\\(A_1^{'} \quad A_2^{'})\begin{pmatrix}L_g^{'}\\L_h^{'}\end{pmatrix}+B_1 l_e^{'}(t)+\varepsilon(t) & ,21\le t\le 29\end{cases}. \tag{4}$$

3 Analysis of the investment environment under the Incremental Distribution Reform of the Park

Accurate load forecasting for industrial parks based on the VAR models can improve the investment decision-making ability of power grid enterprises. The impact of the liberalisation of the investment business of incremental distribution grids on the investment of grid enterprises is mainly in the following areas.

(1) Investment scale. After the liberalization of the incremental distribution business, the total amount of investment will change significantly after the introduction of competition, whether in the "stock area" or the "blank area". The new power grid reform will affect the investment in power grids, as it is difficult for grid enterprises to make a uniform investment in power grids as a whole. To achieve sustainable development of the company, it must be required to go to continuously raise funds and make investments in the power grid. Grid enterprises to raise funds should focus on investment risk, in the face of social capital competition, grid enterprises need to set a more reasonable cost management model, to avoid the total deviation of the investment scale is too large, affecting business operations.

(2) Investment benefits. Incremental distribution grid reform can revitalise some of the stock assets, thus improving the efficiency of the use of grid assets and reducing the duplication of grid construction and wasteful investment. However, for the stock area to launch the incremental distribution investment business, there will be multiple investment entities. However, according to the actual situation, the stock part of the distribution network operation is still responsible by the power grid enterprises, so that the cost in the case of not reduced, the stock of the incremental part still has a certain competition, making the power grid enterprises in this part of the benefit is affected, it is difficult to be guaranteed.

(3) Investment focus. Incremental distribution grid investment should be combined with local government development planning, land planning, industrial planning and other sciences in an organic way, which makes incremental distribution grid investment no longer a simple act of grid construction, but should be developed together with local gas supply, heating, distributed power supply, electric vehicles and other public services to meet all kinds of access needs.

(4) Investment risk. Under the traditional mechanism, the power grid company has monopoly in the distribution area, often in the investment without considering the economy, more concerned about the technicality, reliability improvement, but after the opening of the electricity sales side, the introduction of social capital competition, power grid enterprises can not be as before only focus on the technical aspects, but should be comprehensive consideration of economic benefits, the development of incremental area investment situation is complex, if not strengthen the risk analysis, and then incremental area investment benefits are not obvious, then it will bring resistance to the subsequent development of incremental distribution business and lose the first opportunity.

4 System dynamics-based linkage of factors influencing grid investment

Investment decisions in grid planning need to be analysed by incorporating multiple types of factors, and investment options are the result of the interaction and coordination of different factors. Therefore, the overall process of grid planning can be described as a dynamically evolving system with the following characteristics.

(1) There is not a single relationship between the various types of factors in grid investment, but rather an interaction and a chain of conduction feedback between them, and ultimately a multi-level chain feedback structure.

(2) In a dynamically evolving system, there are complex mechanisms of action based on a single linear feedback for each factor. In a particular situation, the feedback mechanism needs to be expressed through a combination of quantitative and qualitative models, a single model is

difficult to meet the requirements of the formulation, the use of system dynamics of the model, the use of table functions and related equations can play a very good representation effect.

(3) In investment planning, the causal relationships between systems are not immediately apparent, and there is a certain lag in the impact of some variables on the whole system. In addition, the relative delays in information sharing, decision analysis and transmission processes of the participants in the investment decision will make the impact of different variables appear with different degrees of lag.

Total electricity demand can be made up of two components: the stock component and the incremental component. The power system reform document clearly proposes to liberalise the incremental distribution grid business, allowing social capital to enter and participate in market competition. The investment planning decision is based on the relationship between the planned investment capacity, the planned incremental assets of the grid enterprises and the actual planned investment capacity, and a system dynamics stack flow diagram is established.

Figure 1. Feedback relationship for investment in the campus grid

5 Analysis of the impact of grid investments in incremental distribution parks

(1) the impact of the profit of the power grid company, after the reform of the power system, the total profit of the power grid company can be composed of three parts: one is the main business income, i.e. transmission and distribution profit, the profit model of the power grid is changed from the traditional purchase and sale price difference to the collection of over-grid fees; two is the guaranteed profit, the power grid needs to undertake the guaranteed power supply service of its power supply business area, mainly for the first production of agriculture and residential users; three is the establishment of power sales company after Power sales profits, the new situation, part of the electricity will be due to competition by other power sales body snatched, the power grid company set up by the power sales company can be directly negotiated with the power plant purchase price, sales side of a number of power sales company into, power sales company has the right to independent pricing, its sales price can be based on their own operating conditions and competitors' electricity price level to determine.

(2) Impact of investment situation. The impact of the grid company's actual investment in the previous year is divided into two main components: firstly, through the transfer to fixed assets, which affects transmission and distribution prices; and secondly, through meeting and improving the supply capacity, which affects the amount of electricity transferred. The impact of these two components ultimately affects the profit for the year. The realisation of this year's profit has a binding effect on the next year's investment plan. The actual investment each year affects the approved permitted return of the grid company through the transfer of fixed assets, and the approved permitted cost through the depreciation of fixed assets, which in turn affects the transmission and distribution price. As can be seen, the annual investment is ultimately recovered through the grid's transmission and distribution profits. However, the focus of the electricity market reform is to strictly control grid profits, so given the government's control actions, the actual investment of the grid company is not reflected in each year's transmission and distribution price; instead, it is gradually reflected in the next regulatory cycle through the accumulation of certain regulatory cycles. As a result, there is a cyclical delay in the recovery of investment.

6 Concluding remarks

This paper analyses the impact of the new situation on the investment efficiency of power grids from the aspects of load forecasting for industrial parks based on the VAR models, transmission and distribution price reform, electricity sales side market reform, incremental distribution grid reform and energy transformation and clean substitution, laying a theoretical foundation for the subsequent determination of the factors influencing the investment efficiency of power grid enterprises and the establishment of the model. In the context of the new situation, grid enterprises pay more attention to investment efficiency management, strive to improve the efficiency of asset utilization in their own operations, win market competitiveness, further optimize the investment structure and enhance the economic benefits of grid operations; in addition, grid enterprises also pay attention to the social and environmental benefits of project investment, ensure the safe and stable operation of the power system and power quality, while vigorously promoting large-scale centralized and distributed clean In addition, grid enterprises also focus on the social and environmental benefits of project investment, ensuring the safe and stable operation of the power system and power quality, while vigorously promoting large-scale centralized and distributed clean energy consumption.

In terms of investment impact, the annual investment by grid companies is eventually recovered through the grid transmission and distribution profits. However, the focus of the electricity market reform is on strict control of grid profits, so given the government's control actions, the actual investment of grid companies is not reflected in the transmission and distribution prices of each year; instead, it is gradually reflected in the next regulatory cycle through the accumulation of certain regulatory cycles. As a result, there is a cyclical delay in the recovery of investment.

Acknowledgments

This work was supported by the Science and Technology Project of SGCC, Research on the competition situation and operation mode of distribution companies in the park (52170018000S).

References

1. Flexible block offers and a three-stage market clearing method for distribution-level electricity markets with grid limits[J]International Journal of Electrical Power and Energy Systems, 2021, 130.
2. Evaluation and Analysis of Total Factor Input and Output of Power Grid Enterprises under the New Situation[J]IOP Conference Series: Earth and Environmental Science, 2021, 680(1).
3. Research on Decentralized Trading Strategy of Electricity Market Based on Blockchain Technology[J]E3S Web of Conferences, 2020, 185
4. Regulating electricity distribution networks under technological and demand uncertainty[J]Energy Policy, 2020(prepublish)
5. Dynamic pricing in electricity and natural gas distribution networks: An EPEC model[J]Energy, 2020, 207
6. ADMM-based market clearing and optimal flexibility bidding of distribution-level flexibility market for day-ahead congestion management of distribution networks[J]International Journal of Electrical Power and Energy Systems, 2020, 123.

 https://doi.org/10.1051/e3sconf/202132901005

Fenton-like oxidation of Hydroxypropyl guar gum catalysed by Cu(II) complex at high pH

Weijun Ni [1], Yubo Lian [2], Yan Wei [3], Bo Zheng [4], Man Liu [5], Le Qu [1], Ling Zhou [1,*]

[1] Xi'an Key Laboratory of Tight oil (Shale oil) Development, Xi'an Shiyou University, Xi'an Shaanxi, China
[2] Xi'an Changqing Chemical Group Co. Ltd, Xi'an, China
[3] Changqing Drilling Company of CCDC, Xi'an, China
[4] Oil Production No.1, Changqing Oilfield Company, Yanan, China
[5] Oil Production No.11, Changqing Oilfield Company, Xi'an, China

Abstract. Large amounts of wastewater containing hydroxypropyl guar gum (HPGG), polyacrylamide (PAM) and carboxymethyl cellulose (CMC) are produced in the process of fracturing. In the work, a Fenton-like system in the presence of H2O2 was employed to degrade wastewater. The effects of oxidant concentration, catalyst dosage, the temperature and pH on the degradation efficiency of the polymer were studied in detail. Results showed that the prepared complex (Cu(II)L) exhibited a great catalytic effects in the range of pH 7-12. It was also found that under the conditions of 45°C and pH=10, when the amount of H2O2 was 5.0% (mass ratio to hydroxypropyl guar gum) and the amount of Cu (II)L complex was 10%, HPGG had a great reduction rate, and its viscosity value can be reduced from 18 to 6.47.

Keywords: Guar gum; wastewater treatment; Fenton-like oxidation.

1 Introduction

It is well known that a large amount of wastewater will be produced in the process of fracturing and oil recovery. Guar gum is often used as a fracturing agent in oil fields due to its high viscosity, and has now become the main source of oil field wastewater [1,2]. It is a galactomannan derived from the seed of a leguminous plant Cyamopsis tetragonolobus, which is an amazingly versatile and efficient high polymer hydrophilic hydrocolloid, economical in use and easy to handle [3,4]. Now, hydroxypropyl guar gum (HPGG), the derivative of guar gum, has been the main water-soluble additive in the oil extraction process [5,6]. What's more, some soluble polymers, anionic carboxymethyl cellulos (CMC) and partially hydrolyzed polyacrylamide (HPAM) as thickeners are also widely in oil field fracturing. Adsorption [7,8], advanced oxidation, biological treatment and so on have been applied to treat the drilling wastewater. The Fenton chemistry, in which the decomposition of hydrogen peroxide is catalyzed by iron to form hydroxyl radicals, was the most effective way for the degradation of various persistent and refractory organic pollutants in wastewater [9-12]. Fenton oxidation has been used for a long time due to its high catalytic activity for organic pollutants in wastewater. However, the process works only in the range of pH equal to about 3, and the remaining trivalent iron needs a separation system, which make it uneconomical and limited in application. An advantage of a Fenton-like catalyst is its characteristics of a wide working pH range (pH 3-7) compared to the conventional Fenton catalyst [13]. In addition to the Fe^{2+}-catalyzed homogeneous Fenton-like processes, Cu^{2+} and Cu^{2+}-complex can also act as the catalyst. The basic mechanism of the Cu^{2+}-catalyzed Fenton-like processes has been reported by some researcher, as described in Eq.1 and Eq.2. Iboukhoulef et al. [14] applied a Fenton-like system $H_2O_2/Cu(II)$ to treat phenolic compounds from olive mill wastewater and 62% removal rate was achieved at 65 min treatment at 50°C. Lee et al. [15] also applied the Cu^{2+}-H_2O_2 Fenton-like process to degrade diclofenac and carbamazepine with UV light. Under the optimum reaction conditions, the removal rates of the two pharmacetutical compounds can reach 47% and 66% within 4 h, respectively. Maekawa et al [16] studied the mineralization of phenol by Fenton oxidation with Cu^{2+} as the catalyst and found that 94% of phenol can be mineralized in dark conditions.

$$Cu^{2+}+H_2O_2 \rightarrow Cu^{+}+HO_2^{\bullet}+H^{+} \quad (1)$$

$$Cu^{+}+H_2O_2 \rightarrow Cu^{2+}+ HO^{\bullet}+OH^{-} \quad (2)$$

In this paper, using an convenient and eco-friendly one-pot two-step method [17], formaldehyde, piperazidine and p-methylphenol are refluxed for 2h in the formaldehyde solution, then the copper chloride solution is added under vigorous stirring to obtain homogeneous copper complex for treating polymer fracturing agent in the process of oil recovery under high pH conditions.

* Corresponding author: 2081617535@qq.com

2 Experimental

2.1 Materials

All reagents used in the experiments are of analytical grade without further purification. HPGG is purchased by Changqing Oilfield, and CMC and PAM are purchased from Environmental Protection Co., Zhengzhou, Henan.

2.2 Catalyst preparation

According to our previous work [18], the "one pot two steps" method is also applied the preparation process of the catalysts. Firstly, p-methylphenol, formaldehyde and piperazine are added to absolute methanol solution and refluxed at 65°C for 2 h to obtain the ligand solution (L). Subsequently, the L solution was added to the copper chloride under continuous stirring, which is named the complex Cu(II)L. The preparation process of homogeneous catalyst was shown in Fig. 1.

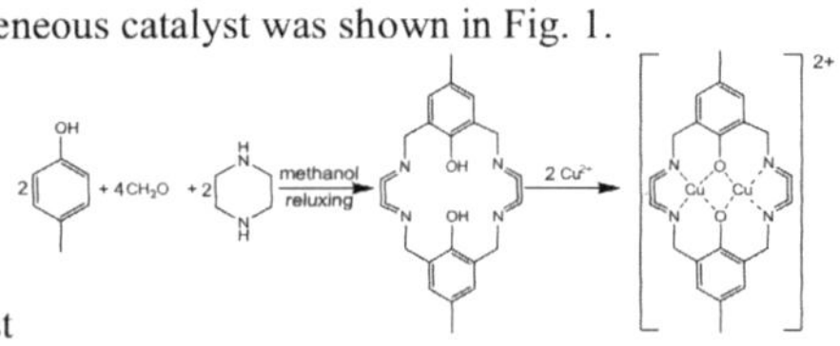

catalyst

Fig. 1 Synthesis of copper complex

2.3 Fenton-like oxidation process

Different polymers were dispersed in distilled water at room temperature and stirred 30 minutes on a magnetic stirrer with a stirring rate of 100 rpm to form 0.3% glue solution, then settled for 12 h to complete the swelling [19]. 5 mL hydroxypropyl guar gum glue, amount of H_2O_2 and catalyst dosage was added in an Ubbelohde viscometer and well-distributed by shaking at specific temperature. At a predetermined temperature, an Ubbelohde viscometer was used to measure the viscosity of the polymer under different degradation times [20]. The initial pH is selected under neutral condition and the pH of further experiment was adjusted by sodium hydroxide solution. In Fenton-like process, H_2O_2 concentration, pH value, catalyst amount, and degradation temperature were studied thoroughly because of their significant effect parameters in oxidation capacity.

3 Results and discussion

3.1 Effect of H_2O_2 dosage

H_2O_2, as the main of HO•, plays a key role in the treatment of organic waste water in the Fenton-like processes [21]. As shown in Fig. 2, the viscosity of HPGG was investigated when H_2O_2 dosage was increased from 1% to 10% (mass ratio to HPGG) under 45℃ at pH 7.0. From the results, the viscosity of HPGG decreased significantly within 15 minutes and then remained stable. The enhancement of HPGG degradation was due to the increase of hydroxyl radicals produced from H_2O_2. Moreover, there was no significant improvement in viscosity reduction rate when H_2O_2 dosage exceeded 5%. This was because the scavenging of hydroxyl of hydroxyl radicals occurred at the high H_2O_2 concentration through Equations (1) and (2), which led to the increase in the number of hydroxyl radicals in the solution [22]. Hence, 5% was chosen as the optimum dosage of H_2O_2 in all subsequent experiments.

$$H_2O_2 + \bullet OH \rightarrow H_2O + \bullet HO_2 \quad (3)$$

$$\bullet HO_2 + \bullet OH \rightarrow H_2O + O_2 \quad (4)$$

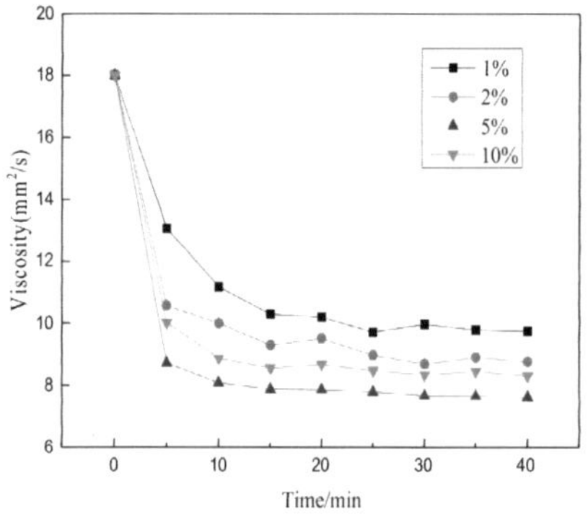

Fig. 2 Effect of concentration of $Na_2S_2O_8$

3.2 Effect of catalyst dosage

Generally, the amount of catalyst is directly proportional to the degradation rate of the polymer, but an excessive amount of catalyst is detrimental to the degradation results [23]. Fig. 3 showed consecutive experiments performed on the same wastewater sample. It can be seen from the figure that after adding the catalyst, the viscosity of HPGG decreases quickly and stabilizes after 20 minutes. Moreover, increasing the catalyst dosage by 1% to 10% can significantly increase the degradation rate of HPGG due to the produce of more hydroxyl radicals generated from H_2O_2. However, when the amount of catalyst reached 15%, the viscosity reduction rate of HPGG began to decrease, which is mainly due to the decomposition of specific free radical generated by H_2O_2. Therefore, 10% was selected as the optimum catalyst dosage for the degradation of hydroxylpropyl guar gum.

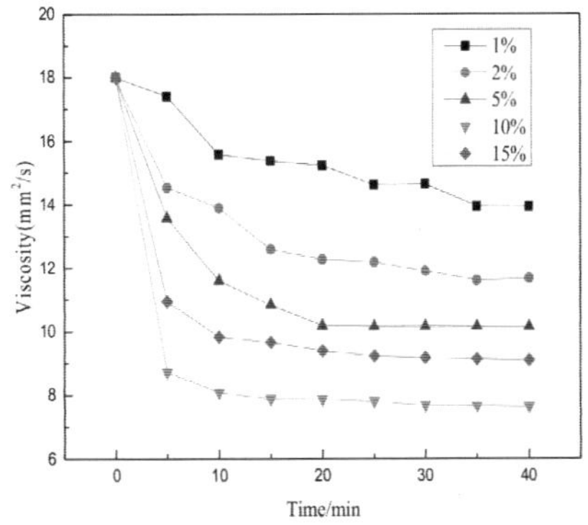

Fig. 3 Effect of dosage of catalyst

3.3 Effect of the temperature

Since the thermal degradation reaction is a non-spontaneous endothermic process, temperature affects the degradation rate of polymers [24]. Considering the application temperature of the catalyst in the oilfield, Figure 8 shows the change of viscosity of HPGG solution at different temperatures under the action of H_2O_2 and catalyst. It can be clearly seen from the figure, the higher the temperature, the faster the rate of oxidative degradation. According to Arrhenius equation, higher temperatures by increasing the reaction rate constant can provide more energy to overcome the reaction activation energy and then accelerate the reaction [25]. When the temperature increased to 45°C, the viscosity of HPGG decreases significantly, and the viscosity value can reach 7.88 after 15 minutes of degradation. Therefore, the following experiment was performed at 45℃.

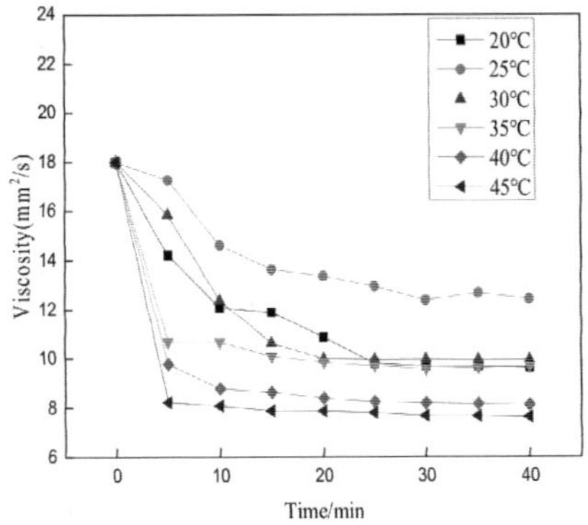

Fig. 4 Effect of reaction temperature

3.4 Effect of the pH

As we all know, conventional Fenton oxidation needs to adjust the pH to about 3 first to ensure the activity of the catalyst. The fracturing fluid is usually alkaline [26], so it requires a lot of cost to pre-treat the wastewater. This paper evaluated the removal effect of the prepared Cu(II) complex catalyzed by H_2O_2 to degrade HPGG under high pH conditions. The solution was adjusted with 0.1M NaOH solution to a pH in the range of 7-12, and the corresponding results are shown in Figure 5. Obviously, the removal efficiency under alkaline conditions is higher than that neutral condition. The highest removal performance reflecting by great viscosity drop of HPGG from 18 to 6.47 was observed at pH of 10. Different from the narrow pH required for conventional Fenton oxidation, using Cu(II)L complex as catalyst, the Fenton-like system had a wide applicable range of pH.

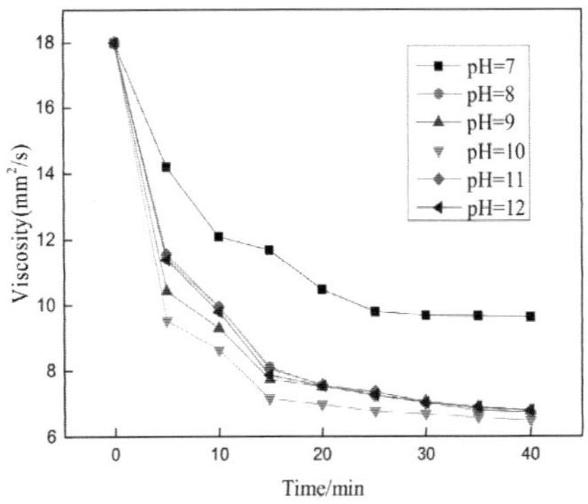

Fig. 5 Effect of pH value on the degradation

4 Conclusions

In this study, we used a Fenton-like system Cu(II) complex/H_2O_2 to degrade hydroxypropyl guar gum in oilfields. At 45°C and pH in a wide pH rage of 7.0 to 12.0, Cu(II) complex, hydroxyl radical and temperature affect the catalytic degradation rate of HPGG. Therefore, the homogeneous Cu(II)L complex/H_2O_2 system has huge application prospects in the treatment of oilfield polymers.

Acknowledgements

This work was financially supported by the grants from Open Fund of Shaanxi Key Laboratory of Carbon Dioxide Storage and Enhanced Oil Recovery (YJSYZX20SKF0008) and Shaanxi Provincial Key Research and Development Program (2019ZDLGY06-03). And we thank the work of Modern Analysis and Testing Center of Xi`an Shiyou University.

References

1. D. Mudgil, S. Barak and B.S.J. Khatkar. Guar gum: processing, properties and food applications-A Review. Journal of Food Science and Technology, 2014, 51(3): 409-418. 12,
2. Y. Tang, H. Ren, P. Yang, et al. Treatment of fracturing fluid waste by Fenton reaction using transition metal complexes catalyzes oxidation of hydroxypropyl guar gum at high pH, Environmental Chemistry Letters, 2019, 17: 559-564.

3. M. Mei, J. Yang, X. Zhang, et al. Cysteine Fe(III) catalyzed oxidation of common polymer used in oilfield by H2O2 in a wide pH range, Russian Journal of Applied Chemistry, 2019, 92(1): 134-139.
4. Babuponnusami, K. Muthukumar. A review on Fenton and improvements to the Fenton process for wastewater treatment, Journal of Environmental Chemical Engineering, 2014, 2(1): 557-572.
5. Y. Tang, L. Zhou, Y. Xue, et al. Preparation of nanoscale zero-valent metal for catalyzed clean oxidation of hydroxypropyl guar gum at neutral pH value, Desalination and Water Treatment, 2020, 197: 328-334.
6. V. Arantes, C. Baldocchi and A.M.F. Milagres. Degradation and decolorization of a biodegradable-resistant polymeric dye by chelator-mediated Fenton reactions, Chemosphere, 2006, 63(10): 0-1772.
7. L. Zhou, M. Slaný, B. Bai, et al. Enhanced removal of sulfonated lignite from oil wastewater with multidimensional MgAl-LDH nanoparticles, Nanomaterials, 2021, 11, 861.
8. Y. Tang, Z. Li, Z. Xu, et al Synthesis of hierarchical MgO based on a cotton template and its adsorption properties for efficient treatment of oilfield wastewater, RSC Advance, 2020, 10: 28695-28704.
9. W. Chu, K.H. Chan, C.Y. Kwan, et al. Degradation of atrazine by modified stepwise-Fenton's processes, Chemosphere, 2007, 67(4): 0-761.
10. Goi, Y. Veressinina, M. Trapido Degradation of salicylic acid by Fenton and modified Fenton treatment, Chemical Engineering Journal, 2008, 143(1-3): 1-9.
11. M. Trapido, N. Kulik, A. Goi, et al. Fenton treatment efficacy for the purification of different kinds of wastewater, Water Science & Technology, 2009, 60(7): 1795-1801.
12. N. Dulova, M. Trapido, & A. Dulov Catalytic degradation of picric acid by heterogeneous Fenton-based processes, Environmental Technology, 2011, 32(4), 439-446.
13. F.L.Y. Lam, A.C.K. Yip, & X. Hu Copper/mcm-41 as a highly stable and pH-insensitive heterogeneous photo-Fenton-like catalytic material for the abatement of organic wastewater, Industrial & Engineering Chemistry Research, 2007, 46(10), 3328-3333.
14. H. Iboukhoulefa, A. Amraneb, H. Kadia Removal of phenolic compounds from olive mill wastewater by a Fenton-like system H2O2/Cu(II)-thermodynamic and kinetic modeling, Desalination & Water Treatment, 2016, 57(4): 1874-1879.
15. H.J. Lee, H. Lee, & C. Lee Degradation of diclofenac and carbamazepine by the copper(II)-catalyzed dark and photo-assisted Fenton-like systems, The Chemical Engineering Journal, 2014, 245: 258-264.
16. J. Maekawa, K. Mae, H. Nakagawa. Fenton-Cu2+ system for phenol mineralization, Journal of Environmental Chemical Engineering, 2014, 2(3): 1275-1280.
17. X. Xiang, L. He, Y. Yang, et al. A one-pot two-step approach for the catalytic conversion of glucose into 2,5-diformylfuran, Catalysis Letters, 2011, 141(5): 735-741.
18. Y. Tang, L. Zhou, Z. Xu, et al. Heterogeneous degradation of oil field additives by Cu(II) complex-activated persulfate oxidation, Environmental Progress & Sustainable Energy, 2020, e13562.
19. Y. Tang, H. Ren, P. Yang, et al. Catalytic oxidation of polymers used in oilfield by metal-1,2-benzenediol complex, Desalination and Water Treatment, 2018, 120: 304-310.
20. H. Gu, X. Tang, R.Y. Hong. Ubbelohde viscometer measurement of water-based Fe3O4 magnetic fluid prepared by coprecipitation, Journal of Magnetism & Magnetic Materials, 2013, 348(12): 88-92.
21. N. Wang, T. Zheng, G. Zhang, et al. A review on Fenton-like processes for organic wastewater treatment, Journal of Environmental Chemical Engineering, 2016, 4(1): 762-787.
22. G. Wei, Y. Li, S. Cai, et al. Photo-Fenton degradation of ethyl xanthate catalyzed by bentonite-supported Fe(II)/phosphotungstic acid under visible light irradiation, Water Science & Technology, 2018, 2017(2): wst2018174.
23. Y. Tang, H. Liu, L. Zhou, et al. Enhanced Fenton oxidation of hydroxypropyl guargum catalyzed by EDTA-metal complexes in a wide pH range. Water Science and Technology, 2019, 79(9): 1667-1674.
24. L.I. Enhui, J. Xinyao, W. Chenge, et al. Degradation Kinetics and Stability of Anthocyanins from Blueberry, Food Science, 2018.
25. A.O. lebuegu, C.P. Ezenwa. Removal of Endocrine Disrupting Chemicals in Wastewater Treatment by Fenton-Like Oxidation, Water, Air & Soil Pollution, 2011, 217(1-4): 13-220.
26. L. Zhou, Z. Xu, J. Zhang, et al. Effective degradation of hydroxypropyl guar gum at wide pH range by heterogeneous Fenton-like using supported zero-valent copper, Water Science and Technology, 2020, 82, 1635-1642.

Study on power load adjustment capacity of a 300MW cogeneration unit in heating season

Jinxu Lao, Wei Zheng, Lingkai Zhu, Junshan Guo, haizhen Lv

Shandong Electric Power Research Institute, ShanDong Province JiNan, 250003 China

Abstract. After the formulation of the "3060" double carbon target, the installed capacity of new energy has increased rapidly and the proportion of power generation has become higher and higher. In the heating season, under the operation mode of "determining electricity by heat" of cogeneration units, in order to meet the consumption of renewable energy and the heating guarantee of people's livelihood, the contradiction with power grid peak shaving is increasing. This paper mainly introduces the load capacity of a 300MW cogeneration unit in the heating season. Through the test method, find out the load capacity of the unit under different heating capacity, which can be used as a reference for power generation enterprises and power grid dispatching.

1 Background

The cogeneration unit uses high-grade energy for power generation, and the low-grade steam after power generation is used for industrial production or heating. It can not only meet the needs of production and life, but also improve the comprehensive utilization efficiency of energy, and realize the policy of "proper distribution, different needs, corresponding temperature and cascade utilization" for energy. The advantage of cogeneration is to turn the hierarchical utilization of energy into reality. The cogeneration [1] unit itself has high energy efficiency, but limited by the operation mode of "determining electricity by heat", it reduces the peak shaving capacity of the power system. In order to meet the consumption of renewable energy and the heating guarantee of people's livelihood, the contradiction with power grid peak shaving [2-4] is increasing. Therefore, it is necessary to find out the actual load capacity of cogeneration units in the heating season and break the restrictions of the original heating working diagram. Taking a 300MW cogeneration unit as an example, its load capacity is obtained through test [5].

2 Equipment introduction

2.1 Overview of steam turbine equipment

The steam turbine is C300/256-16.7/0.39/537/537subcritical, one intermediate reheat, single shaft, two cylinders and two exhaust, condensing steam turbine, with eight stages of non regulated regenerative steam extraction. Rated power 300MW, rated main steam flow 900.9t/h; The maximum continuous output is 321mw and the maximum steam flow is 979t / h. Table 1 shows the main design parameters of steam turbine.

Table 1 Main parameters of steam turbine under rated extraction conditions

Name	Company	THA	Rated heating	Maximum heating
Main steam flow	t/h	900.9	950.5	950.5
Main steam pressure	MPa	16.67	16.67	16.67
Main steam temperature	℃	537	537	537
Reheat pressure	MPa	3.179	3.319	3.319
Reheat temperature	℃	537	537	537
Feed water temperature	℃	272.4	275.6	275.6
Heating extraction capacity	t/h	/	350	550
Heating extraction pressure	MPa	/	0.39	0.39
Heating extraction temperature	°C	/	246.3	246.3

2.2 Overview of boiler equipment

The boiler is SG-1025/17.4-M851subcritical, intermediate primary reheat, natural circulation and coal-fired drum boiler. The positive pressure direct blowing four corner tangential combustion is adopted, the main steam temperature is regulated by two-stage water spray desuperheating, and the reheat steam temperature is regulated by flue gas baffle and water spray

* Corresponding author: 1686844029@qq.com

desuperheating. Table 2 shows the main design parameters of the boiler.

Table 2 Main design parameters of boiler

Name	Company	numerical value
Rated evaporation	t/h	1025
Superheated steam outlet pressure	MPa	17.5
Superheated steam outlet temperature	℃	540
Reheat steam flow	t/h	842.05
Reheat steam inlet pressure	MPa	3.86
Reheat steam inlet temperature	℃	326.7
Reheat steam outlet temperature	℃	540
Exhaust gas temperature -- before correction	℃	135.6
Exhaust gas temperature - corrected	℃	128.3

2.3 Introduction to heating system

The heating steam extraction of the four units in the plant is the exhaust steam extraction of intermediate pressure cylinder, and there is no industrial steam. The steam extraction of the steam turbine is realized by adjusting the pressure through the butterfly valve of the connecting pipe of the medium and low pressure cylinder. The heating steam extraction pipeline is arranged in a unit system. The steam extraction of the unit is sent to the corresponding heat network heater through the steam extraction pipeline to heat the circulating water of the heat network, and the circulating water pump of the heat network is used for external heating. There are 2 initial heating stations in total, including 4 heating network heaters and 0 heat network circulating pumps in the initial heating network station of phase I; The initial station of phase II heat supply network has 5 heat supply network heaters, 9 heat supply network circulating pumps, all of which are electric pumps. The return water of the heat supply network first enters the initial station of the phase II heat supply network. After the pressure head is raised by the circulating water pump of the heat supply network, it enters the phase I and phase II heat supply network heaters respectively. The temperature is heated from 60 ℃ to about 120 ℃ and sent out.

3 Test process

3.1 Maximum output test when heating extraction steam is 150 t/h

Gradually increase the main steam flow to 950t/h, close to the maximum steam inlet of 979t/h, and then bring the unit heating extraction flow to 150t/h. at the same time, the outlet water temperature of the heating network meets the requirements of the current thermal power company. During the test, the heating extraction flow was 148.50t/h, and the average active power of the generator was 289.48MW.

3.2 Minimum output test when heating extraction steam is 150 t/h

First, gradually reduce the unit load to about 171MW. At this time, the intermediate exhaust pressure is 0.09MPa, close to the alarm value of 0.08MPa, but the heating extraction steam flow is higher than the required 150t/h. adjust the heating extraction steam flow to 150t/h by turning down the heating quick closing valve. Because there is a logical relationship between the heating butterfly valve of the connecting pipe of the medium and low pressure cylinder when the heating quick closing valve is fully open, The electrical load of the unit cannot be reduced again by adjusting the opening of the heating butterfly valve of the connecting pipe of the medium and low pressure cylinder, but can only be maintained in the current state. During the test, the heating extraction steam flow of the unit was 142.33t/h, and the average active power of the generator was 171.03MW.

3.3 Maximum output test when heating extraction steam is 250 t/h

Gradually increase the main steam flow to 950t/h, close to the maximum steam inlet of 979t/h, and then bring the heating extraction flow of the unit to 250t/h. During the test, the heating extraction flow is 250.06t/h, and the average active power of the generator is 276.92MW.

3.4 Minimum output test when heating extraction steam is 250 t/h

Gradually reduce the unit load to about 150MW. At this time, the heating extraction flow of the unit is about 250t/h, and the load reaches 50% of the rated capacity required by the dispatching agreement. During the test, the heating extraction steam flow of the unit is 250.93t/h, and the average active power of the generator is 150.02MW.

3.5 Maximum output test when heating extraction steam is 350 t/h

Gradually increase the main steam flow to 950t/h, close to the maximum steam inlet of 979t/h, and then bring the heating extraction flow of the unit to 350t/h. During the test, the heating extraction flow was 346.87t/h, and the average active power of the generator was 262.41MW.

3.6 Minimum output test when heating extraction steam is 350 t/h

Firstly, bring the heating extraction flow of the unit to 350t/h, gradually close the steam inlet butterfly valve of the low-pressure cylinder, reduce the steam inlet of the unit and maintain the heating extraction flow unchanged. When the unit load is reduced to about 203MW, the opening of the steam inlet butterfly valve of the low-pressure cylinder is 16.04%. Due to the jamming of the butterfly valve, the butterfly valve cannot be closed continuously, and the test is started under the current state. During the test, the heating extraction steam flow of the unit is 348.60t/h, and the average active power of the generator is 203.02MW.

3.7 Output test at heating extraction 440 t/h

Firstly, bring the heating extraction flow of the unit to 440t/h. at this time, the steam inlet butterfly valve of low-pressure cylinder has been closed to the minimum opening (it can only be reduced to 16% due to valve jamming), meeting the requirements of minimum electrical load test; Meanwhile, the main steam flow is 950t/h, close to the maximum main steam flow of 979t/h, meeting the requirements of the maximum electric load test. During the test, the heating extraction flow was 439.11t/h, and the average active power of the generator was 233.10MW.

4 Test result

4.1 Difference between actual output and design output

Table 3 Comparison between actual output and design value of unit

Heating extraction capacity (t/h)	Upper limit of output (MW)		Lower limit of output (MW)	
	Design value	Test value	Design value	Test value
145.42	298.85	289.48	197.20	171.03
250.49	281.39	276.92	177.47	150.02
347.74	265.23	262.41	159.26	203.02
439.11	250.04	233.10	188.29	233.10
145.42	298.85	289.48	197.20	171.03
250.49	281.39	276.92	177.47	150.02
347.74	265.23	262.41	159.26	203.02
439.11	250.04	233.10	188.29	233.10
145.42	298.85	289.48	197.20	171.03

4.2 Limiting factors of unit output

The maximum electric load test of the unit is limited by the maximum steam inlet of the unit. During the test, the main steam flow is about 950t/h, which is close to the maximum steam inlet of the unit of 979t/h.

In the minimum electric load test, when the heating extraction steam flow is 347.74t/h and 439.11t/h, the steam inlet butterfly valve of low-pressure cylinder has been closed to the minimum opening (it can only be reduced to 16% due to valve jamming). In order to maintain the heating extraction steam flow, it is impossible to continue to reduce the unit load; When the heating extraction flow is 250.49t/h, it reaches 50% of the rated capacity required by the dispatching agreement. In order to ensure the safety of the unit, the load of the power plant will not be reduced; When the heating extraction steam flow is 145.42t/h, in order to ensure that the heating extraction steam flow reaches the test value and maintain the middle exhaust pressure greater than 0.08MPa, the measures of reducing the heating quick closing valve shall be taken during load reduction. Due to the logical relationship that the heating butterfly valve of the connecting pipe of the medium and low pressure cylinder can be adjusted only when the heating quick closing valve is fully open, the unit load cannot be further reduced by adjusting the opening of the heating butterfly valve, Can only maintain the current state.

4.3 Heating working condition diagram

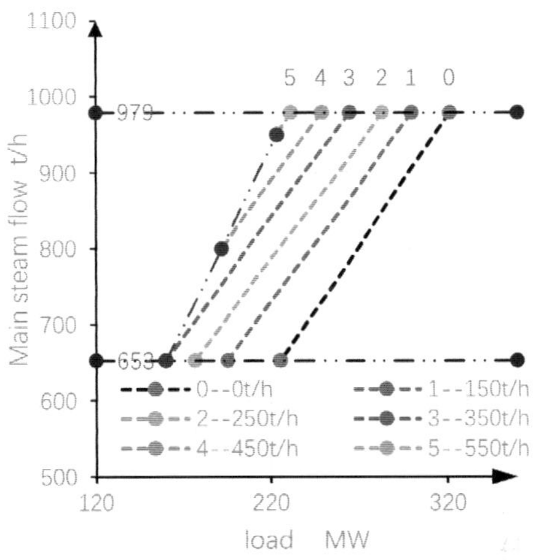

Fig. 1 Design heating condition of unit

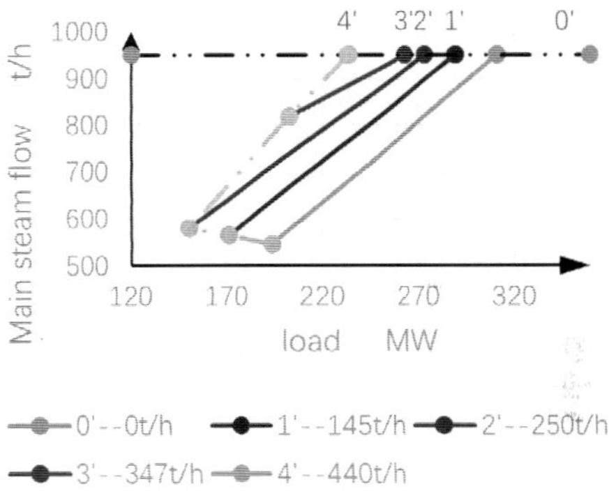

Fig. 2 Actual heating condition of the unit obtained from the test

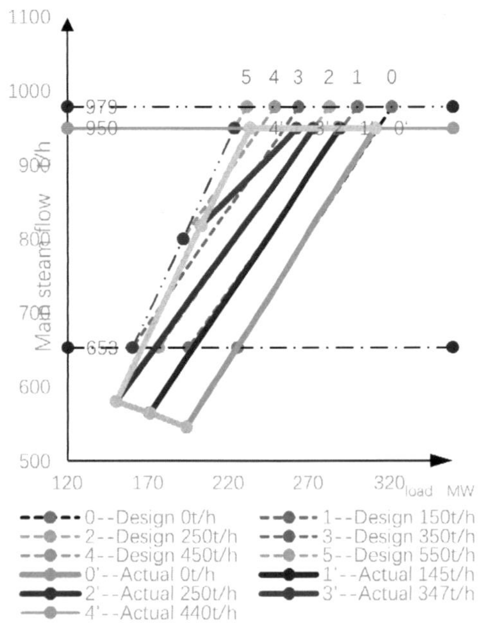

Fig. 3 Comparison between actual and designed heating conditions of the unit

It can be seen from the above table and the heating working condition diagram that the maximum output of the test is slightly lower than the upper limit of the design output, mainly because the main steam flow of the test is about 950t/h , which does not reach 979t/h under the design working condition; The minimum output of the test is lower than the lower limit of the design output when the heating extraction flow is 145.42t/h and 250.49t/h, and higher than the lower limit of the design output when the heating extraction flow is 347.74t/h and 439.11t/h. When the heating extraction steam flow is 347.74t/h and 439.11t/h, the lower limit of load is higher than the design output, mainly because the opening of butterfly valve of connecting pipe of medium and low pressure cylinder is about 16%, which does not reach the minimum steam inlet of low pressure cylinder.

4.4 Upper and lower limit fitting formula of output

According to the heating working condition diagram, the upper and lower limit formulas of output are fitted, as shown in the table below.Where, x is the heating extraction capacity; y is Unit load.

Table 4 upper and lower limit fitting formula of unit design output

Steam extraction interval	Upper limit fitting formula of output	Lower limit fitting formula of output
0≤x<350	y = -0.1662x + 323.02	y = -0.1878x + 224.51
350≤x≤550		y = 0.3176x + 48.824

Table 5 upper and lower limit fitting formula of actual output of unit

Steam extraction interval	Upper limit fitting formula of output	Lower limit fitting formula of output
0≤x<250.49	y = -0.1876x + 320.98	y = -0.3055x + 222.47
250.49≤x≤439.11		y = 0.4416x + 42.69

5 Conclusion

After long-term operation and heating transformation[6] of cogeneration units and the needs of heat users, and the current difficult situation of peak regulation of thermal power units is not considered in the early design stage, the original design heating working condition diagram can not reflect the real electric load adjustment range of the unit; At the same time, in the heating season, the contradiction between power grid peak shaving and cogeneration units continues to deepen. Under the new situation of vigorously developing new energy, cogeneration units are bound to break through the operation range of the original heating working condition diagram. Through tests, find out the actual electrical load adjustment range of the unit, remove the margin left for safe operation, and reflect the real peak shaving level of the unit.

Acknowledgements

This work was supported by the science and technology project of Shandong Electric Power Research Institute (ZY-2021-25), which carries out the research on the heating and peak shaving capacity of different types of cogeneration units in Shandong power grid.

References

1. Dawei Huang, Zheng Huang, Weichun Ge, et al. Analysis on electric power flexible regulation capacity for combined heat and power units in heat-led mode[J]. Automation of Electric Power Systems, 2018,42(24):27-35+52+36-39.
2. Haiying Dong, Lei Fang, Kun Ding, et al. Peak regulation strategy of csp plants based on operation mode of cogeneration[J].Acta Energiae Solaris Sinica,2019,40(10):2763-2772.
3. Yongqi Liu, Hongpeng Zhang, Qun Li, et al. Design and practice of electricity peak-shaving assistant service market in northeast power grid[J]. Automation of Electric Power Systems, 2017, 41(10):148-154.
4. Guangcai Zhang, Ke Zhou, Fen Lu, et al. Discussions on deep peaking technology of coal-fired power

https://doi.org/10.1051/e3sconf/202132901006

plants[J]. Thermal Power Generation, 2017, 46(09):17-23.

5. Xuhui Zhang, Zhonghua Zhao, Fuxing Cui, et al. Experimental Research on Deep Peak Regulation Characteristics of 1030 MW Ultra Supercritical Unit[J]. Shandong Electric Power, 2021,48(02):58-62.
6. Fuxing Cui, Daning You, Xuhui Zhang, et al. Research on Increasing Flexibility of Thermal Power Units by Flue Gas-water Heat Exchanger[J]. Shandong Electric Power, 2021,48(05):54-58+64.

 https://doi.org/10.1051/e3sconf/202132901007

Mathematical model and influence factor analysis for productivity prediction of fractured-vuggy carbonate reservoir when drilling into cavern reservoir

Yufeng Gong [1,*], Yan Fu [2], Xinyi Zhao [3], Kaifeng Wen [4], Tao Zhang [4], Jie Yan [2], Chengquan Li [5], Chengyong Li [1]

[1] School of Energy, Chengdu University of Technology, Chengdu, China
[2] Engineering Technology Research Institute, PetroChina Southwest Oil & Gas Field Company, Chengdu, China
[3] China National Petroleum Corporation Southwest Oil and Gas Field Branch Chuanzhong Oil and Gas Mine, Suining, China
[4] No.2 Gas Production Plant, PetroChina ChangQing Oilfield Company, Yulin, China
[5] Development Division, Southwest Oil and Gas Field Branch, China National Petroleum Corporation, Chengdu, China

Abstract. The seam hole type carbonate didn't touch type drill in the cave reservoir oil and gas resources is rich, but due to the large difference of unused reservoir capacity, capacity of main control factors and the evaluation model is not clear, so the reservoir is difficult to reasonable and effective development, in order to improve the cave reservoir development effect, find the factors influencing the cave reservoir capacity of the master,This article through to carry out the cave type carbonate reservoir capacity model test, the cave reservoir productivity prediction model is set up and analyze the factors affecting capacity, studies have shown that the wellbore storage coefficient and skin factor of cave type carbonate reservoir had a greater influence on the vertical well productivity, this paper research content for the cave type of carbonate reservoir reserves producing has a certain guiding significance.

1 Preface

Carbonate reservoirs are widely distributed in China, with 29.3×10^8t of proven petroleum reserves, which are indispensable for oil and gas exploration and development. Buried hill fracture-vuggy-type reservoirs are rich in untapped reserves. However, fracture-vuggy-type carbonate reservoirs are complex and diverse in their reservoir space. Fracture-vuggy carbonate reservoirs cannot be developed reasonably and effectively because of the strong heterogeneity of assemblage relationship and seepage field distribution and large difference in reservoir productivity.

In order to realize stable and high production and effective development of fracture-vuggy carbonate reservoir, this paper, based on the basic principle of seepage mechanics, establishes the productivity prediction model of fracture-vuggy carbonate reservoir and analyzes the influencing factors of productivity. The research shows that the effects of wellbore reservoir coefficient, skin coefficient, reserve-volume ratio and interflow coefficient on single well productivity of vertical well in fracture-vuggy carbonate reservoir are analyzed, and the reliability of mathematical model established in this paper is verified by field data. The research results of this paper can provide support for reasonable and effective reserve production evaluation technology, and it is of great significance for stable and high production and enhanced oil recovery of fracture-vuggy carbonate reservoirs.

2 Mathematical model of cavernous seepage through drilling

2.1 Establishment of physical model

Suppose a circular reservoir, outer boundary distance r_e(m), formation thickness h(m);The reservoir presents a compound distribution of two zones. Zone 1 is a karst cave directly connected to the wellbore, and Zone 2 is a fracture-vuggy zone. The radius of hyperosmotic zone is r_1(m).The low permeability zone is $k_2(10^{-3}\ \mu m^2)$, and the porosity is ϕ_2 .The formation fluid is slightly compressible, and the flow through the formation satisfies Darcy's law. The viscosity of the fluid is μ(mPa s), and the comprehensive compressibility coefficients of zones 1 and 2 are C_{t1} and C_{t2} (MPa^{-1}).The pressure of each reservoir and the cave is the original formation pressure p_i(MPa).The wellbore radius is r_w(m), the production rate is q (m^3/d), and considering the well storage and skin effects, the wellbore storage coefficient is C(m^3/MPa), and the skin coefficient is S. The influence of gravity and capillary force is not considered.

* Corresponding author: gongyufeng1995@stu.cdut.edu.cn

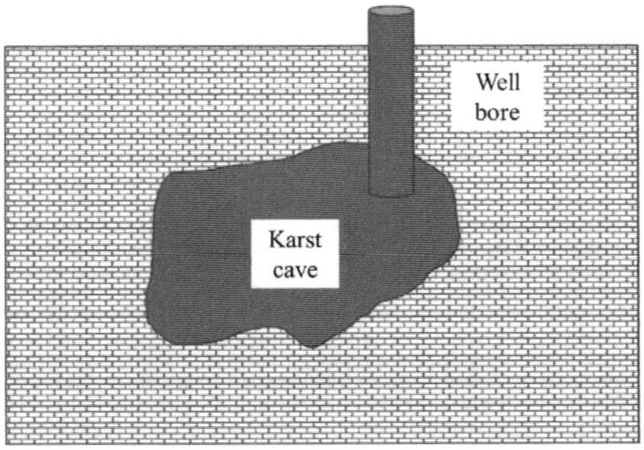

Figure 1. Physical model of cavernous reservoir.

2.2 Establishment of mathematical model

The dimensionless definitions of dimensionless pressure, dimensionless time, channelling coefficient, fracture-cave to matrix storage capacity ratio and fracture-cave storage capacity ratio were made:

$$\psi_{jD}=\frac{k_f h}{1.842\times10^{-3}qB\mu}\left(\psi_i-\psi_j\right),j=f,m \tag{1}$$

$$t_D=\frac{3.6k_f t}{\left(\phi_f\mu C_f+\phi_m\mu C_m\right)r_w^2} \tag{2}$$

$$\lambda=\alpha\frac{k_m r_w^2}{\mu};\omega=\frac{\phi_f C_f}{\phi_m C_m+\phi_f C_f} \tag{3}$$

$$\omega_v=\frac{\phi_v C_v}{2\left(\phi_f C_f+\phi_v C_v\right)};r_D=r/r_w \tag{4}$$

The dimensionless equations are as follows:

$$\frac{1}{r_D}\frac{\partial}{\partial r_D}\left(r_D\frac{\partial\psi_{fD}}{\partial r_D}\right)+\lambda\left(\psi_{mD}-\psi_{fD}\right)=\omega\frac{\partial\psi_{fD}}{\partial t_D} \tag{5}$$

Matrix diffusion equation:

$$\lambda\left(\psi_{mD}-\psi_{fD}\right)+\left(1-\omega\right)\frac{\partial\psi_{mD}}{\partial t_D}=0 \tag{6}$$

Initial conditions:

$$\psi_{jD}\left(r_D,t_D=0\right)=0,j=f,m \tag{7}$$

The outer boundary conditions of the banding:

$$\frac{\partial\psi_{fD}}{\partial r_D}\bigg|_{r_{eD}}=0 \tag{8}$$

Cave area equation:

$$\psi_{fD}\bigg|_{R_D}-\psi_{wD}=SR_D\frac{\partial\psi_{fD}}{\partial r_D}\bigg|_{R_D} \tag{9}$$

$$\frac{V_D}{24A_D}\omega_v\frac{d\psi_{wD}}{dt_D}=R_D\frac{\partial\psi_{fD}}{\partial r_D}\bigg|_{R_D}+1 \tag{10}$$

2.3 Establishment of Laplace mathematical model

According to the definition of Laplace transform, the dimensionless equation is transformed into a Laplace space expression:

Diffusion equation:

$$\frac{1}{r}\frac{\partial}{\partial r_D}\left(r_D\frac{\partial\overline{\psi}_{fD}}{\partial r_D}\right)+\lambda\left(\overline{\psi}_{mD}-\overline{\psi}_{fD}\right)=\omega s\overline{\psi}_{fD} \tag{11}$$

$$\lambda\left(\overline{\psi}_{mD}-\overline{\psi}_{fD}\right)+\left(1-\omega\right)s\overline{\psi}_{mD}=0 \tag{12}$$

Initial conditions:

$$\overline{\psi}_{jD}\left(r_D,s\right)=0,j=f,m \tag{13}$$

Closed outer boundary conditions:

$$\frac{\partial\overline{\psi}_{fD}}{\partial r_D}\bigg|_{r_{eD}}=0 \tag{14}$$

Cave area equation:

$$\overline{\psi}_{fD}\bigg|_{R_D}-\overline{\psi}_{wD}=SR_D\frac{\partial\overline{\psi}_{fD}}{\partial r_D}\bigg|_{R_D} \tag{15}$$

$$\frac{V_D}{24A_D}\omega_v u\overline{\psi}_{wD}=R_D\frac{\partial\overline{\psi}_{fD}}{\partial r_D}\bigg|_{R_D}+\frac{1}{u} \tag{16}$$

General solution:

$$\overline{\psi}_{wD}=Ak_0(r_D\sqrt{u})+BI_0\left(r_D\sqrt{u}\right) \tag{17}$$

External boundary conditions:

$$\frac{\partial\overline{\psi}_{wD}}{\partial r}\bigg|_{r_{De}}=0 \tag{18}$$

Substitution:

$$-\sqrt{u}Ak_1\left(r_{De}\sqrt{u}\right)+BI_1\left(r_{De}\sqrt{u}\right)\sqrt{u}=0 \tag{19}$$

Solution:

$$A=\frac{I_1\left(r_{De}\sqrt{u}\right)B}{k_1\left(r_{De}\sqrt{u}\right)} \tag{20}$$

The general solution is as follows:

$$\overline{\psi}_{wD}=aBk_0\left(r_D\sqrt{u}\right)+BI_0\left(r_D\sqrt{u}\right) \tag{21}$$

Substitute into the cave equation:

$$aBk_0\left(r_{DV}\sqrt{u}\right)+BI_0\left(r_{DV}\sqrt{u}\right)-\left(\overline{\psi}_{wD}+\frac{R_{DV}\lambda}{u}\right)= \\ SR_{DV}\left[-aB\sqrt{u}k_1\left(r_{DV}\sqrt{u}\right)+B\sqrt{u}I_1\left(r_{DV}\sqrt{u}\right)\right] \tag{22}$$

$$\frac{V_D}{24A_D}\omega_v\frac{d\psi_{wD}}{dt_D}=R_D\frac{\partial\psi_{fD}}{\partial r_D}\bigg|_{R_D}+1 \tag{23}$$

Solution:

$$B=\frac{\overline{\psi}_{wD}+R_{DV}\lambda/u}{ak_0\left(r_{DV}\sqrt{u}\right)+I_0\left(r_{DV}\sqrt{u}\right)-SR_{DV}\sqrt{u}\left[I_1\left(r_{DV}\sqrt{u}\right)-ak_1\left(r_{DV}\sqrt{u}\right)\right]} \tag{24}$$

Make:

$$b=ak_0\left(r_{DV}\sqrt{u}\right)+I_0\left(r_{DV}\sqrt{u}\right)-SR_{DV}\sqrt{u}\left[I_1\left(r_{DV}\sqrt{u}\right)-ak_1\left(r_{DV}\sqrt{u}\right)\right] \tag{25}$$

There are:

 https://doi.org/10.1051/e3sconf/202132901007

$$B = \frac{\overline{\psi}_{wD} + \frac{R_{DV}\lambda}{u}}{b} \tag{26}$$

Substitute into the karst cave equation:

$$\frac{V_D}{24A_D}\omega_v u\overline{\psi}_{wD} = R_{DV}\left[-a\sqrt{u}\frac{\overline{\psi}_{wD} + R_{DV}\lambda}{b}k_1\left(r_{DV}\sqrt{u}\right) + \frac{\overline{\psi}_{wD} + R_{DV}\lambda}{b}\sqrt{u}I_1\left(r_{DV}\sqrt{u}\right)\right] + \frac{1}{u} \tag{27}$$

Solution:

$$\overline{\psi}_{wD} = \frac{b + \lambda R_{DV}^2\sqrt{u}I_1\left(r_{DV}\sqrt{u}\right) - \lambda a R_{DV}^2\sqrt{u}k_1\left(r_{DV}\sqrt{u}\right)}{u\left[b\frac{V_D}{24A_D}\omega_v u + aR_{DV}\sqrt{u}k_1\left(r_{DV}\sqrt{u}\right) - R_{DV}\sqrt{u}I_1\left(r_{DV}\sqrt{u}\right)\right]} \tag{28}$$

Cave type capacity response model:

$$q = \frac{2\pi k_f h}{B\mu\psi_{fD}}(p_i - p_f) \tag{29}$$

3 Mathematical model of cavernous seepage through drilling

It can be seen from the seepage pattern diagram of vuggy carbonate reservoir that the seepage pattern of this type of well can be divided into three categories: karst cave radial flow, fracture linear flow and boundary quasi-steady flow.

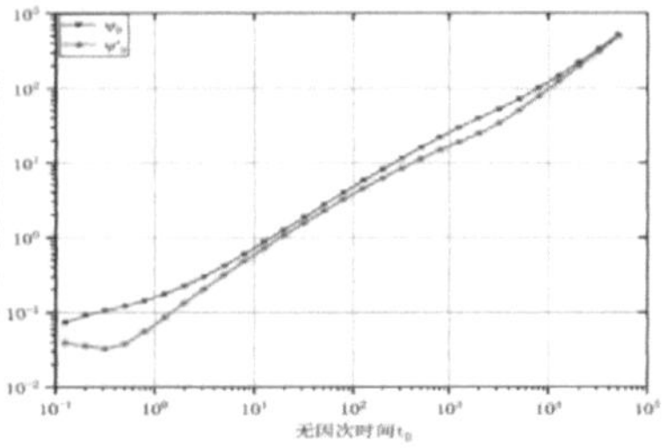

Figure 2. The seepage model of cavern-fracture carbonate reservoir.

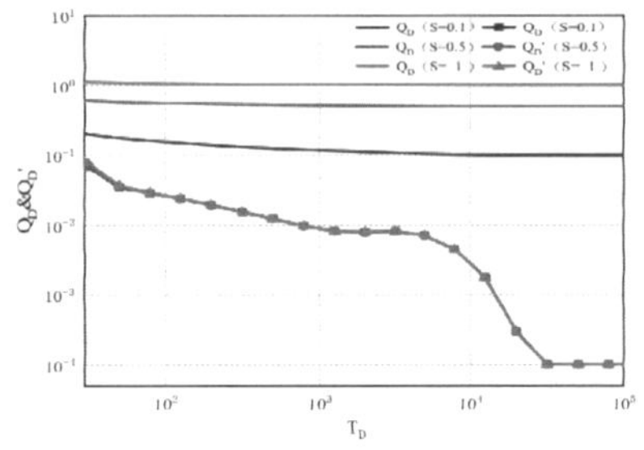

Figure 3. Relation diagram of influence of skin coefficient on productivity of cave-fracture carbonate reservoir (log-log coordinates)

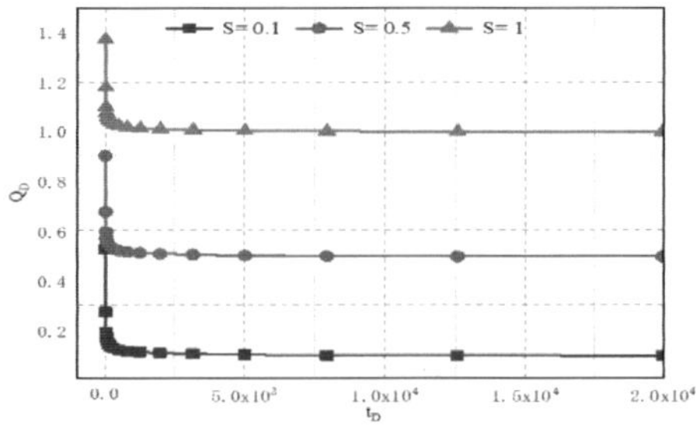

Figure 4. Effect of skin coefficient on productivity of vugge-fracture carbonate reservoir (Cartesian coordinates)

Fig. 3 is the skin factor of hole-seam type carbonate reservoir capacity log-log influence diagram, Fig. 4 is the skin factor of hole-seam type carbonate reservoir capacity cartesian coordinate effect diagram, skin factor of hole-seam type mainly embodies the influence of carbonate reservoir production at the beginning of the row in phases, the skin factor, the greater the production decline rate is lower, The larger the value of dimensionless yield curve is, the derivative curve of middle and late yield basically coincides.

Fig. 5 are seam wellbore storage coefficient for the hole - type carbonate reservoir capacity log-log influence diagram, Fig. 6 seam wellbore storage coefficient for the hole - type carbonate reservoir capacity cartesian coordinate effect diagram, wellbore storage coefficient of seam hole - type of carbonate reservoir production effect is mainly embodied in the stage of high rank of the former, wellbore storage coefficient, the faster the yield decline rate is, the larger the dimensionless yield curve value is, and the middle and late production curves basically coincide with each other without obvious influence.

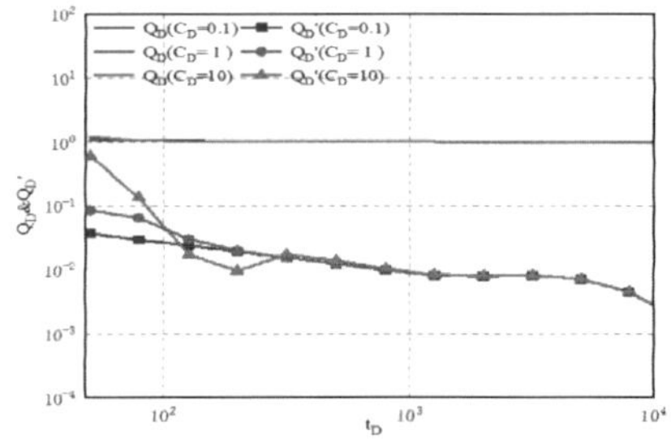

Figure 5. Influence of wellbore reservoir coefficient on productivity of cave-fracture carbonate reservoir (log-log coordinates)

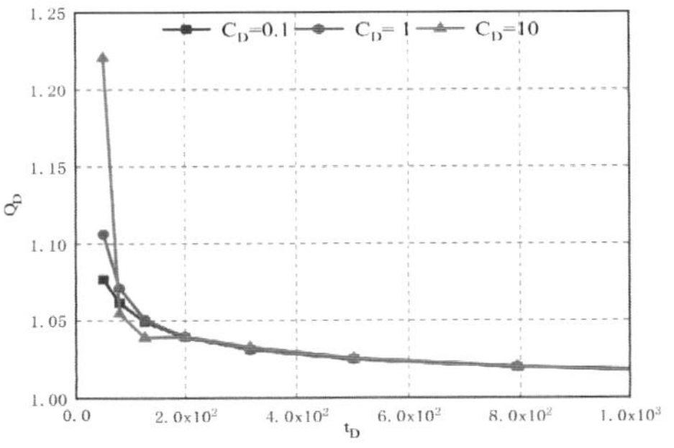

Figure 6. Relation diagram of influence of wellbore reservoir coefficient on productivity of cave-fracture carbonate reservoir (Cartesian coordinates)

4 Conclusion

This paper to carry out the buried hill type seam hole carbonate reservoir geological model based on simplified, starting from the basic principle of seepage flow mechanics, establish buried seam hole type carbonate reservoir productivity prediction model and analyze the factors affecting capacity, through Z6 block of single well productivity verification on the spot of this paper to establish the mathematical model of reliability is higher. The results of influence factor analysis show that the larger the storage capacity ratio is, the greater the dimensionless yield curve value is. The higher the interflow coefficient, the greater the dimensionless production curve value. The larger the wellbore reservoir coefficient is, the faster the production decline rate is. The larger the skin factor, the higher the dimensionless yield.

References

1. Xia Tian, Study on productivity evaluation of buried hill reservoir [D]. Northeast Petroleum University, 2015.
2. Wen Huaizhong, Evaluation of development effect and optimization of production allocation of YM32 carbonate reservoir [D]. Southwest Petroleum University, 2014.
3. Wan Yizhao, Liu Yuewu. Three-dimensional discrete fracture-cavity numerical well test model for fracture-cavity reservoirs [J]. Chinese Journal of Theoretical and Applied Mechanics, 2015,47(06):1000-1008.
4. Liu Jian, Liu Yuetian, Liang Wuquan, et al. Study on out-of-step injection and production of fractured oil reservoirs in carbonate buried hill [J]. Science, Technology and Engineering, 2013(36):49-54.
5. LIN Xing, Study on numerical simulation method of buried-hill fractured-vuggy carbonate reservoir [D]. Southwest Petroleum University, 2014.
6. Feng Jinde, Study on seepage theory and reservoir engineering application of fractured low permeability reservoir [D]. China University of Petroleum (Beijing), 2007.
7. Xu Tonghao, Geng Shaoyang, Li Yao, Hu Gang.Productivity evaluation model of buried-hill fractured-vug carbonate reservoir and analysis of influencing factors [J]. Petrochemical Industry Application, 2020,39(09):40-46
8. Kang Bo, Simulation and application of multi-scale reservoir flow in deep karst condensate gas reservoir [D]. Southwest Petroleum University, 2020.

Preparation of organic polymer flocculant and its application in drilling fluid

Ya Shi [1], Wenyu Ji [2], Xiwen Jia [3], Le Qu [4], Jie Zhang [1], Rui Zhou [1], Chen Gang[1], Yongfei Li [1]

[1] Shaanxi Province Key Laboratory of Environmental Pollution Control and Reservoir Protection Technology of Oilfields, Xi'an Shiyou University, Xi'an, 710065, China
[2] Xi'an Changqing Chemical Group Co. Ltd, Xi'an Shaanxi, 710068, China
[3] Oil Production No.8, Changqing Oilfield Company, CNPC, Xi'an, 710000, China
[4] Xi'an Key Laboratory of Tight oil (Shale oil) Development, Xi'an Shiyou University, Xi'an, 710065, China

Abstract. A flocculant (DES0-0) was synthesized by epichlorohydrin and diethylamine, followed by adding crosslinking agent such as ethylenediamine, diethylenetriamine and triethylenetriamine to obtain the crosslinked polymer DES0-1, DES0-2, DES0-3. The flocculation effect of DES0-1, DES0-2, DES0-3 and PAM was studied by measuring the volume, transmittance and particle size of the clarified solution after flocculation. The results showed that the flocculation effect of triethylenetetramine crosslinked polymer (DES0-3) and PAM was better. Meanwhile, the flocculation effect was more optimistic in the strong alkali environment.

Key words: Epichlorohydrin; Diethylamine; Flocculant; Crosslinking agent.

1 Introduction

Drilling fluid was often used in modern oil and gas well drilling operations. In drilling operations, drilling fluids provided a series of important functions, such as preventing formation fluids from entering the wellbore, removing drill cuttings, suspending drill cuttings during drilling pauses, and keeping the drill bit cool and clean[1-5]. Due to the rapid development of the petroleum industry, the composition of drilling wastewater had become more and more diversified in recent years. At present, the harmless treatment of drilling fluid waste is only carried out from physical, chemical, biological and other aspects that including coagulation, adsorption, reverse osmosis, chemical oxidation and electrolysis, and biological treatment. It has been widely used in drilling wastewater treatment[6,7]. In conventional operations, drillingfluids were always recirculated by removing useless solid pollutants because they were costly and hazardous if directly discharged to the environment[8-10]. However, due to solid contaminants will increase the viscosity and friction of the drilling fluid, reduce the rate of penetration (ROP), and thereby reduce the performance of the drilling fluid ,the most important task of drilling fluid recycling is to remove the useless solid phase in the return fluid[11]. Flocculation is an effective and economical method to treat water and wastewater[12,13]. Natural and synthetic polymers can make colloidal suspensions unstable and have been found to be widely used as flocculants [14]. Water-soluble synthetic polymers have the disadvantages of non-biodegradability and high cost. Therefore, there is an increasing demand for efficient, inexpensive, and environmentally friendly flocculants for water and wastewater treatment. ndustrial flocculants are sometimes not suitable for oilfields due to their poor water solubility, small molecular weight, low charge density, and high concentration[15]. Polyamines are quaternary ammonium cationic polymers that have been used as flocculants and charge neutralizers in the pulp and mining industries. A number of studies have found that polyamines are effective in a wide pH range are easy to handle and immediately dissolve in aqueous systems[16,17]. Moreover, the cationic polymer neutralizes the charge by compressing the electric double layer, adsorbs and then forms a bridge, and the removal rate is high[18,19]. In this paper, ethylenediamine and epichlorohydrin with two-step polymerize to polyamine (DES0-0), using ethylenediamine, diethylenetriamine and triethylenetetramine as crosslinking agents to synthesize crosslinked polymer DES0-1, DES0-2 , DES0-3 mix with PAM. After flocculation , the flocculation was evaluated by measuring in these aspects which and the particle size diameter of the floc and the volume of the supernatant and the light transmittance.

https://doi.org/10.1051/e3sconf/202132901008

2 Materials and methods

2.1 Materials

Rereagents involved in the experiment, such as epichlorohydrin, diethylamine, ethylenediamine, diethylenetriamine and triethylenetetramine (analytical purity), were purchased from Chengdu Kelong Chemical Reagent Factory. Polyacrylamide (molecular weight is more than or equal to 3 million) were provided by Shanghai Shanpu Chemical Co., Ltd.

2.2 Preparation of polyamine salt and flocculant

Adding sodium oleate and epichlorohydrin with a molar ratio of 1:1 in a round-bottom flask, 1% potassium iodide was used as a catalyst and ethanol was used as a solvent under refluxing for 2 h to obtain product 1. In another round bottom flask, epichlorohydrin and diethylamine with a molar ratio of 1:1 was added based on ethanol as the solvent, and then transfered product 1 into this flask, continued to reflux for 2 h to evaporate the solvent ethanol to obtain the final product. After dripping, the mixture was heated at 80°C and keept the reaction for 240 min to obtain the polyamine.

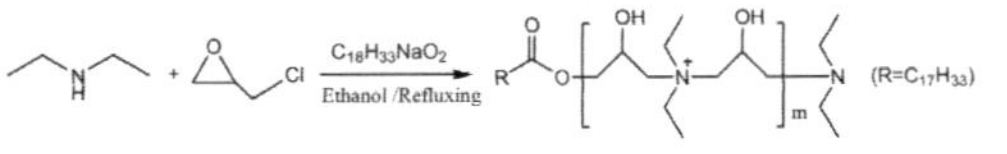

Figure 1 Synthesis mechanism

2.3 Crosslinking reaction

After adding the resulting polyamine salt medicine, a certain proportion of cross-linking agent was added. The reaction was carried out for 5 h kept with 60℃ until there was no pungent odor in the reactant to obtain the DES0-1, DES0-2, DES0-3.

2.4 Synthetic viscosity measurement

The viscosity of the polymer was measured using a 0.8-0.9 mm ubbelohde viscometer. The time of synthetic stream passing through capillary tube was measured at room temperature, followed by the corresponding kinematic viscosity was converted into standard viscosity according to formula.

2.5 Laser particle size determination method

Flocculant was added to the drilling fluid base slurry for flocculation, then transfered an appropriate amount of suspension to measure the particle size and distribution of flocs with laser diffraction particle size analyzer.

2.6 Base slurry configuration

The 4% bentonite was added to a certain amount of water under stirring with certain speed for 2 h, followed by age for 24 h.

2.7 Flocculation performance measurement

After the flocculation by adding flocculant to drilling fluid base slurry, an appropriate amount of suspension was transfered, followed by centrifugation. The volume of the supernatant was measured and the light transmittance was measured with a visible spectrophotometer.

2.8 Characterization of flocculants

A certain amount of synthetic products were analyzed by Fourier infrared spectrometer. The mixture consisting of sample and potassium bromide with a ratio of 1:100 (mass ratio) was put in mold, followed by tablet compressing.

3 Results and discussion

3.1 Optimization of flocculant synthesis conditions

3.1.1 Screening of reactant molar ratio

Table 1 and Figure 2 reflected the viscosity of the synthesized compound under different molar ratios. It suggested that when the molar ratio of epichlorohydrin to diethylamine was 1.5:1, the kinematic viscosity and dynamic viscosity of the compound reach the maximum, implying that the molecular weight of the composite is the largest. The larger the molecular weight, the better the flocculation effect. Therefore, the molar ratio of reactants is 1.5:1.

Table 1 Compound related data in different molar ratios

The molar ratio	Kinematic viscosity m^2/s	Dynamic viscosity /mpa.s
1:1	1.800	2.03
1:2	2.590	2.94
1:1.5	5.025	9.52
1.5:1	12.750	15.58
2:1	10.800	13.50

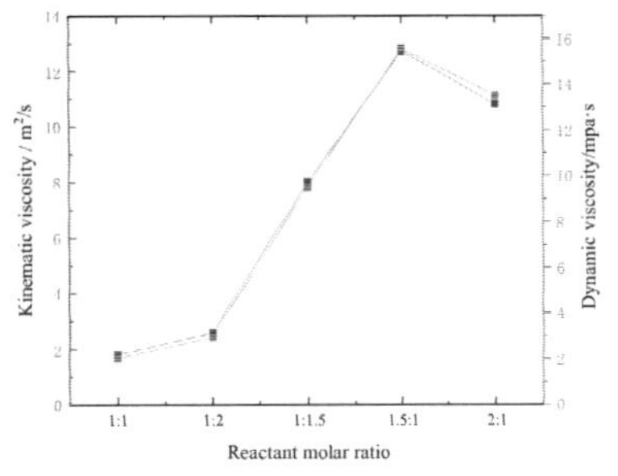

Figure 2 The influence of reactant molar ratio on product viscosity

3.1.2 Screening of measurement standard for crosslinking agent

Table 2 and Figure 3 showed that the relevant data of the product after crosslinking with a 3% crosslinking agent and a reactant with a molar ratio of 1.5:1. It illustrated that the composition containing triethylenetetramine had the highest viscosity. However, epichlorohydrin and amines were used to synthesize polymers under the condition of a certain amount of crosslinking agent, the greater of the cationicity which the smaller the amine group contained in the crosslinking agent structure. The flocculation had a great relationship with the content of cations, so ethylenediamine was selected as the content determination standard.

Table 2 Data related to the composition with different crosslinking agents

Type of crosslinking agent	Kinematic viscositym^2/s	Dynamic viscosity /mpa·s
Ethylenediamine	10.76	13.00
Diethylenetriamine	12.26	15.23
Triethylenetetramine	14.63	19.74

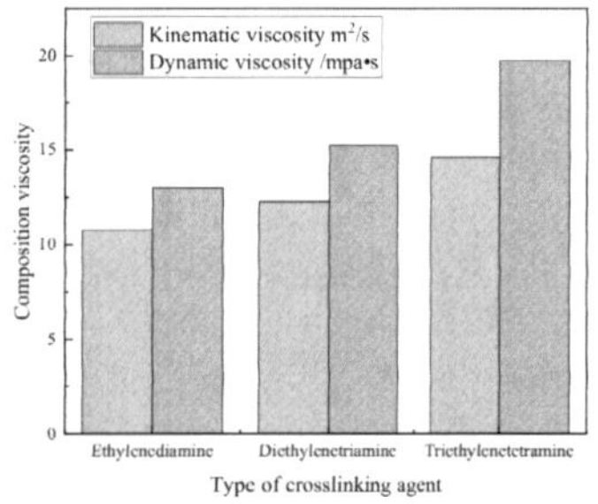

Figure 3 The effect of crosslinking agent on polymer viscosity

3.1.3 Determination of the dosage of crosslinker

Dethylamine crosslinking with reactants which the molar ratio was 1.5:1 were seleced to explore the influence of the content of the cross-linking agent on the viscosity of the synthetic products. A blank which added a certain amount of distilled wate was set during the measurement, and then set five different dosages of 0%, 1%, 2%, 3%, 4% to determine the optimum dosage of crosslinking agent. It can be seen intuitively from Table 3 and Figure 4 that the viscosity of the composition containing 4% of the crosslinking agent was relatively large, and the viscosity of the 3% and 4% organic polymer composition was not very different. Comparing with the cationicity,due to it was found that 3% of the polymer had a higher cationicity, a polymer with a content of 3% was selected.

Table 3 Data related to the composition with different content of cross-linking agent

Crosslinking agent content	Kinematic viscositym^2/s	Dynamic viscosity /mpa.s
0%	6.30	7.70
1%	7.01	8.47
2%	8.36	9.78
3%	10.99	13.23
4%	12.64	15.29

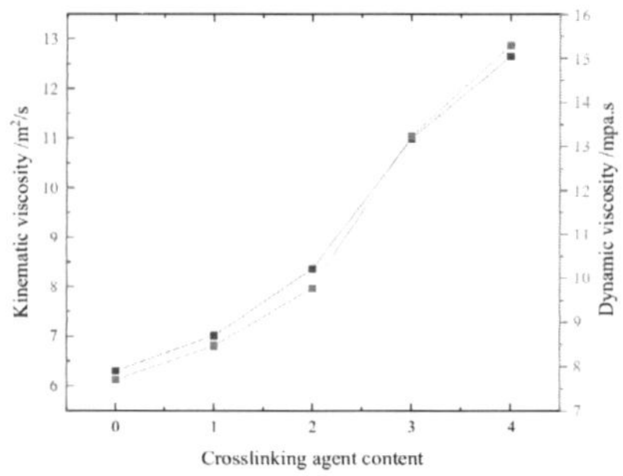

Figure 4 The influence of crosslinking agent content on polymer viscosity

3.2 Evaluation of flocculation effect of flocculant and PAM

The mixture which had different concentrations of flocculant and 70ppm PAM was added to the drilling fluid base slurry. The flocculation effect on the drilling fluid base slurry was investigated by analyzing the volume and permeability of the centrifugal liquid and the particle size diameter of the flocs. The experimental results were followed.

Table 4 DES0-0 and PAM compound flocculation effect

DES0-0	0	0.10 %	0.15 %	0.20 %	0.25 %	0.30 %
Volume of supernatant/ mL	152.0	152.0	150.0	153.0	157.0	157.0
Transmittance /%	99.5	79.4	90.0	96.4	99.0	100.0
Particle size diameter /μm	503.1	144.6	148.2	150.5	152.5	281.0

Table 5 DES0-1 and PAM compound flocculation effect

DES0-1	0	0.4%	0. 5%	0.6%	0.7%
Volume of supernatant/mL	151.8	130.2	128	140	139.8
Transmittance /%	98.8	81.8	85.3	91.6	98.1
Particle size diameter /μm	463.5	151.3	153.2	154.6	189.7

Table 6 DES0-2 and PAM compound flocculation effect

DES0-2	0	0.4%	0. 5%	0.6%	0.7%
Volume of supernatant/mL	151.8	145	157	148	151
Transmittance /%	98.8	94.2	94.3	96.7	95.9
Particle size diameter /μm	463.5	147.5	151.7	158.5	169.7

Table 7 DES0-3 and PAM compound flocculation effect

DES0-3	0	0.10%	0.2%	0.25%	0.30%
Volume of supernatant/mL	153	126.1	137.2	145.1	146.5
Transmittance /%	98.5	90.8	90.9	95.1	98.9
Particle size diameter /μm	489.1	147.5	150.6	153.8	156.5

With the continuous increase of the mass fraction of flocculant was added into the drilling fluid, the high molecular chain will have the effect of net capture and adsorption bridging on the bentonite. Meanwhile, due to the rich positive charge contained on it will neutralize the negative charge on the end face of the bentonite, which will have the effect of pulling and sinking the solid suspension and the transmittance and the particle size diameter of flocs are also increasing. The DES0-3 had a better effect with PAM.

3.3 Flocculation effect of flocculant in different pH environment

The pH value of the drilling fluid base slurry was adjusted by hydrochloric acid and sodium hydroxide. The experimental results of the flocculation effect on the drilling fluid base slurry at different pH were investigated by analyzing the volume and permeability of the centrifuged liquid and the particle size diameter of the flocs. The experimental results were followed.

Table 8 Flocculation effect of DES0-0 under different pH conditions

pH	4	5	6	7	8
Volume of supernatant/mL	173.0	177.0	175.0	160.0	160.0
Transmittance /%	99.4	100.0	99.5	99.0	99.0
Particle size diameter /μm	114.4	125.3	112.6	106.1	183.2

Table 9 Flocculation effect of DES0-1 under different pH conditions

pH	4	6	7	9	11
Volume of supernatant/mL	145.1	146.1	134.2	134.0	143.0
Transmittance /%	97.4	99.9	86.9	94.1	91.8
Particle size diameter /μm	103.2	182.2	141.3	146.1	133.2

Table 10 Flocculation effect of DES0-1 under different pH conditions

pH	4	6	7	9	11
Volume of supernatant/mL	170.0	171.1	172.5	160.3	168.6
Transmittance /%	98.8	99.9	99.9	88.8	95.1
Particle size diameter /μm	153.3	196.4	191.3	156.3	163.2

Table 11 Flocculation effect of DES0-1 under different pH conditions

pH	4	6	7	9	11
Volume of supernatant/mL	155.1	152.0	152.2	151.1	159.5
Transmittance /%	87.2	96.8	97.5	93.6	97.5
Particle size diameter /μm	110.3	92.5	93.1	112.5	176.4

It can be seen from table 8,9,10,11 that the flocculation effect of the DES0-1, DES0-2 and PAM is the best when pH is 6, while the flocculation effect of the products synthesized with DES0-3 and PAM is the best when pH is 11. This is because the crosslinked products of diethylamine and triethylenetetramine as crosslinking agents can dissociate more cations which enhance the electrostatic attraction under alkaline conditions, so that the adsorption performance is improved, and the flocculation effect is also improved. However, due to the total negative charge of the bentonite particles itself is small, the total negative charge of the base slurry itself is small cause to the adsorption capacity for cations is small. When the pH was too high, the cations dissociated from the flocculant will appear to be excessive. On the contrary, the particles will be repelled due to the positive charge, which is not conducive to agglomeration.

3.3.1 Infrared analysis

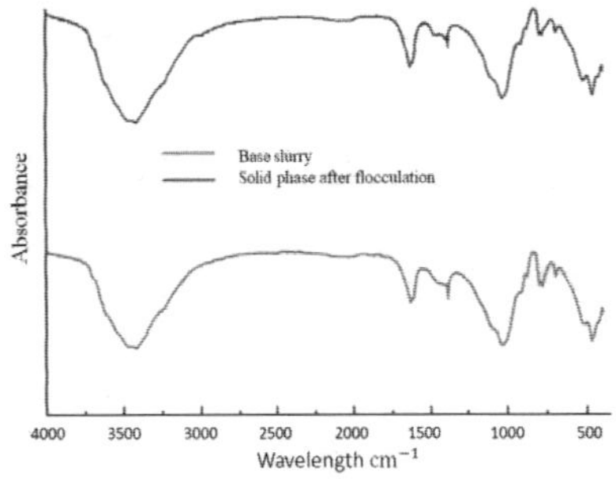

(a) DES0-0

https://doi.org/10.1051/e3sconf/202132901008

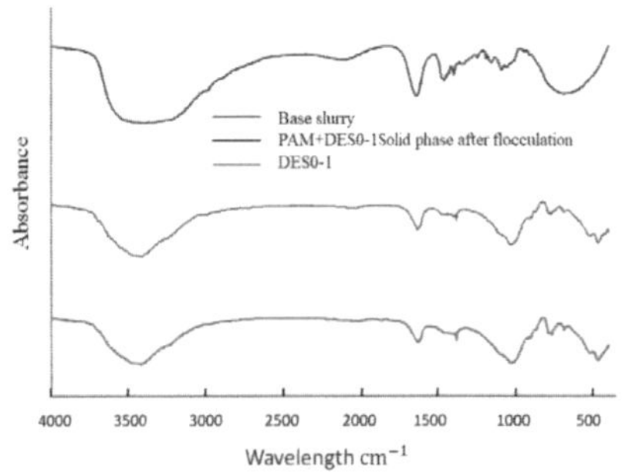

(b) DES0-1

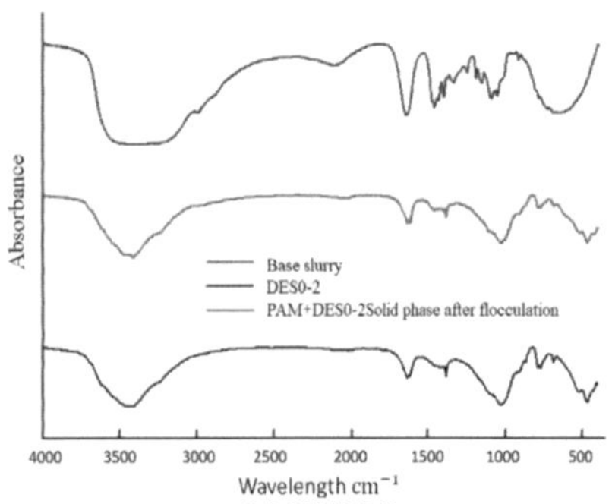

(c) DES0-2

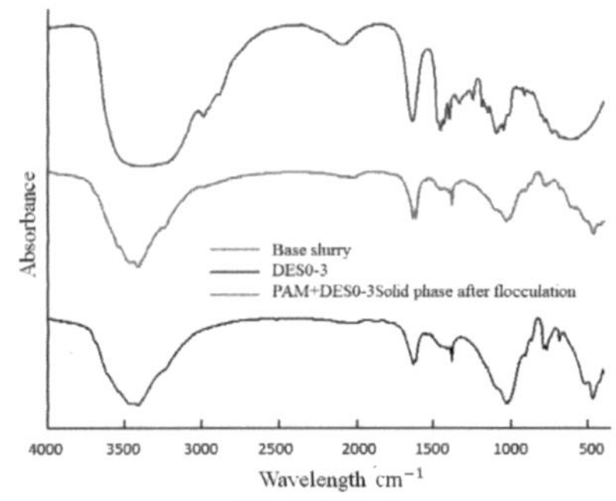

(d) DES0-3

Figure 5 FTIR spectra of polymer

It can be concluded from the above infrared spectra that the special peak shape of the ternary epoxy in the molecular structure of epichlorohydrin had disappeared. 978cm^{-1} was the characteristic absorption peak in the molecular structure of quaternary ammonium salt. A series of sharp peaks from 2800cm^{-1} to 3050cm^{-1} were the stretching vibration peaks of methyl CH_3 - and methylene CH_2 -, which indicates that cationic quaternary ammonium salt had been formed. The absorption peak of water molecular structure appeared at about 3300 cm^{-1}. Compared with the infrared characteristic peaks of the base slurry and flocculant, the main characteristic absorption peaks of the base slurry and flocculant were shown in the infrared spectrum of the solid phase after flocculation and indicating that the flocculant had been adsorbed on the base slurry and played the role of adsorption flocculation.

4 Conclusions

In this paper, a series of crosslinked polymer such as DES0-1, DES0-2, DES0-3, was synthesized via adding ethylenediamine, diethylenetriamine and triethylenetriamine to a flocculant (DES0-0) obtained by epichlorohydrin and diethylamine, followed by flocculation evaluation. The results suggested that the relative molecular mass of triethylenetetramine (DES0-3) was higher than ethylenediamine (DES0-1) and diethylenetriamine (DES0-2). The polymer chain had a great effect on the net capture, adsorption and bridging of bentonite, and a pull-sinking effect on suspended solids. Meanwhile, it was found that the cross-linked product of triethylenetetramine as a cross-linking agent will not dissociate too many cations. The excess cations would cause the particles to be positively charged and charge repulsion, which is not conducive to agglomeration. Therefore, the triethylenetetramine cross-linked polymer can achieve better results when it is compounded with PAM under alkaline conditions.

Acknowledgments

The work was supported financially by Shaanxi Provincial Key Research and Development Program (2019GY-136), Key Scientific Research Program of Shaanxi Provincial Department of Education (21JY035) and Youth Innovation Team of Shaanxi University. And we thank the work of Modern Analysis and Testing Center of Xi`an Shiyou University.

References

1. Fink J. Petroleum engineer's guide to oil field chemicals and fluids (Second Edition) [M]. 2015.
2. Hart PJ, Snyder RK. Method for on-site treatment of oil and gas well waste fluids: US, US7022240 B2[P]. 2006.
3. Landis CR, Collins RP, Anderson EA, et al. Novel injection flocculation and compression dewatering unit for solids control and management of drilling fluids and methods relating thereto: US, US20130015141 A1[P]. 2013.
4. Nagham Amer Sami. Effect of magnesium salt contamination on the behavior of drilling fluids[J]. Egyptian Journal of Petroleum,2016, 25(4): 453-458.
5. Ball AS, Stewart RJ, Schliephake K. A review of the current options for the treatment and safe disposal of drill cuttings[J].Waste Manag Res, 2012, 30(5): 457-473.
6. Wang Y, Wang T, Yan X. Discussion on harmless treatment technologies of drilling fluid waste in oil and gas fields[J]. IOP Conference Series Earth and Environmental Science, 2020, 446: 052037.

7. Wang JP, Chen YZ, Ge XW, et al. Gamma radiation-induced grafting of a cationic monomer onto chitosan as a flocculant[J]. Chemosphere, 2007, 66(9): 1752-1757.
8. Jiang G, Peng S, Li X, et al. Preparation of amphoteric starch-based flocculants by reactive extrusion for removing useless solids from water-based drilling fluids[J]. Colloids & Surfaces A Physicochemical & Engineering Aspects, 2018, 558: 343-350.
9. Elnenay AEMH, Nassef E, Malash GF, et al. Treatment of drilling fluids wastewater by electrocoagulation[J]. Egyptian Journal of Petroleum, 2016, 26(1): 203-208.
10. Hamed SB, Belhadri M. Rheological properties of biopolymers drilling fluids[J]. Journal of Petroleum Science & Engineering, 2009, 67(3-4): 84-90.
11. Peng, Shuanglei, Jiang, et al. Flocculation of submicron particles in water-based drilling fluids by CMC-g-DMDAAC[J]. Journal of Petroleum Science & Engineering, 2018, 162: 55-62.
12. Yang WY, Qian JW, Shen ZQ. A novel flocculant of Al(OH)3-polyacrylamide ionic hybrid[J]. Journal of Colloid & Interface Science, 2004, 273(2):400-405.
13. Tripathy T, Singh RP. High performance flocculating agent based on partially hydrolysed sodium alginate–g–polyacrylamide[J]. European Polymer Journal, 2000, 36(7): 1471-1476.
14. Aly AA. Preparation and evaluation of some anionic starch derivatives as flocculants[J]. Starch-Strke, 2010, 58(8): 391-400.
15. Jing Z, Hong Z, Wang F, et al. Preparation of a new inorganic–organic composite flocculant used in solid–liquid separation for waste drilling fluid[J]. Chemical Engineering Journal, 2011, 171(1): 350-356.
16. Joo DJ, Shin WS, Kim YH, et al. Effect of polyamine flocculant types on dye wastewater t reatment. Separation Science and Technology,2003, 38(3), 661–678.
17. Yue QY, Gao BY, Wang Y, et al. Synthesis of polyamine flocculants and their potential use in treating dye wastewater[J]. Journal of Hazardous Materials, 2008, 152(1): 221-227.
18. L Besra and D.K Sengupta and S.K Roy and P Ay. Polymer adsorption: its correlation with flocculation and dewatering of kaolin suspension in the presence and absence of surfactants[J]. International Journal of Mineral Processing, 2002, 66(1-4): 183–202.
19. Avci D, Mol N, Dagasan L. New cationic polyelectrolytes for flocculation processes of baker's yeast waste water[J]. Polymer Bulletin, 2002, 48(4-5): 353-359.

Measurement of the coordination effect of polyphenols used in oil field by colorimetric method

Wen Tian [1], Huanyu Liu [2], Xiaoyu Han [3], Rui Zhou [1], Le Qu [4], Jie Zhang [1], Sanbao Dong [1]

[1] Shaanxi Province Key Laboratory of Environmental Pollution Control and Reservoir Protection Technology of Oilfields, Xi'an Shiyou University, Xi'an, 710065, China
[2] The Shale Oil Development Department of Longdong Area, Changqing Oilfield Company, Qingyang, 745100, China
[3] Gas Production No.2, Changqing Oilfield Company, CNPC, Yunlin, 719000, China
[4] Xi'an Key Laboratory of Tight oil (Shale oil) Development (Xi'an Shiyou University), Xi'an Shiyou University, Xi'an, 710065, China

Abstract. Polyphenols are widely used in drilling fluid, cement slurry and water treatment in oil fields. The paper is based on the characteristics of the abundant phenolic hydroxyl groups in plant polyphenol compounds that can form stable complexes with iron ions, and the characteristics of phenolic compounds forming blue-violet to brown complexes with iron salts, and the color of iron salts itself for light yellow, a color method is designed to evaluate the quality of polyphenols, which provides a simple and reliable method for the quality analysis of industrial polyphenols.

Keywords: Phenolic compounds; quantitative analysis; coordination; colorimetric

1 Introduction

Polyphenols are a kind of renewable energy with wide distribution, variety and large quantity. Its rich phenolic hydroxyl groups give it a series of unique chemical properties, which not only have the effect of reducing viscosity and fluid loss in drilling fluids, but also have a certain anti-collapse effect [1,2]. Polyphenol compounds can form a protective film on the surface of steel to prevent further corrosion, so these compounds also have a certain corrosion inhibition [3,4]. It is precisely because of its rich phenolic hydroxyl structure that polyphenol compounds can react with high-valent ions such as iron ions and aluminum ions to form stable complexes [5], thereby forming a protective film to prevent steel corrosion. However, the ortho-para phenolic hydroxyl group in the phenolic hydroxyl structure of plant phenols is easy to be oxidized to quinone structure, resulting in the decline of its coordination ability [1]. Based on the characteristics of phenolic compounds forming blue-violet to brown complexes with iron ions, the color of iron ions themselves is light yellow. When iron ions are gradually added, the color of the mixed solution system gradually darkens (the absorbance increases). When the ion reaches the saturation coordination, its increase will be significantly weaker, that is, an inflection point is generated [6-8]. From this inflection point, the amount of coordinated phenolic hydroxyl provided by plant phenols can be calculated according to the conventional coordination form of phenolic hydroxyl and trivalent iron ion 6:1. Therefore, the absorbance change of the complex formed with iron salt is used to evaluate the effective phenolic hydroxyl content in plant phenolic materials, and as an index to evaluate their quality.

2 Experimental

2.1 Materials

$FeCl_3$ was purchased from Tianjin Komiou Chemical Reagent Co., Ltd. Phenol was supplied from Zhengzhou Zhongtian Experimental Instrument Co., Ltd. Hydroquinone was purchased from Tianjin Third Chemical Reagent Factory. Phloroglucinol was purchased from Ouhai Chemical Reagent Factory in Wenzhou City, Zhejiang Province. Tannin was purchased from Yancheng Ourui International Trade Co., Ltd. Sodium Lignosulfonate was purchased from Beijing Jiahemu Technology Co., Ltd. Alkali lignin was purchased from Luohe East China Lignin Co., Ltd.

2.2 Method

In this experiment, for monophenols, first prepare 50ml of the same molar concentration of ferric chloride solution and polyphenols solution in a volumetric flask, and then take 10ml volumetric flasks, numbered 1# to 10#. Add 3ml phenol solution into 10 volumetric flasks in sequence, and add 0 ml, 0.3ml, 0.35ml, 0.4ml, 0.45ml, 0.5ml, 0.6ml, 0.75ml, 1ml and 1.5ml ferric chloride solution in sequence. Fix the volume with industrial ethanol and

 https://doi.org/10.1051/e3sconf/202132901009

measure its absorbance at 510nm (the maximum absorption wavelength of ferric iron). If the absorbance value is between 0.2-0.7, the measured data is valid, The best coordination ratio can be obtained by fitting two straight lines from the scatter trend diagram and finding the intersection point. For bisphenols and triphenols, using the same method, the volume of ferric chloride solution increases the corresponding multiple of phenolic hydroxyl group under other conditions. or gallic acid and tannic acid with known molecular weight, prepare a solution with appropriate molar concentration according to the previously known trend line of polyphenols, draw a similar trend line, and then calculate the best ratio of the substance. For sodium lignosulfonate and alkali lignin with unknown molecular weight, use the solution with the same mass concentration as ferric chloride and adopt the above method, Draw two straight lines to calculate the trivalent iron ion equivalent, that is, the mass of alkali lignin that can effectively coordinate with each gram of ferric chloride, which has achieved the purpose of quantitative analysis.

3 Results and discussion

3.1 Mechanism study

At the wavelength of 510 nm, different absorbance values were obtained after the color development of ferric chloride and fixed concentration phenol in different ratios. The molar ratio of ferric ion to phenolic compounds was taken as the X axis and the corresponding absorbance was taken as the Y axis. In certain range, iron ions coordinate with phenolic hydroxyl groups of phenols, the absorbance increases with the increase of iron ion concentration, and the color of the solution system gradually deepens [9]. After exceeding a certain concentration, because there is no coordinated phenolic hydroxyl group, continue to increase the concentration of iron ion, the increase range of absorbance of the system becomes weak, and only the color of iron ion itself is increased [10]. Calculate the linear relationship between the two intervals. The X value at the intersection is the best coordination molar ratio between phenols and ferric ions, and the concentration of iron ions (C_{Fe}, mol/L) can be found here. The ratio of the concentration to the phenol concentration (Cp, g/L) is multiplied by the coefficient of iron ion saturation coordination 6, to obtain the coordination index M, $M=(C_{Fe}/C_p)\times 6$ (mol/g). In this experiment, products with high purity, such as chemical pure and analytical pure products, are used as control experiments to obtain their coordination index and compare it with that of industrial products. The closer they are, the higher the quality is, so their quality can be evaluated [11].

3.2 Coordination of monophenol with ferric chloride

The color development of phenol and ferric ions is shown in Figure 1. It can be seen from the figure that the intersection value is X=0.19, Y=0.49, that is the molar ratio of ferric chloride to phenol is 0.19, and the same molar ratio of phenol to ferric chloride is 5.4. In a word, when the molar ratio to ferric chloride is 5.4, it is the effective coordination ratio.

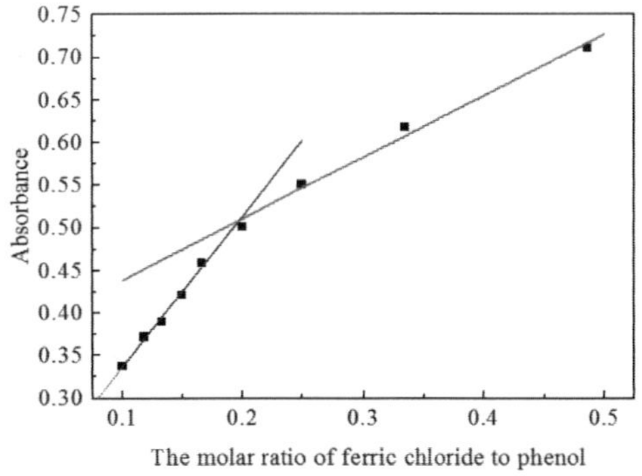

Fig. 1 Relationship between molar ratio of ferric chloride to phenol and absorbance

3.3 Coordination of bisphenol with ferric ion

The color development of hydroquinone and iron ion is shown in Figure 2. It can be seen that the intersection value is x = 0.36 and y = 0.41, the molar ratio of ferric chloride to hydroquinone is 0.36 and the molar ratio of hydroquinone to ferric chloride is 2.8. In other words, when the molar ratio of hydroquinone to ferric chloride is 2.8, a saturated coordination ratio is formed, indicating that it is difficult for both hydroxyl groups in the molecule to participate in coordination.

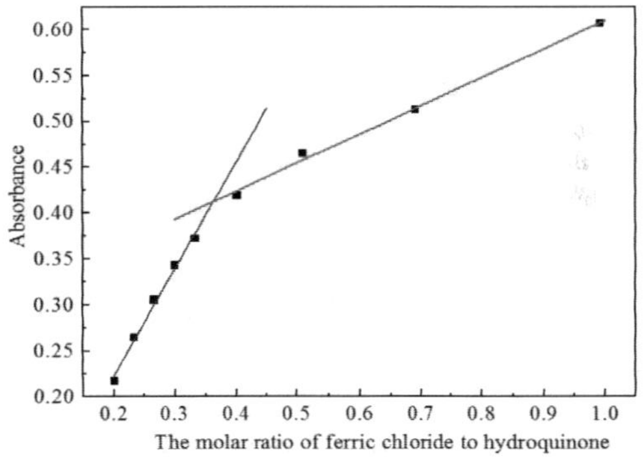

Fig. 2 Relationship between molar ratio of ferric chloride to hydroquinone and absorbance

3.4 Coordination of triphenol with ferric ion

The color development of phloroglucinol and iron ion is shown in Figure 3. The intersection point can be seen from the figure: x = 0.54, y = 0.36, that is when the molar ratio of ferric chloride to phloroglucinol is 0.54 or the molar ratio of phloroglucinol to ferric chloride is 1.8, that is, when the molar ratio of phloroglucinol to ferric chloride is 1.846, an effective coordination ratio is formed and close to the theoretical value of 1:2, indicating that the three hydroxyl groups in the molecule are difficult to participate in coordination.

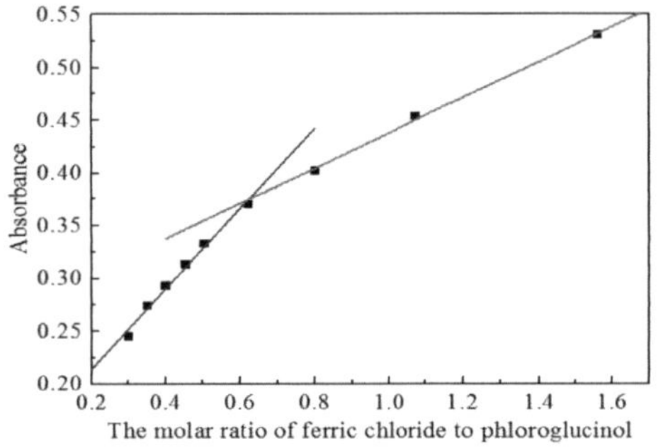

Fig. 3 Relationship between molar ratio of ferric chloride to phloroglucinol and absorbance

3.5 Evaluation of coordination ability of polyphenol materials with unknown quality

Based on the above research, industrial tannic acid is used for coordination reaction with iron ions. From Fig. 4, it can be seen that the coordinates of the intersection point are: x=2.23, y=0.32, that is the molar ratio of ferric chloride to tannic acid is 2.23. The value shows that 6 mol of phenolic hydroxyl and 1 mol of ferric chloride are effectively coordinated. Therefore, the phenolic hydroxyl group that can effectively coordinate with ferric ion in 1mol tannic acid is $2.23 \times 6 = 13.4$mol.

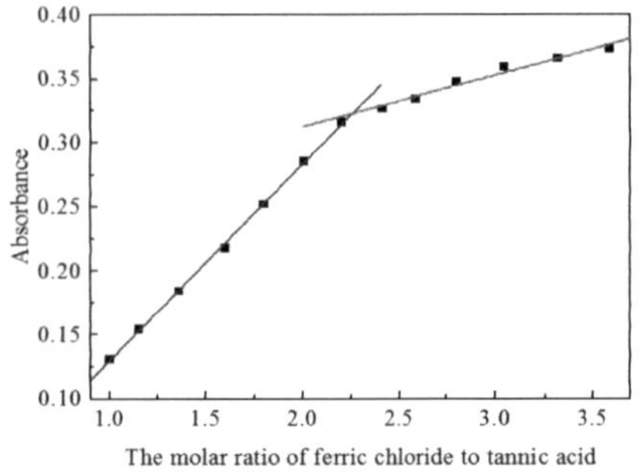

Fig. 4 Relationship between molar ratio of ferric chloride to tannic acid and absorbance

Industrial sodium lignosulfonate is used to coordinate with iron ions. The color development results are shown in Figure 5. The intersection coordinates are x = 0.09, y = 0.47, that is the mass ratio of ferric chloride to sodium lignosulfonate is 0.09, and the mass ratio of sodium lignosulfonate to ferric chloride is 10.8, which shows that 1 g of ferric trichloride can effectively coordinate with 10.8 g of sodium lignosulfonate.

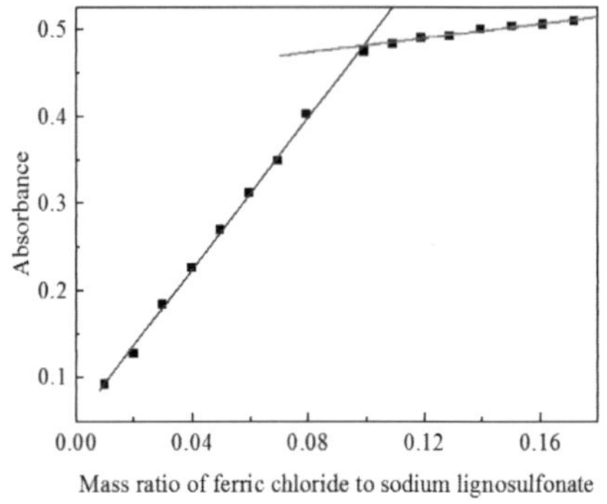

Fig. 5 Relationship between mass ratio and absorbance of ferric chloride and sodium lignosulfonate

Industrial alkali lignin and iron ions are used for color development, as shown in Figure 6. The coordinates of the intersection point are x=0.06, y=0.52, that is, the mass ratio of ferric chloride to alkali lignin is 0.06, and the mass ratio of alkali lignin to ferric chloride is 15.8, then 1 g of ferric chloride can be compared with 15.8 g of alkali lignin is effectively coordinated.

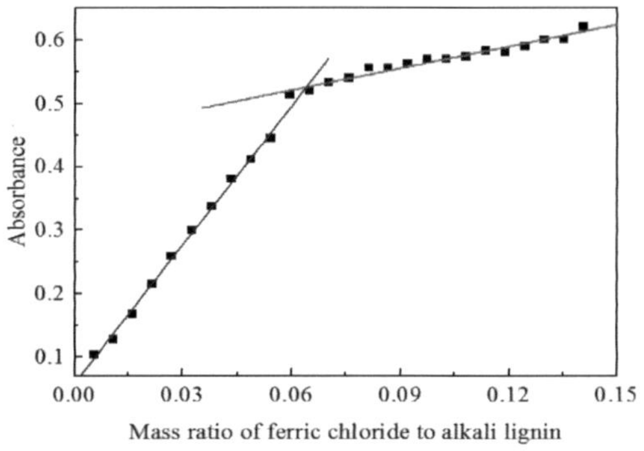

Fig. 6 Relationship between mass ratio and absorbance of ferric chloride and alkali lignin

4 Conclusion

On the one hand, through the study of the coordination ratio of monophenols, bisphenols and triphenols to ferric chloride, it is found that the effective coordination numbers of these three phenolic compounds are close to the theoretical value, but they are less than the theoretical value, on the other hand, the method can be used to quantitatively analyze the coordination ability of industrial tannic acid, sodium lignosulfonate and alkali lignin with iron ions, which provides a simple and reliable method for the quality analysis of industrial polyphenols.

Acknowledgments

The work was supported financially by Shaanxi Provincial Key Research and Development Program (2019GY-136), Key Scientific Research Program of Shaanxi Provincial Department of Education (21JY035)

https://doi.org/10.1051/e3sconf/202132901009

and Youth Innovation Team of Shaanxi University. And we thank the work of Modern Analysis and Testing Center of Xi`an Shiyou University.

References

1. Qi Wang, Enqi Liu. Research status of plant polyphenols[J]. *Journal of Shanxi Agricultural Sciences,* 2009, 37(1): 92-94.
2. Zhe Yu, Zhexiong Jin. Application and research progress of plant tannins[J]. Heilongjiang Medicine Journal, 2015, 28(1): 20-23.
3. Gang Chen, Peng Wang, Weichao Du, Jie Zhang. Progress in study and application of waste mud disposal technologies[J]. *Drilling Fluid & Completion Fluid,* 2020,37(1): 1-8.
4. Jie Zhang, Yongfei Li, Yuntian Jing, et al. Research on Influencing Factors of Clean Treatment of Waste Drilling Fluid in Chad[J]. *Advances in Fine Petrochemicals,* 2018, 19(1): 1-4.
5. Jie Zhang, Fan Zhang,Weichao Du, Gang Chen. Study on the preparation of clean fracturing fluid flowback fluid with strong inhibitory drilling fluid[J]. *Complex Hydrocarbon Rsserviors,* 2019, 12(2): 67-72.
6. Aiyong Qiu, Junhai Liu, Haijun Zhang. Extraction and application of plant polyphenols[J]. *Cereals & Oils,* 2003, (6): 10-11.
7. Jianlong Dong, Zili Wang, Yuanyuan Mao,et al. Research on detection and removal methods of sulfide in oilfield wastewater[J]. *Environmental Protection and Technology,* 2018, 24(3): 1-5.
8. Fangling Qin, Yue Fan, Qian Wang, et al. Study on the conditions of UV/H_2O_2/ferric oxalate complex system for the treatment of glycan lignin drilling waste fluid[J]. *Journal of Northwest Uniersity (Natural Science Edition),* 2015, 45(6): 893-898.
9. Gang Chen, Min Pang, Yunjuan Mo, et al. Study on Application of pomegranate peel extract in oilfield water treatment[J]. Chemical Engineering of Oil & Gas, 2013, 42(1): 83-85.
10. Jie Zhang, Li Liu, Li Zhang, Gang Chen. Study on electrical conductivity characteristics of plant phenol drilling fluid treatment agent[J]. *Complex Hydrocarbon Reserviors,* 2013, 6(4): 1-3.
11. Xiongxiong Liu, Jie Zhang, Weichao Du, Gang Chen. Research on quantitative analysis method of industrial lignin[J]. Technology & Development of Chemical Industry, 2019, 48(8): 44-46.

Preparation and performance evaluation of polyvinyl alcohol grafted polyacrylic acid for plugging

Jiangbo Liu [1, 2], Yi Ren [3], Qin Zhang [4], Xiangyun Wang [5], Weichao Du [5, *]

[1] Oil & Gas Technology Research Institute Changqing Oilfield Company, X i'an, Chian
[2] National Engineering Laboratory of Tight Oil & Gas Field Exploration and Development, Xi'an, Chian
[3] Nanpu Operating Area, Jidong Oilfield, Tangshan, Hebei, China
[4] Changqing Drilling Branch, CNPC Chuanqing Drilling Engineering Company Limited, Xi'an, Chian
[5] Xi'an Key Laboratory of Tight oil (Shale oil) Development (Xi'an Shiyou University), Xi'an Shiyou University, Xi'an, 710065, China

Abstract. Well-leakage is a very difficult problem in the process of drilling at home and abroad. At present, it mainly relies on the cumulative effect of plugging slurry to implement plugging regardless of the number of times and costs. Commonly used plugging slurry and plugging technology are difficult to guarantee the success rate and scientific of plugging malignant leakage, which has an extremely bad impact on the development of the petroleum industry. In this study, With polyvinyl alcohol and acrylic acid as main agents, a plugging material of acrylic acid grafted polyvinyl alcohol was prepared, and the influence of the composition of the reaction system on the gel state of the system was studied. Through a series of single-factor experiments, the best formula for preparing the plugging material was determined, polyvinyl alcohol and acrylic acid (specific gravity 1:3) + crosslinking agent (0.1wt%) + initiator (0.3wt%). The gel plugging agent using above formula to prepare has good gel strength and suitable gel time, and has a good application prospect in solving the problem of lost circulation.

1 Introduction

During drilling operations, when the pressure of the drilling fluid column exceeds the formation pressure and the fractured formation is encountered, the drilling fluid can easily enter the pores and fractures of the formation, leading to the occurrence of lost circulation accidents [1-3]. As a more complicated accident in petroleum exploration and development, it needs to spend a lot of resources to deal with it. If it is not handled properly, it will lead to vicious leakage incidents and cause great environmental pollution problems. Adding plugging materials is a common treatment method in leakage accidents [4]. The plugging materials will have physical and chemical effects under the conditions provided by the formation, forming a network space structure, and forming a sealing layer at the formation cracks, thereby controlling leakage and ensuring stability Well wall [5]. The plugging materials include conventional plugging materials such as walnut shells, and chemical plugging materials that use the static force of the polymer at the interface and the intermolecular forces to bond the compounds to achieve the purpose of plugging [6]. In petroleum exploitation, Chemical gel is often used as a plugging material. When the plugging agent enters the formation, under the action of a certain temperature and pressure, various components interact [7]. Retention, accumulation, bridging, and cracking occur Filling and reinforcement, thereby improving the cementing and sealing pressure-bearing capacity of the gel in the leakage formation, achieving the purpose of sealing the leakage layer [8]. Among them, polymer gel plugging agent is an important branch of chemical gel plugging agent, compared with general plugging materials. The advantages are obvious, with: (1) low density is beneficial to the protection of oil and gas reservoirs; (2) gelling time is controllable, and the controllable time range is large, which is convenient for it to accurately reach the missing layer for plugging; (3) plugging The leaking slurry has strong fluid loss performance, and the formation of filter cakes can help stabilize the well wall; (4) It can be used in various applications and can block micro-fractures and relatively large fractures; (5) The gel plugging agent has good compatibility with the formation [9-11]. It will not affect the formation environment. And for large fractures in complex formations, traditional materials are prone to false plugging, and when polymer gel plugging agent is used to plug the leaky layer, polymer molecules and water When the molecules are in contact, the internal chemical bonds will be connected to each other [12]. A gel with a special three-dimensional network structure and high viscoelasticity is formed, and the number and types of intermolecular chemical bonds determine the shape and performance of the polymer spatial structure.

* Corresponding author: duweichao@xsyu.edu.cn

2 Experimental

2.1 Materials

Polyvinylamide was purchased from Shandong Yuhuang New Energy Technology Co., Ltd. Polyvinyl alcohol was purchased from Tianjin Zhiyuan Chemical Reagent Co., Ltd. Acrylic acid was purchased from Luoyang Haohua Chemical Reagent Co., Ltd. N. N-methylene bisacrylamide was purchased from Tai'an Changxing Engineering Materials Co., Ltd.

2.2 Gel Preparation

Preparation of 10% polyvinyl alcohol solution: 15g polyvinyl alcohol and 150ml distilled water were added into a three neck flask, which was put into a water bath pot and stirred at 95 °C until polyvinyl alcohol was completely dissolved. A certain amount of acrylic acid was added into the beaker, and 40% sodium hydroxide solution was slowly added in the cold-water bath to neutralize the acrylic acid, so that the neutralization degree of acrylic acid reaches the preset value. An appropriate amount of polyvinyl alcohol solution was added to the neutralized acrylic acid solution, and then the crosslinker N, N-methylene bisacrylamide, initiator ammonium persulfate and sodium sulfite were added to the mixed solution in turn. After stirring the mixed solution evenly, it was put into a water bath for heating, and the gelation of the reaction was observed.

3 Results and Discussion

3.1 Effect of Polyvinyl alcohol concentration on gel-forming condition

The composition of the gelling system is as follows: the concentration of polyvinyl alcohol was 7%, 8%, 10%, 11%, 12% and 13% respectively, the mass ratio of polyvinyl alcohol to acrylic acid was 1:3, the neutralization degree of acrylic acid was 60%, the dosage of initiator (ammonium persulfate and sodium sulfite) was 0.3% of the mass sum of polyvinyl alcohol and acrylic acid, and the dosage of the crosslinker (N,N-methylene bisacrylamide) was 0.1% of the mass sum of polyvinyl alcohol and acrylic acid, and the reaction temperature was 100 °C. The gel-forming conditions of different systems were shown in Table 1.

Table 1 The gel-forming conditions of system with different Polyvinyl alcohol concentration

Polyvinyl alcohol concentration	Gelation time (h)	Gel strength
7%	7	I
8%	7	I
10%	6.5	I
11%	4.5	I
12%	5.5	I
13%	5.5	I

From Table 1, the gelation time of each system was 7 h, 7 h, 6.5 h, 4.5 h, 5.5 h and 5.5 h respectively. When the concentration of polyvinyl alcohol in the reaction system was no more than 10%, the gelation time of the reaction system was relatively short. However, when the concentration of polyvinyl alcohol increases in the reaction system, the gelation time of the reaction system gradually decreased. The reason was that when the dosage of polyvinyl alcohol in the reaction system was small, the contact probability between polyvinyl alcohol and the active radical produced by the initiator was small, which made the reaction slower and the gelation time of the reaction system longer. When the dosage of polyvinyl alcohol was large, the chance of collision between polyvinyl alcohol molecules and active radicals was significantly improved, which speeded up the reaction rate of the reaction system. When the concentration of polyvinyl alcohol was higher than 12%, the gelation time of the gel system was no longer affected by the amount of polyvinyl alcohol. The gel strength of each reaction system reached grade I, and 11% polyvinyl alcohol concentration was selected as the best concentration for subsequent experiments.

3.2 Effect of initiator on gel-forming condition

The composition of the gelling system was as follows: the concentration of polyvinyl alcohol was 11%, , the mass ratio of polyvinyl alcohol to acrylic acid was 1:3, the neutralization degree of acrylic acid was 60%, and the dosage of the crosslinker (N,N-methylene bisacrylamide) was 0.1% of the mass sum of polyvinyl alcohol and acrylic acid, the initiator systems were ammonium persulfate (oxidation), sodium sulfite (reduction), ammonium persulfate-sodium sulfite (oxidation-reduction), the reaction temperature was 100 °C. The gel-forming conditions of different systems are shown in Table 2.

Table 2 The gel-forming conditions of system with different initiator systems

initiator system	Gelation time (h)	Gel strength
ammonium persulfate	4	I
sodium sulfite	Not gelled	D
ammonium persulfate-sodium sulfite	2	I

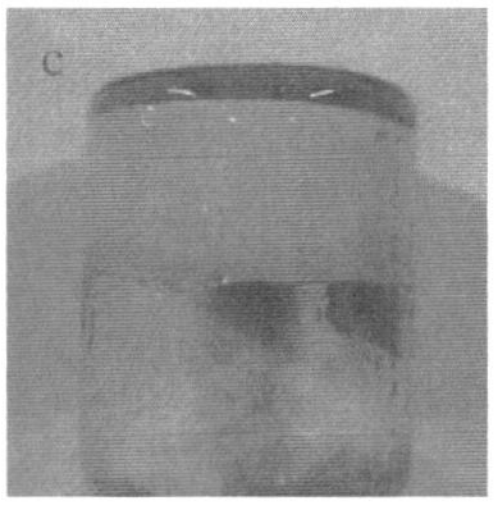

Figure 1 The gel-forming morphology of systems with initiator systems: (a)ammonium persulfate, (b)sodium sulfite, (c)ammonium persulfate-sodium sulfite

Ammonium persulfate ($((NH_4)_2SO_4)$) is an inorganic oxidizing initiator, which is used for aqueous solution polymerization and emulsion polymerization. When the temperature is higher than 60 ℃, ammonium persulfate may decompose effectively, and the decomposition product is free radical, which can effectively initiate the polymerization reaction. Sodium sulfite (Na_2SO_3) is usually used as a reducing initiator. According to the experimental data in Table 2, when ammonium persulfate was used as initiator, the gelation time of the reaction system was longer, which was 4 hours, and the strength reached grade I. When the initiator was sodium sulfite, the reaction system was basically not gelled. When ammonium persulfate sodium sulfite was used as initiator, the gelation time of the reaction system was 2 hours, the gelling time was appropriate and the strength was excellent. Therefore, ammonium persulfate- sodium sulfite was used as initiator for subsequent experiments.

3.3 Effect of initiator concentration on gel-forming condition

The composition of the gelling system was as follows: the concentration of polyvinyl alcohol was 11%, , the mass ratio of polyvinyl alcohol to acrylic acid was 1:3, the neutralization degree of acrylic acid was 60%, the dosage of the crosslinker (N,N-methylene bisacrylamide) was 0.1% of the mass sum of polyvinyl alcohol and acrylic acid, and the dosage of the initiator (ammonium persulfate-sodium sulfite) was 0.1%, 0.3%, 0.4%, 0.6% of the mass sum of polyvinyl alcohol and acrylic acid, the reaction temperature was 100 ℃. The gel-forming conditions of different systems are shown in Table 3.

Table 3 The gel-forming conditions of system with different initiator concentration

initiator concentration (%)	Gelation time (h)	Gel strength
0.1	1.5	I
0.3	1.0	I
0.4	1.0	I
0.6	1.0	I

Figure 2 The gel-forming morphology of systems with different initiator concentration: (a)0.1%, (b)0.3%, (c)0.4%, (d)0.6%.

It can be seen from Table 3 that when the amount of initiator was 0.6% of the mass sum of the two monomers, the crosslinking reaction rate was too fast and the gelation time was too short. When the amount of initiator was 0.1%, 0.3% and 0.4% of the mass sum of polyvinyl alcohol and acrylic acid, the gelation time of the reaction system was 1 hour. The reason was that when the amount of initiator was relatively small, there were relatively few active free radicals produced by the initiator in the reaction system, so that the polymerization rate was slow. As shown in Figure 2.2, compared with the reaction system with other concentrations of initiator, the system with initiator concentration of 0.3% of the sum of the mass of two monomers had the greatest gel strength. The reason was that the concentration of living radicals in the reaction system affected the crosslinking density and affect the gel strength.

3.4 Effect of neutralization degree of acrylic acid on gel-forming condition

The composition of the gelling system was as follows: the concentration of polyvinyl alcohol is 11%, , the mass ratio of polyvinyl alcohol to acrylic acid is 1:3, the neutralization degree of acrylic acid is 0, 60%, 92.5%, 97.5%, the dosage of the crosslinker (N,N-methylene bisacrylamide) is 0.1% of the mass sum of polyvinyl alcohol and acrylic acid, and the dosage of the initiator (ammonium persulfate-sodium sulfite) is 0.3% of the mass sum of polyvinyl alcohol and acrylic acid, the reaction temperature is 100 ℃ . The gel-forming conditions of different systems were shown in Table 4. According to the table, when the neutralization degree of acrylic acid was 0%, the gelling time of the reaction system was too fast. When the neutralization degree was 60%, when the neutralization degree of acrylic acid is 92.5% and 97.5%, the gelation time of the reaction system was long, and the gel strength reached I grade. The main reason is that when the neutralization degree of acrylic acid was relatively small, the concentration of acrylic acid in the reaction system is high, and the internal reaction of acrylic acid was easy to self-react, resulting in the acceleration of chemical reaction and the final gelation time of the gel system. When the neutralization degree of acrylic acid was large, the concentration of -COONa generated by neutralization reaction between acrylic acid and sodium hydroxide was too high, which had an effect on the rate of chemical reaction and made the system of gelation time too long. Therefore, 60% is the best neutralization degree of crosslinking reaction.

Table 4 The gel-forming conditions of system with different neutralization degree of acrylic acid

Neutralization degree (%)	Gelation time (h)	Gel strength
0	0.5	I
60.0	1.0	I
92.5	5.0	I
97.5	Not gelled	C

Figure 3 The gel-forming morphology of systems with different neutralization degree of acrylic acid: (a) 0, (b) 60%, (c) 92.5%, (d) 97.5%.

4 Conclusion

In this paper, a resin plugging agent was synthesized by solution polymerization with acrylic acid and polyvinyl alcohol as polymerization monomers, ammonium persulfate and sodium sulfite as redox initiator and N, N-methylene bisacrylamide as crosslinking agent. At the same time, the experimental conditions were optimized by controlling variable method. The synthesis process was as follows: the specific gravity of acrylic acid and polyvinyl alcohol was 4:1, the dosage of initiator was 0.3% of the mass sum of acrylic acid and polyvinyl alcohol, the dosage of crosslinking agent was 0.1% of the mass sum of acrylic acid and polyvinyl alcohol, the neutralization degree of AA was 60%, and the water bath was set at 70 ℃. The plugging agent prepared by the above experimental conditions has excellent gel strength and suitable gelation time, and has a good application prospect in solving the well leakage problem.

Acknowledgements

The work was supported financially by Shaanxi Provincial Key Research and Development Program (2019GY-136) and Youth Innovation Team of Shaanxi University. And we thank the work of Modern Analysis and Testing Center of Xi`an Shiyou University.

References

1. Y. S. Wang, H. Zhang, J. F. Fu. New progress in chemical plugging technology [J]. Science and Technology Vision, 2015, 1(1): 9-10.
2. J. W. Fang, Z. K. Lu, Z. He. Application of chemical gel plugging agent pressure-bearing plugging technology in Shunbei 3 well [J]. Drilling Fluid and Completion Fluid, 2017, 34 (6): 13-17.
3. X. C. Cao, B. R. Li, Y. Qin. Research and application of polymer gel plugging agent [J]. Contemporary Chemical Industry, 2015, 44(11): 65-67 .
4. X. P. Wang, Y. X. Zheng, C. X. Zhang. Preparation of P(AA-AMPS) water-absorbent resin and determination of salt resistance in high temperature environment [J]. New Chemical Materials, 2019, 47(3): 185-188.
5. X. R. Li, Y. Zhang, N. Yang. Performance evaluation of pre-crosslinked gel microsphere plugging agent [J]. Oilfield Chemistry, 2016, 33(4): 581-584.
6. J. Y. Mou, J. J. Shao, B. Y. Lu. Plugging performance of acid fracturing temporary plugging agent for fractured-vuggy reservoirs [J]. China Science and Technology Papers, 2016, 11(3): 258-262
7. Y. Y. Liu, Y. H. Zhang, H. Zhang. Experimental study and field application of high-strength fiber composite temporary plugging agent [J]. Petrochemical Technology, 2017, 24(8): 101-104.
8. Wilson A. Thermally stable fluid-loss pill performs better than guar-base gels [J]. Journal of Petroleum Technology, 2013, 65(11): 134-136.
9. P. Gamage, J. P. Deville, J. Shemian. solids-free fluid-loss pill for high-temperature reservoirs [J]. SPE Drilling & Completion, 2014, 29(1): 125-130.
10. C. L. Hu, P. X. He, G. C. Jiang. Self-degradable leak prevention and plugging system for workover operations [J]. Acta Petrolei Sinica, 2016, 37(2): 150-157.
11. R. J. Liu, D. Lu, D. Wang. Research progress of in-situ intercalation polymerization to prepare

nanocomposites [J]. Shandong Chemical Gong, 2016, 45(15): 52-54.

12. H. Jia, H. Chen, B. Chen. Leak prevention mechanism and application of elastic liquid rubber plug for well repair [J]. Acta Petrolei Sinica, 2018, 39(3): 349-356.

https://doi.org/10.1051/e3sconf/202132901011

Research on the non-point source pollution characteristics and water environmental capacity within small river basin of agricultural areas a case study of the small watershed near Gan village, Tanxia town

Liya Yang [1,2,3,*], Hao Zhou [1], Liyuan Dai [1,3]

[1] Wanjiang University of Technology, Maanshan, 243031, China
[2] Ma'anshan Engineering Technology Research Center of Engineering Monitoring and Safety Monitoring China
[3] Maanshan Engineering Technology Research Center for Water Resources Efficient Utilization in Hilly Region China

Abstract. A small watershed of a tributary of the Gantang River was chosen as the study object. The water environment capacity and the amount of non-point source pollutants entering the river were calculated by combining field measurements and survey data. The results reveal that the water quality in the basin exceeds the standard severely, with 23.12t of TN, 1.64t of TP, 58t of CODMn, and 12.05t of NH3-N exceed during the plentiful period; The watershed exceeded the TN standard by 14.3t, TP by 1.64t with 17.489t of CODMn remaining in water environmental capacity and 0.71t of NH3-N exceed during the drought period. Finally, suggestions for the prevention and control of the water environment in the watershed were proposed in response to the pollution characteristics, which are of great significance for improving the water environment of the agricultural area.

Keywords: Non-point source pollution, water environmental capacity, water quality analysis.

1 Introduction

Non-point source pollution is the main cause of surface water pollution, with agricultural non-point source pollution contributing the most. Due to its large spatial and temporal variation, the uncertainty and randomness of pollutants and emission pathways, and the complexity of the composition and process, it has become a major source of pollution to the surface water environment. Non-point source pollution is well recognized by humans, but not well researched, and even less quantitative study. For the quantitative research need in order to evaluate its impact, the current research is dominated by non-point source pollution, while water environment capacity research of rivers is also relatively small. In this research, a small river basin in a tributary of the Gantang River in Tanxia Town, Lingchuan County, is selected as the research object. Based on the actual local situation and the characteristics of agricultural non-point source pollution, the main pollutants are identified as TN, TP, NH_3-N and COD_{Mn}, and combined with the water quality monitoring results, suitable parameters are selected for the calculation of the water environment capacity, the amount of pollutants entering the river, as well as the achievement of control targets and the amount of pollutants to be reduced, is intended to provide a method and reference basis for the management of non-point sources and water pollution control.

2 Overview of the research watershed

Tanxia Town belongs to the northwestern part of Lingchuan County, Guilin City, Guangxi Zhuang Autonomous Region, and the watershed within the area is a tributary of the Gantang River. The study of small watershed is located near Gan Village, Tanxia Town, with a river length of about 8 km and a total watershed area of about 12 km^2. the soil in the research area has a low organic matter content, contains less nitrogen, phosphorus and potassium, and has low fertility. The vegetation cover in the area is well established, reaching over 70%. The watershed is located in the central subtropical monsoon climate zone, with an average annual rainfall of 2011.1 mm, with rainfall mainly concentrated from March to August and the most rainfall for the year in May. November to January is the drought period and April to June is the plentiful period.

The watershed is a typical agricultural area with no industrial pollution sources or large-scale farms. Local agriculture is mainly based on citrus and rice cultivation, and farming is dominated by swine and poultry. Agricultural surface sources are the main source of pollution in the river. Two monitoring sections were set up in the upper and lower

* Corresponding author: 527615613@qq.com

reaches of the river basin, 1300m apart, with no tributaries flowing between the two sections. Based on the characteristics of agricultural surface pollution, four monitoring items were identified, including TN, TP, COD_{Mn} and NH_3-N. Six months of field sampling and monitoring were carried out in accordance with the monitoring technical standards of the "Analytical Methods for Water and Wastewater Monitoring". The monitoring results showed that the concentrations of TN and COD_{Mn} in the water body of this river section far exceeded the pollution concentration values for Class III water in the "Environmental Quality Standards for Surface Water". The results of the experiment monitoring are shown in Table 1.

Table 1 Tanxia town watershed monitoring results (mg/L)

	Time	TN	TP	COD_{Mn}	NH_3-N
Upstream monitoring cross-sections	Plentiful period	4.58	0.1	7	0.14
	Drought period	5.93	0.13	15.8	0.82
Monthly average pollutant levels	Plentiful period	4.39	0.09	7.3	0.13
	Drought period	5.88	0.12	17.3	0.99

3 Computational model of the water environment capacity

3.1 Watershed monitoring objectives

Through the field monitoring results, it could be seen that the pollutants in the basin are mainly TN, TP, COD_{Mn} and NH_3-N. To make the total water environment in the basin effectively controlled, then the amount of pollutants entering the river must be controlled. In this paper, TN, TP, COD_{Mn} and NH_3-N will be selected as the calculation factors of the water environment capacity from the problem of total pollutant control. Based on the field measurement data, and the surrounding monitoring information and field survey results, the water environment capacity of this small watershed is calculated.

3.2 Computational model

The calculation of the water environment capacity requires consideration of the dilution effect of the water body and the biochemical decontamination effect. Only the impact of non-point source pollution on the environmental capacity is considered in the river. For rivers with predominantly non-point source pollution, it could be assumed that pollutants enter the river in a uniform form along the river, then the river water environmental computational model can be expressed as:

$$W = 365 \times 86.4 \times Q_0 \times \left[C_S \times \exp\left(\frac{KL}{86400u} \right) - C_0 \right]$$

Where: W is the water environmental capacity of the river section; Q0 is the average flow rate of the river section; CS is water quality standard concentration in the basin; C0 is upstream incoming pollutant concentration; K is overall attenuation coefficient; u is average flow rate of the river section; L is length of the river section. Stream flow and cross-sections are based on measured data. The cross-sectional flow rate in the basin during the drought period is 0.01m/s; the cross-sectional flow rate in the basin during the plentiful period is 0.08m/s.

2.2.1 Overall attenuation coefficient

A section of the river basin with a length of 400m, a straight channel, stable water flow, no tributaries in the middle and no outfalls was selected and sampling points were placed upstream and downstream to monitor pollutant concentrations.

$$K = 86400 \times \ln \frac{C_1}{C_2} \times \frac{v}{l}$$

Where: K is the pollutant attenuation coefficient; C_1 and C_2 are the pollutant concentrations at the upstream and downstream sections of the river; l is the distance between the upstream and downstream sections; and v is the average velocity of the water. Due to the small amount of monitoring data and large errors, the pollution attenuation coefficient was combined with the actual local conditions and empirical formulas, and the final calculation of the overall attenuation coefficient is shown in Table 2

Table 2 Overall attenuation coefficient

K	TN	TP	COD_{Mn}	NH_3-N
Plentiful period	0.732	1.821	-0.725	1.281
Drought period	0.018	0.173	-0.196	-0.407

3.3 Sewage discharge quantity

The people income of Tanxia town is mainly agriculture, forestry, livestock farming and fishing, etc. There are no industrial factories in the area. The survey mainly includes the agricultural population, the livestock breeding quantity, agricultural fertilizers and pesticides quantity. In this paper, we will calculate the quantity of pollutants entering the river from three aspects: pollution from agricultural life, livestock breeding and fertilizers and pesticides on farmland.

The questionnaire investigation of the local population and the monitoring data of the water quality around the toilets in the watershed, in which the content of pollutants contained in human excretion of faeces and urine is used to determine the discharge coefficients through the regulations of the National Health Bureau study, and the visit to seven survey points in the study area. The survey shows that the actual population living in the area is about 3,600 people, the cultivation area is about 12,000 mu, about 400 hogs are kept, and the total number of chickens and ducks is about 200. Referring to the calculation method in the 《Fertiliser Manual》 and combining the results of the Guilin part of the

2016 National Pollution Source Census, the regional pollutant emissions are shown in Table 3 below.

Table 3 Agricultural domestic pollution emission statistics

Item	TN	TP	COD_{Mn}	NH_3-N
Agriculatural domestic sewage entering the river quantity (g/a)	2102.4	578.16	78840	9198
pesticides and chemical fertilizers entering the river quantity (kg/a)	8280	432	36000	7200
livestock manure entering the river quantity (g/a)	5756	1573	33530	2358
pollutants entering the river quantity (t/a)	11.15	1.22	77.02	11.42

3.4 Calculation of water environment capacity

The estimation of the water environmental capacity took into account the amount of non-point source pollution entering the river along the river as the background value of pollutants in the water bodies of the small watershed, reflecting only the impact of non-point source pollution on the water bodies of the watershed, and the results of calculating the amount of water environmental capacity entering the river in this section and the remaining water environmental capacity are shown in Table 4.

Table 4 Water environment capacity for various pollutants

Time		TN	TP	COD_{Mn}	NH_3-N
Plentiful period	Water environment capacity	-11.972	-0.630	18.902	-0.630
	Remian	-23.122	-1.847	-58.118	-12.050
Drought period	Water environment capacity	-3.150	-0.630	94.509	10.710
	Remian	-14.300	-1.847	17.489	-0.710

4 Conclusions and recommendations

4.1 Conclusion

Through the study of the characteristics of the water environment capacity of the small watershed near Gancun in Tanxia Town during the plentiful period and the drought water period, the monitoring results show that the pollutant TN and COD_{Mn} concentrations are higher than the water quality standard of category 3. According to the calculation results of the water environment capacity, the water environment capacity of the small watershed near Gancun is almost always exceeded throughout the year, with an excess of about 23 tonnes of TN, 1.64 tonnes of TP, 58 tonnes of COD_{Mn} and 12 tonnes of NH_3-N during the plentiful period, and an excess of about 14 tonnes of TN, 1.64 tonnes of TP, 0.7 tonnes of NH_3-N and 58 tonnes of COD_{Mn} remaining during the drought water period.

From the above research content, it can be concluded that the water environment capacity of the basin shows a large difference in the remaining water environment capacity during the dry season and the wet season. Only the TP water environment capacity remains stable throughout the year. TN, COD_{Mn} and NH_3-N all have a higher residual capacity in the drought period than in the plentiful period. The area to which the basin belongs has higher summer temperatures, rainfall and amounts of pesticides and fertilisers required for crop growth, considering that the increase in pollutant concentrations should be significantly related to rainfall and temperature.

4.2 Recommendations

Rural agricultural non-point source pollution seriously restricts the development of the agricultural economy and the improvement of the local ecological environment. To eradicate environmental pollution problems, the starting point is to cut off the source of water environment pollution. The discharge of production and domestic sewage should be treated accordingly, first of all, sewage will be directly introduced into sewage treatment plants, rather than directly dumped such as rivers in farmland; the application of chemical fertilizers and pesticides in farmland should also be based on the actual local situation, the local soil survey sampling, according to local conditions, the choice of suitable fertilizer products, fixed-point quantitative application of fertilizers, rather than a single excessive application of fertilizers. Introduce advanced planting techniques, develop ecological agriculture, encourage farmers to use efficient and non-polluting green fertilisers and organic fertilisers, develop and promote new pesticides with high efficiency, low toxicity, low residue and easy decomposition; for pollution from livestock farming, it is recommended to realise the separation of humans and animals, implementing a pond and three changes, promoting four-in-one and other college agro-ecological models, harmless treatment and comprehensive use of livestock manure, developing clean farming. In addition, it is recommended to develop clean aquaculture, and attach importance to environmental protection requirements in areas such as circle structure, manure cleaning, and source matching.

Acknowledgments

This work was financially supported by scientific research projects of Maanshan Engineering Technology Research Center for Water Resources Efficient Utilization in Hilly

https://doi.org/10.1051/e3sconf/202132901011

Region (WREU202102) , scientific research projests of Ma'anshan Engineering Technology Research Center of Engineering Monitoring and Safety Monitoring (2020gcjc02) and The Program of Cultivating University Top-Notch Talents (gxyg2021244) fund.

References

1. Xiang Meng, Gu Shixiang, Gao Rong, Xie Chongbao, Cui Yuanlai. Investigation of farmland non-point Source and Countermeasures of Agricultural Water Saving and Emission reduction in Qilu Lake Basin [J]. China Rural Water Conservancy and Hydropower, 2021 (08): 127,132.
2. Hu Yawei, Ma Qiang, Zhai Laizheng, Jin Xiaohui. Calculation of agricultural non-point source pollution load in Yellow River diversion irrigation area based on SWAT [J/OL]. Ecological environment monitoring of the three Gorges: 1-22 [2021-10-03].
3. Zang Mengyuan, Li Ying. Study on load estimation and Control Countermeasures of Agricultural non-point Source pollution [J]. Shandong Agricultural Sciences, 2021, 53 (02): 142-147.
4. Gong Shifei, Ding Wuhan, Xiao Nengwu, Guo Yuanping, Ye Qingsong, Wang Wei, Li Hu. Characteristics of agricultural non-point source pollution in typical watershed of core water source area of Danjiangkou Reservoir [J]. Journal of Agricultural Environmental Science, 201951 38 (12): 2816-2825.
5. Sun Qian. Improvement and application of stochastic simulation method of groundwater flow based on T-PROGS [D]. China University of Geosciences (Beijing), 2019.
6. Wu Sheng, Zhang Junsen, Zhang Donghong, Guan Fei, Li Xiaogang, Ma Youhua. Research Progress on Evaluation and Comprehensive Control of Agricultural non-point Source pollution in small Watershed [J]. Environmental pollution and Prevention, 2018 Magi 40 (06): 710-716.
7. Ji Dingyu, Wang Qingan, she Hongying, Wang Dongrui, Zhang Xiang. Study on Agricultural non-point Source pollution characteristics and Water Environmental capacity of small Watershed in Hilly area of Central Sichuan-- A case study of Huanglaxi Watershed [J]. Journal of Water Resources and Water Engineering, 2011 Ji 22 (04): 81-84.

https://doi.org/10.1051/e3sconf/202132901012

Sub-23 nm Solid Particle Number Emission Characteristics for a Heavy-duty Engine Fuelled with Compression Natural Gas

Xiaowei Wang [1], Lin Zhang [1], Mingda Wang [2,*], Xiaojun Jing [1] and Xuejing Gu [2]

[1] CATARC Automotive Test Center (Tianjin) Co., Ltd., Tianjin 300300, China
[2] Chinese Research Academy of Environmental Sciences;State Environmental Protection Key Laboratory of Vehicle Emission Control and Simulation, Beijing 100012, China

Abstract: Cold and hot WHTC (World harmonized Transient-State Cycle) were separately run on the engine test bed for a heavy-duty natural gas engine fuelled with high calorific value natural gas and low calorific value natural gas. The particle emissions including PN10 (number of particles above 10nm) and PN23 (number of particles above 23nm) were measured. The results show that the transient emission of PN10 and PN23 have basically the same trends. The weighted specific emission of PN10 is 21.6 times of PN23. Cold start PN emissions account for a relatively large proportion. Fuel property has a great influence on the PN emissions of natural gas engines. The increase of carbon-containing fuels such as methane and ethane in the fuel gas will lead to a rapid increase in PN emissions especially in the cold start process. It is extremely important to strengthen the control of sub-23nm particulate emissions for heavy-duty natural gas engines.

Key words: Heavy duty natural gas engine, sub-23 nm particle, particle number, emission

1 Introduction

According to the "China Mobile Source Environmental Management Annual Report (2021)" released by the Ministry of Ecology and Environment [1], China's motor vehicle particulate matter emissions will reach 68,000 tons in 2020, of which diesel vehicle particulate matter emissions account for more than 90%. The emission of particulate matter is harmful to the environment and human health. The environmental hazard of particulate matter mainly stems from its optical properties. The particles absorb and scatter light to reduce visibility and cause haze weather [2]. Compared with atmospheric particulate matter with larger particle size, the particulate matter emitted by motor vehicles has a small particle size with a long residence time in the atmosphere and a long transportation distance. The specific surface area of particles is larger, and it is possible to adsorb more toxic and harmful substances. Fine particles have stronger penetrating power and can penetrate into the human respiratory system and even the brain tissue [3].
The regulation of "Limits and Measurement Methods for Pollutant Emissions from Heavy Duty Diesel Vehicles (China's Sixth Phase)" [4] not only sets limitation on particulate matter quality, but also requires to count the particle number with a particle size between 23nm-2.5μm (PN emission). The limitation of PN emission is 8×10^{11}#/kWh at the World Unified Steady State Cycle (WHSC) and 6×10^{11}#/kWh at the World Unified Transient Cycle (WHTC), respectively.
However, laboratory tests reveal that particles in sub-23nm region should not be neglected [5-9]. Studies have shown that 30%-50% of GDI (gasoline direct injection) engine particulate matter, and 50%-100% of PFI (port fuel injection) engine particulate matter, are not controlled under the current particle definition with a particle size between 23nm-2.5μm. Both European Union and China has set up working group to consider whether the emission regulations should be revised so that particles with smaller size can also be controlled and monitored .
The study on sub-23 nm particles is mainly concentrated on vehicles and engines fuelled with gasoline or diesel, but less attention to engine fuelled with compression natural gas. Due to the price advantage of natural gas, heavy-duty trucks equipped with gas engines are becoming increasingly popular. The statistics shows that the cumulative sales of natural gas heavy duty trucks from January to September 2020 is 116500 in China with a year-on-year increase of nearly 29%. Natural gas is considered to be a clean fuel. Reports have pointed out that the HC emission of natural gas vehicles can be reduced by 40%, the CO emission can be reduced by 50%, and there is no soot emission [10]. In fact, although the particle emission quality from natural gas engine is very low, its emissions contain a large number of small particulate matter, with a geometric average diameter of only 30nm [11], indicating that the PN emission

* Corresponding author: wangmingda@vecc.org.cn

https://doi.org/10.1051/e3sconf/202132901012

especially the fine particle need to be attached great importance. We have investigated the sub-23 nm solid particle number emission characteristics for a heavy-duty engine [12], so in this paper, we used a 10.4 L heavy-duty natural gas engine to run cold and hot WHTC on the engine test bed separately. The particle number emissions were measured by using a AVL particle counter which can simultaneously measure PN10 (number of particles with size above 10nm) and PN23 (number of particles with size above 23nm). The PN emission characteristics of heavy-duty natural gas engine with two kinds of natural gases were obtained and analyzed.

2 Experimental Setup

2.1 Test equipment

The tests were carried out on the engine test bed with a full-flow dilution sampling system (CVS). The particle number emission was measured by a particle counter developed by AVL company. The exhaust gas first goes through a volatile particle remover (VPR) to oxidize and evaporate hydrocarbons and store sulfides so as to minimize the impact on particle number counting accuracy. Then the particles in the exhaust is counting by light scattering.

The engine speed and torque were measured by AVL electric dynamometer. Other equipment used in this test includes air flow meter, intake air temperature conditioning, etc. The main test equipment is summarized in Table 1.

Table 1 Test equipment

Equipment name	Equipment Type and Manufacturer
AC Dynamometer	INDY S50-4/3001-1BV-1
Test bed control system	AVL PUMA Open V1.5.3
Intake air temperature conditioning	AVL Air Conditioning System 2400
Air flow meter	ABB SensyFlow P
Gaseous emission measurement	AVL AMA i60
Full-flow dilution sampling system	AVL CVS i60
Particle counter	AVL APC^{plus} ADVANCED with AVL CPC option for sub-23nm
Particle quality sampling equipment	AVL PSS i60

2.2 Tested engine

The tested engine was a heavy-duty natural gas engine with a displacement of 10.4 liter with a rated power of 235 kW and a maximum torque of 1450 Nm. The specific parameters of the engine can be seen in Table 2.

Table 2 Main parameters of tested engine

Parameter	Value
Engine capacity	10.4 L
Bore×Stroke	128 mm×135 mm
Compression ratio	12
Rated power/speed	235 kW/1900 rpm
Maximum torque/speed	1450 Nm/1200 rpm
Idle speed	600 rpm
Emission Control Technology Route	EGR+TWC
Emission Standards	China VI

2.3 Fuels

In this paper, two types of compression natural gas were used in accordance with the requirements of the regulations. The first is a low calorific value natural gas (NG1) which is a mixture of 85.45% methane and 14.54% nitrogen by volume fraction. The high calorific value natural gas (NG2) is a mixture of 94.11% methane, 3.5% ethane, 1.2% CO_2 and 1% nitrogen.

2.4 Test plan

In this study, WHTC was chosen as the test cycle. Except the hot WHTC (WHTC-H), the WHTC including engine cold start (WHTC-C) was also run to investigate the particle number emission characteristics under cold condition. The impact of fuel type on particle number emission was also considered in the test plan.

3 Results and Discussions

3.1 PN10 and PN23 results under WHTC

Table 3 shows the emissions of PN10 and PN23 and the weighted emissions under the hot and cold WHTC cycle. Among them, PN10 and PN23 are the measured values from the second particle counter. The weighted average PN emission result is calculated by the following formula.

$$PN_{i,weighted} = \frac{(0.14 \times PN_{i,cold} + 0.86 \times PN_{i,hot})}{(0.14 \times W_{act,cold} + 0.86 \times W_{act,hot})}$$

Among then, the subscript i represents PN10 or PN23, the subscript cold and hot represents the cold WHTC and hot WHTC, W represents the cycle work.

Table 3 PN specific emission results

	WHTC_C(#/kW.h)	WHTC_H(#/kW.h)	Weighted WHTC(#/kW.h)
PN23_NG1	4.04×10^{9}	1.07×10^{9}	1.48×10^{9}
PN10_NG1	8.71×10^{10}	2.31×10^{10}	3.2×10^{10}
PN23_NG2	6.97×10^{10}	9.54×10^{8}	1.06×10^{10}

 https://doi.org/10.1051/e3sconf/202132901012

PN10_NG2	1.50×10^{12}	2.06×10^{10}	2.27×10^{11}

For these two fuels used, the weighted WHTC brake specific emission results of both PN10 and PN23 are lower than the regulation requirement of 6×10^{11} #/kWh. The weighted emission of PN23 of NG1 is 1.48×10^{9} #/kWh, and the weighted emission of PN10 is 3.2×10^{10} #/kWh. The weighted WHTC brake specific emission of PN10 is 21.6 times that of PN23. Under the cold WHTC, the brake specific emission of PN23 of NG1 is 4.04×10^{9} #/kWh, and the brake specific emission of PN10 is 8.71×10^{10} #/kWh. For WHTC-H, the brake specific emission of PN23 of NG1 is 1.07×10^{9} #/kWh, and the brake specific emission of PN10 is 2.31×10^{10} #/kWh. Under the hot and cold WHTC cycle, the brake specific emission of PN10 is 21.6 times that of PN23. For PN23, the brake specific emission of WHTC-C is 3.78 times that of WHTC-H. For PN10, the brake specific emission of WHTC-C is 3.77 times that of WHTC-H. The weighted brake specific emission of PN23 of NG2 is 1.06×10^{10} #/kWh, and the weighted brake specific emission of PN10 is 2.27×10^{11} #/kWh. The weighted brake specific emission of PN10 is 21.5 times that of PN23. For WHTC-C, the brake specific emission of PN23 of NG2 is 6.97×10^{10} #/kWh, and the brake specific emission of PN10 is 1.5×10^{12} #/kWh. For WHTC-H, the brake specific emission of PN23 of NG2 is 9.53×10^{8} #/kWh, and the brake specific emission of PN10 is 2.06×10^{10} #/kWh. For both WHTC-C and WHTC-H, the brake specific emission of PN10 is 21.5 times that of PN23. For PN23, the brake specific emission of WHTC-C is 72.99 times that of WHTC-H. For PN10, the brake specific emission of WHTC-C is 72.64 times that of WHTC-C .

It can be seen that PN10 are much higher than those of PN23 for heavy-duty natural gas engines. The main reason is that the particle size of natural gas engines is smaller. Thiruvengadam et al found that heavy-duty natural gas engine with equivalent ratio combustion equipped a three-way catalytic converter has an obvious nuclear mode peak at 10 nm in the particle size distribution [13]. Alanen et al found that the exhaust particles of natural gas engines have a peak size of less than 10nm in diameter under steady-state conditions [14]. Hou Xianjun et al also found that the particulate matter of natural gas fuel mainly exists in the form of nuclear state [15]. All these shows that it is more urgent to control the emission of smaller particles for natural gas engines.

3.2 PN emission characteristics under WHTC

As shown in Figure 1, PN10 and PN23 exhibit a very consistent trend. The highest emission peak appears at the cold start phase. This is mainly due to the low exhaust temperature and the low catalytic efficiency of the three-way catalytic converter. There are two larger peaks at 700s and 1200s. Among them, the engine is running from deceleration to idling at 700s, which causes decreases in exhaust temperature and catalytic efficiency. At 1200s, the engine was running from acceleration to deceleration, and the air-fuel ratio control was unstable during the process, which results in incomplete combustion and increased PN emission. In the last stable high-speed stage of the cycle, the emissions of PN10 and PN23 are lower. The main reason is that the exhaust temperature is high at this time, and the particle is more likely to be oxidized.

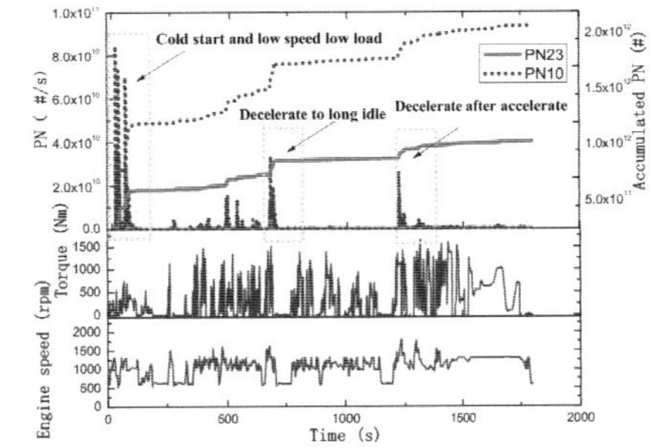

Figure 1 PN transient and accumulated emission under cold WHTC for NG1

The transient emissions and cumulative emissions of PN10 and PN23 of hot WHTC with NG1 fuel is presented in Figure 2. The trend of PN10 and PN23 is still very consistent. In the low-speed and low-load stage, the generation of particulate matter is less. Additionally, because the engine is running in a hot state, the efficiency of the three-way catalytic converter is higher. Therefore, compared with the cold state WHTC, there is no obvious peak at this stage. The largest PN emission peak appears in the acceleration to deceleration phase of about 1200s.

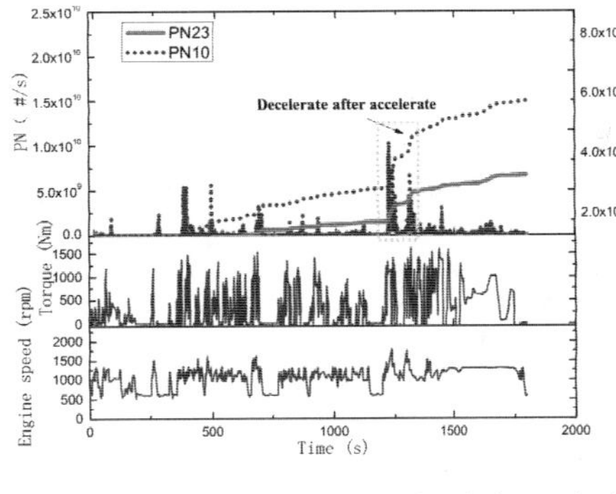

Figure 2 PN transient and accumulated emission under hot WHTC for NG1

PN10 and PN23 emission characteristics under WHTC-C and WHTC-H with NG2 fuel are seen in Figures 3 and 4. Similar to NG1 fuel, the trends of PN10 and PN23 are very analogous. In Figure 3, the largest peak occurs during the cold start phase, and the larger peak occurs during the deceleration to long idle and acceleration to deceleration phases. However, compared with NG1 fuel, the PN emissions of NG2 fuel during the cold start phase are very serious, especially for PN10. The cumulative emissions results show that more than 95% of PN emissions are produced at this stage.

https://doi.org/10.1051/e3sconf/202132901012

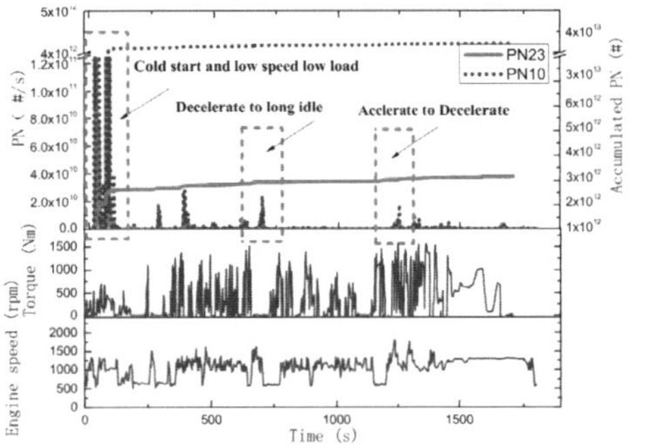

Figure 3 PN transient and accumulated emission under cold WHTC for NG2

There are fewer particles in the low-speed and low-load phases in Figure 4, and the larger particle peaks appear in the deceleration to long idle and acceleration to deceleration stages.

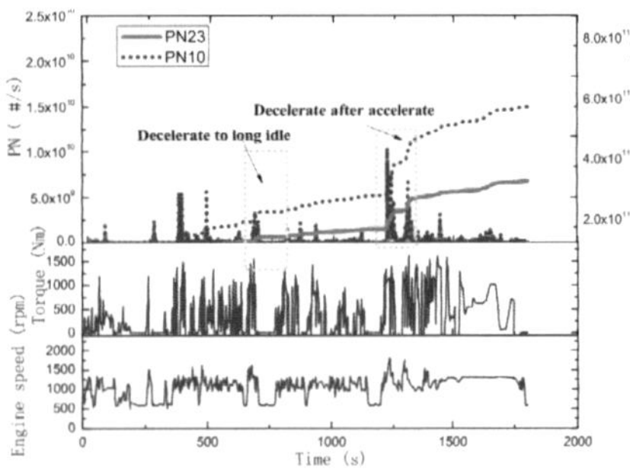

Figure 4 PN transient and accumulated emission under hot WHTC for NG2

From the above analysis, it can be seen that PN emissions are very serious when the exhaust temperature has not reached the optimal working temperature of the three-way catalytic converter for this natural gas. When the exhaust temperature rises, operating conditions become the main influencing factor of PN emissions. In the process of acceleration and deceleration, poor control of the air-fuel ratio will result in a rich mixture in some areas of the combustion chamber, thereby increasing the emission of particulate matter. In addition, fuel has a great influence on the PN emissions, and we will conduct a detailed analysis in the next section.

3.3 Effects of fuel on PN emissions

Figure 5 shows the PN brake specific emission of the two different fuels in the hot and cold WHTC. In the cold WHTC, the brake specific emissions of PN10 and PN23 of the high calorific value NG2 fuel are 17.25 times and 17.2 times that of the low calorific value NG1 fuel, respectively. In the hot WHTC, the brake specific emissions of PN10 and PN23 of NG2 fuel are 0.89 times that of NG1 fuel. The fuel property has little effect on the PN emission in hot WHTC but extremely impact in cold WHTC. Combined with Figure 3, it can also be seen that the impact of NG2 fuel on PN emissions is mainly due to the cold start stage. Amirante et al. used a mixture of methane and propane in different proportions to explore the effect of fuel on emissions, and found that the addition of propane would lead to an increase in the number of particles in the size range of 5 to 30 nm [16]. McTaggart-Cowan et al. found that the addition of ethane and propane will lead to an increase in soot particulate matter emissions [17]. In this paper, the main components of NG1 fuel are methane and nitrogen, while NG2 fuel contains a certain amount of ethane in addition to the higher content of methane. The increase in carbon-containing fuels has led to an increase in PN emissions.

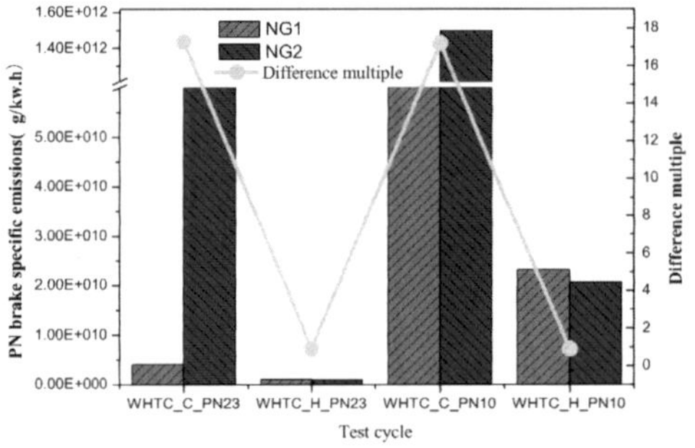

Figure 5 Brake specific emissions of PN10 and PN23 with NG1 and NG2.

Since the impact of NG2 fuel on PN emissions is mainly derived from the cold start phase of the cold WHTC cycle, the other pollutants and exhaust temperature of the two fuels during the first 150 seconds in the cold WHTC are compared, as shown in Figure 6 . It can be seen from Figure 6 that there is basically no difference in the exhaust temperature of the two fuels. The temperature rises to close to 300°C in less than 100s. The total hydrocarbon (THC) and methane (CH4) emissions of NG2 fuel are slightly higher than that of NG1 fuel, but the difference is not significant. However, the carbon monoxide (CO) emission of NG2 fuel is much higher than that of NG1 fuel, and the nitrogen oxide (NOx) emission of NG1 fuel is higher than that of NG2 fuel. Therefore, it can be concluded that for the NG2 fuel, the incomplete combustion in the cold start stage is more serious than that of NG1 fuel, so the CO emission is much higher. This also explains why the PN emission of NG2 fuel during the cold start stage is much higher than NG1 fuel.

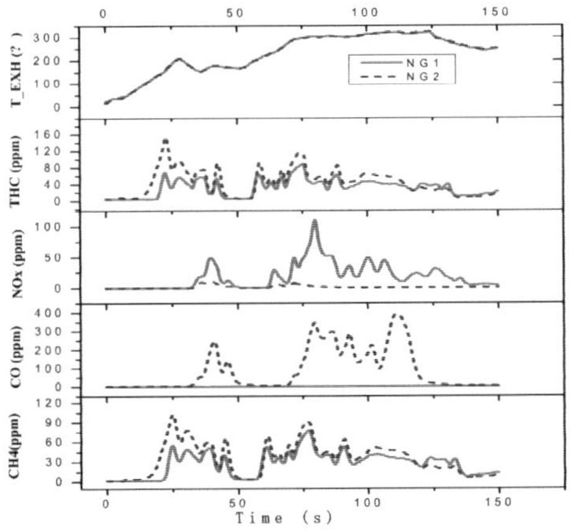

Figure 6 Emissions and exhaust temperature of engine with NG1 and NG2.

4 Conclusion

The heavy-duty natural gas engine was run in cold and hot WHTC, and the results showed that:
1) For heavy-duty natural gas engines, the emissions of PN10 are much higher than those of PN23. The weighted PN10 brake specific emission are 21.6 times that of the PN23. It is necessary to strengthen the control of the sub-23 nm PN emissions of natural gas engines.
2) The transient emission trend of PN10 and PN23 for a natural gas engines are basically the same. PN emissions at cold start stage are very serious. When the exhaust temperature rises, PN emissions will increase during the acceleration and deceleration process.
3) Fuel has a great influence on PN emission for natural gas engines. The increase of carbon-containing fuels such as methane and ethane in the fuel will lead to a rapid increase in PN emissions during the cold start process.

References

1. China mobile source environmental management annual report in 2021 — the situation of vehicle emission[J]. Ministry of Ecology and Environment of the People's Republic of China, 2021.
2. Nel A, Xia T, Madler L, et al. Toxic potential of materials at the nanolevel [J]. Science, 2006, 311(5761):622-627.
3. Oberdorster, G., Sharp, Z., Atudorei, V., et al. Translocation of Inhaled Ultrafine Particles to the Brain[J]. Inhal Toxicol., 2004, 16(6-7):437-45.
4. GB17691-2018 Limits and measurement methods for emissions from diesel fueled heavy-duty vehicle (Chinese Ⅵ) [S]. Ministry of Ecology and Environment of the People's Republic of China, 2018.
5. Giechaskiel B., Vanhanen J, Väkevä M., eta al. Investigation of vehicle exhaust sub-23 nm particle emissions[J]. Aerosol Science and Technology, 2017 51(5):626-641.
6. Hou Yuxu, Ge Yunshan, Wang Xin et al, Emission characteristics of Sub-23 nm Particle from a Turbocharged Direct-Injection Gasoline Engine[J]. Chinese Internal Combustion Engine Engineering, 2021, 42(4):24-29.
7. Stéphane Z, Mickaël L, Loïc R, et al. Measurement of Sub-23 nm particles emitted by gasoline direct injection engine with new advanced instrumentation [C], SAE 2019-01-2195.
8. Leach, F., Lewis, A., Akehurst, S., Turner, J. et al., Sub-23 nm Particulate Emissions from a Highly Boosted GDI Engine[C], SAE 2019-24-0153.
9. Khan, M.Y., Peng, W., Scott, N., Lin, Y. et al., Heavy-Duty Engines Exhaust Sub-23 nm Solid Particle Number Measurements[C], SAE 2021-01-5024.
10. Elia D, Riccardo A, Giuseppe C, et al. Evolution of Soot Particle Number, Mass and Size Distribution along the Exhaust Line of a Heavy-Duty Engine Fueled with Compressed Natural Gas[J]. Energies 2020, 13 (15) :3993.
11. Hajbabaei M, Karavalakis G, Johnson K C, et al. Impact of natural gas fuel composition on criteria, toxic, and particle emissions from transit buses equipped with lean burn and stoichiometric engines. Energy, 2013, 62: 425–434.
12. Xiaowei Wang, Xiaojun Jing, Xuejing Gu, et al. Future Regulationbased Particle Number Emission Characteristics for a Heavyduty Diesel Engine, IOP Conference Series: Earth and Environmental Science, 2021
13. Thiruvengadam A, Besch MC, Yoon S, et al, Characterization of particulate matter emissions from a current technology natural gas engine[J]. Environ Sci Technol. 2014 Jul 15;48(14):8235-42.
14. Alanen J, Saukko E, Lehtoranta K,et al, The formation and physical properties of the particle emissions from a natural gas engine[J]. Fuel, 2015 162:155–161.
15. Hou Xianjun, Xiong Na, Wang Youheng, et al., Particulate Matter Emission Characteristics of Gasoline/Compressed Natural Gas Dual-Fuel Engine[J]. Chinese Internal Combustion Engine Engineering, 2021, 42(3):74-80.
16. Amirante R, Distaso E, Di Iorio S,et al. Effects of natural gas composition on performance and regulated, greenhouse gas and particulate emissions in spark-ignition engines[J]. Energy Convers. Manag. 2017, 143:338–347.
17. McTaggart-Cowan GP, Rogak SN, Munshi S R, et al The influence of fuel composition on a heavy-duty, natural-gas direct-injection engine[J]. Fuel 2010, 89: 752–759.

Evaluation of main factors affecting fractured horizontal well productivity of tight sand gas reservoir in Sichuan Basin

Linqing Liu [1], Shuo Zhai [2,*], HaiTao Li [1], Jing Wang [1]

[1] Exploration and Development Research Institute of Petro China Southwest Oil & Gas Field Company, Sichuan, China
[2] Chengdu University of Technology, Sichuan, China

Abstract. Hydraulic fracturing horizontal well is the key technology of tight gas development. After fracturing, the gas well shows the characteristics of great difference in production energy and many factors affecting production capacity. Therefore, taking 10 fractured horizontal wells in JQ gas field in Sichuan Basin as an example, comprehensively considering the influence of geological and engineering factors, and based on the statistical analysis of gas well productivity and the laws of various index parameters, 17 geological and engineering parameters were selected, and the main factors controlling productivity were studied by grey correlation method. The results show that the tight gas fracturing horizontal well can obtain high production, which is jointly determined by geological and engineering factors. The five main control factors affecting the productivity of tight gas fracturing horizontal well are as follows: proppant type, preflush, ϕ·h. Sand volume and reservoir volume.

Keywords. Tight gas, fractured horizontal well, productivity, grey correlation method, main control factors.

1 Introduction

The world is rich in tight gas resources, with geological reserves of about 201×1012m3. With the progress of hydraulic fracturing technology, the production of tight gas is higher and higher, which occupies a more and more important position in the world energy pattern. However, The porosity, permeability and abundance of tight gas are lower than those of other gas reservoirs, hydraulic fracturing technology improves the productivity and economic benefits of tight gas wells, but there are many factors affecting productivity and it is difficult to evaluate productivity.

However, traditional theoretical methods and numerical simulation methods are difficult to evaluate and analyze the factors affecting production capacity. Based on the statistics of productivity influencing factors of 10 fractured horizontal wells in JQ gas field, Sichuan Basin, 17 engineering and geological parameters are analyzed by grey correlation method, and the main factors affecting the productivity of fractured horizontal Wells, namely productivity controlling factors, are obtained. Through this study, we have a better understanding of development sweet spots, sweet spots, fracturing and development mode formulation in tight areas.

2 Productivity characteristics and influencing factors of gas wells

Jinqiu gas field in Sichuan Basin is dominated by fractured horizontal wells, and the length of horizontal section is between 400-1200 meters. Multiple clusters of perforations and continuous sand addition process are adopted in the fractured section of horizontal wells. The productivity of Jinqiu gas field varies greatly. The minimum absolute open flow rate of single well is $1.14\times 10^4 m^3/d$ and the maximum is $214.05\times10^4 m^3/d$. The relationship between geological parameters and open flow rate of gas well and engineering parameters and open flow rate of gas well are obtained according to the productivity data obtained from the test in the early stage of gas well production.

As shown in Figure 1, the absolute open flow of gas well has a strong positive correlation with reservoir thickness, $\phi\cdot h$, reservoir volume and pressure, so the physical properties of reservoir have a great impact on the productivity of gas well.

* Corresponding author: 2020050127@stu.cdut.edu.cn

https://doi.org/10.1051/e3sconf/202132901013

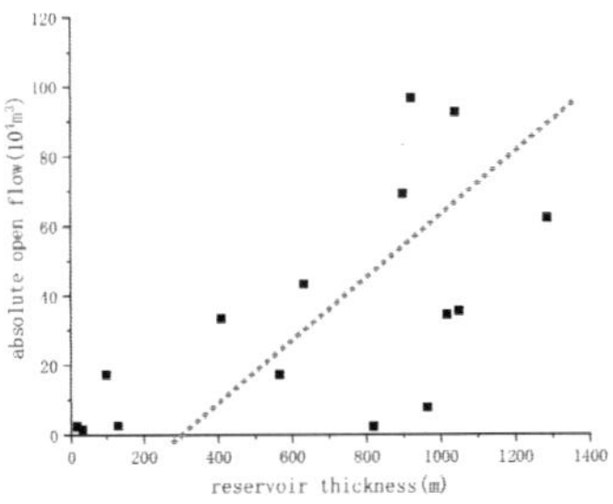

a. Relationship between absolute open flow and reservoir thickness

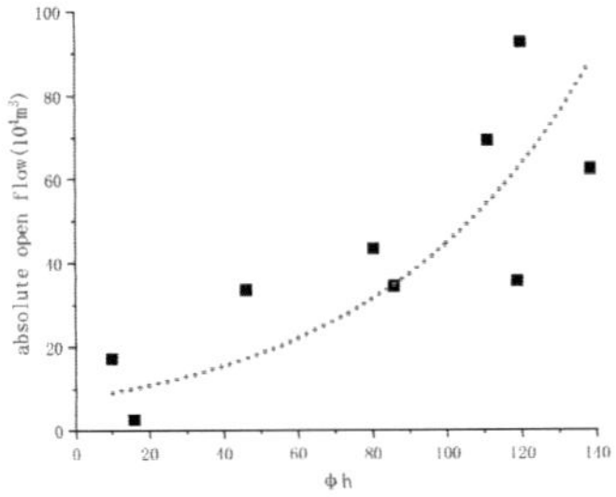

b. Relationship between absolute open flow and ϕ·h

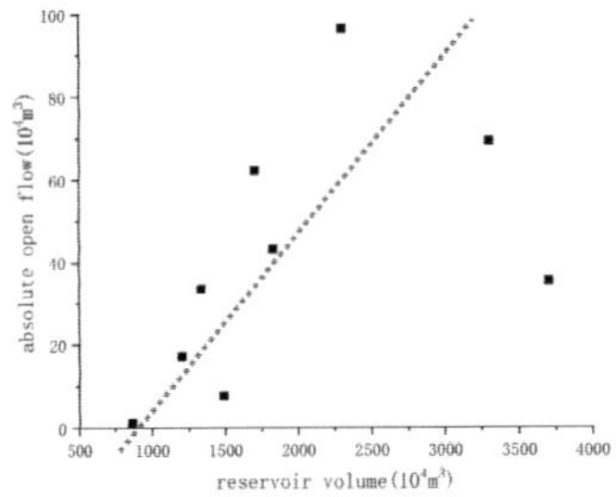

c. Relationship between absolute open flow and reservoir volume

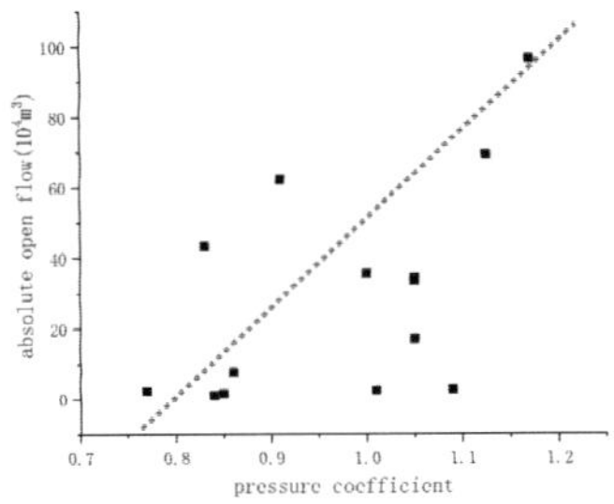

d. Relationship between absolute open flow and pressure coefficient

Figure 1. Relationship between absolute open flow of gas well and geological parameters

As shown in Figure 2, engineering parameters also have a certain positive correlation with absolute open flow of gas wells, and its influence trend is more complex than the relationship between geological parameters and absolute open flow.

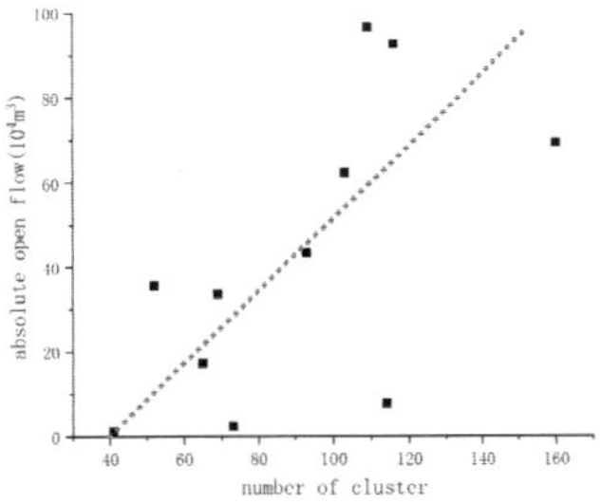

a. Relationship between absolute open flow and perforation number of cluster

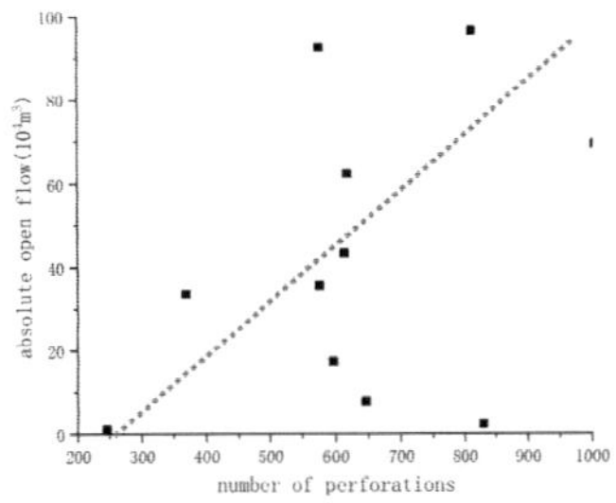

b. Relationship between absolute open flow and number of perforations

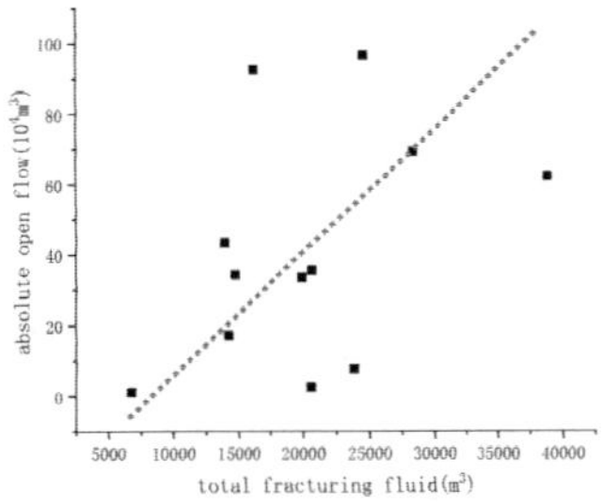

c. Relationship between absolute open flow and total fracturing fluid

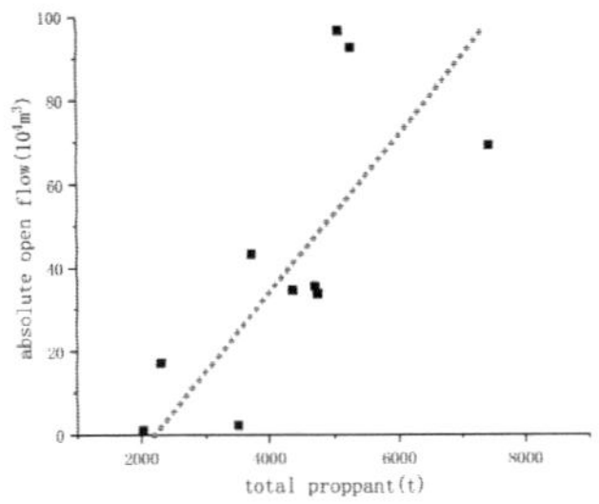

d. Relationship between absolute open flow and total proppant

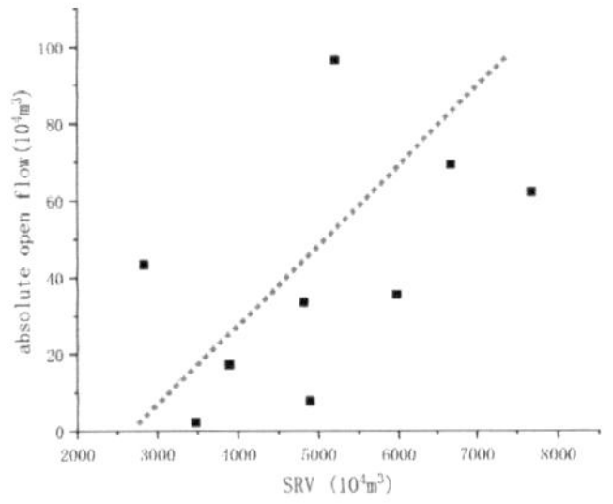

e. Relationship between absolute open flow and SRV

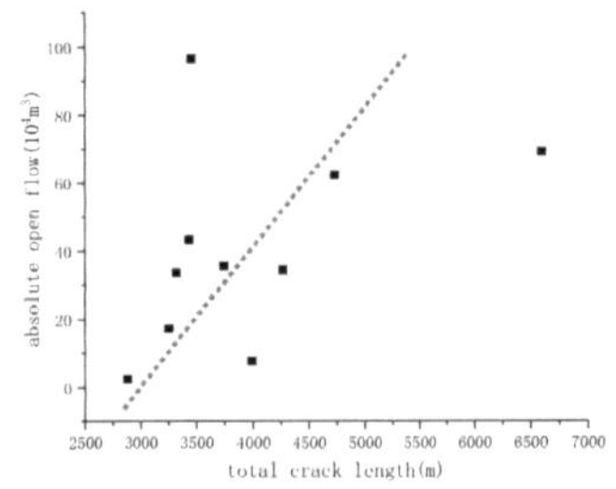

f. Relationship between absolute open flow and total crack length

Figure 2. Relationship between absolute open flow of gas well and engineering parameters

According to the statistical relationship between geological and engineering parameters and absolute open flow of gas wells, the productivity of fractured horizontal wells in tight gas reservoirs is comprehensively affected and restricted by many factors. It is very important for gas well drilling and gas reservoir efficient development to clarify the main controlling factors of productivity.

3 Grey correlation analysis

Grey correlation analysis method is an analysis method based on multi-factor comparison to get the main control factors, and the strength of correlation between factors is represented by grey correlation degree.The closer the change of grey correlation degree between two factors is, the stronger the correlation is. The geological parameters analyzed include porosity, permeability, water saturation, energy storage coefficient, pressure coefficient, reservoir thickness and reservoir volume; Engineering parameters are: number of clusters, number of perforations, amount of preflush, amount of sand carrying fluid, total amount of fracturing fluid, 70/140 proppant dose, 40/70 proppant dose, total amount of proppant, fracture length and SRV. The process is as follows:

(1) The invalid water injection ratio is selected as the reference sequence, and the geological and production parameters are the comparison sequence, expressed as （1）;

$$\begin{cases} X_0 = \{X_0(1), X_0(2), X_0(3), \ldots\ldots, X_0(n)\} \\ X_i = \{X_i(1), X_i(2), X_i(3), \ldots\ldots, X_i(n)\} \\ X_m = \{X_m(1), X_m(2), X_m(3), \ldots\ldots, X_m(n)\} \end{cases} \quad (1)$$

Where,X_0 - reference sequence; X_i - comparison sequence; j=0,1,2…,m; k=1,2,…,n。

(2) In the process of grey correlation analysis, the influence of dimension and weight relation among all factors should be comprehensively considered. All factors should be unified in the same level and equal weight condition for calculation and analysis, and the dimensionless data should be carried out through the initial value method. First, select the optimal sequence and record it as $\{x_0'(1), x_1'(1), \text{L}\ x_n'(1)\}$. The calculation formula is:

$$Y_j(k) = \frac{X_j(k)}{X_j'(1)} \quad (2)$$

Where, $Y_j(k)$ is the dimensionless sequence.

(3) Take the difference sequence:

$$\Delta_{0i}(k) = |Y_0(k) - Y_i(k)| \quad (3)$$

Where, $\Delta_{0i}(k)$ - absolute difference.

(4) Find the maximum and minimum difference of two levels:

$$\Delta_{\max} = \max_i \max_k |Y_0(k) - Y_i(k)| \quad (4)$$

$$\Delta_{\min} = \max_i \max_k |Y_0(k) - Y_i(k)| \quad (5)$$

Where, $\Delta_{\max}$-maximum difference of two stages; $\Delta_{\min}$-minimum difference between two levels.

(5) The correlation coefficient is calculated as follows:

$$\xi_{0i}(k) = \frac{\Delta_{\min} + \rho\Delta_{\max}}{\Delta_{0i}(k) + \rho\Delta_{\max}} \quad (6)$$

Where, ρ-Resolution coefficient. Its function is to magnify the difference between correlation coefficients numerically, $\rho \in (0,1)$, usually 0.5, ρ The smaller the value, the greater the difference between the correlation coefficients.

https://doi.org/10.1051/e3sconf/202132901013

4 Analysis results of main control factors

According to the data of various influencing factors of fractured horizontal wells in JQ gas field, geological and engineering parameters are processed based on grey correlation analysis method, and the correlation degree comparison diagram of all factors of open flow can be obtained. As shown in FIG. 3, $\phi \cdot h$ has a great influence on reservoir volume and open flow. Among engineering factors, 70/140 mesh proppant, SRV, preconditioning fluid and open flow rate are the most important. Therefore, engineering factors are more influential than geological factors.

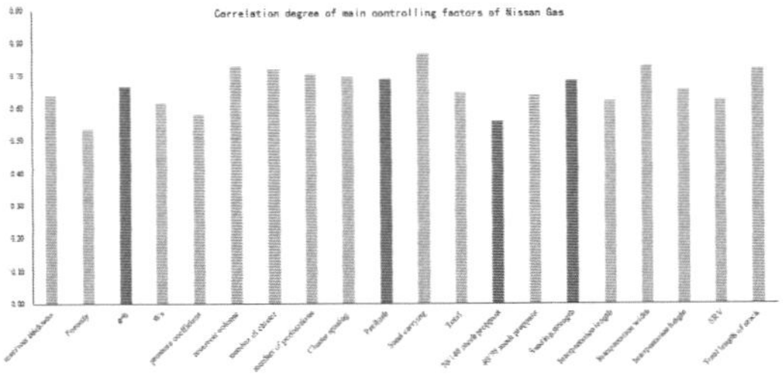

Figure 3. Main control factors of absolute open flow

5 Analysis results of main control factors

(1) There are 17 factors influencing the productivity of fractured horizontal Wells in tight gas. Based on the grey correlation analysis of these influencing factors, five main controlling factors are obtained, which are proppant type, pre-fluid, $\phi \cdot h$, sand volume and reservoir volume in order.
(2) Whether tight gas wells can obtain high production is determined by geological and engineering factors, which complement each other. High reservoir quality is the material basis for shale gas wells to obtain high production, and high engineering quality is its technical guarantee.

References

1. Fu Jianbin. Productivity prediction method of fractured wells in tight gas reservoir considering multi factors[J]. Fault-Block Oil & Gas Field, 2021,28(02):156-161.
2. Luo Hongwen, Li Haitao, An Shujie, et al. Analysis of Influencing Factors of Temperature Profile of Fractured Horizontal Wells in Tight Gas Reservoirs[J]. Special Oil & Gas Reservoirs, 2021,28(04):150-157.
3. Duan Wenshen. Exploration potential and favorable direction of tight gas in Middle Jurassic Xiashaximiao Formation in Sichuan Basin[J]. Petroleum Geology & Experiment, 2021,43(03):424-431+467.
4. Deliang Zhang, Liehui Zhang, Yulong Zhao et al. A composite model to analyze the decline performance of a multiple fractured horizontal well in shale reservoirs[J]. Journal of Natural Gas Science and Engineering, 2015, 26(C)
5. Liu Zixiong, Li Xiaonan, Wang Jinwei, et al. Study on method of sweet spot interval identification of tight gas reservoir in horizontal well[J]. China Petroleum Exploration, 2021,26(03):117-125.
6. Liang Xue, Shao-Hua Gu, Xie-Er Jiang, Yue-Tian Liu, Chen Yang. Ensemble-based optimization of hydraulically fractured horizontal well placement in shale gas reservoir through Hough transform parameterization[J]. Petroleum Science, 2021,18(03):839-851.
7. HU Zubiao, ZHANG Jianqin, WANG Qingchen, et al. Water Base Drilling Fluid Technology for Ultra-Long Horizontal Drilling in a Tight Gas Well in Changqing Oilfield[J]. Drilling Fluid & Completion Fluid, 2021,38(02):183-188.
8. TANG Mingming, LU Shuangfang, LIANG Hongru, YAN Bihui, SHEN Shan. Study on the Fracturing Parameters and Productivity of Horizontal Wells[J]. Acta Geologica Sinica(English Edition), 2015, 89(S1):321-323.
9. LIU Naizhen, ZHANG Zhaopeng, ZOU Yushi, MA Xinfang, ZHANG Yinuo. Propagation law of hydraulic fractures during multi-staged horizontal well fracturing in a tight reservoir[J]. Petroleum Exploration and Development, 2018,45(06):1129-1138.
10. WANG Lei, LIU Yingming, WANG Caizhi, FAN Yiren, WU Zhenguan. Real-time forward modeling and inversion of logging-while-drilling electromagnetic measurements in horizontal wells [J]. Petroleum Exploration and Development, 2021, 48(01): 159-168.

Research on new transmission network planning considering adjustable comprehensive resources

Zhenghua Chen, Junjie Cao, Shengzhi Lu, Yezhang Cao, Shanshan Zheng, Xiaonan Yang

State Grid Jiangsu Electric Power Co., Ltd. Yangzhou Power Supply Company China

Abstract. With the large-scale renewable energy grid integration, the problems of traditional transmission network planning methods, such as high investment, low utilization rate and difficulty in implementation, have gradually emerged. Therefore, in order to improve the level of new energy consumption, effectively improve the operation efficiency of power grid and optimize the investment benefit, this paper proposes a new mode of power grid planning based on adjustable integrated resources. This paper focuses on two key planning methods: new energy confidence capacity assessment methods and load forecasting methods considering demand side resources.Then the paper takes a city-level 220 KV power grid as an example, applies a new power grid planning method to carry out planning, and verifies the effectiveness of the proposed planning model and method.

1 Introduction

Transmission grid planning is an important part of power system planning. Reasonable and effective transmission grid planning can ensure the healthy development of the power system[1].However, the uncertainty of the power system has also increased due to the increasing proportion of grid-connected renewable energy sources[2]. Therefore, it is particularly important to consider new grid planning with adjustable integrated resources such as renewable energy.

At present, domestic and foreign scholars are also doing a lot of related research on how to carry out new power grid planning.The literature [3] analyzed the challenges and changes brought about by the development of new energy to traditional power transmission technologies, and puts forward the key technologies for the application planning of DC transmission grids.The literature [4] proposed the transmission network planning theory with multi-scene technology, robust optimization technology and collaborative planning technology as the core, and prospected the future transmission network planning methods. In reference [5], HRPTD-20, a transmission network planning and testing system considering transmission and distribution coordination, was constructed under the high proportion of renewable energy access, and the rationality and effectiveness of the test system was verified by comparative analysis. The literature[6] analyzed the important role of load management in power grid planning and concluded that load management was more cost-effective than traditional power grid expansion. The literature[7] proposed a new reliability-oriented mixed integer linear programming method for power networks, and demonstrated that the optimization of the location and capacity of power plants in the power system can reduce the average unsupplied energy cost. In literature[8], a grid planning model for renewable energy access system embedded with unit combination and network structure optimization was established, which provides an important auxiliary decision-making basis for power grid planning.

2 Traditional power grid planning model

In recent years, the demand for electricity has been accelerating, and the grid planning and design process has also been continuously improved.When planning and designing traditional power grids, the main purpose is to ensure the optimal situation of the safe and coordinated operation of the system, reduce the system cost in the life cycle as far as possible, and then improve the quality of the whole planning process to meet the peak load and set the planning constraints.

However, with the large-scale grid connection of renewable resources, the continuous development of energy storage technology, and the gradual promotion of demand response measures, there is a new demand for power grid planning. The influence of adjustable comprehensive resources on power transmission network planning is also increasing, which can be divided into the following three aspects: (1) Large-scale renewable energy grid integration increases the random fluctuation of power flow and planning targets in transmission network planning; (2) After the demand response resource mechanism is introduced, the demand response resources that can be saved on the demand side to replace the power transmission and distribution resources are also included in the power planning, so that the resources on both sides of supply and demand can be utilized more efficiently and economically; (3) When planning the power grid,

* Corresponding author: 594842366@qq.com

reasonable allocation of the location and capacity of energy storage can change the temporal and spatial characteristics of load and renewable energy power, thus changing the transmission performance of the power grid and solving the problem of short-term overload of transmission lines.

There are some following problems in the current traditional power grid planning mode :(1) renewable energy is not considered in the power balance planning, and the resulting power balance planning scheme is prone to imbalance in the implementation process, resulting in serious abandonment of wind and light by renewable energy, resulting in resource waste; (2) There are many factors affecting load change, and the fluctuation of power grid load is large. As an important premise and foundation of power grid planning, the accuracy of power load prediction is usually not high; (3) Due to various reasons, the overall urban development plan cannot be well implemented, and the nature of land use in some regions often changes, resulting in uncertainty of load growth, low accuracy of power grid planning and poor adaptability of planning schemes.

Therefore, it is particularly important to take adjustable comprehensive resources into account in power grid planning. By establishing a new optimal dynamic planning mode of power grid based on adjustable comprehensive resources, the operation efficiency of power system can be improved, the absorption of renewable energy can be promoted, the phenomenon of abandoning wind and light can be reduced, and the power grid can operate more safely and stably.

3 A new transmission network planning mode based on adjustable comprehensive resources

To build a new type of grid based on adjustable integrated resource planning to build new type of power system, to adapt to the economic and social development planning, and user's growing electricity demand as the basic goal, adjustable integrated resource characteristic is taken into account in planning and its impact, reasonable arrangement of all levels of power grid construction, promote the power grid network frame structure optimization, to achieve efficient operation of the power grid to At the same time, improve the reliability of power supply, reduce power loss as much as possible, and ensure the sustainable development of power grid.Specific ideas are as follows:

(1)Take meeting electricity demand as the fundamental task

Through the continuous improvement of grid reliability and power quality, to maximize the social demand for electricity, and strive to achieve on the premise of realizing social benefits, the development of enterprises, improve the grid, improve the goal of income.

(2) Consider the characteristics and impacts of adjustable comprehensive resources

In view of the "new energy and energy storage" output combination can provide relatively stable power supply, is conducive to the balance of power, the adjustable integrated resources of renewable energy and energy storage resources, should be included in the power balance points area, according to its expected output condition, the methods of confidence capacity assessment, renewable energy and energy storage equivalent to the conventional unit, Incorporate into transmission network planning. In addition, for the demand response resources in the adjustable comprehensive resources, load prediction should be carried out considering the demand response resources, and the load prediction results should be further used as a reference for transmission network planning.

(3) Select a reasonable time section to carry out safety check

Considering the fluctuation and uncertainty of renewable energy output, the nodes with the lowest output stability and minimum output are selected as the time section according to the output curve to carry out power grid security check.

(4) Demand response cost should be considered in the economic evaluation of transmission network planning scheme

The demand response cost should be considered in the economic evaluation of transmission network planning scheme to measure the effectiveness and economy of the new planning mode more effectively.

(5) Promote the coordination between the superior power grid and the subordinate power grid

Subordinate power grid and superior power grid as two ladder units, superior power grid planning as guidance, responsible for macro guidance to subordinate units; As the basis, the subordinate power grid planning is responsible for the execution and implementation of basic tasks.

Among them, when considering the characteristics of adjustable comprehensive resources, the available capacity of renewable energy is determined by using confidence capacity assessment and other methods according to its expected output, and the renewable energy and energy storage are equivalent to conventional units and incorporated into the transmission network planning. For the demand response resources in the adjustable comprehensive resources, load prediction is carried out considering the demand response resources to ensure the accuracy of load prediction, and the load prediction results are further used as a reference for transmission network planning.

4 Confidence Capacity Evaluation of Renewable Energy

Among new energy sources, uncertain energy sources such as wind power generation and solar photovoltaic power generation have great fluctuations in their output due to the meteorological environment, resulting in great differences between their actual output and initial installed capacity. New energy capacity of confidence can be summarized new energy influence on abundance degree of power system capacity, not only has the intermittent new energy and conventional unit installed capacity of q can compare at the same level, and make the uncontrolled, with volatility and randomness of new energy in traditional power planning analysis and calculation of possible. Therefore, the confidence capacity assessment of renewable energy used in the planning of new power grid is as follows.

4.1 Evaluate the confidence capacity of a single intermittent power supply

In this paper, the method used to calculate the confidence capacity assessment is equivalent conventional unit capacity, and the unit with random outage rate is used to replace the renewable energy.
Calculate the confidence capacity of a single intermittent power supply
Assume that the system contains conventional unit capacity Cr and intermittent power supply to be evaluated. Take wind power as an example, its capacity is Cw, and assume equivalent unit capacity is Ck, which have the same reliability parameters as conventional units. f{C} represents the reliability index of the system when the total installed unit capacity is C; f{Cr,Cw} represents the reliability of the system when the conventional unit capacity is Cr and the wind turbine capacity is Cw; f{Cr,Ck} represents the reliability of the system when the conventional unit capacity is Cr and the equivalent unit capacity is Ck respectively. If there is:

$$f\{C_r, C_w\} = f\{C_r, C_k\} \tag{1}$$

Then the wind farm confidence capacity Cv of installed capacity Cw is defined to be equal to equivalent unit capacity Ck. f{C} can take different power systems in the calculation of reliability index, such as system load time expected loss, power shortage probability, etc., as an evaluation standard of system reliability.

4.2 Assess the confidence capacity of the wind-solar-storage combined power generation system

In the sequential Monte Carlo simulation with time sequence characteristics considered in this paper, if there is a power difference and the available capacity of energy storage equipment cannot meet the difference, the current hour is short of power hours, and the LOLE short of power hours of the system can be obtained after accumulation, which can be expressed as:

$$f_{LOLE} = P[Y_{dc}(t-1) + P_h(t) + P_m(t) < P_L(t)]T \tag{2}$$

When the wind and landscape storage combined power generation is carried out, the total output power of the wind and landscape storage combined power generation system is set as:

$$P_a = P_h + Y_{dc}(t-1) \tag{3}$$

Depending on the load and unit state, different formulas are used to calculate the EENS of the system. When the load P_L is greater than the maximum value $P_{a\,max}$ of the combined power generation of scenery storage, there is:

$$f_{EENS} = P_L - \int_{P_{a\min}}^{P_{a\max}} P_a \cdot f_{P_a}(P_a)dP_a \tag{4}$$

When $P_{h\,min} \leq P_L \leq P_{h\,max}$,

$$f_{EENS} = \int_{P_{a\min}}^{P_L} (P_L - P_a) \cdot f_{P_a}(P_a)dP_a \tag{5}$$

5 Load forecasting considering demand-side resources

In order to better compare the economics of traditional power grid planning and new power grid planning that considers adjustable integrated resources, on the basis of traditional load forecasting, the power saving effect of energy-efficient demand-side resources and the peak-shaving of load-type demand-side resources are superimposed. Grain filling effect, realizing regional maximum load forecasting considering demand-side resources.

5.1 Annual electricity load forecast

When the demand-side resources are not considered, the forecast of the annual electricity load. As shown in formulas (7) and (8):

$$q_1^0 = q_2 \cdot (1+\varphi)^T \tag{6}$$

$$P_1^0 = q_2^0 / \beta_0 \cdot H \tag{7}$$

In the formula, it represents the total power consumption in the target year without considering the demand side resources, MWh; is the initial annual power consumption, MWh; is the average annual growth rate of power consumption; T represents the forecast period, year; represents The maximum load in the target year without considering the demand side resources, MW; is the average load rate; H represents the total effective time in the forecast period, in hours.

5.2 Load forecast correction

(1) The maximum load forecast revision considering the energy-saving effect of energy-efficient resources
Assuming that the type of energy efficiency resource owned by this type of user is I, the calculation method of the energy saving that can be achieved in the calculation area is as follows:

$$q_{n,e} = q_{n,1}^0 - \Delta q_{n,e}, n \in N \tag{8}$$

Among them, represents the power saving brought by energy efficiency resources of the nth user in the area, MWh; it is the total power consumption after the effect of the energy efficiency resources in the coverage area of the substation.
(2) The maximum load forecast correction considering the effect of peak shaving and valley filling of load resources
When this type of user responds to multiple load resources, the superimposed effect is as follows:

$$\Delta P_n^r = \sum_j^J \alpha_j \cdot \Delta P_{n,j}^r \qquad n \in N, j \in \Omega_J \tag{9}$$

$$P_n^D = P_n^e - \Delta P_n^r, n \in N \tag{10}$$

Among them, represents the total number of load resources owned by this type of user; represents the probability of this type of user responding to the load

 https://doi.org/10.1051/e3sconf/202132901014

resource, which is related to the resource type. This parameter can be counted by monitoring the response of a user to the load resource in real time; represents coverage The maximum load, MW, of the nth type of user in the area after being affected by all demand-side resources.

(3) Regional comprehensive maximum load forecast revision considering the role of multiple demand-side resources

Since the types and numbers of users are different, the time node when the maximum load occurs when calculating the comprehensive maximum load in the coverage area may not be exactly the same. After the load simultaneous rate is introduced and adjusted, it can be obtained:

$$P^D = \alpha_n \left(\sum_{n=1}^{N} P_n^r \right), n \in N \quad (11)$$

$$P_{\max}^D = \alpha_n P^D \quad (12)$$

$$\Delta S = R\left(P_1^0 - P_{\max}^D\right) \quad (13)$$

Among them, represents the maximum load of multiple types of users in the coverage area of the substation under the comprehensive action of demand-side resources, MW; represents the simultaneous load rate, which is used to adjust the maximum load between different types of power users; N is the total number of user types; is the area The comprehensive maximum load is as follows, MW; represents the substation capacity that can be avoided or postponed after taking into account the resources on the demand side, MVA; R represents the capacity-to-load ratio.

6 Example analysis

6.1 Basic data

This paper uses a municipal 220 kV power grid as a pilot, and uses the scale and output of new energy and energy storage as basic data. The planning scenario is set on the basis of the load and power level of the original grid plan, and the demand side response is considered. The impact of resources on the grid load and the impact of new energy power stations and energy storage power stations on power output. The settings of the five scenarios are shown in Table 1:

Table 1 Scenario setting

	Energy storage configuration ratio	Demand-side resources ratio
Scenario 1	10%	10%
Scenario 2	5%	10%
Scenario 3	15%	10%
Scenario 4	10%	12.5%
Scenario 5	10%	7.5%

Carry out confidence capacity assessment and load forecast for different scenarios according to 3.1 and 3.2.

6.2 Planning scheme analysis

Perform 500kV power balance analysis and 220kV capacity-load ratio analysis for municipal power grids. According to the 500kV and 220kV substation capacity requirements under different scenarios, check and optimize the necessity and construction sequence of the original power grid project to form 5 different scenarios of power grids List of planned capacity items. The original power grid planning and the number of capacity projects under the five scenarios and the demand for substation capacity are shown in Table 2,and the planned capacity project arrangements for the power grid under each scenario are shown in Table 3.

Table 2 Number of power grid planning projects under different scenarios

	Number of items	500kV	220kV
Original plan	14	3	11
Scenario 1	6	1	5
Scenario 2	6	1	5
Scenario 3	5	0	5
Scenario 4	2	0	2
Scenario 5	7	1	6

Table 3 Capacity of power grid planning projects under different scenarios

	Substation capacity	500kV	220kV
Original plan	5230	3250	1980
Scenario 1	1090	250	840
Scenario 2	1150	250	900
Scenario 3	840		840
Scenario 4	360		360
Scenario 5	1330	250	1080

It can be seen from the table that the number and capacity of power grid planning under the five scenarios have been significantly reduced. The largest number of planned projects is 7 in scenario 5, the least is 2 in scenario 4, and the 14 in the original planning report. In comparison, the number has more than doubled. Among them, the capacity of the power grid planning project in scenario 5 is reduced from the original plan of 5230 to 1330, which is a reduction of 3900; the capacity of the power grid planning project in scenario 4 is reduced to 360, which is a reduction of 4870. In addition, the capacity of Scenario 1 grid planning project is reduced by 4140 compared with the original plan, Scenario 2 is reduced by 4080, and Scenario 2 is reduced by 4390.

6.3 Evaluation of planning schemes

The economics of power grid planning schemes under 5 different scenarios are evaluated. The operating period is

30 (years), the fixed asset conversion rate is 95%, the grid operation and maintenance rate is 2%, and the load reduction compensation price is 10 (yuan). /kw), the discount rate is 8%, and the annual adjustable resource utilization cost mainly considers the demand response cost. The economic benefits of each scenario are calculated and compared using the total investment during the planning period and the life cycle cost indicators. The results are shown in Table 4:

Table 4 Economic Comparison of Each Scenario

Unit: Million

Planning scenario	New load during planning period (MW·h)	Construction investment	Unit investment to increase supply load (MW·h/yuan)	Total life cycle cost
Original plan	1989	1479.41	134	1762.2591
Scenario 1	1794	400.41	448	531.2785
Scenario 2	1794	435.41	412	572.9701
Scenario 3	1794	359.69	499	482.7732
Scenario 4	1745	199.69	874	317.1267
Scenario 5	1843	455.41	405	584.3219

Comparing the total cost of the five set scenarios in the table with the original planning report during the planning period, it can be found that the total cost of the five scenarios considering comprehensive adjustable resources during the planning period is lower. Therefore, the comprehensive adjustable resources are included in the transmission grid planning. It can reduce the grid investment cost during the planning period to a certain extent. In addition, the investment efficiency of each planning scenario considering adjustable comprehensive resources has been significantly improved. In terms of increasing the supply load of unit investment, the most obvious improvement in investment efficiency is in Scenario 4, followed by Scenarios 3 and 1, and Scenario 5 is the worst.

In summary, the planning idea of considering comprehensive adjustable resources is conducive to improving the economics of power grid operation. Scenario 4 with the highest demand response participation rate is the most economical, which can bring higher investment efficiency and greater economic benefits to power grid companies.

7 Conclusion

This paper proposes a new power grid planning model based on adjustable integrated resources, and takes a municipal power grid as an example to apply new grid planning methods to carry out planning. The planning results show that adjustable integrated resources such as renewable energy, energy storage, and demand response are in place. Power grid planning has a certain substitution effect on traditional power. Among them, the available potential of demand response resources and the impact of incorporating them into power grid planning are greater. Therefore, this article puts forward the following related suggestions for the planning of the transmission grid: (1) Focus on the alternative role of adjustable resources in grid planning, and use it as an alternative resource on the power generation side for comprehensive planning; (2) In the future, power transmission In the process of network planning, the incentive space of the power market mechanism should be tapped, and the incentive role of the power market should be considered in the power grid planning; (3) The power system planning problem based on big data can be analyzed and processed by the collected power big data to reduce the impact Dependence on the physical model.

Acknowledgements

This work was financially supported by State Grid Jiangsu Electric Power Co., Ltd. Yangzhou Power Supply Branch Management Consulting Project "Research on Transmission Grid Planning Based on Adjustable Integrated Resources".

References

1. Liu Shuai, Kong Liang, Liu Zifa, Li Yuwen, Chen Yixuan. Transmission grid planning method based on deep reinforcement learning [J]. Electric Power Construction, 2021, 42(07): 101-109.
2. Liu Dundun, Cheng Haozhong, Liu Jia, Zeng Pingliang, Zhang Jianping, Lu Jianzhong. Overview and prospects for robust planning of transmission grids[J]. Power System Technology, 2019, 43(01): 135-143.
3. Lu Jingjing, He Zhiyuan, Zhao Chengyong, et al. Key technologies and prospects for HVDC grid planning[J]. Automation of Electric Power Systems, 2019, 43(2): 182-191.
4. Liu Lu, Cheng Haozhong, Wu Yaowu, et al. Research progress and prospects of transmission grid planning methods for high proportion of renewable energy[J]. Automation of Electric Power Systems, 2021, 45(13): 176-183.
5. Li Aowei, Liu Lu, Cheng Haozhong, et al. Transmission grid planning and testing system considering transmission and distribution coordination under high proportion of renewable energy access[J]. Automation of Electric Power Systems, 2021, 45(11): 19-27.
6. Wintzek Patrick et al. Application and Evaluation of Load Management Systems in Urban Low-Voltage Grid Planning[J]. World Electric Vehicle Journal, 2021, 12(3): 91-91.

https://doi.org/10.1051/e3sconf/202132901014

7. Ghayoor Fatemeh and Ghannadpour Seyed Farid and Imani Din Mohammad. Bi-objective robust optimization for reliability-oriented power network planning by considering distributed generation effects: A case study in Iran[J]. Sustainable Energy, Grids and Networks, 2021, 26(prepublish): 100455.
8. Wang Weichen, Zhang Tianyu, Xuan Wenbo, etc. Power grid planning method considering unit commitment and network structure optimization[J]. Proceedings of the CSU-EPSA, 2021, 33(2): 108-115.

Research on Capacity Configuration and Optimal Operation of Microgrid Energy Storage under Demand Management

Wensheng Li [1], *Donglei Sun* [1], *Yajin Li* [2], *Xian Wang* [1], *Rui Liu* [1], *Dayang Yu* [2]

[1] Economic Research Institute of State Grid Shandong Electric Power Co., Ltd., Jinan 250022, China
[2] School of Electrical Engineering, Shandong University, Jinan 250061, Shandong, China

Abstract. In order to reduce the energy consumption cost, considering the influence of the correlation of photovoltaic output, load demand and peak valley TOU price in different periods on the optical storage capacity configuration, the system operation strategy and capacity configuration principle are determined based on the peak valley TOU price and energy storage charge state information and the actual size of photovoltaic output and load.Integrated with the historical load, time of use electricity price, state of charge, charge discharge power and other constraints of the energy storage system, the minimum objective function is utilized to optimize the output power of energy storage.The effectiveness of the model is verified by an example. The sensitivity analysis shows that different demand tariff rules affect the user demand declaration strategy. The energy storage system planning selects the light storage combination with appropriate capacity according to the demand tariff rules and the change of energy storage investment cost, which has practical engineering value.

Keywords: Capacity configuration; energy consumption cost; demand electricity price; optimal scheduling.

1 Introduction

The dispatching control of energy storage and peak electricity price on the user side of microgrid can effectively reduce the user peak load and reduce the capacity electricity charge. For the power grid, it can bring the reduction of peak shaving cost caused by the reduction of load fluctuation and the reduction of system capacity caused by the reduction of maximum load. However, limited by the cost of energy storage capacity, under the price policy and market environment, how to balance the power consumption cost and energy storage investment of users, minimize the power consumption cost of microgrid users, and realize the energy storage evaluation of industrial large users with demand management and peak valley arbitrage need to be further studied.

Scholars have studied the optimal allocation and investment economy of energy storage applications. Reference [9] studies the cost and return on investment model of user side battery energy storage in the whole life cycle. Reference [10] establishes the investment, operation and maintenance cost of energy storage power station and the income model under different applications, and considers the charge and discharge depth and service life of battery. Document [11] proposed the operation mode of the battery energy storage system on the user side of the industrial park, considered the charge discharge operation under different energy storage strategies, and established the income and cost model of energy storage operation, but only carried out peak valley price difference arbitrage without considering the maximum demand control. Reference [12] establishes a month before and day before two-stage model for energy storage optimization, which takes the minimum total electricity charge and the maximum peak valley arbitrage as the objective function to calculate the charge and discharge power of output energy storage, but does not consider the situation that the number of cycles will affect the battery loss.

To sum up, the main source of income of energy storage investment on the user side is the saved user electricity bill, but most models do not fully consider the impact of parameters, performance characteristics and life cycle of energy storage system on its cost and income value. In order to ensure the best economic benefit in the operation of energy storage, an energy storage evaluation model is established with the minimum comprehensive cost of users, to evaluate the economy of users installing energy storage and optimize the energy storage configuration; The annual and intra day optimization models of energy storage are constructed. In the annual optimization, determine the monthly maximum demand reported by the user; In the intra day optimization, the daily energy storage charge and discharge power is optimized based on the maximum demand constraint. At the same time, this paper further analyzes the return on investment and user power cost of energy storage under different sensitive parameters, and verifies the economy of the model.

2 Demand management

Under the two-part electricity price, the electricity charge is mainly composed of basic electricity charge and electricity charge. The demand electricity charge belongs to one of the ways of basic electricity charge. The demand management is to reasonably control the user load so that the maximum power does not exceed the maximum power of contract electricity. Therefore, by making full use of the flexibility of energy storage power regulation, discharging at the peak of power consumption and controlling the maximum demand value of load, so as to reduce the basic electricity charge is an effective means to improve the income.

The basic electricity charge to be paid by the user is the maximum demand reported in the current month multiplied by the unit price (different in different regions). On this basis, when the actual maximum value exceeds 1.05 times the approved value, the part exceeding the approved value will be doubled. The electricity charge model per unit demand is:

$$\tilde{c} = c(x,\tilde{D}) = \begin{cases} p_1 x + p_2(\tilde{D}-kx), kx < \tilde{D} \le C \\ p_1 x, \tilde{D} \le kx \end{cases} \tag{1}$$

Where, C is the transformer capacity; $\tilde{D}$ is the actual maximum demand; x is the maximum demand, which shall not be lower than a certain distribution transformer capacity; k is the threshold of punitive price, k = 1.05 according to the current regulations; p1 and p2 are the demand price of the actual maximum demand within 1.05 times the reported maximum demand and the demand price of the part exceeding 1.05 times the reported maximum demand respectively. According to the current regulations, p2 / p1 = 2.

3 Energy storage capacity configuration

3.1 Minimum objective function of user power cost

The objective function of energy storage capacity allocation is to minimize the user's power consumption cost in the research period T, that is, the sum of electricity charge and demand charge minus the minimum demand response income, which can be expressed as:

$$\begin{aligned} \min C &= \sum_{t=1}^{T}(C_1 + C_2 + C_3)\cdot \Delta t + C_4 \\ C_1 &= (P_{lt} - P_{vt} + P_{ch,t} - P_{dis,t})\cdot p_t \\ C_2 &= P_{vt}\frac{C_{pv}}{P_{pv}\cdot H \cdot N_{pv}} \\ C_3 &= P_{dis,t}\frac{C_{BESS}}{E_{BESS}\cdot n_{bess}} \\ C_4 &= \alpha \cdot P_{demand}\cdot T_n \end{aligned} \tag{2}$$

Where, C is the total power consumption cost of microgrid users; C_1 represents the power purchase cost of microgrid users from the power grid; C_2 represents the cost of photovoltaic power generation; C_3 represents the energy storage and discharge cost. T is the number of periods in a research cycle and the duration interval of each period; C_4 indicates the electricity demand of the user; R represents the benefit of user participation in demand response.

Where, P_{lt} represents the power grid load in period t; P_{vt} is the photovoltaic power generation power in period T, $P_{ch,t}$、$P_{dis,t}$ which respectively represents the charging power and discharge power of energy storage battery in period t, and p_t is the time-of-use price of power grid.

Where, C_{pv} represents the overall photovoltaic investment cost (yuan), P_{pv} represents the photovoltaic installed capacity (kw), H represents the maximum utilization hours of photovoltaic year, and N_{pv} represents the photovoltaic operation life.

Where, C_{BESS} represents the overall investment cost of energy storage (yuan), E_{BESS} represents the rated capacity of energy storage configuration (KWh) and n_{bess} represents the number of energy storage cycles.

Where, P_{demand} represents the maximum demand of users and α is the electricity charge per unit demand.

3.2 Constraint condition

Capacity of the previous time + charging and discharging power of the current period * time = capacity of the current time.

$$E_t = E_{t-1} + \eta_{ch}\cdot P_{ch,t}\cdot \Delta t - \Delta t \cdot P_{dis,t} / \eta_d \tag{3}$$

Where, E_t and E_{t-1} represent the remaining capacity of the battery in t period and t-1 period, and η_{ch} η_d are the charging efficiency and discharge efficiency of the energy storage battery respectively.

2) The inequality constraint of the remaining battery power in period t is:

$$E_{\min} \le E_t \le E_{BESS} \tag{4}$$

Where, E_{BESS} represents the rated capacity of energy storage configuration and $E_{\min}$ is the lower limit of energy storage capacity.

3) The charging and discharging power of energy storage shall not exceed the limit.

$$\begin{aligned} 0 \le P_{ch,t} \le P_{\max} \\ 0 \le P_{dis,t} \le P_{\max} \end{aligned} \tag{5}$$

Where, $P_{\max}$ is the maximum charge discharge power limit of the energy storage system.

Regardless of the grid connection of energy storage, the configuration of energy storage is completely used to absorb photovoltaic and reduce user load. Therefore, the maximum charge and discharge power of the energy storage system only needs to meet the power imbalance within a certain period T within the assessment cycle t, that is:

$$P_{\max} \le \left| \max_{t\in[1,T]} [P_{lt} - P_{vt}] \right| \quad (6)$$

4) Maximum demand constraint
Guided by peak valley electricity price and demand response compensation, energy storage may form new load spikes, or even exceed the original maximum demand, resulting in the increase of demand electricity charge. Therefore, this paper introduces the maximum demand constraint.

$$\max(P_{lt} - P_{vt} + P_{ch,t} - P_{dis,t}) \le 1.05 P_{demand} \quad (7)$$

The above model is a dynamic programming problem. Equations (3) to (7) constitute the constraints of the objective function, including the constraints of battery power, charging demand, charging power, etc. In the model, $P_{ch,t}$、 $P_{dis,t}$, E_{BESS} , P_{demand} are decision variables, and other parameters are known quantities.
The above model is a dynamic programming problem. This paper establishes the model in Python and calls the optimization function library to solve it.

3.3 Evaluating indicator

This paper uses the return on investment index of energy storage in the whole life cycle to evaluate whether users are suitable for installing energy storage.

$$R = B / C_{BESS} \quad (8)$$

Where, R represents the return on investment, B represents the return on energy storage, that is, the electricity saved before and after energy storage, and C_{BESS} represents the overall investment cost of energy storage (yuan).

4 Analysis of case

4.1 Basic data

Taking the power consumption data of an industrial user as an example, the typical daily power load is shown in Figure 1.
According to the available photovoltaic sites, the maximum installed capacity of photovoltaic is 4.5342mwp. According to the calculation, the typical output curve of photovoltaic is shown in Figure 2. The annual photovoltaic power generation is 5562mwh and the annual utilization hours of photovoltaic are 1235.9h. The photovoltaic investment cost is 3500 yuan / kW, and the operation and maintenance cost is not considered.

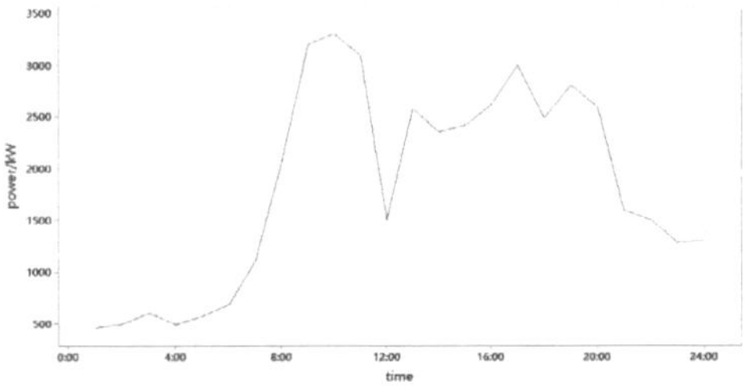

Fig. 1. Typical daily load curve

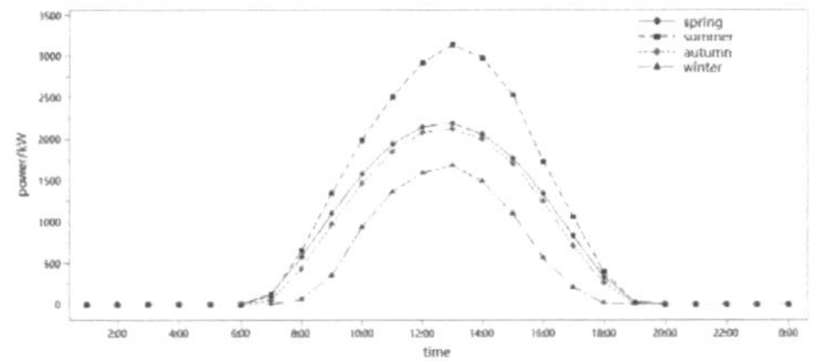

Fig. 2. Typical output curve of four seasons photovoltaic

The peak valley time division of Shandong Province is shown in Table 1. The time of use electricity price is: peak electricity price of 1.0102 yuan / kWh, peak electricity price of 0.8948 yuan / kWh, normal electricity price of 0.6062 yuan / kWh and valley electricity price of 0.3177 yuan / kWh. The maximum demand electricity price of Shandong power grid is 38 yuan / kW · month.
The type of energy storage battery is lithium iron phosphate, and the battery price is 900 yuan / kWh. The battery cycle life is 1500 times, and the battery charge / discharge efficiency is 92%.

4.2 Simulation analysis

Combined with the actual load power data of the user, this paper calculates the optimal energy storage capacity configuration of the user, considers the impact of lithium iron phosphate battery charge and discharge on capacity loss, and uses the objective function in Section 2.1 to evaluate the economy of energy storage after participating in demand management.
The scheduling interval is 1 hour. Assuming that the output power of photovoltaic and load is constant within 1 hour, the fluctuation of four typical days in spring, summer, autumn and winter represents the change of one year, and 8760 hours of the whole year are analyzed in combination with the days of different typical days.
When energy storage participates in peak valley arbitrage under TOU price, the optimal operation scheme is solved. Take the load data from June 29 to July 28 for the test, and set the equivalent charge and discharge times M = 2. Figure 3 shows the load curve comparison before and after typical daily energy storage participating in peak valley arbitrage. The abscissa is the sampling point with a sampling interval of 15min. Solve the model to obtain the energy storage operation strategy with the minimum electricity charge. The optimal configuration of energy

storage is 2347.5kwh and the reported maximum demand is 2235.5kw.

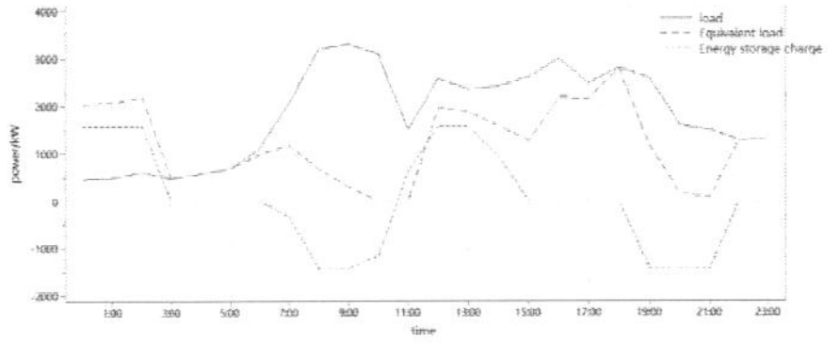

Fig. 3. Typical daily energy storage in spring and equivalent load curve after peak valley TOU price optimization

It can be seen from the figure that under the optimized energy storage charge and discharge strategy, the energy storage is attracted by the price difference between peak and valley. It is selected to charge and discharge in valley and peace periods and peak periods. By reducing the peak load, the final power consumption curve is smoothed as much as possible while earning the income from peak cutting and valley filling.

It can be seen from the analysis in Figure 4 that the rated power of energy storage and the monthly comprehensive cost of users tend to decrease first and then increase, while the expected income is on the contrary. This is because after the energy storage is increased to a certain capacity, the growth rate of monthly electricity saving is not enough to offset the growth rate of energy storage cost, resulting in the increase of comprehensive cost and the decrease of expected income. It can be seen intuitively from the figure that when the energy storage power is 998kw, the comprehensive cost of users is the lowest and the expected income is the largest.

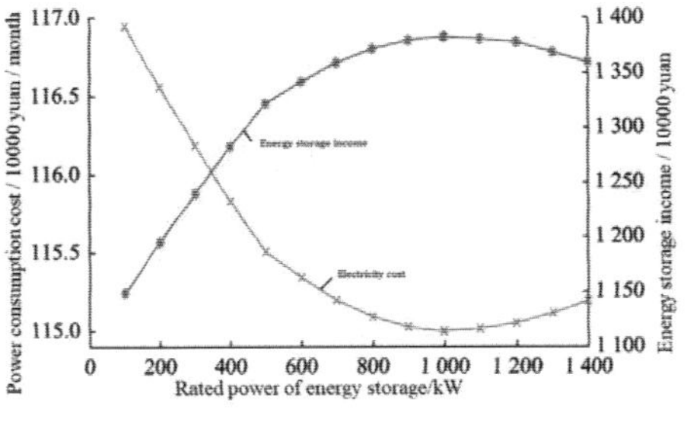

Fig. 4. Impact of energy storage allocation on users' economic benefits

4.3 Sensitivity analysis

Under different demand electricity prices, analyze the declaration of maximum demand. As shown in Figure 5, the four demand electricity tariff curves are the electricity tariff curves when P1 = [42,39,33,30] from top to bottom. The optimal solution of the maximum contract demand is marked at the bottom of each curve. It can be seen from the figure that the change of unit demand electricity price can not affect the user's contract maximum demand declaration strategy, but can only affect the user's demand electricity fee expenditure, and this impact is in direct proportion to the unit demand electricity price.

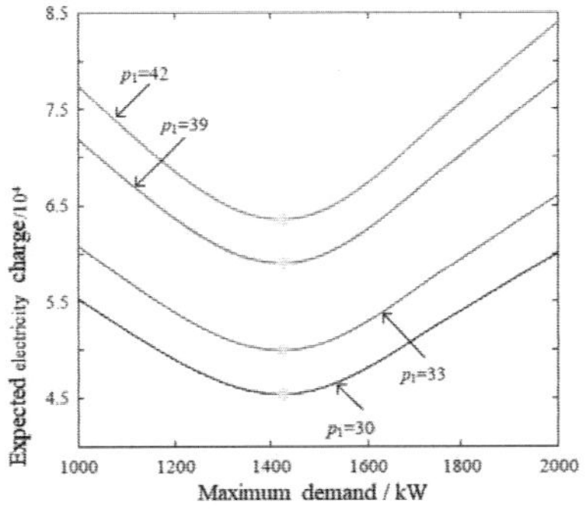

Fig. 5. Demand tariff curve under different demand tariff

In the current regulations, the part where the actual maximum demand exceeds 1.05 times of the contract maximum demand will be doubled, that is, the threshold of punitive charge is k = 1.05. In this section, users in the same case are simulated under the same other conditions and only the punitive charging threshold K changes, so as to intuitively show the impact of the change of punitive charging threshold on users' demand declaration. The simulation results are shown in Figure 6.

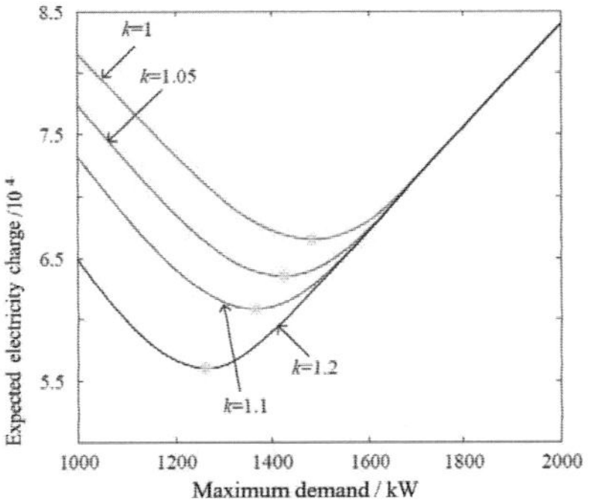

Fig. 6. Demand electricity charge curve under different penalty rates

The four demand tariff curves are the tariff curves of k = [1,1.05,1.1,1.2] from top to bottom. The optimal solution of the maximum contract demand is marked at the bottom of each curve. It can be seen from the figure that the adjustment of punitive charge threshold K can affect the declared value of users and the minimum value of electricity demand. The greater the punitive charge rate, the smaller the declared value of the user's maximum contract demand, and the smaller the corresponding minimum demand electricity charge.

5 Conclusion

Taking the minimum comprehensive cost of users as the objective function, this paper establishes an energy storage evaluation model. The example shows that it is suitable to install energy storage for large industrial users with load peak and large peak valley difference in the daytime; For users with relatively flat load curve, it is not suitable to install energy storage.

A two-stage energy storage optimization model is established in this paper. In the pre month optimization, the minimum total electricity charge of users is taken as the objective function to determine the reported maximum demand. In the intra day optimization, the maximum peak valley arbitrage of users is taken as the objective function to optimize the energy storage charge and discharge power and further improve the economic benefits of users. At the same time, through sensitivity analysis, this paper discusses the impact of the adjustment of demand tariff rules on users' declaration strategy and demand tariff from the perspective of power companies, and discusses the changes of demand tariff rules parameters including unit demand tariff and punitive charge threshold, which provides valuable conclusions and suggestions for the formulation and adjustment of demand tariff rules of power grid companies.

Acknowledgment

This work was supported by the science and technology projects of Economic Research Institute of State Grid Shandong Electric Power Co., Ltd. (SGSDJY00JPJS2000227).

References

1. CHEN Changqing, CHEN Changqing.Research on grid energy storage and photovoltaic coordination planning with flexible load[J].Power System Protection and Control, 2021,49(4):169-177
2. LI Jianlin, MA Huimeng, HUI Dong. Present development condition and trends of energy storage technology in the integration of distributed renewable energy [J] . Transactions of China Electrotechnical Society, 2016, 31(14) :1-10.
3. ZHANG Mingxia, YAN Tao, LAI Xiaokang, et al. Technology vision and route of energy storage under new power grid function configuration[J]. Power system technology, 2018, 42(5) :1 370-1 377.
4. BARTON J, INFIELD D. Energy storage and its use with intermittent renewable energy[J]. IEEE Transactions on Energy Conversion, 2004, 19(2) :441-448.
5. FAN Haifeng, YU Zhipeng, LIU Wenlong, et al. A demand assessment approach of energy storage for participating in fast frequency regulation of power system[J]. Electric Power Engineering Technology, 2019, 38(2) :18-24.
6. XU Haihua, WANG Xudong, ZHU Xingyang, et al. Research on the optimal configuration of energy storage in user side integrated energy system[J]. Power Demand Side Management, 2020, 22(2) :13-20.
7. LI Hua, WANG Simin, GAO Jie. Research on energy storage control strategy of wind solar storage combined generation system based on tracking plan [J]. Electrical & Energy Management Technology, 2019(4) : 71-78.
8. GUO Li, XUE Guiyuan, WU Chen, et al. Research on economic evaluation of energy storage applied to load shift [J]. Power Demand Side Management, 2019, 21(5) :31-34.
9. XUE Jinhua, YE Jilei, TAO Qiong, et al. Economic feasibility of user-side battery energy storage based on whole-life cycle cost model[J]. Power System Technology, 2016, 40 (8) : 2 471-2 476.
10. XIONG Xiong, YANG Rengang, YE Lin, et al. Economic evaluation of large scale energy storage allocation in power demand side[J]. Transactions of China Electrotechnical Society, 2013, 28(9) : 224-230.
11. JIA Xuecui, LI Xiangjun, WANG Huanling, et al. Re - search on consistency assessment method for energy storage battery based on operating data fusion[J]. Distribution & Utilization, 2017, 34(4) : 29-35.
12. CHEN Lijuan, WU Tiantian, LIU Huibo, et al. Demand management based two- stage optimal storage model for large users[J]. Automation of Electric Power System, 2019, 43(1) :194-200.

 https://doi.org/10.1051/e3sconf/202132901016

Construction and application of comprehensive early warning system for coal and gas outburst in Xinjing Coal Mine

Yunlong Zou [1, 2, 3, *]

[1] State Key Laboratory of the Gas Disaster Detecting, Preventing and Emergency Controlling, Chongqing, 400037, China
[2] Chongqing Research Institute of China Coal Technology and Engineering Group Crop., Chongqing, 400037, China
[3] Chongqing University, Chongqing, 400044, China

Abstract. In order to further strengthen the prevention and control of coal and gas outburst in Xinjing Coal Mine, the online comprehensive analysis and early warning index system and rules of coal and gas outburst suitable for Xinjing Coal Mine were studied. Based on the corresponding early warning computer system and guarantee mechanism, a comprehensive early warning system for coal and gas outbursts in Xinjing Coal Mine was established, realizing real-time intelligent early warning of outburst dangers in working faces. The system realizes the standardization and dynamic management of outburst prevention information at the working face, as well as the real-time dynamic update and sharing of outburst prevention information, which improves the efficiency of mine outburst prevention management and the level of mine safety. The actual application of the system in Xinjing Coal Mine shows that the system can provide an effective reference for the comprehensive early warning of outbursts in other outburst mines of Yangquan Coal Industry Group.

1 Introduction

Coal and gas outburst poses a serious threat to the safety of mine production [1-2]. The primary link of coal and gas outburst prevention is to judge the possibility and intensity of outburst based on the precursor information of the outburst [3]. A certain predictive index can only indirectly and partially reflect a certain aspect of the outstanding risk, and there is no early warning of the outstanding risk and its severity from multiple directions and perspectives, which has considerable limitations [4-8]. At the same time, with the development of computer technology and information technology, the informatization of coal mine safety management will also be an irreversible development trend [9-13]. Therefore, with the help of computer technology and information technology, a coal and gas outburst warning system suitable for the specific conditions of the mine is established to realize the information management of mine safety and the early warning of mine outburst disasters. This will be one of the important technical means for coal mines to effectively prevent and control outstanding disasters and ensure coal mine production safety in the future.

At present, the coal and gas outburst disasters in the mines of Yangquan Group are serious, and the outburst has become the main factor restricting the safe and efficient production of the outburst mines of the group. Taking the Xinjing Coal Mine of Yangquan Coal Group as the pilot unit, aiming at the problems existing in the current gas prediction and forecasting in the mine, comprehensive information such as gas geology, gas emission, and roof pressure in the outburst mine is integrated, there is an urgent need to enhance the mine's ability to prevent coal and gas outburst disasters and improve the safety management level for Xinjing Company to, so as to promote the application of experience to other outburst mines of Yangquan Coal Group.

2 Research on early warning index system of coal and gas outburst

The research is the prerequisite and foundation for the implementation of the entire early warning system project. It is the development of early warning system software and provides a technical basis for the realization of comprehensive, intelligent, advanced and dynamic early warning of coal and gas outbursts in coal mines [14-15]. The system mainly focuses on the research of comprehensive early warning technology [16]. The index system framework of the entire early warning system (as shown in Figure 1) mainly includes six aspects: gas geology, mining impact, daily prediction, gas emission, mine pressure, and outburst prevention measures. Under the guidance of this framework, examine the specific early warning indicators and rules suitable for Xinjing Company, and determine the critical value of each early warning indicator and its relationship with the degree of outstanding danger, and finally establish a prominent early warning indicator system that is in line with Xinjing Company's actual situation.

* Corresponding author: 2010152@cqccteg.com

https://doi.org/10.1051/e3sconf/202132901016

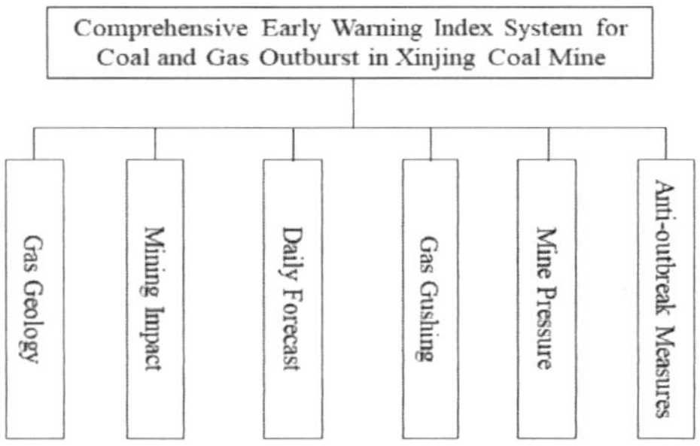

Figure.1 Framework of comprehensive early warning indicator system for coal and gas outburst in Xinjing Coal Mine

Early warning rules should include status warning rules and trend warning rules. Among them, the state early warning rules are the rules set to reflect the current state of the work face with outstanding hazards, and the trend early warning rules are mainly the rules formulated to reflect the changing trend of the prominent hazards in the front of the work face.

Xinjing's outstanding early warning rules are divided into 6 categories, a total of 57, including 11 gas geological categories, 26 mining impact categories, 7 daily forecast categories, 4 gas emission categories, 6 mine pressure categories, and 3 outburst prevention measures. In the production process, the value of each early warning indicator reflects the state of the early warning elements of the working face [3].

3 Construction of a comprehensive early warning system for coal and gas outbursts

The coal and gas outburst early warning computer system is a large-scale computer program based on the combination of GIS, C/S and B/S. Through the outburst early warning module, the system analyzes the coal and gas outburst early warning data in the comprehensive early warning data server, and sends out early warning information according to the analysis results [17]. According to the characteristics and related requirements of the outburst prevention and control work in Xinjing Coal Mine, the comprehensive early warning system for coal and gas outburst in Xinjing Coal Mine includes 7 subsystems and 1 outburst early warning website. Seven of the subsystems are mainly geological measurement management system, gas geological dynamic analysis system, gas outburst dynamic management and analysis system, gas emission dynamic analysis system, mining progress management system, mine pressure monitoring and early warning system, and gas outburst comprehensive early warning platform [3].

The stable operation of the coal and gas outburst early warning system is the prerequisite for accurate early warning. After the software adaptability adjustment and preliminary system test are completed, the system functions are gradually improved and the overall operation is stable. The specific performance is as follows:

(1) The hardware of the early warning system runs reliably. During the inspection, the early warning server and client computers were operating completely normally, and there were no machine failures such as crashes or abnormal shutdowns. After on-site testing, the UPS power supply can guarantee the normal operation of the early warning server for 20 minutes in the event of an abnormal power failure. The SMS publishing platform can promptly and accurately publish the warning result information.

(2) The early warning system software responds quickly. After many tests, the various subsystems of the early warning system open the local mine map generally within 10s. When the office network is normal, the time to open the server mine map directly is within 15s. Data synchronization and early warning analysis can generally be completed within 5 minutes under normal network conditions.

(3) The error rate of the entire early warning system is low. When the user operates and maintains the early warning system in accordance with the "Xinjing System Early Warning System Job Responsibility" and related operational requirements, the early warning system can calculate early warning indicators in a timely and accurate manner and release the early warning results in a timely manner. The system runs continuously and stably for longer than 3 months.

4 Application effect analysis

4.1 Follow-up inspection of daily early warning effect at work face

According to the relevant information, during the inspection period, there were two major outburst dangers in the 3# Coal South Eighth Lane in Lunan Second District, one occurred on February 2, and the K1 value was predicted to reach 1.28mL/g•min1/2, the thickness of the soft layer is 0.8m.

The trend pre-alert for the zero o'clock shift on January 30 is "orange", and the trend warning for the shift at four o'clock on February 1 will be upgraded to "red", and the state warning for the eight o'clock shift on February 2 will be upgraded to "danger", and the trend warning will be "orange". The K1 forecast on duty exceeded the standard, and the triggering factors of this early warning process were all gas gushing (Figure 2). The early warning results were accurate and had good forecasting advancement (the K1 value was 2 days ahead). In addition, on February 2nd, the thickness of the soft layer exceeded the warning threshold of 0.3m.

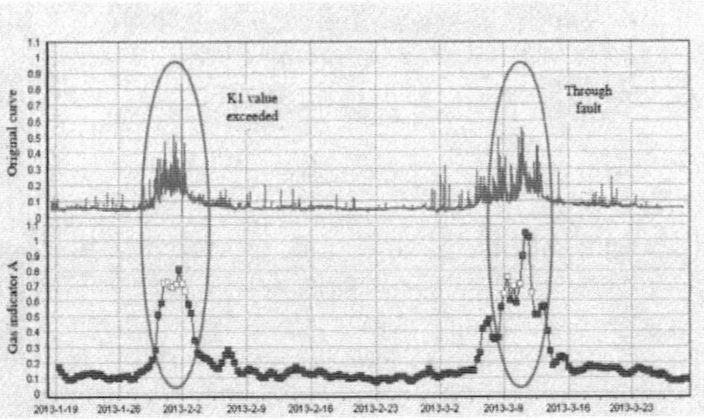

Figure.2 Early warning results of gas emission in Nanba main lane

The pre-warning of the eight o'clock shift on March 7 was "orange". And at 8 o'clock on March 10, the trend warning was upgraded to "red" and the status warning was "threatening", indicating that the working face has entered the structural influence range and the structure will be revealed soon. By the start of the zero o'clock shift on March 9, it gave an "orange" warning from the gas gushing angle, and then the zero o'clock shift was upgraded to a "red" warning on March 11. This situation shows that under the action of geological structure, gas gushing is abnormal. In this early warning process, two aspects of gas geology and gas gushing were carried out to warn the outstanding hazards in front of the work. The law of occurrence of outstanding hazards was completely consistent with theoretical research and on-site conditions. This further shows that the early warning system has good advancement and reliability.

4.2 Analysis of typical daily warning results

The daily forecast of No. 8 Coal East No.1 Main Lane in Lunan No.1 District exceeded the standard at 8 o'clock on July 4th. K1 reached 1.49mL/g•min1/2, and the K1 value was too large, indicating that there is a greater risk of outburst in front of the work. The trend warning result of the zero o'clock shift on July 2 was "orange", and the status warning result was "normal". The trend warning result of the eight o'clock shift on July 2 remained "orange", and the status warning result was upgraded to "threatening". However, because the predicted K1 value of the working face did not exceed the standard (the index value is 0.35mL/g•min1/2), and there is no obvious gas dynamic phenomenon, the working face continued to move forward until the zero o'clock shift on July 3, the early warning result is upgraded to "red".

On the same day, the K1 value still did not exceed the standard (the index value was 0.35 mL/g•min1/2), and there was no dynamic phenomenon, so the tunneling operation was still carried out according to the normal procedure until the K1 value of the eight o'clock shift on July 4 exceeded the standard. The work face stopped work and construction measures were taken. Subsequently, the results of the trend warning and status warning of the shift at 8 o'clock on July 5 became "normal".

The triggering factor of this alarm was abnormal gas gushing, and it issued a prominent hazard warning 2 days ahead of Xinjing's daily forecast of K1 value, indicating that the early warning is well advanced and can leave a sufficient safety distance for the implementation of anti-outbreak measures. At the same time, from July 1st to July 3rd, the gas emission index showed a gradually increasing pattern of change (Figure 3), which was in full compliance with the general law of "zone and zone" distribution of prominent hazards. After July 4th, with the gradual elimination of the prominent danger of the working face, the warning level has also been reduced, which fully shows that the warning results have high reliability.

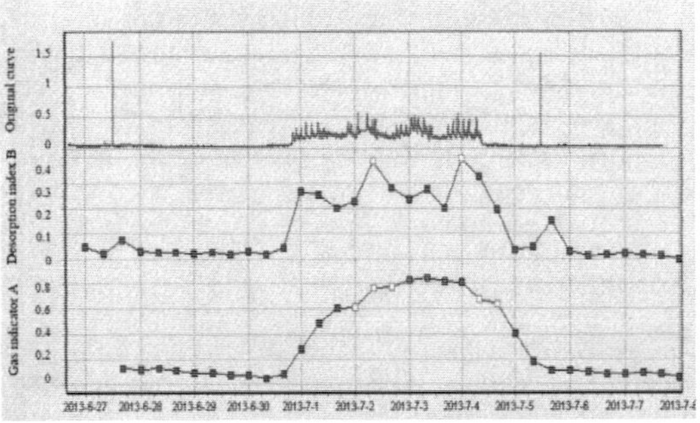

Figure.3 Early warning results of Dongyi main lane

The early warning result of Nanba main lane is shown in Figure 4. It can be seen from Figure 4 that the trend warning result of the early warning system at the eight o'clock shift on July 28 was "orange", and the trend warning result of the zero shift was upgraded to "red" on July 29, and the K1 value of the eight o'clock shift predicted that the K1 value exceeded the standard on July 30. This time, the early warning system has advanced the K1 value by 2 days from the angle of gas emission to provide early warning of the prominent danger in front of the Nanba main lane excavation work.

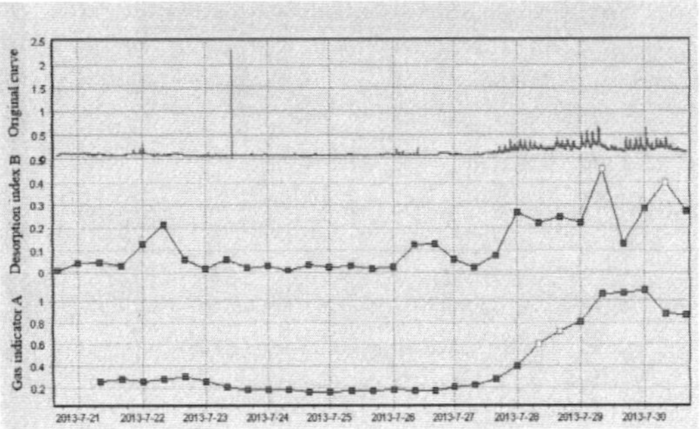

Figure 4. Early warning results of Nanba main lane

5 Conclusion

(1) Based on the actual site data and data of Xinjing Coal Mine, comprehensively analyzed the impact on the outburst hazard of the mining face of Xinjing Coal Mine, and established consideration of gas geology, mining impact, daily prediction, gas emission, mine pressure and Six factors including anti-outburst measures are applicable to Xinjing's coal and gas outburst online

comprehensive analysis and early warning indicator system and rules.
(2) According to the characteristics and related requirements of the outburst prevention and control work of Xinjing Coal Mine, a comprehensive early warning system for coal and gas outbursts of Xinjing Company was established. The system mainly includes the geological survey management system, the gas geological dynamic analysis system, and the outburst prevention dynamic management and analysis.
(3) The comprehensive early-warning system for coal and gas outburst constructed in Xinjing Coal Mine has been inspected on site application effects and showed that the software and hardware of the comprehensive early-warning system can ensure the continuous and stable operation of the early-warning system, as well as timely and accurate release of early-warning results.

Acknowledgments

The study was supported by the Special Fund Project for Safety Production in Chongqing (Coal Development) (2020B182[5985]20015501), Chongqing Research Institute Independence Project (202001120).

References

1. Si Hu, Zhao Jiannan, Hu Qianting. Analysis of causes of coal and gas outburst accidents based on big data theory[J]. Journal of Xi'an University of Science and Technology,2018,38(04):515-522+537.
2. Jiang Anfei, Sun Dongling, Liu Yanbao, et al. Study on the effect of gas pressure on the propagation of coal and gas outburst shock wave [J]. Mining Safety and Environmental Protection, 2021, 48(02): 18-22.
3. Zhao Xusheng, Sun Haitao, Cao Jie, et al. Applications of online integrated system for coal and gas outburst prediction A case study of Xinjing Mine in Shanxi China [J]. Energy Science and Engineering,2020,8(6): 1980-1996
4. Zhao Xusheng, Ma Guolong, Qiao Wei, et al. Coal and gas outburst warning index system based on fault tree analysis [J]. Mining Safety and Environmental Protection, 2019, 46(03): 37-43.
5. Ning Xiaoliang. Research status and development trend of early warning technology for coal and gas outburst[J]. Industry and Mine Automation, 2019, 45(08): 25-31+37.
6. Cao Kang, Li Zhonghui, Yu Desheng, et al. Research on precursor characteristics and comprehensive early warning of coal and gas outbursts in driving face [J]. Coal Science and Technology, 2020, 48(11): 147-152.
7. Song Dazhao, He Xueqiu, Qiu Liming, et al. Research on dynamic real-time monitoring and early warning technology of regional and local outburst hazards[J]. Coal Science and Technology, 2021, 49(05): 110-119.
8. Xu Xuezhan. Continuous prediction technology and application of coal and gas outburst based on gas emission characteristics [J]. Energy and Environmental Protection, 2021, 43(01): 1-7.
9. Pu Yang. High-yield and high-efficiency mine coal and gas outburst far and near field collaborative prediction technology[J]. Coal Mine Safety, 2020, 51(07): 70-74+80.
10. Ning Xiaoliang. Coal and gas outburst warning technology based on big data[J]. Mining Safety and Environmental Protection, 2020, 47(04): 51-56.
11. Long Nengzeng, Yuan Mei, Wang Guanliang, et al. Research on real-time early warning of coal and gas outburst based on data mining[J]. China Mining Industry, 2020, 29(11): 88-93+99.
12. Yao Yahu. Application of multi-parameter outburst warning technology based on gas geological characteristics in new coal mines [J]. Mining Safety and Environmental Protection, 2021, 48(03): 94-99.
13. Cheng Xiaoyang, Pu Yang, Song Zhiqiang. Application of outburst warning system based on gas monitoring data source in Xiayukou Coal Mine [J]. Mining Research and Development, 2021, 41(07): 161-165.
14. Niu Jufen, Cheng Wuyi. Construction of early warning index system for coal and gas outburst based on time dimension[J]. Coal Engineering, 2012(05): 96-98.
15. Tan Guowen. Research on the determination method of coal and gas outburst warning indicators based on gas emission characteristics in Wangjiazhai Coal Mine [J]. Mining Safety and Environmental Protection, 2015, 42(03): 34-37.
16. Che Yuheng. Application of multi-source information intelligent early warning system for coal and gas outburst [J]. IOP Conference Series: Earthand Environmental Science, 2021,781(4): 042027.
17. Ning Xiaoliang. Construction and application of comprehensive early warning system for coal and gas outburst in Wangjiazhai Coal Mine[J]. Energy Technology and Management, 2014, 39(02): 5-7.

Sedimentary micro-phase and reservoir display characteristics in shale Oil

Cai Xiao [1, 2]

[1] The Key Laboratory of Unconventional Oil & Gas geology, CGS; China
[2] SINOPEC East China Oil and Gas Company Exploration and Development Research Institute China

Abstract. Sedimentary microphilic and its exhibition feature is of great significance for identifying the distribution of remaining oil. This paper studies the development of geological characteristics of Zhidan Oil Zone, mineralogue and scanning electron microscopy experiments, and obtains the sedimentary micro-phase and reservoir display characteristics. The results show that combined with logging data, the core phase analysis, logging phase analysis, single well phase analysis, etc. Taking the leading edge of the Delta, the microphase type such as the river, the diversion bay and the underwater natural embankment is developed under water, due to the lifting of the lake, form a number of invested, plus, replenishment, in which water The downstream river is facilitating the advantages of the reservoir. The sedimentary phase and the angle effect of the zone are dominated by 4 + 5, chang 6 oil and gas, showing the characteristics of the rock sex reservoir. The type of reservoir is a rock sex reservoir. The main development of sandstone is impaired and the sandstone lens is a rock sex reservoir. The reservoir is an important influencing factor in the reservoir. The area has no uniform oil and water interface, lacking edge, bottom water.

Keywords: Zhidan oil area; extended group; deposition; reservoir exhibition.

1 Introduction

Understanding the sedimentary environment, deposition microforming and exhibition feature of the middle and later reservoirs, which have important practical significance for identifying the distribution of remaining oils, and the post-injection well network design is important. Through the system collection and organizing the core, recording of the study area And a large number of logging materials, the formation is detailed, and there is an important research significance to further clarify the feature of the reservoir. Zhidan oil plant 3043 well area is located in the northeastern part of Zhidan County, with an area of about 33km2 and 723 in total wells. Among them, there is 166 injection wells, 557 oil mining well; current average single well Nissan liquid 0.84m3, Nissan oil 0.37 T, 48% water; in which the water moisture is ≥80% oil well 40, due to water flooding, high water stops lying in the well 70; the main oil layer is 4 + 5 and chang 6 oil layer group, and some areas simultaneously develop length 3 and lower Combined oil layer group. This paper studies the development of geological characteristics of Zhidan Oil Zone, mineralogue and scanning electron microscopy experiments, and obtains the sedimentary micro-phase and reservoir display characteristics.

2 Method and experiment

2.1 Sedimentary microphase division

Combined with the recording phase, the logo phase and the seismic phase analysis are the main methods and means of the current sedimentary microformation analysis. Conventional deposition microphase analysis mainly includes a combination analysis of rock sex, core deposition characteristics, heavy mineral vertical and horizontal planar display characteristics, ancient water flow analysis, etc. At present, the system binds to the sedimentary microphase analysis technology, and the deposition microformation method of the logo phase, the log phase and the seismic phase has been widely used. From the perspective of longitudinal research accuracy, it is carried out in a sand group or Sedimentary microplasma analysis of single sand layer is the largest formation unit of oil and gas release exploration. From a specific deposition microphase planar view technique, for the same deposition microformation study formation unit (such as sand group or single sand layer), a well point conventional deposition microformation, recording phase and logging phase analysis The starting point determines the deposition microphase type, and the study results of the study of sandstone thickness, the planar extrapolation

 https://doi.org/10.1051/e3sconf/202132901017

is determined to determine the deposition micro phase boundary on the plane. When planned, the sedimentary micropomposition is selected as the oil and gas reservoir relationship is a superior phase, indicating a percentage of different deposition microspices in longitudinal direction. The deposition microphase type facilitating the development of the oil and gas is then determined by the deposition microphase type analysis belonging to the oil reservoir to be known. On the basis of the former research work, the core of the center of the work area was observed and the core phase analysis was formed, forming a comprehensive understanding of the deposition of 4 + 5, chang 6 oil layers in the 359 well region. Its research methods are as follows (Figure 1):

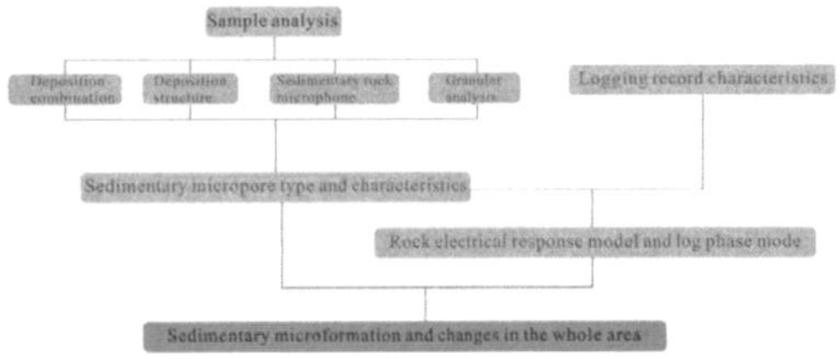

Figure 1 Sedimentary microphlic research flow chart

2.2 Scanning Electron Microscopy (SEM)

The scanning electron microscope experiment was carried out at China University of Petroleum (Beijing). The model used was FEI Quanta 200. The sample was observed by argon ion polishing before the observation. In the experiment, the energy type of the shale was combined with the energy spectrometer to determine the pore type in the shale. The resolution is 2 nm.

3 Results and discussion

3.1 Rock phase sign and rocky characteristics

The rock phase is the sum of the rocky characteristics formed in a particular deposition environment, the same rock represents the product under the same deposition conditions. By analyzing the characteristics of different rocks, structures, deposition structures, types, etc., can reflect the hydraulic conditions of different abrasive depositions, and restore the original deposition environment. According to the core observation, the zone develops 8 major rock types.

(1) Block layer: mainly gray sandstone (Figure 2), single layer thickness is small, bottom is a mutation contact, the layers are not obvious, the up transition is the parallel layer, and it is slightly thinner. The rhyme rhythm is rotated back, reflecting the rapid accumulation under strong water conditions, mainly in the river in the front of the Delta.

(2) Parallel layer: mainly gray fine sandstone (Figure 3, Figure 4), a single layer thickness, better sorting. The thickness of the tattoo layer is between 0.2 ~ 0.5 cm, which consists of parallel flat continuous or intermittent texture, and the texture is displayed by char. Reflects the water-motion conditions of water flow. It is mainly seen in the leading water under the front of the Delta. .

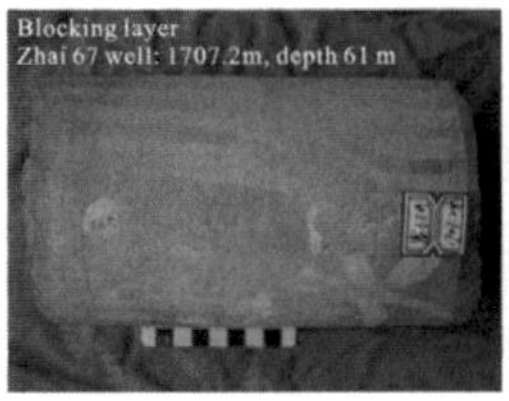

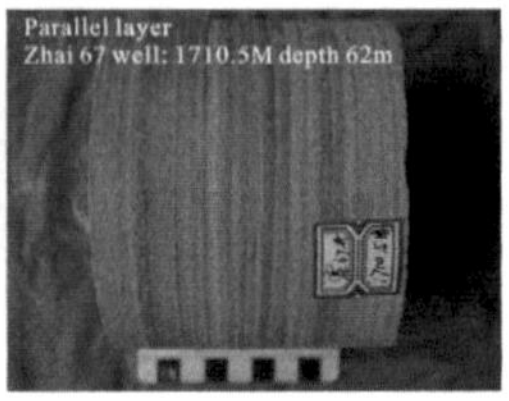

Figure 2 Block layerFigure 3 Parallel layer

(3) Interlaced layers: Mainly gray fine sandstone (Figure 5), a single layer thickness, better sorting. Composed of a series of slanting layers, the texture is displayed by charging, reflects strong hydraulic conditions, mainly seen in the front of the Delta frontal water.

(4) Horizontal layers: Mainly developed in mudstone, mudstone, muddy powder sandstone. Its texture is thin and parallel to each other, indicating that in low flow of low energy, forms a suspension substance deposition, and is common in diversion bay deposition.

(5) Deformation Hierarchical: Common and Sandstone and fine sandstone, belonging to the layers of layers in the layer. Due to the liquefaction effect in the deposited layer or the shear effect of the water, it is common in the deposition of the shunt.

(6) Sand pattern layer: there is a sand-shaped migration. Mainly in the sandstone, muddy powder sandstone, is a multi-layered small small interleaving layer, and the lower interface of the layer is micropro-waveform, the tattoo is irregular, intermittent or continuous, and the fine layer is inclined to one side and Under the convergence, see fine plant debris, carbonia and rich mica tablets, and often with parallel layers, plate-shaped staggery and small interlaced layers. It is mainly formed in an environment with weak hydrodynamic conditions, such as the Delta Division Diversion Bay Deposition Environment.

(7) Cycloma: Commonly in the sand mud, multiple development in powder ~ fine sandstone, muddy powder sandstone and sand mudstone, fine layer is curved, sometimes clearly lenticular, corrugated The mud layer and the sand layer are unregulated to each other in front of the Delta premium sand deposition.

(8) Flush surface structure: It is a deposition structure that is common in the study area. It is a surface of strong water to form unevenness in the surface of the dodel. It is shown in the mudstone, and the sandstone is stuck in sandstone. And more mud and tear muddy fragments. The flushing surface is generally developed at the bottom of the river. When there are multiple discarded rivers to overlap each other, there can be seen a plurality of mudble layers.

According to the observation and statistics of the sheet and the surface electron microscope, the length of the Zhidan Oilfield 3043 well regions extended 4 + 5 and chang 6 reservoir rock types are mainly light gray, gray, gray brown, gray green for powder - fine granules and Medium - fine grain chute. The ingredients are main, the stone is the stone, the dark mineral is small, the selection is good, the mud gel, the subprogramme, and the

 https://doi.org/10.1051/e3sconf/202132901017

looseness. The overall mineral maturity is low, the structure is high, and the rocky characteristics of the formation of a strong sandstone reservoir.

3.2Sedimentary type

According to the above deposition phase identification method, the predecessor is taken from the large number of geological studies in the region, and the sedimentary microforming of the length of 4 + 5 and 6 in the region is divided. The length 4 + 51 in the area is the Delta Plain Asia, and the natural embankment, the natural embankment, the river is deposited; the length of 4 + 52, chang 6, the period is the leader of the Delta, subdivided into the water Divide the river, underwater natural embankment, the split bay deposited microphones. According to regional data, core observation and logging data analysis can be divided into the leading edge deposition system in the Delta, and the detailed division is shown in Table 1.

1. Underwater shunt

The research area underwater shunt river sand body is an important reservoir, and the water shifts in the water is more straightforward. The more it is more smaller. At the bottom of its deposition, there is often an unevenly scored surface, and there are more mud and mudstone tore. In the rock section, the river props are multi-rhythm. The single sand layer has a small thickness, which exhibits a tendency to gradually thinner from the main river to the river wing, and the uniform block shape of the positive rotary or top mutation is generally present. The obvious mud rock compartment, but the particle size distribution is unstable, mostly medium, fine, abrasive thin layer interactive body, and local muddy strip development. The natural potential curve is a bell or box shape, represents the deposition of the side, and the plus. The natural gamma curve curve is strong and abnormal, the microelectrode is large, and the permeability is good. The sound speed curve is low, dental, but has a peak. The resistivity curve is a medium amplitude wave.

Table 1 3043 Well Region chang 4 + 5, chang 6 Deposition System Division

Layout		phase	Subphor	Microphone
Chang4+5	Chang4+5^{1}	Delta	Delta plain	Distilted river, natural embankment, river
	Chang4+5^{2}		Delta front edge	Underwater shunt Underwater natural embankment Diversion Bay
Chang6	Chang6^{1}			
	Chang6^{2}			
	Chang6^{3}			
	Chang6^{4}			

2, divide the river

It is part of the grid in the Delta plain, with the characteristics of general river deposition, mainly laminated fine sandstone, laminated, laminated sandstone, regular rhythm. The sandstone sorting and molarity is good, and the debris particles are circular or rounded. Generally, the bottom is medium ~ fine sand sand, often containing the sediments such as mud, plant stems, and the upwards of sanding, muddy powder sandstone and powder sand mud. The sanding surface is seen in the bottom of the sand, and the sand layer has a slot or plate-shaped interlaced hierarchy and a wave interlaced layer. The cross section is a top bottom convex or double convex lens shape, and the sand body is often sinking in a downstream mudstone layer, wherein the outermost thickness and thickening are thinned to both ends. The natural potential curve is mostly a box-shaped negative abnormality, and there is a bell shape, a bell-shaped box and a box-shaped ~ funnel is negative.

3, natural embankment

On the land on the land. It is usually seen in the longitudinal section and is located on both sides of the shunt main river bed, and its morphology is controlled by the main river channel. The lithology is mostly a sandstone and muddy powder sandstone, and the thin layer is thin sandstone and mudstone. Cyber layer, sand grain interleaving layer, horizontal hierarchy and water wave marks are more common. The natural potential curve and the natural gamma curve are low in a lower center finger, which is a plurality of funnel or step-shaped stacks, and from the main river road to the river wing, the bell shape is gradually increased, the curve shape gradually became irregular zigzag, curved amplitude Small.

3.3 Reservoir type

The sedimentary phase and the angle effect of the zone are dominated by 4 + 5, chang 6 oil and gas, showing the characteristics of the rock sex reservoir. The type of reservoir is a rock sex reservoir. The main development of sandstone is impaired and the sandstone lens is a rock sex reservoir. The reservoir is an important influencing factor in the reservoir. The area has no uniform oil and water interface, lacking edge, bottom water.

(1) Sandstone tilt rocky reservoir

Such reservoirs are widely distributed, in the horizontal length of 4 + 5, the chang 6 oil layer group Delta frontal water is divided into the shunt of the river sand body in the upper direction of the bay mudstone, and the reservoir superficial direction sandstone is destroyed. Gravity is a dense powder sandstone, mudstone. The reservoir sandstone side flaps and the upper direction of the reservoir (mudstone) are formed on the plane.

(2) Sandstone lens body rock sex reservoir

This type of reservoir is more developed in this region, mainly with mud mud rock, sandstone, and sandstone body generally lenticular output, and the reservoir feature is coincident with sand body development, generally a lens-shaped distribution.

4 Conclusion

(1) On the basis of mound heart observation, combined with logging data, the core phase analysis, log finding analysis, single well phase analysis, etc., the logging phase recognition mode is established, and the 3043 well region length 4+ is obtained. 5. chang 6 oil layer group mainly based on the leading edge of the Delta, and the

microformation of the river, the diversion bay and the underwater natural embankment is developed. Due to the rise and fall of the lake, a number of invested, plus, The replenishment is rotated back, wherein the water split river is facilitating the advantages of the reservoir. The research section leads 4 + 5 and chang 6 reservoir rock types are mainly long-stone sandstone, and the debris ingredients are mainly stone and quartz, followed by black microconuts and stings. The debris particles are mainly in line contact and point-line contact, and local visible points are in contact. The cementation type is mainly pore cement, thin film cement, film-porous cement, and increase-pore cement, a small amount of embedded cement. The pore type is mainly granulated hole and a variety of dissolves, a small amount of crack holes.

(2) The sedimentary phase and the angle effect of the region are dominated by 4 + 5 and chang 6 oil and gas, showing the characteristics of the rock sex reservoir. The type of reservoir is a rock sex reservoir. The main development of sandstone is impaired and the sandstone lens is a rock sex reservoir. The reservoir is an important influencing factor in the reservoir. The area has no uniform oil and water interface, lacking edge, bottom water.

Acknowledgments

Thanks to the Funding support by Study on the microscopic pore structure and adsorption capacity of organic matter and clay minerals in shale (Number: DD20160178-03-02) .

References

1. Yang F, Ning Z F, Hu C P, et al. Characterization of microporous pore structures in shale reservoirs[J]. Acta Petrolei Sinica, 2013, 34(2): 301-311.
2. Bustin R M, Bustin A M M, Cui A, et al. Impact of shale properties on pore structure and storage characteristics[C]//SPE shale gas production conference. Society of Petroleum Engineers, 2008.
3. Jiao K., Yao S., Wu H., etc. Research progress on characterization methods of shale gas reservoir pore system [J]. Geological Journal of China Universities, 2014(01): 151-161.
4. Zhou S, Yan G, Xue H, et al. 2D and 3D nanopore characterization of gas shale in Longmaxi formation based on FIB-SEM[J]. Marine and Petroleum Geology, 2016, 73: 174-180.
5. Dung Le T, Murad M A. A new multiscale model for methane flow in shale gas reservoirs including adsorption in organic nanopores and on clay surfaces[C]//International Symposium on Energy Geotechnics (1st: 2015: Barcelona). Universitat Politècnica de Catalunya. Departament d'Enginyeria del Terreny, Cartogràfica i Geofísica, 2015.
6. Wei M, Zhang L, Xiong Y, et al. Nanopore structure characterization for organic-rich shale using the non-local-density functional theory by a combination of N 2 and CO 2 adsorption[J]. Microporous and Mesoporous Materials, 2016, 227: 88-94.
7. Misch D, Mendez-Martin F, Hawranek G, et al. SEM and FIB-SEM investigations on potential gas shales in the Dniepr-Donets Basin (Ukraine): pore space evolution in organic matter during thermal maturation[C]//IOP Conference Series: Materials Science and Engineering. IOP Publishing, 2016, 109(1): 012010.
8. Ju B, Wu D. Experimental study on the pore characteristics of shale rocks in Zhanhua depression[J]. Journal of Petroleum Science and Engineering, 2016, 146: 121-128.
9. Yang, S, Wang, J, Liu, G, Li H. (2019). Horizontal well of shale gas complex fracturing casing failure mechanism. , 944, 898-902.
10. Xiong J, Liu X, Liang L. Experimental study on the pore structure characteristics of the Upper Ordovician Wufeng Formation shale in the southwest portion of the Sichuan Basin, China[J]. Journal of Natural Gas Science and Engineering, 2015, 22: 530-539.
11. Leng, Aolin, Liu, Zihan, Xing, Guangyuan, et al. China's Investment Incentive Strategy for Shale Gas Development[J]. Emerging Markets Finance and Trade, 2019(6):1-14.
12. Guo, Xue Li, Li, Jun, Zeng, Yi Jin, et al. Analytical Method to Evaluate Casing Stress during Multi-Fracturing for Shale Gas Wells[J]. 2019, 944:1050-1060.
13. Wang M, Yang J, Wang Z, et al. Nanometer-Scale Pore Characteristics of Lacustrine Shale, Songliao Basin, NE China[J]. PloS one, 2015, 10(8): e0135252.
14. Wan Y, Pan Z, Tang S, et al. An experimental investigation of diffusivity and porosity anisotropy of a Chinese gas shale[J]. Journal of Natural Gas Science and Engineering, 2015, 23: 70-79.
15. Curtis M E, Cardott B J, Sondergeld C H, et al. Development of organic porosity in the Woodford Shale with increasing thermal maturity[J]. International Journal of Coal Geology, 2012, 103: 26-31.
16. Fishman N S, Hackley P C, Lowers H A, et al. The nature of porosity in organic-rich mudstones of the Upper Jurassic Kimmeridge Clay Formation, North Sea, offshore United Kingdom[J]. International Journal of Coal Geology, 2012, 103: 32-50.
17. Lowell S, Shields J E, Thomas M A, et al. Characterization of porous solids and powders: surface area, pore size and density[M]. Springer Science & Business Media, 2012.
18. Clarkson C R, Solano N, Bustin R M, et al. Pore structure characterization of North American shale gas reservoirs using USANS/SANS, gas adsorption, and mercury intrusion[J]. Fuel, 2013, 103: 606-616.

The Progress of Offshore CO_2 Capture and Storage

Peilin Liu [1], Xueyuan Wang [1,*], Wenfeng Chen [1], Rong Hu [1], Xiaohan Li [2]

[1] Engineering Research & Design Department, Offshore Oil Engineering Co., Ltd, 300451, Tianjin, China
[2] Offshore Division, Offshore Oil Engineering Co., Ltd, 300451, Tianjin, China

Abstract. With the development of the offshore oil and gas fields, more and more offshore oil and gas fields are found to have high carbon dioxide. In addition, as peaking carbon dioxide emissions and carbon neutrality were written into the government work report for the first time, the correct separation and emission of CO_2 have become a key issue that needs to be solved by offshore oil and gas fields. In this paper, we studied two CO_2 separation methods suitable for offshore platforms and the current status of CO_2 offshore storage and application. Moreover, the development of offshore carbon dioxide storage application was investigated in detail, and the technical characteristics and application prospects of CO_2-EOR and CO_2 replacing combustible ice were analysed and discussed. This paper analyses the challenges and countermeasures of offshore CO_2 storage from many aspects. It provides a theoretical reference for future CO_2 treatment in offshore oil and gas fields.

1 Introduction

With the development of the offshore oil and gas fields, more and more offshore oil and gas fields are found to have high carbon dioxide (CO_2), and the carbon dioxide content of some offshore field is up to 40%[1]. Especially in offshore gas fields, the high content of CO_2 have a potential corrosion risk to the upper structures and pipeline. But CO_2 cannot be emitted directly into the atmosphere, it is considered to be a major contributor to the greenhouse effect and climate change. Therefore, the correct separation and emission of CO_2 have become the key problem to be solved by offshore oil and gas fields.
At the fourth Plenary Session of the 13th CPC Central Committee (2021), peaking carbon dioxide emissions and carbon neutrality were written into the government work report for the first time, which pointed out that peaking carbon dioxide emissions should be reached before 2030 and achieve carbon neutrality before 2060[2]. The latest assessment report of Intergovernmental Panel on Climate Change (IPCC) also shows that to achieve the global temperature limit of 1.5 ℃, global CO_2 emissions will need to be reduced by about 45% in 2030 compared to 2010, and "net zero" emissions will need to be achieved in 2050[3]. Carbon capture and storage (CCS) is a methodology to separate CO_2, then to store the CO_2, commonly originating from power generation, industrial processes, and from CO_2 gas fields. IPCC report[4] has indicated that in the absence of CCS implementation, the required total cost to mitigate global climate change may escalate up to 138%. CCS has been recently known to play a vital role in global climate change. Therefore, the further development of CO_2 separation and capture technology in offshore oil and gas fields conform to the basic requirements of the country in the field of energy. Meanwhile, as the carbon-rich matrix, the successful recovery and reuse of CO_2 can also solve the fuel supply problem of the offshore drilling platform and make preparation for the next step of offshore CO_2 storage.

2 The technology of CO_2 capture

At present, the technologies used for CO_2 capture mainly include adsorption capture technology, absorption capture technology, membrane separation technology, chemical cycle combustion capture technology, cryogenic, electrochemical capture and so on[5]. However, due to the complex and compact structure, small space and strict weight requirements of the offshore platform, on-site CO_2 separation on the offshore platform required that the CO_2 separation device can be installed in the limited space of the platform and the weight of the platform can bear the weight range. Considering the characteristics of offshore platform and various separation technologies, membrane separation technology and supergravity chemical absorption separation technology are suitable for offshore platform.

2.1 Membrane separation technology

Membrane separation is a kind of technology which uses the different permeability of gases to realize the separation. The gas separation membrane is mainly composed of polymeric materials and permeates the membrane by gas dissolution and diffusion[6].
The CO_2 membrane separation system used in offshore platforms consists of pre-treatment and membrane separation system (Figure 1). The pre-treatment part mainly includes gas-liquid separation system and heating

* Corresponding author: wangxy138@cooec.com.cn

system. Under a given pressure, CO_2 has better permeability than CH_4. It leaves through the membrane to the low pressure side as permeable gas, and most of CH_4 remains in the high pressure side as residual gas. The residual gas is enriched in CH_4, and the permeated gas is enriched in CO_2.

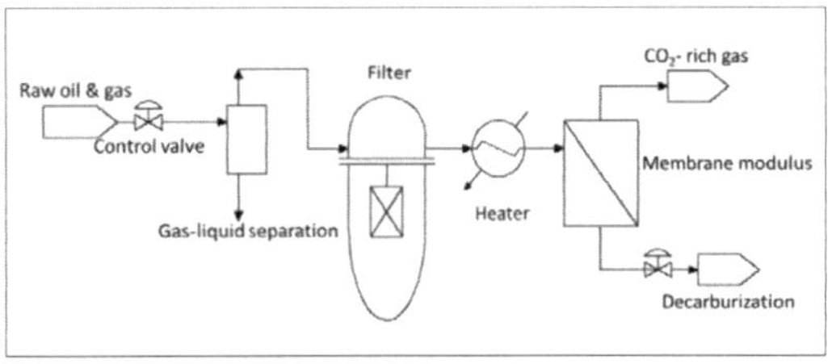

Figure 1. Process diagram of CO2 removal by first membrane separation

In order to reduce the loss rate of hydrocarbon and improve the purity of CO_2, two-stage membrane separation technology can be used. The process of two-stage membrane separation and decarbonisation device is shown in Figure 2. The membrane separation part is divide into two stages, and the raw gas is separated into the first stage membrane separator M-01. Decarbonized gas (residual gas) is obtained at the high first pressure side, and CO_2-rich gas (permeable gas) is obtained at the low pressure side. The primary permeable gas enters the compressor and compresses to a certain pressure, then enter the secondary membrane to separate M-02. The residual gas returns to the inlet of raw gas, and the CO_2-rich permeable gas is discharged and collected uniformly[7].

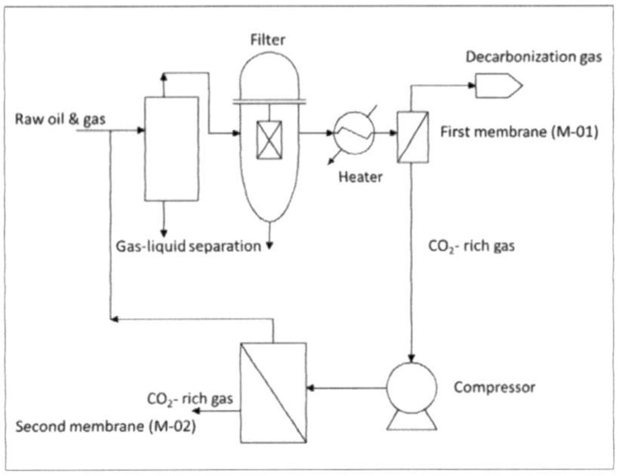

Figure 2. Process diagram of CO2 removal by second membrane separation

Membrane separation has the advantage of simple process, less equipment, less floor space, less operation and maintenance workload and low system energy consumption. Its greatest advantage lies in the combination of flexible membrane, easy to expand.

2.2 Supergravity chemical absorption separation technology

Supergravity removal CO_2 technology is a comprehensive CO_2 removal technology combining chemical absorption and supergravity technology. It generate a gravitational field of 10~1000g by high speed rotating to greatly strengthen gas-liquid mass transfer during CO_2 capture and improve CO_2 capture effect. The supergravity device equipment was shown in Figure 3[8]. CO_2 supergravity chemical absorption technology is to use rotating packed bed (RPB) instead of the vertical static tower, make a fully contact between gas and liquid in RPB. Mass and heat transfer are carried out under the condition of high dispersion of liquid phase, rapid updating of surface and strong disturbance of phase interface, so that the chemical reaction process of adsorbent and CO_2 in mixed gas is strengthened, so as to realize the technology of CO_2 and natural gas separation[9]. Meanwhile, the efficient mass transfer effect of RPB is beneficial to reduce the size of the equipment, reduce the occupying space, reduce the investment cost of the equipment, and the operation of RPB is flexible, easy to assemble into skid-mounted equipment, and easy to move, all of these performance are far superior to the traditional tower equipment [8].

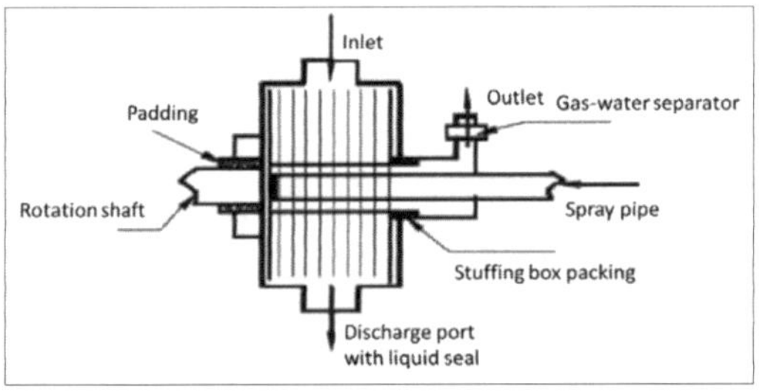

Figure 3. Schematic diagram of supergravity device equipment

Absorbent commonly used for CO_2 absorption include monoethanolamine (MEA), diethanolamine (DEA), triethanolamine (TEA), ammonia, diethylene glycol amine, methyldiethanolamine (MEDA), etc.. However, the focus of current research on supergravity chemical absorption technology is to select effective CO_2 absorbent. It can react with CO_2 quickly and efficiently to form intermediate compounds in a supergravity environment. MEA is a kind of absorbent which has been studied more, but its combination with supergravity action needs further experimental verification and optimization. In addition, the development of new absorbent is also the focus of current research.

The process of supergravity absorption technology is complicated, and the energy consumption of the moving equipment is high, the stability needs to be improved. Meanwhile, the matching selection of the supergravity machine with CO_2 absorbent is being studied and tested, and further screening tests are needed to improve the separation efficiency and meet the requirements of industrial use.

3 Offshore carbon storage and application

In addition to conventional CO_2 separation and reprocessing through industrial processes, offshore CO_2 can be sequestering directly in subsea or used as a driving force for enhanced oil and gas recovery or gas hydrate exploitation.

3.1 Offshore CO_2 storage

The CO_2 produced by offshore gas field can be buried on land or subsea. From the engineering and economic point, it is ideal to inject CO_2 on land or near shore. Generally, before the exploration of the offshore gas field, the appraisal well and seismic data were limited on the sea, the geology surrounding the gas field area is clear, including the type of structure, reservoir-cap conditions. Therefore, if inject on land or near shore, relevant exploration needs to be carried out to determine the thickness, occurrence and distribution range of high permeability salt water layer and cap layer. In consequence, CO_2 in offshore gas field is more suitable for deep saline water storage mode.

According to the method of carbon dioxide capture, the geological storage in salt water layer can be divided into primary capture and secondary capture. Primary capture is the sequestration of carbon dioxide directly blocked by geological traps. Secondary capture means that carbon dioxide is dissolved in salt water and chemically reacts with salt water and reservoir minerals to produce new minerals and achieve permanent storage of carbon dioxide. According to the reservoir site, it can be divided into geological traps and monocline structures.

In 1996, the Sleipner Gas field[10] in the North Sea of Norway started the CO_2 geological storage on offshore salt water layer for the first time, and proposed that the injection and storage of CO_2 in the underground aquifer must meet the following physical and chemical conditions[11]: (1) The top of the aquifer is at least 800 m below the seafloor. Below this depth, the hydrostatic pressure will exceed the critical pressure of CO_2, which will exist as a supercritical fluid, facilitating the storage of more CO_2. And the deep aquifer is mostly saline water with high salinity, which cannot be used as drinking water and will not affect groundwater resources. (2) A waterproof or weakly permeable layer shall be covered above the aquifer to ensure the separation of the aquifer from shallow potable water and surface water to prevent possible contamination of water quality due to the release of CO_2. (3) The aquifer is near the injection well, the surrounding area should have maximum porosity and permeability. (4) The permeability of the entire storage area should be relatively low to ensure long-term CO_2 separation in the aquifer. (5) Injection point should be as close as possible to CO_2 emission source, so as to reduce CO_2 transport costs.

Considering the geological characteristics of offshore gas fields in China, the following conditions should be met for CO_2 offshore storage: (1) The layers must be relatively stable in distribution. Shallow mudstone is not good for sealing as cap layer and does not conform to the requirements of supercritical temperature and pressure. Therefore, it is ideal for cap layer depth to be at least 500 m. (2) the most suitable burial depth is 600~1000 m, where the temperature is about 25~30℃ and the pressure is 6~12 Mpa, meeting the requirements of supercritical temperature and pressure of carbon dioxide. The thickness of sandstone is more than 100 m, with high porosity and high permeability. The porosity is more than 30%, and the permeability is more than 500×10-3 μm.

CO_2 offshore sequestration is similar to the land, also includes carbon dioxide capture, purification, transportation, injection and other processes, among which the mode of CO_2 transport at sea is the key to determine the sequestration scheme. There are two options for carbon dioxide transport by sea: one is by ship, the other is by subsea pipeline. When using ship transportation, in order to make the ship effective load, reduce the transportation cost, it is advisable to liquefied carbon dioxide for transportation, which requires the installation of liquefaction device in the platform. Pipeline transportation can be carried out in three phases, gas, liquid and supercritical state. The actual conditions of heat tracing, insulation, construction, pressure, operation and maintenance and investment of submarine pipelines should also be considered.

3.2 Offshore CO_2 storage used for enhanced oil and gas recovery (EOR)

Offshore CO_2-EOR can be seen as a way to facilitate offshore storage. Over the past four decades, CO_2-EOR technology has been in operation onshore[12], particularly in North America, but the large amount of CO_2 produced by offshore has not been used on a large scale. According to the report, the key obstacle of offshore CO_2-EOR projects are the investment in facility modifications, loss of revenue during the modifications, emission of CO_2, uncertainty in reservoir performance, and lack of transportation infrastructure. In order to prevent the release of CO_2, Norway has been put offshore CO_2 storage into practice since 1996[10], and Brazil is also working on a project related to offshore CO_2-EOR. The world's first project to implement offshore CO_2-EOR was launched in 2011(so-called Lula-pilot) and a second one in 2013 (Lula-NE).

In the CO2-EOR process, CO2 is injected into an oil reservoir under high pressure. Oil displacement by CO2 injection relies on the phase behaviour and properties of the mixture of CO2 and oil, which are strongly dependent on reservoir temperature, pressure and oil composition. There are two main types of CO2-EOR processes[13]: Miscible CO2-EOR is a multiple contact process involving interactions between the injected CO2 and the reservoir's oil. Figure 4 provides a one-dimensional schematic showing the dynamics of the miscible CO2-EOR process. Immiscible CO2-EOR occurs when insufficient reservoir pressure is available or the reservoir's oil composition is less favourable (heavier) [14].

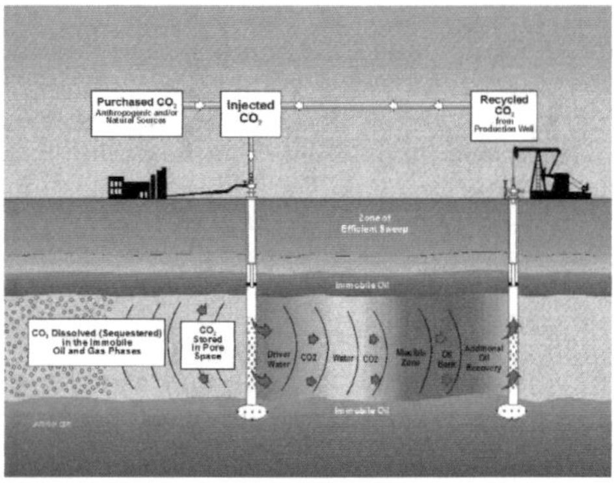

Figure 4. The principles of miscible CO2-EOR

In a CO_2-EOR operation floodable hydrocarbons (mainly oil), CO_2 and brine are produced to surface are production well. The elements involved in a typical offshore CO_2-EOR are indicated in Figure 5[15]. The production mechanisms are principally the same in onshore and offshore CO_2-EOR setting, and future CO_2-EOR operation offshore will mimic and utilise technologies known from the more matured onshore business. However, the following challenges remain for offshore CO_2-EOR implementation: (1) Due to the limited space and weight of offshore platforms, large equipment cannot be placed. (2) Offshore oil and gas Wells tend to be directional and farther away than onshore Wells. (3) Compared to onshore fields, offshore oil fields have typically achieved higher recovery rates prior to the use of CO_2-EOR. (4) On the sea, carbon dioxide must be transported by ship or subsea pipelines, both of which incur additional costs compared to onshore solutions. (5) The reservoir dynamic management model is inconsistent with that of onshore oil and gas fields. All of these challenges and differences will result in higher investment and operating costs of offshore CO_2-EOR [15].

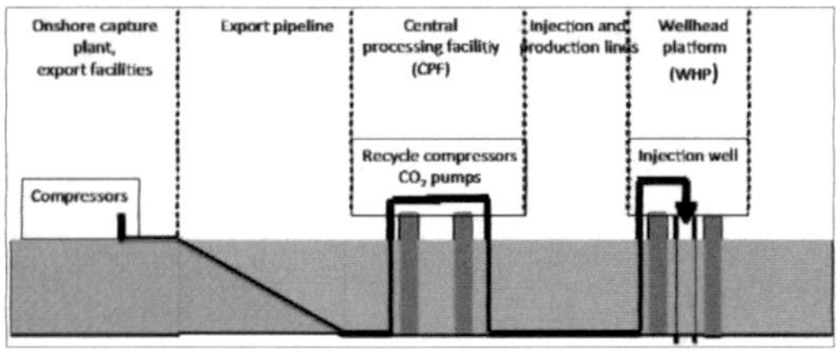

Figure 5. Schematic diagram of offshore CO2-EOR project facilities

3.3 Offshore CO_2 storage used for combustible ice exploitation

Based on CO_2 geological storage technology and CO_2 replacement gas hydrate technology, the co-combustion ice mining technology is a new technology that comprehensively uses CO_2 and obtains multiple benefits. It is also a useful complement to Carbon Capture, Utilization and Storage (CCUS).

Ebinuma and Ohgaki first proposed the idea of CO_2 replacement for combustible ice[16]. After long term research and demonstration, CO_2 replacement of combustible ice has been proved to be highly feasible in terms of dynamics and thermodynamics, and its replacement reaction is shown in Equation (1)[17]. Compared with other exploit technologies, the biggest advantage of CO_2 replacement of combustible ice is that it can maintain the stability of the original reservoir structure and reduce some geological disasters such as earthquakes. The schematic diagram of replacement exploitation is shown in Figure 6[18].

$$CH_4 \cdot nH_2O + CO_2\,(g) \rightleftharpoons CH_4\,(g) + CO_2 \cdot nH_2O\ (n \geqslant 5.75)\quad (1)$$

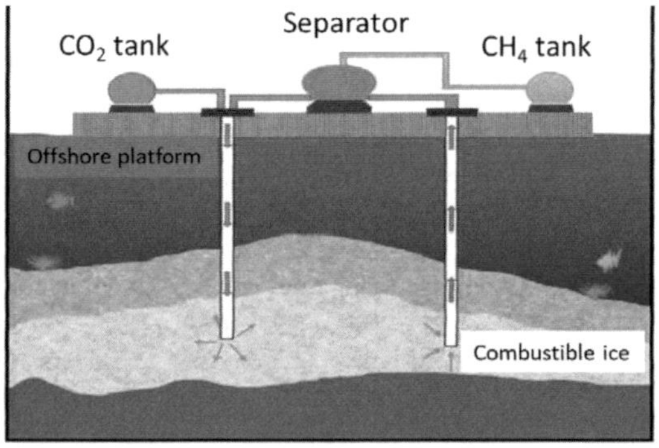

Figure 6. Schematic diagram of CO2 replacement of combustible ice

The research shows that if liquid CO_2 is injected into combustible ice reservoir, the hydrophilicity of CO2 is better than CH4, and the pressure required to generate CO_2 hydrate is lower than that required to maintain the stability of combustible ice at the same temperature. Therefore, in a certain pressure range, natural gas hydrate (NGH) will decompose into water and CH_4, CO_2 will combine with water generated by decomposition to form hydrate and remain stable, so CO_2 will drive away CH_4. In this displace process, the formation of CO_2 hydrate is exothermic, while the decomposition of combustible ice is endothermic. It can be seen from Equations (2) and (3) that the exothermic is greater than the endothermic, so this process can also occur spontaneously [19].

$$nH_2O + CO_2\,(g) \rightarrow CO_2 \cdot nH_2O\ (\Delta H_f = -57.98\ \text{kJ/mol})\quad (2)$$

$$CH_4 \cdot nH_2O \rightarrow CH_4 + nH_2O\ (\Delta H_f = 54.49\ \text{kJ/mol})\quad (3)$$

The technology of CO_2 storage combined with combustible ice exploit is beneficial in many aspects. The combustible ice reserves are huge, which is undoubtedly an important follow-up energy source in the future. The CO_2 offshore storage combined with combustible ice exploit can accelerate the development of combustible ice, which is of great significance to alleviate the current situation of energy shortage in the world. Although CO_2 storage combined with combustible ice have broad prospects, it still faces some challenges. Firstly, the replacement of CH_4 by CO_2 is not complete, and the water cages wrapped with CH_4 molecules in combustible ice have different sizes. CO_2 molecules can only enter large holes, but not small ones, resulting in the retention of CH4

 https://doi.org/10.1051/e3sconf/202132901018

molecules in small holes. For this problem, it is necessary to combine other exploit methods to fully release CH_4. Secondly, the rate in CO_2 replace CH_4 of combustible ice is low, and the reaction rate will be lower and lower as the replacement reaction progresses, which requires to find conditions or catalysts to accelerate the replacement rate.

4 Conclusion

In summary, with the large scale development of offshore oil and gas fields, how to deal with CO_2 emission is a major problem and challenge that needs to be treated seriously at present. A series of factors such as investment cost, platform space and system energy consumption should be considered for CO_2 processing in offshore platforms. Subsea geological conditions and CO_2 gathering and transportation should also be considered in the process of further CO_2 offshore storage. China is a Marine big county whose the sea area is vast and salt water layer is all over the ocean, which is the favourable condition for the implementation of CO_2 offshore storage. In addition, further application of CO_2 offshore storage can be considered, such as offshore CO_2-EOR and combine with combustible ice exploitation. This paper provides theoretical reference for further offshore CO_2 capture and storage.

References

1. L. Ma, Y.H. Liu, Silicon Valley, 7,12 (2014)
2. Y.Z. Sun, X.F. Guo, Y. Ding, Q, Liu, Modern Chemical Industry, 39, 1 (2019)
3. Report on the Work of the Government, (2021)
4. N.R. Sukor, A.H. Shamsuddin, T.M.I. Mahlia, M.F. M. Isa, Processes, 8, 350 (2020)
5. X.Y. Wang, T. He, J.H. Hu, M. Liu, Environment Science Nano, 8, (2021)
6. Z.W. Meng, Y. Liu, S.K. Ren, Low Temperature and Specialty Gases, 33, 05 (2015)
7. Y. Jin, G.C. Wang, Z.J. Li, Chemical Engineer, 1, (2008)
8. J.Y. Liu, F.Q. Li, X.F. Zhang, Energy and Energy Conservation, 10, (2013)
9. W.L. Huang, B.L. Zhang, H.B. Liu, N. Mao, Modern Chemical Industry, 37, 03 (2017)
10. IEAGHG Report (2016)
11. W. Zhang, Y.L. Li, Environmental Pollution & Control, 12, (2006)
12. R.M. Erick, D.K. Olsen, J.R. Ammer, W. Schuller, Society of Petroleum Engineers, 154122, (2012)
13. CSLF, (2017)
14. IEAGHG Report (2009)
15. S.G. Goodyear, M.P. Koster, K.A. Marriott, A. Paterson, A.W. Sipkema, I.M. Young, Society of Petroleum Engineers, 144939, (2011)
16. K. Ohgaki, K. Takano, H. Sangawa, T. Matsubara, S. Nakano, Journal of Chemical Engineering of Japan, 29, 3 (1996)
17. F.Y. Jin, Y. Guo, W.F. Pu, Y.B. Liu, P. Lian, Journal of Southwest Petroleum University(Science & Technology Edition), 35, 03 (2016)
18. K. He, Modern Chemical Industry, 38, 04 (2018)
19. E.M. Yezdimer, P.T. Cummings, A.A. Chialvo, The Journal of Physical Chemistry A, 106, 34 (2002)

E3S Web of Conferences **329**, 01019 (2021) https://doi.org/10.1051/e3sconf/202132901019
GESD 2021

Multi-stage Depressurization Analysis of Subsea Production System Based on LedaFlow Software

Wenfeng Chen, Pengpeng Ju, Xueyuan Wang *, Dong Wang, Feilong Liu

Engineering Research & Design Department, Offshore Oil Engineering Co., Ltd, 300451, Tianjin, China

Abstract. This paper briefly summarizes that during the depressurization operation in subsea pipeline for maintenance or prevention of hydrate formation, because the flare knock-out drum pump is designed for ambient temperature, and cannot deal with the low temperature fluid in the depressurization process. Therefore, the low temperature fluid needs to be heated in the knock-out drum to meet the temperature requirement of pump. In order to avoid the volume of low temperature fluid exceeding the surge volume of the flare knock-out drum, the concept of multi stage depressurization is innovatively proposed in this paper, which meets both the requirements of pump temperature and the surge volume of flare drum. Taking a gas field project in the South China Sea as an example, the simulation analysis of depressurization temperature and pressure is carried out by using different size of the blowdown valve. It provides a good reference for the similar project in future.

1 Introduction

Subsea production systems (SPS) are widely used in deep-water and marginal oilfield development[1]. The water depth of a gas field project in the South China Sea is 1500m, using the subsea production system and semi-submersible platform (SEMI) development program, the natural gas is exported after drying and dehydration treatment in SEMI, and the condensate oil is stored in SEMI for regular export after qualified dehydration treatment. The SPS consists of seven production wells (W1-W7), with relatively scattered well locations, using three subsea manifold (E1~E3) and double subsea pipeline development scheme. Oil & gas wells output pipelines are connected to subsea manifold through jumpers, and production fluid are connected to SEMI for processing[2]. The schematic diagram is shown in Figure 1.

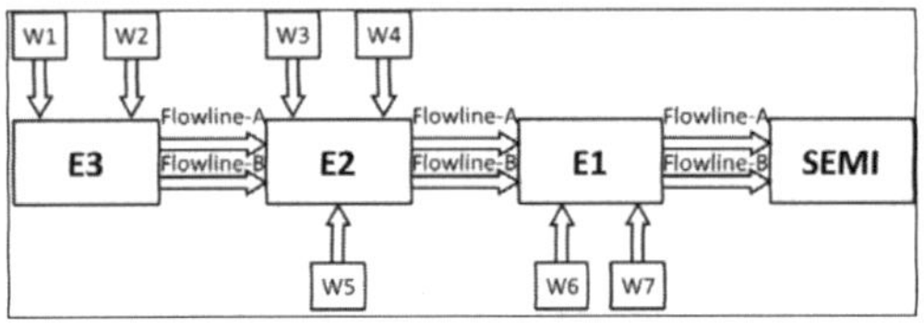

Figure 1. Schematic diagram of SPS development in target gas field

When hydrate is generated in the subsea pipeline and SPS will block pipelines and affect production, depressurization operation should be carried out on the platform to reduce the operating pressure in the subsea pipeline and make it lower than the hydrate dissolution pressure, so as to ensure the safety of production and operation.

2 Depressurization process

Hydrate formed in gas field is a kind of appearance and mechanical properties are similar to ice substance, which is generated by methane, ethane, propane, nitrogen, carbon dioxide, hydrogen sulfide and other gas composition with water under high pressure and low temperature conditions, it will reduce the flow of the pipeline area, or even blocking pipe, and possibly as high-speed flow of gas, damage to the valve in the pipeline, elbow and other parts[3]. After hydrate formation, domestic and foreign projects take remedies such as depressurization, methanol injection and heating. In the process of depressurization, in order to prevent the pressure difference between the two sides of the hydrate from being too high and causing danger in the high-speed operation of the pipeline, the pressure on both sides of the hydrate should be relieved at the same time. For this deep-water project in the South China Sea, the blowdown valve at the outlet of the double sea pipe can be released at the same time.

The platform outlet valve is closed, and the SPS continues to produce until the subsea tree reaches the shut-off pressure. At this time, the pressure corresponding to the outlet of the subsea pipe is 308 bar (1 bar = 105 pa), which is the initial discharge pressure. The high-pressure production fluid in the subsea pipeline is discharged through the blowdown valve at the outlet of the subsea pipeline on the platform, and the discharged gas goes into

* Corresponding author: wangxy138@cooec.com.cn

the flare system for processing. The discharged low-temperature fluid goes into the flare dispensing tank for storage and is sent back to the production process by booster pump. HP flare knock-out drum (TS-V-3401) min design temperature is -99 °C, surge volume is 10 m3, but HP flare K.O drum pumps (TS-P-3401A/B) min design temperature is ambient temperature (AMB). The discharged low-temperature fluid can be heated to AMB by the electric heater (TS-EH-3401) in the tank and transported to the process flow. The processing capacity of SEMI platform flare system is 1176×104 Sm3/d. the maximum gas volume during pressure depressurization should not exceed the processing capacity of flare. All parameters in the depressurization process should meet the requirements of the above design parameters[4]. The depressurization process of the subsea pipeline is shown in Figure 2.

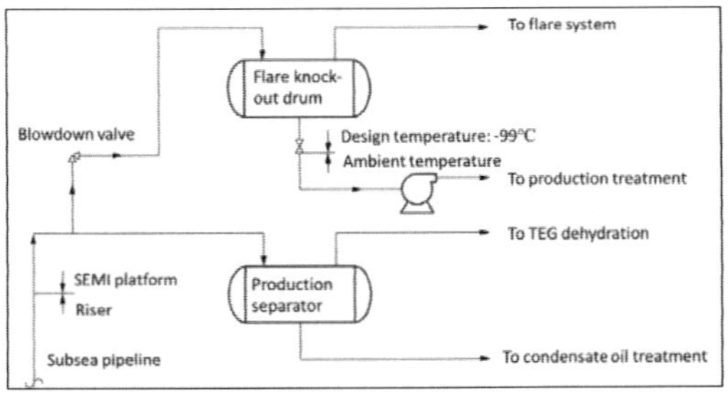

Figure 2. The depressurization process of the subsea pipeline

3 Simulation method

The initial condition for flowline blowdown was obtained with the integrated wellbore and flowline simulation model. The valves at flowline outlets were closed while the subsea chokes were kept the original opening. When equilibrium of system was established, the pressure at subsea wellheads reached the Shut-In Tubing Head Pressure. The subsea chokes were subsequently closed.

The reservoir gas composition of Year 2020 was used for the minimum temperature analysis, and higher gas system would result in higher drop in temperature. The flowing conditions were simulated with all production forecast flowrates. Consequently, the predicted minimum temperature is more conservative (lower), and the predicted depressurization time is shorter than those with production forecast flowrates.

The back pressure at the outlet of the blowdown valves was assumed to be 3 bar. Two blowdown cases were performed for the performed for the blowdown valve size of 0.5 and 1 inches.

4 Simulation results with target gas field flowlines

During the depressurization operation, the low temperature fluid will be carried out by gas from subsea pipeline into the flare knock-out drum storage, because of the booster pump and the production processing designed for AMB, discharged low temperature fluid cannot be pump directly into the production processing system for treatment. The liquid must be stored in the knock-out drum to heat up until its temperature increases to AMB, when it can be pumped to the production processing system for treatment. According to the requirements of this working condition, if the conventional direct discharge method is adopted, the storage capacity of the knock-out drum will be 47.2 m3 at a minimum, which is much larger than the current surge volume of 10 m3. Therefore, based on the direct depressurization simulation of conventional projects, the multi-stage depressurization is creatively proposed. The surge volume of the flare knock-out drum is taken as the limit condition, and the pressure of the subsea pipeline is reduced by multi stage depressurization, so as to provide the operator with enough time for low temperature fluid treatment, that is, to dissolve the hydrate generated. Thus, the operation problems caused by the normal temperature design of the booster pump can be avoided.

The simulation results for depressurization of target gas field flowlines in Year 2020 with 0.5 inch diameter blowdown valves are as follows.

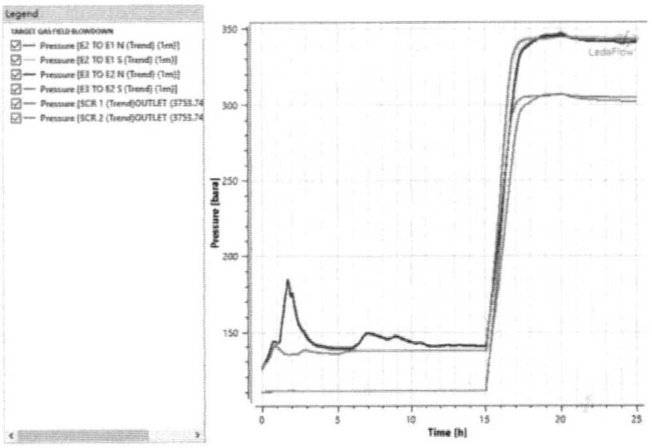

Figure 3. Pressure at target gas field subsea wellheads and flowline outlet prior to depressurization

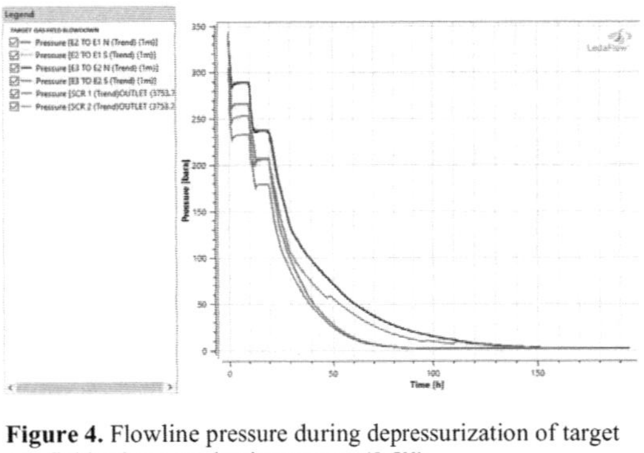

Figure 4. Flowline pressure during depressurization of target gas field subsea production system (0.5'')

https://doi.org/10.1051/e3sconf/202132901019

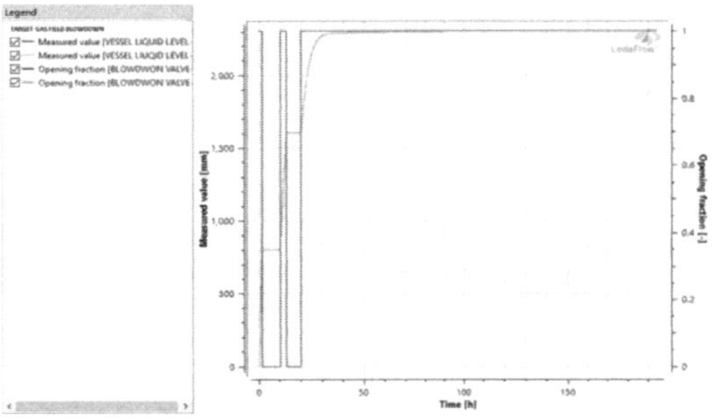

Figure 5. Volume of low-temperature liquid from flowlines and open/close status of valves during depressurization of target gas field subsea production system (0.5")

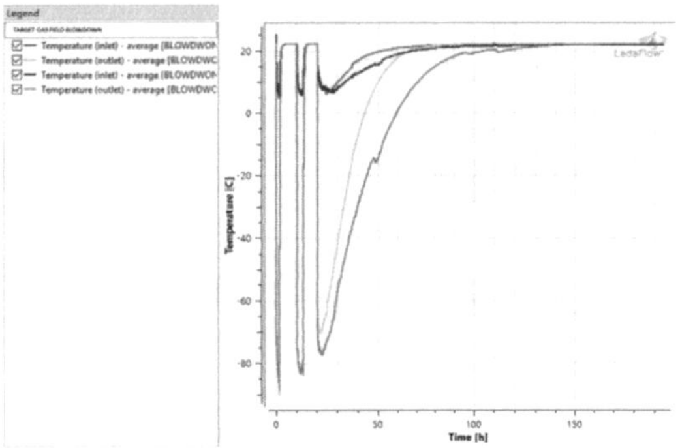

Figure 6. Low temperature during depressurization of target gas field subsea production system (0.5")

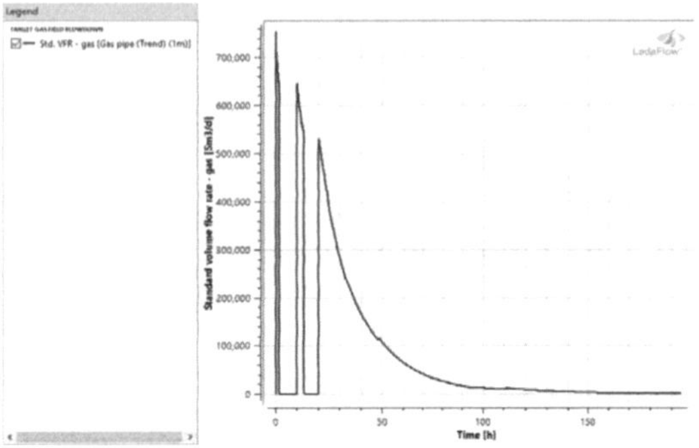

Figure 7. Standard condition gas flowrate during depressurization of target gas field subsea system (0.5")

It can be seen from Figure 3 to Figure 7, when 0.5 inch blowdown valve was used for depressurization of target gas field flowlines in 2020, the flowline outlet pressure was reduced from 304 bar to 232 bar during the first stage of depressurization. The E3-E2 N & E2-E1 N flowline inlet pressures were reduced from 341 bar to 289 bar. During the second stage of depressurization, the flowline outlet pressure was reduced to 179 bar, and the E3-E2 N & E2-E1 N flowline inlet pressures were reduced 237 bar. During the third stage of depressurization, the flowline outlet pressure was reduced to 3 bar, and the E3-E2 & E2-E1 flowline inlet pressures were reduced 3.4 bar. It took about 139.2 hours for the flowline outlet pressure to reduce to the specified back pressure.

During the first stage depressurization, the low temperature liquid from the flowlines reached the volume of 10 m^3 in 1.5 hours, and reached the volume of 10 m^3 in 3.4 hours during the second stage. During the third stage, the volume of low temperature liquid stopped increasing after reaching 8.9 m^3. Therefore, the total volume of low temperature liquid from all stages of depressurization was 28.9 m^3.

The lowest temperature downstream of the blowdown valve was -90.5°C during the first stage depressurization, -84.5°C during the second stage depressurization and -77.3°C during the third stage depressurization, all within the low temperature design limits for the piping downstream of the blowdown valve and the knock-out drum of the high pressure flare system.

The maximum gas flowrate at the outlet of the blowdown valve was 75.3×10^4 Sm^3/d during the first stage, 64.4×10^4 Sm^3/d during the second stage and 53.1×10^4 Sm^3/d during the third stage. Since the flare system on the SEMI platform is design for 1176×10^4 Sm^3/d, the gas flare during depressurization is within the flare system capacity.

The simulation results for depressurization of target gas field flowlines in Year 2020 with 1 inch diameter blowdown valves are as follows.

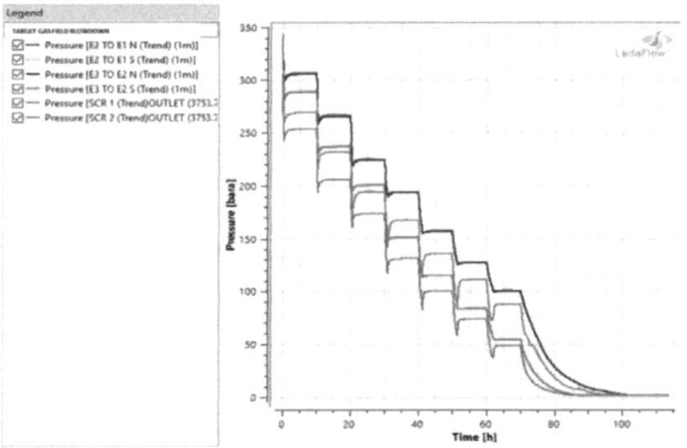

Figure 8. Flowline pressure during depressurization of target gas field subsea production system (1'')

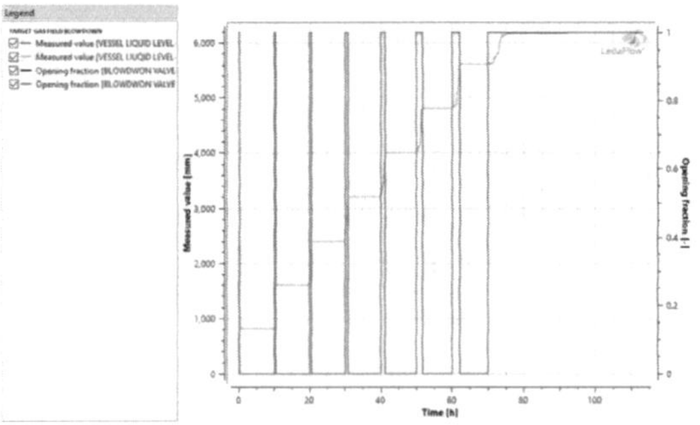

Figure 9. Volume of low-temperature liquid from flowlines and open/close status of valves during depressurization of target gas field subsea production system (1")

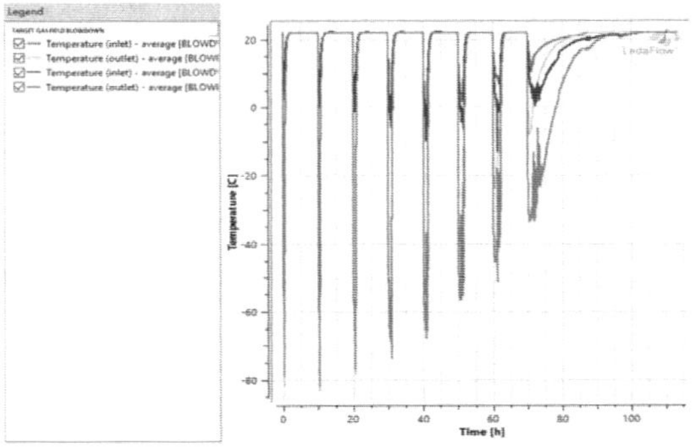

Figure 10. Low temperature during depressurization of target gas field subsea production system (1")

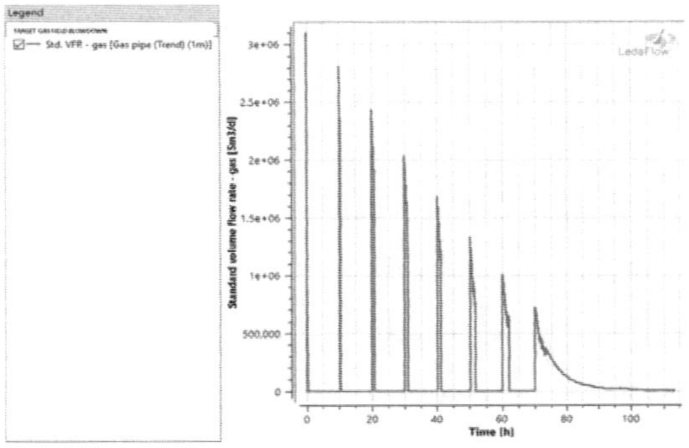

Figure 11. Standard condition gas flowrate during depressurization of target gas field subsea system (1")

It can be seen from Figure 8 to Figure 11, when 1 inch blowdown valve was used for depressurization of target gas field flowline in 2020, the flowline outlet pressure was reduced from 304 bar to 254 bar, and it takes 8 stages to reduce to 3 bar. The E3-E2 N & E2-E1 N flowline inlet pressures were reduced from 334 bar to 307 bar and takes 8 stages to reduce to 3.4 bar. It took about 91.2 hours for the flowline pressure to reduce to the specified back pressure.

During the first stage depressurization, the low temperature liquid from the flowlines reached the volume of 10 m^3 in 0.28 hours, and the last stage, the volume of low temperature liquid stopped increasing after reaching 7.7 m^3. Therefore, the total volume of low temperature liquid from all stages of depressurization was 77.7 m^3.

The lowest temperature downstream of the blowdown valve was -83.2°C during the all stages depressurization, all within the low temperature design limits for the piping downstream of the blowdown valve and the knock-out drum of the high pressure flare system.

The maximum gas flowrate at the outlet of the blowdown valve was 310×10^4 Sm^3/d during the first stage, the following stages are 281, 243, 203, 168, 133, 100 and 72×10^4 Sm^3/d, respectively. Since the flare system on the SEMI platform is design for 1176×10^4 Sm^3/d, the gas flare during depressurization is within the flare system capacity.

Table 1 summarizes the simulation results for depressurization of target gas field with 0.5 inch and 1 inch blowdown valves.

Table 1. Summary of target gas field depressurization simulation results

BDV Size	in	0.5	1
Number of depressurization	/	3	8
Total Volume of low temperature liquid form flowlines during depressurization	m^3	28.9	77.7
Minimum temperature at the outlet of blowdown valve	°C	-90.5	-83.2
Maximum gas flowrate at the outlet of blowdown valve	10^4 Sm^3/d	75.3	310
Time for the flowline outlet to be depressurization to 10 bar	h	90.8	83.3
Time for the flowline outlet to be depressurization to 3 bar	h	139.2	91.2
Minimum temperature at the upstream of blowdown valve	°C	4.5	-12.9

From the above analysis results, 0.5 in and 1 in blowdown valve have their own advantages, and the operator can determine the size of the blowdown valve according to the concerned depressurization parameters.

5 Conclusion

In summary, the depressurization analysis for regular project, the high pressure fluid in the subsea pipeline is simulated with directly discharged by blowdown valve. Under the condition of the discharge gas flow does not exceed the processing capacity of the flare treatment, the low temperature flow does not exceed the capacity of the flare knock-out drum pump, and the low temperature at the outlet of the blowdown valve does not exceed the minimum design temperature of the depressurization system, then the appropriate size and time of the blowdown valve are determined. However, this method does not take into account the condition that the flare knock-out drum pump is designed at AMB, which cannot transport low temperature fluid in time, and the power of the heater in the flare knock-out drum does not meet the load requirements.

In the design process of the South China sea project, the multi stage depressurization concept was proposed innovatively. Under the condition of the flare knock-out drum surge volume, to provide operators with enough time for handing low temperature fluid. The contradiction

https://doi.org/10.1051/e3sconf/202132901019

between low temperature fluid transport and the minimum design temperature of booster pump or other equipment was well solved, which provided a reference for the development of similar projects in the future.

References

1. L.Song, S.G. Yang, B.L. Liu, Ocean Development and Management, 30, 06 (2013)
2. Z.D. Huang, X.M. Li, R.J. Shen, J. Luo, N. Li, China Offshor Oil and Gas, 26, 01 (2014)
3. W.F. Chen, P.P. Ju, P.L. Liu, H. Zhang, X.Y. Zhou, T. Deng, Ocean Engineering Equipment and Technology, 7, 02 (2020)
4. Transient Simulation Study Report for Target Gas Field Flowlines, (2018)

Research and application of L-shaped ground well gas drainage technology in mining area

Junhui Fu [1, 2, *]

[1] Chongqing Research Institute Ltd. of China Coal Technology and Engineering Group, Chongqing, China
[2] Shandong university of science and technology, Qingdao, China

Abstract. By law mining fracture morphology analysis, based on coal face gas flow characteristics, obtain the L-shaped surface well layout area in the mining area under the influence of mining. Analysis of overlying strata "three zones" distribution, the calculation method of the height of overburden rock "three zones" designed in mining area ground well structure proposed safety drainage system construction plan. In Sihe Mine was L-shaped ground mining area well engineering practice, the average gas production at a concentration of 80%, with an average drainage net amount 22 000 m3 / d, a total of more than 300 days running, drainaging 6.5 million CBM m3. After the implementation of drainage work surface corner gas content fell 46.5%, to effectively improve production efficiency, for the surface treatment gas play a key role. Successfully removed Mining Face Gas for safe production constraints, to protect the mining of coal mine safety, and achieved good social and economic results, with good promotional value.

1. Introduction

At present, some domestic coal mines are faced with the confusion of high difficulty and high pressure of gas control in the working face. Surface coalbed methane wells can alleviate the pressure of gas control in coal mines [1-3]. After long-term exploration and research by researchers, the purpose of using surface wells in mining affected areas to control gas overrun in the upper corner of coal face has been basically achieved [4-5].Zhang[6] et al. presented that the coal mine gas (coalbed methane) is a high-quality clean source of energy and a good industrial, chemical, power generation and residential fuel. Zhou[7] et al. presented that the coalbed methane drainage is quite important for the safe production in coal mines and the utilization of gas resources.It is proven significantly important to study the coalbed methane drainage in the mining area. DNWhittles[8] et al. conducted a research on the borehole boring casing failure with the coal mining process in the collapsed area. The damage effect of casing and the effect of lithological parameters were analyzed. Pierre-Jean[9] of France utilized an experimental method to study the failure characteristics of the surrounding rock, pointing out that the principal stress on the borehole wall has an important effect on the borehole wall stability. Song[10] et al. studied the quantitative relationship between the geological borehole failure characteristics and the geostresses, initially obtaining the relationship among different geostresses and deformations of the ground drills[11].

Considering that the gas extraction range of CBM vertical surface wells is limited and the mining area covered is small, it is necessary to consider adding a section of horizontal hole on the basis of vertical surface wells to form the CBM extraction mode of "L" surface wells, and study the surface wells in the "L" mining affected area through reasonable optimal layout. In order to achieve the purpose of wide coverage and excellent gas drainage effect, so as to improve the gas drainage efficiency, alleviate the gas control pressure of the mining face, reduce the gas overrun risk of the mining face, and enhance the safety guarantee of high yield and high efficiency of the coal mine.

2. Horizon analysis of horizontal section of L-shaped surface well in mining area

2.1 Distribution law of mining fractures

After the coal seam is mined out, the original stress balance around the goaf is damaged, resulting in the redistribution of stress, resulting in the deformation, rotation and fracture of the rock stratum, which develops from bottom to top to the surface. When there is a key layer in the overburden, it may develop to a key layer and stop. Large fractures are generated in the rock stratum below the key layer. The fractures can be clearly seen through similar simulation tests, as shown in Figure 1. It can be seen from the figure that after pressure relief, the rock stratum above the cut hole and working face will rotate under the action of gravity, resulting in vertical fractures [12].

* Corresponding author: 289649796@qq.com

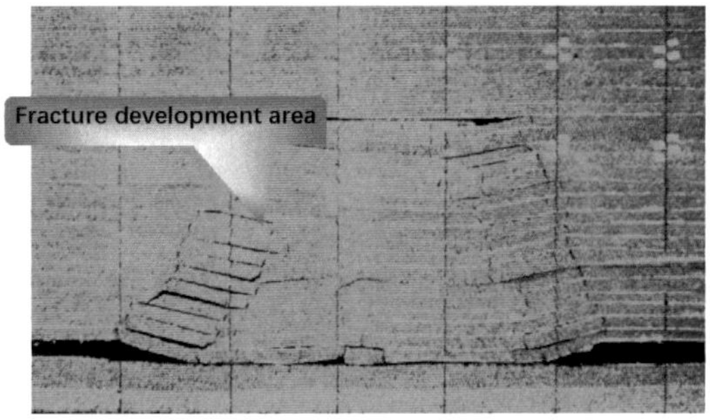

Figure1. Distribution pattern of overburden fractures

Starting from the opening of the cut hole, with the advance of the working face, the mining fracture continues to develop. The mining fracture in the middle of the goaf is the most developed, which is the first stage of mining fracture development. When the goaf area reaches a certain value, it enters the second stage of mining fracture development. At this time, the mining fracture in the middle of the goaf tends to be compacted, and there is a connected mining fracture development area around the goaf, that is, the mining overburden fracture "O" ring. The mining fracture "O" ring can be maintained for a long time and is the storage space and flow channel of pressure relief gas.

The flow of coal seam pressure relief gas is a continuous process: first, in the form of diffusion, the gas flows from the coal body without cracks to the surrounding cracks; In the second step, in the form of seepage, the gas flows along the fracture to the drainage borehole, and the mining fracture is the channel for gas flow. Obviously, arranging the drainage boreholes in the area where the fractures are developed and can be maintained for a long time is conducive to the flow of pressure relief gas into the drainage boreholes. It can be seen from Figure 1 that the "O" ring shape can be obtained by cutting the ellipsoidal polishing surface with the horizontal plane within the crack height, as shown in Figure 2. The "gas River" formed by the "O" ring is mainly reflected in the separation fractures. There is more gas storage space in the separation fractures, and the mining vertical fractures are more used as the gas flow channel between rock layers. The "O" ring fracture space of mining fracture will exist for a long time and will be used as the storage space for pressure relief gas. However, with the advance of the working face, the "O" ring will continue to extend along the strike direction of the working face.

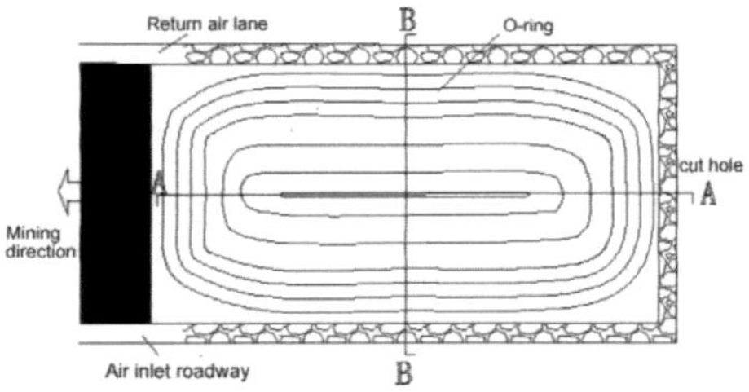

Figure 2. O- ring shape of stope overburden

2.2 Distribution characteristics of "three zones" of stope overburden

By combining the actual geological conditions and mining technical conditions of Sihe mine, we appropriately modify the empirical formula of "three zones" height given by predecessors, so as to analyze the distribution characteristics of "three zones" suitable for Sihe mine.

(1)Height of caving zone

Suppose that the primary mining thickness of the coal seam is m, after the mining area reaches a relatively large area, the height of the caving zone is h, and the crushing expansion coefficient of the caving rock block is K. When the volume of the caving rock block is equal to the sum of the mining space and the space of the caving rock layer itself, all the mining space disappears. At the same time, the overlying non caving rock layer is supported by the caving rock block, and the caving process stops. As the falling rock blocks are stacked and compacted, the overlying non falling rock stratum bends and sinks, resulting in cracking and damage. Considering that when the overburden is not a hard roof, the overburden will generally sink continuously during the occurrence and development of overburden caving, which reduces the volume of caving rock block required to fill the mining space and will lead to the reduction of caving height. The falling height can be expressed as:

$$H_m = \frac{M-\Delta M}{(K-1)\ cos\alpha} \tag{1}$$

Where,Hm is the height of caving zone, ΔM is the subsidence of overburden before caving, M is the mining height, and K is the crushing expansion coefficient of caving rock block; α Is the dip angle of coal seam.

(2)Fracture zone height

According to "GB12719-91 code for hydrogeological and engineering geological exploration of mining area", and according to the condition that the overburden of Sihe mine is medium hard rock stratum, the calculation formula of water diversion fracture zone is as follows:

$$\mathrm{H_t} = \frac{100\times M}{3.3n+\ .8} + 5.1 \tag{2}$$

Where, Ht is the height of water conducting fracture zone, M is the mining height, and N is the number of coal layers.

(3)Bending subsidence zone

According to the definition of bending subsidence zone, bending subsidence zone is a continuous and integral moving rock stratum from the top boundary of fracture zone to the surface. But it is at least a function of time. After full mining, the curved subsidence zone must develop to the surface. If the mining is not sufficient, the curved subsidence zone may stop when it develops to a key layer of overburden.

3. Wellbore structure design of L-shaped surface well in mining area

Based on the analysis of drilling deformation and failure characteristics and in combination with the characteristics of gas surface wells in the mining area, the surface wells in the mining affected area are arranged approximately

 https://doi.org/10.1051/e3sconf/202132901020

0.2 ~ 0.4L away from the return air side (L is the length of the working face). The well body structure needs to be determined according to the characteristics of overburden movement. Generally, the well body structure of three openings is adopted. The first opening is designed to be about 20m below the bedrock, and the casing adopts J55 API casing, The second spud is arranged from below the surface to the upper part of the fracture zone, N80 casing with good shear resistance is adopted, and the third spud is arranged with open holes, as shown in the figure.

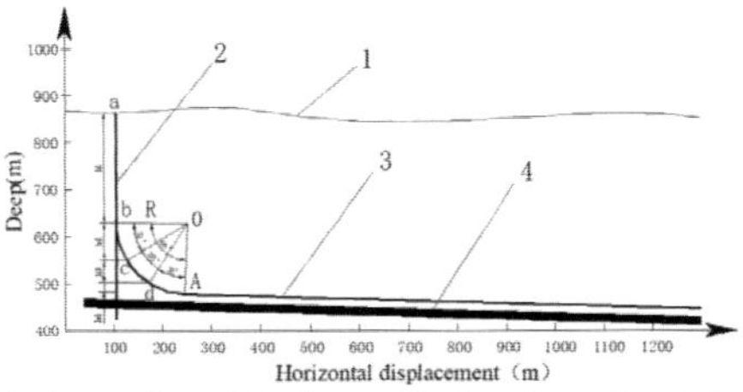

(1 is the surface, 2 is the vertical well section, 3 is the horizontal section, and 4 is the coal seam.)

Figure 3. Schematic diagram of L-shaped surface well in mining area

4. Construction of safe extraction system

According to the effect and purpose of ground coal seam test, five types of equipment are mainly required to be purchased: safety monitoring and control equipment, pumping and mining basic equipment, pumping and mining data acquisition equipment, fire-fighting equipment and power supply and power auxiliary equipment. The layout of safety pumping system is shown in the figure. At the same time, carry out the research on the matching of pumping equipment and process adaptability, including calculating the pipe diameter wear resistance of mining fracture field, casing and pumping pipeline, selecting the water ring vacuum pump installed on the ground, arranging explosion-proof and data monitoring equipment, continuously monitoring pumping data, and evaluating the adaptability of pumping equipment in combination with the changes of pumping data.

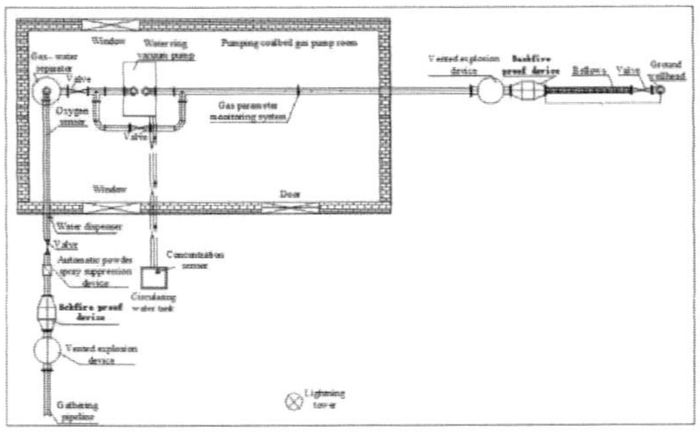

Figure 4. Layout of surface borehole coalbed methane drainage system

5 Engineering application

The test working face of L-shaped surface well in the mining area is 3313, the panel name, the working face elevation is 394 ~ 466m in the east third panel, the ground elevation of 3303 fully mechanized mining face is 788 ~ 960m, and the working face elevation is 405 ~ 486m. This working face is a fully mechanized mining face with one-time mining height. The mining thickness is about 6.13m, the coal seam dip angle is 0~12 °, the average dip angle is 6 °, and the strike length is 1239.88m, The working face is 301.50 m long. The coal seam structure is simple, and the L-shaped surface well in the mining area is arranged 45m near the return air side. The absolute gas emission during mining in the working face is 29.73m3/min.

It can be seen from Figure 5 that continuous extraction is realized in the mining affected area and goaf of the coal seam, and the drilling structure is intact; The average produced gas concentration is 80%; The average net extraction volume is 20000 m3 / D, the cumulative operation is more than 300 days, and the extraction of coalbed methane is more than 6.5 million m3.

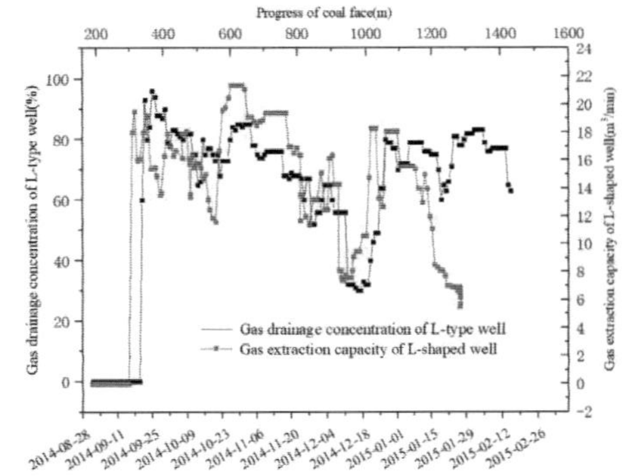

Figure 5. Relationship between gas purity, gas concentration and distance from coal mining face to well

LocationAfter the implementation of drainage, the gas concentration in the upper corner of the working face decreases by 46.5%, as shown in Figure 6. Effectively improve production efficiency and play a key role in gas control in this face. Successfully eliminate the restriction of goaf gas on safety production of working face.

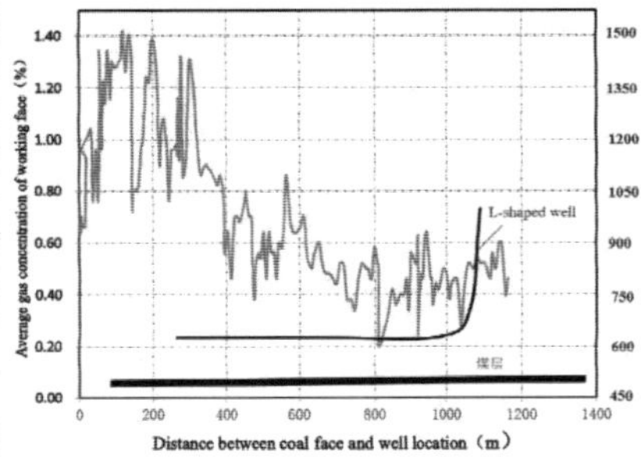

Figure 6. Change of gas concentration in working face after drainage

6 Conclusion

1) Through the analysis of the mining fracture shape law and according to the gas flow characteristics of the coal mining face, the basic principle of L-shaped surface well layout in the mining area under the influence of mining is obtained: it should be arranged near the return roadway side and in the "O" ring.
2) The distribution characteristics of "three zones" of overburden in stope are analyzed, and the calculation method of "three zones" height of overburden is put forward, which provides a basis for the horizontal horizon height of L-type surface wells in mining area.
3) The well structure of L-shaped surface well in mining area is designed, and the construction scheme of safe pumping system is put forward.
4) The engineering practice of L-shaped surface well in mining area is carried out in Sihe mine, which has achieved good gas drainage effect, solved the problem of difficult gas control for coal mining face, ensured the safety of coal mining, achieved good social and economic effect, and has good popularization value.

Acknowledgments

The word is supported by Chongqing innovation leading talent support program (cstccxljrc201911), Special fund for technological innovation and Entrepreneurship of Tiandi Technology Co., Ltd (No. 2020-TD-MS012). Shanxi Province Unveiling Bidding Project (20201101001), and major science and technology Project of Shanxi province (20201102001).

References

1. J. Zhang, D.l. Sun, X.C. Huang, Experimental study on gas drainage technology in mining area by surface drilling [J]. Mining safety and environmental protection, 007,34 (1),pp.1-5
2. R. F. Li, Y.P. Liang, J. Zhang. Influencing factors of gas extraction efficiency in goaf by ground drilling [M]. Journal of coal, 2009, 34 (7),pp.942-946
3. L. Guo, J.H. Fu, R. F. Li, Application of underground combined gas extraction technology in Yuecheng mine [J]. Mining safety and environmental protection, 2013, 40 (2). pp.59-61
4. F.B. Zhou, T.Q. Xia, Y.K. Liu, et al. Calculation model of gas flow in pressure relief coal seam and goaf by surface drilling [M]. Journal of coal, 2010,35 (10): 1639-1643
5. H.T. Sun. Study on deformation and failure mechanism of surface drilling under the influence of mining [D]. Chongqing: Chongqing University, 2008
6. Seedsman RW. Geotechnical sedimentology - its use in underground coal mining. Int J Rock Mech Min Sci 2001;45(2):147–53.
7. N._Spiezia, V. A. Salomoni, C. E. Majorana, Plasticity and strain localization around a horizontal wellbore drilled through a porous rock formation', International Journal of Plasticity 78 (2016) 114-144
8. D.N.Whittles, I.S.Lowndes, S.W.Kingman, C.Yates, S.Jobling. The stability of methane capture boreholes around a long wall coal panel[J]. International Journal of Coal Geology, 71 (2007) 313–328.
9. Pierre-Jeam. Laboratory investigation of rock fracture around boreholes[J], America: University of California, 1990.
10. I.S. Song. Borehole Breakouts and Core Disking in Westerly Granite: Mechanisms of Formation and Relationship to In Situ Stress [D].America: University of Wisconsin-Madison, 1998.
11. Rostami J, Elsworth D, Watson R Study of borehole stability for gas boreholes in longwall mining areas. 2012.Pennsylvania State University, State College, Pennsylvania
12. J.H. Fu.Research on gas drainage technology and application of surface wells under the influence of mining [J], China coalbed methane, 2014, 11 (6): 36-40

Application of Green Power Generation Technology for Distributed Energy

Qiang Fu [1,*], Chengxi Fu [2], Peng Fu [3] and Yuke Deng [4]

[1] Guangdong Key Laboratory of Electric Power Equipment Reliability, Electric Power Research Institute of Guangdong Power Grid Co., Ltd., Guangzhou, China
[2]School of Engineering, City University of Hong Kong, Hong Kong, China
[3] School of Management, Jilin University, Changchun, China
[4]School of Electrical Engineering, Guangxi University, Nanning, China

Abstract. This article discusses and analyzes the technical strengths and weaknesses of the green power generation that can be used for distributed system (power generation) power generation, for instance, solar power generation, wind power, hydrogen fuel cells, biomass power generation, and small gas turbines. The key to the discussion is to apply the technical distributed power generation of solar power stations. In addition, it also discussed the use of "focusing solar power generation high-temperature solar thermal power conversion system software" technical completion of distributed system power distribution. Low-cost, high-temperature solar thermal power generation is selected as the power generation solution medium, the power generation is technically low consumption and high-efficiency, the volume and power generation methods are conveniently equipped, the stability is high, and the economic development is environmentally friendly.

1 Introduction

Distributed Generation (DG) or Distributed Resource (DR) is generally defined as the composition of small and medium-sized diesel generators or distributed power sources and energy storage technology, which are connected around the customer side to meet some end-user customers' requirements[1]. Their storage capacity is generally small, ranging from tens of kilowatts to hundreds of thousands of kilowatts. Common forms of Distributed Generation include some environmentally friendly new energy, such as gas, hydrogen power, solar power and wind power. Distributed Resource can use a variety of new energy technologies. Generally speaking, the Distributed Resource system is a way of immediately connected to the 380 V or 10 kV distribution network system for grid connected operation or free operation.

In addition, We can sort the DR into the following categories [2,3].

Power generation system software that takes this kind of technology as the key technology, such as gas turbines, natural gas vehicle engines or gas turbine reciprocating vehicle engines, small turbine engines, etc.

Hydrogen fuel cell power generation system software includes ammonium hydrogen sulfate fuel cell (PAFC), proton exchange membrane hydrogen fuel cell (PEMFC) [4], molten rock carbonate hydrogen fuel cell (MCFC) and solid metal oxide hydrogen fuel cell (SOFC) .

Photovoltaic battery power generation system software

Wind turbine system software [5]

Bio-energy power generation system software [6,7]

1.1 Socioeconomics

Compared with traditional large and medium-sized thermal power plants, the Distributed Resource power generation system software has the advantages of less project investment, less risk, smaller floor space, and shorter capital construction cycle time. The DR power generation system software is built around the load core, and there is no need to build expensive transmission and distribution Internet and power distribution stations. It saves land resources, and the load is consumed locally, which greatly reduces the line construction cost and line loss of long-distance power transmission.

1.2 Ecological environment protection

Using gas or hydrogen, solar power and wind power, due to reduce fuel use or no need for fuel, can reduce the emission of harmful substances, thereby alleviating the working pressure of natural recovery of the environment. Many nearby principles can reduce the basic construction of large-space, long-distance, and high-voltage transmission lines. It not only reduces the electromagnetic radiation generated by the high-voltage transmission line, but also reduces the land resources and route aisle of the high-voltage transmission line. And It can also reduce the broadscale cutting down of trees, which is beneficial to ecological environment protection.

* Corresponding author: 13926402226@163.com

1.3 The diversification of the main uses of electric energy

Distributed Resource can use various alternative electric energy sources, for example, green resources (gas), new resources (hydrogen) and renewable resources (wind and solar power). In the meantime, the process of converting fuel energy into electric energy, power generation system will produce a lot of heat energy,which is usually abandoned by traditional power plants. Distributed Generation can play a dual role of power supply and heating in residential or commercial areas by establishing cogeneration devices, improve fuel utilization efficiency and reduce pollution. Therefore, this is also a new method to deal with energy shortage issues.

1.4 Reliable switching power supply

DR is transformed into system software separately from each other. In the event of a problem in the power grid, DR can ensure the energy supply of its customers and prevent some devastating adverse effects. Meanwhile, a lot of Distributed Resource power generation system software is included in the power grid and can be maintained when Distributed Resource cannot operate normally because of its own matters. Thus, the Distributed Resource power generation system software built around us and connected to the large power grid can further improve the overall energy supply, especially the operational safety and stability of the power supply system.

1.5 Solutions for power distribution in remote areas

It is very difficult to distribute power from the power grid in remote areas and areas with sudden load growth. Therefore, the use of separate system software for solar, wind, and bio-energy power plants may be the best approach to avoid a large amount of capacity expansion cost of transmission and distribution equipment.

2 Categories and characteristics of the power generation function of Distributed Resource power generation systems

2.1 Small turbocharged power generation system software

The output power of each small turbine engine ranges from 0.2 to 500 kW even can be expanded to 1 MW. Its main technical principle are using radial flow centrifugal impeller automobile engine (centrifugal turbocharger and multi-stage centrifugal fan) and heat recovery cycle. The key to small turbocharged diesel generators is fuel gas, methane gas, automotive gasoline and diesel engines. It has the advantages of high stability, long service time, low noise, light weight, small output capacity, clean energy, rich materials, and low consumption. This is also the most widely used of Distributed Resource system[8].

2.2 Wind turbine system software

Among all kinds of energy, wind energy is relatively simple to use. It is different from coal, oil and natural gas. It needs to be excavated and transported to the boiler equipment of thermal power plant for combustion; It is also different from hydraulic energy. A dam must be built to drive the water turbine; Nor does it require expensive devices and protective equipment like atomic energy. The utilization of wind energy has a broad future because of its simplicity and convenient installation. Especially in coastal islands, mountainous areas and plateau areas where there is a lack of hydraulic resources, fuel and inconvenient transportation, there is high-speed wind, which is a very valuable energy. If it can be used to generate electricity, it will be quite beneficial to the life and production of local people

2.3 System software of solar power station

Solar energy is the most smart and practical technology of all renewable energy sources. The techncal nature of solar power generation includes solar thermal power generation [9] and solar cell power generation[10]. Solar thermal power generation technology is an important project of solar energy utilization. It gathers solar energy, heats working medium and drives steam turbine to generate electricity. The optimal capacity range for general applications is one. At present, solar thermal power generation can be roughly divided into three categories: solar focusing power generation, solar pool power generation and solar chimney power generation. According to the form of receiving system, solar focusing power generation can be divided into three types: slot line focusing system, tower system and disc system. Solar cell power generation technology directly converts light energy into electric energy, and generates DC electric energy according to the photovoltaic effect of solar cell semiconductor materials. A solar cell usually has an area of less than one square centimeter, can only produce a voltage of about 0.5V, and the output power is very limited, usually less than 1W. Therefore, in order to output more power, according to the actual demand, many cells can be assembled into solar cell modules, and can also be connected in series or in parallel.

The characteristics of solar power generation are: inexhaustible resources, no air pollution, not subject to regional restrictions, and very convenient. Photovoltaic cells can be used as a separate power supply system, and can also be connected to the power grid. However, if the solar power station system software is successfully applied to the DR system software, some problems must be dealt with. Such problems include: how to further control the cost, how to solve the problem of low solar light than energy, how to solve the change of radiant power within 24 hours, and how to solve the problem of power engineering stability. In the new approach clearly presented below, most of these issues have been dealt with.

2.4 Hydrogen fuel cell power generation system software

Hydrogen fuel cell [11] immediately uses radon gas obtained from various fuels (gas, gas and crude oil), reacts with O_2 in the air with the help of electrolyte solution, and produces water together with electricity. For the most part, it converts mechanical energy into electromagnetic energy. It immediately converts the mechanical energy into electromagnetic energy, instead of the traditional thermal power plant system software, which converts the mechanical energy of the fuel into heat and then into mechanical kinetic energy, so it is simple and efficient. The chemical energy is then converted into electromagnetic energy. Because there are too many conversion processes, the overall power generation is high and the efficiency is not high.

The hydrogen fuel cell has the advantages of high heat transfer efficiency, zero emission, no noise, no vibration, application of water saving, stable and reliable driving force output, and strong ability to solve load changes. However, if hydrogen fuel cells are widely used in the DR power generation system software, the problem of increased project investment costs can be dealt with. The installation cost of hydrogen fuel cell is 3000-4000 USD per KW. Therefore, more efficient new technologies must be used to improve the level of manufacturing, expand the scale of production, and then improve the level of automation technology..

3 Use the new "Focus Solar Power High Temperature Photothermal Power Conversion System Software" to complete DR power distribution

3.1 Technicality of solar power station

Solar power is inexhaustible, and it is not easy to cause environmental pollution. It is the optimal power source for large-scale public applications. Therefore, scientific researchers have been exploring ways to use the sun. So far, this technology has only been used to produce ultra-low temperature and low-taste electric energy, such as boiling water. Aiming at high temperature and high quality electric energy, especially electric energy production and manufacturing, Because the technological development is relatively limited, the product cost is high. In addition, the technology cannot be compared with traditional fossil fuels. There are several solar power generation technologies, the following part will briefly introduce their realization in detail.

3.1.1 The technicality of solar power generation..

This technology has already experienced a development trend for more than a century. So far, there have not been all important improvements. This technology still uses semiconductor materials for photoelectric conversion, such as photovoltaic cells and crystalline silicon. The cost of large-scale power generation is much higher than that of non-renewable resources. It is not possible to meet the needs of large-scale power generation in today's economic situation. Therefore, the process seems to have difficulties in replacing or filling the traditional power engineering manufacturing level [12]..

3.1.2 The technical nature of solar thermal power generation.

This technology has been used on a large scale. Its crucial issue is tracking manipulation. Because of the diversification and precision requirements of its tracking, it is impossible to reasonably reduce the cost of the lighting module. At the same time, because the conversion of CSP is generated in the open indoor space, there are many damages caused by light reflection and thermal damage. CSP has high conversion efficiency and low efficiency, and its rationality is far less than traditional (electric energy production and manufacturing) methods..

3.2 Focusing solar power high temperature CSP conversion system software

High-temperature CSP materials can be produced at low cost to generate electricity [14]..

3.2.1 Key criteria.

The sun can be focused based on the reflective surfaces on the three sides. The concentrated sun reaches the juxtaposed mirror area of the glare focus point and can be concentrated on the bathroom mirror. There is a small round hole on the focal point of the dazzling mirror on the heat storage safety channel. The focusing lamp can go into the hole to complete the conversion of photothermal power generation. According to the basic principles of heat [15], this method is similar to the heat conduction standard in boldface, which can complete the conversion of high-efficiency CSP. Fuqing in the safe passage is used as a material for heat conduction and heat storage. Its technical basic principles have been recognized in the operation process equipment, and the tracking control system has been grasped to achieve high-precision and reasonable regulations. The Fuqing substance can be heated to above 800℃ after passing through the reflective surface three times. It can be connected to the current motor to meet the requirement of becoming larger. Figure 1 shows the photos of the proposed machinery and equipment.

Figure 1. The high-temperature solar thermal power conversion system software is completed according to the focus solar power generation

3.2.2 Technical characteristics.

The heat storage capacity is not restricted, and the construction is convenient and flexible. The system software can be easily connected to generators of various capacities according to different methods. It can be used together with steam turbine or gas turbine power generation. Its geometric figure is relatively low in height, which is convenient for shock resistance. Each module has the same posture, so it is easy to operate. It has high stability and low cost. There are no special or bureaucratic capitalist regulations to implement this technology, and at the same time it does not require a lot of electrical equipment. Even if processing machinery and equipment are used, the facility can be completed at low cost.
There is a certain degree of thermal radiation damage when focusing on the three reflective surfaces, so the kinetic energy transmission, transformation and utilization rate are high.
All components can be mass-produced, thoroughly meeting the necessity of mass application.
The middle heat transfer medium (salt) can be heat treated to a temperature above the change. High-temperature storage can be completed at the same time, and high-end heat can be given at the same time. The annual operating time of generator equipment has been greatly improved. The heat exchanger standard in the heat storage safety channel is stable, and the super large total area in the heat storage safety channel can ensure the timely and flexible adjustment of the motor load. This will further help improve the comprehensive rationality of generator equipment.

3.2.3 Estimated development trend.

At this stage, the investment per KW of the system software and the cost of wind power are very high. However, its power generation capacity is much higher than wind power generation. At the same time, the selectivity of engineering construction sites is far better than wind power generation. Therefore, its rationality is better than wind and solar power generation. With the further improvement of the production process, the cost can be the same as the basic coal chemical industry.
The results obtained at this stage indicate a good vision for the direction of development. Nowadays, there are a variety of gas turbine machines and equipment with different volumes on the sales market. They have given the flexibility to select special KW-manufactured small turbine engines and the established system software according to the application. In addition, the entire process of scientific research and development will be completed without project investment exceeding 10 million US dollars. After all the development and design of the model enterprise are carried out, the method will be comprehensively improved, and the electric energy production process will be more cleaned up.

4 Conclusion

Distributed power generation (DPG) technology is a new technology. It can develop along with the rapid development of renewable resources technology. Because of the concept of natural environment, DR technology must be more and more critical for analysis. The reform and innovation of the power system will promote the development trend of distributed power technology.
In the current stage of work, "Focus Solar high-temperature solar thermal power conversion system software" is discussed as an uncertain distributed system resource (DR) technology. The high-temperature photo thermal power generation material for generating electricity can be produced at low cost. It has the advantages of low cost, high conversion efficiency, volume distribution, flexible and reliable power generation methods, economic development, and environmental protection. This technology is a new type of emerald green power generation technology. Through further improvement of soundness, it is hoped that a new direction will be clarified for the development trend of green energy..

References

1. Zheng Jianchao, Current Status and Prospects of Power Technologies at the Frontier. Electric Power, 32(10): pp1-9. 1999
2. M. Begovie, A. Pregelj, A. Rohatgi. Impact of Renewable Distributed Generation on Power Systems. the 34th Hawaii International Conference on System Sciences, ppl-10,2001.
3. Wang Hui, Qiu Yuchang, LI Rubiao," Development of Distributed Power Generation Technology." Northwest Chιna Electrιc Power; pp 33(3).,2005.
4. H.-C. Chen, P.-H. Chen, ontroller design of proton exchange membrane fuel cell for sustainable energy. Energy Education Science and Technology Part A 28(1):pp 271-284, 2011.
5. C. Sevim, D. Kultur, S. Basak, Wind power grid connection characteristics for Turkey. Energy Education Science and Technology Part A 26(2): pp155-164, 2011.
6. S. Acar, A. Ayanoglu., Higher heating values (HHVs) of biomass fuels. Energy Education Science and Technology Part A 28(2): 749-758, 2012.
7. HAN Shuqi, LI Wenxin, CHEN Chong, et al. Modeling and Control of Controllable D-PMSG Based on Hybrid Energy Storage of Wind Power Hydrogen Production and Supercapacitor[J]. Guangdong Electric Power, 2019, 32(5): 1-12.
8. Ding Ming, Wang Min., Ributed generation technology. Electric Power Automation Equipment 24(7): pp31-36,2004.
9. C. H. Wang, Y. Chen, J. Huang., Thermal and electrical mechanism of thermoelectric generation chip and its performances in solar energy thermal

power system. Energy Education Science and Technology Part A, 28(1): pp331-338,2011.

10. B. Yelmen, T. Ustuner, M. Ustuner, Determining the potential of solar energy and benefiting from this potential using photovoltaic system in Turkey. Energy Education Science and Technology Part A 27(2): 347-358, 2011.
11. Larminie, James. Fuel Cell Systems Explained, Second Edition. SAE International, ISBN 0-7680-1259-7, 2003.
12. O. Guler, S. Eyigun., Turkey solar potential and viability of grid connected solar photovoltaic power plant in different regions. Energy Education Science and Technology Part A, 28(2): pp879-890,2012.
13. SU Wei, ZHONG Guobin, XU Kaiqi, et al. Review of Evaluation Method for Economy of Energy Storage Technology[J]. Guangdong Electric Power, 2019, 32(1): 29-35.
14. Carlo Ombello, The world's first molten salt concentrating solar power plant. http://www.guardian.co.uk/environment/2010/jul/22/first-molten-salt-solar-power, 2010 06 22.
15. QIAN Zeyu, ZANG Haixiang, CHENG Lexiang, et al. Research on Allocation for Hybrid Energy Storage System Capacity Based on FDM[J]. Guangdong Electric Power, 2018, 31(6): 46-53.

E3S Web of Conferences **329**, 01022 (2021)
GESD 2021
https://doi.org/10.1051/e3sconf/202132901022

Research on the Interactive Response Relationship between Price and Load Based on Market Elasticity

Pengcheng Zang, Zhijun Xu, Yan fei Su and Yingjun Li

Inner Mongolia Electric Power (Group) Co., Ltd. Ulanchabu Electric Power BureauUlanchabu, China

Abstract: With the rapid development of local economy, the frequent occurrence of extreme climates, and the increase in the scale of new energy installations, short-term peak loads have repeatedly reached new highs, and the characteristics of the power system "double high, double peaks, and double-sided random" will be further highlighted, making the power grid supply and demand Balance faces greater challenges. In this context, the role of electricity demand response has become more prominent. Therefore, based on the principle of market elasticity, this paper studies the interactive response relationship between electricity price and load, establishes a demand response model based on price elasticity, and verifies demand response projects based on actual cases. The effect of peak clipping and valley filling.

1 Introduction

There according to the principles of economics, the elasticity of electricity market includes the self-elasticity of electricity and the cross-elasticity of electricity. The self-elasticity is used to measure the impact of current price changes in a single period on electricity demand, while the cross-elasticity is used to measure the impact of price changes in a multi-period on electricity demand. At present, domestic and foreign scholars have done a lot of research on demand response model based on market elasticity.

Taking into account the price elasticity of demand for consumer electrical equipment, literature [1] proposed a method for evaluating the price elasticity of residential electricity demand based on the characteristics of equipment power consumption. According to the definition of demand price elasticity, the use characteristics and self-elasticity of residential electrical equipment were analyzed. Cross-elastic correlation, combined with the classification of equipment adjustability associated with electricity consumption characteristics, constructs a residential electricity price response behavior analysis model based on equipment electricity adjustment fuzzy inference;Reference [2] focuses on how to automate and intelligentize the load on the user side Participate in demand response, design and implement a user-side automatic demand response system based on load decomposition to realize user participation in price-based demand response;Literature [3] studies the demand price elasticity matrix and proposes a method based on elasticity effect weights. Improve the price elasticity matrix model; in order to improve the operability of demand response potential evaluation; literature [4] proposed a quantitative evaluation method for demand response potential considering the statistical characteristics of load power consumption. This method is based on the principles of econometrics to obtain electricity demand-price elasticity Coefficient, quantify the electricity price level and the power demand under the incentive policy;literature [5] uses the user's demand price elasticity matrix for power commodities to establish a peak-valley time-of-use electricity price mathematical model;literature [6] establishes a reasonable time-of-use price and compensation Price, based on the electricity price elasticity coefficient, establishes an electricity demand response model, and uses the price signal to transfer part of the electricity consumption during the peak load period to the load trough period, achieving peak-shaving and valley-filling;literature [7] according to demand price elasticity Matrix (PEMD), proposes an improved PEMD model based on elastic effect weight; Literature [8] proposes a demand-side management evaluation model of distribution system based on the concept of elasticity and time-of-use pricing.

Research on the interactive relationship between load and electricity price based on market elasticity

Analysis of Electricity Market Elasticity

The price elasticity of electricity demand refers to the relative change in the amount of electricity demand caused by the relative change in the price of electricity, that is, the ratio of the rate of change of the user's electricity consumption to the rate of change of the corresponding price within a certain period of time. According to economic principles, a typical demand-price relationship curve is shown in Figure 1.

* Corresponding author: 595910434@qq.com

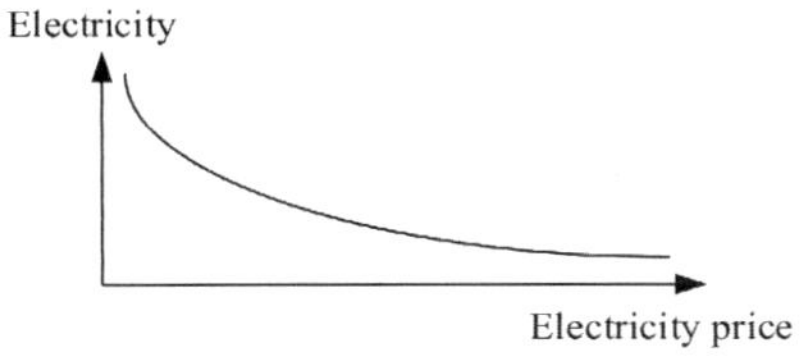

Figure 1. Typical demand price relationship curve

Under normal circumstances, when the price of electricity rises, the demand for electricity will decrease accordingly. The response to the electricity demand triggered by the electricity price, that is, the price elasticity of demand represented by the elastic coefficient, is shown in equation (1):

$$\varepsilon = \frac{\Delta q}{q_0}\frac{p_0}{\Delta p} \tag{1}$$

In the formula, Δq and Δp respectively represent the relative q_0 increase of the original load and the original p_0 electricity price.

Because in real life, the electricity demand of most users in a certain period of time is not only affected by the electricity price of that period, but also related to the electricity price of an adjacent time. In the multi-period response, the load elasticity coefficient can be divided into the self-elasticity coefficient and the cross elasticity coefficient according to the response time period. The self-elasticity coefficient represents the change in user load caused by the electricity price in the current period, and the cross-elasticity coefficient represents the impact of the electricity price in other periods on the user's electrical load. The coefficient of self-elasticity and cross-elasticity are as follows:

$$\varepsilon_{ii} = \frac{\Delta q_i}{q_i}\frac{p_i}{\Delta p_i} \tag{2}$$

$$\varepsilon_{ij} = \frac{\Delta q_i}{q_i}\frac{p_j}{\Delta p_j} \tag{3}$$

In the formula, ε_{ii} represents the self-elasticity coefficient, ε_{ij} represents the cross-elasticity coefficient, and the subscripts represent the i-th and j-th periods, respectively. From the self-elasticity coefficient and the cross-elasticity coefficient, the elasticity matrix E of electricity load and electricity price is obtained as follows:

$$E = \begin{bmatrix} \varepsilon_{11} & \varepsilon_{12} & \cdots & \varepsilon_{1n} \\ \varepsilon_{21} & \varepsilon_{22} & \cdots & \varepsilon_{2n} \\ \vdots & \vdots & & \vdots \\ \varepsilon_{n1} & \varepsilon_{n2} & \cdots & \varepsilon_{nn} \end{bmatrix} \tag{4}$$

The interactive relationship model between electricity load and electricity price

The interactive relationship model between electricity load and electricity price in the period 1~n can be expressed as follows:

$$\begin{bmatrix} \frac{\Delta q_1}{q_1} \\ \frac{\Delta q_2}{q_2} \\ \vdots \\ \frac{\Delta q_n}{q_n} \end{bmatrix} = \begin{bmatrix} \varepsilon_{11} & \varepsilon_{12} & \cdots & \varepsilon_{1n} \\ \varepsilon_{21} & \varepsilon_{22} & \cdots & \varepsilon_{2n} \\ \vdots & \vdots & & \vdots \\ \varepsilon_{n1} & \varepsilon_{n2} & \cdots & \varepsilon_{nn} \end{bmatrix} \begin{bmatrix} \frac{\Delta p_1}{p_1} \\ \frac{\Delta p_2}{p_2} \\ \vdots \\ \frac{\Delta p_n}{p_n} \end{bmatrix} = E \begin{bmatrix} \frac{\Delta p_1}{p_1} \\ \frac{\Delta p_2}{p_2} \\ \vdots \\ \frac{\Delta p_n}{p_n} \end{bmatrix} \tag{5}$$

Thus, the energy load of each period after the price-based demand response transformation can be obtained as:

$$\begin{aligned} q_t' &= q_t + \Delta q_t \\ &= q_t[1 + E(t,t)\frac{\Delta p_t}{p_t} + \sum_{\substack{i=1 \\ i\neq t}}^{n} E(t,i)\frac{\Delta p_i}{p_i}] \end{aligned} \tag{6}$$

Then the load change in this period is:

$$\Delta q_t = q_t[E(t,t)\frac{\Delta p_t}{p_t} + \sum_{\substack{i=1 \\ i\neq t}}^{n} E(t,i)\frac{\Delta p_i}{p_i}] \tag{7}$$

According to the current situation of demand response in various provinces and cities in China and the research experience of many scholars, the price elasticity coefficient of demand during peak, peak, flat, and valley periods can be obtained as shown in Table 2-1. The self-elasticity coefficient of electricity demand price during valley period is -0.15 , Which shows that the change in electricity price during the valley period is negatively correlated with the change in the electricity demand during that period, and when the electricity price in the valley period increases by 1%, the user's power demand in the valley period decreases by 0.15%. The cross elasticity coefficient of the electricity demand price during the normal period to the valley period is 0.03, indicating that the change in electricity price during the normal period is positively correlated with the change in the electricity demand during the valley period, and when the electricity price in the normal period increases by 1%, the user's power demand in the valley period decreases by 0.03% .

Table 1. Price dlasticity coefficient of electricity demand by users

ε	Valley	flat	peak	sharp peak
Valley	-0.15	0.03	0.06	0.04
flat	0.03	-0.15	0.02	0.06
peak	0.06	0.02	-0.15	0.03
sharp peak	0.04	0.06	0.03	-0.15

 https://doi.org/10.1051/e3sconf/202132901022

2 Example analysis

This section selects the time-of-use electricity price before the implementation of the demand response and the peak electricity price after the implementation of the demand response in a province, as well as the forecasted load data of a typical summer day in 2025 (as shown in Figures 2 and 3).

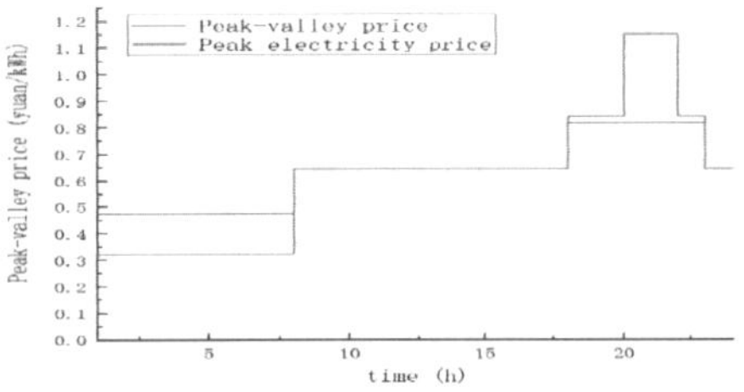

Figure 2. Electricity price before and after the implementation of demand response

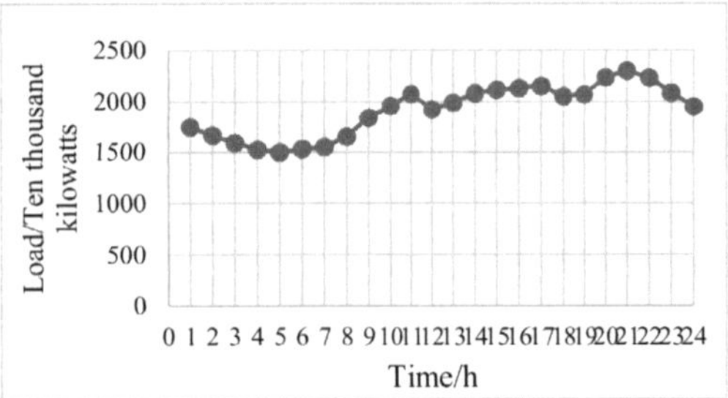

Figure 3. Typical daily load curve in the summer of 2025

The interactive relationship model between electricity load and electricity price in time period 1~24 is calculated as follows:

$$\begin{bmatrix} \frac{\Delta q_1}{q_1} \\ \frac{\Delta q_2}{q_2} \\ \vdots \\ \frac{\Delta q_{24}}{q_{24}} \end{bmatrix} = \begin{bmatrix} \varepsilon_{11} & \varepsilon_{12} & \cdots & \varepsilon_{1,24} \\ \varepsilon_{21} & \varepsilon_{22} & \cdots & \varepsilon_{2,24} \\ \vdots & \vdots & & \vdots \\ \varepsilon_{24,1} & \varepsilon_{24,2} & \cdots & \varepsilon_{24,24} \end{bmatrix} \begin{bmatrix} \frac{\Delta p_1}{p_1} \\ \frac{\Delta p_2}{p_2} \\ \vdots \\ \frac{\Delta p_{24}}{p_{24}} \end{bmatrix} = E \begin{bmatrix} \frac{0.4759625-0.47179}{0.47179} \\ \frac{0.4759625-0.47179}{0.47179} \\ \vdots \\ \frac{0.6452625-0.64371}{0.64371} \end{bmatrix} = \begin{bmatrix} 0.02768352 \\ 0.02768352 \\ \vdots \\ 0.0024118 \end{bmatrix} \quad (8)$$

Assuming that users who implement time-of-use electricity prices account for 50%, and users who do not implement time-of-use electricity prices account for 50%, from equation (8), the energy load of each period after the price-based demand response transformation can be obtained as:

$$\dot{q_t} = q_t + 0.5 q_t \frac{\Delta q_t}{q_t}$$

$$= \begin{bmatrix} 1749 \\ 1665 \\ \vdots \\ 1948 \end{bmatrix} + 0.5 \begin{bmatrix} 1749 \times 0.02768352 \\ 1665 \times 0.02768352 \\ \vdots \\ 1948 \times 0.0024118 \end{bmatrix} = \begin{bmatrix} 1773 \\ 1688 \\ \vdots \\ 1995 \end{bmatrix} \quad (9)$$

Specifically, according to formula (1) ~ formula (9) to calculate the load of each period after the implementation of the time-of-use electricity price, as shown in Table2. and Figure4.shown.

Table 2. Loads in each period after the implementation of time-of-use electricity

Time period		Original load(Ten thousand kilowatt)	Load after implementation(Ten thousand kilowatts)	load change(Ten thousand kilowatts)
Valley	1	1749	1773	24
	2	1665	1688	23
	3	1591	1613	22
	4	1526	1548	22
	5	1500	1521	21
	6	1535	1556	21
	7	1553	1575	22
	8	1656	1679	23
flat	9	1842	1886	44
	10	1956	2003	47
	11	2073	2123	50
	12	1923	1969	46
	13	1989	2036	47
	14	2079	2129	50
	15	2114	2164	50
	16	2131	2182	51
	17	2152	2203	51
	18	2049	2098	49
peak	19	2068	2083	15
	20	2234	2250	16
sharp peak	21	2300	2171	-129
	22	2230	2104	-126
peak	23	2084	2099	15
flat	24	1948	1995	47

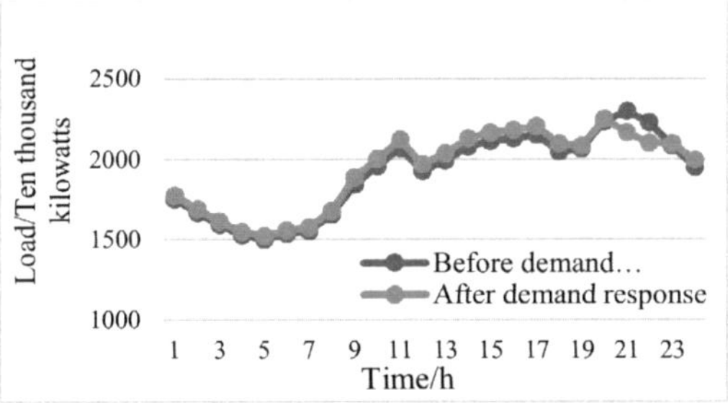

Figure 4. Changes in load in each period before and after the implementation of time-of-use electricity prices

According to Figure 4, the implementation of time-of-use electricity prices can achieve the effect of peak shaving and valley filling. When the peak-to-valley price difference is 0.665 yuan, a total of 2.55 million kilowatts of load will be transferred during peak periods to flat-valley periods, and the total load during valley periods will increase by 1.78 million kilowatts. The peak cut percentage of the highest load before the implementation of time-of-use electricity prices can reach 5.61%. The percentage of grain filling at the lowest load reached 1.40%.

3 Conclusion

From this study, the implementation of time-of-use electricity prices has played a role in peak-shaving and valley-filling. When the peak electricity price is raised by

0.01 yuan, the peak-to-valley price difference increases to 0.675 yuan. During the peak period, a total of 2.63 million kilowatts of load is transferred to the flat valley period. The load during the valley period increases by 1.83 million kilowatts, and the peak cut percentage of the highest load can reach 5.81% , The percentage of grain filling at the lowest load reached 1.43%. When the peak electricity price was increased by 0.02 yuan, the peak-to-valley price difference reached 0.685 yuan. During the peak period, a total of 2.71 million kilowatts of load was transferred to the flat valley period. The load during the valley period increased by 1.89 million kilowatts. The peak load of the highest load before the time-of-use electricity price was implemented. The percentage can reach 5.99%, and the percentage of valley filling with the lowest load reaches 1.48%. Therefore, increasing the peak-to-valley spread can improve the peak-shaving and valley-filling effect.

Acknowledgments

This work was financially supported by Science and Technology Project of Inner Mongolia Electric Power Group (Limited) Company (WD-ZXZB-2021-SC0402-0304): Research on Key Technologies of Coordinated Control of Ulanqab Regional Power System.

References

1. Xu Guangda, Zhang Li, Liang Jun, Sun Donglei, Zhao Long, Han Xueshan. Evaluation of residential electricity demand price elasticity based on equipment power consumption characteristics[J]. Automation of Electric Power Systems, 2020, 44(13): 48-55.
2. Sun Zhiqing, Wang Shouxiang, Zhou Kai, Liu Tianyu. User-side automatic demand response system based on load decomposition[J]. Journal of Electric Power System and Automation, 2016, 28(12): 64-69+88.
3. Liu Haotian, Chen Jin, Zhu Xi, et al. A peak-valley time-of-use electricity price incentive strategy for residents based on price elasticity matrix[J]. Power System Protection and Control, 2021, 49(5): 116-123. DOI: 10.19783/j.cnki.pspc.200527.
4. Li Zhangyun, Wang Gang, Ding Maosheng, et al. Demand response potential evaluation considering the statistical characteristics of load power consumption[J]. Chinese Science and Technology Paper, 2017, 12(5): 529-536. DOI:10.3969/j. issn.2095-2783.2017.05.010.
5. Sun Wei, Li Shan. Research on time-of-use electricity price in power demand side management [J]. Electrical Application, 2008,27(6):63-66. DOI:10.3969/j.issn.1672-9560.2008.06.019 .
6. Sun Qian, Peng Jianchun, Pan Juntao, et al. Congestion management for multi-period power demand response[J]. Power System Technology, 2010, 34(9): 139-143.
7. Wang Lei, He Jianjian, Zhou Mei, et al. Research on multi-frequency control of power grid peak-valley time-sharing load based on demand response[J]. Electronic Design Engineering, 2020, 28(11): 5-8, 13. DOI: 10.14022 /j.issn1674-6236.2020.11.002.
8. Juan Carlos Galvis, Agostino Costa. Demand Side Management Using Time of Use and Elasticity Price[J]. IEEE Latin America Transactions, 2016,14(10):

Research on Formaldehyde Release Rate of Car mats

Song Zhao[a], Yanling Wang, Lei Wang, Yanan Lu, Xueli Han, Jin Liu*

Hebei Academy of Product Quality Supervision and Inspection, Shijiazhuang 050200, Hebei China
[a] 930575554@qq.com

Abstract: The environmental test cabin method was used to simulate the actual use environment of car mats to collect test gas, The formaldehyde content in the collected gas was measured by ultraviolet spectrophotometer, and the formaldehyde release rate was calculated. The results indicated that formaldehyde was released in all 30 batches of samples at a rate ranging from 0.02 to 0.13mg/ m2 (m2•h).

Keywords: Car mats, formaldehyde release rate, environmental test cabin, ultraviolet spectrophotometer

1 Introduction

As people's living standards are getting higher and higher, the number of cars in China is increasing year by year, and cars have gradually become the means of transportation for people's daily travel[1]. When people pay attention to the convenience of travel, they often ignore the serious consequences of environmental pollution in small cars, which poses a serious health threat to drivers and passengers[2]. There are many reasons for the primary pollution of the automotive interior environment. As an important automotive interior product, the quality of the automotive floor mat directly affects the automotive interior environment.

Inferior floor mats emit toxic and harmful gases, which is one of the main causes of air pollution in the vehicle. At present, the VOC detection method for the air in the vehicle has become mature[3]. GB/T 27630-2011 guidelines for air quality evaluation in passenger cars also set corresponding requirements for the limits of eight volatile organic compounds such as benzene in the air in the vehicle, but for the car floor mats, There is no corresponding national standard. Formaldehyde is a colorless and soluble gas with strong pungent smell, which mainly comes from the inferior components and adhesives of automobile footmats. Formaldehyde can enter the human body through the respiratory tract. It has the characteristics of long-term, latent and hidden. It is a protoplasm poison. When combined with protein, it will lead to the imbalance of nervous system and immune system. It is also a carcinogen. Excessive inhalation of formaldehyde will produce adverse reactions such as fainting, fatigue, muscle weakness, syncope, and even death[4-5].

Hebei Province is a large province for the production of automobile floor mats. The production is mainly concentrated in Xinle City of Shijiazhuang and Nangong city of Xingtai. It is mainly small enterprises and even family workshops. The product quality is uneven. The problems of product quality and safety are reflected in the use safety and air pollution in the car. In recent years, the relevant national departments have not conducted supervision and random inspection on the automobile foot pad products, and the product quality situation of the industry is not clear. Therefore, it is urgent to study and monitor the product quality of automobile foot pad products.

In this study, the environmental test chamber is used to simulate the actual use environment of automobile foot pad products, collect the test gas, test the formaldehyde content in the collected gas by UV-Vis spectrophotometer, calculate the formaldehyde release rate, and discuss the use risk of automobile foot pad products.

* Corresponding author: szjyliujin@126.com

2 Experimental

2.1 Materials

In this study, the environmental test chamber is used to simulate the actual use environment of automobile foot pad products, collect the test gas, test the formaldehyde content in the collected gas by UV-Vis spectrophotometer, calculate the formaldehyde release rate, and discuss the use risk of automobile foot pad products. Car floor mats are common in the market, including latex type, silk ring type and large surrounding type (Fig. 1). The samples are purchased from the auto fittings (supplies) business premises registered in Hebei Province with the business license, and the scale of the business premises is small. The sample unit is set, and the sampling quantity is 1 set / batch, a total of 30 batches. Among them, there are 15 batches of emulsion type, 11 batches of wire loop shape and 4 batches of large enclosure

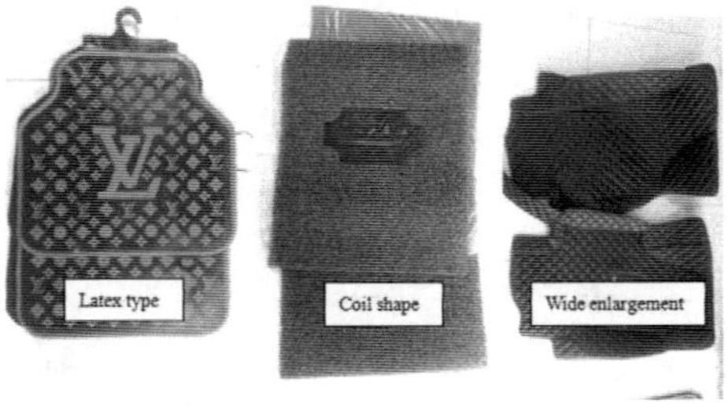

Figure 1 Sample category

2.2 Experimental method

Prepare the sample with the material / cabin load ratio of 0.4cm2/m3, cut the surface artificially and cover the ground with aluminum foil. The sample shall be built into the environmental test chamber for testing 1h after preparation. Place the sample in the middle of the environmental test chamber (air temperature 60 °C, relative humidity 5%, air exchange rate 1 time / h, air velocity on the sample surface 0.2m/s), the emission surface is horizontal and upward, so that the air flow passes through the time surface evenly, and quickly close the cabin door of the environmental test chamber to start the test. The time of putting into the environmental test chamber is 0, and the air in the chamber is sampled and analyzed after 24 hours of balance.

Put 5ml absorption solution into the bubble absorption tube, sample at the flow rate of 0.5l/min, and the gas production volume is 10L. Transfer all the sample solution into a 10ml colorimetric tube, wash the absorption tube with a small amount of absorption solution, combine to make the total volume of 5ml, add 0.4ml ammonium ferric sulfate solution, shake well and place for 15min. At the wavelength of 630nm, the absorbance of each tube solution was measured with a 1cm cuvette and water as the reference. At the same time, 5ml of unsampled absorption solution was used as blank control. Determine the formaldehyde content according to the standard curve.

Formaldehyde emission(Ci), $C_i = \frac{(m - m_0) \times 1000}{1000 \times V_0}$

Formaldehyde concentration in sample under standard state(Csi),mg/m3, $C_{si} = C_i \times \frac{101.3}{P} \times \frac{t + 273}{273}$

Formaldehyde release rate (EFI),mg/(m2*h), $EFi = \frac{C_{si} \times V \times ACH}{S}$

Sample volume(V0), L
Working pressure of environmental test chamber (P), kPa
Ambient cabin temperature(t), °C
Sample preparation area(S) ,m2
Sample preparation area and volume in environmental chamber(V), m3
Ventilation rate of environmental chamber(ACH),times/h
Formaldehyde quality of sample blank(m0),μg
Formaldehyde mass of sample solution(m),μg

3 Results and Discussion

Table 1 is the statistics of formaldehyde emission rate test results of 30 batches of car foot mats.

Table 1 Test results of formaldehyde release rate

Name		Latex type	Wide enlargement	Coil shape
Result rangemg/(m²·h)	Measured value	0.02 ~ 0.11	0.05 ~ 0.12	0.03 ~ 0.13
	average value	0.05	0.08	0.06
Total averagemg/(m²·h)			0.06	

It can be seen from table 1 that the detected value range of formaldehyde release rate of the sample is 0.02~0.13mg/(m2·h), and the average value is 0.06 mg/(m2·h). The formaldehyde release rate of different types of automobile footrest samples is slightly different. Among them, the formaldehyde release rate of large surrounding samples is the highest, reaching0.08 mg/(m2·h). The formaldehyde release rate of latex samples is lower than the average value, which is0.05 mg/(m2·h).

Table 2 Body effects of different formaldehyde concentrations

concentration (mg/m³)	effect
0.05	EEG changes
0.06	Eye stimulation threshold
0.06~0.22	Olfactory stimulation threshold
0.12	Upper respiratory tract irritation threshold
0.16	Increase in respiratory diseases
1.0	tissue damage
6.0	Pulmonary irritation
60.0	pulmonary edema
120.0	death

As we all know, formaldehyde has a great impact on human health. According to the data given in the Application Manual of national environmental health standards (Table 2) [4], formaldehyde concentration exceeding 0.05 (mg/m³) can cause harm to human body. According to the actual measurement, the interior space of an ordinary class B car is 3.5m3, the service area of car foot pad is 1.5m², and the car is exposed to the sun for 6 hours, The concentration range of formaldehyde in the vehicle caused by the test sample alone is 0.05 ~ 0.33 mg / m³, which can cause harm to human body (Table 3).

Table 3 Calculated formaldehyde concentration in the vehicle after exposure for 6 hours

Serial number	category	Formaldehyde concentration(mg/m³)	Serial number	category	Formaldehyde concentration(mg/m³)
1	Latex type	0.23	16	Wide enlargement	0.31
2	Latex type	0.10	17	Wide enlargement	0.23
3	Latex type	0.05	18	Wide enlargement	0.13
4	Latex type	0.10	19	Wide enlargement	0.21
5	Latex type	0.26	20	Coil shape	0.15
6	Latex type	0.15	21	Coil shape	0.21
7	Latex type	0.08	22	Coil shape	0.13
8	Latex type	0.08	23	Coil shape	0.15
9	Latex type	0.08	24	Coil shape	0.13
10	Latex type	0.15	25	Coil shape	0.08
11	Latex type	0.21	26	Coil shape	0.10
12	Latex type	0.05	27	Coil shape	0.13
13	Latex type	0.13	28	Coil shape	0.08
14	Latex type	0.13	29	Coil shape	0.33
15	Latex type	0.28	30	Coil shape	0.08

The sources of formaldehyde in automobile floor mats are complex, mainly caused by poor materials, processing solvents and adhesives. The formaldehyde emission rate of latex type automobile foot pad is low, which is related to its material and processing technology. Latex type automobile foot mats are made of a single material, and most products are made of integrated injection molding or pouring molding. High quality latex materials will not add processing additives containing formaldehyde in the processing process, and will not involve the use of adhesives. Therefore, the formaldehyde release rate of this type of samples is lower than that of other types. Silk ring and wide enlargement car mats are not made of single material. Coil shape mats are generally composed of bottom layer and silk ring, which are bonded with adhesive; Wide enlargement type foot mats are generally of multi-layer structure, and most of the layers are made of adhesive bonding or sewing. Therefore, the formaldehyde emission rate of these two types of automobile foot mats is high.

4 Conclusion

In order to simulate the real use environment of automobile foot pads, this risk monitoring adopts the environmental test chamber method to detect the toxic and harmful gases in automobile foot pads. This inspection method can better restore the real use environment of automobile foot pads, and has no restrictions on the type of samples. The results showed that formaldehyde was released from 30 batches of samples, and the release rate ranged from 0.02 to 0.13 mg / (m² • h).

References

1. Hangzhou Ming, Huang shunmin, Weng Wenxiang, Yue Gaodong, Yang pan. Study on volatile organic compounds in automobile floor mats [J]. Chemical management, 2017,(05):196+231.
2. Jia Xiaobo, Liu Rongde, Li Jing. Research and development of PVC special materials for automobile footrest [J]. PVC, 2008(08):21-22+33.
3. Cai Baixue, Luo Yuanyuan, Li Xiaolong, Zhang Li, Wang Hao. Risk assessment based on anti-skid performance of automobile footrest [J]. Journal of Chongqing Jiaotong University (NATURAL SCIENCE EDITION), 2018,37(04):128-132.
4. Zhang Chuanzhen, Zhang Jipeng, Zhao Hong, Shi Jin. Study on the harm of automobile interior pollutants to human health [J]. Journal of Qingdao University (Engineering Technology Edition),2014,29(01):105-110.
5. Li Dong, ye Jihong, Li Chaojun, Dai Danfeng. Harm and prevention analysis of environmental pollution in cars [J]. Management observation,2015(13):58-59+62.

Forecasting method of clean energy development potential considering sector coupling

Yu Zhu [1], Weijie Shen [2,*], Changhao Zhang [3], Yangyang Ge [1], Danting Cai [3]

[1] State Grid Liaoning Electric Power Research Institute, Shenyang 110006, China.
[2] School of Economy and Management, North China Electric Power University, Beijing 102206, China.
[3] School of Electrical and Electronic Engineering, North China Electric Power University, Beijing 102206, China.

Abstract. Sector coupling includes not only the coupling within energy sectors such as electricity, heat and gas, but also the coupling between energy sector and transportation sector, construction sector and industrial sector. This article introduces a method for measuring the development potential of clean energy based on sector coupling. First, the analysis model of electric energy substitution potential is constructed, the analysis object is determined, and the analysis object is quantified. On the basis of the definition of clean energy development potential, an IPAT model for electric energy substitution is constructed to realize the comprehensive evaluation of clean energy. Secondly, based on Markov theory to realize the sub-path calculation of the development potential of clean energy. Finally, based on the IPAT model and the decoupling theoretical model, this paper sets up three different alternative scenarios to make a more comprehensive prediction and analysis of the medium and long-term clean energy development potential, and applies the introduced measurement methods to the calculation of China's clean energy development potential. The results show the effectiveness of the algorithm in the calculation of the development potential of clean energy based on sector coupling.

1 Introduction

Energy is the foundation of human survival and development, an important driving force for the industrial revolution, and a strong support for social and economic operations [1].The literature [2, 3, 4] all pointed out that the development and utilization of clean energy is the current development trend of the world's energy industry. In the literature [5], it is pointed out that China will adopt relevant measures to accelerate the development and utilization of clean energy. Literature [6] pointed out that under the current situation, increasing the proportion of renewable energy use and achieving clean energy is the general trend, and there is huge potential for cross- sector energy coupling.

Sector coupling includes not only the internal coupling of electricity, heat, gas and other energy industries, but also the coupling between the energy sector and the transportation, construction, and industrial sectors. Transportation systems, building energy, and industrial systems are all important loads of the energy system. With the development of the transportation system in the direction of electrification and intelligence, promoting the integrated and coordinated development of energy systems and electrified transportation systems will be the inevitable trend of economic efficiency, safe and stable operation and sustainable development of cities. As the construction sector develops towards green and low-carbon, realizing the coupled development of the energy-building sector is a necessary means to improve energy efficiency and build green buildings. Optimizing and system integration in a larger scope and scale is of positive and far-reaching significance. With the development of industrial production in the direction of electrification and automation, the interaction between industrial systems and transportation systems has become increasingly significant. Promoting the integrated and coordinated development of the energy system and industry is the trend of the coordinated development of the industrial system and the energy industry.

2 Forecasting method of clean energy development potential considering sector coupling

2.1 The basic process of potential calculation

To conduct a mid- and long-term analysis of the development potential of clean energy, we must first determine the object of analysis, quantify the object of analysis, and construct relevant basic models. Since the development potential of clean energy cannot be measured solely by the amount of electricity demand, the amount of electric energy substitution is first defined to quantify the progress of the development of clean energy. Build an IPAT model on the development of clean energy based on the definition of the development potential of

* Corresponding author: shenweijie233@163.com

https://doi.org/10.1051/e3sconf/202132901024

clean energy. Then introduce the decoupling theory as the basis of multi-scenario construction and parameter setting. Finally, the scenario analysis method and the econometric analysis method are combined to realize the prediction and analysis of the development potential of clean energy. The basic process of potential analysis is shown in Figure 1.

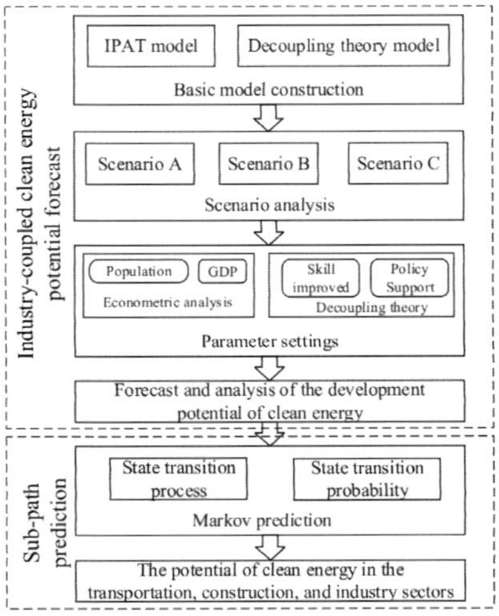

Figure 1. Flow chart of analysis of the development potential of clean energy.

2.2 Model construction of the development potential of clean energy based on IPAT

In order to realize the quantitative calculation of the development potential of clean energy, the amount of electric energy substitution is defined in this article as an objective basis for analyzing the development potential of clean energy. Set the base year TB. If the terminal energy consumption pattern is maintained at the base year level, the proportion of electric energy in the terminal energy is the same as that in the base year, and the increase in electric energy consumption in year t compared to the baseline electric energy consumption is defined as the amount of electric energy substitution:

$$D_{e,t} = (Y_{e,t} - \frac{Y_{e,T_B}}{Y_{T_B}}.Y_t) \quad (1)$$

In the formula: $D_{e,t}$ is the amount of electric energy substitution; $Y_{e,t}$ is the actual electric energy consumption in year t; Y_t is the total terminal energy consumption in year t.

Barry Kemenler, Paul Ralph Ehrlich, and JP Holder jointly proposed the IPAT model in the 1970s[7].The IPAT model is a well-known formula that uses population (P), affluence (A), and technology level (T) to evaluate the pressure of the environment. Its expression is:

$$I = P \times A \times T \quad (2)$$

The establishment of the IPAT model is to study the impact of population on environmental changes and reflect the relationship between changes in the environment affected by the three factors of population, wealth and technology. Electric energy substitution is also affected by the comprehensive effect of population, economic growth and technological level. Population growth, economic improvement and the development of related technologies will all promote the substitution of electric energy for terminal energy. Among them, the technological progress of electric energy substitution can be reflected in the intensity of electric energy consumption and the proportion of electric energy in the terminal energy consumption. As the technologies related to electric energy substitution become more mature, the intensity of electric energy consumption and the proportion of electric energy in terminal energy consumption will increase accordingly.

Therefore, by decomposing the above relational expressions, the expression of the IPAT model for electric energy substitution can be obtained:

$$C_t = P_t \times A_t \times (Y_e / G)_t \times (D_e / Y_e)_t \quad (3)$$

In the formula: Ct represents the total amount of electric energy replacement in year t; P_t represents the total population in year t; A_t represents the per capita GDP level in year t; D_e represents the amount of terminal electrical energy replacement; Y_e represents the terminal electrical energy consumption; G represents the GDP value; $(Y_e /G)_t$ represents the intensity of terminal power consumption in t; D_e/Y represents the ratio of electric energy substitution to the total terminal electric energy consumption, that is, the electric energy terminal substitution structure.

Comparing the electric energy substitution situation in year t with the base year, suppose the population growth rate is α, the average annual growth rate of per capita GDP is β, and the increase rate of terminal power consumption intensity is γ, which represents the degree of electrification of terminal energy consumption, that is, the degree of technological progress. Use δ to indicate the rate of change of the proportion of electric energy substitution in the terminal's electric energy consumption, that is, the rate of change of the electric energy terminal's substitution structure.

Since the government-led industrial structure optimization and adjustment and the consumption-oriented role of clean energy are the main factors that affect the change of the terminal energy substitution structure, and the terminal clean energy mainly refers to electric energy, so δ can be regarded as the strength of the government's support for electric energy substitution. In the model, in year t, the relationship between various factors about the base year is as follows:

$$Pt = P0 \times (1+ \alpha) \quad (4)$$

$$At = A0 \times (1+ \beta) \quad (5)$$

$$(Ye / G)t = (Ye / G)0 \times (1 + \gamma) \quad (6)$$

$$(De / Y)t = (De / Y)0 \times (1 + \delta) \quad (7)$$

In the formula: Pt and P0 are the total population in year t and the base year respectively; At and A0 are the GDP per capita in year t and the base year respectively; De is the replacement amount of terminal electric energy; Ye is the consumption of terminal electric energy.

Therefore, the change in the total amount of electric energy substitution in the base year in year t can be expressed as:

$$C = C \times [(1+\alpha) \times (1+\beta) \times (1-\gamma) \times (1+\delta)]t \quad (8)$$

In the formula, Ct and C0 are the total electric energy substitution in year t and the base year, respectively.

2.3 Basis for multi-scenario setting

Since the increase in the proportion of terminal power consumption can promote economic growth, that is, economic growth and the degree of power substitution have a certain correlation, which can be measured by the decoupling theory. The degree of decoupling of development can be used as an important evaluation basis for technological progress and policy guidance. χ indicates the degree of decoupling. Here, the rate of change of GDP per capita λ and the rate of change of electric energy substitution μ are used to express the degree of dependence between economic growth and terminal electric energy substitution:

$$\chi = \mu / \lambda \quad (9)$$

When the degree of decoupling >1, it indicates that the growth rate of electric energy substitution is greater than the economic development speed, the speed of electric energy substitution is fast, and the economic growth depends on the degree of electric energy substitution, and it is in the development stage of not decoupling. When the degree of decoupling=1, the growth rate of electric energy substitution is lower than the economic growth rate or consistent with the economic development rate. Economic growth does not completely or rarely depends on the degree of electric energy substitution, and the process of economic growth and electric energy substitution is in a decoupling stage.$(1+\beta)/(1+\gamma)\times(1+\delta)$ >1 indicates that the speed of economic development is faster than the process of electric energy substitution, and the implementation of electric energy substitution is highly dependent on economic development. At this time, the degree of technological advancement and government support has made little effort to promote the substitution of electric energy, and decoupling from economic development cannot be achieved. $(1+\beta)$ / $(1+\gamma)\times(1+\delta)$>=1 indicates that the speed of economic development is basically the same as the process of electric energy substitution, and the popularization and application of electric energy substitution is not completely dependent on economic development. At this time, technological progress is the main factor to promote the substitution of electric energy, and decoupling from economic development can be achieved. (1+β) / (1+γ)×(1+δ) <1 indicates that the speed of electric energy substitution is greater than the speed of economic development, and the promotion of electric energy substitution no longer depends on economic development. At this time, in addition to the impact of technological progress on the development of electric energy substitution, the support of relevant policies for electric energy substitution also promotes electric energy substitution, and the absolute decoupling from economic development has truly been realized.

2.4 Basis for sub-path calculation

There are many coupling interaction variables between the energy sector and transportation sector, construction sector, industry sector and other sectors, and there is a certain degree of time and space correlation. Existing research is mostly based on the integrated system composed of energy and a single sector, such as the integrated system of the energy and transportation sector. The integration with multiple sectors will make the complexity of the model and the difficulty of solving it drastically increase, so it is necessary to adopt a better method to deal with changes in complex systems. The Markov method [8] is a probabilistic method for predicting the state of the system. It can predict the state of the system at various times in the future based on the current state of the system, and can effectively deal with the problem of calculating the potential of clean energy in the transportation sector, construction sector, and industrial sector routes under the coupling of sectors.

(1) State, state transition process.

In Markov prediction, state refers to a certain result of a certain event at a certain moment (or period).Generally speaking, as the researched event and its predicted target are different, the state can be divided in different ways. For example, there are "child", "juvenile", "youth", "middle-aged", "old age" and other states in the population system forecast.

The state transition process refers to the transition from one state to another during the development of an event. The state transition that the development of the system changes with the change of time, or the relationship between state transition and time, is called the state transition process, referred to as the process.

(2) State transition probability and state transition probability matrix.

In the development process of the system, the possibility of starting from a certain state and transferring to another state at the next moment is called the state transition probability. According to the definition of conditional probability, the state transition probability $P(E_i \to E_j)$ from state E_i to state E_j is the conditional probability $P(E_j / E_i)$, namely:

$$P(E_i \to E_j) = P(E_j / E_i) = P_{ij} \quad (10)$$

Assuming that a certain predicted system has n possible states, mark P_{ij} as the state transition probability from state E_i to state E_j, and make a matrix:

$$P=\begin{bmatrix} P_{11} & P_{12} & \cdots & P_{1n} \\ P_{21} & P_{22} & \cdots & P_{2n} \\ \cdots & \cdots & \cdots & \cdots \\ P_{n1} & P_{n2} & \cdots & P_{nn} \end{bmatrix} \quad (11)$$

Then P is the state transition probability matrix.

(3) State probability.

The state probability $\pi_j(k)$ represents the probability that the event is in the state E_i at the kth moment after k times of state transitions under the condition that the initial state is known. According to the nature of probability, there are obviously:

$$\sum_{j=1}^{N}\pi_j(k)=1 \quad (12)$$

Starting from the initial state, the state transition process of reaching the state E_j after k times of state transitions can be regarded as first reaching state E_i $(i=1,2, \ldots, n)$ after $(k$-1) state transitions, and then reaching state E_j through one state transition from E_i. According to the no aftereffect of Markov process and the Bayes conditional probability formula, there are:

$$\pi_j(k)=\sum_{i=1}^{n}\pi_i(k-1)P_{ij}(j=1,2,\ldots n) \quad (13)$$

If the row vector $\pi_j(k)=[\pi_1(k),\pi_2(k),\ldots\pi_n(k)]$ is recorded, the recursive formula for successively calculating the state probability can be obtained from the formula (13):

$$\begin{cases} \pi(1)=\pi(0)P \\ \pi(2)=\pi(1)P=\pi(0)P^2 \\ \cdots \\ \pi(k)=\pi(k-1)=\pi(0)P^k \end{cases} \quad (14)$$

In (14), $\pi(0)$ is the initial state probability vector.

According to the above analysis, if the initial state of an event at the 0th moment is known, the probability of the event being in various possible states at the kth moment after k state transitions can be obtained by the recursion formula (14) (i.e. $\pi(k)$). Thus, the state probability prediction of the event at the kth moment is obtained, that is, the potential calculation of the sub-path of clean energy in the transportation sector, construction sector and industrial sector under the coupling of sectors is realized.

3 Case analysis

In order to make a more comprehensive prediction and analysis of the development potential of clean energy in the medium and long term, this paper sets three different alternative scenarios based on the IPAT model and the decoupling theoretical model to provide support for the analysis of the development potential of clean energy based on different paths.

Among them, scenario A is based on the existing electric energy substitution development, and does not consider the effect of technological progress and policy support on electric energy substitution, that is, electric energy substitution and economic development are not decoupled. Scenario B introduces the promotion of technological substitution on the basis of the development of existing electric energy substitution, and realizes the decoupling of electric energy substitution and economic development. Scenario C Increases the driving force of policy support for electric energy substitution on the basis of scenario B, and realizes the complete decoupling of electric energy substitution and economic development. Predict the parameters of population, economic development, electric energy substitution technology level and related policy support under scenario A, scenario B and scenario C.

1) Population parameters. With reference to the predictions and judgments of the "National Population Development Strategy Research Report", "Population and Social Development Report 2021" and other relevant plans, and a comprehensive analysis of the changes in China's future population growth rate, the population growth rate from 2021 to 2025 is expected to be 2.4‰,the population growth rate from 2026 to 2030 is 2.2‰, the population growth rate from 2031 to 2035 is 1.3‰, the population growth rate from 2036 to 2040 is 1.3‰, the population growth rate from 2041 to 2045 is 3.1‰, and the population growth rate from 2046 to 2050 is 3.1‰. Set population parameters based on the above data.

2) Economic parameters. Economic growth has a certain degree of influence on terminal power consumption. According to the forecast of the Chinese Academy of Social Sciences "China's Economic Situation Analysis and Forecast Blue Book in 2021", China's economic growth rate in 2016 will be between 6.6% and 6.8%. By 2030, the economic growth rate will gradually slow down due to structural issues such as population aging. At the same time, the Economist Intelligence Unit report pointed out that the global economic growth in 2050 will not decline sharply compared with its historical level, but economic growth will slow down after 2030. Based on the actual GDP in 2021, it is estimated that 5.4% in 2021-2025, 4.1% in 2026-2030, 3.8% in 2031-2035, 3.75% in 2036-2040, and 3.5% in 2041-2045. From 2046 to 2050, it is 3.4%.

3) Technical level. Assume that in scenario A, the development of electric energy substitution related technologies stagnates, the terminal electric energy demand remains at the current level, and the terminal electric energy consumption intensity remains unchanged. The electric energy substitution process is not decoupled from economic development due to the lack of technological advancement. The application field of electric energy substitution technology will be expanded on a large scale in 2021. Small coal-fired boilers are gradually being replaced by electric boilers, electric boilers + heat pumps, electric boilers + gas boilers, etc. Heating technologies such as carbon crystals, electric heating films, and heating cables are gradually being used in daily life. Related technologies such as military port shore power and railway electrification have also been applied on a large scale. It is assumed that electric energy

https://doi.org/10.1051/e3sconf/202132901024

substitution technology develops in coordination with economy; the demand for terminal power increases greatly with the development of power substitution technology; the consumption intensity of terminal electric energy increases, and the growth level is consistent with the economic development level of each stage, so that the decoupling degree of electric energy substitution and economic development is approximately 1. In scenario C, the technical level of electric energy substitution at all stages is consistent with scenario B.

4) Policy support intensity. It is assumed that in scenarios A and B, the government does not provide policy support and related interference in the work related to electric energy substitution, and the electric energy terminal substitution structure maintains stable development. Without government intervention, electricity substitution and the economy have not been decoupled or completely decoupled, and the degree of decoupling is not greater than one. Scenario C is based on China's work plans and goals related to electric energy substitution, increasing the government's support for electric energy substitution, enhancing the ability of electric power terminal substitution, changing the substitution structure, and greatly increasing the development potential of clean energy.

Since 2015, the government has issued 226 support policies to encourage electric energy substitution, guiding the society to actively choose electric energy, and gradually eliminating high-polluting and low-efficiency energy use methods. Since electric energy substitution can respond well to the green development requirements of China's 13th Five-Year Plan, and has been written into The State Council's Action Plan for Air Pollution Prevention and control and other important documents, many provinces and cities have also included the work of electric energy substitution in their 13th Five-Year Plan. Therefore, it is expected that the government's promotion of electric energy substitution will enter a peak period in the next 5 years. Thanks to the national policy guidance and the active efforts of the State Grid Corporation of China, in 2015, compared with 2014, the amount of electricity replacement increased by nearly 25 billion kW-h, with a growth rate of 0.49. It is expected that policy support will continue to increase significantly from 2016 to 2020 until the level of technological progress related to electric energy substitution stabilizes, the field of electric energy substitution has been fully expanded, and the structure of electric energy terminal substitution has stabilized. It is estimated that after 2040, the work of electric energy substitution can be carried out well without government support. The economy of electric energy substitution makes the society actively choose electric energy as the main energy consumption of the terminal.

In addition to the promotion of electric energy substitution by technological and economic development, increase the government's encouragement and support for electric energy substitution related work, so that electric energy substitution and economic development are completely decoupled, and the development of electric energy substitution does not need to rely on economic development.

According to the decoupling theory, the population, per capita GDP, technological progress and policy support parameter settings under scenario A, scenario B and scenario C are used to predict the development and change of the end-use energy structure, as shown in the following table. Among them:α is the annual growth rate of the population;β is the annual growth rate of GDP per capita;γ is the growth rate of the level of electric energy substitution technology; σ is the rate of change of policy support.

Table 1. Basic parameters of scenario analysis.

Year	Scenario	α/%	β/%	γ/%	σ/&
2016-2020	A	0.35	6.00	0.00	0.0
	B	0.35	6.00	5.71	0.0
	C	0.35	6.00	5.71	4.5
2021-2025	A	0.24	5.40	0.00	0.0
	B	0.24	5.40	5.54	0.0
	C	0.24	5.40	5.54	4.1
2026-2030	A	0.21	4.10	0.00	0.0
	B	0.21	4.10	4.93	0.0
	C	0.21	4.10	4.93	3.2
2031-2035	A	-0.13	3.80	0.00	0.0
	B	-0.13	3.80	4.30	0.0
	C	-0.13	3.80	4.30	2.6
2036-2040	A	-0.13	3.75	0.00	0.0
	B	-0.13	3.75	3.43	0.0
	C	-0.13	3.75	3.43	2.1
2041-2045	A	-0.31	3.50	0.00	0.0
	B	-0.31	3.50	3.21	0.0
	C	-0.31	3.50	3.21	1.8
2046-2050	A	-0.31	3.40	0.00	0.0
	B	-0.31	3.40	2.84	0.0
	C	-0.31	3.40	2.84	1.4

Table 2. China's total electricity consumption by sector.

Power consumption	Total (kWh)	Industry sector (kWh)	Construction sector(kWh)	Transportation sector(kWh)
2020year	77091	51530	1061	1822
2019 year	74866	50698	991	1752
2018 year	71508	49094	887	1608
2017 year	65914	46052	789	1418
2016 year	61205	42996	725	1251
2015 year	58020	41550	698	1125
2014 year	57829	42248	721	1059
2013 year	54203	39236	675	1000
2012 year	49762	36232	608	915
2011 year	47000	34691	571	848
2010 year	41934	30871	483	734
2009 year	37032	26854	421	617
2008 year	34541	25388	367	571
2007 year	32711	24290	309	531
2006 year	28588	21267	271	467
2005 year	24940	18521	233	430
2004 year	21971	16424	202	449
2003 year	19031	14169	179	406
2002 year	16465	12402	154	303
2001 year	14723	10944	154	309

Through calculations, the development potential of clean energy under sector coupling will show a substantial growth trend before 2030 under scenario A, scenario B, and scenario C, as shown in Figure 2. Among them, the growth rate of clean energy development potential under scenario A is smaller than scenario B and scenario C. This is because scenario A does not have technological progress and related policies to promote the replacement of electric energy. In scenario C, due to the increase of government support for electric energy substitution related work, the growth rate of clean energy development potential under this scenario is greater than that in scenario B. Between 2030 and 2040, the development potential of clean energy will tend to grow steadily. Until 2045, as the alternative field is close to saturation, the development potential of clean energy will gradually stabilize and there will be no substantial growth.

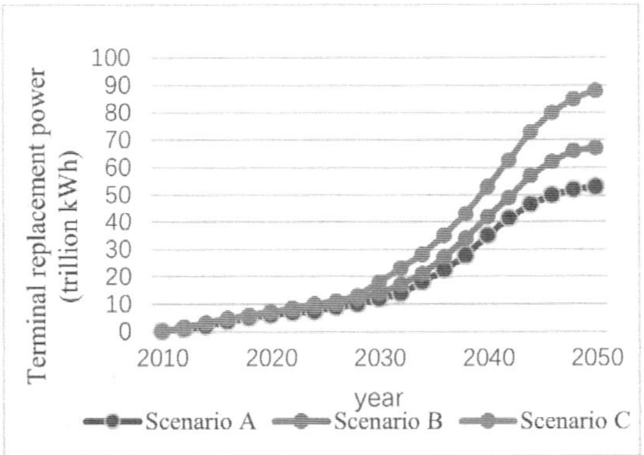

Figure 2. Forecast of clean energy development potential under various scenarios.

Compared with scenario A, the electric energy substitution process under scenario B and scenario C is significantly greater than scenario A. The increased policy support in Scenario C has a particularly significant effect on the promotion of electric energy substitution, and it is much higher than in Scenario B, which only relies on technological progress to promote electric energy substitution. In Figure 4, in 2020, 2030, 2040 and 2050, Scenario B will reach clean energy development potential of 480 million kWh, 1.71 billion kWh, 4.39 billion kWh and 6.5 billion kWh, respectively. Compared with the clean energy development potential under Scenario A, the alternative electric energy is increased by 8.13%, 13.96%, 18.43% and 22.05%, respectively. In Scenario C, due to increased policy support, the development potential of clean energy will grow more rapidly. Compared with scenario A, this scenario will increase the alternative electric power by 520 million kWh, 2.03 billion kWh, 5.55 billion kWh, and 8.61 billion kWh, respectively, and the increase rate reaches 17.83%, 35.52%, 49.87% and 61.76%.

Next, calculate the state transition probabilities in 2030, 2040, and 2050 under different scenarios based on Markov theory, and the results are shown in Table 3. Combining the above prediction results of clean energy development potential, the state transition probability matrix is used to calculate the clean energy development potential of the transportation sector, construction sector and industry sector under different scenarios in 2030, 2040 and 2050. The calculation results are shown in Figure 3.

Table 3. Probability of state transition of sector path for clean energy development.

	Industry sector			Construction sector			Transportation sector		
	Scenario A	Scenario B	Scenario C	Scenario A	Scenario B	Scenario C	Scenario A	Scenario B	Scenario C
2030 year	0.617	0.598	0.571	0.015	0.017	0.018	0.029	0.031	0.034
2040 year	0.551	0.542	0.536	0.017	0.019	0.020	0.035	0.037	0.038
2050 year	0.499	0.481	0.477	0.020	0.023	0.025	0.041	0.046	0.051

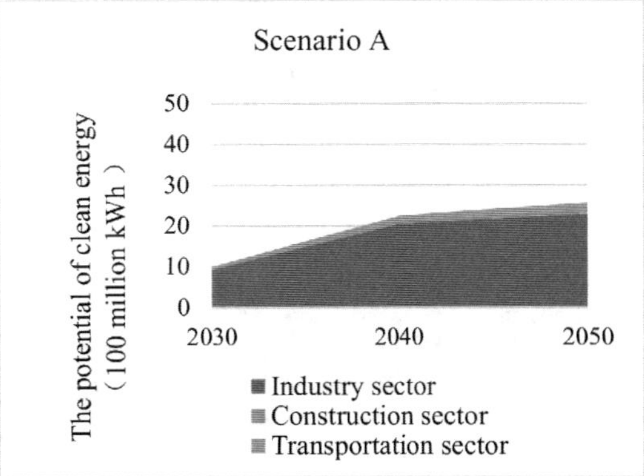

(a) Estimation results of clean energy potential of sub-path under scenario A.

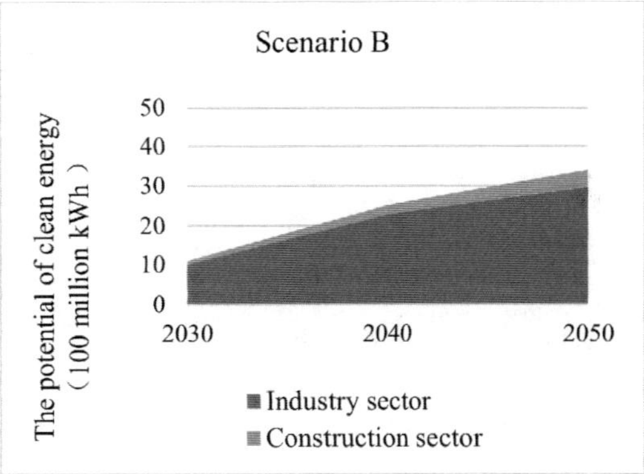

(b) Estimation results of clean energy potential of sub-path under scenario B.

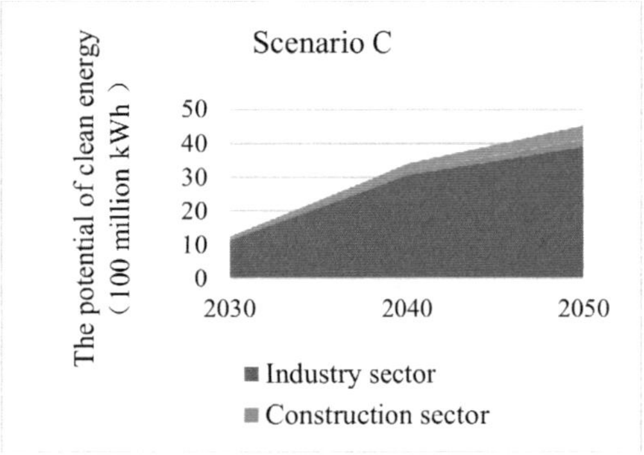

(c) Estimation results of clean energy potential of sub-path under scenario C.

Figure 3. Calculation of the development potential of clean energy in the industry, construction and transportation sectors under various scenarios.

It can be seen from Figure 3 that, horizontally, the development potential of clean energy has shown a substantial growth trend under the influence of factors such as policy support and technological progress. In scenario A, the clean energy development potential of the industry, construction, and transportation sectors will reach 926 million kWh, 23 million kWh, and 44 million kWh respectively in 2030; they will reach 2.0387 million kWh, 80 million kWh, and 155 million kWh respectively in 2040; they will reach 2.295 billion kWh, 94 million kWh, and 179 million kWh respectively in 2050.Compared with scenario A, the clean energy development potential of the industry, construction, and transportation sectors under scenario B will increase by 9.82%, 26.08%, and 20.45% respectively in 2030; increase by 11.6%, 26.8%, and 20% respectively in 2040; increase by 29.9%, 55%, and 51.22% respectively in 2050.In scenario C, the clean energy development potential of the industry, construction, and transportation sectors will reach 12.33%, 24.56%, and 29.03% respectively in 2030; they will reach 34.7%, 42.8%, and 39.3% respectively in 2040; they will reach 31.15%, 43.76%, and 46.6% respectively in 2050.

4 Conclusion

From a vertical perspective, in the current stage of clean energy development, the main alternative areas are concentrated in the industry sector, and electric energy replaces scattered coal, coal-fired boilers and other areas with relatively mature technologies; secondly, in the transportation sector, electric motors replace internal combustion engines, realizing electric vehicles instead of fuel vehicles; finally, in the construction sector, it is mainly reflected in the combination of cold and hot spots and smart energy management systems. It is recommended to introduce subsidy policies related to clean energy development. On the one hand, it can improve the economics of electric energy substitution, and effectively guide the society to choose low-carbon, high-efficiency clean energy to replace traditional energy. On the other hand, it is recommended to increase the research and development of electric energy substitution related technologies in transportation, industry, construction and other sectors to accelerate the improvement of the technical level, so as to enhance the development potential of clean energy.

Acknowledgement

This study is funded by the science and technology project of SGCC (Research on key technologies and modes of efficient operation of integrated energy system considering P2X and industry coupling, SGZJSX00XTJS2001473)

References

1. Fang Guochang, Wang Li, Gao Zhengye. Discussion on the coupling mechanism of energy internet and energy transition and identification of influencing factors[J]. Coal Economic Research, 2020, 40(11): 10-17.
2. Gao Zhiyuan,Zhang Tao,Zhao Lei,Zhang Jing,Hu Yuou,Cao Yang.Design of Priority Substitution Rules for Clean Energy Marketization and Related Analysis[J].Electric Power Automation Equipment, 2021,41(08):105-110.
3. Wang Huan,Ma Bing,Jia Lingxiao,Yu Yang,Hu Jiaxiu,Wang Wei.The role of key minerals in the clean energy transition under the goal of carbon neutrality, supply and demand analysis and recommendations[J/OL].Chinese Geology:1-19 [2021- 10-29].

   http://kns.cnki.net/kcms/detail/11.1167.P.20211025.1627.002.html.
4. Jiao Yuping, Cai Yu. An Analysis of China-Latin America Clean Energy Cooperation in the Context of Energy Transition[J/OL].Latin American Studies:1-14[2021-10-29].http://kns.cnki.net/kcms/detail/11.1160.C.20211015.2133.002.html.
5. China will accelerate the development and utilization of clean energy[J].Henan Science and Technology, 2021,40(09):2.
6. Hu Ming. Promote energy coupling development and break through industry energy barriers[N]. China Energy News, 2020-12-14(004).
7. Xingjunxia,Chen Yujie,Yao Hongyi.Analysis of Qinghai Province's Economic Growth and Energy Consumption Based on IPAT Model[J].Bohai Economic Outlook,2020(05):90-91.
8. Masao Nagasawa. Markov Processes and Quantum Theory[M].:2021-07-15.

https://doi.org/10.1051/e3sconf/202132901025

Efficient resourcezation of sulfanilamide wastewater and A study on this reaction

Yingying Tian [1, 2], Kai Wen [2] and Dezhan Chen [1, *]

[1] College of Chemistry, Chemical Engineering and Materials Science, Shandong Normal University, Jinan 250014, P. R. China
[2] College of Resource and Environmental Engineering, Shandong Agricultural Engineering University, Jinan 250100, P. R. China

Abstract. Sulfanilamide wastewater which is come from the dyestuff plant is characterized by a high content of hardly biodegradable dissolved, high acidity, and organic matter. The dyestuff plant`s wastewater has become a water pollution of the world's attention. Therefore, efficient removal of sulfanilic acid and hydrochloric acid from wastewater before drainage is of great importance. This paper designed a new process engineering containing a recovery of sulfanilic acid and hydrochloric acid, at the same time which can cycle the mother liquor without any wastewater. The experimental results showed that, using this new process engineering is better than that in traditional methods.

1 Introduction

Wastewater from dyestuff plant has a negative impact on environment and water resource. In particular, great concerns arise from sulfanilamide wastewater because it is characterized the high acidity, high concentrations of organic matter, and vast emissions[1]. Many dye factory use waste carbide slag to make neutralization treatment wastewater containing acid ,then go through a solid-liquid separation process to treatment the wastewater[2]. Furthermore, some studies on using a two-phase anaerobic system sulfanilamide wastewater treatment[3] or the adsorption of resin adsorption method is adopted to aminobenzene sulfonic acid to biochemical treatment process[4]. This treatment effect is obvious, but such process`loading is small, cycle is long, which does not apply to sulfanilamide factory mass production, and there are still a lot of sewage discharge[5]. So exploration with economic benefits, being suitable for mass production, and having the environmental benefits of wastewater treatment process is particularly important. This article through to study on sulfanilamide wastewater containing acid, put forward a new post-treatment technology of waste acid water route, recycled the economic value of products, and circulate the mother liquid, basically achieved zero emissions. New process mainly steamed out of hydrochloric acid by using ammonium chloride solid-state metathesis reaction. By means of the adjustment of the temperature and solvent, precipitation in aminobenzene sulfonic acid, and effective use of SO42 - in the waste water, such as the production of ammonium sulfate sulfate products, the feasibility of this technology has a large-scale production.

2 Experimental part

2.1 Reaction principle

Sulfonamide waste acid solution, see table 1 for main components.

Table 1. Main components of sulfanilamide waste acid solution (%)

H_2S O_4	HA c	HC l	 amide	Sulfanil
35~4 0	0.8 4	0.3 7		2~3

Through the component analysis ,the experimental objectives were established as follows: recovery of sulfanilic acid and hydrochloric acid,a large number of SO_4^{2-} in waste acid water were further synthesized into sulfate products, and the mother liquor produced was recycled.

* Corresponding author: tavis_cn@sina.com

2.2 Technological process

The waste acid water (calculated by sulfuric acid content of 36%) and the solid ammonium chloride are mixed in the mixing kettle at molar ratio 1:1, after mixing, it was heated and evaporated in the evaporation kettle, the pressure in the kettle was maintained at 0.05 MPa, distillate occured when the temperature of bottom residue reached 85°C,after heating and distilling for one hour, the temperature of bottom residue continued to rise, distillate gradually disappeared, distillation was completed. More than 90% hydrochloric acid and distillation residue mainly containing ammonium bisulphate can be recovered.

$H_2SO_4 + NH_4Cl \rightarrow NH_4HSO_4 + HCl$

NH_4HSO_4 crystals were precipitated from the distilling residue, which can be refined to produce NH_4HSO_4 products; it also can add the mother liquor to the distilling residue to produce $(NH_4)_2SO_4$ products by adding ammonia water to adjust the pH value of the system, and the generated ammonium sulfate mother liquor is the mother liquor.

$NH_4HSO_4 + NH_3 \cdot H_2O \rightarrow (NH_4)_2SO_4 + H_2O$

Based on the property of slight solution of sulfanilic acid in cold water, the temperature of the system was adjusted to about 20 °C, and the crystal of sulfanilic acid was separated and refined.

3 Results and discussion

3.1 Process conditions for evaporation of hydrochloric acid

According to the azeotropic distillation principle of water and HCl, hydrochloric acid in acid wastewater was recovered by adding solid ammonium chloride into acid wastewater for stirring and distillation. Under normal pressure, the boiling point of ammonium bisulfate is 490°C, the decomposition temperature of sulfanilic acid is 280°C, and the azeotropic temperature of water and HCl is 110°C[6]. Under normal pressure, the azeotropic system of water and HCl can be formed at the temperature of 110°C without taking out ammonium bisulfate and sulfanilic acid.

3.1.1 Distillation temperature

Under normal pressure,the majority of hydrochloric acid can be steamed by controlling the distillation temperature at 110-140°C;under reduced pressure, distillation temperature varies with pressure. The pressure is -0.05MPa, the relationship between distillation time of wastewater containing acid, tower bottom and tower top temperature, Cl^- concentration in bottom residue are shown in figure 1:

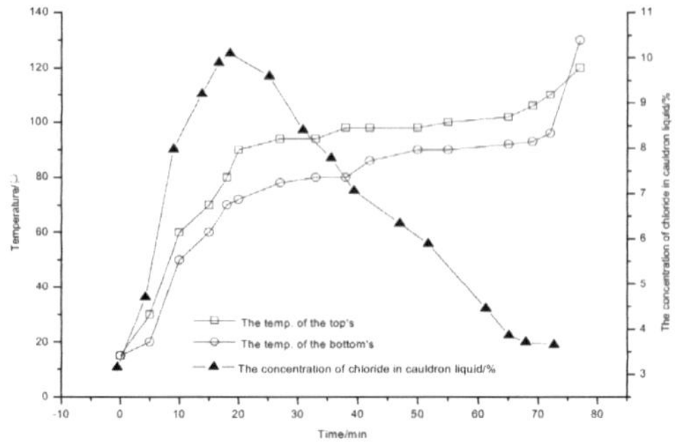

Figure 1. Relation curves of distillation time,kettle bottom and kettle top temperature, Cl- concentration in bottom residue

3.1.2 Distillation pressure

Pressure has a great influence on the evaporation rate of hydrochloric acid. Compared with normal pressure distillation, reduced pressure distillation can increase the evaporation rate of hydrochloric acid by about 20%, as shown in table 2. Through the experiment, the hydrochloric acid can be effectively evaporated by stirring and evaporation under the control of pressure in the range of -0.05 to -0.06MPa.

Table 2. Under different pressures,evaporation rate of hydrochloric acid at the same distillation time (%)

Pressure/MPa	ordinary pressure	-0.02	-0.04	-0.06
Hydrochloric acid evaporation rate/%	68	75	88	93

https://doi.org/10.1051/e3sconf/202132901025

3.2 Ammonia is neutralized with ammonium bisulfate to produce ammonium sulfate

The distilling residue contains a large amount of NH_4HSO_4. When the temperature is lower than 70°C, NH_4HSO_4 can crystallize and be sold as a product; it also can add the mother liquor for dissolution and add ammonia water for neutralization to produce ammonium sulfate products.

Refer to the process of producing ammonium sulfate by direct neutralization of sulfuric acid and ammonia, the content of sulfuric acid in the system is 36%, in order to obtain qualified ammonium sulfate products, the pH value of the system needs to be adjusted. When ammonia water is added to the mother liquor containing 40% sulfuric acid, crystals are precipitated when the pH of the system approaches [7]. Through the experiment, the pH value of the system was adjusted from 6.5 to 8.5 by adding ammonia water to monitor the precipitation amount of ammonium sulfate. When the pH of the system is 8±0.5, the maximum amount of ammonium sulfate can be crystallized out, and the ammonia taste of the product is small.

3.3 Separation of sulfanilic acid

Through analysis, the organic matter in waste acid is in the form of sulfanilic acid. N-acetylsulfanilic acid is soluble in water, it slowly hydrolyzes into acetic acid and sulfanilic acid in aqueous solution, the hydrolysis can be accelerated by heating. Sulfanilic acid is slightly soluble in cold water(Under normal pressure, at 10°C, the solubility of sulfanilic acid is 0.8g/100g water), solubility increases with temperature [8], and this property can be used to separate sulfanilic acid.

Hydrogen ions on the sulfonic acid group of sulfanilic acid in ammonium sulfate mother liquor at different pH values were detected by ^{1}H-NMR, it is concluded that the sulfanilic acid solubility in the mother liquor of ammonium sulfate under different pH value (see figure 2), and the best conditions for separation of sulfanilic acid is ammonium sulfate mother liquor pH is about 2, the content of sulfanilic acid is the least in solution. By adjusting the pH value of the system to about 2, 83% of sulfanilic acid in waste acid water can be recovered.

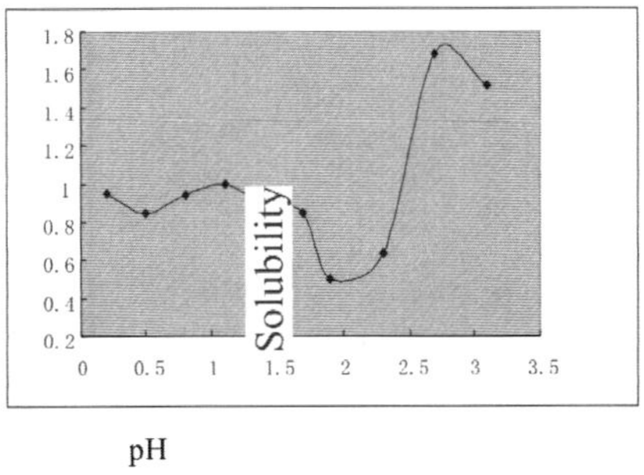

Figure 2. Solubility of p-aminobenzenesulfonic acid in ammonium sulfate mother liquor under different pH

4 Conclusions

The new process can steam hydrochloric acid and recover sulfanilic acid from waste acid water by adding NH_4Cl solid to sulfonamide wastewater,using SO_4^{2-} in acid wastewater to produce sulphate products to improve the economic value of sulfonamide wastewater so that sulfanilamide wastewater reach the resource treatment, at the same time,mother liquor is recycled in the system, and the whole processing process basically reaches zero discharge. This conforms to the requirements of modern chemical industry for economic, environmental and social benefits. Compared with the traditional waste acid liquor treatment, the new process has a great improvement, and has great market promotion value.

References

1. Shi Jing, Wang Hongwu, Ma Luming.Research status of sulfa wastewater treatment in China [J]. Journal of liaoning institute of technology,2006,26(5) : 326-328
2. R.M. Baena-Nogueras, E. González-Mazo, P.A. Lara-Martín.Degradation kinetics of pharmaceuticals and personal care products in surface waters: photolysis vs biodegradation Sci. Total Environ., 590 – 591 (2017), pp. 643-654,
3. Gao Jun, DIAO Litong, Li Xinjun.Study on the treatment of acid containing wastewater from sulfa production [J]. Industrial water treatment, 1998,18 (5) : 17-18
4. Ren Nanqi, Liu Yanling, Liu Min.Experimental study on the treatment of sulfa wastewater by two-phase anaerobic system [J]. China water & wastewater, 2001,17 (2) : 8-11
5. Sun Yue, Zhu Zhailian, PAN Bingcai.Treatment of sulfa wastewater by resin adsorption [J]. Environmental protection of chemical industry, 2003,23 (1) : 9-13
6. P. Chakraborty, M. Mukhopadhyay, S. Sampath, B.R. Ramaswamy, A. Katsoyiannis, A. Cincinelli, D. Snow Organic micropollutants in the surface riverine sediment along the lower stretch of the transboundary river Ganga: occurrences, sources and ecological risk

assessment, Environ. Pollut., 249 (2019), pp. 1071-1080

7. Liu Guangqi, Ma Lianxiang, Liu Jie.Chemical chemical physical properties data manual (inorganic volume) [M]. Chemical industry press, 1998,411,477
8. Chen Shi, Meng Cong, Li Wu. Chemical safety and environmental protection, 2007,33 (2) 43-46. (in Chinese
9. Liu Guangqi, Ma Lianxiang, Liu Jie.Chemical chemical properties data manual (organic volume) [M]. Chemical industry press, 1998,26

Research on Energy Storage Cost Model in Distributed Energy System Environment

Shuo Yin [1], Zhe Chai [1], Xing Chen [1], Meng Yang [1], Yao Lu [1, *], Kyle Georgiou [2]

[1] State Grid Henan Economic Research Institute, Zhengzhou, China
[2] University of Waterloo, Canada

Abstract. With the "dual carbon" goal proposed and the direction of building a new power system dominated by new energy sources clarified, the energy storage industry has entered the fast lane of accelerated development. However, due to the imperfect marketization mechanism and the lack of effective power market ancillary service policies, it is difficult to 00EFfectively recover the cost of energy storage construction, which restricts the development of the energy storage industry to a certain extent. Therefore, it is urgent to clarify the development trend and cost recovery mechanism of typical types of energy storage under the distributed energy system, and clarify the impact of the operating cost of the electrochemical energy storage system. This paper systematically analyses the domestic and foreign energy storage investment channelling mechanisms, clarifies the impact of energy storage cost system operation on costs, and helps the development of the energy storage industry.

Keywords: Electrochemical energy storage; Cost of investment; Grooming mechanism.

1 Preface

Energy storage investment cost diverting mechanism. From the perspective of international practices, energy storage applications in the United States, Europe, Australia, Japan and other regions are relatively active, mainly distributed energy storage, and mainly support the development of energy storage system through incentive policies and market mechanism design, to help users reduce electricity prices, increase the proportion of renewable energy utilization, and improve power quality or disaster recovery capabilities. From the perspective of domestic policies, the status of energy storage in the energy system and industrial development is gradually established. At the local level, local policies have been introduced to promote the development of the energy storage industry in terms of grid operation, auxiliary services, demand-side response, and subsidy mechanisms. In general, energy storage should use market-based methods to recover costs as much as possible.

2 The impact of electrochemical energy storage system operating costs

The influencing factors of the charge-discharge cycle cost of electrochemical energy storage equipment are mainly related to DOD and discharge efficiency. First discuss the relationship between cycle cost and DOD. It should be clear that the relationship between cycle cost and DOD is related to the initial state of the SOC and the end state of the SOC cycle of the electrochemical energy storage system. For example, there are two situations: ① SOC is discharged from 100% to 40%; ② SOC is discharged from 80% to 20%. Although the DOD of the depth of discharge of the two types of cases is 60%, the cycle cost in the case of type ② is obviously higher.Therefore, this example also confirms from the side that one of the variables in the operation model of the energy storage system is SOC instead of DOD.

The second factor that affects the cycle cost of an electrochemical energy storage battery is the high current intensity in the state of charge and discharge. The following formula reflects the discharge efficiency of the battery through the DCR variable, and the physical meaning is expressed as the SOC change per unit time:

$$DCR_t = \frac{SOC_{t-1} - SOC_t}{T} = \Delta SOC_t \quad (1)$$

The discharge efficiency of an energy storage battery can be represented by "C". For example, 1C means that the battery is fully discharged within 1 hour, 2C means that the battery is fully discharged within half an hour, 3C means that the battery is fully discharged within 20 minutes, and so on.In order to accurately establish the relationship between the loss model of the electrochemical energy storage battery and the DOD and discharge efficiency, the following Figure 1 shows the relationship between the cycle period of the lithium ion battery and the DOD (at a given temperature and discharge efficiency):

* Corresponding author: luyao@ha.sgcc.com.cn

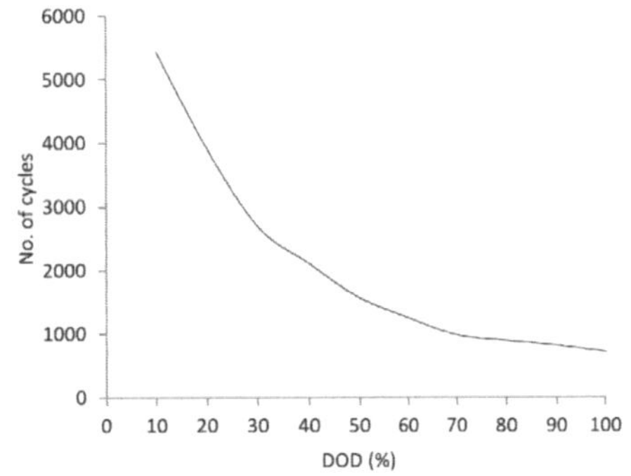

Figure 1 The relationship curve between the cycle period of lithium battery and DOD

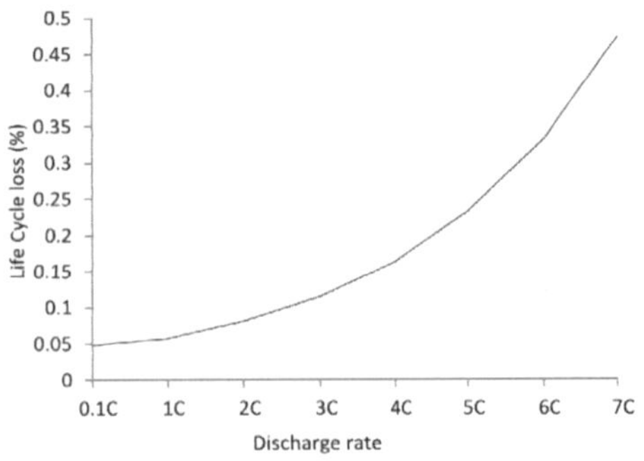

Figure 2 Lithium battery cycle life loss and discharge efficiency curve

Through the relationship curve in Figure 1, the mathematical expression of the relationship between the cycle period of the lithium battery and the DOD can be obtained:

$$L(D) = \gamma D^{-\omega} \tag{2}$$

In the above formula, γ, ω is a parameter of the mathematical relationship between the number of cycles and the DOD in the case of a lithium battery 1C discharge efficiency.

Similarly, the relationship between cycle life loss and discharge efficiency of a lithium battery is shown in Figure 2. It can be found that in the range of discharge efficiency below 3C, the relationship between cycle life loss and discharge efficiency is close to linear.Combining the information in Fig. 1 and Fig. 2, the relationship curve between the cycle period and the DOD under different discharge efficiencies can be obtained, as shown in Fig. 3.

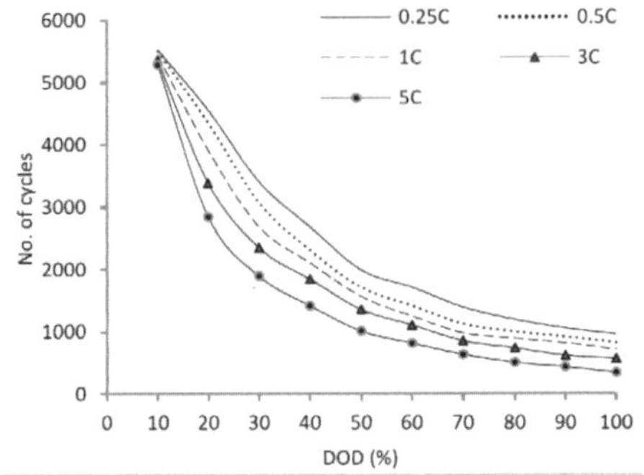

Figure 3 Cycle period under different discharge efficiency

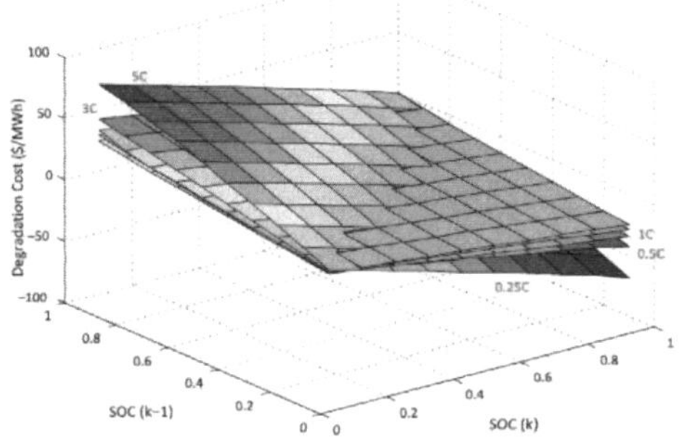

Figure 4 Cycle cost under different DOD and discharge efficiency Relationship curve with DOD

Considering the SOC for the start and stop of the energy storage battery, the mathematical expression of the cycle cost at a certain discharge efficiency is as follows:

$$C1_t = \frac{C^B}{B_{cap}^E \eta^2 \gamma}\left[(1-SOC_t)^{\omega} - (1-SOC_{t-1})^{\omega}\right] \tag{3}$$

In the formula, C^B is the construction cost of energy storage, B_{cap}^E is the rated capacity of the energy storage system, and η is the cycle efficiency of the energy storage cycle. Based on the above formula, for different discharge efficiencies and different values of SOC_t and SOC_{t-1}, the cycle cost of the energy storage battery can be represented by Figure 4.Note that the cyclic cost function is a non-linear function, which is difficult to apply to the market optimization model. Linear simplification processing is required for the cyclic cost model. The linear simplification results are as follows:

$$C1_t = aSOC_t + bSOC_{t-1} + cDCR_t + d \tag{4}$$

In the formula, a, b, c, d are the coefficients of the battery cycle cost model.

3 Incentive Grooming Mechanism

Incentive policies in the United States are mainly geared towards user-side distributed energy storage. The United States has formulated a self-generation incentive plan. For energy storage equipment, the installed power subsidy will be provided at the standard of US$2/W before 2016. After 2016, the power generation subsidy will be granted according to the installed capacity. There are five stages. The subsidy amount for the first stage is 0.5 to 0.6 USD/Wh, and thereafter, each stage will be reduced by 0.05 USD/Wh.

Table 1. The subsidy standards for the five rounds of subsidies in the 2017 edition of the SGIP manual

Energy storage project	First round /$·(W·h)	Second round /$·(W·h)	Third round /$·(W·h)	Forth round /$·(W·h)	Fifth round /$·(W·h)
Large-scale energy storage (> 10kw)	0.5	0.4	0.35	0.3	0.25
Obtain ITC's large-scale energy storage	0.36	0.29	0.25	0.22	0.18
Home energy storage (≤10kw)	0.5	0.4	0.35	0.3	0.25

Germany's incentive policies are mainly for distributed optical storage. Its subsidy policy lasted for 6 years. Since 2013, the subsidy policy for photovoltaic storage has mainly provided 30% of the investment amount of household energy storage equipment, and provided low-interest loans to units or individuals purchasing photovoltaic energy storage equipment. The photovoltaic energy storage system grants a subsidy of up to 600 Euros/kW.

Table 2. List of subsidy policies of German optical storage.

release time	Subsidy object	Condition requirements	Subsidy standard
March 2013	Energy storage system supporting household photovoltaic	The peak power of photovoltaics is below 20kW; only up to 60% of the photovoltaic power generation can be sent to the grid; the energy storage system has a 7-year warranty.	30% of subsidies for the installation of energy storage systems; newly installed photovoltaic energy storage systems are given a subsidy of up to 600 Euros/kWp.
March 2016	The object of funding is a stationary battery energy storage system (not a photovoltaic device) that matches photovoltaic equipment, and only one battery energy storage system that matches photovoltaic equipment can be subsidized	The photovoltaic system with energy storage battery must be installed after December 2012, and the peak power cannot exceed 30kWp; the power fed back by the photovoltaic system to the grid cannot exceed 50% of the peak power; the system service life must be at least 20 years; the battery system must Have a 10-year warranty; installers must have relevant qualifications.	Depending on the year of application, The proportion of subsidies provided by banks is also different; 2016/3/1-2016/6/30: The loan subsidy ratio is 25%; 2016/7/1-2016/12/31: The loan subsidy ratio is 22%; 2017/1/1-2017/6/30: The loan subsidy ratio is 19%; 2017/7/1-2017/9/30: loan subsidy ratio 16%; 2017/10/1-2017/12/31: loan subsidy ratio 13%; 2018/1/1-2018/12/31: loan subsidy 10%;

Japan's incentives are mainly for residential use of energy storage systems. Energy storage incentive policies are mainly used to alleviate the grid management challenges brought about by a large number of distributed solar energy, which has also increased the demand for battery energy storage systems. Since 2018, the Ministry of Economy, Trade and Industry of Japan has invested about 98.3 million U.S. dollars to provide 66% of the cost subsidy for families and businesses that install lithium electronic batteries. So far, Japanese government subsidies can account for 40-50% of the entire battery retail price.

4 Market-oriented Governance Mechanism

In the United States, FM ancillary services are the main form of energy storage participating in the market. The Federal Energy Regulatory Commission (FERC) issued Decree No. 841 in 2018, requiring all electricity markets to allow energy storage to participate in the wholesale electricity market, ancillary service market, and capacity market. From a practical point of view, the FM auxiliary service market is currently the most important form of participation in energy storage. 76% of electrochemical energy storage in the United States participates in power frequency modulation auxiliary services. The California Independent Operator (CAISO) has the most outstanding performance in this regard, and can participate in the day-ahead FM market through the non-generation resource model (NGR).

In Germany, a higher frequency market bidding model is used to achieve revenue stacking. The German Federal Power Grid Administration has optimized the bidding time and the minimum bid size for the second and third frequency modulation. Starting from July 2018, the bidding time for the second frequency modulation and the third frequency modulation has been changed from weekly to daily. The supply time period has also been changed from "2 segments per day, 12 hours per segment" to "6 segments per day, 4 hours per segment". High-frequency bidding and a shorter service supply window allow energy storage capacity to participate in more target markets, and can more effectively stimulate aggregate energy storage capacity to achieve revenue stacking.

In Australia, independent ancillary service providers can directly participate in the market. The Australian Energy Market Commission (AEMC) issued the "National Electricity Modification Rules" to open ancillary services to new market participants, namely, market-oriented ancillary service providers other than large power generation companies. Market participants can provide ancillary services at one location, or aggregate loads or units from multiple locations to provide services. Since the implementation of this rule in 2017, it has greatly increased the opportunities for energy storage to participate in the Australian power ancillary service market.

5 Concluding remarks

The United States, Europe, Japan and other countries generally introduced direct fiscal and taxation measures such as initial investment subsidies, tax reductions and exemptions. At the same time, they also actively guide energy storage to participate in power market transactions, and "two-pronged" support energy storage development. Domestic energy storage diversion is mainly through government regulation and diversion, including the successive introduction of local policies to promote the development of the energy storage industry in terms of grid operation, auxiliary services, demand-side response and subsidy mechanisms.

The core products of energy storage are electric and electricity. From the perspective of pricing mechanism, its price policy is related to function positioning and operation mode. The diversification of its functions and operating models determines the diversification of its investment recovery channels and pricing diversification. In principle, energy storage should adopt market-based methods as much as possible to recover costs.

Acknowledgments

This work was supported by the Science and Technology Project of SGCC, Research on the competition situation and operation mode of distribution companies in the park (52170018000S).

References

1. Analytical method for estimating leakage of reservoir basins for pumped storage power stations[J], Bulletin of Engineering Geology and the Environmen, 2021. PP 1-14.
2. Safety warning of lithium-ion battery energy storage station via venting acoustic signal detection for grid application[J], Journal of Energy StorageVolume 38, 2021.
3. Competitive model of pumped storage power plants participating in electricity spot Market——in case of China[J], Renewable EnergyVolume 173, 2021. PP 164-176.
4. Atomic layer deposition in the development of supercapacitor and lithium-ion battery devices[J], CarbonVolume 179, 2021. PP 299-326.
5. Use of Energy Storage in Low Thrust Spaceflight[J], Critical Asian StudiesVolume 30, Issue 1. 2012.
6. Design and Test of Controller in Power Conditioning System for Superconducting Magnetic Energy Storage[J],ICPE(ISPE) symposium 2007. PP 966-970.

Investigation and Water quality Analysis of Agricultural non-point Source pollution in Tanxia Town

Liya Yang [1,2,3,*], Hao Zhou [1], Liyuan Dai [1,3]

[1] Wanjiang University of Technology, Maanshan, 243031, China
[2] Ma'anshan Engineering Technology Research Center of Engineering Monitoring and Safety Monitoring China
[3] Maanshan Engineering Technology Research Center for Water Resources Efficient Utilization in Hilly Region China

Abstract. Taking the small watershed in Tanxia town as an example, the classified investigation of agricultural non-point source pollution in the tributaries of Lijiang River was carried out, the sources of local agricultural non-point source pollution were analyzed, and the surface water environmental quality standard and Nemero index method were used to evaluate the water quality of the basin. The results show that the water quality of the basin is poor and the pollution exceeds the standard all the year round, and the source of agricultural non-point source pollution mainly depends on the pollution of agricultural chemical fertilizer. Improving the use of local chemical fertilizer and rational application of chemical fertilizer are the key factors to improve local agricultural non-point source pollution.

Keywords: Non-point source pollution, Chemical fertilizer, Tanxia Town.

1 Introduction

The amendment to the Clean Water Act of the United States in the last century defines non-point source pollution as a source of pollution that enters surface and underground water bodies in a wide, dispersed and trace form. With the rapid development of economy, non-point source pollution has become the main source of agricultural pollution. As a large agricultural country, the severe situation of non-point source pollution in rural areas is not optimistic. Due to the abuse of chemical fertilizers and pesticides, agricultural non-point source pollution can't be controlled reasonably and effectively.

At present, the situation of water pollution in the Lijiang River basin is not optimistic. Agricultural non-point source pollution directly affects the ecological environment and drinking water safety in the Lijiang River basin. Taking Tanxia Town as an example, this paper studies the current situation of agricultural non-point source pollution in the tributary basin of the Lijiang River, provides a powerful tool for the study of non-point source pollution, has potential application value, and is of great significance to the construction of ecological civilization..

2 Background

Tanxia Town is located in the northwest of Lingchuan County, 18km from the urban area of Guilin, the river basin belongs to the Lijiang River basin, the river runs through many villages in the town, the full-length 8km, the basin area 12km^2. Tanxia Town is located in the mid-subtropical monsoon climate zone, with four distinct seasons, abundant rainfall and hot rain in the same season, which is conducive to the growth of all kinds of crops. The average annual rainfall is 1941.5mm and the annual average temperature is 18.7℃. The content of organic matter in the soil in the study area is less, the content of nitrogen, phosphorus and potassium is less, and the fertility is low. The regional vegetation cover is good, reaching more than 70%. As an important grain production base in Guangxi, the town has 3320 hectares of cultivated land and 2.3 million mu of orchards.

2.1 Sources of agricultural non-point source pollution

The population income of Tanxia town is mainly agriculture, forestry, animal husbandry, fishery and so on. There are no industrial factories in this survey area. The sources of non-point source pollution in the basin are mainly concentrated in agricultural pollution, and the contents of this survey mainly include rural population, the amount of agricultural chemical fertilizer, insecticidal dose, livestock breeding, the content of pollutants in the basin and so on.

Through the collection of data, on-the-spot investigation and household inquiry, nine villages in the research basin of Tanxia Town were investigated, and the basic situation of local farmers' population, planting, breeding and

* Corresponding author: 527615613@qq.com

domestic sewage discharge in 2013 were investigated by questionnaire.

2.2 Rural domestic sewage

The discharge of domestic sewage from rural residents and the excretion of local residents are the main sources of non-point source pollution in rural areas. This survey counted the number of local residents and monitored the water quality around the residential toilets, and calculated the total load of rural domestic sewage as shown in Table 1. Among them, the content of pollutants in human feces and urine is determined by the research regulations of the National Health Bureau, which stipulates that the emission coefficient of total nitrogen is 0.584kg per person per year, and the emission coefficient of ammonia nitrogen is 0.16kg per person per year. The average discharge of feces and urine per person per day is 250g and 1000g respectively.

Table 1. Statistical data unit of rural domestic pollution

Project	TN	NH_3-N
Feces	730kg	200 kg
Urine	2920 kg	800 kg

2.3 Pesticides and fertilizers

The main crops in the study area are rice and citrus. This survey visited seven villages in the study area, according to the results of field investigation, combined with the results of the 2016 national survey of pollution sources in Guilin, to determine the application amount of pesticides and chemical fertilizers per unit area in the basin, as shown in Table 2 below.

Table 2. Application rate of pesticides and chemical fertilizers

Position	Pesticide(kg/mu)	Chemical fertilizer (kg/mu)	
		Compound fertilizer (kg/mu)	Urea(kg/mu)
Gan Village	0.07	100	60
Jiang family	0.07	80	60
Huangbai	0.05	110	70
Cha futian	0.08	100	75
Shangliyuan	0.1	100	50
Xialiyuan	0.095	100	50
Xinqiao	0.1	150	75
Average	0.08	105.7	62.9

The rice in the study area is two seasons a year, the amount of compound fertilizer per season is 10~20kg/mu, three times per season, each time decreasing 3~5kg per mu, urea application rate is 5~10kg/mu, fertilizing twice per season, each time decreasing 1~2kg per mu. Citrus is fertilized about 3 times a year, the amount of compound fertilizer per mu is about 10kg, and the amount of urea per mu is about 5kg.

2.4 Pollution caused by livestock faeces

The livestock and poultry found in the study area are mainly pigs, chickens and ducks. In the process of breeding, the manure of livestock and poultry is an important source of agricultural non-point source pollution. A small amount of livestock and poultry manure can be used as organic fertilizer in agricultural production and is absorbed by crops. if the scale of livestock and poultry breeding is large, a large amount of manure is bound to cause some pollution to the environment. Precipitation scours the untreated livestock manure, making the livestock manure accumulated around the farm flow into the water body with the water flow, causing pollution to the river.

The excretion of feces and urine of livestock and poultry is related to many factors, such as animal species, growth cycle, feed and so on, but the effect on the result is not obvious. With reference to the calculation method in the Fertilizer Handbook and combined with the data, Table 3 was selected as the pollutant parameter of livestock and poultry manure in the study area.

Table 3. Average content of pollutants in livestock and poultry faeces per

unit kg/t:

Project	Pig manure	Pig urine	Chicken manure	Duck manure
TN	5.88	3.3	9.84	11
NH_3-N	3.1	1.4	4.78	0.8

2.5 Water pollution.

According to the location of the watershed in the study area, this paper selects Gan village outlet, Chaibutian, Huangbai, Xinqiao North, Huangbai Village Trading Market, Donggang Canal, Shangliyuan and Guanli as water quality monitoring points. The water sample data of 8 months from September 2013 to April 2014 are tested, the outliers are removed, and the average value is taken. The test results are as follows:

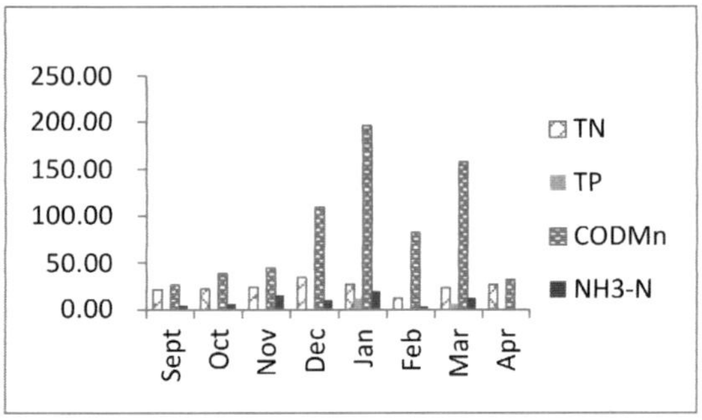

Fig. 1 Monthly average concentration of pollutants

The Environmental Quality Standards in Surface Water and the Sanitary Standard of Water for domestic Application by the Ministry of Health in 2001. According to the function and standard of water area, surface water is divided into five types. The functions of the five types of water are decreasing in turn, of which Class I II and Class III are mainly suitable for drinking water, IV water can only be used as general industrial water, and Class V water is only suitable for general landscape waters. The source of water in the basin has drinking water requirements, so the environmental quality standards of surface water in the basin should meet the requirements of III or above.

Table 4. Environmental Quality Standards in Surface Water

Project(mg/L)	Class I	Class II	Class III	ClassIV	Class V
TN	0.2	0.5	1.0	1.5	2.0
TP	2	4	6	10	15
COD_{Mn}	0.02	0.1	0.2	0.3	0.4
NH_3-N	0.15	0.5	1.0	1.5	2.0

3 Evaluation of water quality.

At present, because the single factor index method is simple to calculate and reflect intuitively as a more common calculation method in water quality evaluation, but its disadvantage is that it can not fully reflect the comprehensive situation of water environment pollution, and the results are not representative. In order to make the evaluation results comprehensive, this paper uses the Nemero comprehensive pollution index method to evaluate the water quality of the river basin, and the improved Nemero comprehensive index method takes into account the impact of pollution factors with lower concentration on the environment. Not only consider the impact of pollution factors with the largest pollution index. This makes the result of the evaluation more accurate. In this paper, the measured concentration of monitoring index and the reference value of III water are selected as the standard concentration. The specific calculation formula is as follows:

$$F_i = \frac{c_i}{s_i}$$

$$P_i = \frac{\sqrt{F_{\max}{}^2 + \overline{F}}}{2}$$

$$\overline{P} = \frac{1}{n}\sum_{i=1}^{n} P_i$$

C_i is the measured concentration, S_i is the standard concentration, P_i is the cross-section Merro index, P_s is the weighted average Nemero index.
The corresponding relationship between Nemero index and water quality grade is shown in Table 5.

Table 5. The corresponding relationship between Nemero index and water quality grade

Water quality category	Class I	Class II	Class III	ClassIV	Class V
Nemero index	P<0.59	0.59≤ P<0.74	0.74≤ P<1.00	1.00 ≤P< 3.50	P≥3.50

According to the measured data, the water quality in this area is calculated as follows:

Table 6. Water quality evaluation results

	Nemero index				Weighted average Nemero index	Water quality category
Month	TN	TP	COD_{Mn}	NH_3-N		
Sept	0.97	0.96	0.46	2.59	1.25	IV
Oct	1.08	0.73	0.77	2.49	1.27	IV
Nov	3.18	3.10	7.35	3.44	4.27	V
Dec	3.21	3.21	2.09	2.76	2.82	IV
Jan	0.00	0.00	0.00	0.00	0.00	-
Feb	0.00	0.00	0.00	3.79	0.95	III
Mar	2.25	1.65	0.00	16.14	5.01	V
Apr	2.42	2.33	1.98	4.21	2.73	IV

4 Discussion.

In recent years, with the attention to the problem of non-point source pollution in China, the domestic non-point source pollution model will also be established. The research on non-point source pollution model is still in its infancy. With the effective control of point source pollution, the treatment of non-point source pollution will improve the problem of agricultural non-point source pollution in China.
According to the Nemero index method, the water quality in November and March is the worst, and the water quality in February is relatively good. The monitoring data show that except for the accident of total phosphorus, the contents of total nitrogen, permanganate index and ammonia nitrogen in surface water are much higher than the standard concentration of surface water. Among them, the change of the content of pollutants in the water body will show an irregular "W" shape with the month.
Generally speaking, the main source of surface water pollution in small watersheds is caused by chemical fertilizers and pesticides in local agricultural cultivation. According to the investigation, the seasonal characteristics of fertilizer application in the watershed were obtained, and considering the delay of the migration and transformation of pollution factors with rainfall and runoff, it was found that there was a certain coincidence between the time of concentrated application of fertilizer and the period when the concentration of pollutants exceeded the standard. it is proved that there is a certain relationship between the content of pollutants in water and the excessive application of local pesticides and fertilizers. The intensity of rainfall and the amount of rainfall also have a certain impact on the concentration of pollutants in the basin.

Because the soil in the basin itself is not fertile, the high demand of local farmers for crop yield has led to a relatively high content of pollutants in the basin. Blind fertilization is not the only way to increase yield. The government should increase the relevant training for farmers in scientific and rational planting and breeding in order to improve the deterioration of the local water resources environment.

Acknowledgments

This work was financially supported by scientific research projests of Maanshan Engineering Technology Research Center for Water Resources Efficient Utilization in Hilly Region (WREU202102) , scientific research projests of Ma'anshan Engineering Technology Research Center of Engineering Monitoring and Safety Monitoring (2020gcjc02) and The Program of Cultivating University Top-Notch Talents (gxyg2021244) fund.

References

1. El Behairy Radwa A.,El Baroudy Ahmed A.,Ibrahim Mahmoud M.,Kheir Ahmed M. S.,Shokr Mohamed S.. Modelling and Assessment of Irrigation Water Quality Index Using GIS in Semi-arid Region for Sustainable Agriculture[J]. Water, Air, & Soil Pollution,2021,232(9):
2. Li Donglin,Zhang Hucai,Chang Fengqin,Duan Lizeng,Zhang Yang. Distribution and health-ecological risk assessment of heavy metals: an endemic disease case study in southwestern China.[J]. Environmental science and pollution research international,2021:
3. Xiang Meng, Gu Shixiang, Gao Rong, Xie Chongbao, Cui Yuanlai. Investigation of farmland non-point Source and Countermeasures of Agricultural Water Saving and Emission reduction in Qilu Lake Basin [J]. China Rural Water Conservancy and Hydropower, 2021 (08): 127,132.
4. Hu Yawei, Ma Qiang, Zhai Laizheng, Jin Xiaohui. Calculation of agricultural non-point source pollution load in Yellow River diversion irrigation area based on SWAT [J/OL]. Ecological environment monitoring of the three Gorges: 1-22 [2021-10-07].
5. You Ruyue, ao Tianqi, Zhu Hong, Gao Danyang, Zhang Xing. Comparative study on evaluation methods of water quality in small watershed [J]. Sichuan Environment, 2021 and 40 (02): 73-81.
6. Zang Mengyuan, Li Ying. Study on load estimation and Control Countermeasures of Agricultural non-point Source pollution [J]. Shandong Agricultural Sciences, 2021, 53 (02): 142-147.
7. Zhang Caili, Lin Lin, Jiang Yi, Lu Kai. Study on the variation characteristics and influencing factors of water quality in Huaihe River Basin (Fuyang section) from 2009 to 2018 [J]. Anhui Agricultural Sciences, 2020,48,23: 95-100,107.
8. Li Su, Yan Zhihong, Xu Dan, Wang Shuqian. Application of improved Nemero Index method in Reservoir Water quality Evaluation [J]. Science, Technology and Engineering, 20205,20 (31): 13079-13084.
9. You Ruyue, ao Tianqi, Zhu Hong, Gao Danyang, Zhang Xing. Comprehensive Assessment of Water Environment pollution in Xiaoanxi Watershed [J]. Hydropower, 2020, 46 (11): 6-10, 64.

Effect and cognition of hole filling measures in ultra-low permeability reservoir block

Ziwei Li*

Third Oil Mine, Daqing Oilfield Company Limited, Tenth Oil Production Plant, Daqing, Heilongjiang 166405, China

Abstract. In this paper, based on the application of water-flooded layer interpretation plate, multi-well comparison, and the combination of dynamic and static analysis methods to optimize the well selection of fill hole wells, and through the analysis of the production of fill hole wells in **block in recent years, we summarize the relationship between fill hole effect and fill hole well injection and extraction well spacing, connection relationship, water content before the measure, and the relationship between fill opening thickness of the patch opening, so as to provide reference for the next step to continue to carry out fill hole work and find fill hole potential.

Keywords: Encrypted wells, hole filling, effect

1 Problem formulation

Over the past few years, a total of 275 encrypted oil recovery wells have been put into production in Block A. Since the old block had been developed with water injection for more than 10 years at the time of encryption, the distribution of subsurface oil and water was complicated and the water flooding condition in some well areas was serious, the layers with the same oil and water layer, a higher degree of water flooding and a higher risk of hole injection explained at the time of encryption well injection were not injected. Of the 275 oil production wells that were encrypted, a total of 90 wells were not completely shot open, with 179 total un-shot layers and a total un-shot thickness of 435.3m. After densification, through the adjustment of the injection and extraction system and the adjustment of the water injection scheme, the subsurface oil and water are redistributed and some of the un-injected layers have a certain potential for hole filling.

2 Application of oil recovery well filling technology to improve the relationship of the injection and extraction and increase the degree of reserve utilization

2.1 Small injection and extraction well spacing, After hole filling to increase the liquid and oil effect becomes better, But there is a risk of rising water content.

Among the 28 wells in the fill hole, 10 wells with 140m injection and extraction distance, 10 wells with 210m injection and extraction distance and 8 wells with 237m injection and extraction distance. The wells with 210m spacing between injection and extraction wells have the best effect after hole patching; the wells with 140m distance between injection and extraction has a large fluid increase after hole patching, but the water content is high, and after water injection adjustment, it has an oil increase effect; the wells with 237m injection and extraction spacing are less effective due to the larger well spacing.
The 140m well spacing fill hole well has good fluid and oil increase effect, but the water content is high, through the water injection adjustment, a certain oil increase effect can be achieved. There are 10 wells with 140m spacing between injection and extraction wells, producing 2.2t of liquid per day, 0.6t of oil per day and 72.7% of water before filling the hole, 9.0t of liquid per day, 1.9t of oil per day and 78.6% of water have been produced in the initial stage after filling the hole, and well increasing 1.3t of oil per day and 5.9 percentage points of water, currently producing 5.9t of liquid per day, 0.8t of oil per day and

* Corresponding author: liziwei@petrochina.com.cn

 https://doi.org/10.1051/e3sconf/202132901028

86.5% of water with a current average single well increasing, the current average daily oil increase per well is 0.2t. Due to the small well spacing of 140m, the increase in fluid after the fill hole is larger, and the rise in water content is also larger, but through the comprehensive adjustment of water injection after the fill hole, a certain effect of oil increase can be achieved. The effect of 210m well spacing of hole patching wells to increase fluid and oil is good. There are 10 wells with 210m well spacing, 8 wells with good effect after filling, with 80% efficiency, and 2 wells with poor effect, which are mainly affected by oil layer interference.

Before the measure, it produced 1.8t of liquid per day, 1.1t of oil per day containing 39.9% of water. In the initial stage after filling the hole, it produced 6.5t of liquid per day, 4.5t of oil per day, and increased 3.4t of oil per day containing 30% water, now it produces 3.1t of liquid per day, 1.6t of oil per day containing 49.5% water, with an average single well validity of 522d and a cumulative increase of 637t of oil. There are 8 wells with 237m of injection and extraction well spacing, of which 5 are effective, with 62.5% measure efficiency. Due to the large distance between wells, the effect of fluid and oil increase is the worst. The average daily oil increase of a single well is 2.0t. In the initial stage after filling the hole, and the average daily oil increase of a single well is 0.2t at present, with an average single well validity of 325d and an accumulated oil increase of 298t.

2.2 The effect is good after filling wells with moderate fluid production and low to medium water content before the measure

According to the relationship between daily liquid production, water content before the measure and the initial daily oil increase after the measure, it can be seen that: both curves are parabolic, when the daily liquid production is less than 1.0t or more than 2.5t, the effect of the measure becomes worse, the liquid volume before the measure is low, which means that the liquid supply capacity of the area is insufficient, and the oil increase after the fill hole is relatively poor, when the liquid volume is high and the water content is higher than 40.0%, the liquid output capacity of the shot hole layer is strong, the degree of water flooding is high, and the new fill open layer. The effect of measures is relatively poor because of inter-layer interference.

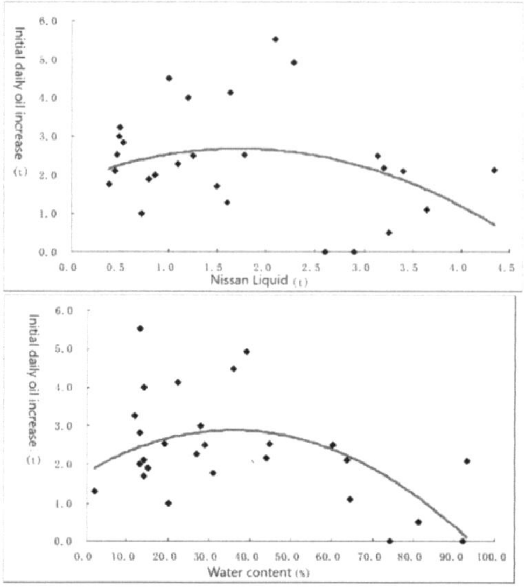

Figure 1 Relationship between daily fluid production and water content before the measure and daily oil gain in the initial period after the measure

Statistical analysis was performed on 28 fill hole wells, which were divided into 3 intervals: fluid production before the measure was less than 1.0t and water content was less than 40.0%; fluid production was between 1.0-2.5t and water content was less than 40.0%; fluid production was more than 2.5t and water content was more than 40.0%.From the statistical results, it shows that there are 10 wells producing 1.0-2.5t with water content less than 40.0%, 8 effective wells, 80.0% measure efficiency, 3.3t daily oil increase in the initial stage, 0.3t daily oil increase at present, 487 days validity, 579t cumulative oil increase; 9 wells producing less than 1.0t with water content less than 40.0%, 7 effective wells, 77.8% measure efficiency, 2.2t daily oil increase at the beginning, 0.5t daily oil increase at present, 405 days validity, 404t cumulative oil increase, 5 wells are still effective at present. The initial daily oil increase of 2.2t, the current daily oil increase of 0.5t, valid for 405 days, cumulative oil increase of 404t, 5 wells are still effective, the type of wells after the fill open water content rose greatly, indicating that although has shot open layer poor fluid supply, low fluid production before the measure, but the fill hole layer may be medium water flooding, a larger increase in fluid, with the effect of oil increase; fluid production greater than 2.5t, water content greater than 40.0% of the 9 wells The effect is good 3 wells, the measure efficiency 33.3%, affected by inter-layer interference, the effect is poor, the initial daily oil increase of 2.2t, currently ineffective, valid for 204 days, cumulative oil increase of 191t.

2.3 Water-driven wells with large reservoirs and good connectivity are effective in filling holes

The larger the water-driven reserves of the fill hole well, the better the results. The 28 wells that were filled with holes increased water-driven reserves by a total of 15.03×104t, with an average of 0.54×104t of water-driven reserves added to a single well

The relationship between water-driven active reserves and initial daily oil gain and cumulative oil gain shows that: the larger the water-driven active reserves, the higher the initial daily oil production and cumulative oil gain of the make-up well. The effect is better if the well is well connected.

The average single well injection pressure is 11.2MPa, daily water injection is 18m3, cumulative water injection is 5.42×104m3. 20 wells have good connection relationship, 13 wells have good effect after taking measures, the efficiency of measures is 65.0%, the average single well connection is 2.3 water injection wells, the average injection and extraction well distance is 210m, the main oil layer is two-way and The average single-well daily oil increase was 2.2t at the beginning after the hole patching, and now the average single-well daily oil increase is 0.5t, the average single-well validity is 401d, and the cumulative oil increase is 454t; 8 wells have poor connection relationship, 5 wells have good effect after the measure, the measure efficiency is 62.5%, the average single-well connection is 1.4 water injection wells, the average injection and extraction well distance is 225m, the main oil layer two-way and above is 22.0%. The average daily oil increase in a single well is 2.5t at the beginning of the period after the hole patching, and the current average daily oil increase in a single well is 0t, with an average single well validity of 257d and an accumulated oil increase of 216t.

2.4 Good results in making up for oil formation thickness and making up for oil recovery wells with large main oil formation thickness

According to the thickness of the fill-opened oil layer, the thickness of the fill-opened main oil layer, the initial daily oil increase and cumulative oil gain relationship, it can be seen that: the thickness of the fill-opened oil layer and the thickness of the fill-opened main oil layer are large, the activated reserves are large, and the fill-opening effect is better.

From the statistical table of the effect of measures, we can see that there are 18 wells with thickness greater than 5.0m, including 12 wells with good effect, 66.7% of the effective measures, 2.3t of oil increase per day at the beginning after filling the hole, 420d of the average single well validity, 468t of cumulative oil increase; 10 wells with thickness less than 5.0m, including 6 wells with good effect, 60.0% of the effective measures The initial daily oil increase was 2.2t, the average single well validity was 279d, and the cumulative oil increase was 273t.

3 Conclusion and Awareness

3.1 In encryption, the small layer that is not shot open can be selectively filled with holes, which can achieve a certain effect of oil increase.

3.2 The effect of filling wells with 210m and 140m injection and extraction spacing is better than that of wells with 237m injection and extraction spacing, but the 140m well spacing is closer and there is a risk of water content rising after filling.

3.3 Priority should be given to wells with fluid volumes <2.5t and medium and low water content for hole filling. For oil recovery wells with high fluid volume and high water content, in order to avoid inter-layer interference affecting the effect of hole filling, the high water-bearing layer in the initial injection opening level should be blocked first to improve the oil enhancement effect of the measure.

3.4 The water drive is more effective in filling holes in wells with large reserves and good connectivity.

3.5 Better results for wells with a large formation thickness and a large main formation thickness

References

1. Li Daopin. Development of low permeability sandstone oil fields [M]. Beijing: Petroleum Industry Press. 1997.
2. XuQi et al. Seepage characteristics of extra-low permeability reservoirs and the law of water drive development [M]. Beijing: Petroleum Industry Press.2019.

 https://doi.org/10.1051/e3sconf/202132901028

3. Lu Baoping. New advances in petroleum engineering technology [M]. Beijing: Sinopec Publishing House. 2014.
4. Lian Qingcun. Oil reservoir engineering [M]. Beijing: Petroleum Industry Press. 2006.
5. Li Daopin. Decision Theory of Efficient Development of Low Permeability Reservoirs [M]. Beijing: Petroleum Industry Press. 2003.
6. Chen Taoping. Petroleum Engineering [M]. Beijing: Petroleum Industry Press. 2000.

https://doi.org/10.1051/e3sconf/202132901029

Characteristics and potential area optimization of Oil layer A reservoir in T Oilfield

Junhong Gao

The Ninth Oil Production Plant of Daqing Oilfield Co., Ltd., Heilongjiang, China Daqing 163000 China

Abstract. Oil layer A of T oilfield has a low development degree. At present, there are 5 medium and high-yield wells in the area, showing a good evaluation and development prospect. In order to understand the reserve potential of this area, the reservoir forming law and reservoir characteristics are further analyzed by analyzing the structure and reservoir development characteristics. The research shows that Oil layer A has two types of reservoirs: one is structural reservoir, which is distributed in the large structural trap near the Aogula-Halahai fault zone in the east of the oilfield. It is characterized by close oil source, short migration path and stable reservoir development. It is a "large and enriched" reservoir. The other type is lithologic reservoir, which is distributed at the sedimentary end of the western structure of the oilfield, and is characterized by distant oil source. After long-distance secondary migration, it is greatly influenced by the development degree and physical properties of the reservoir, but it is a "small but fat" reservoir with high oil saturation. On the basis of clarifying reservoir types, characteristics and evaluation ideas, this paper carries out reserve optimization classification, and selects three potential areas: block Y1 and block Y2 which are mainly controlled by structure and block Y3 which is controlled by lithology.

Keywords: T oil field, Oil layer A, Reservoir characteristics, Potential area optimization.

1 Introduction

T Oilfield is structurally located on two secondary structural units, Longhupao-Da'an Terrace and Qijia-Gulong Sag in the western part of the central depression area in the northern Songliao Basin. There are many longitudinal oil-bearing layers, from top to bottom, they are A, Saertu, Putaohua, Gaotaizi, Qing 1st Member and Fuyu oil layer. Oil layer A has the characteristics of low degree of development, shallow burial, relatively simple oil-water relationship, and high abundance of oil and gas reserves, showing good evaluation and development prospects. Previous studies mainly focused on the Sapu and Gaofu oil layers in the lower part [1-6]. Due to the sparse fracture on the top surface of Oil layer A, single formation occurrence, large difference in sand body type, scale and thickness, the scale of Oil layer A reservoir is relatively small. Therefore, previous studies on Oil layer A are relatively few, which restricts the upgrading and effective production of Oil layer A reserves in this area. In this paper, by systematically analyzing the structure and reservoir development characteristics, the reservoir forming rules and reservoir characteristics are further deeply analyzed, and then the evaluation ideas are determined. Three potential blocks of "big but rich" and "small but fat" reservoirs are found, which are used step by step according to the reservoir implementation degree. It has certain guiding significance for the exploration and development of Oil layer A of T oilfield and similar oilfields.

2 Geological characteristics research

2.1 Structure background

The structural pattern of the top surface of Fuyu-Oil layer A in the T oilfield is roughly the same, from deep to shallow, the overall structure is a monoclinic high in the northwest and low in the southeast. The structure is relatively flat, and the structure changes from complex to simple. The Aogula-Halahai fault zone and the Bayanchagan fault zone, which run through the entire area in the north-south direction, are developed on the monoclinic. Many local structures are formed under the background of the fault zone. These fault blocks and micro-amplitude structures are Favorable location for oil and gas accumulation [7].

The overall structure of the top surface of Oil layer A is relatively simple, and it is still a monoclinal structure. It is still the inherited development of two fault zones that control the structural pattern, but the number of associated small faults is reduced. Many structural traps are developed along the fault zone. Gulong sag is located to the east of Aogula Halahai fault zone in the East. The structural dip angle becomes larger and the structural axis is obvious. The formed structural traps are large, the trap amplitude is relatively large and the number is large. Three large structural traps are developed along the eastern fault zone, with an area of 0.8 ~ 3.5 km2 and a closing range of 20 ~ 30m. Bayanchagan fault in the west has a small development scale, gentle structure, few local structures and small trap amplitude and area. Two relatively large structural traps are developed, with a trap area of 0.4 ~ 0.9 km2 and a closure amplitude of 10-15m (Figure 1).

2.2 Sedimentary and reservoir characteristics

Oil layer A is dominated by the delta sedimentary system in the northeast, experienced three large-scale regresses, and developed three reverse cycles. The reservoir is dominated by estuary bar and sheet sand deposits, and the predelta facies develops to the southwest. During the deposition of the A5 small layer, the T area was in the pre-delta-semi-deep lake subfacies. During the deposition of the A4 small layer, the delta front advanced west and south, developed delta front estuary bar and sheet sand deposits, and developed front mud microfacies at the northwestern edge of the block (Figure 2). During the deposition of the A3 small layer, the T area was in the pre-delta-semi-deep lacustrine facies. During the deposition period of A2 sublayer, it is in the delta front and pre delta deposition, and sheet sand bodies are developed, which are mainly distributed in the northeast and south of the block, and the front mud microfacies are mainly distributed in the northwest of the block (Figure 3). During the sedimentary period of A1 sublayer, it is in pre delta semi deep lake facies, mainly developing front mud microfacies [8]. During the evolution process of lacustrine transgression and lake regression, the distribution of the various small sedimentary facies belts controls the development range of the reservoir. The A4 and A2 small layers are in the front subfacies of the northern delta sedimentary system, depositing estuary bars and sheet sand sand bodies, providing good storage space for Oil layer A oil and gas reservoirs in the T oilfield.

On the whole, the distribution of sandstone in Oil layer A is stable, with sandstone thickness of 2-1 ~ 4m and single sandstone thickness of 1-4m. Sandstone is mainly distributed in the eastern part of the work area, with thickness of 6-12m generally, and sandstone development layers are generally between 2-6 layers, of which 2-4 layers are the main ones. Sandstone is thinned along the upward dip direction of stratum, and the northwest and southwest are areas with low sandstone thickness. Vertically, sandstone is mainly developed in small layers A2 and A4. The penetration rate of small layer A2 sandstone is 63.2%, and that of small layer A4 sandstone is 88.5%. The average core porosity of Oil layer A is 20.3%, and the air permeability is 61.7 × 10-3 μ m2, medium porosity and medium permeability reservoir.

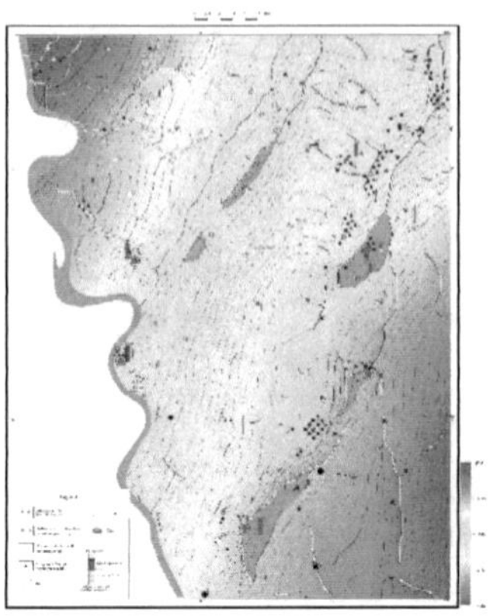

Figure 1 Oil layer A top surface structural trap diagram

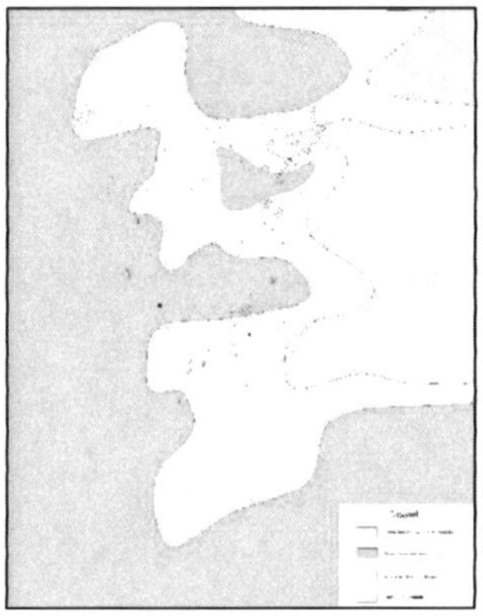

Figure 2 Plane distribution diagram of microfacies of A2 small layer deposition

Figure 3 Plane distribution diagram of microfacies of A4 small layer deposition

2.3 Analysis of reservoir types, characteristics and reservoir forming laws.

2.3.1 Reservoir forming analysis

T oilfield is adjacent to the oil-generating center of Gulong sag. The effective source rocks in the west of Daqing Placanticline are mainly distributed in Qijia-Gulong area, and the first member of Qingshankou Formation is the main source rock in the middle and shallow layers. The analysis of source rocks and aromatic hydrocarbon types of crude oil in A, Sarthou, Putaohua, Gaotaizi and Fuyu oil layers in T area shows that the crude oil in Oil layer A mainly comes from the source rocks of the first member of Qingshankou Formation.

The effective thickness of Oil layer A in T oilfield is mainly between 1 and 4m, mainly in the east and west. The east along the Aogula-Halahai fault zone is north-south, distributed in a belt shape, and the effective thickness is generally between 2 and 3 m. The western part is effectively developed at the tipping point of the sandstone or where the thickness of the sandstone becomes thinner (Figure 8). It is mainly developed in the A4 small layer, the effective drilling encounter rate of the A4 small layer is 29.9%, and the effective drilling rate of the A2 small layer is 8%.

The medium and high-yield wells in Oil layer A of T oilfield are mainly distributed along the Aogula Halahai fault zone in the East. It can be seen that the structure controls the reservoir. The sedimentary microfacies of Oil layer A are estuarine bar and sheet sand. Seismic reservoir prediction [9] and drilling data show that the reservoir in the east of the oilfield is developed stably. The estuarine bar and sheet sand body are blocked by faults in the structural updip direction, forming local structural traps. The monoclinic structure in the East is inclined to the oil generation center of Gulong depression, and the structural traps on it are favorable parts for oil and gas accumulation, The oil and gas generated by the lower source rock migrated to the trap along the fault zone and formed a reservoir nearby (Figure 4).

The western part of the oilfield is located in the prodelta facies, which is located at the sedimentary end in the structural updip direction. The sand body is thinned or even pointed out along the structural updip direction. The oil and gas discharged from the lower source rocks are transported to Oil layer A through faults. Because of the sparse faults on the top of the Oil layer And the single occurrence of the stratum, it is difficult to form large-scale effective structural traps. At the western end of the sedimentation, due to the influence of the sedimentary environment, the physical properties of the reservoir at the edge of the sand body become poor, and it is easy to form a lithologic oil reservoir trap (Y3 block). In the lithologic trap formed in the area where the thickness of the sandstone becomes thinner (Figure 5). However, in areas with particularly developed sand bodies, the physical properties of the reservoirs are good. Due to the multi-phase activities of faults, it is easy to cause secondary migration of oil and gas, so the degree of oil and gas enrichment in the areas with particularly developed sand bodies is low or not enriched [10] .

2.3.2 Reservoir types and characteristics

Oil layer A reservoir types mainly include structural reservoir and lithologic reservoir. Structural reservoir has the characteristics of near oil source, short migration path and stable reservoir development. Lithologic reservoir has the characteristics of far oil source, long migration path, great influence by reservoir development and physical properties, but high oil saturation.

The structural trap formed in the Aogula-Halahai fault zone in the east is a favorable place for oil and gas accumulation. The high part of the structure in the trap produces pure oil, which transits to the low part to be the same layer of oil and water and the reservoir type is structural reservoir (Figure 6). This type of reservoir is close to the oil-generating center of the Gulong sag, with a large structural dip and an obvious structural axis. The formed structural traps are relatively large, and the range of the traps is relatively large. The fault zone provides channels for oil and gas migration and a large-area developed estuary. Dams and sheet-shaped sand bodies provide storage space for oil and gas. The oil test and production test proved that the reservoirs in this area have strong liquid supply capacity, large liquid supply radius, and a long distance from the oil-water transition zone. It is a large and enriched oil reservoir.

The sand body in the west is thinned and even pinched out along the structural updip direction. The high part of the structure in the sandstone updip direction produces pure oil, and the oil-water layer transits to the water layer from the same oil-water layer to the low part. The reservoir type is lithologic reservoir (Figure 7). There is no relatively large fault zone at the edge of this kind of reservoir to provide a channel for oil and gas migration. After the oil and gas are primarily migrated to Oil layer A through far large faults, they will undergo long-distance secondary migration and be stored in lithologic reservoir traps due to the deterioration of reservoir physical properties due to the influence of sedimentary environment. Therefore, the reservoir scale is relatively small. However, oil test proves that this kind of reservoir is a small and fat dessert reservoir with strong liquid supply capacity and large liquid supply radius.

 https://doi.org/10.1051/e3sconf/202132901029

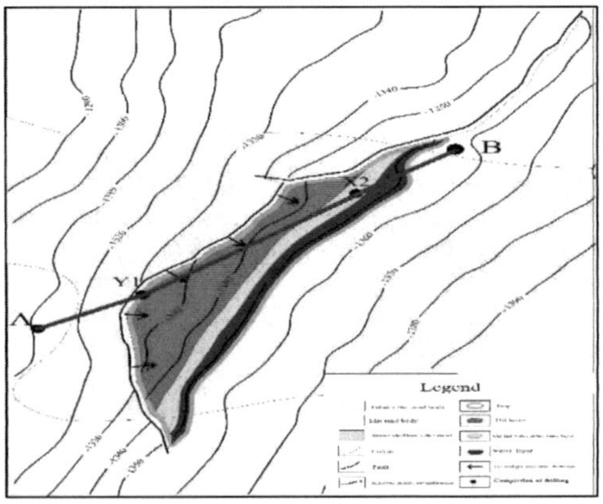

Figure 4 Structural oil reservoir formation model diagram

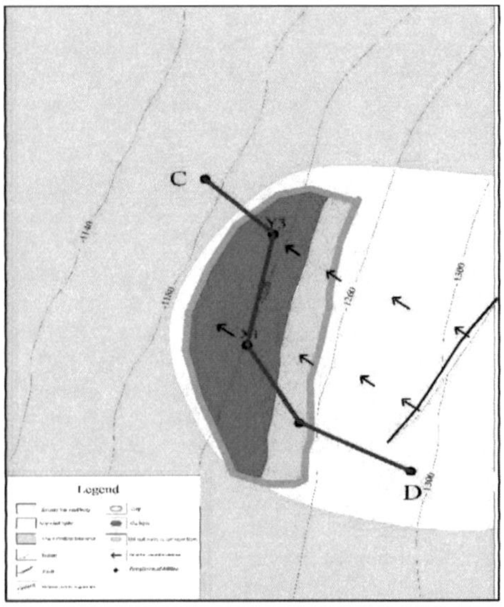

Figure 5 Lithological oil reservoir formation model diagram

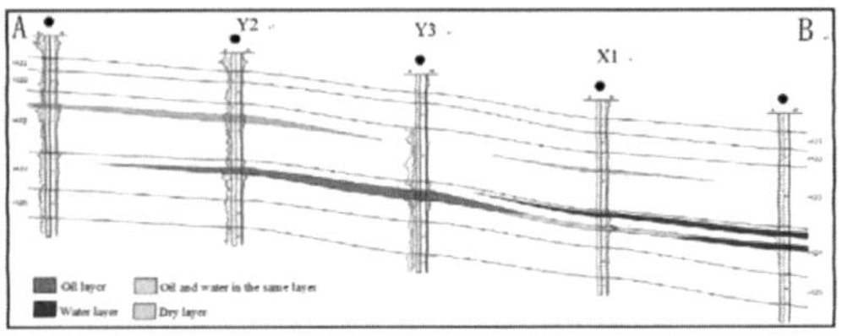

Figure 6 Oil layer A typical structural oil reservoir profile (AB)

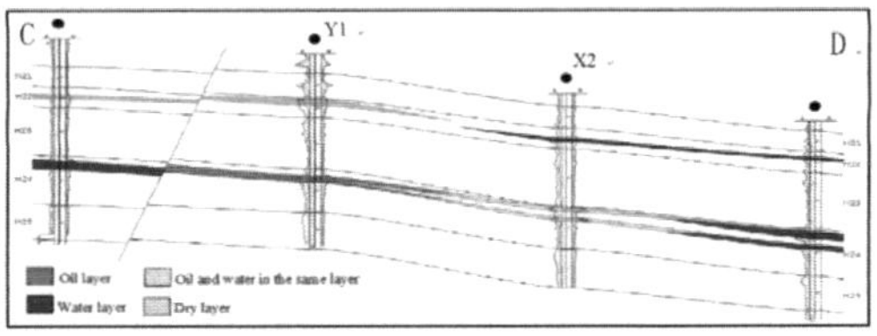

Figure 7 Oil layer A typical lithological reservoir profile (CD)

3 Evaluation ideas and optimization of potential areas

3.1 Evaluation ideas

According to the geological understanding of Oil layer A in the T oilfield, the evaluation ideas are determined: First, look for structural oil reservoirs formed by structural traps near the eastern Aogula-Halahai fault zone. The second is to look for lithologic oil reservoirs formed in the sandstone updip pinch-out or sandstone thinning area near the Bayanchagan fault zone in the west. Based on this, the potential area is optimized, the potential area is classified according to the degree of well control, the certainty of oil-water interface and the degree of reservoir implementation, and step-by-step production is considered. The block with confirmed reservoir reserves is class I area, and development wells are directly deployed. The block with low degree of reservoir implementation is class II area. Evaluation wells are deployed to further implement the reserve scale.

3.2 Potential area optimization

According to the above evaluation ideas, three potential blocks of Oil layer A are selected: the structural high part with large structural traps developed along the eastern fault zone, blocks Y1 and Y2 with good reservoir physical properties, and block Y3 controlled by lithology in the west (Figure 8). According to the development scale of the reservoir, the degree of well control, the physical properties of the reservoir, the thickness of the reservoir, and the oil test production, the potential areas of well deployment are classified (Table 1).

Block Y1 is located in the structural trap of Aogula Halahai fault zone, with a trap area of 3.5km2 and a closure amplitude of 30m. After drilling, it is confirmed that the thickness of sand body is large, and the oil test production and well control degree are high. Well Y1 in the high part of the structure produces pure oil, and well X2 in the low part of the structure produces oil and water together. The oil-water boundary is determined. It is a class I area, and development and production is given priority.

Block Y2 is located in a small structural trap in the Aogula-Halahai fault zone, with a trap area of 0.21km2 and a closure range of 5m. There is no well control in the trap. Well Y2 in the lower part of the trap is tested for oil. It is a water-bearing industrial oil layer. Because the trap is located in the high part of the structure and the reservoir has good physical properties, it is a favorable structural control potential area, but There is no well control in the trap, so it is determined to be a Class II area. Deploy evaluation wells in the high part of the structure to further realize the reserve potential.

Y3 block is a lithologic reservoir in the western sandstone updip pinchout area. Although the oil production of exploration and evaluation wells is relatively high, it is a Class II area due to its thin sandstone and low degree of well control. The evaluation wells are deployed in the

https://doi.org/10.1051/e3sconf/202132901029

structure of this area to further confirm the lithologic pinchout zone.

Table 1 Reserve potential classification table of Oil layer A in T oilfield

Potential block	Reservoir type	Position	Trap area (km2)	Structure amplitude (m)	Main layer sandstone/effective thickness (m)	Test oil output (t)	Oil-water boundary	Reserves Category
Y1	Structural reservoir	Aogula-Halahai fault zone	3.5	30	3.6/2.2	12.48	Determine	Type I
Y2	Structural reservoir	Aogula-Halahai fault zone	0.25	5	No well control	/	Uncertain	Type II
Y3	Lithologic reservoir	Updip pinch-out zone of western sandstone	/	/	3.4/1.8	8.4	/	Type II

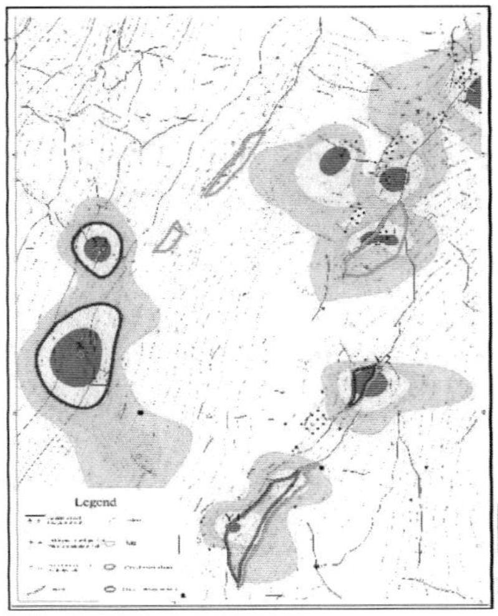

Figure 8 Effective thickness and reserves evaluation diagram of Oil layer A

4 Conclusions

The Oil layer A reservoir of T oilfield is mainly deposited by delta front estuary bar and sheet sand. The oil layer is mainly developed in the A4 small layer. The eastern part is controlled by structure and the west is controlled by lithology.

Oil layer A reservoir types are structural reservoir and lithologic reservoir. Structural morphology and reservoir development control the distribution and enrichment of oil and gas. The "large and rich" structural reservoir is distributed near the Aogula Halahai fault zone in the east of the oilfield, and the "small and fat" reservoir is distributed in the updip pinch out of the sandstone in the West and the small circle near the fault zone.

The structural reservoir in the east is close to the oil-generating center of the most important hydrocarbon-generating depression in Songliao Basin, which is rich in oil source supply. The Aogula-Halahai fault zone communicates with the source rocks, providing a channel for oil and gas migration, and forming a large structural trap to provide storage space for oil and gas, so the oil and gas are accumulated nearby, so the reservoir scale is relatively large.

The lithologic oil reservoirs in the west are located at the end of structurally up-dip deposition. Because there is no relatively large fault zone at the edge of the reservoir to provide channels for oil and gas migration, oil and gas can migrate through long distances and be stored in high parts of the structure due to storage. The deterioration of layer physical properties forms lithologic reservoir traps, and the reservoir scale is relatively small.

According to the above research, Y1 and Y2 blocks near the eastern fault zone and Y3 block at the updip pinch out of sandstone in the West are preferred, which are divided into two types according to their different characteristics for step-by-step production.

References

1. Li Yihe. Study on hydrocarbon accumulation regularity and target optimization of Putaohua oil reservoir in Talaha syncline [D]. Northeast Petroleum University, 2017: 75-96.
2. Liu Gang. Stratigraphic sequence and sedimentary facies of Putaohua, Gaotaizi and Fuyu oil in Talaha-Qijia area [D]. Zhejiang University, 2018: 65-72.
3. Liu Xin. Study on hydrocarbon accumulation regularity of Pu taohua oil reservoir in Talaha-Changjiaweizi area [D]. Northeast Petroleum University, 2013: 46-69.
4. Yang Yufang, Zhong Jianhua. Re-understanding of the sedimentary and accumulation characteristics of the first member of Yao formation in The Aigujia area of Taraha, northern Songliao Basin [J] . Modern geology, 2010: 301—305 .
5. Xu Chengwu . Study on reservoir geology of Salao oil layer in Talaha area, northern Songliao Basin [D].Daqing Petroleum Institute, 2004:54-58 .
6. Lian Xiaoliang . Study on reservoir sedimentology of Putaohua oil layer in Changjiaweizi area, Talaha [D] . Jilin University, 2007:43-45
7. Ji Qingsheng,Han Ling. New Proved Oil Reserves Report in Block Y, T Oilfield[R]. 2006: 13-16

 https://doi.org/10.1051/e3sconf/202132901029

8. Ma Fujian. Study on fine stratification of Heidi Miao oil layer in Telaha-Yingtai-Da 'an area, northern Songliao Basin [D]. Daqing Petroleum Institute, 2003:24-26
9. Liu Yuanji.Study on reservoir comprehensive evaluation inTarahasi area,Daqing[D].Zhejiang University, 2011:52-60
10. Yao Zhongwen,Wan Shuangshuang. Accumulation mechanism of Aogula-Talaha fault zone in Songliao Basin. [R]. Mineral rock, 2017: 66-73.

https://doi.org/10.1051/e3sconf/202132901030

Discussion on Fine Characterization Technology of Single Sand Body in Fluvial Reservoir

Mei Xu

Geological Institute of the Fourth Oil Production Plant of Daqing Oilfield Co., Ltd, Daqing, Heilongjiang, 163511, China

Abstract. Fuyu oil layer of an oilfield is an ultra-low permeability oil layer. The remarkable characteristics of this oil layer are poor physical properties, low single well production, 2-4 layers of sand body longitudinally, large change frequency of plane upper layer and continuous characteristics. In order to realize effective oil layer development and ensure effective reservoir utilization. According to the poor physical properties of this oilfield, this paper analyzes the microscopic and macroscopic responsiveness of the reservoir, ensures effective permeability and thickness of permeable sand body, quantificationally depicts the plane distribution of permeable sand body, and effectively determines the distribution range of the reservoir according to the description of single sand body, thus providing a good technical means for subsequent scale production.

Keywords: Rivers, Reservoir, Description of single sand body, Sand seepage characterization.

1 Introduction

The structure of an oil field is located in the syncline axis and East Wing of the central depression area of the basin. The main purpose of development is Fuyu oil layer, which is characterized by poor reservoir physical properties and low single well productivity, which is very unfavorable to the development of the oil field [1-3]. Therefore, in order to ensure the productivity and realize the utilization of oilfield resources, it is necessary to strengthen the use of fine characterization technology of single sand body in fluvial reservoirs, strengthen the research on related rock-electricity relationship, realize quantitative characterization of permeable sand body, promote the promotion of local resource development potential, and lay a foundation for the improvement of economic benefits in this area.

2 Define reservoir permeability

The productivity of an oil well in an oil field is affected by many factors, among which the most important ones are sand microfacies, reservoir physical properties and oil-bearing property of oil layers. The related research on microscopic throat characteristics shows that the reservoir has different characteristics even at different throat positions of the same hole position, and there are significant differences between throat radii of different permeability levels, while the differences between pore radii are relatively small. Therefore, strengthening the control of throat radius is of key significance to enhance the permeability of reservoirs, and it is also the only factor that can be controlled. By strengthening the analysis of the relationship between throat radius and drainage pressure, the throat radius can be kept less than 0.24 μ M, the displacement pressure can increase significantly [4-5]. If there is a good linear relationship between throat radius and permeability, the throat radius can be kept at 0.24 μ M, the permeability can reach 0.2md, so it is necessary to scientifically control the throat radius, so as to effectively improve the development potential of the reservoir. Therefore, when studying reservoir parameters, fully realize the integration of linear data, grasp the relationship between core and electrical logging curve, reasonably calculate permeability, realize the image description of sand body, clarify the reservoir distribution, and provide an effective basis for subsequent oilfield exploitation.

3 Quantitative characterization of permeable sand bodies in ultra-low permeability reservoirs

The information characteristic of ultra-low permeability reservoir is thin interbedded sand and mudstone, and the shale content of this layer is relatively high, which has a great impact on the characterization of permeability. Therefore, in the characterization, it is necessary to use the traditional method to separately respond to the acoustic time difference and express the properties of reservoir objects. Therefore, in the implementation of quantitative characterization, it is necessary to use the relative value of natural gamma to fully reflect the reservoir lithology, so as to scientifically calculate the permeability. In practical research, it is necessary to give priority to strengthening the analysis of natural gamma relative value on acoustic time difference, so as to scientifically correct mudstone and reconstruct the permeability calculation method of acoustic time difference. Because the shale content of the reservoir in the study area is obviously higher than that in other areas, and the calcium content is also higher, it is necessary to use the relative values of acoustic time difference and natural gamma ray respectively for calculation. Attention should also be paid to the high shale content or large deviation under other circumstances. If the acoustic difference under high shale content is large, it may be due to the influence of high calcareous condition to a great extent, and the relative value of natural gamma ray at this time will be higher than the normal level, so the shale content calculated by using the relative value of acoustic time difference and natural gamma ray will be higher, and the permeability value will become larger.Therefore, in the detailed theoretical calculation, it is necessary to follow the relevant ideas and correct the amount of mudstone according to the relative acoustic value, so as to reconstruct the acoustic time difference curve, reduce the impact of mud on mining and achieve good permeability. Firstly, the core depth homing value can be calculated according to the actual situation of mining, so as to ensure that the influence on acoustic time difference can be corrected on the basis of calculating shale content. Because the mudstone time difference δ Δtsh is larger than the sandstone skeleton time difference δ TMA, the time difference δ T is larger in terms of the measured results in muddy reservoirs, so the time difference δ Δt e after correction can be reflected in the following formula:

$$\triangle te=\triangle t-Vsh\ (\triangle tsh-\triangle tma)$$

$$Vsh=\frac{2^{GCUR\cdot\triangle GR-1}}{2^{GCUR-1}}$$

$$\triangle GR=\frac{GR-GR_{\min}}{GR_{\max}-GR_{\min}}$$

Vsh——Clay content, f;
△GR——natural gamma relative index, f;
GCUR——Hilchie index, no unit;
GR——The natural gamma value of the target layer, API;
GRmax——the natural gamma value of pure mudstone, API;
GRmin——The natural gamma value of pure sandstone, API.

When analyzing acoustic data, it is necessary to analyze the porosity and corrected acoustic moveout of the core to analyze the core permeability, so as to establish a scientific fitting model and construct the porosity and permeability formula. Subsequently, through the mudstone correction of acoustic moveout, the impact on calcareous and mudstone is minimized to ensure the improvement of correlation fitting coefficient. Of course, comparative analysis can be carried out in this process to ensure the orderliness of multi-well evaluation results. By using this method, the effectiveness of permeability can be more scientifically ensured, the coincidence degree of core analysis can be ensured, the microscopic and macroscopic seepage capacity of the reservoir can be deeply embodied in detail, and the expression of correlation can be realized. If the bellowing radius is > 0.24μ m and k > 0.2md, it can represent the production limit of ultra-low permeability reservoirs in this oilfield. In this process, using direct software for analysis can ensure the information extraction in the reservoir with permeability > 0.2md, and then draw the distribution map under the control of corresponding lithofacies to ensure the distribution characteristics of sand body plane, so as to effectively describe and identify the permeability.

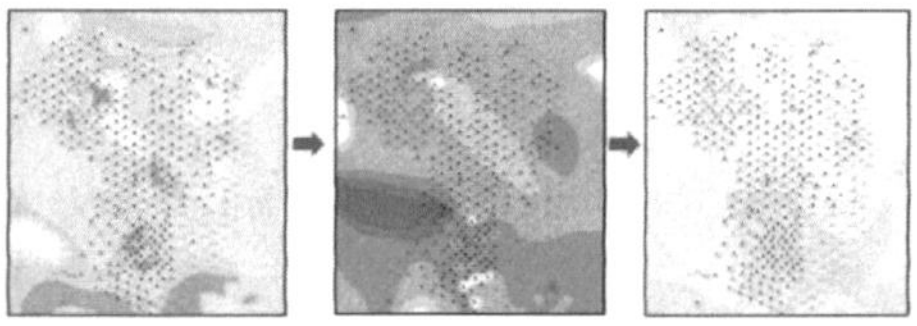

Fig. 1. Depiction map of permeable sand body in an oilfield

4 Analysis of fine characterization of single sand body in reservoir

Fuyu oil layer in this oilfield belongs to multi-stage superimposed channel sand body deposition, which is characterized by strong heterogeneity. In practical research, only by perfect single-stage channel characterization can the longitudinal superimposed relationship of reservoirs be better ensured and the stability of reservoir boundaries be ensured. First, the electrical logging curves can be used to analyze the morphological and rhythmic characteristics. If conditions permit, Fuyu oil layer can be divided into 22 stages of single sand bodies in detail, and then the structure of sand bodies can be identified according to core and logging data. Of course, the division and correlation of multiple unit strata can also be realized by means of facies control and skeleton section closure. Then, according to the characteristics of electrical logging curve of single sand body, the identification attributes of different standards are established [6-7]. In this process, it is necessary to analyze various single-stage channel analysis methods such as "morphological difference", "rhythm difference" and "amplitude difference" according to the dense well pattern data, study the distribution state of well control area, guide the distribution of sedimentary microfacies in low well control area, complete the plane depiction of sedimentary microfacies of single sand body in multiple periods, and control the width of channel sand body in the range of 500 ~ 1000m.

5 Identification and application of braided river single sand body

Shallow sandy braided river, because the river channel is often changed, in the underground braided river reservoir, only the last braided river channel before abandonment can be identified. During the flood period, the river channel will be submerged, so a single river channel will be formed. In the falling period, it will emerge from the water surface and be reflected in the channel in the form of braiding, and then in the dry season, the river channel will become abandoned, but this river channel is usually very narrow. Therefore, the braided river of this style usually needs to be deposited in the way of "wide dam and narrow channel".

5.1 Plane combination of braided and beach dams

In order to accurately grasp the distribution of core beach dam and braided channel, it is necessary to clarify the differences between them, and then set obvious identification marks to distinguish core beach dam and braided channel.

5.1.1 Filling style of braided river channel

By observing the river outcrop and the core of coring wells in the study area, it is used to distinguish different filling styles: Sandy filling, argillaceous semi filling and argillaceous filling. Because of its different hydrodynamic performance, it shows a coarser vertical aggradation sand body. The vertical binary structure of the sand body is not very clear, but it still has strong positive rhythm characteristics, which is reflected in the bell shape on the curve.

5.1.2 Identification and combination of beach dam

For the identification mark of core beach dam, it is necessary to observe the vertical sequence and the relative depth of sand top. After confirming the relevant marks, it is necessary to scientifically divide the braided flow zone sand body and clarify the quantitative scale of typical core beach dam. Through the analysis of global satellite photos and the analysis of factors such as the width, length and width of Xintan dam, it is found that the relationship between them is positive correlation, so as to provide a good geological basis for subsequent oil exploitation.

6 Conclusion

To sum up, through the discussion of fine characterization technology of single sand body in fluvial reservoir and the combination of micro and macro, it can be more clear that the throat radius is greater than 0.24μm and the permeability is greater than 0.2mD, which is the lowest permeability reservoir development limit for oilfield exploitation. In this process, if mudstone is used for correction, a physical parameter model can be established by zoning, the technical process of quantitative characterization of permeable sand bodies can be described in detail, and scientific identification and characterization of high permeability strips can be done well. At the same time, through the anatomy of dense well pattern, we can establish a perfect river channel identification method, more clearly define the spatial distribution law of sedimentary sand body of Fuyu superimposed river channel sand body, analyze the identification and application of braided river single sand body through the characterization of permeable sand body and fine characterization of single sand body, effectively use reservoir identification technology, and clarify the development potential of the established area.

References

1. Lan Xue. Fine description technology of single sand body in fluvial tight reservoir[J]. Petroleum Knowledge, 2019(4): 38-39.
2. Zhang Pengkai, Yue Dali, Rong Chengrui. Application of "single sand body" modeling method based on configuration interface in fluvial facies reservoir configuration modeling [C]. The 15th National Paleogeography and Proceedings of the Academic Conference on Sedimentology. 2018:1-2.

3. Chen Bo, Duan Dongping, Liu Yinghui, et al. Seismic sedimentology interpretation and sedimentary evolution analysis of middle-deep fluvial sand bodies in Xihu Depression[J]. China Offshore Oil and Gas, 2019, 31(3): 117-126.
4. He Yuhang, Wang Qingyu, Zhu Wei, et al. Prototype geological model of single sand body in underwater distributary channel based on dense well pattern [J]. Daqing Petroleum Geology and Development, 2018, 37(3): 43-48. Chinese.
5. Zhu Jinli. Quantitative description of remaining gas distribution and potential tapping strategies for strong heterogeneous tight sandstone gas reservoirs: Taking the northern old area of Block Su 11 in Sulige Gas Field as an example [J]. Natural Gas Industry, 2020, 40(11) :89-95.
6. Li Bing. Identification method of thin-bed sand connectivity[J]. West China Prospecting Engineering, 2019, 31(7): 71-75.
7. Li Hongye. Research and application of single sand body fine characterization technology [J]. China Chemical Industry Trade, 2020(24):248-249.

Structural model splicing method based on well seismic combination

Dan Sun

Geological Research Institute of the Fourth Oil Production Plant of Daqing Oilfield Co. Daqing, Heilongjiang 163511 China

Abstract. The use of small blocks with model-building results for overall structural model splicing can unify fault understanding, effectively integrate existing small block model results, and form a unified and continuous overall model. This article focuses on the structural models of two small blocks in the central and western parts of a certain area, and studies the method of splicing structural models at the sedimentary unit level to lay the foundation for the precise tapping of the potential of the fault edge.

Keywords: Fine modeling of Fault; Construction model splicing.

1 Introduction

At present, an oilfield has entered into an extraordinarily high water-bearing period, the remaining oil distribution is scattered, and the edge of the fault has become a favorable area for residual oil tapping. Special well designs such as high-angle wells are designed to be close to the fault, so it is important to finely characterize the three-dimensional spatial spread of the fault and pinpoint the spatial location of the fault. Since the well-seismic combination of tectonic studies was carried out in an oilfield in 2007, it has deepened the understanding of tectonic features in three-dimensional space, achieved full coverage of well-seismic combination in the whole area, and finely delineated the sedimentary unit-level stratigraphic model, with a total of 18 three-dimensional fine tectonic models for small blocks. The drawbacks of the sub-block tectonic model have been found in practice in two main areas: firstly, the incomplete data of the fault model wells at the boundaries do not provide highly accurate cross-block model results, which limits the application of the results; secondly, the irregular shape of the generated fault polygons does not provide high accuracy in the delineation of the upper and lower plates. The well seismic based tectonic model splicing technique solves the drawbacks of the block tectonic model in the articulation area, and the seamless splicing of the tectonic model deepens the understanding of the overall tectonic features and plays an important role in the overall dredging work of the fault margin.

2 Research ideas and advantages of constructing model splices

2.1 Research ideas for constructing model splices

Tectonic model stitching requires a combination of information to accurately describe all geological formations within a range. The small block models are stored as multiple independent model files and only a single file can be manipulated and it is not possible to splice two model files. The traditional method of constructing a structural model is point-> line-> surface->body, the process of constructing a model is a reverse process from body->surface->line->point, first the results of the existing cell block structural model, into the database, the database for stitching fusion, and then the overall construction of the model. The splicing of structural models is a process of turning the parts into a whole. It requires a certain understanding of the structure in each small block. The modeling idea is based on the integration and unified management of the structural modeling data (Figure 1). The main stages can be divided into the following 3 phases: (1) Establishing the overall model work area. (2) Fusing the fault model results of each sub-block to establish the overall fault model. (3) Fusing the results of the layer model of each sub-block to establish the overall layer model.

 https://doi.org/10.1051/e3sconf/202132901031

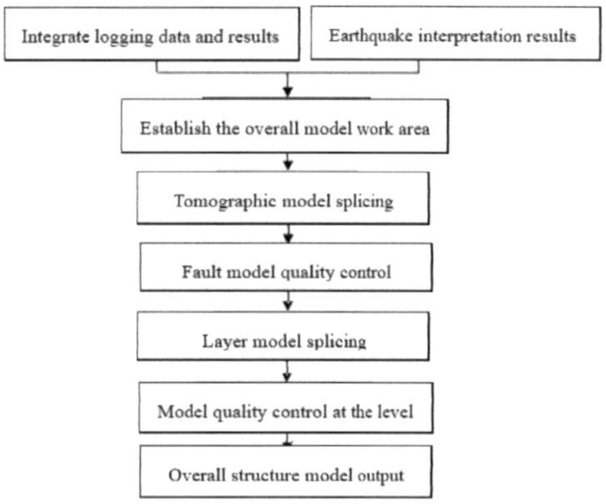

Figure 1 Constructing the model stitching process

2.2 The advantages of constructing model splices

Small block model stitching is a process of fusing data from zero to whole, with quality checks on well data combined with seismic data, which modelers can stitch to suit their needs and the computing power of their computers. This way the accuracy of the model is not compromised and the efficiency of building models of large areas is increased. Advantages of small block model stitching:
(1) High modeling efficiency. Because the small block model has been adjusted, it is only necessary to process the overlapping parts of multiple blocks for fine-tuning. The adjustment effort is only one third of that of conventional modelling and the modelling time is one fifth of that of conventional modelling.
(2) Improve the accuracy of boundary tectonics. Both the fault model and the stratigraphic model are derived from the original sub-block model. For the understanding of inter-block boundary tectonics, the well data and seismic interpretation results can be used during the splicing process to establish an overall unified 3D fine tectonic model to further understand the overall tectonic features.

3 Key techniques for constructing model splices

3.1 Fault model stitching and fusion technology

3.1.1 Fault model data volume fusion

The fault model data volume is extracted from the small block structural model, and the simulation is re-simulated under the control of the well break point data, and the fault is re-understood to ensure that the accuracy of the processed fault model is consistent with the original model. The fusion of the fault model data volume is divided into three categories: i. Inherited: the fault within the work area, extracting the fault model data and applying it directly to the fault model; ii. Reconstructed: the fault as a whole is at the boundary of two sub-blocks, using well seismic combined with the data to recalculate the fault model; iii. Spliced: the same fault is divided between different sub-blocks, extracting the fault model data between each sub-block separately. (see Figure 2)

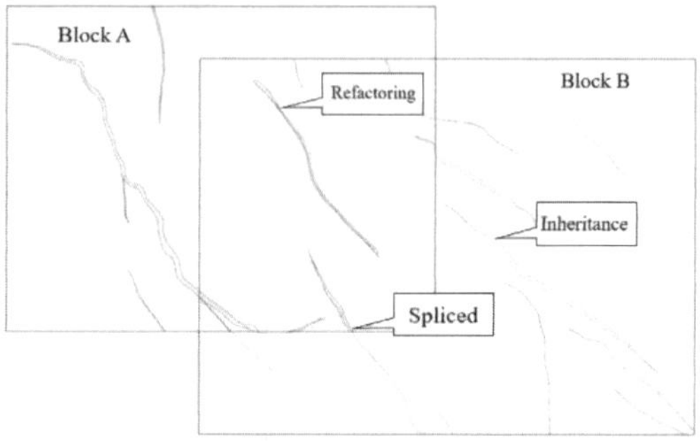

Figure 2 Schematic diagram of the fault splicing of two cell blocks

3.1.2 Handling of fault intersection relations

The conventional method for handling fault intersection relations is manual and manual adjustment, which is very laborious and time consuming. The fault model uses an "improved binary tree algorithm" when dealing with fault intersection relations, which is a new algorithm different from the traditional software fault modelling algorithm (pillar-based method \strict binary tree method). This method uses the division of the fault line to define the range of the fault, avoiding the influence of modeling methods based on pillar and fault upper and lower walls. One-click adjustment of the cut-off relationship between faults can automatically handle the intersection of faults and faults, making editing fast and convenient. It not only improves the efficiency of fault modeling, but also improves the accuracy of fault model (Figure 3).

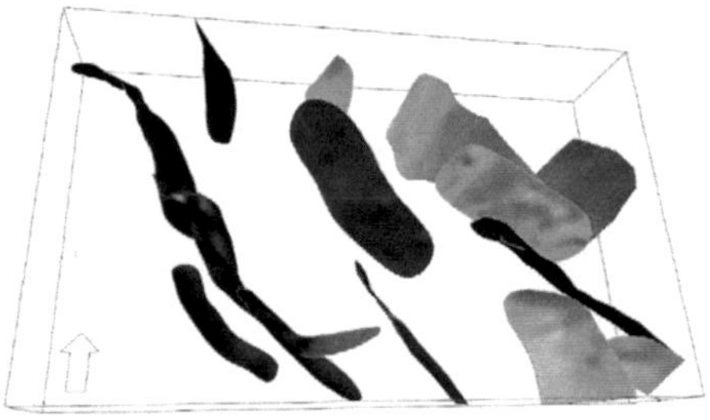

a Fault cross-cutting relationship before processing

https://doi.org/10.1051/e3sconf/202132901031

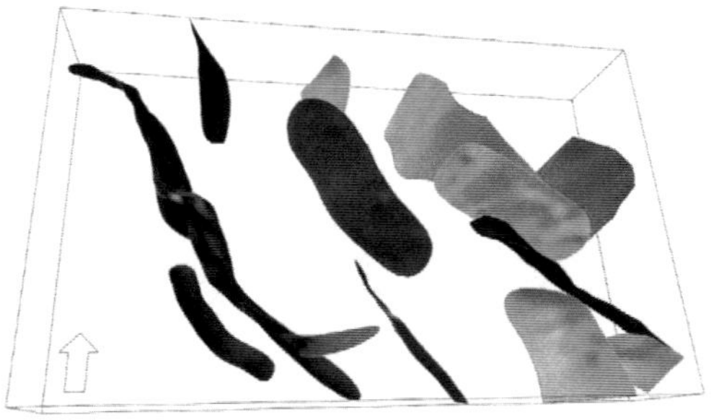

b Fault cross-cutting relationship after processing

Figure 3 Effect picture before and after processing of fault intersection relationship

3.2 Layer model splicing and fusion technology

The layer model is to establish a set of skeletal grid system within a given closed modeling boundary after the fault is accurately established. Therefore, when the structure changes at the boundary of the small block, the local layer model reconstruction method is used to reconstruct the horizon model of the overlap area: Using the boundary fault as the boundary, filter out the layer model data of 1-2 well spacing on each side of the fault (e.g. Figure 4 delete the data in the red border of block A and delete the data in the green border of block B), merge the layer model data of the two blocks into a new layer data, complete the layer model construction, complete the stitching of the layer model and unify the construction understanding.

In this paper, a hierarchical structure modelling technique is used to build a layer model, i.e. a layer model with seismic interpretation data (top surface model of the formation group) is built first, and the top surface model of each formation group is built using the drilling stratification data and the fused layer model data, and under the constraints of this model, the model of small layers or sedimentary units is built. Since the second-level structure is completed on the basis of the upper-level structure, the structure plane data except the overlap area is inherited from the small block model, so the hierarchical structure modeling can reduce the workload of small-level structure adjustment, and can Keep the structure trend consistent from layer to layer.

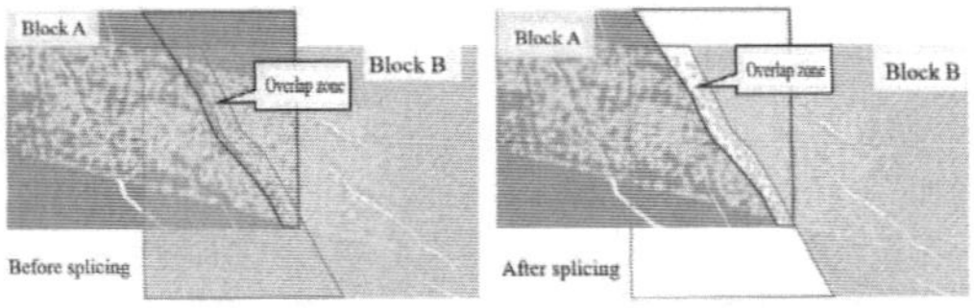

Figure 4 Schematic diagram of the splicing of the horizon model of the two small blocks

4 Applications

In order to effectively use the results of the block well seismic modelling, a three-dimensional tectonic model based on the combination of wells and seismic was developed for the central and western part of a district that already had the results of the block tectonic model, with a total of 1,583 wells (648 wells in the central part of a district and 935 wells in a district) in the central and western part of a district with a seismic work area of 16km2 and 8 faults in the Saputo high objective. During the model splicing process, the breakpoint combination rate was 95.5%, and the error of the tectonic surface was less than 2‰ (Figure 5). After the splicing process, the faults and layers of the tectonic model are spatially unified, and the results of the existing block model are effectively integrated to form a unified and continuous overall model, which improves the accuracy of the identification of boundary faults.

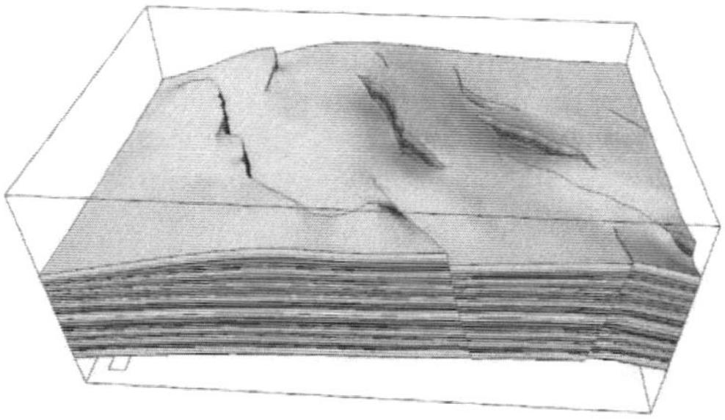

Figure. 5 Results of tectonic model splicing in the west-central part of a district

5 Conclusion

(1) The stitching technology of the construction model unifies the boundary fault recognition, effectively integrates the results of the existing cell block model, inherits the model data of the cell block, greatly reduces the model adjustment work, improves the efficiency of modeling, and saves a lot of time for the subsequent attribute modeling and tapping work.

(2) The establishment of the overall tectonic model can output the tectonic map of any layer, and any tectonic profile, and the continuous and consistent spreading of tectonic and fault features in three-dimensional space over a large area, which lays the foundation for the study of the overall tectonics and evolution of the whole region and the local fine tectonic features, and will greatly improve the effectiveness and efficiency of the application of tectonic research results.

References

1. Jiang Yan, Li Jie, et al. Well Seismic Combined Integral Tectonic Modeling Techniques in the Daqing Changyuan Saltu Oilfield. Daqing Petroleum Geology and Development, 2015,34(1): 151-156.
2. Liu Lei, Jin Li, et al. Research Methods for Developing Seismic Tectonic Modeling Techniques. Highland Earthquakes, 2009, 21(3):46-51.
3. Zhang Yang, Wang Qi, et al. Tectonic Modeling and Applications for Complex Fracture Block Reservoirs. Liaoning Chemical Industry, 2012, 4(5):512-514.
4. Kong Hong. Application of RMS geological modeling software in MTZ area. Journal of Jianghan Petroleum Institute, 2003,25:9-10
5. Chong Renjie, Yu Xinghe. Key Points for Quality Control of 3D Geological Modeling of Reservoirs. Frontiers in Marine Geology, 2011,27(7):64-69.

Application of well-seismic reservoir prediction technology in oilfield

Xiaona Zhang

Key Research Room, Institute of Geology, Fourth Oil Production Plant, Daqing Oilfield Co., Ltd., Daqing, Heilongjiang, 163511, China
zhangxiaona@petrochina.com.cn

Abstract. This study mainly takes the thick oil layer in Lamadian Oilfield as an example. This oil layer is mainly developed in the area dominated by fluvial sediments, and the reservoir sand body space changes frequently. If we only rely on the traditional single well logging data, it is difficult to directly and effectively predict the changes of the whole sedimentary sand body, and the spatial configuration relationship between river boundary and other sedimentary microfacies can not be effectively predicted. In order to effectively improve the understanding of cross-well sand bodies, it is necessary to effectively combine well seismic with reservoir prediction technology, give full play to the advantage of high resolution of well seismic in the longitudinal direction, and predict its specific adaptability and application according to different situations, so as to better improve the prediction accuracy of reservoir sand bodies and provide important information data for the rational adjustment of cross-well encryption and the formulation of effective measures to tap potential. Based on this, this article deeply analyzes the application value of well-seismic combined reservoir prediction technology in oil fields.

Keywords: Well earthquake, Oilfields, Reservoir prediction technology.

1 Introduction

In this study, taking the area of 17km2 in the second area of Labeibei as an example, there are 1162 wells in total, and the distribution range of sand bodies is very wide in fluvial sedimentary environment. In the vertical direction, there are three development modes, namely, a large set of thick laminated sand bodies, a thin interbedded sand and mud model, and an isolated lenticular sand body with continuous distribution of river channels, interwoven strip distribution of river channels (branch-lump sand bodies) and a single strip distribution mode of river channels. After four times of infill development, the overall distribution of both underground reservoirs and rivers in the study area is very complex. It is necessary to find new ideas and technical methods in continuous research and exploration, and only one step to strengthen the understanding and prediction of cross well sand bodies.

2 Seismic reservoir prediction

Seismic reservoir prediction can be roughly divided into two aspects, one is seismic correlation analysis, the other is attribute extraction. When extracting seismic attributes, different extraction methods need to be adopted according to the different meanings represented by each attribute in general. The correlation between seismic attributes and sand body can be directly analyzed by extracting its specific attributes, sand body distribution and sand body thickness. Simply speaking, it is the analysis of intersection diagram, and then the reservoir conditions are predicted by using the above-mentioned relationship. In this study, the pressure difference-flow method was mainly used to determine the starting pressure gradient of reservoir water flooding and emulsion flooding. The measurement can directly prove that the emulsion has significant application value in reducing seepage resistance and reducing the starting pressure gradient. At the same time, it can also provide important information and theoretical support for the displacement of oil reservoir emulsions. When the microemulsion flow passes through the small pores, the surfactants in the system can be directly adsorbed in the gaps, which can not only reduce the tension of the solid-liquid interface, but also effectively reduce the seepage resistance in the boundary layer. In addition, ionic surfactants can effectively ionize the solution to generate charged particles. These so-called charged particles will directly enter the rear of the boundary layer, and will interact with ions in the boundary layer. Under the interaction, they will play the role of compressing and diffusing the electric double layer. At the same time, due to this effect, the boundary layer will gradually become thinner. Once the boundary layer is lower than the standard range, the actual overflow area in the pores will gradually increase. Based on this, microemulsion system can effectively reduce the gradient of starting pressure, and promote the seepage resistance to decrease gradually.

3 Well-seismic reservoir description

Taking the second area of Labeibei as an example, this paper mainly describes how well earthquakes combine with reservoirs, and the concentration of alkali and surfactant has certain influence on the interfacial tension of ternary composite system. In the whole experimental concentration range, the interfacial tension between the ternary composite system and crude oil will gradually decrease with the increasing concentration of alkali and surfactant [1]. When the concentration of alkali is greater than 0. 2% and the concentration of surfactant is greater than 0. 9%, the tension between the ternary composite system and the crude oil will achieve ultra-low interfacial tension to a large extent. Taking the second area of Labeibei as an example, this paper mainly describes how well earthquakes combine with reservoirs, and the concentration of alkali and surfactant has certain influence on the interfacial tension of ternary composite system. If the concentration of surfactant is above 0. 05%, the interfacial tension can reach the order of 10 ~ 2 Mn / m, but within the concentration range of the actual test, the interfacial tension does not reach ultra-low interfacial tension [2]. Therefore, by analyzing the construction of data centers in the international, domestic and petroleum fields, it is found that the construction of enterprise data centers can realize the development of integrated business, which can become the main way to enhance the competitive strength and position of enterprises in society, and at the same time provide important information and data for the production, management and decision-making of enterprises [3]. Therefore, by analyzing the application of information in domestic and foreign enterprises and institutions, it can be found that the construction of data center can better promote the process of enterprise informatization, realize the unified collection and storage of each core data, and realize the sharing of information resources.

4 Verification of coincidence rate of well-seismic reservoir prediction technology

In order to better explore the advantages of the combination of well earthquake and reservoir prediction technology applicable to the target layer in this study area,In this study, three methods were used to predict the inter-well reservoir conditions,This method mainly includes single amplitude attribute, multi-attribute neural network fusion, geostatistical inversion and prediction,At the same time, it is divided into two levels: sand layer group and sedimentary unit. Generally, each sand layer group includes four sedimentary units, and some sand layer groups contain three sedimentary units,The general thickness of single layer of deposition unit is about 4m [4-5]. The single attribute prediction is mainly based on the well-point position and the correlation of the well seismic data to rationalize its specific attributes and slice conditions, so as to use it as the final choice, that is, the root mean square amplitude attribute. Use this attribute along the entire distribution trend of the layer slices to predict the distribution of the sandbody plane. At the same time, it is necessary to focus on the existence of the Saertugong oil layer group (hereinafter referred to as the sl oil layer group) and the SII oil layer group in this study. The seismic reflection axis of the top interface is strongly interfered with. At the same time, the wavelet decomposition and reconstruction technology is used to predict, which significantly improves the reflection effect of the overall target layer [6].

Geostatistical inversion usually adopts JAS (n) software to reconstruct the density curve with the participation of acoustic wave and resistivity, which can greatly improve the accuracy of wave impedance calculation, and then through the physical analysis of well seismic rock [7-8]. Determine the lithology sensitivity curve, seismic attributes and lithology division sensitivity values, and then use the natural potential pseudo-wave impedance inversion constraint to effectively combine the well seismic and geostatistics co-simulation technology process, which can be used to a large extent To ensure the accuracy of the response effect, improve the effectiveness of the inversion effect and the vertical high resolution [9].

By analyzing the application effects of these three different seismic prediction techniques in post-reservoir post-inspection wells (see Table 1),Compared with the single attribute technique, the multi-attribute technique has significantly improved the prediction coincidence rate of the later mudstone and 3m thick sandstone, which can reach 90. 01% and 88. 58% respectively. By analyzing the data, we can directly clarify the specific distribution of this part of the reservoir. For the sedimentary unit level with an average stratum thickness of 4m, in the process of actual prediction, seismic inversion technology, that is, post stack geostatistical inversion, can significantly improve the prediction accuracy of thin sandstone with a thick bottom of less than 3m and the prediction value of thin reservoir can be significantly improved compared with seismic attribute technology [10-21].

Table 1. Statistical table of coincidence rate of seismic prediction for different sandstone thicknesses

Statistical unit	Earthquake prediction technology	Seismic prediction coincidence rate of thickness mudstone/%	Seismic prediction coincidence rate of sandstone with different thickness/%.		
			0-2m	2-3m	>3m
SII1-3 (Sand group level)	Rms amplitude attribute	78. 5	28. 5	44. 4	75. 3
	Multi-attribute neural network fusion.	90. 9	33. 2	49. 8	88. 5
SII1+2b (Sedimentary unit level)	Rms amplitude attribute	63. 2	27. 7	39. 6	78. 5
	Multi-attribute neural network fusion.	83. 7	32. 4	46. 2	88. 4
	Post-stack geostatistical inversion	82. 5	66. 7	78. 2	88. 7

5 Conclusions

In conclusion, by effectively combining the selected seismic attributes with single well sedimentation, a complete sedimentary facies result map of the target horizon can be made. By analyzing the renderings, we can directly evaluate and analyze the connectivity between wells, and effectively identify the ineffective circulation layer with low efficiency. According to its analysis and evaluation results, the development of oil and water wells can be accurately subdivided and adjusted, and at the same time, the changes in the injection-production relationship between wells where the sedimentary facies diagram has changed can be accurately identified.

References

1. Wei Feng, Xue Guoqing, Xiong Cong, et al. Classification and comprehensive evaluation of facies-controlled reservoirs in sparse well areas combined with well-seismic: Taking the Zhujiang Formation of the WC-A Oilfield in the Pearl River Mouth Basin as an example[J]. Journal of Oil and Gas Technology, 2020, 42(2): 17-26.
2. Wei Feng, Xue Guoqing, Xiong Cong, et al. Classification and comprehensive evaluation of facies-controlled reservoirs in sparse well areas combined with well-seismic——Taking Zhujiang Formation of WC-A Oilfield in Pearl River Mouth Basin as an example[J]. Journal of Oil and Gas Technology ,2020,42(02):17-26.
3. Ma Yuliang, Zheng Huan, Xie Ruijie, et al. Several main problems and processing methods of well seismic data in reservoir prediction[J]. Science Technology and Engineering, 2019, 19(33): 132-138.
4. Li Suhua, Hu Hao, Zhu Lan, et al. Identification and prediction of bioclastic shoal karst reservoirs in the Maokou Formation in Yuanba area, northern Sichuan[J]. Geophysical Prospecting for Petroleum, 2021, 60(4): 584-594.
5. Li Jianxiong, Gu Yuemin, Dang Huqiang, et al. A method for combined well-seismic reservoir prediction in high-density well pattern development areas: Taking the JN2 block of the SL basin as an example[J]. Petroleum Geophysical Prospecting, 2011, 46(3):457-462.
6. Zhao Weiping. Application of combined well-seismic attribute analysis technology in deep-water turbidite reservoir prediction[J]. Journal of Engineering Geophysics, 2016, 13(2): 213-220.
7. Guo Jigang, Nie Zhikun, Mao Chuanlong, et al. Application of multi-attribute fusion method in two-dimensional seismic coal seam prediction: Taking the two-dimensional work area of Dacheng Uplift as an example[J]. Computing Techniques for Geophysical and Geochemical Exploration, 2020, 42(6)): 732-742.
8. Yang Feng, Nie Hui, Chen Hongtao, et al. Carbonate prestack reservoir prediction based on broadband seismic data: Taking Block B in the eastern margin of the Caspian Basin as an example[J]. Geophysical Prospecting for Petroleum, 2019, 58(4) :555-562.
9. Zhu Jianbing, Wang Xingmou, Feng Deyong, et al. The method and application of fluvial reservoir prediction based on bidirectional cyclic neural network[J]. Geophysical Prospecting for Petroleum, 2020, 59(2): 250-257.
10. Ma Manhong. Research on the salvage technology of falling objects in oilfield wells[J]. Chemical Management, 2013(10): 114-114.
11. Fan Longwen. Discussion on the technology of salvaging complex falling objects underground[J]. Henan Science and Technology. 2013(5)
12. Wu Qi. Handbook for Downhole Operations Engineers Beijing: Petroleum Industry Press. 2002:215-253
13. Ren Fade. Discussion on the method of salvaging falling objects in oilfield downhole operations [J]. Petrochemical Technology. 2016, 23 (8): 260
14. He Wang, Xie Liyun, Wei Ziyu. Evaluation of engineering geological suitability of long-distance natural gas pipeline project crossing area——Comment on "Oil and Gas Pipeline Integrity Evaluation Technology" [J]. Xinjiang Geology, 2019, 37(4): Post 4 .
15. Wang Qingguo, Guo Min. Research on the influencing factors of gas pipeline integrity based on ISM and AHP[J]. Gas and Heat, 2019, 39(4): Post-insertion 1- Post-insertion 4.
16. Ma Bin, Jiang Feng, Ma Xuqing. Integrity evaluation technology of defective pipelines[J]. Gas and Heat, 2019, 39(5): 32-39.
17. Cui Ruixue, Wang Lin, Zhao Deyin, et al. Research on technology development of pipeline integrity management platform[J]. Contemporary Research in Chemical Industry, 2020(11): 108-110.
18. Jin Jian, Zhu Xueshan, Liu Yanyang, et al. Research on big data integration technology methods for oil and gas long-distance pipelines[J]. Oil and Gas Field Surface Engineering, 2020, 39(2): 77-81.

19. Jiang Zhongan, Zheng Dengfeng, Zeng Fabian, et al. Research on the safety management model of oil and gas pipelines based on hazard theory [J]. Journal of Hunan University (Natural Science Edition), 2021, 48(4): 56-65.
20. Jia Shuai. The application of intelligent information technology in the integrity construction of the Beijing-Tianjin Second Oil Pipeline[J]. Oil and Gas Field Surface Engineering, 2021, 40(2): 97-102.
21. Hou Lina, Li Yuanpeng, Yan Wei, et al. Construction and fusion method of oil and gas field pipeline and station integrity management system[J]. Oil and Gas Field Surface Engineering, 2021, 40(3): 63-69.

Analysis on the Important Application of Geophysical Exploration Technology

Dan Yu

Liaohe oilfield exploration and Development Research Institute, Panjin, Liaoning 124000 China

Abstract. Geophysical prospecting technology is widely used in current geological exploration, metal exploration, oil and gas resources exploration, etc. This technology has many advantages such as low cost, high efficiency and good accuracy of survey results, and plays an important role in many industries. At present, the level of geophysical prospecting technology in China is constantly improving, and its application effect is also constantly improving, so it is necessary to clarify the key application points of this technology. Therefore, this paper will conduct in-depth research and Analysis on the important applications of geophysical exploration technology, elaborate the basic concepts and specific applications of geophysical exploration technology, and put forward some reasonable opinions and measures, in order to further improve the application level of this technology.

Keywords: Geophysical survey technology; Basic concepts; Specific application; geological prospecting; Optimization measures

1 Introduction

Geophysical exploration technology is widely used, which can accurately obtain the specific information of the explored area, so as to determine the actual situation of the earth structure. It has good application effect in the fields of construction engineering and metal exploration, and can provide scientific information and data support for subsequent development and construction. Geophysical exploration technology is a general term of technology, which includes the application of many specific technologies, and different technologies have corresponding principles and characteristics. In order to further promote the application level of geophysical exploration technology, this paper analyzes its specific application in many fields.

2 Brief introduction to geophysical exploration technology

Geophysical exploration technology is a professional category of geology. It takes the content of physics as the technical basis, uses the measurement and observation of the change and distribution of physical field, completes the exploration of the earth's constituent elements, various material structures existing in space and their evolution process, and can analyze the change law to obtain the geological structure, geological conditions and At the same time, it plays an important role in natural disaster monitoring, prediction and response. Because there are some differences in density, elasticity, conductivity, radioactivity, thermal conductivity and magnetism among the different rock media that make up the crust, and the differences will cause local changes in geophysical fields, by controlling the distribution and changes of these differences, the goal of geophysical prospecting can be achieved. Compared with other geophysical exploration technologies, geophysical exploration technology has many advantages, such as portable equipment, low cost, high investigation efficiency and wide working space, and is usually used in combination with drilling technology [1].

Geophysical exploration technology is usually based on the physical properties of rock strata, such as rock permeability, rock density, rock conductivity, rock elasticity, rock thermal conductivity and rock radioactivity. Common exploration methods include gravity exploration method, magnetic exploration method, electrical exploration method, seismic exploration method, geothermal exploration method and nuclear exploration method. According to the spatial location and area where geophysical exploration technology is applied, it can be divided into ground geophysical exploration method, aerial geophysical exploration method, marine geophysical exploration method and drilling geophysical exploration method, etc. According to different exploration objects, it can be divided into metal geophysical exploration method, petroleum geophysical exploration method, coal geophysical exploration method, hydrogeophysical exploration method and engineering geological geophysical exploration method.

DanYu@petrochina.com.cn

3 Analysis of specific methods of geophysical exploration technology

Analysis of gravity exploration method

Gravity exploration is the most common technical form in geophysical exploration technology. Through precise measuring equipment, the geological strata and the gravity change and density of existing ores are explored and measured, so that the basic situation in the exploration area can be obtained. It has good application effect in the exploration of mineral resources, and can divide faults, magma and rock mass, which is an important basis for mineral prediction [2]. Before the application of gravity exploration method, the density data of rock must be clear before the specific situation can be identified. The following table shows the density values of several common rocks when the method is applied.

Table 1. Density values of common rocks

Serial number	Name	Density (g/cm3)
1	Basalt	2.6—3.3
2	Granite	2.4—3.1
3	Quartzite	2.6—2.9
4	Gneiss	2.4—2.9
5	Mica schist	2.5—3.0

3.1 Magnetic prospecting analysis

Magnetic exploration method is a technical way to analyze the changes of different magnetic fields by using magnetic equipment and taking the magnetic differences existing in natural ores or rock masses as exploration technology. It has a good application effect in geological problems and mineral resources exploration. The magnetic survey method requires simple operating equipment, small size and high work efficiency, and can ensure the accuracy of survey results, making outstanding contributions in the fields of geological survey and resource exploration in China.

3.2 Analysis of seismic survey method

The application of seismic survey method is mainly to detect the medium density elastic wave group in different geological strata, survey the signals and radio waves in the strata through equipment and instruments, and obtain the detailed information of the strata by manually activating the earthquake. It has good results in the fields of coal resources exploration, oil and gas resources exploration and so on. According to the practical application of seismic survey method, taking its application in coal resources exploration as an example, the three-dimensional seismic model built by high-resolution imaging technology can intuitively show three-dimensional images of geological conditions, thus providing a basis for subsequent exploration work.

3.3 Analysis of electrical survey method

The electrical prospecting method is a geophysical prospecting technique with good application effect in recent years, which mainly uses the ground electrical method, aviation electrical method and direct current method to realize the exploration of geological strata. In the process of application, DC method has the highest accuracy, so it has a good application effect in hydrological survey. However, this method has higher requirements for topographic environment, so the survey environment must be treated accordingly before application. In addition, the electrical survey method also includes transient electromagnetic method, which can complete the underground electromagnetic transmission without connecting the return line to form electromagnetic pulse, and then combined with corresponding instruments to obtain various information of geological layers. The following table shows the classification and application of electrical prospecting methods.

Table 2. Classification and application of electrical prospecting method

Category	The nature of field	Method name	Application
		Natural electric field method	Looking for technical sulfide deposits and stratigraphic mapping of yellow titanium mineralization
Direct current method		Induced polarization method	Looking for gold sulfide deposits, groundwater exploration and oil exploration
	Natural field	Charging method	Foundation engineering exploration and groundwater exploration
AC electric method		Resistivity method	Explore the Karst Development Zone, determine the fluctuation of bedrock and determine the thickness of weathered layer
	Artificial field	Frequency electromagnetic sounding	Large depth geological survey

4 Specific application analysis of geophysical prospecting technology in different fields

Geophysical survey technology has been applied in many fields at present. According to different application objects, different geophysical survey technologies are adopted, and the comprehensive survey results are remarkable, which can effectively guarantee the accuracy of survey results. The following are the specific applications of geophysical survey technology in different fields.

4.1 Application of in deep metal mineral resources exploration

The deep part belongs to the second depth space. In the depth range of 500-2000m, it is difficult to investigate because of its deep burial depth. Therefore, it is necessary to use corresponding geophysical exploration technology to help the investigators determine the thickness of sedimentary cover structure and weathered layer, and establish deep geophysical inversion model, so as to determine the deep structural environment of metal mining area, At the same time, the deep lithology mapping can be carried out to determine the ore bearing horizon. According to the formation and distribution of the technical mining area, there is a certain relationship between granite body and basic ultrabasic intrusive rock mass. Therefore, geophysical exploration can be used to map the deep lithology in the metal mining area to help the investigators determine the abnormal field distribution and shape of rock mass with different physical attributes. Taking a deep metal mineral resources development project as an example, in the course of investigation in this project, through the technical means of controlled-source audio magnetotelluric emission and based on audio gross electromagnetic method, the working principle of controlled-source audio magnetotelluric method is to introduce harmonic current in a certain audio range to the deep underground, observe the frequency value of the audio, and compare the frequency changes, and finally complete the exploration of deep mineral resources [3].

For some deep concealed ore bodies with obvious physical differences from the surrounding rock, the prospecting work can be carried out directly through aviation and ground geophysical exploration methods, which can realize high-precision geophysical exploration. For example, the aeromagnetic method and electromagnetic survey method of large-scale helicopter carried out by the aerospace remote control center of the Ministry of land and resources in Daye iron mine area. According to the detailed interpretation results, more than 5 layers of iron ore bodies were found in the depth of 720~770 meters, and the accumulated thickness of iron ore bodies exceeded 14 meters. Magnetite, pyrrhotite and chalcopyrite were among the found ores.

4.2 Application analysis of in nonferrous metal resources exploration

Generally, nonferrous metals are deeply buried in areas with deep underground depth, so corresponding exploration methods must be adopted to find nonferrous metal minerals. As an important natural resource, nonferrous metals have an important impact on China's industrial production and other aspects. In recent years, the development of China's nonferrous metal exploration industry has been promoted by geophysical exploration technology to a certain extent. The basic principle of geophysical exploration is to identify the physical characteristics of rock strata and lithology in the crust. According to the identification results, it can be clear whether there are non-ferrous metal resources in this area. It is a simple and effective exploration method with fast exploration efficiency and low exploration cost. It is the most basic method in non-ferrous metal exploration. Geophysical exploration technology applied in nonferrous metal exploration is extremely important for the development of China's resource exploration industry. Having more advanced exploration methods means that it can improve the exploration efficiency of nonferrous metal resources and find out more nonferrous metals and related mineral resources distributed in China, which is of great significance for the implementation of China's resource strategy and ensuring national security [4].

4.3 Application analysis in engineering geological exploration

Before the development of road engineering, in order to clarify the geological conditions of the construction area, scientific methods must be used to obtain geological information. The application of geophysical exploration technology has good results and can help the construction unit obtain sufficient and accurate geological information. Taking a road construction project in Q city as an example, the road construction needs to build a tunnel. In order to ensure the safety of tunnel construction, after full demonstration by the designers, it is decided to adopt the micro motion survey technology in geophysical survey technology. Micro-motion detection technology makes use of the basic principle that there are weak fluctuations on the earth's surface. The weak fluctuations on the earth's surface come from the basic activities of nature and human beings. These weak fluctuations will spread outward in the form of vibration and contain a certain amount of energy, which mainly propagates in the form of surface waves. By surveying the energy surface waves, the corresponding information in geological structures can be obtained. The following figure shows the equipment structure and operation schematic diagram of micro-motion detection technology.

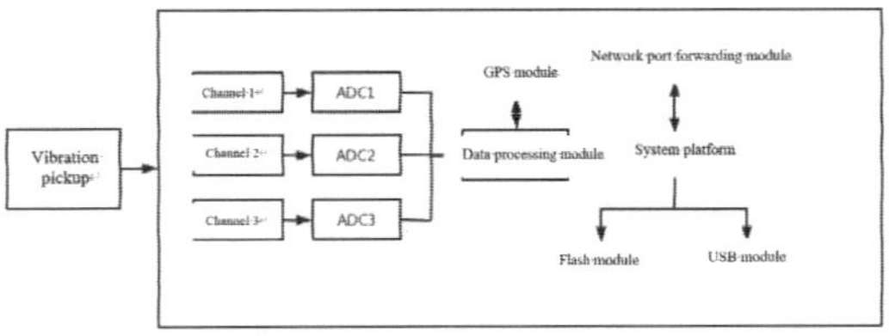

Figure 1. Structure and operation principle of micro motion detection technology equipment

In this project, the basic application process of micro motion geophysical survey technology is: survey station → micro motion observation waveform recording → Rayleigh wave phase velocity dispersion curve → surface wave phase velocity contour map. The main function of the observation station is to collect the signal of micro motion wave, and apply the set algorithm to extract the detected signal from multiple survey stations and record it, then draw it into the Rayleigh wave velocity phase velocity dispersion curve by a specific algorithm, and then draw the contour map directly. According to the contour map, the change of address structure in the detected area can be truly reflected, And all kinds of information in the address structure, the results are more accurate. In this tunnel project, two survey lines are set up along the north-south line of the tunnel, with 70 observation points of the north line survey array and 80 observation points of the south line survey array, thus obtaining the specific geological structure information of the proposed tunnel area, which mainly includes: (1) The fretting wave propagation speed in the tunnel body is relatively slow on both sides near the entrance and exit of the tunnel, and it can be judged from the drilling data that the tunnel near the entrance has passed through silty clay, residual clay and clay. (2) According to the application of combined drilling technology, it can be explained that the middle part of the tunnel mainly passes through the base rock, but there is a certain gap in the velocity of the bottom rock. The relatively complete rock is in the area with high velocity and the relatively broken rock is in the area with low velocity. The change result of surface wave velocity can explain that the continuity of geological structure in the tunnel is poor, It shows that the tunnel body is affected by the nearby fault structure.

5 Conclusions

In this paper, the basic concepts of geophysical survey technology are described in detail, and its various application methods are analyzed. Finally, the specific applications of geophysical survey technology in deep metal survey, nonferrous metal survey and geological survey of road engineering are analyzed, hoping to play a certain reference and help role for the development of related fields and industries, and at the same time promote the continuous improvement of geophysical survey technology level.

References

1. Luo Ji, Tang Wenqian. Analysis on the important application of geophysical exploration technology[J]. China Chemical Industry Trade, 2019, 11(015):85-85.
2. Tong Zhikui. Application analysis of geophysical exploration technology in oil and gas exploration and development[J]. 2021(2015-23):173-173.
3. Long Zhiyuan. Geophysical prospecting technology methods and application research analysis[J]. Electronic Components and Information Technology, 2020, 39(09):118-119.
4. Ruan Jue. Application analysis of integrated geophysical exploration technology in geothermal exploration[J]. Encyclopedia Forum Electronic Journal, 2019, 000(017):784-784.
5. Yu Zhitong, Tian Jian, Xin Zhixiang. Research on the development status and practical application of geophysical exploration technology[J]. World Nonferrous Metals, 2019, 53(002):0058-0058

 https://doi.org/10.1051/e3sconf/202132901034

Analysis of Fracturing Effect of Recovered Wells in Class II Oil Reservoirs and Practices to Safeguard Fracturing Result

Lexiao Lin

Second Oil Production Plant, Geology Brigade, Three Production Room, China

Abstract: In recent years, as the polymer drive of the main class I oil reservoir has been transferred to the subsequent water drive development, the class II oil reservoir has gradually become the main target of the three times oil recovery development, compared with the class I oil reservoir, the class II oil reservoir has the characteristics of relatively poor oil reservoir development and strong non-homogeneity. With the effect of the second-class oil layer successively, the fracturing target is mainly the second-class oil layer, but the fracturing effect becomes worse year by year, and it is difficult to produce in the next step. Therefore, it is important to optimize well selection for fracturing. In this paper, we analyze the effect of fracturing wells in Class II reservoirs in recent years and summarize the factors affecting the effect of fracturing in Class II reservoirs and the practices to ensure the effect of fracturing in Class II reservoirs, so as to provide a reference basis for the selection of wells for Class II fracturing in the future, which is of certain significance.

Keywords: Class II reservoir, fracturing, residual oil, degree of control.

1. Basic information on fracturing in previous years

Extraction well fracturing as one of the main production tools for oilfield development, the second-class oil layer has entered the peak of effectiveness one after another and has become the main fracturing target. The oil recovery rate is rising, but the fracturing effect is getting worse year by year, and it is difficult to go up to production next since 2005. The oil recovery rate has been rising, but the fracturing effect has become worse year by year, making the next step to production difficult. 2018, through optimizing the fracturing selection, the average single well fracturing of the extraction wells increased oil by more than 6t per day. In terms of fracturing scale, the proportion of fracturing in the late injection block reached more than 70%. Based on the analysis of the fracturing oil increase effect, the proportion of "high efficiency wells" can be effectively increased by optimizing the well selection conditions and implementing measures to ensure the fracturing effect.

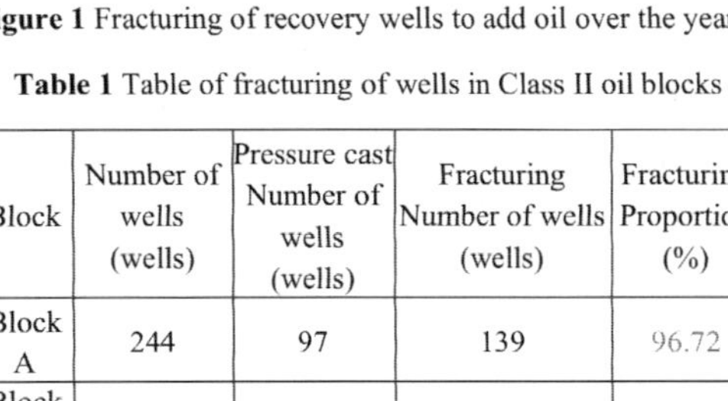

Figure 1 Fracturing of recovery wells to add oil over the years

Table 1 Table of fracturing of wells in Class II oil blocks

Block	Number of wells (wells)	Pressure cast Number of wells (wells)	Fracturing Number of wells (wells)	Fracturing Proportion (%)
Block A	244	97	139	96.72
Block B	197	70	122	97.46
Block C	316	125	101	71.52
Block D	315	124	120	77.46
Block E	248	78	40	47.58
Total	1320	494	522	76.97

* Corresponding author: linlexiao@petrochina.com.cn

2. Summary of experience with fracturing effects

Based on the fracturing results of Class II reservoirs in previous years, the experience is summarized in terms of "factors affecting the fracturing effect of recovery wells" and "practices to ensure the fracturing effect of recovery wells".

The analysis concluded that the effect of fracturing on the oil increase in the recovery well is mainly related to factors such as "the remaining oil in the fractured well, the degree of control, the timing of fracturing, and the injection status of the well area". The approach to ensure the fracturing effect of the extraction well mainly includes "pre-fracturing training, timely post-fracturing adjustment and optimization of the fracturing process".

2.1 Analysis of factors affecting the effectiveness of fracturing of recovery wells

2.1.1 Good fracturing effect in the remaining oil-rich well area.

According to the oil increase effect of different fractured wells, when the water content before injection is less than 96%, the oil increase effect of fracturing is better, and the lower the water content before fracturing, the higher the initial oil increase after the measure. From the oil increase of wells with different water saturation before fracturing, the oil increase effect of fracturing is better when the water saturation is less than 56%, and the lower the water saturation, the higher the initial oil increase after the measure. In addition, the oil increase classification of fracturing wells with different water flooding thickness ratios shows that when the high-water flooding thickness ratio exceeds 40%, the oil increase is relatively low due to the low remaining oil, high water content saturation and relatively high-water content before injection.

Therefore, when selecting wells for oil fracturing, try to select wells with a high-water flood thickness of less than 40% and wells containing less than 96% water before injection.

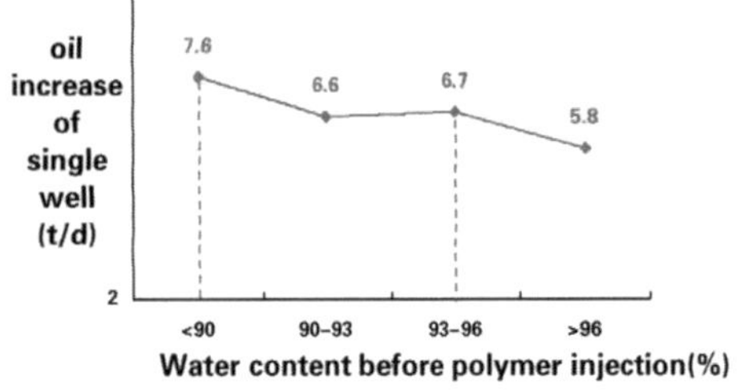

Figure 2 Oil enhancement curves for fractured wells with different pre-injection water contents

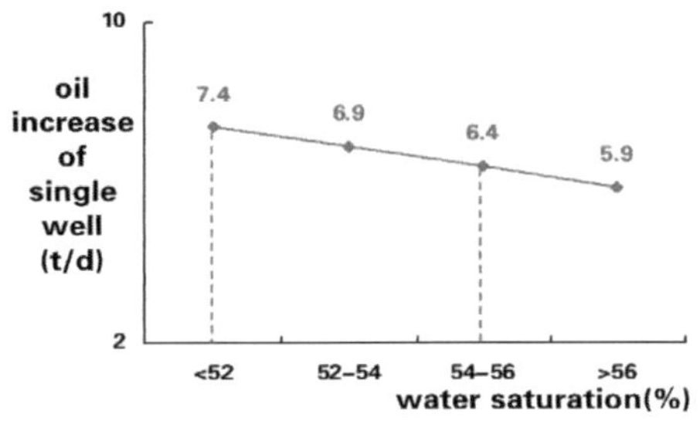

Figure 3 Oil enhancement curve for fractured wells with different water saturation

Table 2 Grading table for oil increase in fractured wells at different high water flood thickness ratios

High flooding Thickness Proportion (%)	Wells	Effective Thickness (m)	Water content Saturation (%)	Water content at different stages			Compare to pre-press	
				Poly injection Aqueous (%)	Before pressing water content (%)	After compression water content (%)	Daily liquid volume (t)	Oil production per day (t)
>40	26	9.7	56.17	93.59	86.07	84.75	35	5.7
20-40	44	6.6	55.59	92.40	85.34	83.25	40	7.1
<20	141	7.5	53.69	91.78	85.11	82.8	36	6.8
Total	211	7.6	54.39	92.15	85.3	83.15	37	6.7

According to the results of the above analysis, the extraction wells are located near the original water-driven injection well rows or wells in the mainstream line of water-driven oil and water wells with poor fracturing effect. According to the statistics of fracturing effect of Class II oil layer in recent years, there are 10 wells affected by the interference of water-driven wells, the water content saturation of these wells is higher than other well areas, the water content before fracturing is greater than 96%, the effect of fracturing to increase oil is not satisfactory, so when conducting fracturing selection avoid water-driven interference wells as far as possible.

Table 3 Comparison of the effectiveness of fractured wells in different areas

Fracturing Wells Location	wells	Effective Thickness (m)	Penetration rate (mD)	Water content Saturation (%)	Pre-pressure situation			Comparison with before and after fracturing		
					Before pressing water content (%)	Liquid production Strength (t/m)	Water content Drop (%)	Daily liquid volume (t)	Oil production per day (t)	Containing water (%)
Water-driven interference	10	9.4	269	55.35	89.04	2.61	7.33	39	3.8	0.7
Fault margin	81	7.8	315	53.41	84.53	3	6.33	34	7.3	-3.42
Injection mining perfection area	120	7.3	273	54.95	85.49	3.56	8.09	38	6.6	-1.7
Total	211	7.6	289	54.38	85.33	3.28	7.41	37	6.7	-2.18

2.1.2 Good fracturing effect during the water content decline period and water content stabilization period.

According to the analysis results of the oil increase effect of fractured wells in different periods in recent years, the oil increase effect of fracturing wells in the declining and stable water-bearing periods is significantly better than that of wells in the non-effective, poorly effective and rebounding water-bearing periods. Based on this experience, 38 wells were fractured in 2018, accounting for 60.32% of the wells fractured, 5.17% more compared to last year. A better oil enhancement effect was achieved.

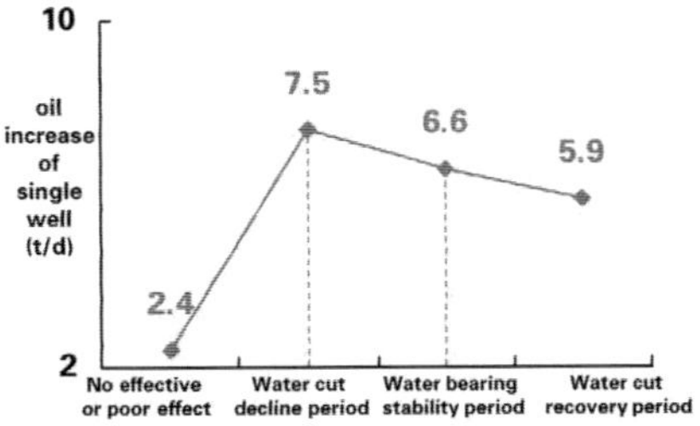

Figure 4 Oil increase curve with different fracturing timing

Table 4 Comparison of fracturing results of extracted wells 2017-2018

	Types	2017			2018			Differential	
		wells	Proportion (%)	Daily oil increase (t)	wells	Proportion (%)	Daily oil increase (t)	Proportion (%)	Daily oil increase (t)
	No effect Poor results	6	4.41	1.8	2	3.17	5	-1.24	3.3
	Declining Period	33	24.26	6.6	28	44.44	7.6	20.18	0.9
	Stabilization period	42	30.88	5.6	10	15.87	6.7	-15.01	1.1
	Rebound period	55	40.44	4.2	23	36.51	4.8	-3.93	0.6
Figure 4 Oil increase curve with different fracturing timing	Total	136	100	5.3	63	100	6.2	0	0.9

Through further analysis, the fracturing effect of the recovery wells with a high degree of effectiveness was good. From the oil increase effect of wells with different water content reduction rate before fracturing, when the water content reduction rate before fracturing exceeds 5%, the oil increase effect is better, and the higher the water content reduction rate, the greater the oil increase by fracturing the wells. Therefore, when selecting wells for fracturing, we should try to select wells with a water content decline and water content stabilization period, while the water content decline rate is greater than 5 percentage points for fracturing.

Table 5 Grading table for fracturing and adding oil in wells with different water content reduction rates

Before fracturing Water content reduction (%)	wells	wells Proportion (%)	Before fracturing				Before and after fracturing comparison		
			Daily liquid volume (t)	Daily oil production (t)	Contains water (%)	Water content Drop (%)	Daily liquid volume (t)	Daily oil production (t)	Containing water (%)
<5	68	32.55	27	3	88.95	1.14	32	5.5	-3.34
5-8	37	17.45	28	3.4	87.77	6.43	43	6.4	-1.6
8-11	40	18.87	25	3.7	85.32	9.44	39	6.5	-1.18
>11	66	31.13	21	4.4	78.44	14.94	37	8.3	-0.75
Total	211	100	25	3.6	85.33	7.41	37	6.7	-2.18

2.1.3 A high degree of control to fracture the recovered well.

According to the oil increase effect of fractured wells with different degrees of control, the oil increase effect of fracturing wells with a degree of control greater than 60% is higher than that of wells with a degree of control less than 60% at different stages, and the proportion of wells with oil increase greater than 6t is also higher, so fracturing wells with a degree of control greater than 60% should be selected for fracturing as far as possible.

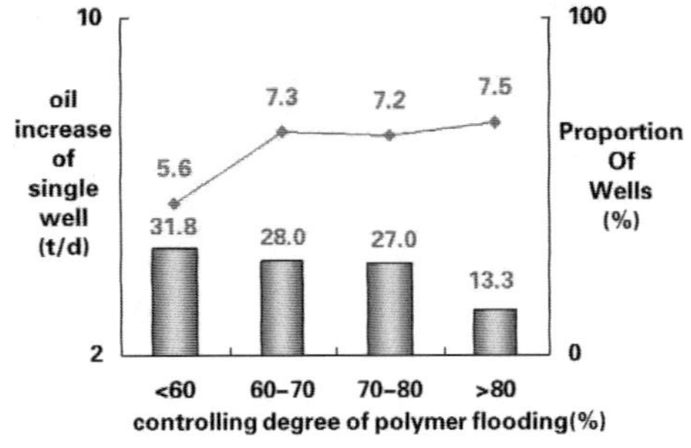

Figure 5 Oil increase curves for fractured wells with different levels of control

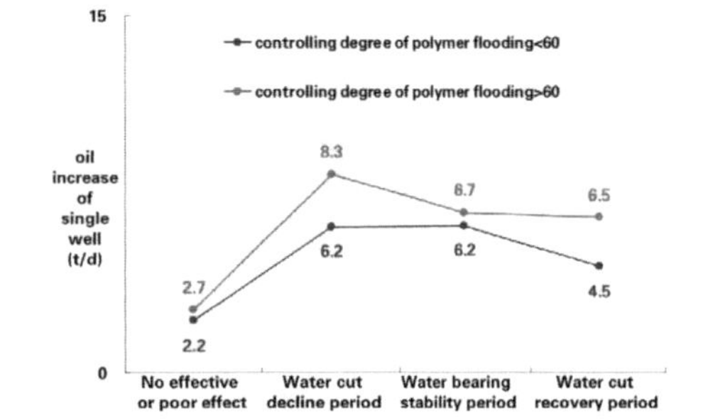

Figure 6 Oil increase curves for different stages of fracturing wells with different levels of control

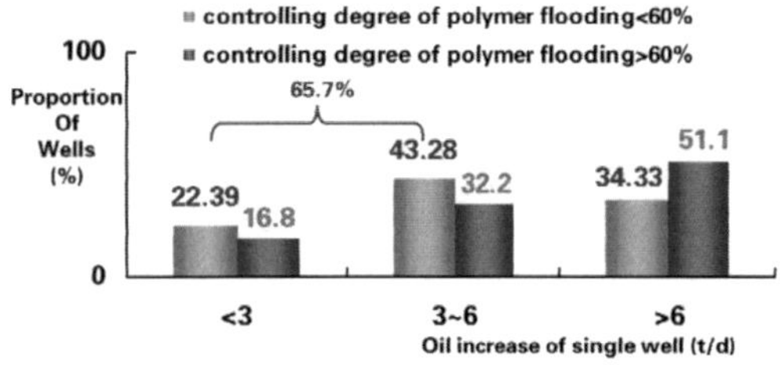

Figure 7 Histogram of the proportion of wells adding oil from fractured wells with different levels of control

2.1.4 Good fracturing of well sets with good well injection in connected wells.

Based on the analysis of fractured well results, among the fractured wells with similar recovery well see-through conditions, connection conditions and well relationships, the well group with good well zone injection conditions had good results. Take two fractured wells in Block D as an example.

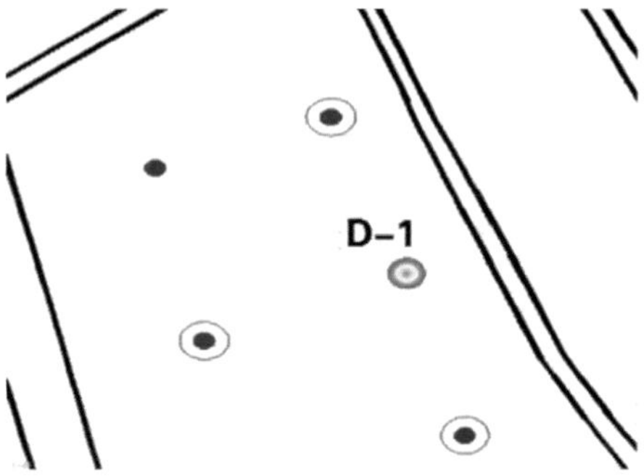

Figure 8 Well location map for well D-1

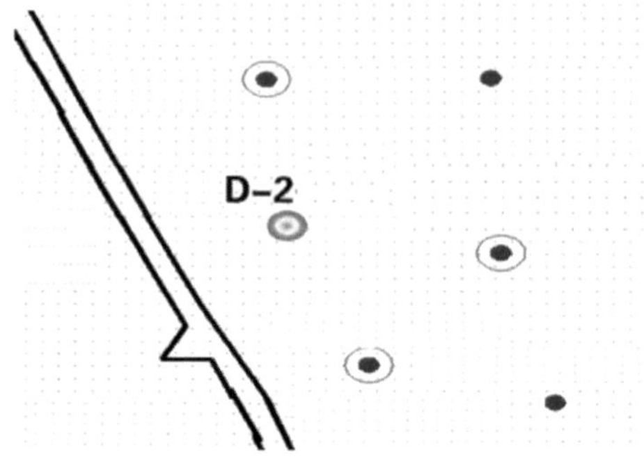

Figure 9 Well location map for well D-2

Table 6 Basic information table for different frac wells

Classification	Fracturing Effect	Extraction wells			Connected injection wells					Compare to pre-press	
		Effective Thickness (m)	Penetration rate (mD)	Poly Drive Control Degree of control (%)	Effective Thickness (m)	Penetration rate (mD)	Allowable differential pressure of injection (MPa)	Injecting Pressure (MPa)	Daily actual injection (m^3)	Daily liquid volume (t)	Daily oil production (t)
D-1well	Bad	8.4	248	79.3	8.7	327	1.42	8.61	130	10	4.4
D-2well	good	6.5	211	79.6	11.8	313	1.64	9.35	153	25	13.3
D Block		7.3	275	63.8	7.3	266					

Both wells are free from water-driven well interference. From the fracturing timing, D1 is fractured during the water-bearing decline period and D2 is fractured during the water-bearing stability period. In terms of effectiveness, both wells have shown good results. Based on the experience of the previous analysis, both wells are suitable for fracturing. However, poor injection in the well zone before D1 pressure, decreasing injection volume leads to a continuous decrease in fluid production and sinkage before pressure in the recovery well, and insufficient fluid supply in the well zone, resulting in a small increase in fluid after fracturing the recovery well. And the injection volume of the D2 well group was reduced at the beginning of the injection period due to the rapid rise in injection pressure, after which only one well in the well area had difficulty in injection after testing and recovered after well washing, and the pre-fracturing sinkage had an improving trend, proving that the injection volume was guaranteed and a better fluid and oil increase was achieved after fracturing.

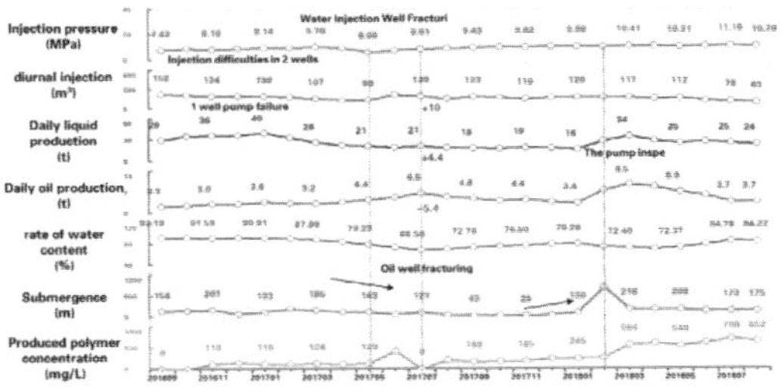

Figure 10 Production curve for well D-1

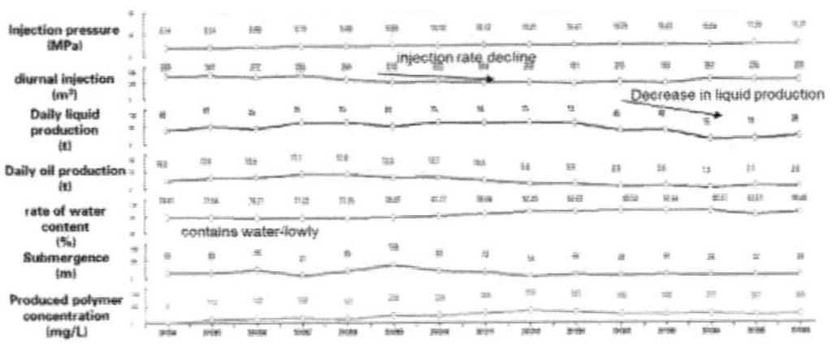

Figure 11 Production curve for well D-2

2.2 Practices to Safeguard the Effectiveness of Fracturing Extraction Wells

2.2.1 Pre-press cultivation to ensure fracturing of the recovery well.

It has been proven that for wells with fracture selection conditions, but poor well zone injection conditions, pre-fracture cultivation can be implemented to ensure that the fractured well increases fluid and oil. Take well A-1 as an example.

The A-1 well is an oil well located in the central area of Block A. The well is dominated by a thin differential layer development and the numerical model oil saturation results indicate that the well is relatively rich in remaining oil.

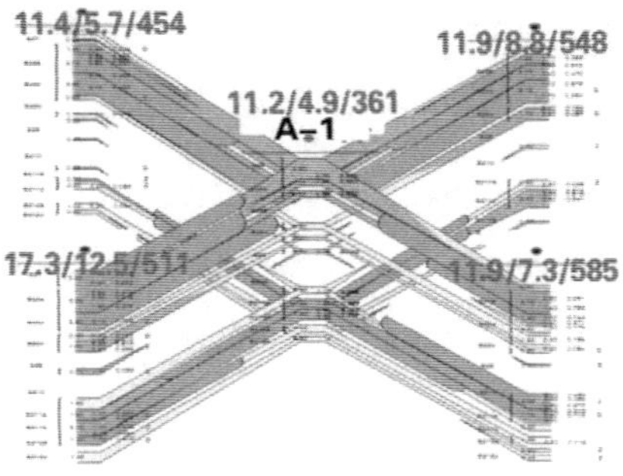

Figure 12 Censored map of the A-1 well group

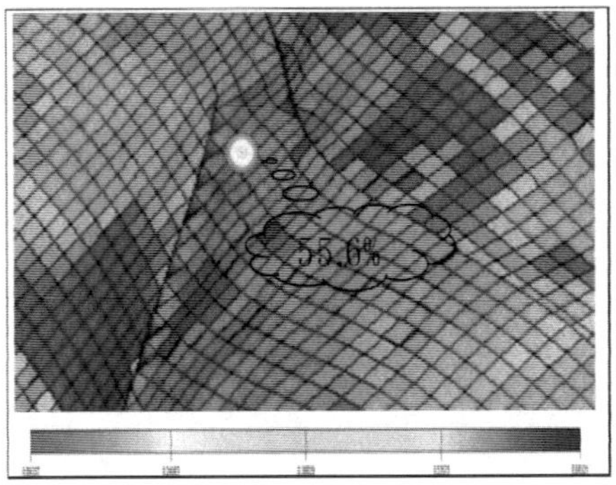

Figure 13 Numerical simulation of oil-bearing saturation in Block A

As can be seen from the production curve, the well has low initial water content and relatively enriched remaining oil. The significant drop in injection volume in the well area is mainly due to the implementation of dissection and controlled injection in three wells in the well area, injection difficulties in one well and operational reallocation, which resulted in a large drop in fluid volume in the recovered well.

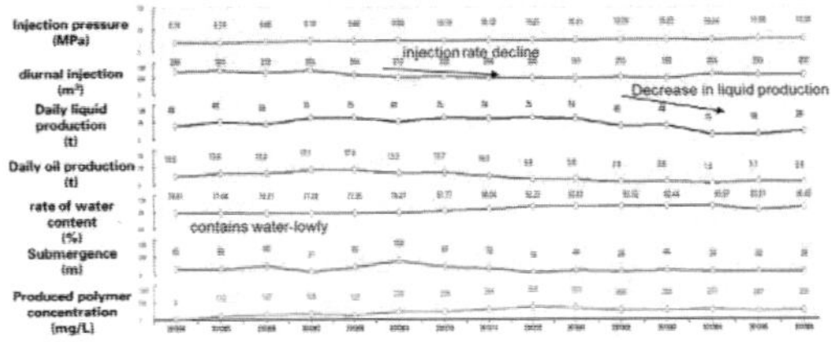

Figure 14 Pre-frac extraction curve for well A-1

In order to guarantee the fracturing effect, the injection wells in the well area were planned to implement programmed fluid withdrawal and fracturing to increase injection in the first five months of the measure, and after ensuring the fluid supply capacity, the fracturing of the extraction wells was implemented, and the daily oil increase reached 13.7t after the measure, and the water content dropped significantly.

Table 7 Comparison of well group A-1 before and after adjustment

Well number	Adjustment Type	Adjustment time	Before adjustment			After adjustment		
			Oil pressure (MPa)	Daily allotment (m^3)	Daily actual injection (m^3)	Oil pressure (MPa)	Daily allotment (m^3)	Daily actual injection (m^3)
A-w1	Fracturing and injection	2013.3	10	80	51	7.31	80	80
A-w2	Program Extraction Solution	2013.6	11.3	60	60	11.9	80	79

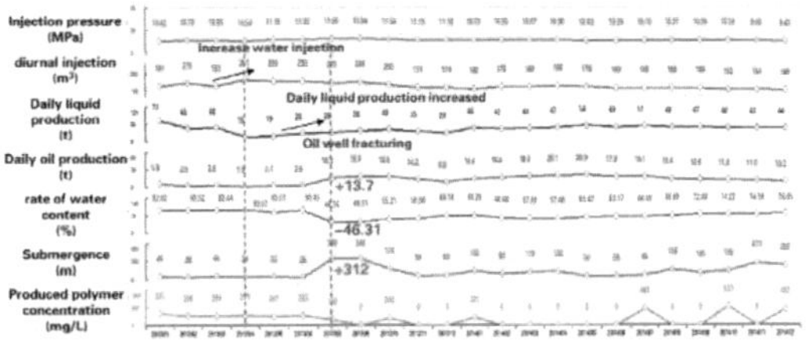

Figure 15 Extraction curve for well group A

2.2.2 Timely adjustment after fracturing to ensure the fracturing effect

After the fracturing program is implemented, the production pressure differential should be amplified in a timely manner according to the dynamic characteristics of the well area to safeguard the fracturing effect and extend the fracturing period as far as possible. Take well D-3 as an example.

The D-3 well is located at the edge of the fault adjacent to Block D and Block C. The well area is connected to a total of three injection wells, two of which belong to Block C and one to Block D. The well is mainly developed with medium and low water flooding, with a water content saturation of 48.5%, which is 7.33% lower than that of the whole area, and relatively rich in remaining oil.

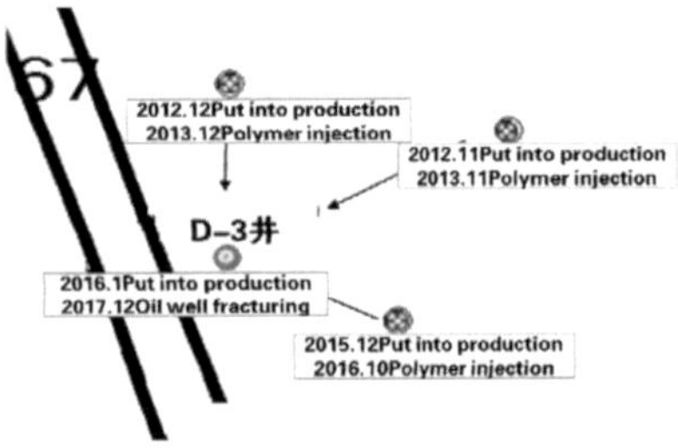

Figure 16 Well location map of the D-3 well group

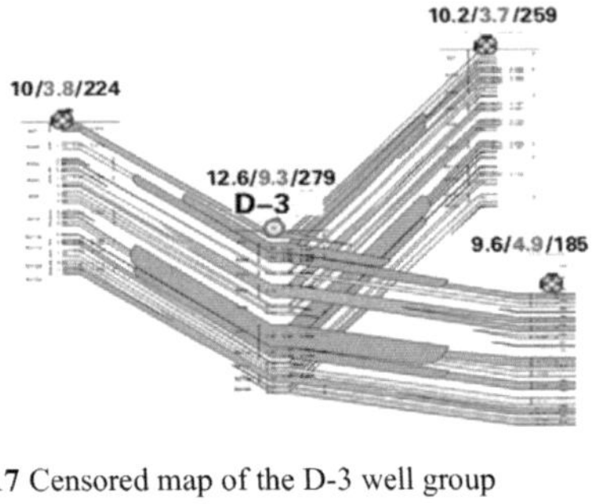

Figure 17 Censored map of the D-3 well group

Influenced by the earlier injection and gathering time in Block C, the well started to see results two months after it was put into production, with a large drop in water content and relatively rich residual oil. After the implementation of fracturing in December 2017, the initial daily fluid increase of 24t and daily oil increase of 14.7t were achieved with good oil increase effect, Two main problems existed at the stage from the production curve, one is that the sinkage has been at the wellhead after fracturing and the production capacity has not been effectively released, and the other is that the fracturing The injection capacity of the well area before fracturing decreased, and the fracturing validity period was difficult to guarantee.

In order to ensure the fracturing effect, firstly, in response to the actual high sinkage, timely implementation of the adjustment of the large participation to enlarge the production pressure difference, the initial period after the adjustment of the participation, the daily fluid increase of 22t, the daily oil increase of 12.6t, to achieve a better fluid and oil increase effect. Second, the timely implementation of fracturing transformation for injection wells in block D of the well area where injection has deteriorated, the injection volume was guaranteed after the measure, and the effect of fracturing the recovery well to increase oil lasted for 7 months, with a cumulative increase of 3,133t.

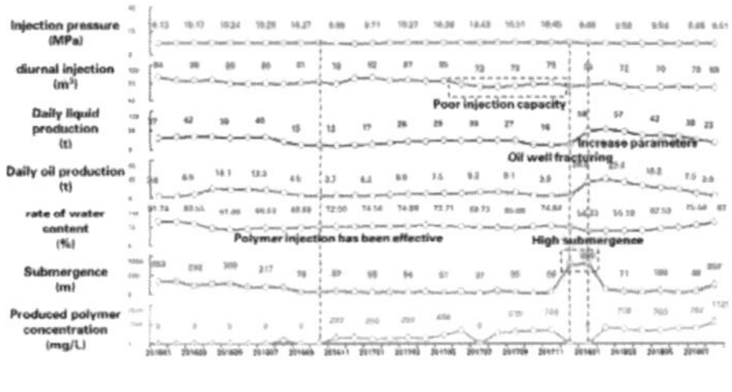

Figure 18 Extraction curve for the D-3 well group

2.2.3 Optimizing the fracturing process to guarantee fracturing results

According to the relationship curve between the fracture penetration ratio and the degree of recovery for Class II reservoirs, when the fracture penetration ratio reaches a certain value, the degree of recovery reaches its highest point, and when the penetration ratio is greater than that value, the degree of recovery begins to decline. This is due to the fact that when the penetration ratio is too high, it tends to cause the injected media to burst in, resulting in a rapid rise in water content. The penetration ratio is determined by the fracture radius. According to the formula for the relationship between sand addition and fracture radius, the amount of sand addition needs to be optimized to keep the fracture penetration ratio around 0.12-0.18 as much as possible to ensure the maximum fracturing effect, so when there is a high water absorption ratio layer in the injection well in the well area, the amount of sand addition should be controlled to prevent the penetration ratio from being too high and causing the injection media to break through.

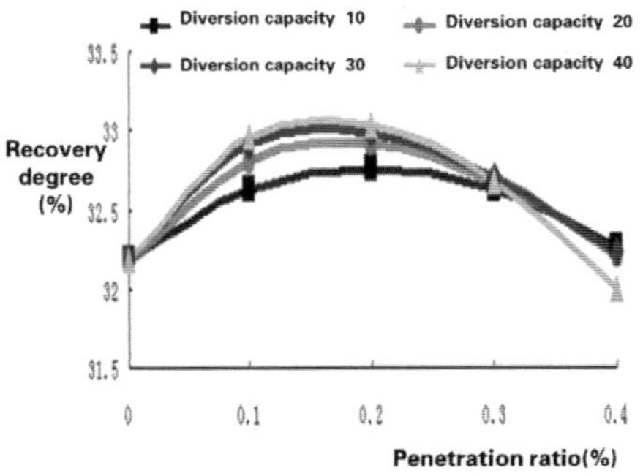

Figure 19 Curve of the effect of fracture penetration ratio on the degree of recovery

Calculation formula of penetration ratio :

$\lambda = r/L$ where: λ - Penetration ratio(%)
r- Crack radius(m)
L- Well distance(m)

The relationship between the crack radius and the amount of sand :

$r=0.0188\chi^2+1.7338\chi+11.114$

式中: r- Crack radius(m)
χ - Sanding amount(m^3)

Block	Well distance (m)	Long cracks (m)	Single layer sanding amount (m^3)	Penetration ratio
A-B	175	18-22	4-7	0.10-0.14
C-E	125	18-22	4-7	0.14-0.19

Figure 20 Equation for penetration ratio and fracture radius versus sand addition

3. A few points of awareness

3.1 The effect of fracturing a well from a Class II formation is highly uncertain and is usually not determined by one factor, but rather by the co-existence of several factors that affect the fracturing effect.

3.2 Based on past experience with fracturing to increase oil, select wells with characteristics such as "rich residual oil, water content in the declining and rebounding phases with a high degree of effectiveness, and strong injection capacity in the well area" for fracturing as far as possible to increase the probability of producing fractured high-efficiency wells.

3.3 According to the well area injection and extraction conditions, adopt pre-pressing cultivation, timely and effective post-pressing tracking and adjustment and optimization of the fracturing process, which can guarantee the fracturing effect of the recovery well.

References

1. Zhou Dehua, Jiao Fangzheng, Ge Jiali. Recent progress in fracture seepage research[J]. Offshore Petroleum, 2004(6)
2. Li Jingcai, Li Lingyun, Yang Shengxian. Establishment and Application of Economic Boundary Plates for Oil Well Fracturing Process Measures [J]. Special Oil and Gas Reservoirs, 2005(6)
3. Sun Fuguang. Analysis of fracturing effect and optimization of process plan for the second-class oil reservoir in Sanan Development Zone. Chemical Engineering and Equipment, 2005(9)

Research and Analysis of Perforating Cable Fishing Technology

Liping Qu

Market development center of oil testing and production branch, China
quliping@petrochina.com.cn

Abstract. Perforation gun and cable are stuck underground, which often occurs in cable perforation. Whether the perforation cable gets stuck properly or not will not only affect the cost of accident treatment, sometimes even affect the success or failure of a well, and sometimes delay the best time of other well layout; Therefore, the unblocking and fishing technology of perforating gun is an important part of perforating operation. The fishing operation of perforating gun sticking can be divided into two categories: one is the fishing of cable, the other is the fishing of perforating gun body. The shape of fish falls into rope and tube. The stuck fishing work must accurately grasp: well condition data, fish falling situation, fishing tool combination, fish head situation, etc. Combined with the process of accident treatment, this paper studies and analyzes the fishing technology of perforating cable and perforating gun.

Keywords: Perforated cable, External hook fishing spear, Internal hook fishing spear, Overshot

1 Introduction

The development of society can not be separated from the support of energy. Therefore, oil production has been widely concerned. The problem of falling objects in wells is a common problem that puzzles oil production. The concept of downhole falling objects in oil fields is broad, mainly referring to a variety of tangible materials falling from oil production wellhead, pipe string, logging and other devices. According to the shape, they can be divided into pipe well falling objects, rope well falling objects, rod well falling objects, etc. These wells must be cleaned up in time, otherwise once the oil layer is blocked, the threat to the entire oil well will be fatal. There are two main methods for cleaning the falling objects in the well. One is to destroy it in the well, but the more extensive method is to use the technique of downhole fishing.

2 The basic strategy of rope salvage

2.1 External hook fishing of movable teeth

The pipe string combination from bottom to top is: movable tooth external hook+adjusting nipple (3 ~ 4m)+hemispherical baffle (diameter less than 2 ~ 4mm of casing inner diameter)+pipe string.

Operation method: when the drilling tool is lowered to 10m above the top of the fish, rotate at the bottom. When the tension meter shows that the hanging weight of the drilling tool decreases, continue to apply force for rotation, and then lift it up to observe the change of the hanging weight and judge the fishing effect.

If the ropes in the well have become a group, the outer hook can be moved when the outer hook cannot be inserted. Because when this kind of external hook is blocked, the movable hook retracts into the hook body, so that the hook body is easily inserted into the rope-like falling mass. When lifting the pipe string, the movable hook is opened by the elastic force of the spring to hang the rope falling objects.

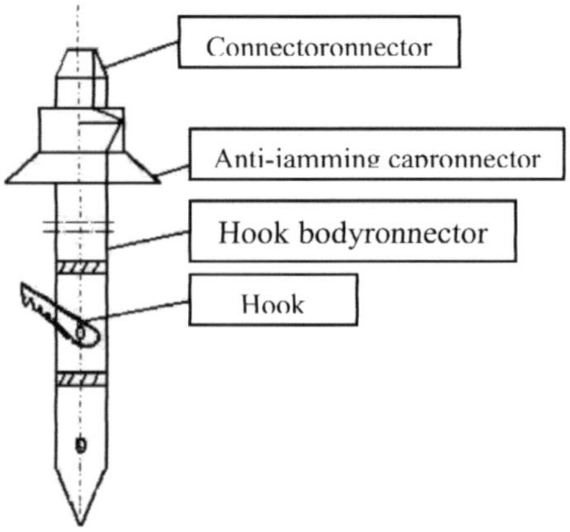

Fig. 1. The elastic force of the spring to hang the rope falling objects.

* Corresponding author: quliping@petrochina.com.cn

2.2 Double arm internal hook fishing

The combination of the pipe string from bottom to top is: double-arm outer hook + adjusting sub (3 ~ 4m) + hemispherical baffle (diameter less than the inner diameter of the casing 2 ~ 4mm) + pipe string.

Operation method: Lower the drilling tool slowly when it is lowered to 10m above the fish top. When the hanging weight drop is observed from the weight indicator, use the methods of lifting and lowering, rotating and changing the direction for fishing. Requirements: anti falling device shall be installed at the wellhead to prevent falling objects under the well; The downhole tools shall be identified by a specially assigned person to ensure that they are solid and reliable.

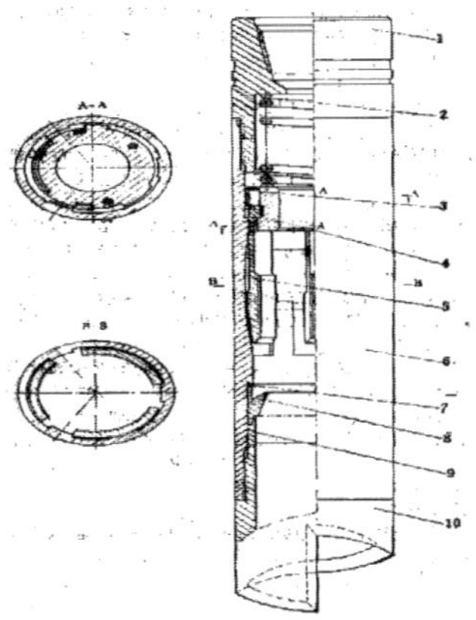

Fig. 2. The downhole tools shall be identified by a specially assigned person to ensure that they are solid and reliable

3 Basic fishing strategy of perforating gun body

3.1 Buckling barrel:

It can not only be used for fishing and make-up, but also release fish, but also circulate well washing fluid. In the fishing operation, the inverted fishing drum is one of the important supporting tools of the inverted tool, and can also be used together with the inverted drill pipe.

Principle of action:(1) Lower the tool to the top of the fish drop, rotate the tool, and introduce the fish drop. When the slip with an inner diameter slightly smaller than the outer diameter of the fish falls into contact with the fish, the cylinder begins to slide down relative to the slip, the spring is compressed, and the limit sits on the lower end face of the upper joint, forcing the slip to expand outwards, and the fish enters the slip. Because of the spring force, the teeth of the slip always cling to the outer surface of the fish. If the drilling tool is lifted up after the lowering is stopped, the cylinder moves upward relative to the slip. As the slip fits with the cone of the cylinder, the clamping force of the three slips on the falling fish increases with the increase of the lifting force; The triangular teeth eat into the outer wall of the falling fish. Continue to raise, and salvage can be realized.

(2)If torque is applied to the drill pipe at this time, it will be transmitted to the slips and the fish through the key on the barrel, so that the fish joint will be loosened, that is, the undercut will be realized.

(3)If fishing and back-off fail, if you want to quit fishing and retrieve the tool, you can relax and knock down the drilling tool to make the slip separate from the cone surface of the cylinder, then turn right, and the inner chamfer at the bottom of the slip enters the included angle of the inward inclined plane, locking the slip on the cylinder and rotating with the cylinder. You can exit the fish by lifting the drill string.

3.2 Retractable slip overshot:

Structure of slip overshot: slip overshot is basically composed of outer barrel and internal components. The same outer cylinder can be used with slips of different sizes to operate fish of different sizes. The outer cylinder is mainly composed of an upper joint, a main body and a shoe guide. The spiral slip is composed of packing, spiral slip and control ring; the basket slip is mainly composed of basket slip and milling shoes.

Working principle: the fishing part of the overshot is screw slip or basket slip. Its external wide serrated thread and internal fishing tooth are left-hand threads. The clearance between the wide sawtooth thread and the cylinder is large, which makes the slip expand and shrink within a certain stroke in the cylinder. As long as an axial pressure is applied, the falling fish can enter the slip. With the insertion of the falling fish, the slip rises and expands. Under the action of elasticity, the slip teeth bite the falling fish. When the drill string is lifted up, the slip moves downward in the barrel, the diameter is reduced, and the falling fish is caught more firmly. Because the threads of the cylinder and the slip are left-handed threads, and their rotation is restricted by the control ring, when releasing the fish, just loosen the slip and rotate the drilling tool clockwise, and the fishing cylinder can be withdrawn from the fish.

Salvage steps:

(1)Circulating fish head: record the circulating pressure of low pump stroke (30 ~ 50 SPM) before fishing. It is estimated that about 0.5 meters above the fish top, the fish top will be washed with large displacement. Flush the fish head with a fixed low pump. When the suspension weight becomes smaller, immediately stop the brake lever. Record the depth of the well when it encounters resistance or when the pump pressure rises, that is, the position of the fish top.

(2)Fishing: slowly lower the drilling tool, pressurize 4.9 to 49 kn, put the fish head into the barrel, confirm the insertion depth, and then upper the drilling tool. If the suspended weight increases, it is proved to catch the falling fish.

(3)Treatment after fishing: After fishing, the fish can be lifted from the bottom of the well by 0.5 to 0.8 meters, and braked twice or three times. If the hanging weight does not drop, it is proved that the fish has been caught before tripping out. If not, the above operations can be repeated.

(4)Tripping out: after successful fishing, due to the control of tripping out speed, it is better to use hydraulic tong to break out. It is strictly prohibited to rotate the rotary table. When the fish is pulled out of the wellhead, clean the overshot with clean water. After the fish falls out of the well, try to avoid dismantling fishing tools at the wellhead to prevent the fish from falling into the well again.

4 Perforating cable and perforating gun fishing examples and fishing technology research analysis

When perforating well zq-52 in 2016, the lifting cable was stuck when the construction was carried out to the ninth shot and the depth was 2574m. Party A decided to use the bottom cutter string to go into the well to cut off the cable lower bottom cutter string. It can't continue to descend down to 2037 meters, and the cable is broken when it is lifted up.

Penetration point of perforating vehicle: 2037m. 537m cable, 4m 102 gun and 3M 102 empty gun are left in the well.

a. The first salvage pipe string (movable tooth fishing spear): down to the 217th, depth: 2084.49m, the hanging weight of the pipe string is reduced from 13t to 8t, the string is rotated 13 times, and the pipe is lifted to the 212th. The weight rose from 13t to 17t, returning to normal tonnage instantly.

Pull out all fishing strings and pull out cable 40m.

b. The second salvage pipe string (live-tooth fishing spear): down the Φ73mm EUE tubing down to the 221st, depth: 2122.76 meters, the hanging weight of the pipe string is reduced from 13 tons to 3 tons (big hook weight) and rotate 15 times forward, slowly Lifting, lifting to the 213th, the hanging weight of the pipe string rose from 13 tons to 15 tons, after 15 minutes of maintaining the hanging weight of 15 tons, slowly raising to 17 tons, the hanging weight returned to normal (13t).

Lift the fishing string and fish out the cable for 50m.

c. The third salvage pipe string (outer hook fishing spear-73mm tubing welding system): down Φ73mm EUE tubing down to the 223rd, depth: 2149.84 meters, the hanging weight of the pipe string is reduced from 19 tons to 10 tons (big hook weight) Spin 20 times, slowly lift up to the 215th, the suspended weight of the pipe string rises from 19 tons to 23 tons, and slowly rises to 23 tons, and the suspended weight returns to normal (19t).

Lift the fishing string and fish out the cable for 20m.

d. Lower the fourth fishing string (external hook Spear - 73mm oil pipe welding): Φ 73mm eue tubing runs down to the 222nd, depth: 2140.28m, string hanging weight is reduced from 19T to 15t, positive rotation is 10 turns; Slowly lower the pipe string, and the hanging weight of the pipe string is reduced from 15 tons to 10 tons, with positive rotation of 20 turns. Slowly lower to the No.223rd pipe, with a depth of 2,149.84m, and the hanging weight of the pipe string is reduced from 10 tons to 7 tons and rotated for 10 times; Slowly lift it up to the No.215th pipe, and the hanging weight of the pipe string rises from 19 tons to 23 tons. Continue to slowly lift it up, and the hanging weight returns to normal.

Lift the fishing string and fish out the cable for 30m.

e. The fifth trip to salvage the pipe string (live-tooth fishing spear): down the Φ73mm EUE tubing down to the 225th, depth: 2168.98 meters, the hanging weight of the pipe string slowly decreases from 19 tons to 3 tons (hook hanging weight). Lift and lower the movable pipe string, down to the 227th tubing, with a depth of 2180.16 meters. The pipe string cannot continue to descend. Slowly lift the pipe string. The suspended weight is increased from 3 tons to 10 tons and rotated 15 times. No change in tonnage (normal hanging weight 19 tons), start the pipe.

Lift the fishing string and fish out the cable for 12m.

f. Lower the sixth fishing string (external hook fishing spear - 38mm oil well pump welding): Φ 73mm EUE tubing was run down to the 227th, and the string hanging weight was slowly reduced from 19T to 16t, with 15 turns of forward rotation; The suspended weight of the pipe string is slowly reduced from 16t to 12t, and the forward rotation is 15 turns; The hanging weight of the pipe string is slowly reduced from 12 tons to 9 tons, and the forward rotation is 15 turns. Slowly lift up the pipe string, and there is no change in tonnage (normal suspended weight of 19 tons), and lift the pipe.

Lift the fishing string and fish out the cable 42m.

g. Lower the seventh fishing string (external hook fishing spear - 38mm oil well pump welding): Φ 73mm eue oil pipe is lowered to the 230th (square entry: 0 m), the hanging weight of the pipe string is slowly reduced from 19 tons to 12 tons, and the forward rotation is 15 turns. The depth is 2200.61 M. The hanging weight of the pipe string decreased slowly from 12 tons to 8 tons (square in 2 meters) and rotated 15 times with a depth of 2,202.61 meters; Slowly lift the pipe string to 227, and after the lifting tonnage (normal hanging weight of 19 tons) rises to 23t, return to normal tonnage (19t), and pull out the pipe.

Lift the salvage pipe string, the hanging weight of the pipe string suddenly rises to 35t at the start and stop 150, and then returns to the normal tonnage of 19t. The pipe string is lifted out, and the outer hook fishing spear (38mm oil pump welding) is pulled off (5.5m tools are left in the well) .

h. Run the fishing string (movable tooth spear) on the eighth trip: lower Φ 73mm eue tubing was run down to the 150th, and the string hanging weight decreased slowly from 10t (normal hanging weight) to 2T, with 15 turns of forward rotation; Slowly lift up the string, restore the normal tonnage (19T) and start the pipe.

Lift the fishing string and fish out the cable for 30m.

The lead mold of the perforating car: the length of 1497.5 meters from the platform.

i. Lower the ninth fishing string (inner hook spear): Φ 73mm eue oil pipe down to the 162nd, depth: 1554.16m. The tonnage of pipe string is reduced from 15t to 14T; Lift the hanging weight of the string to 16t; Pipe lifting.

Trip out the fishing string and fish out the cable for about 343m (the perforation marks are marked, and the perforation personnel judge that it is broken from the cable head).

j. The next tenth salvage pipe string (windowed overshot): Lower the Φ73mm EUE tubing down to the 268th, and the hanging weight of the pipe string decreases slowly from 22 tons (normal hanging weight) to 20 tons and rotates forward for 5 times, and slowly raises the highest tonnage. After rising to 30t, the normal tonnage (30t) will be restored, and the pipe will be started.

The fishing string was lifted and no tools were fished out. Punch the lead mold with the perforating car, and the depth is 2570m. The fish top is shown as a 38mm pump cylinder.

k. Run the fishing string on the 11th trip (reverse fishing barrel): lower Φ 73mm eue tubing is lowered to the 268th, with a depth of 2572.16m. The string hanging weight is slowly reduced from 22T (normal hanging weight) to 21t, and the string is rotated forward for 2 turns; Slowly lower the string to 12t and suddenly rise to 16t (fish jacking tool); Continue to slowly lower the string to 4T; Slowly lift up the string, go up 4m, the tonnage rises to 38t, and suddenly falls back to 22T (normal tonnage); Slowly lower the pipe string for 1m, and if there is resistance indication, start the pipe.

Fishing string: pull out 38mm pump spear tool.

l. run the 12th fishing string (retractable slip overshot): Lower the Φ73mm EUE tubing down to the 269th, depth: 2574m, the suspended weight of the tubing string decreased slowly from 22 tons (normal suspended weight) to 4t; slowly lift the tubing string, no tonnage change; continue to slowly lower the tubing string to 12t and suddenly rise to 16t (fish top into the fishing bucket); continue to slowly lower the pipe string to 4t; slowly lift the pipe string, the hanging weight of the pipe string rises to 40t, suddenly drops to 22t (normal tonnage), and the pipe is started.

Fishing string: successfully fished out the perforation gun body.

Research and Analysis on fishing technology for the fishing process of the well:

(1) The upper part of all fishing tools into the well is equipped with a Φ118mm spacer ring to prevent the cable from running up.

(2) The length of the movable tooth fishing spear is 1.75m, the cable is inserted into the positive rotation, the pipe string of the pressurized part is suspended, and a small amount of cable is gathered, so the efficiency value of the movable tooth fishing spear is reduced.

(3) The outer diameter of the outer hook fishing spear (Φ73mm tubing welding) is too large compared to the movable tooth fishing spear, and the gap between the spear and the casing is narrow, which is not conducive to cable gathering and cable fishing.

(4) External hook fishing spear (welded by 38mm oil well pump) Φ 38mm oil well pump welded, anchor claw Φ 19mm sucker rod, oil well pump and sucker rod toughness fishing cable are completely OK. However, when the sucker rod anchor claw is planted and welded on the oil well pump body, there will be a thin weakness. When the cable is wound on the upper part of the fishing spear, the lifting of the pipe string will produce a clutching phenomenon, the hanging weight of the pipe string will increase significantly, and then it will be pulled off from the weak point of the fishing spear.

(5) Window-opening fishing barrel is suitable for fishing irregular fish, and the broken fishing spear has an anchor claw, which is not conducive to opening the window and completely entering the fish; The window fishing barrel is too thin to apply heavy load.

(6) The inverted bailer has different application ranges, and the slip in the bailer used in this well is suitable for Φ For the fishing of 73mm oil pipe, 21mm thick tooth block is welded on the slip through processing, which is completely suitable for the fishing of external hook fishing spear (welded by 38mm oil well pump).

(7) When the fishing barrel approaches the fish top, it is best to rotate slowly so that the fish top can completely enter the fishing barrel; Increasing the tonnage may lead to the destruction of the top of the fish, making it impossible to salvage in the next step.

5 Conclusions

(1) Scientific analysis and correct fishing ideas are very important for fishing with multiple falling objects.

(2) For the fishing of flexible falling objects such as cables, the position of fish top shall be judged as far as possible to prevent excessive lowering and compaction of cables or ineffective construction.

(3) For objects with different outer diameters in the well, fishing tools of corresponding sizes should be selected.

(4) According to the actual situation, the effective selection of corresponding technologies and methods can effectively save time in the construction process and successfully fish the falling objects. Whether it is pipe or rope falling into the well, a good combination of methods can always salvage it effectively.

References

1. Ma Manhong. Research on the salvage technology of falling objects in oilfield wells[J]. Chemical Management, 2013(10): 114-114.
2. Fan Longwen. Discussion on the technology of salvaging complex falling objects underground[J]. Henan Science and Technology. 2013(5)
3. Wu Qi. Handbook for Downhole Operations Engineers Beijing: Petroleum Industry Press. 2002:215-253
4. Ren Fade. Discussion on the method of salvaging falling objects in oilfield downhole operations [J]. Petrochemical Technology. 2016, 23 (8): 260

Data Model and Application of Pipeline Integrity Management

Bingtao Dai

Langfang CNPC Lang Wei Engineering Project Management Co., Ltd., Langfang, Hebei, 065000 China

Abstract. Data is an important foundation and premise to ensure pipeline integrity management. Using data model to manage and utilize data is the key to implement pipeline integrity management, which can not only ensure the safety of oil and gas transportation, but also promote the stable development of China's social economy. Based on this, this paper deeply analyzes the development status of oil and gas pipeline integrity management data model, and makes an in-depth exploration on the establishment and application of oil and gas pipeline integrity data model by comparing various data models, combined with the current situation of oil and gas pipeline integrity management, and with the help of the advantages of apdm model, such as data set division and spatial information management.

Keywords: Oil and gas pipeline, Integrity management data model, Establishment process.

1 Introduction

For a long time, oil and gas pipelines have undertaken important energy transportation tasks. However, because oil and gas products belong to high-risk transportation media, and the geological conditions of pipeline route areas are complex, and it is difficult to maintain in some areas of the route, the safety of oil and gas pipelines has become one of the topics focused on by countries all over the world. At the same time, oil and gas pipeline engineering involves many disciplines, and has the characteristics of many pipeline basic data, complex information types and sources, and dynamic real-time update of operation information. Therefore, the rationality and accuracy of data processing mechanism can directly affect the subsequent identification and evaluation results. In this situation, it is necessary to build a more scientific and reasonable data model with the help of the pipeline integrity management concept and implementation criteria proposed by API and ASME. So as to further strengthen the data management and utilization of oil and gas pipelines and ensure the safe operation of oil and gas pipelines.

2 Development status of data model of pipeline integrity management

The certainty of data model can provide guarantee for data sharing and information integration. For oil and gas pipelines, the composition of pipeline system is complex (see Figure 1), and there are many kinds of data information involved. Including specific pipeline facilities, operation and maintenance, internal and external inspection, hydrogeology and spatial data, etc. Therefore, it is necessary to establish a data model with reasonable construction, strong pertinence and wide application range to manage and display the data. At present, the existing pipeline models mainly include ISAT, Smallworld, ISPDM, PODS, APDM, etc. Among them, Integrated Spatial Analysis Technology (ISAT) was an earlier oil and gas pipeline data model in 1977, which specifically includes pipeline facilities, crossing and related pipeline events, and has been basically replaced by PODS. The Smallworld data model integrates applications based on the Long-distance Pipeline Management Solution (GTO). The GTO has also become part of the Smallworld software solution. Industrial Standardized Pipeline Data Management (ISPDM) is designed according to the relevant standards of the European pipeline industry sector, not only integrating data warehouse and Web technology. At the same time, multiple modules such as inspection, maintenance process and risk are also introduced, which can provide users with the function of three-dimensional modeling [1].

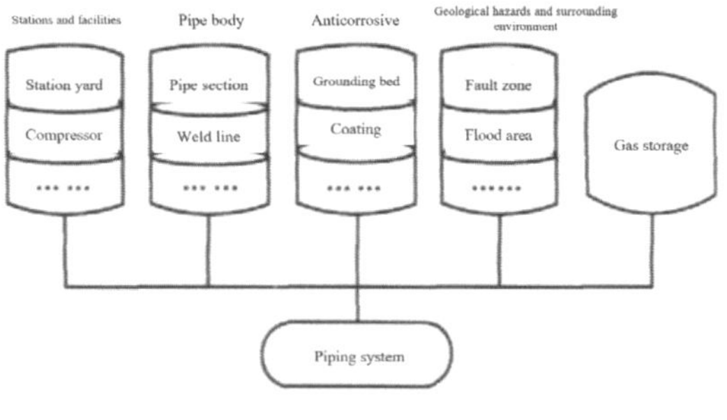

Figure 1 Composition of piping system

https://doi.org/10.1051/e3sconf/202132901036

Pipeline Open Data Standard (PODS), which covers all pipeline management services, is a relational data model specially designed for the application management of oil and gas pipeline technology system. ISAT is the main design source of this model, which can provide a completely open database modeling platform for society. At the same time, in the process of the gradual development of pods model, the design of database standardization, pipeline business expansion and geographic information application has also been further improved, and the connectivity and robustness of the system have been significantly enhanced, which has been recognized and applied by the majority of pipeline operators. For example, in pods6.0, the model has covered 30 classification modules such as environmental information, pipeline repair, internal and external detection and geographical entities, including more than 700 data tables and more than 5800 data items.

ArcGIS pipeline data model (apdm) is a pipeline data model developed on the basis of Geodatabase technology of ESRI company. Among them, apdm5.0 model specifically covers 9 data classification sets, such as cathodic protection, centerline and hierarchy, instrument reading, equipment, operation, integrity and event support [2], which can greatly enhance the flexibility of data template operation. At the same time, operating enterprises can flexibly formulate and implement models according to specific business requirements, and can also improve the models by modifying and adding existing elements, so as to realize the interoperability between database management and pipeline business as soon as possible. In addition, the advantages of APDM are mainly reflected in the process of connecting pipe spatial data with attribute data, and it also supports the 3D analysis function related to GIS in specific applications. At present, there have been a large number of successful application cases of APDM at home and abroad, and this model has been recognized by the pipeline industry. This paper will focus on analyzing and researching the integrity data management model based on APDM.

3 Establishment of data model of pipeline integrity management

At present, the data set of oil and gas pipeline integrity data model is very complex, mainly including centerline and hierarchy, facilities along the line, basic geography, operation, pressure, cathodic protection, crossing and third-party damage, integrity detection and evaluation, identification and risk evaluation of high consequence areas, efficiency evaluation, etc. In total, there are more than 190 items in the pipeline data sheet, including more than 60 items referencing and modifying the original apdm structure sheet and more than 130 items creating new sheets. Therefore, this model not only completely retains the core classes of apdm, but also expands and modifies the optional classes. According to the specific requirements of integrity management, it focuses on the construction of relevant data structures such as occupation, high consequence area identification and risk evaluation, cross and third-party destruction, integrity detection, evaluation and efficiency evaluation [3].

Affected by many factors, such as poor natural environment, unreasonable urban planning, third-party damage, complex interleaving and parallel of municipal facilities and pipelines, oil and gas pipeline failure accidents occur frequently, and most of them occur in economically developed and densely populated areas. The data set of "Occupy, Intersection and Third Party Destruction" in the oil and gas integrity data model is based on the infringement data in APDM, which adds records of many common pipeline hazards such as occupy, illegal drilling and excavation destruction, and further expands the classification of municipal facilities, such as fire-fighting pipelines, sewage pipelines, cable optical fibers and railways, etc., constructs a global attribute domain, and clearly marks the elements of municipal facilities lines.

ASME and API have made clear provisions on the relevant requirements for the original data of potential influencing factors in high consequence areas. Among them, pipeline risk management (see Figure 2 for the specific process) and evaluation, as a key link of oil and gas pipeline integrity management, should comprehensively analyze and consider the identification of risk sources, selection of evaluation models, rules for pipe section division, failure consequences, risk factor tree level and other factors. At the same time, it is also necessary to carry out detailed analysis and improvement on the high-consequence areas, sections of high-consequence areas, the scope of high-consequence areas and risks in operation categories in APDM supervision category, refine the classification indicators of identification standards, and divide the main risk factors into design and construction, corrosion, excavation damage, operation and maintenance, deliberate damage and natural disasters, etc., so as to provide comprehensive and reliable basis for risk segmentation, such as soil type, mileage, regional grade, anticorrosive coating and station.

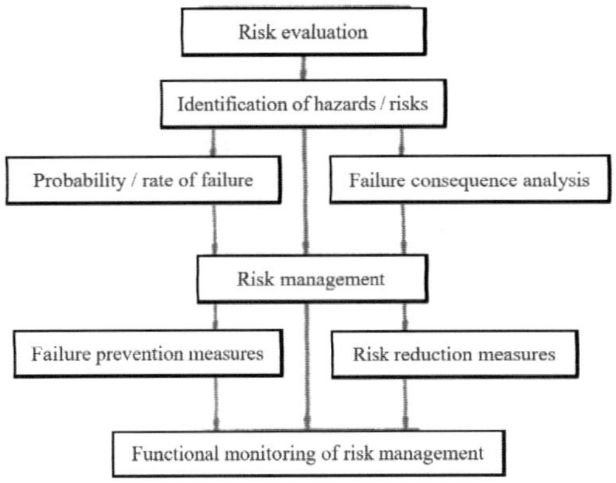

Figure 2 Risk management process

At this stage, the most intuitive way to distinguish and analyze pipeline defects is pipeline inspection. At the same time, the accuracy of pipeline inspection data directly affects the credibility and quantification of the integrity evaluation, providing a valuable reference for the maintenance of oil and gas pipelines. For APDM, the detection categories specifically include anomalies, closely spaced potential measurement readings, anomaly clusters and detection ranges, but there is no clear division of defect types [4]. Therefore, it is necessary to build the "Integrity Testing and Evaluation" data set, refine the internal testing data, classify and mark the pipe body dents, manufacturing defects, metal losses and weld defects (longitudinal, circumferential and spiral), increase various evaluation information such as corrosion stress analysis, defect analysis and durability analysis, and also include the input of the results of external testing methods such as DCVG technology of corrosion protection layer damage detection and ground magnetic leakage.

In addition, as an important hub for the development, storage and transportation of offshore oil and gas fields, the operating environment of submarine pipelines is more complicated, and the detection and maintenance is more difficult, and the requirements for environmental data in integrity management are more stringent, requiring detailed data records for tidal level period, geological sampling-static exploration coordinates and other aspects [5]. In the application, the design objects of apdm are only aboveground pipeline system and buried pipeline system, so there is no in-depth thematic expansion for submarine pipeline and offshore facilities. Therefore, it is necessary to establish a submarine pipeline data set to objectively and accurately predict the specific trend of soil foundation evolution by collecting relevant environmental information, such as tide level, water depth, ocean current, soil mechanics index, sediment concentration and seabed topographic coordinates. The requirements for facilities along the submarine pipeline and cathodic protection The point elements on the center line are mainly located by mileage, the offline point elements are mainly located by the conversion of GPS coordinates (x.y, z) or the distance from the nearest reference target, while the line elements are mainly It locates through the interval of known mileage [7]. The facilities along the pipeline include valves, girth welds, fixed piers, anti-corrosion coatings and hydraulic protection, etc., as well as various activities.

4 Implementation and application of data model for oil and gas pipeline integrity management

Geodatabase spatial database lays a good modeling foundation for lidm model, and still uses the behavior and relationship in apdm abstract classes. As the object class, relation class and feature class in APDM, the core class can clearly point out the expression rules of elements along the oil and gas pipeline model and centerline features with the help of unified coordinate system types and reference attributes [6].

Station location, three pipelines and PS datum point are the core elements of the model. In specific applications, GIS focuses on the absolute positioning of pipeline positioning points, and other relative positioning points can be easily calculated after the station values are clear and quantified. Columns are represented by line feature classes, and there are three sub-classifications (subtypes), including level, mileage pile and continuity. The central point elements are located mainly by mileage, while the offline point elements are located mainly by the conversion of GPS coordinates (x.y, z) or by the distance from the nearest reference target, while the line elements are located mainly by the known mileage interval [7]. Facilities along the pipeline specifically include various elements such as valves, girth welds, fixed piers, anticorrosive coatings and hydraulic protection, as well as various activities. For example, elements such as cathodic protection, line change, pipe change, monitoring and emergency can be superimposed on the center line of oil and gas pipeline, and a dynamic linear reference corresponding to the abstract relationship can be established. The level of pipeline events is mainly divided by the pipe network and subsystems. Among them, the pipe network is divided by creating geometric networks, and the subsystems are divided by thematic classification and plane topology. However, in the oil and gas pipeline integrity management data model, the identification attribute of any event is unique. Therefore, it should be clear that the feature class or object class does not depend on the classification. Therefore, it should be distinguished by main and foreign keys to enhance the relationship between feature class and object class.

The application of oil and gas pipeline integrity management data model to oil and gas pipeline integrity management has the following advantages: first, Geodatabase spatial database technology can realize the integrated management of pipeline construction, spatial data and operation data, and greatly improve the accuracy of data entry and editing through attribute verification and global attribute domain setting, Reduce operational errors. Secondly, through the expression of topological space, we can define the characteristics of environmental elements, event elements and pipeline elements, and at the same time, we can also define the relationship among the elements, so that the elements can make corresponding response according to the specific changes of the specified elements, and effectively realize the dynamic cascade of pipeline information [8]. Third, in the process of oil and gas pipeline integrity management, the relationship between data collection classification and level undertaking should be improved, so as to enhance the purpose of data collection, improve the rationality of evaluation process, and lay a good foundation for thematic data extraction, data warehousing and the construction of integrity data warehouse. Fourth, because the data model of oil and gas pipeline integrity management has strong expansibility, it can meet the long-term application needs of oil and gas pipeline integrity management.

5 Conclusions

To sum up, the integrity management of oil and gas pipelines is very complicated, with strong overall situation and high data quality requirements. Therefore, only by building a unified data model and formulating reasonable data collection standards can we ensure the smooth development of the integrity management of oil and gas pipelines. At the same time, in order to further improve the effectiveness and standardization of oil and gas pipeline data management, in addition to establishing lidm data model, it is also necessary to improve the existing pipeline data classification methods and collection forms, actively improve the construction methods of data set data structures such as infringement, identification, evaluation and submarine pipelines, and deeply explore the methods and methods of oil and gas pipeline integrity data management, So as to ensure the safe operation of oil and gas pipelines.

References

1. He Wang, Xie Liyun, Wei Ziyu. Evaluation of engineering geological suitability of long-distance natural gas pipeline project crossing area—— Comment on "Oil and Gas Pipeline Integrity Evaluation Technology" [J]. Xinjiang Geology, 2019, 37(4): Post 4.
2. Wang Qingguo, Guo Min. Research on the influencing factors of gas pipeline integrity based on ISM and AHP[J]. Gas and Heat, 2019, 39(4): Post-insertion 1- Post-insertion 4.
3. Ma Bin, Jiang Feng, Ma Xuqing. Integrity evaluation technology of defective pipelines[J]. Gas and Heat, 2019, 39(5): 32-39.
4. Cui Ruixue, Wang Lin, Zhao Deyin, et al. Research on technology development of pipeline integrity management platform[J]. Contemporary Research in Chemical Industry, 2020(11): 108-110.
5. Jin Jian, Zhu Xueshan, Liu Yanyang, et al. Research on big data integration technology methods for oil and gas long-distance pipelines[J]. Oil and Gas Field Surface Engineering, 2020, 39(2): 77-81.
6. Jiang Zhongan, Zheng Dengfeng, Zeng Fabian, et al. Research on the safety management model of oil and gas pipelines based on hazard theory [J]. Journal of Hunan University (Natural Science Edition), 2021, 48(4): 56-65.
7. Jia Shuai. The application of intelligent information technology in the integrity construction of the Beijing-Tianjin Second Oil Pipeline[J]. Oil and Gas Field Surface Engineering, 2021, 40(2): 97-102.
8. Hou Lina, Li Yuanpeng, Yan Wei, et al. Construction and fusion method of oil and gas field pipeline and station integrity management system[J]. Oil and Gas Field Surface Engineering, 2021, 40(3): 63-69.

Research and Practice of CO_2 Flooding Technology in Ultra-low Permeability Reservoir of XX Oilfield

Wenxiu Zhou

Geological Research of the Tenth Oil Production Plant of Daqing Oilfield Co., Ltd., Daqing, China

Abstract. CO_2 flooding technology in ultra-low permeability reservoir is the key technology in modern oilfield technology development, which plays a key role in oilfield exploitation of ultra-low permeability reservoir. In this paper, CO_2 flooding technology for ultra-low permeability reservoir in oilfield is analyzed and studied. The main characteristics of ultra-low permeability reservoir in oilfield are briefly analyzed, and the development of CO2 flooding technology is discussed. At the same time, the application of CO_2 flooding technology in ultra-low permeability reservoir of XX Oilfield is studied and summarized.

Keywords: Oil field, Ultra-low permeability, Oil reservoir, CO_2 flooding technology.

1 Introduction

Oil is an important resource used in the development of modern society, and it has a good application in many fields of social production. However, oil resources are non renewable resources. The increasing demand for modern oil resources leads to the depletion of oil reserves in the modern world. And the exploitation of oil resources has also entered a more difficult stage. The development of low and medium permeability reservoirs in oil fields is difficult, which has become an urgent problem to be solved in modern oil field exploitation. In the research process of modern oilfield exploitation technology, relevant experts put forward that CO2 flooding technology can improve the development efficiency of ultra-low permeability reservoirs and play a vital role in the comprehensive exploitation of modern petroleum resources.

2 Brief analysis of ultra-low permeability oil reservoir in oilfield

Ultra-low permeability reservoir specifically refers to the ultra-low permeability oil and gas fields encountered in the process of oilfield development in China, which is an important part of oilfield development in China. Ultra-low permeability reservoir has the following characteristics.

First of all, the ultra-low permeability oil field has the characteristics of high reserves. Because of its relatively difficult oil field exploitation, the exploitation and development efficiency of low permeability oil field is relatively low, and its own oil field reserves are relatively large. According to relevant data, the reserves of ultra-low permeability reservoirs account for more than 1 / 3 of the development reserves of undeveloped oilfields in China, and the reserves of low permeability layers account for about 50% of the total discovered reserves. In the future, low permeability reservoir will be an important working area of oil resources in China.

Secondly, the development of low permeability reservoir has the characteristics of low permeability. According to the relevant data, the development permeability of low permeability reservoir is 0.3 ~ 10 mD, and the oil and gas permeability of the oilfield is relatively small, which affects the actual oilfield development efficiency. In addition, in the current process of oilfield exploitation in my country, the exploitation depth of oilfields with ultra-low permeability de-reservoir is above 3000m, which causes exploitation difficulties.

At present, ultra-low permeability oil reservoirs are the key target of my country's oilfield development. The reasonable exploitation of oilfields in low permeability reservoirs is conducive to the exploitation of my country's petroleum resources. In the actual process of oil resource exploitation, CO2 flooding can improve the development efficiency of ultra-low permeability reservoirs, ensure more effective oil field development and maximize the oil field exploitation effect.

3 Principle analysis of CO2 flooding technology

Now all countries in the world are studying the CO2 flooding technology of ultra-low permeability reservoirs. For example, after the SACROC block of Snyder Oilfield in the United States is put into production, CO2 flooding technology can be adopted. Its principle is mainly to improve the fluidity of crude oil by using CO2. After the

 https://doi.org/10.1051/e3sconf/202132901037

application of CO2 method, the volume of crude oil will expand, so as to realize the oil-gas mixing, and finally realize the good development of the oil field, and finally ensure the better oil-gas development effect.

The CO2 flooding technology in SACROC block of Snyder Oilfield in the United States mentioned above is the earliest CO2 flooding technology applied for patent. In the process of oil exploitation technology research in our country, many oilfields have also started the research of CO2 flooding technology. In the research of oil displacement technology, CO2 is used as oil displacement agent to improve the fluidity of oil in local reservoirs, so as to realize the convenient control of oil exploitation, and also to maximize the efficiency of oil exploitation. In the research of modern CO2 oil displacement technology and methods in China, there are still some problems in the current oil displacement technology. In the research of oil displacement technology, it is found that the technology of ultra-low permeability reservoir still has some problems, such as high miscibility pressure, difficult miscibility and low oil displacement efficiency. Therefore, in the research of modern oil production technology in China, we should pay more attention to the improvement of production technology, Ensure more active and rational oil exploitation and improve oil exploitation efficiency.

4 Research and analysis of CO2 flooding technology in XX Oilfield

4.1 Expounding of CO2 flooding technology in XX oilfield

XX oil field is an important oil field in my country. In its oilfield development technology research, CO2 flooding technology research is an important technology in the region, which plays a key role in oilfield development. In the technical exploitation of the XX oilfield, certain improvements have been made to the CO2 flooding technology currently used in my country. The miscibility of this technology has reached ≥1, permeability>0.5mD, formation crude oil viscosity <12mPa·s, formation crude oil density <0.876 2g/cm3, and oil layer depth>2000. At the same time, in the application of this technology, the lower limit of the physical properties of CO2 flooding in the reservoir is considered. When the reservoir permeability is too low (<0.5 mD), the starting pressure of CO2 flooding will increase significantly, and it is difficult to achieve effective displacement under conventional pressure difference.

4.2 Evaluation of CO2 flooding technology in XX Oilfield

In this research process, the application of CO2 flooding technology in ultra-low permeability reservoir of XX Oilfield is studied. The laboratory test and evaluation standards are established, aiming at the extraction function, phase characteristics, miscible pressure test and oil displacement efficiency of CO2 flooding technology in ultra-low permeability reservoirs. The following is a summary of the specific experimental research.

(1)Analysis of phase characteristics of CO2 flooding technology

During the phase characteristics research, No.1 well in XX Oilfield was selected for experimental analysis. In the process of analysis, the PVT analyzer with high temperature and high pressure was used for experimental research, and it was found that there was a certain relationship between CO2 solubility and pressure in the formation. The greater the pressure, the greater the solubility. The results show that formation crude oil has a strong solubility for CO2, and the higher the pressure, the greater the solubility of CO2 in crude oil. As shown in Figure 1.

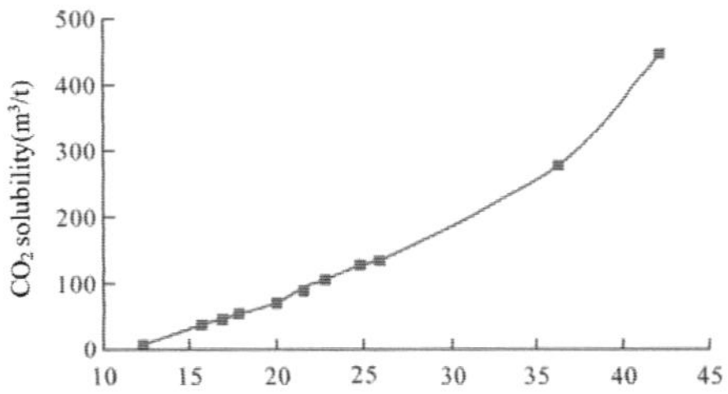

Figure 1 Curve relationship between CO2 solubility and pressure

(2)Analysis of CO2 crude oil extraction

The main principle of oil displacement by CO2 is to analyze the extraction of crude oil by CO2. In the process of local test analysis, PVT instrument is also used for actual analysis. In the process of instrumental analysis, carbon dioxide has a good extraction effect on hydrocarbons in petroleum. Hydrocarbon is the light component of petroleum resources, and the oil in reservoir with reduced permeability of light components is closer to the oil in stratigraphic oil field, thus facilitating the effective development of petroleum resources. Through the good development of oil resources, ensure its rational application. Moreover, the extraction of carbon dioxide can also realize the dynamic miscible state of crude oil, thus facilitating the exploitation of crude oil. Figure 2 shows the change of pure hydrocarbon components in equilibrium liquid phase.

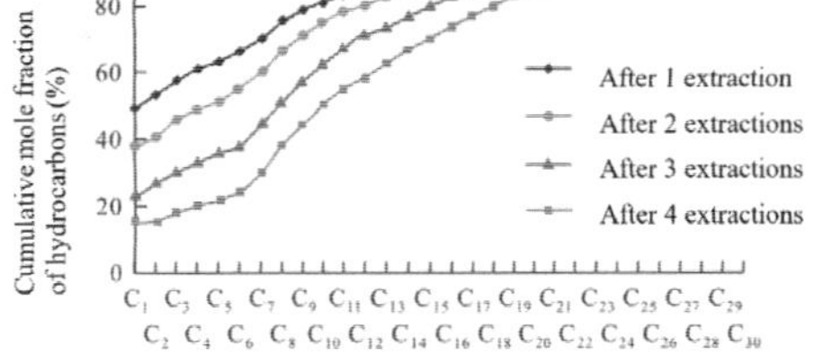

Figure 2 Changes of pure hydrocarbon components in the equilibrium liquid phase

(3) Analysis of the minimum miscible pressure of CO2 and crude oil

The minimum miscible pressure of CO2 and crude oil is a very key link in oil displacement technology, which is very key to the reasonable determination of modern oil displacement parameters. At the same time, in the process of technical research, relevant experts proposed to use the thin tube test method to carry out the minimum miscible pressure test. In the research, the minimum miscible pressure of six oilfields in XX Oilfield is studied. The data research shows that the average miscible pressure of six oilfields can reach about 30MPa. Compared with CO2 flooding technology abroad in the same period, the miscible pressure is generally below 12MPa. Foreign oilfields have high content of light components, relatively low viscosity and density, which is convenient for exploitation. Therefore, in the actual research of oilfield technology, China's oilfield exploitation should pay attention to the effective control of the minimum miscible pressure of crude oil to improve the oilfield exploitation effect.

(4) Study on CO2 oil displacement efficiency

In the research process of this paper, it is very important to study the oil displacement efficiency of the position. To a certain extent, it is related to the working effect of oil displacement efficiency, ensuring that the application of oil displacement efficiency is more reasonable, and also improving the oil displacement effect to the greatest extent. In the study of oil displacement efficiency, this paper designed a physical simulation device, mainly aiming at the water displacement efficiency and CO2 displacement line.

①At first, the oil displacement efficiency of water flooding is 33%, and then the oil displacement efficiency of CO2 miscible flooding is 85%.

②Secondly, in the test, the oil displacement efficiency of the water flooding method after the direct CO2 miscible flooding method can reach 80%.

③In the process of oil displacement efficiency analysis, CO2 oil displacement and water oil displacement are directly mixed for oil displacement, and the oil displacement efficiency can reach 81%.

④In the process of analysis of oil displacement efficiency, direct water flooding is used for oil displacement, and its oil displacement efficiency is about 47%.

Through the study of the above analysis methods, it can be found that the oil displacement method using CO2 and water flooding has the highest efficiency, which can reach more than 81%.

(5) Influence of produced steam reinjection on CO2 flooding

In the process of CO2 oil displacement, CO2 gas will also be produced in the process of oil displacement. In the actual research, it is found that the increase of carbon dioxide content also has a certain influence on the oil displacement efficiency. Therefore, during the experiment in this paper, the influence of carbon dioxide on oil displacement efficiency is analyzed. Fig. 3 shows the influence analysis of carbon dioxide content on oil displacement efficiency. Through analysis, it can be found that the higher the carbon dioxide content, the higher the oil displacement efficiency. The carbon dioxide content reaches 100%, and the oil displacement efficiency can reach more than 80%.

Through the above-mentioned experimental research, it is found that in the research of CO2 flooding oilfield ultra-low permeability reservoir technology, the higher the CO2 content, the better the oil displacement efficiency. At the same time, the combined flooding of water flooding and CO2 flooding has higher oil displacement efficiency. The carbon dioxide has a good extraction effect on the pure hydrocarbon components in petroleum, and can also improve a good oil displacement effect.

5 Summary of key points of CO2 flooding technology in ultra-low permeability reservoir of XX Oilfield

The above research shows that the oil displacement technology of ultra-low permeability reservoir in XX Oilfield has good oil displacement effect. The following is a summary of the key points of oil displacement technology in its oilfield. To ensure more reasonable oil displacement in the oilfield, the following is a detailed technical summary and analysis.

(1) CO2 Huff and Puff Selection Well

CO2 huff and puff well method in ultra-low permeability reservoir can effectively improve the development efficiency of ultra-low permeability reservoir, and in the actual oilfield exploitation process, we should pay attention to adopting necessary well selection principles.

①According to the relevant data, when the porosity reaches 10-20%, the reservoir reserves reach the highest value, but the effect of CO2 huff and puff in the well is relatively low. The main reason is that the porosity of 10-20% is too large, which leads to the low efficiency of carrying residual oil during CO2 spit-out, which affects the actual reservoir exploitation.

②Reservoir pressure is also one of the important indexes affecting well selection for CO2 huff and puff. It also plays a very important role in oilfield development. In actual research, the reservoir pressure of CO2 huff and puff well selection is divided into less than 10MPa, 10-15 Mpa, 15-20 Mpa, 20-25 Mpa and greater than 25 Mpa. However, it is found that when the reservoir pressure is 10-25 Mpa, the CO2 huff and puff recovery effect is the best, and the oilfield development efficiency is relatively high.

③In the process of oil exploitation, relevant research shows that the CO2 huff and puff well selection effect is better when the water content is 50%, which plays a very important role in oil exploitation in low permeability reservoirs [1].

(2) Optimization of injection mode

In the application process of CO2 oil displacement technology, it also includes a good design of its oil displacement technology to ensure that the oil displacement work is more reasonable. At the same time, in the process of oil displacement, it is required to

https://doi.org/10.1051/e3sconf/202132901037

optimize the injection mode. At present, in the process of oil exploitation in CO_2 low permeability reservoir, single-line development, continuous water injection, continuous gas injection and gas-water alternate injection are mainly selected for oil exploitation. In the course of this study, the injection modes of different oil fields are analyzed, and the above four different injection modes are mainly established during the test. In the analysis of injection mode, the single elastic development injection mode is adopted. The oil recovery efficiency is 9%, the oil recovery efficiency of continuous water injection mode is more than 17%, the oil recovery mode of continuous gas injection is 27%, and the oil recovery of steam water alternation can reach about 30%. Through the comparative analysis of test data, it is found that using the method of gas water alternation for recovery control can maximize the recovery effect and ensure that the implementation of oil recovery is more reasonable [2].

(3) Optimization analysis of injection production parameters

In the application of CO_2 flooding technology, the design of injection production parameters is very key, especially the control of injection production speed is directly related to the injection production efficiency. Therefore, in the research in this paper, the above test is also used to analyze the injection production rate of CO_2 flooding technology in an oil field. In the course of the test, the test comparison was mainly carried out for the injection-production speeds of 15t/d, 20t/d, 25t/d, and 30t/d. Through the comparison of the test results, it is found that in the actual test analysis process, the gas injection rate is 20t/d, the oilfield recovery efficiency can reach 32.5%, and the injection-production efficiency of 15 t/d is Only about 30%. Therefore, during the implementation of the petroleum gas injection process, attention should also be paid to the control of the gas injection rate, which requires a gas injection rate of 20t/d [3].

(4) Application of injection-production structure

In the application of CO_2 flooding technology, the effective design of the surface injection-production structure is not included. Through the rational design and application of the injection-production structure, the injection-production effect is ensured to be better. In the whole oilfield exploitation process, the application of CO_2 flooding technology mainly includes hydraulic anchor, backwashing valve, tie-back barrel, packing device, multifunctional main steam valve, main steam layer device and so on. By well controlling the injection-production process, the application of oil production technology can be ensured to be more reasonable, and the oil injection-production effect can be improved to the greatest extent [4].

In addition, in this research process of mining technology, in order to improve the application effect of CO_2 flooding technology, the optimization of oil production string has been completed in the actual technical application. Multi-functional oil production string is used for design. In the process of comprehensive design of the project, the optimized design of oil production string is used to improve the oil production effect. In addition, in the oil production process of XX oilfield, we should also pay attention to the effective control of pipeline corrosion and prevent pipeline corrosion, so as to improve the oil production effect and prevent the impact of corrosive substances on carbon dioxide. In this paper, an economical and reliable anti-corrosion scheme of carbon steel + anti-corrosion coating and FRP is optimized, and a modified imidazoline composite high-efficiency corrosion inhibitor is designed and synthesized. Through the rational use of this preservative, the anti-corrosion effect is optimized [5].

6 Conclusions

In this paper, the author analyzes and studies the CO_2 flooding technology of oilfield ultra-low permeability reservoirs. The article summarizes the key points of the application of the technology, and also summarizes the main influencing factors of the CO_2 flooding technology. It is hoped that it will be helpful to the application of CO_2 flooding technology in ultra-low permeability reservoirs.

References

1. Yang Yong. Research and practice of CO2 flooding technology in ultra-low permeability reservoirs in Shengli Oilfield[J]. Petroleum Geology and Recovery Efficiency, 2020, 27;142(01):17-25.
2. Qian Kun, Yang Shenglai, Dou Hongen, et al. Micro-displacement characteristics of different CO2 injection methods in ultra-low permeability reservoirs[J]. Xinjiang Petroleum Geology, 2020, 41(2): 5-5
3. Lin Qianguo, Zhang Shaojun, Wang Xiangzeng, et al. Establishment of CO2 flooding and storage monitoring system for low and ultra-low permeability reservoirs[J]. Journal of Safety and Environment, 2019(2): 11-11
4. Yang Yong. Research and practice of CO2 flooding technology in ultra-low permeability reservoirs in Shengli Oilfield[J]. Petroleum Geology and Recovery Efficiency, 2020, 27(1): 9-9
5. Hou Pengfei. Study on the effective thickness interpretation standard of Fuyang oil layer CO2 flooding in Block Shu101 of Yushulin Oilfield[D]. Northeast Petroleum University, 2019.

Analysis of the Influence of Gas Distribution on Flora Structure and Oil Displacement Effect During Microbial Oil Displacement

Chenying Wang

Geological Research Institute of the Tenth Oil Production Plant, Daqing Oilfield Co., Ltd., Daqing, Heilongjiang, 163001, China

Abstract. In the context of the rapid development of my country's social economy, the total demand for energy in my country's industrial production continues to increase, of which petroleum energy is an important type of energy required in industrial production. At present, China's petroleum energy mainly depends on imports, so in order to promote the productivity of China's petroleum industry, we must carry out comprehensive innovation of oil displacement technology. Microbial oil displacement is an important technical means, which needs to accurately grasp the factors affecting the oil displacement effect, so as to optimize and innovate the technology. Therefore, this paper will deeply study and analyze the influence of gas distribution on flora structure and oil displacement effect in the process of microbial oil displacement, and put forward some reasonable opinions and measures to further promote the improvement of oil production technology.

Keywords: Keywords microbial oil displacement, gas distribution, flora structure, oil displacement effect, influence analysis

1 Introduction

Microbial oil displacement technology is a commonly used technology in oil production at present. Compared with other oil displacement technologies, microbial oil displacement has better effect and can effectively improve oil displacement efficiency and quality. The application of microbial oil displacement technology will be influenced by many factors. Among them, gas distribution has a direct influence on the structure of flora. Therefore, it is necessary to clarify the specific way and effect of gas distribution on it, so as to optimize and innovate microbial flooding according to the influence. Therefore, this paper takes the crude oil produced in H oilfield as a sample through experiments, and makes a comprehensive analysis of its specific influence effect.

2 Overview of microbial oil displacement technology

After more than 20 years of development, microbial oil displacement technology has been widely used in the field of oil production. According to the different modes of microbial oil displacement, it can be divided into endogenous microbial oil displacement technology and exogenous microbial oil displacement technology. Because the endogenous microbial flooding can directly utilize the microorganisms in the oil reservoir as the object, the application cost is low, and the workflow of microbial flora fermentation can be saved. At the same time, it can effectively deal with the inadaptability of microbial flora to oil reservoir [1]. Endogenous microbial oil displacement technology can activate aerobic microorganisms in oil fields, making them a component of oil displacement. Therefore, attacking the oxygen needed by the development of aerobic microorganisms is the basis of activating aerobic microorganisms. In related studies, the effects of different gas distribution on the structure of flora were studied. It was found that with the extension of culture time, oxygen would be consumed gradually, and the development of flora was consistent. Some researchers found that gas distribution can directly promote microbial oil displacement in the process of oxygen biological consumption and non biological consumption experiments.

3 Experimental research

3.1 Experimental materials

The simulated oil field in this experiment is the reservoir condition of H city oil field. The crude oil sample produced by this oil field is used in physical simulation, which is dehydrated and degassed crude oil. The viscosity of this crude oil sample is 1886mPa·s at 50°C. The water sample in the laboratory is the injection water used in the production of the oilfield, with a salinity of 9980mg/L. Ions include: Cl^-4568mg/L, HCO_3^-855mg/L, SO_4^{2-}

106mg/L, Mg^{2+} 104mg/L, Ca^{2+} 165mg/L. NA^+ + K^+ 4083mg/L. According to the oil and gas industry standards, the reservoir conditions of this oil field meet the requirements for microbial growth tolerance. Glucose, K_2HPO_4、H_2SO_4、CH_3COOH and analytical reagents are common types in the market. Peptone is produced by a Beijing biological company. The trace element liquid is self-made in the laboratory. The activation system for physical simulation application is: 0.25% glucose, 0.2%K_2HPO_4, 0.05% peptone and 0.05% trace element solution. Quartz sand is produced by Zhonglong Petroleum Corporation. The DNA extraction kit was produced by a bioengineering company in China [2].

The core displacement simulation experiment equipment is produced by a Chinese petroleum instrument company. The gas chromatograph model is GC-14B. The phase contrast microscope model is BX50. The high-throughput sequencer is a general-purpose equipment produced in other countries.

3.2 Experimental method

(1) Experimental method design of physical flooding

The experimental method design of physical oil displacement includes:(1)Fill the core with quartz sand according to the following configuration, and measure the core permeability. (2) Pumping vacuum saturated formation water and calculating the original water saturation. (3) Make irreducible water from saturated crude oil, and calculate the original water saturation. (4) Place the core at 60°C for one week, and displace 3PV by water flooding once, and record the oil production, water production and displacement pressure changes in detail. (5) The microbial nutrient solution 0.3PV is sequentially mixed with gas according to different volume ratios of 1:1, 1:5 and 1:10. After the gas distribution is completed, it is injected into the core tube and cultured for 15 days. (6) Secondary water flooding 3PV, recording the changes in oil production, water production and displacement pressure during the period, and calculating the oil displacement and water cut.

Table 1 Quartz sand configuration of sand-filled core

Mesh	Particle size	Particle fraction
30—40	590μm—420μm	8.7%
40—60	420μm—250μm	12.1%
60—80	250μm—178μm	31.2%
80—100	178μm—150μm	27.7%
100—150	150μm—99μm	13.8%
200—300	75μm—57μm	3.2%

The figure below shows the structure of the device used in the core displacement experiment.

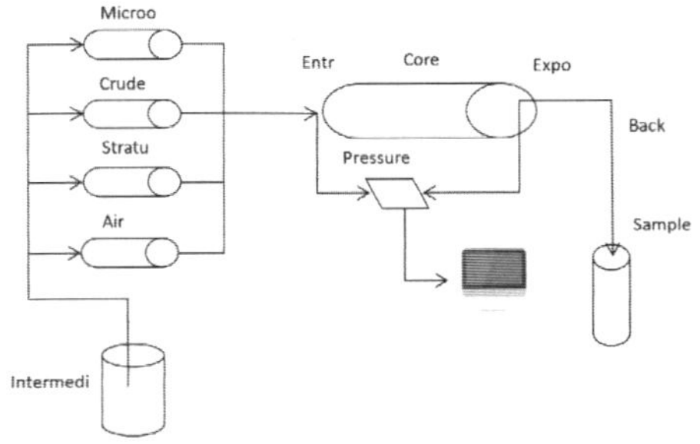

Figure 1 The structure diagram of the device used in the core flooding experiment

(2) Gas distribution

According to the pore volume of the core, the volume of nutrient solution required under the condition of 0.3PV is calculated, and the air quantity required under the normal pressure is calculated according to the liquid-gas ratio value designed in the experimental scheme, and the air quantity is calculated according to the setting of different core liquid-gas ratios. The middle container is used for gas distribution, the lower layer is designed as nutrient solution, and the upper layer is designed as air, which is mixed and injected into the core tube, so as to complete the injection test with different gas distribution [3].

(3) Molecular biology measurement of output fluid

In the process of physically simulating core tube displacement, 20mL of produced liquid is placed in a centrifugal tube, centrifuged at 4°C and 10,000 rpm for 10 minutes, then the bacteria are collected and the supernatant is removed. DNA samples of flora were collected by DNA extraction kit, and then dissolved in water by aseptic method. DNA samples were detected by electrophoresis to ensure sufficient collection, and finally high-throughput sequencing was performed.

(4) Measurement of small molecule acid in produced liquid

In this experiment, the measurement of small molecules in the output liquid adopts the appearance method, and the tested samples and standard samples are analyzed one by one by gas chromatography analysis method to obtain the concentration of small molecules acid in the output liquid. Specifically, it includes: (1) standard sample configuration. The acetic acid standard sample and distilled water were mixed to obtain small molecular acid solutions with different concentrations. (2) Pretreatment of produced liquid samples. Use 10mL produced liquid sample, observe the oil-water separation after standing, collect the supernatant after complete separation, and use 0.45μm microporous filter membrane, add sulfuric acid in the ratio of 1:1 to the filtrate, and then refrigerate for one day. (3) Chromatographic analysis. Chromatographic analysis was carried out on standard samples and tested samples to obtain the concentration of small molecular acids in the produced solution.

4 Results and discussion

4.1 Influence of different gas distribution on displacement effect of physical simulation

The results of physical simulation oil displacement with different gas distribution are shown in the following table.

Table 2 The results of physical simulation flooding with different gas distribution

Core	Pore volume	Saturated oil	Displacement efficiency of one-time water flooding	Liquid-gas ratio	Displacement efficiency of secondary water flooding	Average displacement efficiency	Displacement improves efficiency
1	244 mL	209mL	49.52%	No gas distribution	6.05%	6.02%	—
2	251 mL	216mL	48.05%	No gas distribution	5.99%		
3	230 mL	215mL	47.51%	1:1	10.72%	10.58%	4.56%
4	230 mL	216mL	46.03%	1:1	10.44%		
5	240 mL	225mL	47.51%	1:5	14.89%	14.79	8.77%
6	231 mL	215mL	50.50%	1:5	14.68%		
7	229 mL	210mL	49.04%	1:10	14.36%	16.66%	8.54
8	238 mL	225mL	50.07%	1:10	14.95%		

From the data in the above table, it can be seen that when different gas distribution rates are used, the displacement effect for the acquisition effect is directly affected. Under the condition of increasing gas distribution, the microbial displacement effect is gradually enhanced. It shows that increasing the gas distribution can activate the aerobic microorganisms in the core and promote the growth of aerobic microorganisms. Under the condition of increasing gas distribution, the increasing rate of displacement efficiency keeps decreasing. When the gas-liquid ratio is 1:5, the displacement efficiency is the highest, reaching 8.77%. Under four different conditions of unbalanced gas and gas-liquid ratio of 1:1, 1:5 and 1:10, the water content decreased by 1.53%, 2.66%, 7.31% and 6.98%. With the increase of gas distribution, the maximum reduction range of water content gradually increases, and the maximum reduction range of water content is at the condition of liquid gas ratio of 1:5. It shows that reasonable gas distribution can activate useful microorganisms, facultative anaerobic and anaerobic microorganisms to act in crude oil and improve displacement efficiency [4].

4.2 Changes of produced fluid flora in the process of physical simulation displacement

Bacterial number change

The changes in the number of bacteria in the produced fluid with different gas distribution rates during the displacement period are shown in the figure below.

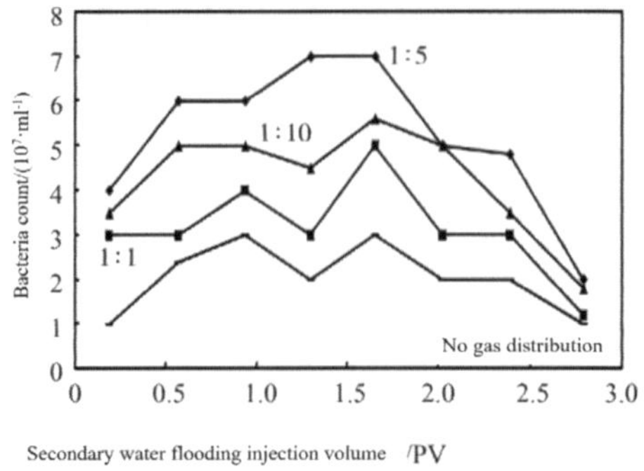

Figure 2 Changes in bacterial counts of produced fluids with different gas distribution rates during displacement

According to the above figure, it can be seen that the number of bacteria in the core produced fluid exceeds 10,000,000 /mL, indicating that the microorganisms in the core are fully activated. The number of bacteria in the output liquid is different in different liquid-gas ratios, and the number of bacteria in the output liquid will increase with the increase of gas distribution, which is the same as the above test results. When the liquid-gas ratio was 1:10, aerobic microorganisms were fully activated, but the activated number of facultative anaerobic and anaerobic bacteria was less than that of aerobic bacteria, so the overall number of bacteria decreased. In the process of core displacement, the number of bacteria increased first and then decreased, indicating that microorganisms gradually moved forward in the rock core. Under the condition of displacement 1-2PV, the number of bacteria is the highest, and the increase of the number of bacteria indicates that microorganisms are fully activated and grow, so as to improve the displacement effect [5].

(1) Changes in produced fluid flora

Under different liquid-gas ratios, there are some differences between microbial community structures. Different gas distribution can activate different types of microorganisms, in which the dominant flora accounts for more than 66% of the total. The following table shows four kinds of functional bacteria after microbial activation under different gas-liquid ratios.

Table 3 Proportion of four different types of functional bacteria

Bacteria	The proportion of different gas distribution			
	No gas distribution	1:1	1:5	1：1
Bacillus	16.99%	44.15%	50.10%	28.31%
Ferribacillus ferrooxidans	47.13%	1.25%	8.59%	0.49%
Pseudomonad	10.05%	18.40%	18.42%	23.17%
Symbiotic corpuscle	5.27%	3.65%	3.02%	38.80%

According to the above table, the proportion of Bacillus in different gas distribution conditions is kept high, and it is the main bacteria genus of microbial oil displacement. After activation, the proportion of Bacillus under different liquid-gas ratios is the highest when the liquid-gas ratio is 1:5, reaching 51.10%, which is consistent with the highest oil displacement effect. Iron-removing bacteria have the characteristics of nitrate reduction and degradation of crude oil, with the highest proportion of 47.13%. With the increase of gas distribution, the proportion is decreasing continuously, which corresponds to its growth and metabolism law. Pseudomonas is a facultative aerobic microorganism, which can metabolize and produce emulsifier to degrade crude oil. Under different gas distribution conditions, the proportion exceeds 10%, and the proportion gradually increases with the increasing gas distribution, which indicates that Pseudomonas has a good response to oxygen. Symbiosis bacteria are microorganisms that coexist with other bacteria and play an important role in maintaining the diversity of biota. Under the condition of liquid-gas ratio of 1: 10, the proportion is significantly higher, indicating that oxygen has a good activation and promotion effect on symbiotic corpuscles.

Under different gas-liquid-gas ratios, the Shannon diversity distribution indices of microorganisms are 1.95, 2.19, 2.22 and 1.66, respectively. With the gradual increase of gas distribution, Shannon diversity index will increase continuously, indicating that oxygen can activate microbial flora. The Shannon diversity index decreased when the gas distribution reached 1:10, which indicated that the aerobic to facultative aerobic microorganisms in the flora increased continuously with the increase of oxygen, but the diversity decreased. From the point of view of finishing trend, the diversity of microbial flora is the best under the condition of liquid-gas ratio of 1: 5, and in this case, the functional microorganisms are mainly Bacillus and Pseudomonas which produce surface or week and emulsifier, and the displacement effect is the best. According to the results of molecular biology analysis, the conclusion that gas distribution can influence the oil displacement effect is verified [6].

5 Conclusions

In the process of microbial oil displacement, injecting different gas distribution can have a direct impact on flora structure and oil displacement effect. The activation of different types of functional microorganisms is different after different air distribution is injected. Among them, the dominant functional bacteria account for more than 66%, and the functional microorganisms are mainly Bacillus and Pseudomonas which produce surfactants and emulsifiers. The concentration of small molecule acids increased after gas distribution, which indicated that oxygen could promote the activation of aerobic microorganisms. Under the condition of liquid-gas ratio of 1:5, the concentration of small molecular acids was significantly increased, and the metabolic activity of microorganisms was improved. With the increasing gas distribution, the microbial displacement efficiency gradually increases and tends to be stable. Under the condition of liquid-gas ratio of 1:5, the displacement effect is the best, reaching 8.77%. Therefore, when adopting microbial oil displacement technology, it is necessary to adjust the gas distribution, and adjust the gas distribution to the best state according to the actual production situation of the oil field, so as to improve the oil displacement effect, promote the level of oil displacement technology and increase the oil field output.

To sum up, this paper studies the influence of different gas distribution on the characteristics of flora and oil displacement effect in the process of microbial oil displacement through experiments. The results show that different gas distribution will have a direct impact on the characteristics of flora and oil displacement effect, and a reliable conclusion is obtained. I hope it can be used as a reference and help for the development of China's petrochemical industry. Based on this conclusion, the specific application of microbial oil displacement technology can be adjusted, so as to promote the improvement of oil displacement technology level and continuously improve the productivity of petrochemical industry. At the same time, innovate other oil displacement technologies to promote the development and construction of petrochemical industry.

References

1. Jing, Shu Qinglin, Sun Gangzheng, et al. The succession of endogenous microorganisms in reservoirs and their influence on oil displacement[J]. Journal of China University of Petroleum: Edition of Natural Science, 2019, 43(001): 7-7.
2. Liu Tao, Wang Lushan, Hu Jing, et al. The influence of gas allocation on the microbial community structure and oil displacement effect in the process of microbial flooding[J]. Oilfield Chemistry, 2019, 36(001):5-5.
3. Sun Gangzheng, Qian Qin, Hu Jing, et al. Research on endogenous microbial flooding laws of Zhan 3 reservoir under different types of activators[J]. Journal of Xi'an Shiyou University: Natural Science Edition, 2019,000(002) :8-8.
4. Yang Xiuqing, Liang Qi, Han Zuoying. The effect of ethanol on the structure of coal geomicrobial flora and gas production pathways in the process of biogas formation[J]. Journal of Shanxi University: Natural Science Edition, 2019, 42(4):10- 10.
5. Song Shemin. Microbial field enhanced oil recovery technology for target dominant bacteria in oil reservoir[J]. Petroleum Drilling & Production Technology, 2019, 41(002):6-6.
6. Hou Zhaowei, Li Wei, Le Jianjun, et al. Research and application of microbial oil recovery technology in Daqing Oilfield[J]. Petroleum Geology and Recovery Efficiency, 2021, 28(002): 8-8.

Research Status of Brackish Water Desalination Technology

Qiaonan Yang [1, 2, #], Can Hu [1, 2, #], Jie Li [1, 2], Xiaokang Yi [1, 2,*], Jie Zhang [1, 2], Zhilin Sun [3]

[1] School of College of Mechanical and Electrical Engineering Tarim University, Alar 843300, China
[2] School of Key Laboratory of Modern Agricultural Engineering Tarim University, Xinjiang Uygur Autonomous Region, Alar 843300, China
[3] Ocean College, Zhejiang University, Zhoushan 310058, China
#These authors contributed equally to this work.

Abstract. The serious salinization of farmland soil and the shortage of water resources in arid areas are becoming more and more serious; In order to solve the problem of shortage of fresh water resources, find a resource that can be converted into fresh water; Brackish water is derived from farmland saline alkali land. Brackish water desalination technology has increased interest in scientists' research. However, brackish water can be used as an effective substitute for freshwater conversion because of its low salt concentration and can also be widely used in many water deficient countries. However, partial or total desalting of brackish water is essential to meet the water quality requirements of production and life. The selection of appropriate water resources desalination technology needs to understand the operation methods and technical means of the existing brackish water desalination process. Proper combination of desalination technologies can improve the efficiency of brackish water desalination. This review provides the latest advances in data, materials and categories that can help select and design requirements for specific applications.

1 Introduction

About 75% of the earth's surface is covered by water, but only 3% is fresh water[1,2]. According to reports, fresh water is stored in glaciers, of which only 5% is easily available[3]. At present, 40% of the world's people live in arid areas or islands lacking fresh water[4]. In addition, the increase of arid areas in the world, the reduction of the adaptability of water resources to climate change, and the over development of brackish water desalination technology. The application of brackish water desalination technology affects the development of environment and economy[5]. In recent years, it is one of the major challenges faced by capitalists in the new century to bring water into the stock market. It is reported that by 2030, three fifths of the world's people will face the problem of shortage of fresh water resources[6]. In view of this problem, the state will formulate friendly policies to solve the major problem of obtaining water resources in arid areas, and ensure the basic guarantee of industrial and domestic water. Fresh water resources are very important for the economic and social development of a country and region. With the growth of the world population and the rapid economic development, the shortage of fresh water resources is becoming more and more serious. Therefore, the use of brackish water desalination technology has become a new research direction for scientists. In response to the increasing demand for fresh water resources, the key solution to avoid the shortage of fresh water supply is to develop new water resources. Desalination of brackish water is a substitute for fresh water because it is low and rich in salinity in areas facing water shortage.

At present, brackish water, as raw material for freshwater production, accounts for 21% of global raw materials. The development of energy-saving and cost-effective seawater desalination process helps to further expand the use of brackish water. At present, desalination of saline alkali water is the most useful method to solve the shortage of water resources. The combination of solar energy and membrane separation technology has been rapid development in recent years, more and more people like membrane desalination method because of simple operation, small footprint, low energy consumption and the continuous improvement of membrane material performance[7]. Membrane desalination technology mainly includes electrodialysis, reverse osmosis, membrane distillation, nanofiltration and forward osmosis. Among them, a most widely used membrane technology mainly includes electrodialysis, nanofiltration and reverse osmosis. Reverse osmosis technology has become the most advanced technology of brackish water desalination technology. However, in low brackish water desalination, RO energy efficiency will be reduced[8,9]. Nanofiltration NF can selectively remove polyvalent ions[10,11], and this technology is applied to hard water softening.

The promotion of brackish water desalination technology is relatively slow. At present, membrane desalination is mainly used, and there are few studies on solar energy and membrane desalination technology. Different brackish water desalination technologies have different process

* Corresponding author: 1571832388@qq.com

characteristics. Single brackish water desalination technology is not suitable for all brackish water. Improper desalination technology may cause damage to equipment, waste of water resources and even pollution. At present, there is a lack of research on the process integration technology of membrane desalination under the action of solar energy. This paper summarizes the research status of this technology, and integrates the research results of different desalination technologies to provide professional and technical personnel with.

2 Main components of brackish water

Brackish water refers to water with alkalinity greater than hardness and contains a large amount of neutral salt with PH value greater than 7. The sources of brackish water mainly include seawater and saline alkali water. Generally, the salt content is 1~25g/L, which is between freshwater(<1g/L) and seawater (1~35g/L)[12,13]. Brackish water is divided into low salt content brackish water, medium salt content brackish water and high salt content brackish water according to salt content, and the salt content is 1-5g/L, 5-10g/L and >10g/L, respectively. The main feature of brackish water is bitter taste, and it is difficult to drink directly. If brackish water is used for a long time, it will lead to gastrointestinal dysfunction and decreased body immunity. Taking the brackish water in Xinjiang, China as an example, according to table 1, the total salt content of brackish water is 6.89g/l, and no heavy metals such as Pb, CD, Cr, Hg and Sn are detected.

Table 1. Composition of brackish water.

Ion Composition of Brackish Water	Ion mass concentration (mg/L)	Percentage of ion mass concentration (%)
Na+	1674.03	26.32
K+	19.47	0.31
Ca2+	434.60	6.83
Mg2+	291.80	4.59
Cl-	2624.48	41.26
SO42-	1315.94	20.69
total mineralization	6360.32	100

3 Development Status of Brackish Water

3.1 Utilization of Brackish Water in Agriculture

The development and utilization of brackish water resources is an effective method to alleviate the shortage of fresh water in arid areas. On the whole, the global brackish water content is as high as $1.45\times10^9 km^3$, accounting for 3% of the total water volume of the earth. After desalination, it can meet the demand for fresh water[14]. Brackish water is widely distributed in areas. Desalination of brackish water can reduce the salinization harm of brackish water to farmland soil and meet the water consumption of agricultural irrigation[15,16]. In normal farmland irrigation, salts in brackish water, such as Na^{2+}, Ca^{2+}, Mg^{2+}, SO_4^{2-}, will provide fertilizer efficiency for crops to a certain extent and save planting costs [17-19].

Brackish water desalination technology has gradually entered the stage of industrial production from small-scale experimental desalination, providing a strong guarantee for drinking water and agricultural and forestry irrigation water, such as Israel, the United States, Italy, France, Austria and other countries [20]. Especially in some countries with extreme water shortage in the Middle East: Saudi Arabia, the United Arab Emirates, Israel, etc., as well as Japan, Spain and the United States in developed countries, there are more and more successful examples of brackish water desalination. The success of brackish water desalination technology has played a grass-roots guarantee role for local industrial water, domestic water, farmland irrigation water and ecological environment[21].

3.2 Utilization of Brackish Water in Breeding Industry

It is completely feasible to cultivate P. chinensis in saline-alkali areas far from the sea area, and its benefit is much higher than that of other varieties[22]. During the whole breeding process, the daily management is the same as that of shrimp farming in coastal China. As long as the salinity of groundwater in saline-alkali land can reach more than 25% and there is good underground freshwater resources, shrimp culture in China can be carried out.

In order to solve the pressure of freshwater aquaculture, worldwide use of brackish water to develop aquaculture. The United States, Israel, India and Australia have begun to use farmland brackish water desalination technology for commercial production of fresh water[23]. At present, Texas, Arizona, Arkansas, Alabama and Florida in the United States have evaluated the cultivation of inland brine shrimp, and the cultivation of Litopenaeus vannamei in Texas began in 1970. Water in Arizona was used to cultivate shrimp in ponds and kandao, and aquaculture wastewater was used to irrigate crops. Australia also carried out small-scale aquaculture production of squatfish using Weishui in southwest China.

4 Research Progress of Brackish Water Desalination Technology

4.1 Membrane Desalination Technology under Solar Energy

The water in the solar distiller is distilled by absorbing the radiant heat of the sun. The latest distillation technology is based on membrane distillation and multistage distillation. In the process of distillation, the energy required to produce 1 ton of fresh water is 30-80kw [24]. In 2011, winter et al. [25] came from Fraunhofer Solar Energy Research Institute. They considered winding the frame on the damaged MD module, with a surface area of 9-14m2. The test results show that the salinity of feed flow and feed water has a significant effect on the boundary of damaged module. In 2012, Chang et al. [26] studied a solar based thin film refining and desalination framework smdds, which uses porous layer refining. The frame is composed of cold liquid indoor regulator, hot temperature storage tank, siphon and energy recovery device. The results showed that the distillate was 0.358kg/d on the day with radiation, and 0.152kg/d on the cloudy day. In 2015, aim et al. [27] introduced a closed desalination system in a manor in Dubai. The desalination system can meet the demand of 260L/d high-temperature water and 18-25 L/d pure drinking water. The desalination unit adopts a pore layer with an opening area of 11.85m^2 to refine the seawater desalination cycle and horizontal plate gas collector. In 2021, Lacroix introduced an innovative thermodynamic reverse osmosis desalination device [28]. The desalination process of this new desalination device uses the heat energy of the sun to pressurize the salt water above its osmotic pressure to desalinate it.

4.2 Electrodialysis desalination technology

Membrane based desalination technology includes osmotic membrane to separate salt molecules from water. The most commonly used membrane separation technologies are electrodialysis and reverse osmosis. In the electrodialysis process, the salt in the solution is discharged through two electrodes through the selective membrane. This technology is suitable for the desalination of brackish water with salt concentration lower than 3g/L^{-1} [29]. Early electrodialysis methods were widely used in brackish water and seawater desalination, chemical industry, wastewater treatment and food industry [30]. With the development of science and technology, the performance of ion exchange membrane used in electrodialysis process and the process structure of electrodialysis device have been continuously innovated and improved, and have also been widely used in new fields such as bioengineering, medicine, flue gas desulfurization and wastewater treatment in coalbed methane, natural gas and oil manufacturing [31]. Electrodialysis technology has high operability in terms of desalting effect in certain environment and treatment of certain types of brackish water. It can meet the requirements of crops for water with different salt content. The technical cost and energy consumption cost can be reduced to a certain extent and has good flexibility.

4.3 Membrane desalination technology

Reverse osmosis technology is a new membrane separation technology, which is widely used in the process of brackish water desalination. In 1992, Rodgers et al. [32] developed a model to apply pressure to the membrane. They studied the effect of this pressure on the permeability of reverse osmosis membrane, indicating that negative pressure membrane pressure may increase the permeability of reverse osmosis membrane and optimize it. In 2005, Kim et al. [33] compared the research results of three-dimensional model and two-dimensional numerical value on the theoretical properties of thin films. After experimental verification, the research results showed the correlation between the theoretical model of thin films and working stability conditions. In 2019, Ali et al. Also introduced the motion model of membrane [34] and used mathematical methods to show that the high-step change of feed mass flow will indeed significantly affect the permeation effect of reverse osmosis membrane. In addition to these models, the reverse osmosis desalination technology is also studied. In 2021, cheddie et al. [35] studied the reverse osmosis process of seawater under wave pressure, and introduced the permeation model of reverse osmosis device. The experimental results show that periodic pressure application can minimize the polarization effect and improve the performance. These work show that dynamic operating conditions affect the performance of RO membrane, and the membrane behavior needs to be modeled in a dynamic way to evaluate these effects.

5 Existing Problems and Solutions

5.1 Problems faced by reverse osmosis technology

Compared with other technologies, reverse osmosis technology has relatively low energy consumption, but the demand for seawater desalination technology is increasing, which means that more and more energy is used. According to the recent and future prediction of estimated energy pollution, in 2045, the global emissions of carbon dioxide and brackish water desalination technology will reach 35 million tons per year [36]. Kuwait and other countries use part of their oil production for brackish water desalination and national power generation, which has a great impact on environmental pollution, depletion of renewable energy, shortage of water resources, more serious destruction of ozone layer and toxic compounds in groundwater [37]. Environmental protection and freshwater resource utilization technology are the common research goals of many scientists, because many scientists provide better ideas for the expansion of reverse osmosis technology.

5.2 Continue to Increase the Promotion of Brackish Water Desalination

After more than 20 years of development, brackish water desalination equipment has been popularized to some extent, but there are still serious drinking water problems in some economically underdeveloped areas. Therefore, the promotion of electrodialysis and reverse osmosis brackish water desalination methods and support policies need to be vigorously promoted and supported by the government.

5.3 Rational development to ensure long-term sustainable use

Brackish water resources are an integral part of the total water resources, which also need reasonable development to ensure long-term use. At present, the total amount of brackish water resources in China is 198 billion m3/a. Brackish water is the main source of drinking water in some areas, so brackish water and freshwater resources need to be planned and exploited. Therefore, in the shallow brackish water distribution area, brackish water desalination should be comprehensively promoted. The exploitation of shallow underground brackish water should be encouraged. On the one hand, it is convenient to take water. On the other hand, the exploitation of shallow underground brackish water will not lead to ground subsidence, and it is conducive to the recovery of groundwater quality and the improvement of ground soil environment. In line with the principle of "light first and then salty, shallow first and then deep," flexible and diverse forms are adopted to provide healthy drinking water resources for the vast brackish water areas.

6 Conclusion

Fresh water can be obtained from brackish water, saline alkali water and seawater. The most suitable technology for obtaining fresh water resources is to use brackish water with relatively low salt content. Through the combination of other technologies, the recovery rate of fresh water can be improved, so as to minimize the generation of wastewater. Solar energy technology and membrane separation technology, electrodialysis method and membrane separation technology have been widely explored for the treatment of brackish water desalination because of the low energy demand. In recent decades, the simplification of device design and operation, the improvement of membrane technology and the development of electrodialysis methods can reduce the cost of brackish water desalination and make these technologies more visible. However, the consumption of membrane and the working environment of membrane are still limited, so membrane needs to be explored in brackish water desalination technology. Understanding these technologies can make up for the deficiency of brackish water desalination technology in research.

Acknowledgments

The project support for this work includes: major science and technology projects of BINGTUAN: 2018AA003, graduate innovation project of Tarim University: TDGRI202039.

References

1. he Chinese Society of Agricultural Engineering, 2008,24(10): 243-247.
2. Q.S. Zhang, Z. Lou. Application of electrodialysis technology in desalination of amino acids, pharmaceutical intermediates, juice and high salinity wastewater . Qingdao International Desalting Conference, 2012.
3. L. Yu. The Application of Electrodialysis Technique in Food and Pharmaceutical Industry. China Environment Management, 2010.
4. Y.C. Xiao, R.Y. Chen, X. Zheng, et al. Preparation of P-mSA/mCS bipolar membrane and its application in electrosynthesis of 2,3,5,6-tetrachlopyridine. CIESC Journal, 2010, 61(2):525-530.
5. J. Zhao, B. Feng. Study on the electrodialysis desalination of amino acid from fermentation broth. Food and Fermentation Industries. 2006,32(9):32-36.
6. O.M. Kattan, H.J. Mengers, et al. On the isolation of single acidic amino acids for biorefinery
7. applications using electrodialysis. Journal of Membrane Science, 2011. 384: 166–175.
8. R. Benjamin, S. Philippe, et al. Application of relaxation periods during electrodialysis of a casein
9. solution: Impact on anion-exchange membrane fouling. ScienceDirect, 2007. 287:41–50.
10. V. Rodgers, R. Sparks. Effect of transmembrane pressure pulsing on concentration polarization. Membr. Sci. 1992, 68, 149–168.
11. S. Kim, E.M. Hoek. Modeling concentC. Chen, Y. Jiang, Z. Ye, Y. Yang, et al. Sustainably integrating desalination with solar power to overcome future freshwater scarcity in China. Glob. Energy Interconnect.2019,2, 98–113.
12. C. Zhao, Q.Q. Gao, Y.H. Zhang, J.J. Song, F.Z. Sun. Research progress of brackish water desalination technology. Water Resources Development Research. 2021,21(08) :61-65.
13. R.X. Hua, Y.Y. Gao, X.O. Feng. Experimental study on desalination of brackish water in plastic film evaporation shed. Gansu Water Resources and Hydropower Technology. 2021,57(06):16-20.
14. Y. Ibrahim, R.A. Ismail, A. Ogungbenro, et al. The sociopolitical factors impacting the adoption and proliferation of desalination: A critical review. Desalination 2021, 498, 114798.
15. A. Tappe. Investors Can Now Trade Water Futures—CNN. Available online: https://edition.cnn.com/2020/12/07/investing/water-futures-trading/index.html (accessed on 12 December 2020).
16. D. Zetland. The role of prices in managing water scarcity. Water Secur.2021,12, 100081.
17. K.A. Wang, .Mechanical properties of waterdesalination and wastewater treatmentmembranes. Desalination, 2017, 401:190-205.
18. T. Qiu, P.A. Davies. Comparison of configurations for high-recovery inland desalination systems. Water 2012, 4, 690–706.
19. A. Drak, M. Adato. Energy recovery consideration in brackish water desalination. Desalination 2014, 339, 34–39.
20. H. Yacubowicz, J. Yacubowicz. Nanofiltration: Properties and uses. Filtr. Sep. 2005, 42, 16–21.

21. W. Chen, M. Jin, P. Ferre, et al. Spatial distribution of soil moisture, soil salinity, and root density beneath a cotton field under mulched drip irrigation with brackish and fresh water. Field Crop. Res. 2018, 215, 207–221.
22. F. Ren, G. Yang, W. Li, et al. Yield-compatible salinity level for growing cotton (Gossypium hirsutum L.) under mulched drip irrigation using saline water. Agric. Water Manag. 2021, 250, 106859.
23. N. Ghaffour, M.M Thomas, L.A. Gary. Technical review and evaluation of the economics of water desalination: current and future challenges for better water supply sustainability. Desalination, 2013,309: 197-207.
24. A. Maurel. Brackish Water Desalination and Other Non-Conventional Processes for Water Supply. Desalination, 2006,4(5): 72-73.
25. Y.F. Song, X.F. Su, J.G. Zhou, B.W. Su, C.X. Gao. Application and research progress of brackish water desalination process with high recovery membrane method. Chemistry. 2015.78(01):79.
26. G.H. Xiao, L.J. Liu, C.W. Zhang, M.Z. Zhao. Experiment on culturing Penaeus chinensis in brackish water in saline alkali land. Hebei Fisheries. 2006(08):30-34.
27. M.S. Abujazar, S. Fatihah, A.R. Rakmi,et al. The effects of design parameters on productivity performance of a solar still for seawater desalination: A review. Desalination, 2016. 385: 178-193.
28. B. Chaouchi, A. Zrelli, S. Gabsi. Desalination of brackish water by means of a parabolic solar concentrator. Desalination, 2007. 217(1): 118-126.
29. Bundschuh, Jochen, et al. Low-cost low-enthalpy geothermal heat for freshwater production: Innovative applications using thermal desalination processes. Renewable and Sustainable Energy Reviews, 2015. 43: 196-206.
30. H. Aburideh, A. Deliou, B. Abbad, et al. An experimental study of a solar still: Application on the sea water desalination of Fouka. Procedia Engineering, 2012. 33: 475-484.
31. G.H. Xiao, L.J. Li,C.W. Zhang, et al. Experiment on Chinese shrimp culture in brackish water in saline-alkali soil. Hebei Fisheries,2006(08):30+34.
32. W.L. Xu, H.Q. Wang, Y.H. Li. Distribution and aquaculture application of saline alkali water at home and abroad . China Fisheries,2021(07):50-53.
33. D. Winter, J. Koschikowski, M. Wieghaus. Desalination using membrane distillation: Experimental studies on full scale spiral wound modules. Membr. Sci. 2011, 375, 104–112.
34. H. Chang, S.G. Lyu, C.M. Tsai, et al. Experimental and simulation study of a solar thermal driven membrane distillation desalination process. Desalination 2012, 286, 400–411.
35. M. Asim, N.T. Uday Kumar, A.R. Martin. Feasibility analysis of solar combi-system for simultaneous production of pure drinking water via membrane distillation and domestic hot water for single-family villa: Pilot plant setup in Dubai. Desalination Water Treat. 2015, 57, 21674–21684.
36. Q.S. Wang, T.J. Ying,C.Y. Zhao, et al. Application of electrodialysis technology on desalination of soybean oligosaccharides solution. Transactions of tration polarization in reverse osmosis processes. Desalination 2005, 186, 111–128.
37. H. Nassrullah, S.F. Anis, R. Hashaikeh, et al. Energy for desalination: A state-of-the-art review. Desalination 2020, 491, 114569.

Commissioning Analysis and Improvement measures for Biomass Circulating Fluidized Bed Gasifier

Shan Gu [1], Maosheng Liu [2,*], Chunxiao He [3], Jicheng Yang [1]

[1] Zhejiang University of Water Resources and Electric Power, Hangzhou 310018, China
[2] China Jiliang University, Hangzhou 310018, China
[3] Pyneo Co., Ltd., Hangzhou, Hangzhou 310012, China

Abstract. This paper presents the debugging test of biomass circulating fluidized bed gasifier in a rice mill factory of Jiangsu Province. The main debugging process of the gasifier system is reported. The typical problems of the gasifier system during the debugging period are analyzed, including the blockage of the primary cyclone separator, blockage of loop seal, bed inventory leakage of blast cap on air distribution plate, feeding tube coking, and large fluctuation of bed pressure. Reasonable improvement countermeasures and implementation plan is provided. The operation characteristics of the gasifier are preliminaries mastered through debugging, which provides reference for the operation and adjustment of the biomass circulating fluidized bed gasifier. These results provide certain guiding significance for the scale application of biomass circulating fluidized bed gasifier.

Key Words. Biomass Circulating Fluidized Bed Gasifier, Debugging Test.

1 Introduction

The development and utilization of biomass energy are of great significance to reduce greenhouse effect, air pollution and energy pressure. Biomass gasification is considered as one of the cleanest biomass utilization technologies. Due to its higher raw material adaptability, higher gas calorific value, higher production capacity, and wider adjustment range, the circulating fluidized bed (CFB) gasifier has a broader application prospect [1-2].

The commercial development of biomass CFB gasifier abroad is relatively mature. The leading countries in the field of biomass gasification abroad are Finland, Germany, the Netherland and so on [3-5]. At present, biomass gasification plants abroad are generally large in scale, high in automation and complex in process.

Research on biomass gasification technology in China started relatively late. After years of development, some progress has been made in the research and application of small gasification power generation, but it is mainly concentrated in fixed bed gasification, with problems such as low efficiency and small scale. At present, biomass circulating fluidized bed gasifier is still in the laboratory stage in China, and there are few reports of its industrial application.

The existing reports are basically applied to the gasification coupling large coal-fired power generation system, such as the 10.8×10^6 W biomass CFB gasification coupling power generation system of Guodian Changyuan Power Plant, and the 20×10^6 W biomass CFB gasification coupling system of Datang Changshan Thermal Power Plant [6]. There are few reports about the commissioning process and operational problems of biomass CFB gasifier.

This paper summarizes the commissioning and operation characteristics of 5 t/h biomass circulating fluidized bed gasifier in a rice plant in Jiangsu Province, which is aiming to provide reference for the large-scale application of this type of gasifier.

2 Equipment Overview

Biomass circulating fluidized bed gasifier is constructed in a rice plant in Jiangsu Province. The section size of dense phase zone in the lower part of gasifier is 1.2 m × 1.2 m, and the furnace wall tilts outward 30° and gradually transits to the dilute phase zone. The section size of dilute phase zone is 1.8 m × 1.8 m, and the total height of gasifier is 18 m. The schematic diagram of the CFB gasifier system are shown in Fig. 1, and main design parameters of gasifier are shown in Table 1. The gasifier adopts square furnace, which is composed of membrane water wall with external lining castable. The bottom of the furnace is provided with fluidized air chamber, and the ignition of light diesel oil under the bed is adopted. Biomass feeding system and quartz sand feeding system are set in front of the furnace. A two-stage adiabatic cyclone separator is arranged behind the furnace, and a loop seal is arranged under the primary cyclone separator

* Corresponding author: liums@cjlu.edu.cn

to make the material flow back to the furnace. The high temperature ash recovery system is arranged under the secondary cyclone separator, and the high temperature ash is recycled after cooling. The high temperature gas extracted from the upper part of the secondary cyclone separator is fed into the gas steam boiler for combustion to produce saturated steam. Rice husk is used as raw material for the gasifier. Table 2 shows the proximate and ultimate analysis of rice husk. Quartz sand is used as bed inventory. Table 3 shows the physical parameters of quartz sand.

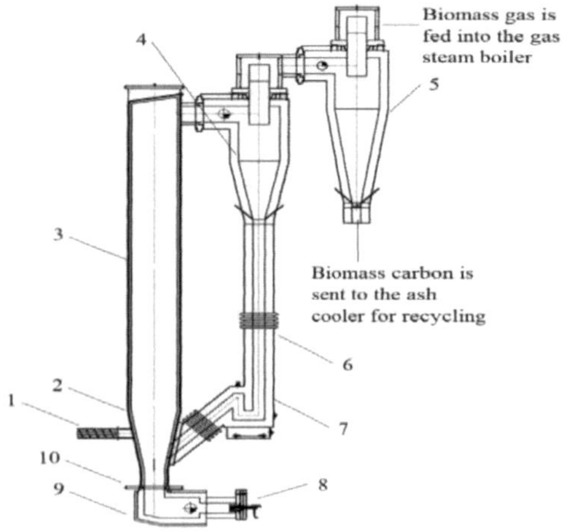

1-biomass feeder; 2-quartz sand feeder; 3-gasifier furnace chamber; 4-primary cyclone separator; 5-secondary cyclone separator; 6-return pipe; 7-loop seal; 8-ignition burner; 9-fluidized air chamber; 10-air distributor

Figure 1. Schematic diagram of biomass circulating fluidized bed gasifier system

Table 1. Main design parameters of biomass circulating fluidized bed gasifier

Item	Value
Gasification mode	Circulating fluidized bed
Design fuel	Rice husk
Fuel consumption	4.6 $t \cdot h^{-1}$ (equivalent to 10% water content, design value, nameplate consumption5 $t \cdot h^{-1}$)
Ignition method	Under-bed ignition by 0# light diesel oil
Bed temperature/℃	700-800
Bed pressure drop/Pa	7 500±100
Fluidizing velocity/ ($m \cdot s^{-1}$)	2-3
Gas production rate/($Nm^3 \cdot kg^{-1}$)	1.6-2.0
Air equivalent ratio	0.23-0.30
Gas calorific value/($kJ \cdot Nm^{-3}$)	5 000-6 300
Gas temperature/℃	<800
Carbon yield/($t \cdot h^{-1}$)	0.5-0.8

Table 2. Proximate and ultimate analysis of rice husk

Mass fraction, %								
Elemental Analysis					Industrial Analysis			
Cad	Had	Nad	Sad	Oad	Vad	Mad	FCad	Aad
39.97	3.87	0.44	0.25	34.90	63.62	7.74	15.81	12.83

Table 3. Physical parameters of quartz sand

Particle size/mm	Average particle size/mm	True density/ ($kg \cdot m^{-3}$)	Packing density/ ($kg \cdot m^{-3}$)	Critical fluidization velocity of cold state/($m \cdot s^{-1}$)	Particle terminal velocity/($m \cdot s^{-1}$)
0-2	1.4	2721	1522	0.62	3.0

3 Debugging Process of Gasification System

The debugging process of gasification system mainly includes three parts: monomer debugging, cold commissioning and hot state debugging. Monomer debugging includes the debugging of valve, bucket elevator, fan and other monomer debugging, such as baking boiler at low temperature and medium temperature, etc. Cold commissioning includes valve transmission and interlocking protection, air pressure test of gasifier and gas steam boiler, cold state aerodynamic field test, and assist system debugging, such as bed inventory system, dust discharge system, ignition system, air and flue gas system, etc. The cold aerodynamic field test is used to investigate the hydrodynamic characteristics of cold furnace and flue gas system. The test shall be conducted after the establishment of flue gas passageway from the gasifier and gas steam boiler furnace, through gas steam boiler outlet to draught fan inlet.

Hot state debugging includes hot state debugging of ignition system and dust discharge system, sodium hydroxide boiling furnace of gasifier and gas steam boiler, baking boiler of gasifier and gas steam boiler at high temperature, steam tightness and safety valve calibration, combustion adjustment test of gas steam boiler, unit start-up and 168 hours trial run. Before the first ignition of the system, all auxiliary systems need to have operational conditions. And rice husk, fuel oil and other resources should be fully prepared and meet the requirements of start-up and trial run of the system.

4 Main Problems and Treatment in Debugging

4.1 The Blockage of Primary Cyclone Separator

In the initial stage of hot state debugging, when the gasifier was still in the combustion state, before turning to the gasification state, several successive blockages occurred in the primary cyclone separator. The separator began to block from the throat of the cone section. The conical section and straight pipe section of the separator had serious ash accumulation. There was obvious ash accumulation in the horizontal flow of the primary cyclone separator.

 https://doi.org/10.1051/e3sconf/202132901040

On the one hand, the bed inventory of the gasifier was sea sand with high alkali metal content. When the temperature of the gasifier is higher than 700 ℃, massive quantity of alkali metal overflow from the bed inventory, resulting in the softening and aggregation of rice husk ash on the inlet flue of the primary separator and the outlet airway of the separator. The agglomerates fall off and block the throat of the separator. With the operation of the gasifier, more and more ash accumulates in the separator. The accumulated ash occurs secondary combustion with the increase of temperature, which makes the ash layer in the separator gradually firm. On the other hand, the exhaust hole in the return pipe was not opened according to the provision when baking boiler. This will result in the failure elimination of water vapor in the castable of the return pipe during the baking boiler. At the early stage of hot commissioning, the residual water vapor in the pipe rises to the primary cyclone separator, which makes the rice husk ash bond into blocks.

To solve this problem, the following measures can be adopted:

(1) Change the bed inventory from sea sand to quartz sand with particle size distribution range of 0-2 mm, to avoid the generation of alkali metal in the bed inventory.

(2) A steam/compressed air blowing device is arranged at the bottom of the cyclone separator cone (see as Fig. 2), and clean the separator periodically. Compressed air is used to clean ash in direct combustion state of gasifier, and steam is used to clean ash in gasification state. A manhole door is provided near the cyclone separator throat. When the return material is found abnormal or cyclone separator is blocked or bridged, priority should be given to cleaning the separator. If this problem still can not be solved, then blowout the furnace and clean dust.

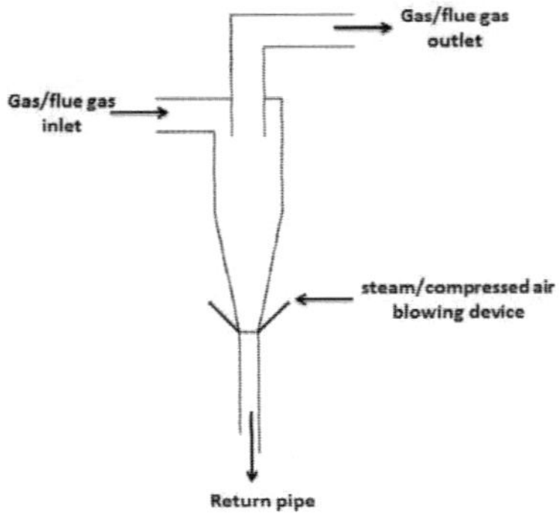

Figure 2. Example of steam/compressed air blowing device of primary cyclone separator

4.2 The Blockage of Loop Seal

After the gasifier was converted from combustion state to gasification state, the output of the feeder was increased to about 40%, and the load of the gas steam boiler reached about 50%. After running for a while, abnormal returning feed occurred. The loop seal under the primary separator of the gasifier was blocked by sintering ash, and the connection inclined pipe of loop seal was obviously agglomerated and blocked, as shown in Fig. 3.

Figure 3. The blockage of loop seal

The reason for the blockage of the loop seal is the secondary combustion of high temperature ash and high temperature combustible gas when meeting the returning air, which makes the loop seal overtemperature and high temperature ash sintering [7]. When superheated steam is used as returning fluidizing medium, no ash coke was generated in the purging part of the returning steam. However, if the returning steam is injected too early or unevenly, there will still be coking and blockage of the loop seal. When the returning steam is injected too early, the hot steam encounters cold materials and precipitates water vapor, and rice husk ash is bonded into blocks under the action of water vapor. When the returning steam is injected unevenly, the rice husk ash will still reburn and caking at the position where the steam is not penetrated.

To solve this problem, the following measures can be adopted:

(1) Air is still used as the fluidization feed-back medium in the gasifier during direct combustion, and the fluidization medium is switched to high temperature superheated steam into the loop seal before gasification. At the same time, maintain the material layer height of the loop seal, and control the appropriate returning air volume or steam volume.

(2) Adding steam nozzles to make injection uniforms. Six groups of steam nozzles are installed at about 200 mm above the loop seal caps, and the blowing angle and direction of steam nozzles are controlled. Insulate the steam pipeline, and change the steam hydrophobic position, to ensure that the returning fluidization medium is steam, rather than precipitated water vapor. Improve the pressure of returning steam, and keep the steam pressure at about 0.4-0.5 MPa.

4.3 Bed Inventory Leakage of Blast Cap on Air Distribution Plate

The blast cap on air distribution plate of the gasifier furnace adopted the mushroom caps with a diameter of 6 mm. However, the bed inventory leakage of ignition air chamber in operation was more serious. In serious case, the bed inventory needed to be replenished more than three times per operation shift, and the furnace needed to be cleaned each shutdown.

Because of the form of direct blowing of the mushroom caps, when the bed pressure fluctuates in the operation of the gasifier, the fine rice husk ash and bed inventory will leak into the fluidization air chamber through the holes of the caps along with the fluctuation of the bed pressure. Moreover, it is inconvenient to repair and replace mushroom caps. The mushroom cap is an integrated structure. The lower part of the cap is welded on the air distribution plate and buried under the refractory materials. When repairing and replacing the caps, refractory materials must be knocked off, leading to the large maintenance cost.

The main measure adopted is to replace the mushroom-shaped caps with the bell-shaped caps, the structure of which is shown in Fig. 4. The connecting pipe of bell caps extends into the top cover of the cap, and the hole of the cap does not directly contact with the bed inventory. Therefore, the leakage problem of mushroom caps can be solved. In addition, because the caps are not fixed in the refractory materials, when replacing the caps, only the top cover of the cap is removed without breaking the refractory materials, which saves time and reduces the maintenance cost [8].

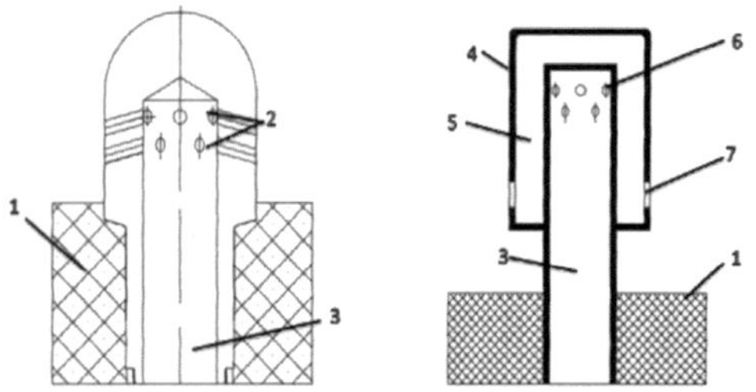

1-refractory material on air distribution plate; 2-hole of mushroom cap; 3- connecting pipe;
4-outer cover; 5-circumferential seam; 6-hole of inner core; 7- hole of outer cover

Figure 4. Structure of blast cap (Left: mushroom-shaped cap, Right: bell-shaped cap)

4.4 The Feeding Tube Coking

During the initial heating stage of hot start-up of gasifier, coking phenomenon occurred several times at the end of feeding pipe near the gasifier wall.

The reason is that the negative pressure control of gasifier furnace is unstable, and the feeding port often runs under positive pressure, leading to the backflow of biomass combustible gas into the feeding pipeline. When condensable gas in the combustible biomass gas meets the cold pipe wall, it condenses and precipitates tar on the feeding pipe, resulting in partial fuel bonding in the pipe. In addition to CO, H_2, CH_4, CnHm, CO_2 and other non-condensable gas, biomass gasification gas also contains some condensable components, commonly known as tar. Tar is gaseous at high temperature, mixed with product gas uniformly, and condensed at low temperature (< 200 °C), which can be easily bonded with water, coke, etc., and blocking pipeline [9]. Tar is inevitable during biomass gasification. With the increase of gasification temperature, tar production in gasification gas decreases rapidly [10]. Therefore, on the one hand, the negative pressure operation of gasifier should be controlled to prevent the reverse channeling of biomass combustible gas. On the other hand, the bed temperature should be strictly controlled to ensure that the gasification reaction is carried out at higher temperatures to reduce tar production. However, high gasification temperature should also be prevented from melting biomass ash and coking in the furnace.

To solve this problem, the following measures can be adopted:

(1) Ensure negative pressure operation of gasifier through operative adjustment. The negative pressure in the furnace at the feeding port should be controlled at about -200 Pa, and the temperature of the gasifier should be controlled at 700-800 °C.
(2) Setting up a steam purging device at the end of the biomass feeder near the furnace wall, periodically purging the end of the feeder.
(3) Adding temperature measuring points in biomass feeding pipe to control feeding pipe temperature. When the temperature in the tube rises, it should be adjusted in time to reduce the negative pressure in the furnace and prevent the reverse channelling of biomass combustible gas.
(4) Controlling the quality of biomass fuel. It is required that the moisture content of the raw material is less than 15% and the particle size is less than 20 mm, so as to prevent the blockage of the biomass feeding system due to the high moisture content and large particles.

4.5 Large fluctuations of bed pressure

In the initial stage of hot state debugging, the bed pressure of gasifier fluctuated greatly for many times. Taking the operation data on April 17, 2020 as an example: the gasifier ignited successfully at 13:28, and gasifier fuel oil was shutdown at 15:13. Bed pressure of gasifier fluctuated violently at 18:30, and stabilized slightly with the increase of bed inventory. Bed pressure of gasifier fluctuated again at 21:30, the bed pressure dropped from 7500 Pa to about 1900 Pa, the highest bed temperature rose to about 1000 °C, and the measures such as reducing fluidization air volume are ineffective. According to the program to stop the feeder at 23:30, the bed temperature dropped to about 600 °C. The bed pressure returned to normal in 7500 Pa at 23:45.

The reason for the violent fluctuation of gasifier bed pressure is that the thickness of the bed inventory is too small and uneven feeding lead to uneven fluidization of bed surface and large swing of bed pressure. Moreover, the mushroom-type blast cap is used on the air distribution plate, with a small empty plate resistance. Under rated fluidization air volume, the empty plate resistance of the blast cap is only about 800 Pa. When feeding or air supply is uneven, it is easy to cause uneven local fluidization of

the air distribution plate. When the thickness of material layer increases, the resistance of bed layer increases, the fluidization of bed surface improves, and the fluctuation of bed pressure improves accordingly. The bed pressure fluctuates again, because the amount of returned material does not match the amount of feed, which leads to the imbalance and severe fluidization in some areas of the bed, and the bed pressure drops sharply. The bed pressure drops to about 1900 Pa, the bed temperature rises to about 1000 °C, because the furnace differential pressure is low, and the material circulation in the furnace is difficult to establish, resulting in high bed temperature. After stopping feeding, the gas precipitation in the furnace is reduced, resulting in the reducing of the amount of feed and bed temperature. Then the feed tends to be stable, and the bed pressure is restored to normal.
To solve this problem, the following measures can be adopted:

(1) Strengthen quality control of biomass fuel to avoid large impurities into the furnace. The height of the starting bed inventory of gasifier should be controlled at about 500 mm, and the coarse particle size should be appropriately increased.
(2) Improve the operation adjustment level, so that the gasifier bed temperature is controlled at 700-800 ℃, and the fluidization air volume is controlled at about 2600-3000 Nm^3/h. Control the feeding speed, the first feeding amount is controlled at about 40 %, and the gasifier is in the combustion state. To change the gasifier from the combustion state to the gasification state, the feed quantity increases rapidly from 40% to 70%, and then the feed quantity increment is controlled at 2-4 %. Increase fuel and air volume sequentially, and keep the bed pressure and bed temperature stability. Timely observer of bed temperature, bed pressure, return valve temperature and pressure changes, predict in advance, timely adjustment, and ensure the stability of gasifier gasification or combustion.

5 Conclusion

Through the implementation and operational adjustment of the above measures, the gasifier system has gradually been upgraded to full load operation. During the subsequent commissioning of the same type of biomass circulating fluidized bed gasifier, the following measures can be adopted to ensure the operation of gasifier under negative pressure: (1) selecting low-alkali metal and anti-caking bed inventory; (2) using inert gas such as steam as fluidization medium; (3) controlling the quality of biomass fuel; (4) enhancing operation adjustment level of operators; (5) controlling of key parameters such as air-fuel ratio, bed temperature, bed pressure, temperature variation rates of returning feed. This paper summarized commissioning experience of 5 t/h biomass circulating fluidized bed gasifier, and provided a series of improvement measures. The results of this paper have certain reference value for the operation and adjustment of biomass circulating fluidized bed gasifier.

Acknowledgments

This work was financially supported by Science and technology project of Water Resources Department of Zhejiang Province fund (Preparation of biomass carbon carrier for ammonia production catalyst by electroreduction of nitrate wastewater).

References

1. G. Mirmoshtaghi, J. Skvaril, P.E. Campana, et al. The influence of different parameters on biomass gasification in circulating fluidized bed gasifiers [J]. Energ Convers Manage. 126 (2016) 110–123
2. Z L Liu, W X Peng, et al. Circulating fluidized bed gasification of biomass for flexible end-use of syngas: a micro and nano scale study for production of bio-methanol [J]. J Clean Prod, 129 (2016) 249-255.
3. D L Granatstein. Case study on Lahden Lampovoima gasification project, Kymijarvi Power Station, Lahti, Finland [R]. IEA Bioenergy: Task 36, 2002.
4. M Tiilikika, M Bolhar-Nordenkampf. Advanced gasification technologies for large scale energy production[C]. Power-Gen Europe 2014, Cologne, Germany, 2014. June 3-5
5. M Bolhar-Nordenkampf, J Isaksson. Operating experiences of large scale CFB-gasification plants for substitution of fossil fuels[C]. 24th European Biomass Conference and Exhibition, Amsterdam, the Netherlands, 2016, June 6-9.
6. S Gu, M S Liu, et al. Operational Characteristics of Biomass Circulating Fluidized Bed Gasifier [J]. Chem React Eng Technol (in Chinese), 37(2021),121-128+138.
7. H LWang. Cause Analysis of Slagging in CFB Boilers and Preventive Measures Thereof [J]. Therm Power Gener (in Chinese), 2007(2):28-30.
8. X Y Ji, X F Lu, et al. Design optimization of the bell type blast cap employed in small scale industrial circulating fluidized bed boilers [J]. Adv Powder Technol, 25(2014) 281–291.
9. Reference to a chapter in an edited book: L L Ma, C Z Wu, L Sun. Biomass Gasification Technology and its Application [M]. Beijing: Chemical Industry Press, 2003
10. C Y Deng, W J Song, et al. Characteristics of Tar Thermal Cracking and Catalytic Conversion during Circulating Fluidized Bed Char Gasification [J]. Energ Fuel, 34(2020):142-149.

 https://doi.org/10.1051/e3sconf/202132901041

Establishment of geology-engineering integration model for shale gas of Wei AH platform in Weiyuan Area

Hai Jiang [1,*], Yang Xiao[2,3], Ziping Liu[1], Zhigang Li[2], Xiang Deng[1] and Longqing Zou[4]

[1] Shale Gas Exploration and Development Project Management Department of CNPC Chuanqing Drilling Engineering Co., LTD, Chengdu, China
[2] School of Energy, Chengdu University of Technology, Chengdu, China
[3] Chengdu Institute of Technology Sunshine Energy Technology Co., Ltd., Chengdu, China
[4] Downhole Operation Company of CNPC Chuanqing Drilling Engineering Co., LTD, Chengdu, China

Abstract: Wei AH platform is a platform composed of 8 horizontal wells in Weiyuan shale gas area. The number of faults in the area is small, and natural fractures are locally developed. Therefore, it is necessary to take volume reconstruction as the core mining technology and establish the geology-engineering integration model for the platform according to the geology-engineering integration concept. On the basis of geological model, combined with one-dimensional rock mechanics profile, the geomechanical model of Wei AH platform is established by means of deep diagenetic theory. Based on geological model and geomechanical model, the well trajectory design model, fracturing design model and reservoir numerical simulation model are modeled and optimized respectively. Finally, it can realize the comprehensive application of geophysical, geological, mechanical and seepage data, and complete the mutual call and matching of all kinds of understanding and models.

Key words: Wei AH platform; Shale gas; Geology-engineering integration; Geomechanical model; Fracturing design model.

1. Introduction

Weiyuan Structure is located in the middle Sichuan gulong flat tectonic zone, Weiyuan to Longnvsi tectonic group [1]. The surface and abdomen structure of well Wei A area is consistent with simple structure, high in northwest and low in southeast, and the axis is near east-west. Wei AH platform is located in the west of well Wei A, where the number and scale of faults are small, and natural fractures developed locally. In order to realize the development of geology-engineering integration in Weiyuan shale gas field, a new connotation of geology-engineering integration with the core technical concept of "volume transformation" is put forward, and the geology-engineering integration model of the field is established [2].

More and more cases about oil and gas exploration and development showed that the technology of geology-engineering integration was an important means to realize the effective exploration and development of complex oil and gas reservoirs [3]. In order to reduce reservoir heterogeneity and modeling uncertainty, Jain et al. applied the idea of geology-engineering integration, combined with high-pressure physical properties, phase percolation data and geological information, to achieve optimal results for unconventional reservoir development [4]. Batch operations on small platforms based on geology-engineering integration technology can effectively save time, increase well drainage area, and ultimately maximize the interests of oil and gas company [5]. Wu et al. introduced the concept of integration of geology and engineering into the Sichuan Basin and the surrounding marine shale gas regions to realize the economic and effective development of shale gas [6]. The integration method of geology and engineering has guided the optimization and adjustment of the entire life cycle of shale gas reservoirs in southern Sichuan, effectively increased the production of single well and EUR, and supported the economic development of shale gas [7].

Based on the geological model, combined with rock mechanics experimental analysis and dynamic rock mechanics parameter from logging, considering the influence of factors such as structure, faults and natural fractures, establishes three-dimensional geomechanics model to simulate the influence of deep reservoir diagenetic environment and three-dimensional stress. Combined with geological model and 3D geomechanical model, simulating the true 3D artificial fracture network based on 3D attribute cube, finally to establish the geology-engineering integration model of Wei AH platform, and provide theoretical basis and technical

* Corresponding author: Jiangh_jx@cnpc.com.cn

support for optimization of fracturing methods and parameters.

2. Brief geological introduction of Wei AH platform and establishment of geological model

2.1 Brief geological introduction of Wei AH platform

The whole area of well Wei A has a trend of steep in north, gentle in south, shallow in west and deep in south. The northwest part is the southeast wing of the Weiyuan structure, which is distributed in a monoclinic manner. The burial depth of the target layer is relatively shallow as a whole. The burial depth of the Wei A is between 2100m and 3900m, with a trend of shallow in north and deep in south, and the overall pattern is similar to the structural map. The surface structure of Wei A area is relatively simple. The strata is steep in the north and gentle in the south. There are no faults. The faults of the Xujiahe Formation, the bottom of the Upper Permian, and the bottom of the Sinian are relatively developed, but the scale of faults is relatively still small. The fault's strike of Wei A area is multi-directional: nearly north-south, north-east, north-west, and nearly east-west.

Most of the blocks in the Wei A well area are buried within 3600m, which is a favorable area for shale gas exploration. The vertical change of the TOC in Weiyuan area is similar and have a good contrast relationship. They all have the characteristics of gradually getting better from top to bottom. Among them, the Longyi 11 layer has the highest abundance of organic matter, which is the most beneficial section for shale gas enrichment. Another condition for shale gas accumulation is the maturity, which is mostly within the range of thermal gas-generating window (Ro is from 0.5% to 2.5%).

2.2 Establishment of geological model for Wei AH platform

Integrating the seismic interpretation, logging interpretation and the fine gas reservoir description of the Wei AH platform, the three-dimensional geological model was established, including a natural fracture prediction model based on ant tracking, which provides a geological basis for the establishment of the geology-engineering integration model.

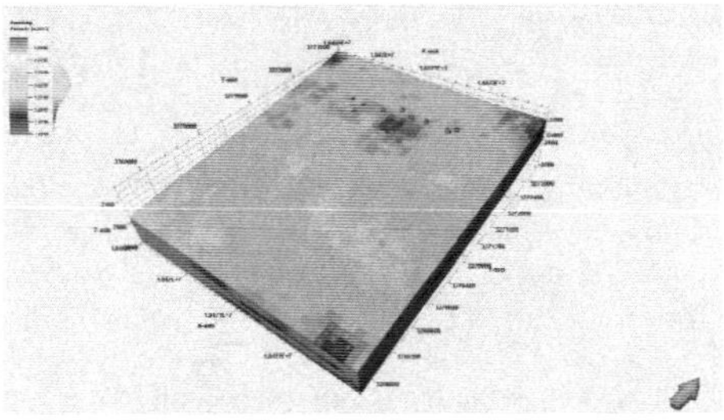

a. Three-dimensional porosity model

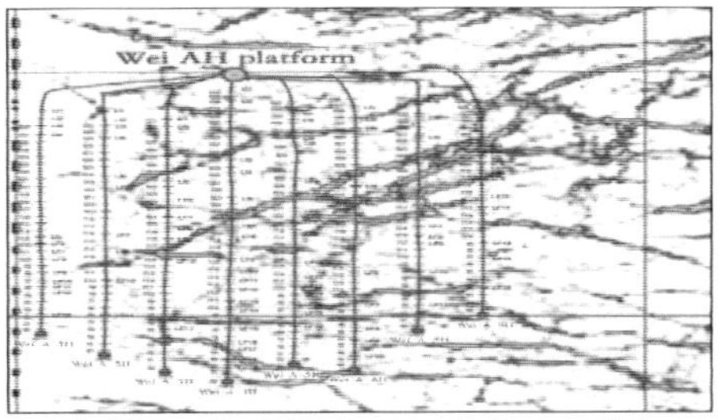

b. The natural fracture prediction model

Figure 1. The three-dimensional geological model of Wei AH platform

3. Establishment of geomechanical model for Wei AH platform

Based on logging interpretation, rock mechanics and acoustic experiments, the one-dimensional rock mechanics profile of the Wei AH well area was established. Synthesized geological model and one-dimensional rock mechanics profile of the Wei AH platform, the deep diagenesis theory was used to consider the influence of overlying, basement and surrounding rock, to simulate diagenetic process by applying circumferential horizontal stress and build the geomechanical model of the Wei AH platform, which provides a geomechanical basis for drilling and fracturing design.

3.1 Initialization of 3D geomechanical model

On the basis of single-well triaxial rock mechanics experiment and acoustic emission experiment data analysis, based on density, longitudinal wave and other logging curve sequences, the Young's modulus, Poisson's ratio, and geomechanical parameters such as horizontal stress are calculated. Through the fitting and correction of the dynamic and static parameters of reservoir rock mechanics, the one-dimensional geomechanical profile models of three wells were obtained.

Loaded the rock mechanics profile of the 8 wells of the Wei AH platform into the logging interpretation attribute body of the geological model, referring to the attribute body modeling process of porosity and permeability, and using Kriging interpolation to generate three-dimensional attribute body of rock mechanical parameters such as Young's modulus, Poisson's ratio, etc., finally to provide a basic model of geomechanical parameters for regional stress simulation.

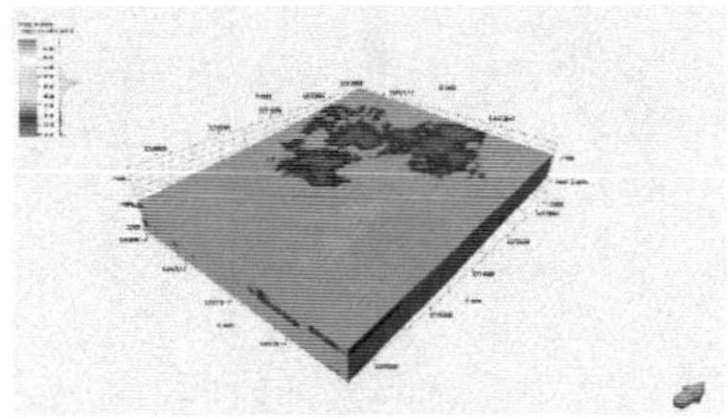

a. Young's modulus model

 https://doi.org/10.1051/e3sconf/202132901041

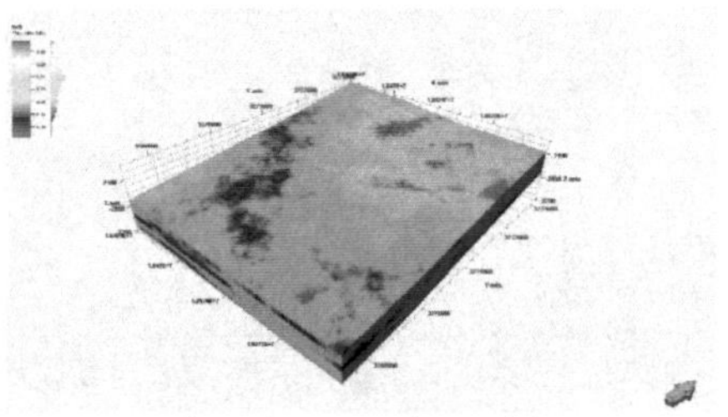

b. Poisson's ratio model

Figure 2. Initialization of 3D geomechanical model

3.2 The deep diagenetic theory and modeling

Different from the gas reservoir production model, the geomechanical model not only needs to consider the gas reservoir, but also the overlying rock, the underlying rock or the rock below the gas reservoir, as well as the lateral or surrounding rock that provides stress boundary conditions. Finite element simulation is widely used in stress analysis of conventional engineering and geomechanics, while finite difference simulation is used in fluid flow analysis. Connected above two types of analysis, it can be described and simulated the coupling characteristics when geomechanics and fluid flow change over time.

In order to simulate the real diagenetic environment of the deep reservoir, the vector displacement field and the continuous displacement compatibility equations were selected, the overlying, basement and surrounding rocks were added to the periphery of the study area, and the equivalent elastoplastic material was filled (Fig. 3a). The equivalent steel plate constitutive equation was selected to eliminate the stress concentration on boundary, and steel plate was added in the peripheral area (Fig. 3b).

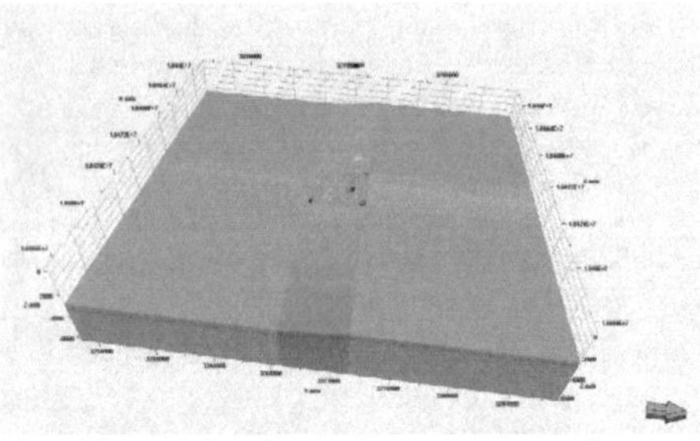

a. To add overlying, basement and surrounding rocks

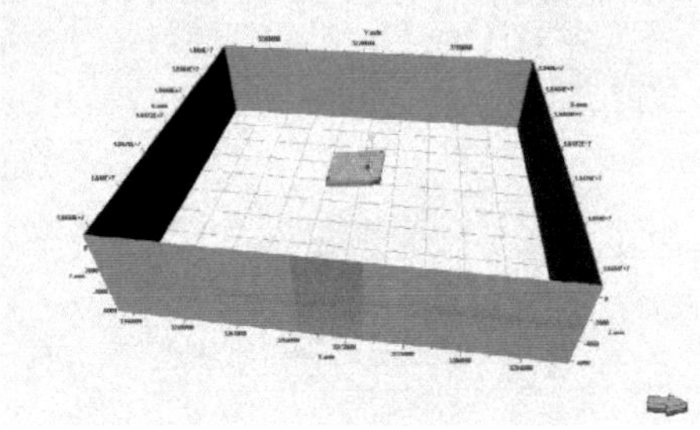

b. To add steel plate

Figure 3. Simulation of deep diagenetic environment

To load the circumferential stress, it is need to determine the current regional stress state and to load the vertical stress according to the rock filling condition, then we should integrate fluid simulation and stress simulation methods, starting from the fracturing changes predicted by the reservoir numerical simulation, to evaluate stress and strain changes caused by production and injection operations. Endowing certain temperature, pore pressure and three-dimensional in-situ stress to the reservoir formation, to simulate the rock diagenesis process, finally the current in-situ stress field distribution model of the research block can be obtained.

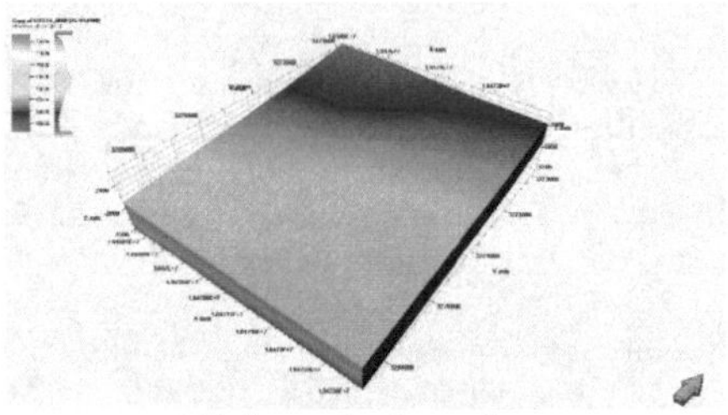

a. Minimum horizontal stress

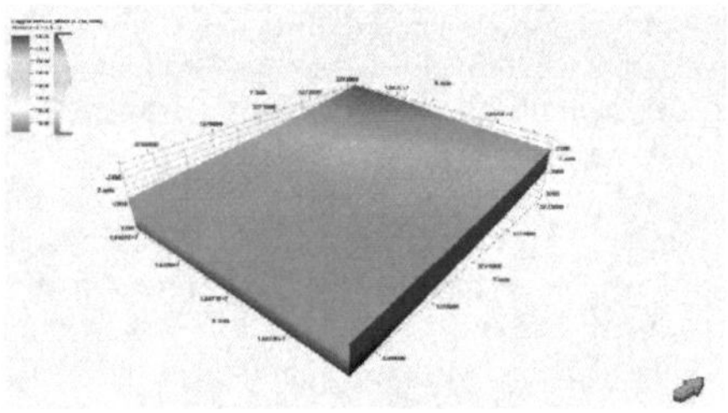

b. Maximum horizontal stress

Figure 4. Horizontal stress distribution map of Wei AH platform

4. Establishment of drilling and completion integration model of Wei AH platform

On the basis of geological and engineering dual sweet spot selection, synthesizing geological and geomechanical model to design well trajectory and fracturing optimization.

4.1 Integrated design for well trajectory

Optimizing the design of horizontal well spacing and row spacing by using the reservoir numerical simulation method. Because the integrated model of fracturing optimization design was considered, therefore, the integrated well trajectory design and fracturing optimization design were actually the process carried out simultaneously and verified by mutual matching. On the basis of geological model, geomechanical model and integrated fracturing optimization design model, geological targets were described and optimized by geological sweet spots such as porosity and gas saturation, and the geological targets were finally set in the

https://doi.org/10.1051/e3sconf/202132901041

geological model. It has laid the foundation for the well trajectory design.

While visualizing the well trajectory design, the engineering sweet spots of horizontal staged fracturing should be considered, especially the geomechanical sweet spots in the geomechanical model. The vertical well section-curved section-vertical well section (or curve section) design could be completed based on the optimization of compressibility evaluation parameters. After setting the well trajectory, using the Ellipsoid of uncertainty method (EOU) to complete the well trajectory correction.

4.2 Integrated design for fracturing optimization

The advantage of integrated fracturing design is to realize seamless connection of data from geophysical prospecting, geology, drilling, completion and reservoir numerical simulation, and to combine three-dimensional attribute models, natural fractures, geomechanics, and even micro-seismic fracture in hydraulic fracturing simulation.

The artificial fracture simulation adopts the UFM model, which can simulate the expansion, deformation and fluid flow of fractures in the complex fracture network, it can also consider the fluid flow in the fracture network and the elastic deformation of fractures. The UFM model can simulate the interaction between hydraulic fractures and pre-existing natural fractures, as well as the expansion of non-planar complex fracturing modes caused by branch fractures which produced by the intersection of hydraulic fractures and natural fractures. In addition to the interaction between hydraulic fractures and natural fractures, UFM also considers the interaction between adjacent hydraulic fracture branches by calculating the "stress shadow" effect of adjacent fractures on each fracture.

a. 3D view of fracturing model

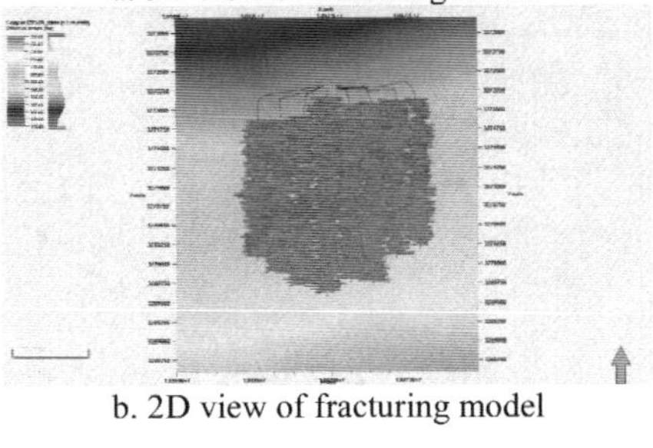

b. 2D view of fracturing model

Figure 5. Integrated fracturing model of Wei AH platform

5. Conclusion

First, the Weiyuan shale gas block has dense reservoirs and complex geological engineering conditions. It is necessary to apply the comprehensive application of geophysical, geological and geomechanical parameters, and introduce the concept of geology-engineering integration to effectively complete the optimization design of drilling, completion and mining method.

Secondly, the geology-engineering integration model of shale reservoirs needs to consider the deep diagenesis process to establish a geomechanical model. Based on the geological model and geomechanical model, the drilling and completion design were completed by synthesizing well trajectory design, fracturing optimization design model and reservoir numerical simulation model. And all the models are nested and mutually verified.

Acknowledgments

This work was financially supported by the National Natural Science Foundation of China (No. 51504042, No.51604043); Subitem of National Science and Technology Major Project (2016ZX05048-001-04-LH); Key Project of Sichuan Provincial Education Department Foundation(18ZA0063).

References

1. Guang Yang, Weizhi Tian, Jiang Lv, et al. Petrological Characteristics of Shale Gas Reservoirs in Long11 Sub-member of Longmaxi Formation in W202 Block of Weiyuan Structure, J. Special Oil & Gas Reservoirs, 2021, 28(2): 34-40.
2. Naizheng Liu, et al. Practice and prospect of geology-engineering integration technology in the efficient development of shale gas in Weiyuan block, J. China Petroleum Exploration, 2018, 23(2), 59-68.
3. Fuhao Zhao, Weian Huang, Rui Yong, et al. Research and application status of geology-engineering integration, J. Oil drilling and production technology, 2021, 43(2): 131-138.
4. A. K. Jain, H. Ahmad, P. K. Choudhary, et al. Integration of Geological, PVT and SCAL Data to Reduce Reservoir Uncertainty of an Unconventional Field in Kuwait, paper IPTC 18008 presented at the International Petroleum Technology Conference, Kuala Lumpur, Malaysia, December 10-12, 2014,
5. Yixiang Lyu. A Project Mode in Developing Oil and Gas on Artificial Islands: Cases Study of EPC, Batch Operation and Extended Reach Drilling, paper SPE 188663 presented at the Abu Dhabi International Petroleum Exhibition & Conference, Abu Dhabi, UAE, November 13-16, 2017.
6. Wu Qi, Liang Xing, Xian Chenggang, et al. Geoscience-to-Production Integration Ensures Effective and Efficient South China Marine Shale Gas Development, J. China Petroleum Exploration, 2015, 20(4): 1-23.

 https://doi.org/10.1051/e3sconf/202132901041

7. Chen Gengsheng, Wu Jianfa, Liu Yong, et al. Geology-engineering integration key technologies for ten billion cubic meters of shale gas productivity construction in the Southern Sichuan Basin, J. Natural Gas Industry, 2021, 41(1): 72-82.

Feasibility study on infilling of horizontal wells with difficult to recover reserves in Fuyu reservoir

Siqi Zheng

School of Resource Exploration Engineering, Daqing Petroluem Institute, Daqing, China

Abstract. Aiming at the inter well drive immobile remaining oil in Fuyu oil layer of block 1 due to the non densification of surface village coverage, taking into account the undeveloped reserves that can not be produced by vertical wells due to being located in the wetland protection area along the river, the feasibility study of layered densification of horizontal wells in the main oil layer with good reservoir development is carried out to explore a new densification mode of horizontal wells in Fuyu tight oil reservoir.

1 Raising questions

The development experience of block 1 shows that infill adjustment can improve the oil production speed and recovery degree of the block. In view of the inter well drive immobile remaining oil caused by the non densification of the surface village coverage, taking into account some of the undeveloped reserves that cannot be produced by vertical wells because they are located in the wetland protection area along the river, the layered design horizontal wells of the main oil layers with good reservoir development are preferred, and a new densification mode of horizontal wells in Fuyu tight oil reservoir is explored.

2 Feasibility study of horizontal well infilling

2.1 The structure of the proposed well area is gentle

Structurally, it is located in the West Wing of nose structure, showing a single slope dipping to the northwest, low in the West and high in the East in the east-west direction, and the stratum dip angle is 1-3.5 °. Based on the drilling data and seismic data, a three-dimensional fine geological model is established by well seismic combination. The structure of the proposed well area is gentle and there is no fault development in the horizontal section.

2.2 The main stratum in the proposed well area is prominent

2.2.1 The main oil reservoirs are developed stably.

Medium and small channel sand bodies are developed in main oil layers A, B and C in the proposed well area, with a width of 400 ~ 800m and a thickness of 2 ~ 5m, mostly distributary plain river deposits; The effective thickness of flattening is 3.2, 2.8 and 4.7m respectively. Small distributary channel sand body and inter river sand are mainly developed in non main oil layers.

2.2.2 High control degree of available reserves.

The continuity of planesA, B and C of the main oil layers is good. Horizontal wells are designed in layers. The average thickness of interlayer A and B is only 1.2m. The control degree of through layer fracturing reserves of horizontal wells is 56.4%. After deducting the influence of narrow channel sand body and inter river sand with an effective thickness of less than 1m, the control degree of available reserves is 73.1% (Table 1).

* Corresponding author: Zhengsq1@petrochina.com.cn

Table 1. Statistical table of effective thickness of reservoir in proposed well area.

Serial number	Well number	Effective thickness of layer A(m)	Effective thickness of layer B(m)	Effective thickness of layer C(m)	Effective thickness of the whole well (m)	Proportion of control thickness in the whole well (%)	Thickness of a and B compartments (m)	Available effective thickness (m)	Proportion of control thickness in available (%)
1	Wel 11	2.8	0	4.4	12.4	58.1		11.1	64.9
2	Wel 12	0	2.1	7.6	18.9	51.3		13.4	72.4
3	Wel 13	2.8	4	3.6	19.4	53.6	1.8	13.6	76.5
4	Wel 14	4.7	2	1.2	19.9	39.7	0.6	12.7	62.2
5	Wel 15	0.6	1.4	0	5	40.0	1.2	4.4	45.5
6	Wel 16	5.6	5	8.3	23.9	79.1	1.2	21.5	87.9
Total/ average average		16.5	14.5	25.1	99.5	56.4	1.2	76.7	73.1

2.2.3 The production proportion of main oil reservoirs is high.

The proportion of liquid production and water absorption of layer A + B + C accounts for about 70% of the total liquid production and water absorption.

3.1 The water flooding risk in the proposed well area is low

3.1.1 Fracture characteristics.

The maximum horizontal principal stress direction of Fuyu reservoir (near east-west direction). The fractures are relatively developed, and the fracture frequency is high, including natural fractures and micro fractures. The multidirectionality of fractures is obvious.

3.1.2 Impact of fractures on water injection development.

The proportion of water flooded oil wells in east-west direction is large. At the initial stage of block 1 development, 300m inverted nine point area well pattern is adopted. Water injection was started one year later. After water injection, the water content of East-West oil wells increased rapidly, and the casing pressure increased significantly after shut in. Four East-West oil wells were flooded five months later, and nine oil wells were flooded one year later, accounting for 81.8% of the number of East-West oil wells and 21.4% of the total number of oil wells.

3.1.3 The effective thickness and water washing ratio of infill wells are low, and the water flooding risk in the proposed well area is low.

The water flooding interpretation results of 48 infill wells in block 1are counted, of which 11 are expanded wells, and the water washing proportion of effective thickness is 12%; 37 wells are infill wells, and the water washing ratio of effective thickness is 15.9%. After being put into operation, the average water content in the initial stage is 26.1%. The risk of water flooding between old wells is low, and the risk of water flooding can be greatly reduced by arranging wells along the fracture direction (near east-west direction).

3.2 Through gradual encryption and adjustment, the development effect is gradually improved

The calibrated recovery factor of block 1 is 13.8%. According to the two algorithms of Tong Xianzhang chart and type a water drive characteristic curve, through gradual encryption and adjustment, the average recovery factor can reach 22.3%.

4 Horizontal well design and productivity prediction

4.1 Design 4 horizontal wells in two horizons

In view of the immovable inter well drive remaining oil caused by the non densification of the surface village coverage, taking into account some of the undeveloped reserves that cannot be produced by vertical wells because they are located in the wetland protection area along the river, the layered design of the main oil layer with good reservoir development is preferred. A total of 4 horizontal wells are deployed, with an oil-bearing area of $1.42km^2$ and geological reserves 27.1×10^4t (Table 2). The orientation of the horizontal well is close to the east-west direction and parallel to the fracture direction. The main stress direction can be used to prevent water flooding and prolong the water breakthrough period; Drilling horizon A_B、C, in which A and B are completed by through layer fracturing, and both small layers can be used; Single horizontal well is selected as well type; The length of horizontal section is 800-1500m.

Table 2. Prediction of controlled reserves of horizontal wells.

Serial number	Well number	Horizon	Single storage coefficient (10^4/m.km^2)	effective thickness (m)	Control area (km^2)	Controlled reserves (10^4)
1	Horizontal well 1	A_B	4.16	4.7	0.28	5.5
2	Horizontal well 2	A_B		6.5	0.22	5.9
3	Horizontal well 3	C		4.9	0.46	9.4
4	Horizontal well 4	C		3.3	0.46	6.3
Total					1.42	27.1

4.2 Production intensity in the proposed well area in combination with the production effect of similar blocks

4.2.1 Determination of oil production intensity of layer A-B.

The physical properties of reservoir in block 2 are similar to those in block 1. The main fracturing intervals are: A and B. the thickness of fracturing sandstone is 8.1m and the effective thickness is 6.8m. Referring to the experience of no fracturing and no oil production in Fuyu reservoir: the oil production intensity of layers A and B is 0.32 t/d.m.

4.2.2 Determination of oil production intensity of layer C.

Select 7 infill wells in Fuyu class II block where the main oil layer is C, other layers are located at the edge of the main sand body or non main sand body, and layer C is all fractured. According to the statistics of the liquid production profile at the initial stage of production, the oil production intensity of layer C is 0.35t/d.m.

4.2.3 Initial oil production intensity of fracture network fracturing in adjacent developed block 3.

The reservoir properties of block 3 and block 1 are similar. In 2014, the average fracturing sandstone thickness of 5 wells put into operation by large-scale fracturing was 19.0m, the effective thickness was 15.2m, the average daily liquid production of a single well was 12.2t, the daily oil production was 6.6t, and the water content was 46.2%. The oil production intensity is 0.43t/d.m.

4.2.4 Determination of oil production intensity in the proposed well area.

Based on the initial oil production intensity of adjacent developed areas and combined with the fracture pattern pressure investment effect, the cumulative injection production ratio in the proposed well area has reached 6.0, the formation energy is sufficient, When tight oil horizontal well volume fracturing is adopted for interval A, B and C (in which A and B are through layer fracturing). The predicted initial oil production intensity of layer A-B、C is 0.3 t/d.m.

4.3 Various methods are used to predict the initial productivity of horizontal wells and improve the accuracy of productivity prediction

Through theoretical formula calculation, the average daily oil production of single well in the initial stage of horizontal well in block 1 is 6.3t/d.

$$q_o = \frac{2n\pi K_h h_o (p_e - p_{wf})}{\mu_o B_o} \left(H_1 + \frac{\beta h}{L} \ln\frac{h_o}{h_f} + \frac{K_h h_o}{K_f c} \ln\frac{h_f}{2r_w}\right)^{-1} \quad (1)$$

$$H_1 = \operatorname{arcch}\left(\frac{\operatorname{ch}\frac{\pi b}{2na}}{\sin\frac{\pi L_f}{2a}}\right) \quad (2)$$

According to the analogy method, the reservoir physical properties of Fuyu oil layer in block 4 and block 5 of Chang 1 are similar to those in block 1. The average effective thickness of the development horizon of 13 East-West horizontal wells is 3m, the average length of the horizontal section is 572m, the average oil production in the 100m horizontal section is 0.5t, and the average daily oil production of horizontal wells in block 1 is predicted to be 6.0t in the initial stage.

Combining the two methods, it is predicted that the average daily oil production of a single well in the initial stage of horizontal well is 6.0t, and the built production capacity is 0.72×10^4t, newly increased recoverable reserves of 6.04×10^4t, initially prepared the well opening plan for horizontal wells in Fuyu oil layer of block 1, and extended the evaluation period under stepped oil prices to 15 years, with an after tax internal rate of return of 6.7%, higher than the industry benchmark rate of return of 6%, which is economically feasible.

5 Conclusion

The artificial fracture is near east-west, and the orientation of horizontal well is parallel to the fracture trend. The principal stress direction can be used to prevent water flooding and prolong the water breakthrough period.

The remaining oil between wells in Fuyu reservoir due to the non densification of the surface village coverage, and the undeveloped reserves that cannot be produced by vertical wells due to being located in the wetland protection area along the river can be effectively produced by layered design of horizontal wells, so as to improve the development effect.

References

[1] Weng Ding Wei, Lei Qun, et al. Fracture network fracturing technology and its field application. Journal of petroleum. 2011,32 (2): 280-284.
[2] Wu Dixiang, Zhang Jifen, Li Hujun, et al. Reservoir physics. Beijing: Petroleum Industry Press, 1994.
[3] Zhang Jifen, Zhao Mingguo, Liu Zhongchun, et al. Fundamentals of enhanced oil recovery. Beijing: Petroleum Industry Press, 1997.

https://doi.org/10.1051/e3sconf/202132901043

Construction Progress of Deep Underground Salt Cavern Gas Storage and Challenges of its Drilling and Completion Technology

Yi Zhang *, Jun Niu, Guoxing Chai Wenqi Ke and Hao Zeng

Petroleum Exploration and Production Research Institute of SINOPEC, Beijing, China

Abstract. With the rapid development of the national economy and the increasing demand for energy, and the acceleration of natural gas storage strategies, underground gas storage has become more and more important in China's oil and gas consumption and energy safety. Underground salt cavern gas storage has many advantages and is very suitable for gas storage and peak shaving。 However, most of the underground salt rocks in China are layered salt beds, and some of the salt beds are deeply buried. This article introduces the development history of deep underground salt cavern gas storage and the technical challenges faced in the process of drilling and completion of salt cavern construction.

1 Introduction

Underground gas storage is a key link in the natural gas industry chain and an important part of the pipeline transportation system. It not only adjusts the balance of supply and demand in the daily natural gas supply, but it also plays an irreplaceable role in an emergency situation in which an accident occurs. Underground gas storage has played a pivotal role in the natural gas storage system of various countries and has formed a relatively complete operating mechanism. Salt cavern gas storage uses dissolved salt mining to store natural gas in underground caves formed by mining salt rock in underground salt layers or salt domes as shown in Figure 1.

Figure 1. Schematic of underground gas storage

It has many advantages, such as flexible gas injection and production, large single well throughput, no leakage of gas storage, and high working gas ratio which is very suitable for gas storage and peak shaving [1, 2]. The underground salt rocks in China are mostly layered salt beds, and some of the salt beds are deeply buried (about 2,000 meters), which makes the construction of salt cavern gas storages face more technical challenges in the process of drilling and completion [3].

* Corresponding author: zhangyi5801@yeah.net

2 Deep Salt Cavern Gas Storage

In the world, the buried depth of the salt cavern gas (oil) storage constructed by the water solution method is generally within 500-1500m. For example, the buried depth of the Huntorf salt cavern in Germany is about 700m, and the buried depth of the Hackberry salt dome cavern in Louisiana, United States is about 800m. At present, the salt cavern gas storages in the world with a depth of more than 1500m mainly include the Harlingen brine production salt mine in the northern Netherlands (2900m), the French Vauvert brine production salt mine (1900m), the Eminence salt cavern gas storage in Mississippi (1800m) in the United States, Hornsea Aldbrough salt cavern gas storage in Yorkshire, UK (1800m), Alberta Strathacona salt cavern gas storage in Canada (1786m), and Hornsea Atwick (1730m) in Yorkshire, UK. The following is a survey of the situation of these salt cavern gas storages.

2.1 Harlingen Brine Salt Mine

The Harlingen brine salt mine is located in the middle area of the Zechstein Salt Basin in the Netherlands, which is part of the Permian Basin in southern Europe. The salt rocks in the salt basin are mainly salt domes and salt pillow structures, which present the characteristics of large thickness and high salt content. For example, the thickness of salt rocks is generally greater than 500m, and the salt content is about 95%. The Harlingen region is adjacent to the northern sea of the Netherlands and belongs to the province of Friesland. There are three gas storages in this area with a vertical buried depth of over 2900m. The gas storage was officially completed and put into operation in 1995. The cave is located in the Zechstein-2 salt rock layer, and the thickness of the salt rock layer is about 500m. The depth of the casing shoe is 2400-2500m, and there is a potash layer with a thickness of about 40m at a position 20-30m away from the casing shoe. The potash layer is mainly composed of carnallite, magnesium sulfate and potassium salt. The dividing line between the gypsum interlayer and the salt rock formation in the Zechstein-2 salt rock formation is relatively obvious, and the thickness of the gypsum interlayer is generally several centimeters to several meters [4].

2.2 Vauvert Brine Salt Mine

Located on the southeastern coast of France, the salt mine was once the main brine production base in France. The salt mine was discovered in 1962. The buried depth of the salt rock is between 1900m and 3000m. The salt content in the salt rock is about 50%. The other insoluble matter content is mainly gypsum and mudstone. Since 1973, the salt rock mining operation has been carried out by the water solution method. The Vauvert Salt Mine has 39 wells for brine production, of which 33 are located in the Parrapon mine and 6 wells are located in the La Galine mine. The distance between the two mines is 2.5km. The volume of a single cavity can generally reach 350,000 square meters to 1.4 million square meters. In the past 40 years, a total of 30 million tons of salt rock have been produced [5].

2.3 Eminence Salt Cavern Gas Storage

The Eminence Salt Cavern Gas Storage is located in the southwestern Mississippi, United States. The gas storage is located in a salt dome-type salt rock formation with a buried depth of between 1725m and 2000m. The salt content in the salt rock formation is about 95%, and the original in-situ stress of the target layer for building the reservoir is about 39 MPa. The first salt cavern gas storage in the area was completed and put into operation in 1970, namely Cavern-1. Cavern-1 was completed in January 1970, and the cavity shape was measured using sonar cavity measurement technology. The cavity height is about 250m and the maximum diameter is about 50m. After it was completed in June of the same year, gas injection and brine removal operations began, and gas injection and brine removal were completed in October of that year. From October to December 1970, the gas storage maintained an operating internal pressure of 6 MPa, after which the pressure increased to 28 MPa. In 1971, the same pressure mode was used for one year, but in the actual operation process, it was found that the injectable gas volume was significantly reduced, and the engineers suspected that the salt cavern volume had contracted greatly. In April 1972, engineers used sonar cavity measurement technology to measure Cavern-1 and found that the cavity bottom was raised by 36m, and the volume was reduced by about 40% compared with the measurement result in January, 1970 [6, 7].

2.4 Hornsea Aldbrough Salt Cavern Gas Storage, Yorkshire

The Hornsea Aldbrough Salt Cavern Gas Storage in Yorkshire, UK, is located on its southeastern coast and provides peak shaving gas for the British National Natural Gas Pipeline. The salt cavern gas storage is composed of a natural gas processing center and 9 caverns, and the roof depth of the gas storage is 1780m-1830m. The earliest time to put the gas storage cavity into operation was June, 2011, and the latest time to put it into operation was November, 2012. The target stratum for the construction of the gas storage is located in the Permian and is a component of the Zechstein Formation. The salt rock formation contains thin interlayers, which alternate with salt rock layers; the main components of the interlayers are dolomite, mudstone and gypsum. According to the results of geological exploration, it can be seen that the Fordon salt rock layer (the target layer for Aldbrough gas storage construction) is rapidly thinning to pinch in the north, and rapidly increasing in the east, with a thickness of more than 300m. In the Aldbrough gas storage area, the thickness of the salt rock formation is about 280 m, the top burial depth is 1660m, and the bottom burial depth is about 1940m [8]

2.5 Strathacona Gas Storage, Alberta

In Alberta, Canada, there are approximately 100 salt caverns used to store natural gas and industrial waste. They are concentrated in Fort Saskatchewan and Strathcona in Edmonton, Alberta (Strathacona). Strathcona is located in the northeast area of Fort Saskatchewan. The Strathacona salt cavern gas storage is built in the Lotsberg formation, which has a buried depth of about 1900m and a thickness of about 220m. The salt rock presents coarse-grained crystals with good transparency and the salt content can reach 99%. The six salt cavern gas storages are operated by Northwest Utilities Company. It is built in a section of salt rock strata with a thickness of about 78m, and the roof of the cavity is about 1786m deep.

2.6 Hornsea Atwick Gas Storage, Yorkshire

Hornsea Atwick gas storage is located in Yorkshire, England, adjacent to Aldrough gas storage. The total working gas volume of the gas storage is about 325×10^6 cubic meters, and the maximum daily gas production volume is 18×10^6 cubic meters. The cavity-building wellbore of Atwick No. 1 gas storage was completed in 1972, the first ultra-deep salt cavern gas storage built by the British Gas Company in Yorkshire. The gas storage is located in the Z2 salt layer. The top of the salt rock has a buried depth of 1770m and a stratum dip of about 2 degrees. It is mainly composed of salt rock, gypsum, polyhalite, potassium carbonate, potash and a small amount of magnesia sulfate. During the construction of the Atwick gas storage, a well encountered a carnallite formation and was close to the fault. The well was eventually abandoned to build a gas storage. Therefore, S-

https://doi.org/10.1051/e3sconf/202132901043

type well trajectories are used in the drilling process in this area later to avoid the adverse effects of this layer. Well 1 is an exploratory well. During the drilling process, the Z2 salt formation was cored. Through analysis of the core, it is found that the 20m thick formation in the upper part of Z2 salt rock contains 22% magnesia sulfate and a small amount of gypsum, and no potash is found. The salt cave in this area is designed as a cylinder with a dome, with a total height of 110m, a diameter of 95m, and a height of 10m of insoluble matter. In order to avoid the adverse effects of carnallite and magnesite rock formations on the cavity creation and airtightness of the gas storage, the casing shoe is designed to be buried at a depth of 20m (distance from the top of the salt rock). The thickness of the salt rock protection layer on the top of the salt cave is 20m, and the distance between the top of the salt cave and the upper layer of carnallite and magnesite is 40m (this part is salt rock) [9]

3 Drilling and Completion Process Challenges

According to the statistics of the International Natural Gas Union (IGU), there are currently 737 underground gas storages in the world. The total working gas volume of underground gas storages is $4110\times10^8m^3$, and the maximum daily gas recovery (PWR) is about $67.7\times10^8m^3/d$. Natural gas consumption annual growth rate has stimulated the global development of underground gas storage. Gas storage construction companies attach importance to basic research such as rock mechanics, string mechanics, and the management of well construction quality.

Gas storage injection and production wells are designed for life according to the principle of no workover for at least 30 years. At present, the Loenhout gas storage in Belgium has achieved this goal [10], and the TIGF gas storage in France has been in operation for more than 50 years. At present, only one well has abnormal casing pressure (except for the plugging well). The Norg gas storage in the Netherlands there was no abnormal casing pressure in the 6 production wells in the salt cavern for 14 years [11].

In order to meet the functions of seasonal peak shaving and emergency gas supply, gas storage injection and production wells need to meet the requirements of large flow injection and production, long life, and high safety. Larger-sized well structures should be adopted as far as possible. The material selection of the production casing and completion string of the gas storage is mainly determined by the quality of the gas injection and production. If there is a corrosion factor, the corresponding anti-corrosion material should be used, and the material selection is relatively conservative. For example, the gas production pipelines of Loenhout gas storage in Europe, Belgium, Yela gas storage and Norg gas storage in the Netherlands are all in diameter of 177.8mm [12].

Under the condition of alternating stress, the cement ring of the wellbore of the gas storage is prone to micro-cracks and micro-annular space, resulting in the risk of gas channeling and casing pressure. In order to improve the cementing quality, the following measures are mainly suggested:

a) In the design of gas storage injection and production wells, the running depth of single-layer casing should be controlled;
b) Pay attention to the cementing quality of each layer of casing and return the cement to the surface;
c) Using ductile expansive cement to reduce the influence of alternating stress on the cement ring;
d) Using self-healing cement technology to fill the micro cracks generated in the later stage;
e) Cementing quality inspection requirements are high, and ultrasonic imaging logging methods (such as IBC) are suggested.

4 Conclusion

Through decades of gas storage research and construction, people have greatly improved their understanding of the quality and requirements of gas storage construction, such as large-sized wellbore, reservoir drilling, horizontal wells, casing threaded gas seal testing, Advanced technologies and concepts such as elastic cement slurry and IBC logging have gradually been accepted by all parties in the construction of gas storage, which has played a very good role in promoting the construction of gas storage. Cementing should be the core of the drilling and completion of gas storage, and great attention should be paid to the wellbore integrity management of injection and production wells and old wells of gas storage. In particular, the quality of cementing of production liners and caps must be guaranteed.

References

1. P. Bérest, B. Brouard, J. G. Durup, Tightness tests in saltcavern wells, J. Oil & Gas Sci. and Tech. 56(5) (2002) 451-469.
2. W.G. Liang , C.H. Yang , Y.S. Zhao, et al. Experimental investigation of mechanical properties of bedded salt rock, J. Inter. Jour. of Rock Mecha. and Min. Sci. 44(3) (2007) 400-411.
3. T.T. Wang, H.L. Ma, X.L. Shi, et al. Salt cavern gas storage in an ultra-deep formation in Hubei, China, J. Inter. Jour. of Rock Mecha. and Min. Sci. 102 (2018) 57-70.
4. P.A. Fokker, G.A.M. Kruse, Salt deformation and subsidence, J. Learned and Appli Soil Mecha. 22, (2002) 57.
5. S.L. Furst, S. Doucet, P. Vernant, et al. Monitoring surface deformation of deep salt mining in Vauvert (France), combining InSAR and leveling data for multi-source inversion, J. Solid Earth, 12(1) (2021) 15-34.

6. K. Allen, Eminence dome-natural-gas storage in salt comes of age, J. Jour. of Petro. Tech, 24(11) (1972) 1299-1301.
7. K. Allen, Eminence--natural gas storage in salt comes of age, J. Soc. Pet. Eng. AIME (1971), 71.
8. T. Beutel, S. Black, Salt deposits and gas cavern storage in the UK with a case study of salt exploration from Cheshire, J. Oil Gas Europ. Maga. 1 (2005) 31-35.
9. D. J. Evans, S. Holloway, A review of onshore UK salt deposits and their potential for underground gas storage, J. Geo. Society, 313(1) (2009) 39-80.
10. J. G. Gallegos, S. Avila, R. Benyon, et al. Experimental evaluation of the performance of humidity analyzers in natural gas under industrial conditions, J. Jour. of Natural Gas Sci. and Eng. 31 (2016) 293-304.
11. B. Orlic, V. Thienen-Visser, G. J. Schreppers, Numerical estimation of structural integrity of salt cavern wells, 50th US Rock Mechanics/Geomechanics Symposium, OnePetro, 2016.
12. G. Yuan, C. Yang, B. Wang, Y. Drilling and completion technologies for the underground gas storage(UGS) in China: A state-of-the-art appraisal, J. Natural Gas Industry 33(2) (2013) 61-64.

Study on carbon emission accounting of warm mix asphalt mixture

Chengqun Wu [1], Wenyu Shi [2, 3], Haoming Shao [2]

[1] Nanjing Vocational Institute of Transport Technology; Nanjing 211188, China
[2] Jiangsu Engineering Technology Research Center for Energy Conservation and Emission Reduction of Modern Transportation; Nanjing 211188, China
[3] School of civil engineering and transportation, Hohai University; Nanjing 210098, China

Abstract: At present, the international community emphasizes reducing carbon emissions, and the national level includes carbon peaking and carbon neutralization as development goals. As a landmark material for energy conservation and emission reduction in road engineering, the research on carbon emission accounting needs to be further refined. Based on the introduction of the basic concept, development process and current situation of carbon emission accounting of warm mix asphalt mixture, this paper focuses on the most novel and recognized energy consumption accounting method and carbon emission calculation method at home and abroad. The research results of carbon emission accounting in the production, transportation and construction of warm mix asphalt mixture are introduced in detail. Finally, the research on carbon emission accounting of this material is evaluated at the present stage and prospected in the future.

1 Introduction

For a long time, the base material often used in road engineering construction in China is hot mix asphalt mixture, but it produces a lot of carbon dioxide in the process of mixing and construction, resulting in environmental damage. The appearance of warm mix asphalt mixture greatly reduces the temperature during mixing and construction and reduces the emission of carbon dioxide. At present, warm mix asphalt technology mainly adopts the way of adding materials to reduce the temperature, generally including organic viscosity reducer method and surfactant method. While ensuring its performance, it can reduce the mixing and compaction temperature to achieve the goal of energy conservation and emission reduction. From the perspective of energy conservation and emission reduction, this paper expounds the basic concept, development history and current situation of carbon emission accounting of warm mix asphalt mixture, analyzes and studies the best carbon emission accounting method at present, and studies the carbon emission accounting of warm mix asphalt mixture from the three aspects of production, transportation and construction.

2 Basic concepts of carbon emission accounting of warm mix asphalt mixture

Firstly, warm mix asphalt mixture is a method that uses some technical means to greatly reduce the mixing temperature of asphalt mixture and hardly change the performance of asphalt mixture, which is called asphalt warm mix technology

Secondly, carbon emission refers to greenhouse gas emissions, such as CO_2, CO and PM (particulate matter), which are all called greenhouse gases. However, since CO_2 is the main body of greenhouse gas emissions and is most well known by the general public, carbon emission can also be called carbon dioxide emission. At present, due to its own gas, carbon emission is vulnerable to the influence of temperature and pressure, generally in terms of mass (kg) Accounting as a company.

Finally, the carbon emission accounting of warm mix asphalt mixture is a kind of research on the specific accounting of carbon emission from the three links of mixture production, transportation and construction.

3 Development and present situation of carbon emission accounting of warm mix asphalt mixture

3.1 Development of warm mix asphalt (WMA)

In 1996, European countries developed warm mix asphalt by using the method of compounding soft asphalt and emulsified asphalt, and paved the test section. In terms of road performance, the effect of warm mix asphalt is comparable to that of hot mix asphalt, but the cost is about 20% higher than HMA [1].

In 2004, the zeolite addition method was introduced in detail in the European Asphalt Parliament. The method of adding admixtures opened another stage of the development of warm mix asphalt technology, and then the research and development of low-temperature organic viscosity reducer was carried out in Europe and Australia. Sasobit was launched in 2005. As a representative product of organic viscosity reduction technology, it has outstanding rutting resistance, which is due to the advantages of other additive products [2].

In 2005, warm mix asphalt technology was introduced and used for the first time in China, and a test road was paved at Badaling Expressway [3]. Subsequently, asphalt foam technology, organic additive method, mineral asphalt method and other related asphalt technologies have been introduced into China.

With the development of warm mix asphalt technology, ten kinds of warm mix additives have been developed, and four main technical methods have been developed: asphalt mineral method, emulsified asphalt warm mix method, organic additive method and foam asphalt warm mix method. In recent years, the research on warm mix asphalt technology has never stopped, and relevant road paving has been carried out. Therefore, its development prospect must be unanimously recognized by all countries and has unlimited potential.

3.2 Development of carbon emission accounting of warm mix asphalt mixture

The earliest Research Institute on carbon emission of warm mix asphalt mixture in China began in 2010, and there are not many such studies and innovative achievements in recent ten years. In 2010, Cheng Ling, Wu Jiazheng and others proved that the warm mixing technology is superior to the hot mixing technology from the perspective of energy conservation and emission reduction by means of measuring energy consumption through experiments [4].

Since then, domestic research has basically focused on three aspects: production, transportation and construction, and the study of carbon emission effect from different types of warm mix agents. So far, the study of carbon emission of warm mix asphalt mixture is relatively comprehensive, but its own impact on energy conservation and emission reduction effect The impact of different construction processes on its carbon emission is still a blank in this field, and further research is needed to fully explain its future development.

4 Study on carbon emission accounting method of warm mix asphalt mixture

4.1 Research on energy consumption accounting method

Among the three main activities of warm mix asphalt mixture, the main energy consumption is electricity, diesel, gasoline and natural gas. These energy consumption can be obtained by indirect or direct measurement of equipment. For example, electric energy can be indirectly obtained by measuring the use of standard coal in thermal power plants; Diesel fuel can be obtained by making specific statistics on the fuel consumption per 100 km of similar models.

According to the research of Yang Bo [5] and others on the quantitative estimation method of energy consumption and emission in the construction stage of asphalt pavement, they adopt the life cycle method (LCA), which is widely used at home and abroad, to quantitatively evaluate the environmental impact of process projects. It has the advantages of systematicness and comprehensiveness, and fully considers the consumption of various materials and energy in each stage of the life cycle of asphalt pavement, So as to realize the compilation and evaluation of input, output and potential environmental impact in the life cycle of a product system. They divided the quantitative estimation methods of life cycle energy consumption in the stage of asphalt pavement construction into actual measurement method, theoretical method and quota method. Among them, the actual measurement method pays attention to field measurement and statistics according to the energy consumption generated by the operation of specific equipment, but its scope of application is relatively limited; The theoretical rule is to calculate the energy consumption according to the energy consumption parameters specified during equipment production, which is representative to a certain extent; The quota method integrates the actual measurement method and the theoretical method, and is based on the project quota managed and improved by the national authoritative department. It is very representative, but it is only applicable to the construction technology and mechanical equipment within the specified range.

4.2 Research on carbon emission accounting based on emission factors

4.2.1 Carbon emission factor.

At present, for the quantitative measurement of carbon emissions, only a few institutions can directly measure the quality of carbon emissions. Most of the other quantitative measurements are quantified and analyzed by emission factors. For most European and American countries, the method of emission factors has long become the basic method for their carbon emission measurement, However, China is only in the field of road vehicle exhaust emission in the field of transportation through quantitative analysis of emission factors. Therefore, at present, the most widely used research method is the emission factor method, which is worthy of reference and innovation by subsequent researchers.

The so-called emission factor refers to the relative value of gas emission generated by unit activity. In the main activity of warm mix asphalt mixture, the carbon emission generated by each ton of mixture is called the carbon emission factor of production activity, and the carbon emission generated by each ton of fuel or other energy consumed by transportation vehicles is called the carbon emission factor of transportation activity, The carbon emission generated by each ton of fuel oil or other energy

https://doi.org/10.1051/e3sconf/202132901044

consumed by mechanical equipment is called the carbon emission factor of construction activities.

4.2.2 Carbon emission factor selection.

The more complex research process is to determine the emission factors, which are also different in different countries. As an accounting study, emission factors given by some authorities as shown in Table 1 are generally used as a reference.

Table 1 emission factor research institutions

Serial number	Organization type	Organization name
1	international body	IPCC (United Nations Intergovernmental Panel on climate change)
2	international body	EPA (National Environmental Protection Agency)
3	international body	EEA (European Environment Agency)
4	international body	Japan Energy Economy Research Institute
5	international body	Chinese Academy of Engineering
6	international body	Energy Research Institute of national development and Reform Commission
7	Urban institutions	Xiangtan Statistical Yearbook (2004)

4.2.3 Research on carbon emission accounting method.

According to the guidance given in the 2006 IPCC guidelines for national greenhouse gas inventories, for greenhouse gas emissions, the general carbon emission calculation method is: emission = activity level multiplied by emission factor. Activity level refers to the quality of carbon emissions per unit activity (kg), and the emission factors are selected according to the data given by major authorities. In addition, the guide also gives guidance on different calculation methods of carbon emission methods in different economic sectors. According to different sources of carbon emission, it is roughly divided into four categories: energy activities, industrial production, agriculture, forestry, land and waste disposal [6].

First of all, the calculation method of carbon emissions from energy activities. This kind of carbon emissions is obviously the main source of greenhouse gas emissions in the world today, because at present, the dependence on fossil fuels such as oil and coal is still at a considerable level in every country in the world. Our daily travel, diet and life, even education and medical treatment are inseparable from energy activities, and energy activities are inevitable According to the 2006 IPCC guidelines for national greenhouse gas inventories, there are three calculation methods that can be used to calculate the carbon emissions from fossil fuel combustion.

Method 1 carbon emission = the quantity of fossil energy burned multiplied by the average emission factor. The quantity of fossil energy burned here shall be selected according to the statistical data given by the region or country, and the emission factor can be used for all relevant greenhouse gases. However, due to the quality difference between gases, when calculating the emission of carbon dioxide, the gas containing "C" shall be considered first As the main body of greenhouse gas, the accuracy of carbon dioxide under this method can be improved to a high level. However, for others, such as methane and NOx, the combustion technology and conditions determine their emission factors. Due to different equipment and different development periods, the technical conditions are significantly different. Therefore, The carbon emissions of such gases are very different based on this method.

Method 2 carbon emission = the quantity of fossil fuels burned multiplied by the average emission factor of a specific country. The statistical method of the quantity of fossil fuels burned here is the same as that of method 1, but the selection of emission factors is not based on the emission factor data provided by IPCC, but based on the localization factors such as domestic fuel combustion characteristics and combustion technology of different countries, Such emission factors are more local and reliable, the calculation results of carbon emissions tend to be reasonable, and have more stability and authenticity for long-term monitoring and calculation of carbon emissions.

Method 3 carbon emission = the quantity of fossil fuels burned under special measurement conditions multiplied by the average emission factor of a specific country. Compared with scheme 2, the quantity of fossil fuels here is not monitored by theoretical method and quota method, but measured according to the detailed emission model or the data given by a single plant. The cost incurred by using this method It is expensive to use and needs to continuously monitor the combustion gas, so it is too redundant for carbon dioxide emission accounting. On the contrary, it is very necessary to monitor other pollutants (such as SO_2, NOx, etc.). In addition, if the flow rate of fuel is difficult to measure or changes greatly, continuous monitoring can also ensure the measurement accuracy.

Secondly, in industrial production, the calculation method of carbon emission. The carbon emission of industrial production is different from the direct combustion of fossil fuels. Its carbon emission is more likely to be the indirect combustion of fossil fuels into a product or a micro chemical reaction, such as chemical industry, metal industry, mining industry, etc.

Taking cement production as an example, limestone containing calcium carbonate ($CaCO_3$) is usually transformed into lime (CaO) by heating or calcination At this time, its by-product CO_2 will be released, and then lime will be mixed into raw materials for reaction to produce clinker, i.e. calcium hydrosilicate. Therefore, the carbon emission here has its specific calculation method: carbon emission = clinker output multiplied by clinker emission factor, and clinker emission factor = lime content multiplied by carbon dioxide relative molecular mass ratio / lime relative molecular mass ratio 。

Furthermore, the calculation method of carbon emissions from agriculture, forestry and land. The carbon emissions from agriculture, forestry and land are mainly the respiration and decomposition of animals and plants, or the combustion of some specific organic substances, such as grassland and forest seasonal fires. The main forms of carbon emissions are CO_2, N_2O and CH_4, in which carbon dioxide comes from the atmosphere and ecosystem; and CH_4 is mainly the interaction between animal manure and soil N_2O is an additional product of nitrification and denitrification.

The calculation of carbon emissions from agriculture, forestry and land is similar to the above two calculation methods. Refer to the 2006 IPCC guidelines for national greenhouse gas inventories , its calculation method is similar to the carbon emission of energy activities, and also has three methods. The activity level data collection method of method 123 is similar to the carbon emission of energy, and the selection of carbon emission factors is also divided into two forms: international general and national specific.

Finally, the carbon emission calculation method of waste disposal. Compared with CO_2 as the main source of emissions from agriculture, forestry and land, the carbon emission of waste is mainly methane (CH_4) The main sources are solid waste landfill, domestic sewage, industrial wastewater and other forms that are prone to produce methane. Generally, the first-order attenuation method is used to calculate the carbon emission of methane gas generated by solid waste landfill, specifically, the solid waste generation multiplied by the solid waste treatment rate multiplied by the methane carbon emission factor.

5 Study on carbon emission accounting of warm mix asphalt mixture in the whole process

5.1 Study on carbon emission accounting in the production process of warm mix asphalt mixture

The Canadian Asphalt Pavement Association once tracked the carbon footprint of a road with a total length of 2.4km. It was found that the production and preparation of mixture accounted for 73% of the total carbon emission, the site transportation accounted for 24% of the total, and the power consumption at the construction site was only 1% of the carbon emission [7]. Therefore, the production process of warm mix asphalt mixture not only produces a lot of carbon emissions, but also the focus of many carbon emission accounting research. In its production process, its energy consumption is related to carbon emission, such as aggregate transportation, warm mix production, aggregate screening, asphalt melting and transportation, mixture mixing, etc. Due to the combustion of fuel and the consumption of electric energy, these links lead to the output and emission of a large number of greenhouse gases, such as N_2O, CO_2, CH_4, etc.

According to the project example [8], as a fast track connecting Northeast China and East China, Tangjin Expressway selected some covered sections and built experimental sections in 2017, using warm mix and hot mix asphalt technology respectively. The test formula used is: 5% of base asphalt, Evotherm warm mix agent, 5% of asphalt, AC-13 asphalt mixture. Both mixtures are mixed in the same mixing plant, and the mixer model and raw material transportation vehicle type are the same. Then, the carbon emissions of aggregate stacking, asphalt heating, aggregate heating and mixture mixing are calculated and analyzed. The test results show that the carbon emissions of site stockpile and mixture mixing are exactly the same, and the biggest difference is in the asphalt heating stage. Due to the addition of warm mixing agent, the melting point of warm mixed asphalt has decreased by 200-300 compared with hot mixed asphalt, which greatly reduces the heat release, thus greatly reduces the carbon emissions and has a wider popularization prospect.

In addition, the research on carbon emission and energy consumption of the overall mixing equipment [9] confirmed that in the links of aggregate and asphalt heating, warm mixed asphalt mixture has an advantage, and the energy consumption and carbon emission of the mixing equipment are relatively small, which can effectively reduce the energy consumption and carbon dioxide emission in the mixing process. In addition, the study also applies to a variety of mixture types (such as AC-10, AC-16, AC-20, SMA-10, OGFC-16, etc.) the carbon emission accounting of mixing equipment is studied. The research shows that the carbon emission of warm mixing is always lower than that of hot mixing for any type of mixture; the energy consumption and carbon emission of the same type of mixture gradually decrease with the increase of nominal particle size; among different mixture types, OGFC is the largest, SMA is the second, and AC is the second Minimum. These conclusions also prove that the current selection of mixture in road engineering is indeed the preferred AC (continuous dense mixture), which can fully reflect the advantages of warm mix asphalt mixture in the production process, and make the development of warm mix asphalt technology become the key direction of road energy conservation and emission reduction.

5.2 Study on carbon emission accounting during transportation and construction of warm mix asphalt mixture

The research on carbon emission accounting during asphalt mixture transportation mainly includes two aspects: one is the carbon emission accounting of raw materials transportation stage, that is, driving trucks to transport raw materials such as aggregate, warm mix, asphalt and mineral powder from their location to the construction site or processing plant; The other is the mixture transportation stage, that is, the carbon emission accounting generated by driving the dump truck to transport the mixture from the mixing plant to the construction site. At this stage, the vehicle types generally include dump truck, truck, powder transport vehicle, asphalt tanker, etc. the fuel used by these vehicles is generally diesel, and the measurement unit is (L). In terms of transportation, the difference is small. Therefore, no relevant research has been carried out in this regard.

The research on carbon emission accounting during asphalt mixture construction includes two aspects: one is the research on carbon emission accounting during paving. Since this stage cannot be continuous and must undergo multi-layer paving, the calculation of energy consumption of paver needs to be recorded carefully for many times; On the other hand, it is the research on carbon emission accounting during compaction. The temperature generated during initial compaction is high, so its carbon emission is higher than that in the other two stages. Therefore, it is necessary to measure the main energy consumption and carbon emission accounting in the initial compaction stage. At present, there are three types of commonly used rollers: drum roller, vibratory roller and tire roller. For pavement construction, asphalt concrete paver is generally selected.

According to the study on pollutant emission during the construction of warm mix asphalt mixture [10], the pollution emission of several asphalt mixtures during construction is compared, and the indoor experiment is supplemented considering the relatively short test section and short sampling time at the construction site. The results show that compared with hot mix asphalt mixture, TOC (organic gaseous matter) emission is reduced by more than 50%, and TPM (total particulate matter) emission is reduced by 90%.

In addition, according to the study on the compaction performance of warm mix asphalt mixture [11], Sasobit warm mix agent, AC-16 mixture and 4.7% matrix asphalt are used for grading composition. The rotating compaction method is used, and the rotating compaction energy index is used as the evaluation criterion to test its compaction performance. The results show that warm mix asphalt mixture is easy to compact, and the lower the compaction work required. Therefore, it is indirectly concluded that in the paving and compaction stage, the mechanical energy consumption of warm mixed asphalt mixture is less, and its carbon emission is also low.

6 Conclusion

(1) The research on energy consumption accounting mainly uses three methods: actual measurement method, theoretical method and quota method. These three methods are basically the three basic frameworks of this kind of research, and their basic contents and basic development have been determined. On its theoretical development, after all, it is only used as a measurement method and does not need more in-depth research. However, as a specific link of measurement, in-depth research can be carried out on specific steps, such as the research and innovation of measurement instruments under the actual measurement method, and so on.

(2) Research on carbon emission accounting, under the guidance of the international organization IPCC on carbon emission accounting for multiple economic sectors, the carbon emission factor method, as an international general method, has been widely recognized in the international community. Undoubtedly, the universality and economy of its accounting have become the two factors most valued by all countries and regions. However, in some specific economic sectors, there are still blank areas of carbon emission accounting, and for some countries, the local average carbon emission factor bank and carbon emission accounting model have not been established. Therefore, these parts can be further improved in future research.

(3) The research on carbon emission accounting in the production process of warm mix asphalt mixture is the most studied aspect. It has the greatest impact on energy conservation and emission reduction, especially in the links of asphalt heating and aggregate heating. However, there is still a lack of a specific evaluation system that can be used to evaluate the advantages and disadvantages of its carbon emission. Such a system should not only consider the reduction of carbon emission, but also put the road performance of the mixture in the same position. In addition, the impact of different types of warm mix agents on carbon emission accounting and refinement of production process need to be further explored.

(4) Research on carbon emission accounting during transportation and construction of warm mix asphalt mixture, there is no obvious gap between these two links and ordinary asphalt mixture, and the economic effect of the research is low. At present, there is basically no specific detailed research. However, with the continuous substitution of fossil fuels by clean energy, such carbon emission research has its development prospect in the fuel substitution of mechanical equipment.

References

1. Li Zhongkai, practice of warm mix asphalt mixture technology in Europe. Sino foreign highway, 2010. 30 (04): 286-291
2. Yang Xiaojuan, Li Shuming and Shi Baohua, technology and application analysis of warm mix asphalt mixture. Petroleum asphalt, 2007 (04): 58-61
3. Li Yulei, research on application technology of warm mix asphalt mixture, Chang'an University, 2017
4. Cheng Ling et al., quantitative study on energy conservation and emission reduction effect of warm mix asphalt paving. Journal of environmental engineering, 2010. 4 (09): 2151-2155
5. Yang Bo, et al., study on quantitative estimation method of energy consumption and emission in asphalt pavement construction stage. Zhongwai highway, 2014. 34 (01): 7-13
6. Cheng Hao, how to calculate carbon emissions -- 2006 IPCC guidelines for national greenhouse gas inventories. China Statistics, 2014 (11): 28-30
7. Wu Junwei, development and application of carbon emission calculation and analysis model of road

 https://doi.org/10.1051/e3sconf/202132901044

engineering. Urban roads, bridges and flood control, 2011 (07):248-250 + 274 + 20

8. Zhang Xu, comparative case study on carbon emissions of warm mix and hot mix asphalt mixtures. China Municipal Engineering, 2018 (05): 92-94 + 113-114
9. Li Shouwei, research on energy consumption and carbon emission of asphalt pavement construction. Highway engineering, 2015. 40 (02): 257-259 + 305
10. Chen Zhipeng, impact analysis of warm mix asphalt mixture construction on environmental protection. Highway and automobile transportation, 2021 (01): 148-150
11. Zou Xiaobin, Sasobit, compaction performance of warm mix asphalt mixture. Fujian communications technology, 2017 (02): 5-7+106

https://doi.org/10.1051/e3sconf/202132901045

Review on Environmental Treatment of Heavy Pollution Industry

Jing Chen [1, 2], Xiaoxue Mu [3]

[1] Nanjing Vocational Institute of Transport Technology, Nanjing, China
[2] Jiangsu Engineering Technology Research Center for Energy Conservation and Emission Reduction of Transportation, Nanjing, China
[3] Business School, Hohai University, Nanjing, China

Abstract. The development of heavy pollution industry has a certain impact on the ecological environment. This paper combs and summarizes the environmental treatment technology of heavy pollution industry from the aspects of air pollution and water pollution. During studing the advantages and disadvantages of various environmental treatment technologies, the authors finds that there are the following problems in the development of environmental treatment technology of heavy pollution industry in China: there is a lack of long-term planning for air pollution environmental treatment; The cost of water pollution control is high and the energy consumption is large. Therefore, relevant environmental governance departments should promote the concept of sustainable development and strengthen the constraints of waste gas and sewage treatment process; Enterprises in heavy pollution industries need to learn and learn from the cutting-edge technologies of environmental governance; Social media positively guide enterprises in heavily polluting industries to protect the ecological environment through public supervision.

1. Introduction

In recent years, the status of heavy pollution industry in China's economic development is increasing day by day, but at the same time, a series of ecological and environmental problems brought by the development of heavy pollution industry can not be ignored. Among them, sewage discharge, global warming, dust storms and haze seriously threaten the natural environment on which human beings depend. The disharmony between the development of heavily polluting enterprises and environmental protection is the main contradiction in China's environmental governance. Only by developing green economy and promoting environmental governance can we fundamentally solve this problem. General Secretary Xi stressed that "green water and green mountains are golden mountains and silver mountains". At present, economic and social development should not only focus on immediate interests, but also concentrate on the sustainable development of future generations. China has entered a new stage of ecological civilization construction, and the environmental management of heavy pollution industry is imminent.

By using the literature research method, this paper comprehensively and deeply did research on the specific application of environmental treatment technology in heavy pollution industry from the aspects of the connotation of heavy pollution industry, the treatment technology and development prospect of air pollution and water pollution. Meanwhile, this study explored the weaknesses of existing research and places for optimization, striving to get suggestions and countermeasures for improving the development of environmental treatment technology in China's heavy pollution industry.

2. Research trends of environmental governance in heavy pollution industry

In 1962, American writer Rachel Carson published a wake-up masterpiece called silent spring, which triggered people's thinking on ecological and environmental problems, and the seriousness of environmental pollution had been displayed in front of people all over the world[1]. This study studied the characteristics and advantages of pollution control technology in relevant fields by sorting out the relevant technical documents of environmental treatment in heavy pollution industries at home and abroad, in order to provide a reliable basis for environmental treatment in heavy pollution industries in China in the future.

2.1 Research on the connotation of heavy pollution industry

There is no unified standard for the definition of heavy pollution industry in academia. Combing the literature of foreign scholars, it is found that the concept of "heavily polluting industry" rarely appears, but the common words are "dirty" industry or pollution intensive industry. Some domestic scholars choose the research object of "heavy pollution industry" based on the division standards stipulated by relevant national departments. The existing literature usually determines which industries belong to "heavy pollution industries" according to the standards such as pollution emission intensity, high and low pollution control cost, pollutant emission and the proportion of wastewater and waste gas emission in the overall industrial emission. The guidance on industry classification of listed companies revised by the CSRC in 2012 had further adjusted and standardized 16 heavily polluting industries on the basis of the division of the Ministry of environmental protection. Heavy pollution industries are divided into 16 sub industries, including thermal power, steel and cement. Its impact mainly includes air pollution and water pollution.

2.2 Research progress of environmental treatment technology in heavy pollution industry

(1) Air pollution and environmental control technology
Air pollution is not just gas pollution. Strictly speaking, air pollution is divided into particulate pollution, gaseous pollution and radioactive pollution[2]. Among them, the air pollutants emitted from the combustion of ore bodies and fuels in heavy industry are one of the main sources of air pollution. Bolen[3] found through research that modern pollution control measures had been adopted in particle plants in Sweden and the United States - circulating fluidized bed with sulfur control, electrostatic precipitator or bag filter with activated carbon and particles in the process of iron ore hardening or pelletization. Liu Huajun and Qiao Liecheng[4] collected the daily ground monitoring data of PM2.5 in Central European cities from 2017 to 2019 by using the pathon technology, quantified the spatial interaction degree of air pollution between China and Europe by using the big data transfer entropy method, and revealed the spatial network structure of the interaction of air pollution between China and Europe from the network perspective. Then they put forward targeted policy suggestions for China EU bilateral cooperation in air pollution control. The identification study of woody biomass samples found that southwest Nigeria had the ability to start large-scale production of these woody organisms. The vigorous investment in the reproduction of these energy crops would promote the use of woody blocks as raw materials for the sustainability of power supply in thermal power plants[5]. These woody fuels brought less air pollutants in the combustion process than traditional fuels, so as to reduce environmental pollution. Zhang Xiao et al.[6] prepared Mn/MCM-41 and IMp-Mn/MCM-41 catalysts by using metal heteroatom substitution and impregnation method, studied the catalytic oxidation performance of toluene in dielectric barrier non thermal plasma (NTP) reactor, and optimized the stability of the catalyst and the parameters of catalytic oxidation conditions. The characterization showed that the catalytic oxidation of toluene in air could be accelerated by the incorporation of manganese into the catalyst in a non thermal plasma system.
The development of heavy industry has brought a lot of air pollution, and coal combustion has brought a lot of sulfur dioxide pollution, especially in the case of serious haze in China, there are more and more inhalable particles and a large number of fine pollutants, which are easy to cause harm to people [7]. Guo Yi [8] studied the technical transformation of boiler dismantling and grid connection and the prevention and control of air pollution, dismantled small coal-fired boilers and incorporated them into large capacity boilers or thermal power plants. Thus, it improved the infrastructure of urban gas for central heating, strengthened the remediation of smoke and dust pollution, and improved the urban greening coverage, so as to achieve environmental protection and emission reduction. Shang Huihui[9] proposed that sulfur dioxide pollution can be reduced through coal desulfurization treatment, that is, coal biological desulfurization technology. Bioleaching used the catalytic enzymes and life activities in microbial strains to release the sulfur in coal. It is a common desulfurization technology. The coal purification technology also belongs to the treatment of pollution. It will greatly reduce the emission of sulfur dioxide in China and effectively alleviate the pressure of air pollution control in China.
(2) Water resources pollution and environmental treatment technology
Wastewater from heavily polluted industries contains organic and inorganic pollutants. Environmental treatment technologies of water resources pollution mainly include biological method, physical method, chemical method and other treatment technologies [10]. In order to study the efficiency of heavy metal adsorbent in industrial wastewater, elboughdiri [11] studied the adsorption behavior of natural zeolite on Cu ions, Pb ions and Cd ions in synthetic metal solution. The research showed that zeolite was suitable for removing Cu, Pb and Cd ions from synthetic wastewater; It is found that the quality of adsorbent, initial solution concentration, initial solution pH, adsorbent particle size and stirring speed affect the removal efficiency of heavy metals in natural zeolite industrial wastewater. Dyes and heavy metals pose a serious threat to drinking water. These pollutants may be harmful or carcinogenic in nature and harmful to humans and different aquatic species. Therefore, it is of great significance to remove heavy metals and dyes and reduce their environmental toxicity.
Domestic scholar Jiang Ting [12] proposed to adopt environmental chemistry technology to treat the pollution of wastewater, waste gas and waste residue. In the process of wastewater treatment, the most widely used ultrafiltration technology is to separate and concentrate wastewater by using microporous aluminum membrane. This technology has the characteristics of high efficiency and high quality. Environmental chemistry technology has great advantages in environmental treatment. Its treatment cycle is short and the investment cost is low. The incidence of secondary pollution is reduced, so the pollution source can be effectively controlled with the result that the pollution treatment effect is greatly improved, and the efficiency and quality of environmental treatment are improved.

3 Problems of environmental treatment technology in heavy pollution industry

(1) Lack of long-term planning for air pollution and environmental control

Reducing pollutant emissions and controlling pollution sources are the main breakthroughs of air pollution and environmental treatment technology. The aggravation of environmental pollution has a direct impact on China's air quality. The government has achieved good results in short-term environmental pollution control through effective haze control measures such as "dust reduction, coal pressure and greening". The environmental governance of heavy pollution industry should promote each other with economic development, but the long-term planning of environmental governance of heavy pollution industry lacks dynamic mechanism and long-term governance planning of sustainable development, resulting in the opposition between environmental governance and economic development. Air pollution control technology needs to be further improved. It is necessary to speed up the development of air pollution control technology from the long-term interests of ecological and environmental protection.

(2) High cost and energy consumption of water pollution and environmental treatment

When carrying out water pollution control, we should scientifically select various ecological restoration technologies, such as biofilm technology, microbial remediation technology and aquatic vegetation restoration technology, according to the requirements of relevant specifications and in combination with the actual situation of water pollution type, pollution scope and pollution degree in order to control and protect the water environment, then promote the sustainable and healthy development of the water environment[13]. Although this kind of efficient and friendly microbial enhancement technology which helps to protect the ecological environment has high application value, it also has defects such as high cost and high energy consumption. Enterprises may abandon this technology because they are unable to pay high fees in sewage treatment, resulting in increased environmental pollution.

4. Analysis on improvement countermeasures of environmental treatment technology in China's heavy pollution industry

4.1 Government level

Relevant environmental treatment department shall, in accordance with the concept of environmental sustainable development advocated by the state and in combination with relevant schedule planning, reasonably arrange the pollution treatment of heavy polluting industries, make internal management orderly, and conduct effective research on the environmental treatment of foreign heavy polluting industries. The government should actively learn from relevant experience and optimize the governance scheme to make it more in line with the actual situation. In addition, we should standardize the order of the national environmental market, strengthen the constraints of relevant systems on relevant waste gas and sewage treatment managers. And the department should severely crack down on the adverse phenomenon of paying attention to interests and ignoring the effect of sewage treatment in carrying out specific work. In view of the unqualified sewage discharged by industrial production enterprises, the government should do a good job in management to promote the rational operation of sewage treatment of industrial enterprises abiding by the national sewage treatment system and rigid regulations.

4.2 Companies level

The development of enterprise treatment technology in heavy pollution industry needs to start from itself, improve the level of technical development, and apply it to promote the improvement of environmental pollution treatment capacity and efficiency of China's heavy pollution industry. In terms of technological innovation, enterprises should actively introduce advanced high-end technology; At the same time, strive to go global, actively grasp the technical learning opportunities of heavy pollution industry and environmental treatment industry, learn skills in communication and exchange, and improve the pollution of air and water waste in China [14].

4.3 Social dimension

The public also plays an important role in the process of environmental governance of heavy pollution industries. The media can strengthen the effective supervision of public opinion, release relevant public opinion through multiple channels such as the Internet, and positively guide enterprises in heavy pollution industries to actively protect the ecological environment and bear the main responsibility of environmental governance. People should actively support environmental protection industries, consume less or do not consume enterprise products with high energy consumption. They should try hard to reduce environmental pollution caused by heavy polluting enterprises with high consumption and high pollution, and promote the development of environmental treatment technology in heavy polluting industries as a result of further promoting the vigorous development of economy and society.

5. Conclusions

On the basis of combing the environmental treatment technologies of heavy pollution industries at home and abroad, this paper studies the environmental treatment technologies of heavy pollution industries from the aspects of air pollution and water pollution; Based on the relevant research at home and abroad, this paper summarizes the advantages and disadvantages of the current development of environmental treatment technology in heavy pollution industry, and puts forward the environmental treatment countermeasures and suggestions suitable for China's national conditions.

Acknowledgments

This work was financially supported by Self financing project of Jiangsu Social Science Foundation (Grant No. 20GLD009).

References

1. He Ke, Zhang Junbiao, Zhang Lu, Wu Xuelian. Interpersonal trust, institutional trust and farmers' willingness to participate in Environmental Governance-Taking the recycling of agricultural waste as an example [J]. Management world, 2015 (05): 75-88
2. Hao Yiren, Meng Chao, Liu kunkun. Analysis on the development of industrial air pollution control technology [J]. China's comprehensive utilization of resources, 2020,38(06): 132-134
3. Bolen, J. Modern air pollution control for iron ore induration[J]. Minerals & Metallurgical Processing, 2014,31(2).
4. Liu Huajun, Qiao Liecheng. Spatial interaction network of air pollution between China and Europe and bilateral cooperative governance -- An Empirical Study Based on big data causal inference technology [J]. Statistical research, 2021,38 (02): 45-56
5. F. B. Elehinafe et al. Identification of the woody biomasses in Southwest, Nigeria as potential energy feedstocks in thermal power plants for air pollution control[J]. Cogent Engineering, 2021, 8(1)
6. Zhang Xiao et al. Catalytic oxidation of toluene in air using manganese incorporated catalyst by non-thermal plasma system[J]. Separation and Purification Technology, 2021, 257
7. Shang Huihui. Application of environmental chemistry technology in environmental pollution control [J]. Chemical design communication, 2019, 45 (09): 236+238
8. Guo Yi. Technical transformation of boiler dismantling and grid connection and control of air pollution [J]. Chemical engineering and equipment, 2020 (11): 280-281+291
9. Shang Huihui. Application of environmental chemistry technology in environmental pollution control [J]. Chemical design communication, 2019, 45 (09): 236-238
10. Men Yonghong, Chen Xiujin, Du Yu, Xu Xuemi. Problems and solutions in industrial sewage treatment in environmental engineering [J]. Management and science and technology of small and medium-sized enterprises (next ten days), 2021 (04): 61-62
11. Noureddine Elboughdiri and Harvey Arellano Garcia. The use of natural zeolite to remove heavy metals Cu (II), Pb (II) and Cd (II), from industrial wastewater[J]. Cogent Engineering, 2020, 7(1)
12. Jiang Ting, Zhu Huijun, Yan Chao, Han Rongrong. Application of environmental chemistry technology in environmental pollution control and protection [J]. Biochemical industry, 2019,5 (02): 117-119
13. Ma Ji. Application of bioremediation technology in water pollution control [J]. Comprehensive utilization of resources in China, 2021,39 (04): 192-194
14. Liu Jiaming. Study on the path of environmental transformation and upgrading of China's heavy pollution industry [D]. Beijing University of chemical technology, 2020

Experimental study on biological dehydration pretreatment of kitchen waste

Yi Lin and Jinliang Sheng*

School of Mechanical Engineering, Tongji University, Shanghai, China

Abstract. Aerobic fermentation treatment technology is one of the effective means of kitchen waste resource treatment at present. Due to the high moisture content of kitchen waste, dehydration pretreatment is needed before aerobic composting.At present, China usually adopts mechanical methods such as forced extrusion for dehydration treatment, which not only has high energy consumption, but also has high content of solid suspended solids (SS) and chemical oxygen demand (COD) in sewage. Sewage is difficult to meet discharge standards.In this paper, biological dehydration technology is proposed to pretreat kitchen waste. Experiments show that biological dehydration pretreatment can dehydrate kitchen waste, effectively reduce the SS and COD values in sewage, which is of great significance to the resource treatment of kitchen waste.

1 Introduction

On July 1, 2019, Shanghai took the lead in implementing the "Shanghai Municipal Household Waste Management Regulations" to classify and treat household waste in four ways . Among them, kitchen waste (wet waste) is mainly organic waste such as fruits, vegetables and leftovers discarded in daily household life, with a high annual output.

Kitchen waste is rich in organic substances and has the characteristics of perishability, high water content and high organic content. After proper treatment, it can be transformed into new resources [1]. Aerobic fermentation treatment technology is one of the effective means of kitchen waste resource treatment at present.

A large number of studies have shown that moisture content is an important factor affecting aerobic fermentation treatment technology, usually controlled at 55%~60%. Therefore, kitchen waste should be pre-treated by dehydration before aerobic composting.

At present, mechanical methods such as gravity, centrifugal force and forced extrusion are usually used for dehydration treatment in China, which not only has high energy consumption, but also high content of solid suspended solids (SS) and chemical oxygen demand (COD) in sewage, making it difficult to meet the sewage discharge standards [2, 3].

In this paper, through the biological dehydration technology of kitchen waste pretreatment, namely to join in the kitchen waste aerobic bacteria, anaerobic bacteria and enzymes and other microbes, degradation of kitchen waste components such as cellulose, starch, protein, make the kitchen waste part of the free water and bound water release, so as to achieve the effect of biological dehydration. At the same time,Microbes generate a lot of heat in the process of metabolism, which also helps dehydrate kitchen waste[4, 5].

2 Experimental materials and methods

2.1 Experimental materials and devices

The raw materials of kitchen waste in this experiment are from Wusi Farm, Fengxian District, Shanghai, and the microbes used are from Sukehan Bioengineering Co., LTD.Biological dehydration experimental device of kitchen waste adopts biochemical reactor of Shanghai Tongjian Yuanda Environmental Protection Mechanical Engineering Co., LTD., as shown in Figure 2, the experimental device consists of biological dehydration chamber, crushing mechanism, stirring mechanism, heating mechanism, drainage tank and so on.

The crushing mechanism for kitchen waste shear crushing, stirring mechanism to kitchen waste and microbial fully mixed, heating device for heating reactor;The effective volume of the biological dehydration chamber was 0.5m3, and 200kg of kitchen waste was injected as experimental raw material.

* Corresponding author: sjl@tongji.edu.cn

Figure 1. Experimental raw material.

Figure 2. Experimental apparatus.

2.2 Experimental methods

In this experiment, microorganism (A) and biological dehydration treatment time (B) were selected as the influencing factors, and the moisture content of kitchen waste (i), PH value of kitchen waste (ii), solid suspended solids SS (iii) and COD (iv) of sewage were used as the investigation indexes.

In the first group, bacillus subtilis was added to decompose cellulose. In the second group, aerobic bacteria, anaerobic bacteria and a variety of enzymes were added to decompose protein, fat and cellulose.

The moisture content and PH value of kitchen waste at 10 hours and 12 hours after material input were tested. SS and COD were measured in sewage samples at 1h, 3h, 6h, 8h, 10h and 12h after material input.The moisture content of kitchen waste is dried and measured by hX-101-00BA4 electric thermostatic drying oven of Itong Century Technology Company; the PH value of kitchen waste is measured by Xima PH 848 test pen; the SS and COD values of experimental sewage samples are measured by Shanghai Kai Lin Environmental Protection Technology Co., LTD.

Figure 3. Electric heating constant temperature drying oven.

3 Results and analysis

3.1 Water content and PH value

The moisture content of the two groups showed an obvious downward trend, as shown in Table 1. Microorganisms effectively degraded the components of kitchen waste, making it release water.

The dehydration effect of the first group was better than that of the second group because the experimental raw materials contained a large number of vegetables with high cellulose content and low protein, starch and other components. The microorganisms added in the first group were mainly Bacillus subtilis, which destroyed cellulose, and could destroy plant cell walls and release water from cells.

The second group contained many proteases and complex strains that destroyed components such as proteins and starches, which were less effective in dehydration than the first group.

When nitrogen sources in kitchen waste are utilized by microorganisms, pH increases.The raw material selected in this experiment contains a large number of vegetables (pakchoi) with a low C/N ratio (C/N) of about 10[6] and rich nitrogen source. Therefore, the PH value of the experimental raw material after 12 hours of microbial action is weakly alkaline.

Table 1. Moisture content and PH value results.

Numble	Moisture content-initial	Moisture content-8h	Moisture content-12h	PH-12h
1	92%	79.14%	72.89%	7.6
2	92%	80.79%	75.94%	7.4

3.2 Experimental sewage SS and COD values

According to the sewage discharge standard, the concentration of more than 60 pollutants is stipulated in the maximum allowable discharge classification standard. However, for the sewage discharged after the dehydration of kitchen waste, its SS and COD values are high and can be discharged up to the standard only after being treated by the sewage treatment plant, which seriously increases the treatment load of the sewage treatment plant.

It can be seen from the data in Table 2 that the data in the control group are SS value and COD value of raw kitchen waste pulverized by the grinder.

Comparing the experimental data set with the mechanical crushing data (SS=18 950 mg/L, COD=43 026 × 103mg/L), it can be seen that the dehydration treatment of kitchen waste by biological dehydration technology compared with mechanical crushing and other treatment methods can effectively and greatly reduce the SS value and COD value in sewage and reduce the burden of sewage plant.

Table 2. SS value and COD value of sewage.

Numble	Element	1h	2h	6h	8h	10h	12h
1	SS mg/L	284	266	288	300	2010	276
	COD mg/L	4.12	4.17	3.46	3.64	5.64	5.77
2	SS mg/L	284	254	2130	2130	2190	305
	COD mg/L	4.74	3.70	9.97	7.10	7.59	5.30

From Figure4, we can see that the suspended solids of four samples are significantly higher than those of the other eight groups. This is because sewage is sampled from the pool at the bottom of the equipment. When the sampling time catches up with the mixing of the mixing equipment or just ends, the sewage at this time is likely to be mixed with a small amount of kitchen waste reactants, and the content of suspended solids in the waste water is high.Controlling sewage discharge time can effectively control the content of suspended solids in sewage.

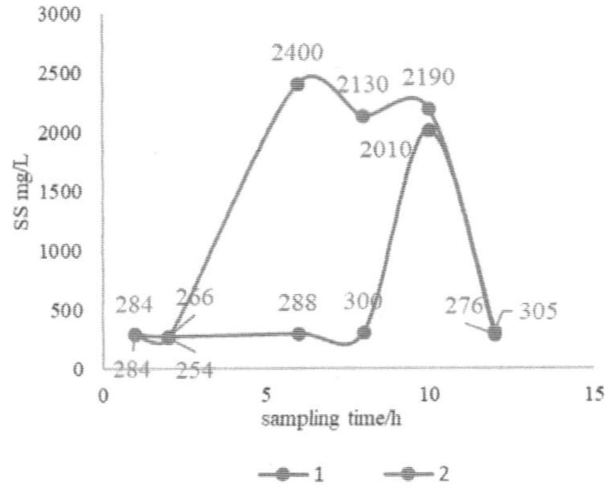

Figure 4. Diagram of SS value changing with time.

As can be seen from Figure 5, COD values of the four groups of water samples are significantly higher than those of other water samples, which is significantly related to the above sewage sampling time.

The solid suspended matter values of the four groups of water samples were also higher, mainly due to the small amount of kitchen waste reactants mixed.

After the exclusion of interference, the value of the remaining sewage suspended solids fluctuates 3 x103 mg/L and 6 x103 mg/L, with no difference of order of magnitude, which is far lower than the COD value of the kitchen waste broken sewage, but still fails to meet the national discharge standard, which may be related to the reaction time.

The duration of action on microorganisms can be days or months, so it is necessary to extend and explore the duration of biological dehydration further.

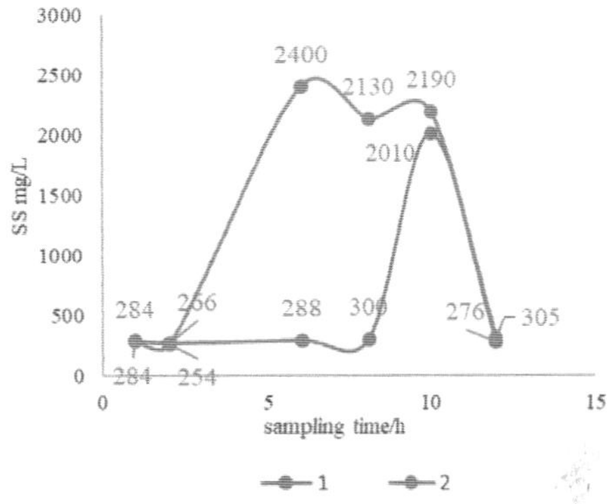

Figure 5. Diagram of COD value changing with time.

4 Conclusion

This paper provides a way to pretreat kitchen waste by using biological dehydration technology, and experiments verify that the biological dehydration pretreatment technology can effectively remove the moisture of kitchen waste and the kitchen waste after reaction is in weak alkaline, which is conducive to the subsequent aerobic fermentation.

In addition, the SS and COD values of the waste water produced by the biological dehydrated kitchen waste are much lower than those of the kitchen waste water produced by the mechanical crushing, which is conducive to the discharge of sewage and effectively reduce the load of the sewage treatment plant.

In order to further explore the application of biological dehydration technology in the pretreatment of kitchen waste, the reaction time can be extended, the sampling time interval can be extended, and the change of moisture content of kitchen waste, SS and COD values in sewage with the reaction time can be further explored, so as to find the optimal reaction time of biological dehydration pretreatment of kitchen waste.

References

1. C Wang, GQ Shen, JJ Pu, Anaerobic fermentation coupled carbonization of food waste and its application prospect in landscape greening, Landscape Architecture, 2020(6)14-18.

2. W Zheng, Current situation and countermeasures of kitchen waste treatment, Energy Conservation and Environmental Protection, 2020 (4)32-33.
3. SR Yu, XM Xie,Discussion on energy treatment and pretreatment of kitchen waste,Energy and Environment, 2021(1)89-90.
4. SY Zhang, JX Liu,JG Geng,Research progress of sludge pretreatment method based on compost , Applied Chemical Industry, 201, 50(7)2008-2013.
5. XJ Liang, SZ Geng, QL Xue, Treatment of food waste by dehydration in situ, Journal of Agricultural Products Processing, 2010(2B)98-102.
6. RQ Li, R Xu, AF Yu, Biodegradation of waste vegetable leaves by microbial fermentation, Bulletin of Soil and Water Conservation, 2019(3) 163-169.

key influencing factors and adjustment countermeasures of recovery in heavy oil reservoir of water flooding

Qingting Wei [1], Guinan Zhen [1], Weilong Wang [1], Jiamin Qin [1], Hongcheng Luo [1], Xuejie Zhao [1], Rui Xue [2,*], Xinbo Wu [2]

[1] Xinjiang Oilfield Co., LTD jiqing operating Area, Karamay 834000, Xinjiang, China
[2] E&D Research Institute of DQ Oilfield, Daqing 163542, Heilongjiang, China.

Abstract: The recovery ratio is an important indicator to evaluate the effectiveness and performance of oilfield development. In order to clarify the key factors of oilfield recovery ratio in the stage of super-high water cut, 9 dynamic indicators of 12 blocks are analyzed in heavy oil reservoir of eastern China in application of reservoir engineering principles and statistical analysis methods. The results show that the four key factors of recovery are the well density, the ratio of producing wells and injecting wells, net cumulative injecting rate and the bottom hole flow pressure, and the weight of each indicator is quantified. In the horizontal contrast, comprehensive forecasting target recovery is compared with the current recovery by the dynamic method to determine according to typical block BXD. Four of important factors are discussed for the previous trend and the countermeasures of according adjustment are put forward. The proposed method and the results have a certain reference value of the theoretical research and practical application in the stage of super-high water cut oil-field.

Key words: Super-high water cut stage; Recovery ratio; Influencing factors; Correlation analysis; Grey relational analysis

1 The introduction

Most of the heavy oil fields in China have entered or are about to enter the ultra-high water cut development stage, so it is urgent to analyze the key factors affecting the recovery efficiency in the ultra-high water cut stage, so as to provide the basis for the adjustment and decision of the next development of the oilfield.There are many factors affecting recovery, and there are many scholars[1-7]The influence factors of recovery efficiency are discussed in depth from geological conditions, development methods and so on.In the following, based on the example of 12 blocks in the ultra-high water cut period of sandstone heavy oil reservoirs in western China, the key influencing factors of oilfield recovery are analyzed by combining objective analysis with subjective analysis.

2 Statistical and quantitative discussion of key factors of influencing recovery

For a lot of factors of influencing the oilfield recovery, they often influence each other. In the analysis of the inluencing recovery factors, several representative indicators could be first selected commonly , focuses on subjective analysis - that is, on the principle of reservoir engineering and practical experience in primarily in the judgment, as the important reference in mathematical methods of quantitative results, at the same time to determine the various factors affecting the synthesis weights of factors influencing the sorting, then appraise the typical block, and the next step adjustment countermeasures are presented.

Taking 9 development indicators related to recovery of 12 blocks in the ultra-high water cut stage (between 90% and 95%) as cases (Tab. 1), the calculated correlation coefficient between two indicators of each block ecovery are shown in Tab. 2.

* Corresponding author: xuerui@cnpc.com.cn

Tab. 1. Indicators of influencing recovery of 12 blocks

Indicator \ block	Block 1	Block 2	Block 3	Block 4	Block 5	Block 6	Block 7	Block 8	Block 9	Block 10	Block 11	Block 12
recovery	43.12	40.56	40.69	38.96	41.90	37.16	36.83	35.56	44.75	34.81	40.14	35.38
Fluid productivity index,	13.04	5.86	9.70	12.92	14.40	6.07	8.16	5.95	5.74	6.12	5.36	2.62
Oil well flow pressure	3.45	2.99	3.41	4.03	3.66	4.48	3.91	3.93	3.01	4.56	3.71	3.72
The intensity of oil recovery	0.15	0.09	0.08	0.11	0.13	0.10	0.18	0.11	0.09	0.06	0.12	0.16
Recovery degree	39.11	32.04	29.22	33.61	33.57	25.52	29.69	22.32	32.33	25.11	30.15	22.65
Oil-well ratio	1.27	2.10	1.29	1.58	1.02	1.88	1.67	1.89	1.56	2.03	1.08	1.81
Cumulative water storage rate	37.90	39.52	41.34	40.41	47.51	39.28	42.49	40.19	42.60	40.99	38.46	35.12
Dynamic pressure difference between injection-production Wells	13.25	17.94	15.00	16.01	13.60	18.04	18.67	15.98	17.42	18.58	17.73	18.62
Injection porosity multiple	1.24	0.73	1.34	1.02	0.99	0.82	0.69	0.59	0.70	0.69	0.79	0.75
Well spacing density	48.03	50.90	49.75	40.07	55.93	32.97	19.24	20.82	37.97	29.14	34.18	18.82

Tab. 2. Correlation coefficient between two factors of influencing recovery

The serial number	*indicators*	*correlation*	*indicators*	*The correlation*
1	Recovery degree	0.7486	Recovery degree	0.8334
2	Net injection percentage	0.7301	Well spacing density	0.7542
3	Well-bottom flowing pressure	0.6876	Injection porosity multiple	0.4791
4	Well spacing density	0.6688	Fluid productivity index,	0.4578
5	The intensity of oil recovery	0.6792	Cumulative water storage rate	0.3310
6	Injection porosity multiple	0.6539	The intensity of oil recovery	0.1139
7	Fluid productivity index,	0.5911	Dynamic pressure difference between injection-production Wells	0.5498
8	Inj. & Pro. Well ratio	0.5744	Oil-well ratio	0.6018
9	Dynamic pressure difference between injection-production Wells	0.5719	Oil well flow pressure	0.7488

After removing the factors affecting the degree of production, the top four indexes calculated by the two methods are the cumulative net injection percentage, well spacing density, well-bottom flowing pressure and Inj. & Pro. Well ratio.The relationship between the recovery degree and recovery is self-evident for the oil field in the ultra-high water cut stage. The greater the recovery degree, the greater the recovery. The well-bottom flowing pressure and Inj. & Pro. Well ratio is not positively correlated with the recovery. In other words,, the smaller the well-bottom flowing pressure and Inj. & Pro. Well ratio are, the higher the recovery will be, which is consistent with theoretical understanding and practical knowing on site.The effects of well pattern density and cumulative water retention rate on recovery in the ultra-high water cut period will be discussed later.Based on the comprehensive analysis and judgment of reservoir engineering principle and field practice, the key factors of recovery in the extra-high water cut period include well pattern density, Inj. & Pro. Well ratio, cumulative net injection percentage and well-bottom flowing pressure. The results of calculating the weight of the several indicators, as shown in Tab. 3:

Tab. 3. Weight of key factors of influencing recovery

category \ index	well-bottom flowing pressure	Inj. & Pro. Well ratio	net injection percentage	*Well pattern density*
Related rights	0.2662	0.2632	0.2792	0.2004
Right of variation	0.1670	0.2895	0.0948	0.4487
Level of power	0.2461	0.3160	0.1917	0.2461
Integrated power	0.2284	0.3005	0.1788	0.2922

According to the statistical results, for the oil field in the ultra-high water cut stage, the most important factor affecting the recovery is the oil-well number ratio, followed by well pattern density and well flow pressure, and the net injection percentage.

On the basis of determining the key factors affecting the recovery efficiency, the following takes BXD block, a typical block of ultra-high water-cut heavy oilfield, as a case for in-depth analysis.

3 BXD recovery efficiency evaluation of typical blocks

(1) discussion of influencing factors

1) Well pattern density

The relationship model between well pattern density and recovery efficiency was proposed by serkachev, a scholar of the former Soviet Union, in the late 1960s according to the statistical results of oilfield data[10]With the increase of well pattern density, the recovery is also increasing. The expression is as follows:

$$E_R = E_D \cdot e^{-a/s} \tag{1}$$

Type:

E_R-- Recovery factor, %;E_D-- Oil flooding efficiency, %; S -- Well pattern density, well /km^2; a -- Well pattern density index.

https://doi.org/10.1051/e3sconf/202132901047

By means of the data of coring Wells since 2012, the flooding efficiency of BXD block is 38%, and the calculated well pattern index is 6.13. The variation trend of the calculated recovery factor with well pattern density is shown in Fig. 1.It is obvious that the recovery rate increases less and less as the well pattern density increases, which is a key factor for the present recovery to be lower than the expected recovery, and the actual recovery will be lower because the well pattern density is much lower than the surface well pattern density because the well pattern density is much lower underground.

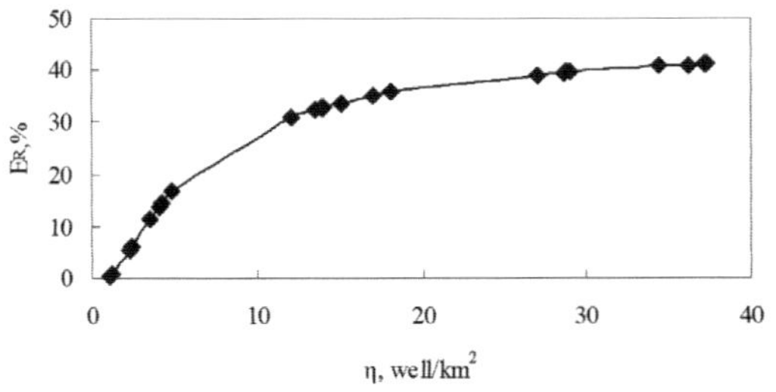

Fig. 1 Recovery curve of BXD block with well pattern density

The variation trend of EOR of BXD block in unit well density under different water cut conditions since the extta-high water cut stage, as shown in Fig. 2.

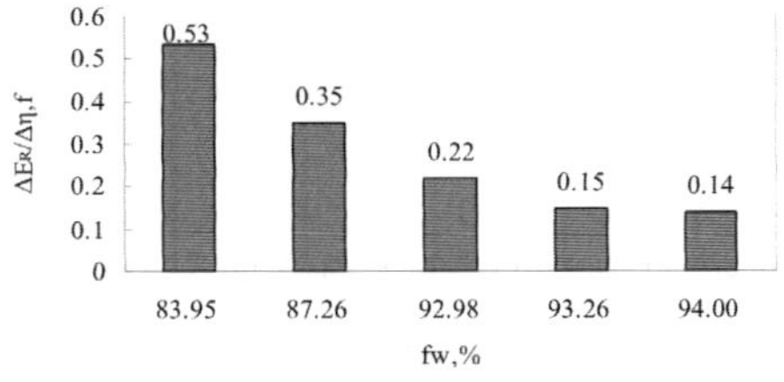

Fig. 2 Enhanced recovery per unit pattern density under different water cut conditions

In the high water cut stage, with the increase of well pattern density, the value of enhanced oil recovery decreases promptly. If the higher recovery is not got before the extra-high water cut stage, even if the well pattern density increases substantially, the increasement of oilfield recovery will decrease remarkably.The variation trend of BXD is that the water cut rises significantly in the initial development stage, which cannot guarantee high recovery efficiency. As a result, the recovery efficiency does not increase significantly even after multiple well pattern infusing and injectable system adjustment. However, infilling adjustment is inevitably a key way of EOR.

2) Inj. & Pro. Well ratio

Reasonable oil-well ratio is a key aspect to guarantee the balancing between injection and production and expand sweep volume for the oil field in ultra-high water-cut stage. It has been verified that injecting water theory and Inj. & Pro. Well ratio have great influence on water injection swept volume and water drive recovery efficiency in waterflooding sandstone oil fields. When well density and water injection multiple invary, with the relative reduction of Inj. & Pro. Well ratio, the number of effective directions of working production well increases, the controlling degree of water dirving increases, and the water injection conformance correspondingly increases. On the contrary, when the Inj. & Pro. well ratio decreases, it decreases and the multi-direction efficiency is lower and lower. Therefor, the sweeping conformance of water injection would decrease correspondingly. In oilfield injection-production pattern adjustment, has a mass of theoretical research and field practice shows that under the condition of the well spacing density rekeys constant, by varying the ratio of oil and water wells numbers to adjust flowing direction, to better the exploiting effect, increasing water flooding recoverable reserves and improve water drive recovery is the purpose of practical (Fig. 3).

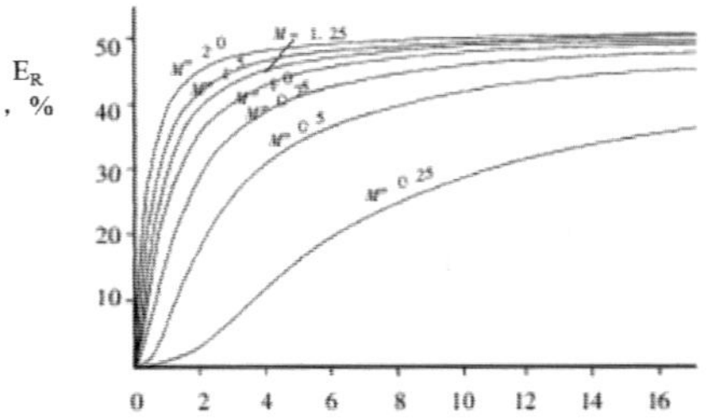

Fig. 3 Relationship curve of well density, Inj. & Pro. well num. ratio and recovery in Shuanghe Oilfield
(From Yang Fengbo, Effect of Inject-production Well Ratio on Water-Flooding Recovery, Xinjiang Petroleum Geology, 1998)

In the case of a constant well pattern density, the lower the Inj. & Pro. Well ratio is, the less the recovery will be. Nevertheless, with the increasing of well numbers, the impact of Inj. & Pro. Well num. ratio on recovery will become smaller and smaller, and the increase of recovery will become smaller.

The variation curve of oil-well ratio in BXD block is shown in Fig.4.Another factor contributing to lower recovery is the gradual adjustment of the oil-well ratio from a high of 2.37 in the 1980s to 1.99 in the 1990s and to 1.53 in the 2000s.

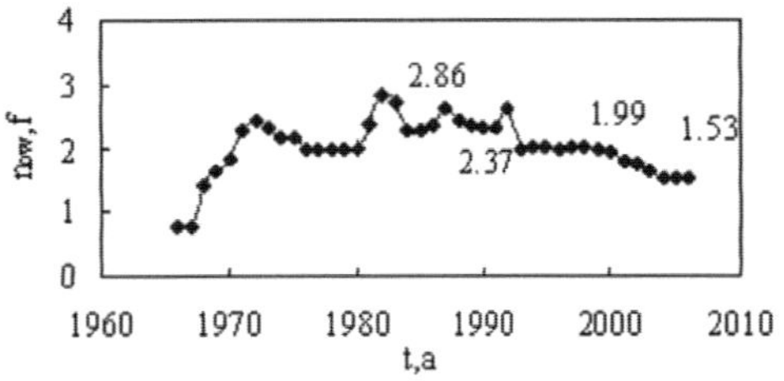

Fig. 4 The changing curve of Inj. & Pro. Well ratio about BXD block

3) net injection percentage

For single-layer oil fields, the recovery factor is the product of the key oil displacement efficiency and sweep volume (Formula 2). The sweep volume coefficient of injected water refers to the difference between the cumulative injection water and cumulative water production divided by the geological reserves, that is, the average oil displacement efficiency of the water-flooded part of the oil layer, also called sweep volume coefficient (Formula 3).

$$E_R = E_D gE_V \quad (2)$$

$$E_v = \frac{W_i - W_p}{N} \quad (3)$$

According to the definition of net injection percence :

$$E_s = \frac{W_i - W_p}{W_i} \quad (4)$$

As for the total injection multiples:

$$\gamma_\phi = \frac{W_i}{N} \quad (5)$$

To obtain:

$$E_v = E_s g\gamma_\phi \quad (6)$$

in a certain range, total injection pore volume ratio is not a great change, thus affecting coefficient is keyly accumulated water rate, conformance with increase with the increase of the accumulated water rate, especially for water drive oilfield, oil displacement efficiency increased under the condition of limited, the key direction of enhanced recovery is to increase the swept volume coefficient.As can be seen from Fig. 5, when the injected pore multiple is certain, the recovery factor increases with the increase of cumulative water storage rate and the increase of conformance.The theoretical relation between cumulative water storage rate and water content is derived[11]. As can be seen from Figure 6, the actual cumulative water storage rate is nearly 6.8 percentage points lower than the theoretical cumulative water storage rate, which is another reason for the low recovery.

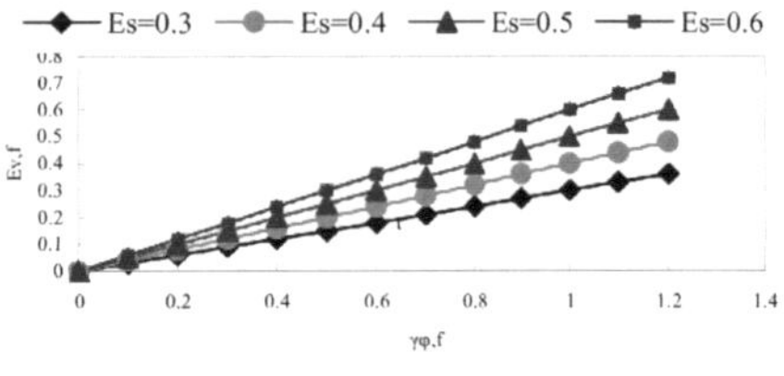

Fig. 5 Variation curves of injection multiple and conformance under different cumulative water storage rates

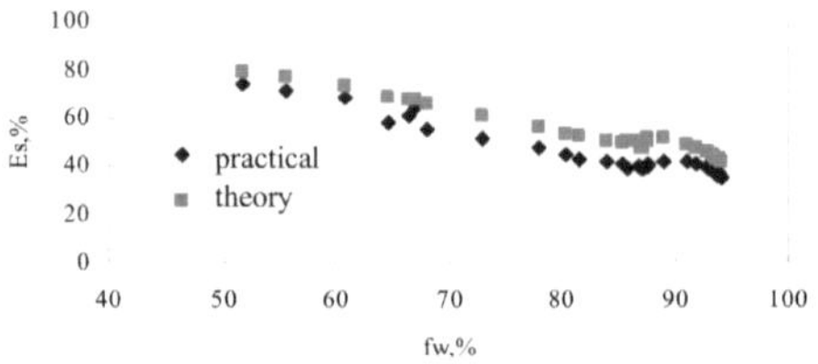

Fig. 6 Variation curve of BD theory and actual water storage rate with water content

4) well-bottom flowing pressure

According to the previous understandings[12], production and recovery theory, there would be a turning corner between well-bottom flowing pressure and recovery, when the well-bottom flowing pressure is about 2.1~3.2MPa, the recovery is the largest.However, the current flow pressure of the block is 3.7mpa, which exceeds the value of 0.4-1.6mpa to maximize the recovery efficiency, which has a approximate influence on the recovery.

In conclusion, one of small recovery of the reasons in BXD block: firstly, the high reservoir heterogeneity (permeability variation coefficient of BXD is 0.5945) caused interally; on the other hand, external factors caused to incorrectly enhanced recovery measurements, it leaded to "the three major contradictions" in particular, determining to encrypt and injection & production system adjustment and keep high flowing pressure. In the early stage of oilfield development, the water cut rapidly rising, which leading to infilling in the middle and late stage. Moreover, the injection-production system adjustment has failed to EOR of BXD, and net injection percentage has been small. To enhance oil recovery of BXD is difficult because of aboving reasons.

(2) development countermeasure

Based on the analysis of the above four aspects, the specific development countermeasures are put forward for BXD.

Firstly, the economic and maxmum well density of BXD at 50\$/ bbl calculated is 60 well/km^2 or so by using the improved Xie's formula, which current well density is 38 well/km^2, enhanced oil recovery by 0.8-1 percentage points through infilling;

Secondly, the recovery of BXD would be improved by 0.1-0.2 percentage points by reducing the production well numbers, and the optimal Inj. & Pro. Well ratio is 1.18 under present conditions.

Thirdly, in terms of improving the net injection percentage, the underground water storage rate can be improved by 0.1-0.2 percentage points by carrying out deep profile control to improve the water absorption profile.

Fourthly, with well-bottom flowing pressure reducing, the fluid of low water cut layers could be producted, and low water cut wells be stimulated to improve the liquid production.

These measurements can validly reduce the declining rate of the oilfield and add oil recovery to 0.3 to 1.6 percent.

4 Conclusion

Through the above research, the following conclusions are drawn:
(1) the paper for the first time put forward from two aspects of straight line and curve of related using the method of correlation coefficient and grey correlation analysis method to analysis the key influence factors of determining recovery rate, the results show that the primary factors influencing the determining oil field recovery is Inj. & Pro. Well ratio, followed by well spacing density and well-bottom flowing pressure, the next is net injection percentage;
(2) The method of weight determination by complex correlation coefficient method and variation coefficient method is analyzed, and the integrated weight calculating method is put forward, which offers theoretical support for the oilfields development of extra-high water cut satage;
(3) The mathematical relation between net injection percentage and sweep volume coefficient is established. It is pointed out that water storage rate can not only be used to evaluate the utilization efficiency of injected water, but also closely related to recovery efficiency.

References

1. Jia Zili, GAO Wenjun, ZHAO Xiaoping, et al. Determination of reasonable well pattern density and injection-production well number ratio in waterflooding oilfield [J]. Xinjiang petroleum geology, 2005,26 (5) : 562-564.
2. Qian Shenhua, Li Yongfu, Yuan Chengzhang. Determination of economic and reasonable well pattern density for oilfield infilling adjustment [J]. Petroleum geology and oilfield development in daqing, 2000,19 (4) : 23-25.
3. Li Zhourong, XUE Zhiguo, Chen Kunlin, et al. Determination method of reasonable well pattern density in later stage of oilfield development [J]. Petroleum geology and recovery efficiency, 2002,9 (3) : 78-80.
4. Guan J A, Cheng Z P. Study on determining parameters of Xie's Formula by Using actual Production Data [J]. Special oil and gas reservoirs, 2003,10 (3) : 57-59.
5. Yang Fengbo. Effect of injection-production ratio on water flooding recovery [J]. Xinjiang petroleum geology, 1998,19 (5) : 410-413.
6. Gao Wenjun, Huang Shibin, Lu Yi. Study on the influence of injection-production well pattern on water flooding recovery [J]. Turpan-hami oil and gas, 2000,5 (2) : 27-30.
7. Peng Changshui, Gao Wenjun, Li Zhengke. Effect of injection-production well pattern on water flooding recovery [J]. Xinjiang petroleum geology, 2000,21 (4) : 315-317.
8. (USA) by R.L. OTT, m. langenecker, translated by zhang zhongzhan.An introduction to statistical methods and data analysis [M].Beijing: Science Press, 2003:35-37.
9. Hu Bozhong, Editor-in-chief.Study on the application of multi-layer production technology in heterogeneous sandstone oilfield [J].Beijing: Petroleum Industry Press, 2000, 315-317.
10. Song Yong.Adjustment technology of water injection development in Daqing oilfield [M]. Beijing: Petroleum Industry Press, 2005, 119-121.
11. Zhang Jifeng, TIAN Xiaodong, Wang Tianzhi, et al.Theoretical relationship between cumulative water storage rate and water cut in water flooding oilfield [J].Xinjiang petroleum geology, 2006,27 (4) : 466-467,477.
12. Gang Wang. Study on the key influencing factors of oil-flood recovery of oilfield in extra-high water cut period, 2020, vol 514, IOP Conference Series: Earth and Environmental Science

A method for determining rational fluid production scale in conglomerate reservoir in the stage of high water cut

Wentao Liu [1], Xiaodong Wang [2], Chaoliang Zhang, Rui Xue, Guifang Zhu [1], Jifeng Zhang [2], Xiaomin Wei [2], Yanan Zhao [2]

[1] Exploration and Development Research Institute of Xinjiang Oilfield, Karamay 834000, China
[2] Research Institute of Exploration and Development, Daqing Oilfield, Daqing 163712, China

Abstract: In view of the problem that water cut rises quickly in conglomerate reservoirs with high water cut, based on the variation law of reservoir water injection and liquid production characteristics, a method for determining the maximum liquid production scale in conglomerate reservoirs with high water cut is established by using reservoir engineering and field statistics.Considering the maximum liquid production capacity at the well end, the maximum injection capacity at the well end and the injecting-production ratio limit to keep the formation energy, a method to determine the maximum liquid production scale in conglomerate reservoir is established to keep the formation energy sufficient.This method ensures that the injection and production ends of the reservoir are coordinated and matched, and the formation energy is sufficient, so that the reservoir development keeps a benign cycle.The research method and conclusion have important reference value and significance for theoretical research and practical application of ultra-high water cut oilfield development.

Key words: high water cut;Injection-production coordination; Liquid withdrawal scale

1 The introduction

The scale of fluid production is a comprehensive index to reflect the level of oilfield development, production management and economic benefit, as well as a key index for planning, scheme design and dynamic analysis.When the oil field enters the high water cut period, the enhanced production with the continuous increase of injection and production scale as the core has become an important measure to improve the oilfield development effect and enhance the recovery efficiency.However, in addition to directly affected by the process equipment and development benefits, the scale of oilfield fluid production is also related to reservoir physical properties, crude oil physical properties and well pattern pattern.At present, there is no quantitative expression of the relationship between reasonable fluid production scale and related parameters such as reservoir characteristics, well pattern and development stage.On the one hand, excessive liquid production scale will lead to a large amount of waste of materials, as well as the rapid rise of water cut and the continuous decline of oil production, while insufficient liquid production scale will lead to the failure to complete the oil production plan and the guarantee of supply demand.At present, the research on the maximum liquid production scale is mainly determined by maximizing the liquid production capacity of the reservoir, that is, the maximum injection pressure at the injection end and the minimum flow pressure at the production end[1~3].However, this method has some shortcomings. It does not take into account the coordination between injection and production, that is, whether injection can meet the demand of production.Therefore, it is necessary to establish a method to determine the liquid production limit in the ultra-high water cut stage to meet the oilfield production demand.

2. Determination method of reasonable liquid production scale in conglomerate reservoir in high water-cut stage

2.1 Reasonable injection and production pressure system

(1) Maximum water injection pressure
The level of water injection pressure is not only related to the injection-production capacity of a well pattern, but also related to the pressure level of the entire injection-production pressure system.Too high injection pressure is easy to cause casing damage of oil and water Wells, which brings great adverse effects to oilfield development.However, the injection pressure is too low to meet the demand of injection water, which will lead to the formation energy reduction.The maximum water

injection pressure can be determined according to the fracturing pressure of the reservoir. Generally, the maximum water injection is about 0.5MPa lower than the fracturing pressure of the reservoir. In this way, casing damage will not be caused, and reasonable injection-production ratio can be guaranteed to meet the needs of the injection water.

$$P_{imax} = \gamma_o \times D_o - 0.5 \quad \text{Type 1-1}$$

P_{imax} Is the maximum water injection pressure of the reservoir, MPa; γ_o Is the fracture pressure gradient of formation, MPa/m; D_o is the medium depth of oil layer, m.

(2) Reasonable formation pressure

Formation pressure is a very important factor affecting the effect of oilfield development, which determines the initiative of oilfield development to a large extent.If the formation pressure is too low, the production cannot meet the production allocation requirements, and the porosity and permeability are easily reduced, which is not conducive to the stability of the reservoir structure.The results of development practice statistics and theoretical research show that the reasonable formation pressure is near the original formation pressure.The development practice statistics and theoretical research results show that the reasonable formation pressure is near the original formation pressure, generally should not be higher or lower than the original formation pressure MPa. ±0.5

$$P_{RR} = P_{Ro} \pm 0.5 \quad \text{Type 1 and 2}$$

P_{RR} Is the reasonable formation pressure of reservoir, MPa; P_{RO} Is the original formation pressure of the reservoir, MPa.

(3) Reasonable oil well flow pressure

The flow pressure of oil well generally decreases with the development of oil field, especially in the high and ultra-high water cut stage, which needs to meet the demand of increasing the drainage amount of oil field.Reservoir water injection pressure, formation pressure and oil well flow pressure are closely related under specific well pattern, oil-well ratio, water-cut stage and reservoir conditions. If two of them are determined, the other one will be determined accordingly.The reasonable flow pressure can be determined according to the injection-production pressure equilibrium intersection method.

$$P_R = P_{iwf} - q_i / J_w = P_{wf} + q_L / J_L + c \quad \text{Type 1-3}$$

P_R Is the current formation pressure of the reservoir, MPa;The coefficient C is determined by the field fitting and is generally between 0 and 1.

When water injection pressure PiwfTake the maximum Pimax, the oil well flow pressure P can be obtainedwfThe minimum value of Pwfmin.

2.2 Maximum injection and production scale

(1) Reservoir water absorption and fluid production index can be calculated according to field dynamic data

In the process of oilfield water injection development, the change of water absorption index and liquid production index is essentially a reflection of the change of seepage resistance near the bottom of oil and water well. With the increasing of water saturation, the seepage resistance keeps decreasing, and the liquid production index increases, while the increase of water absorption index is small.Theory research and development practice shows that high, determining phase conglomerate reservoir, the water injectivity index increase slowly, water injectivity index growth is far lower than the fluid productivity index growth, high water cut stage after production Wells have certain solution space, and the injection well injection has reached saturation, the reservoir of oil phase has been replaced by a water phase.Therefore, the water injection system needs to be strengthened after the high water cut stage.

$$J_L = q_L / (P_R - P_{wf}) \quad \text{Type 1-4}$$

J_LIs liquid production index, m^3/ (d. MPa);q_LIs the daily production fluid of a single well, m^3/ d;

$$J_i^* = q_i / P_i^* \quad \text{Type 1-5}$$

J_i^*Is the apparent water absorption index, m^3/ (d. MPa);q_iIs single well daily water injection, m^3/ d;P_i^*Is the wellhead water injection pressure, MPa.

(2) The maximum scale of water injection and liquid production can be calculated according to the injection-production pressure limit

$$Q_{L\text{-}omax} = n_o \times J_L \times (P_{RR} - P_{wfmin}) \quad \text{Type 1-6}$$

$Q_{L\text{-}omax}$Is the maximum daily liquid production scale at the end of the well, m^3;n_oIs the number of oil Wells, mouth;

$$Q_{imax} = n_w \times J_I^* \times P_{imax}^* \quad \text{Type 1-7}$$

Q_{imax}Is the maximum daily water injection scale at the well end, m^3;n_wIs the number of Wells, mouth.

2.3 The injection-production ratio limit for the oil field to maintain or restore the formation energy

Injection-production ratio is a comprehensive index to measure the replenisation of underground energy and the degree of cavitation. It is a comprehensive index to represent the balance of injection-production in the process of oilfield water injection development, and to reflect the relationship between liquid production, water injection and formation pressure. It is an important basis for the planning and design of oilfield water injection.Reasonable injection-production ratio is an important guarantee to maintain reasonable formation pressure, so as to make the oilfield have strong liquid and oil production capacity, reduce invalid energy consumption, and obtain high oil recovery efficiency.Field dynamic statistics is mainly used to determine the injection-production ratio to maintain or restore formation energy for a long period of oilfield development.Considering the field practice, the injection-production ratio should be kept above 1.0 due to the influence of formation void, injection water overflow and liquid discharge measures.According to the recent production performance data, the variation characteristics of injection-production ratio and formation pressure/total pressure difference are analyzed according to the actual curve of injection-production ratio and formation pressure/total pressure difference. The reasonable injection-production ratio under the condition of maintaining formation energy is determined according to the change trend or inflection point of the two.According to the curve of injection-production ratio and formation pressure variation of typical conglomerate reservoirs in western China, the reasonable injection-production ratio of this reservoir is about 1.1.

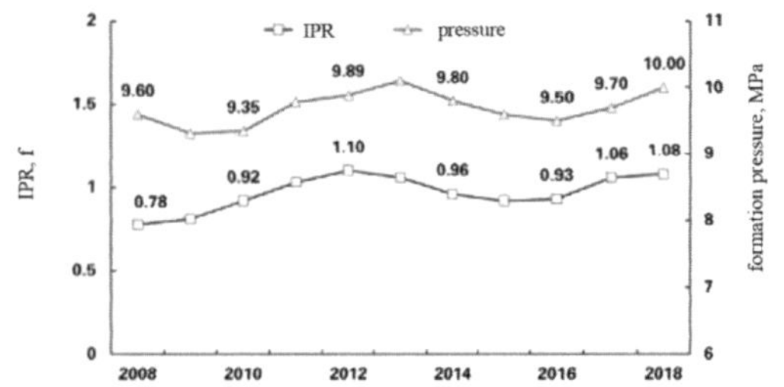

Fig. 1 Variation curve of inject-production ratio and formation pressure in B2D conglomerate reservoir

2.4 Determine the maximum fluid production scale based on the coordination between injection and production ends

The ratio between the maximum water injection scale and injection-production ratio at the well end of the reservoir is defined as the maximum liquid production scale at the well end. If the liquid production scale exceeds the maximum water production scale, the formation energy of the reservoir will be reduced and the oil layer will be degassing, which is not favorable for improving the development effect and enhancing the oil recovery.The maximum fluid production scale based on the well end and the minimum value based on the well end are taken as the maximum fluid production scale of the reservoir.

$$Q_{L\text{-wmax}} = Q_{i\max} / IPR \quad \text{Type 1-8}$$

$Q_{L\text{-wmax}}$Is the maximum daily liquid production scale of the reservoir based on the well end, m^3;IPR is the injection-production ratio, dimensionless;

$$Q_{LR} = \min\left(Q_{L\text{-o}\max}, Q_{L\text{-w}\max}\right) \quad \text{Type 1-9}$$

Q_{LR}Is the maximum daily liquid production scale of the reservoir, m^3.

3. Instance application

3.1 Geological and development characteristics of B2D conglomerate reservoir

(1) Geological features

B2D conglomerate reservoir is located in a basin in western China, with flat surface and convenient transportation.The reservoir develops five sand formations from bottom to top.Subdivided into 11 small layers and 17 single sand layers.The main reservoir is gravelly braided river deposit, mainly developed braided channel and core beach microfacies.The average porosity and permeability of the reservoir are 18.5 and 130.5mD, respectively. It is a medium porosity and medium permeability reservoir.

(2) Mining characteristics

The B2D conglomerate reservoir was put into development in 1959. By December 2018, the recovery degree of the reservoir was 40.63%, the comprehensive water cut was 78.1%, and the recovery rate was 6.75%.According to the change of liquid production rate in B2D oilfield, in the low water cut stage (water cut less than 20%), the oilfield is in the upper production stage, and the liquid production rate rises rapidly.In the middle water cut stage (water cut is between 20% and 60%), the liquid production rate is relatively stable.Into high water cut stage (water cut is more than 60%), oil field in the stage of high and stable yield, liquid withdrawal rate rising fast, but since 2013, as the growth of the scale of liquid withdrawal, a rapidly rising gas-oil ratio, single well to reduce oil and water well casing damage, seriously affecting the effects of oilfield exploitation of improvement and orderly operation of production.

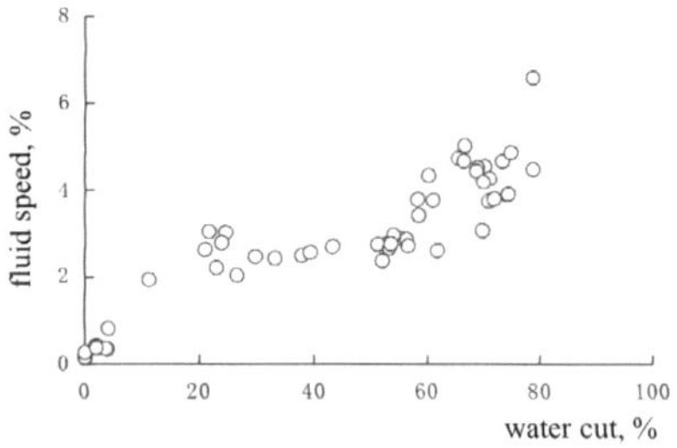

Fig. 2 Variation curve of fluid production velocity in B2D reservoir

3.2 Determination results of reasonable liquid production scale

Taking B2D conglomerate reservoir as an example, statistical calculation is carried out according to the above calculation method.According to the field statistics and reservoir engineering methods, the reasonable injection-production ratio is increased to 1.05, the water injection pressure is increased by 1.6MPa, the reasonable formation pressure is restored to 0.9MPa and remains near the original formation pressure, and the average daily liquid production scale is increased by 80 m^3After the adjustment, the liquid volume was stable, the formation pressure recovered to 10.0MPa, and the liquid production scale was 358 m^3/d rises to 437m^3/d, it is estimated that the recovery rate will be increased by about 1.2%, which can produce great economic benefits and effectively guide the adjustment of reservoir development.

Table 1. Calculation table of maximum liquid production scale of J1b reservoir in conglomerate reservoir in east Qidong 1

parameter	unit	The current value	Maximum/reasonable	Target adjustment difference
Well control reserves	10^4T/mouth	4.01	/	/
The average water cut	%	78.11	/	/
The well number	mouth	33	/	/
Number of Wells	mouth	20	/	/
Fluid productivity index,	m^3/ (d. MPa)	2.1	/	/
Water injectivity index	m^3/ (d. MPa)	2.5	/	/
Wellhead water injection pressure	MPa	8.2	9.8	+ 1.6
Reservoir formation pressure	Mpa	9.7	10.6	+ 0.9
Bottom hole pressure of oil well	Mpa	4.1	4.3	+ 0.2
The water flooding scale	m^3/d	410	490	+ 80
Injection-production ratio	f	0.88	1.05	+ 0.17
Liquid withdrawal scale	m^3/d	358	437	+ 79

4. Conclusion

Based on reservoir engineering and field statistics, a method for determining the maximum fluid production size of conglomerate reservoir in high water-cut stage is established.Considering the maximum liquid production capacity at the well end, the maximum injection capacity at the well end and the injecting-production ratio limit to keep the formation energy sufficient, a method to determine the maximum liquid production scale of the reservoir is established.This method ensures that the injection and production ends of the reservoir are coordinated and matched, and the formation energy is sufficient, so that the reservoir development keeps a benign cycle.This method avoids the previous shortcomings, such as formation energy reduction, casing damage of oil and water well and degassing of formation crude oil, which only consider the production capacity of well end but fail to fully consider the injection capacity of injection end.

References

1. Ji Bingyu, LI Yanxing. Main technical countermeasures for enhancing oil recovery in high water cut stage of la, sa and xing oilfields [J]. Petroleum geology & oilfield development in daqing,2004,23(5):47-53.
2. Zhang Xin-zheng, ZHANG Lie-hui, Xiong Yu. Petroleum geology & oilfield development in daqing,2005,24(3):48-50.
3. Wang Zhijun, Liu Xiuhang, Dong Jing.Quantitative evaluation method and adjustment countermeasures for injection-production adaptability of oilfield block in late period of high water cut [J]. Petroleum geology and oilfield development in daqing, 2005,24 (6) : 51-53.

Economic feasibility evaluation of shale gas stratified development in Chongqing area

Chengxin Yin [1], Jun Zhou [2,*]

[1] Chengdu Aeronautic Polytechnic, Chengdu, 610100, China
[2] Sinopec Research Institute of Petroleum Engineering, Beijing, 100101, China

Abstract. Shale gas as a clean green energy has been rapidly developed in China. At present, the vertical production rate of shale gas in Chongqing area is low, and it is difficult to take into account the whole high-quality shale segment with thickness of tens of meters only by one well pattern. In this paper, the internal rate of return (IRR) is used to calculate the impact of different costs and exploitation conditions on the development effect of two shale gas layers. The results show that shale gas development will be profitable only when the cost per well is controlled at around 70 million and production is provided at more than 100 million cubic meters. Domestic producers should further reduce the cost of shale gas development, increase the productivity of single well and maximize the benefit of domestic shale gas development by means of integrated innovation and technological optimization.

1 Introduction

Shale gas is a new clean natural gas resource, which has gained rapid development in North America, China and other countries and regions. After breakthroughs and large-scale development of shale gas in the United States, shale gas accounts for about 40 percent of the country's annual gas production. In China, due to environmental protection and winter heating, the demand for natural gas is increasing. However, the increment of geological reserves and exploitation degree of conventional natural gas is limited, and the dependence of domestic natural gas is increasing, so it is of great value to improve the development effect of domestic shale gas.

At present, the vertical production rate of shale gas in Chongqing area is low, and it is difficult to take into account the whole high-quality shale segment with thickness of tens of meters only by one well pattern. The shale gas plays in the United States have been developed in three to four vertical well patterns. Therefore, from the perspective of economic benefits, this paper discusses the feasibility of realizing three-dimensional development of shale gas in vertical two sets of well patterns, and provides theoretical basis for the next large-scale commercial development of shale gas in Chongqing.

2 Domestic shale gas development status

With the shale gas revolution underway in the United States, China stepped up its efforts to explore and develop shale gas from around 2010. According to the U.S. Energy Information Administration (EIA), China has 31.6 trillion cubic meters of recoverable shale gas, ranking first in the world. According to the latest survey and calculation by the Ministry of Natural Resources, the recoverable reserves of shale gas in China are 21.8 trillion cubic meters, still higher than the geological reserves of the United States.

At present, shale gas fields with relatively large scale development are mainly distributed in Sichuan Basin. Sinopec shale gas development achieved a breakthrough in Fuling, Chongqing in 2012, thus kicking off the prelude of domestic shale gas development. Chongqing Fuling Shale Gas Field is the first national shale gas demonstration zone, with a cumulative gas production of 200 cubic meters in 2018. It is the first shale gas development zone with commercial exploitation value after shale gas development in the United States. At the same time, CNPC has carried out large-scale exploration and development of shale gas in Changning and Weiyuan areas of Sichuan province. At present, the shale gas resources in Weiyuan block are about 620 billion cubic meters, and the recoverable capacity is more than 260 billion cubic meters.

3 Production prediction model of shale gas well developed by stratified system

The fracture system and two-phase seepage mechanism formed after fracturing of shale gas Wells are complicated. In order to simulate the production process of shale gas horizontal well after multi-stage fracturing as accurately as possible, the dual media model was used to assign values to shale matrix system and natural fracture system respectively, including porosity and permeability, relative permeability, etc (Figure 1).

* Corresponding author: zhou_sripe2015@126.com

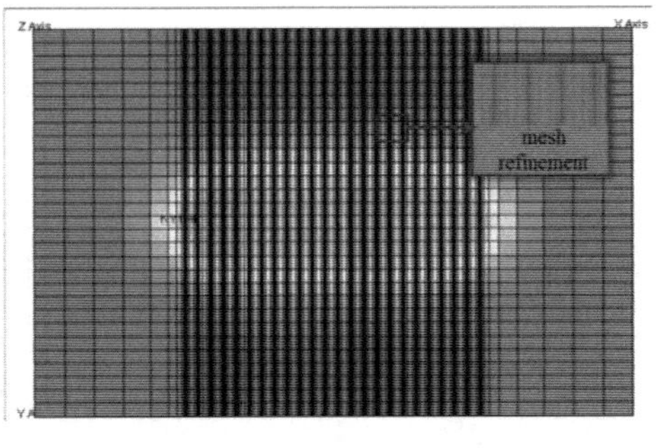

Figure 1. Fracture network fracturing model for shale gas.

The optimized daily production is shown in Figure 2. Taking the 15-year cumulative production as the target, the daily production changes under different reservoir characteristics are calculated according to the distribution of Wells in two development layers. In terms of artificial fracture parameters, the number of fracture stages is 15-20, the length of semi-supported fracture is 200-250m, and the conductivity is 3-4D.cm.

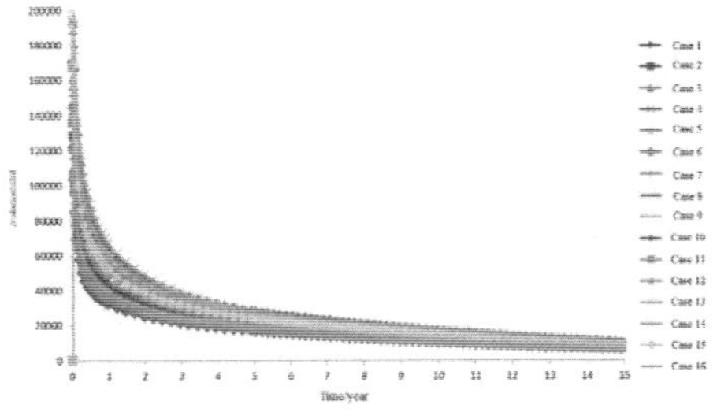

Figure 2. The change of daily output of orthogonal design scheme with time.

Through numerical simulation, it can be concluded that the average initial productivity of the well is 20×10^4 m^3/d. Considering the rapid decline rate of shale gas production and development, according to the production decline model, it is calculated that the cumulative production of shale gas in the two development layers is about 1.5-2.1×10^8 m^3 in 15 years.

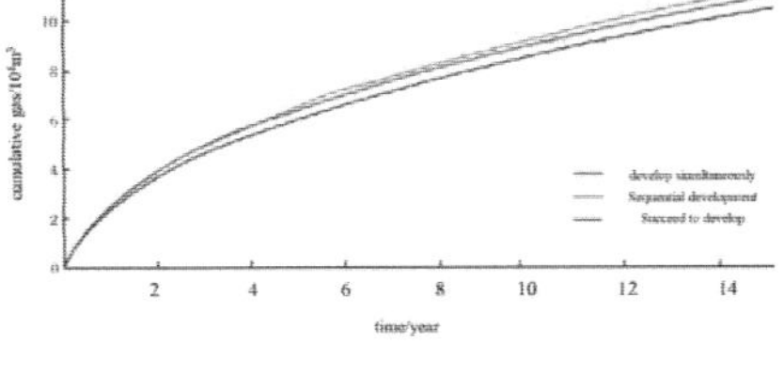

Figure 3. Cumulative gas production under different production modes.

In order to quantify the degree of interference caused by the production of the upper and lower well patterns and explore the effect of sequential production on the mitigation of inter-well interference, three production modes were simulated: single well succession, two Wells successively and two Wells synchronously. A unified production system of $8\times104m^3$/d was adopted for 15 years, and the EUR changes of 15 years were simulated (FIG. 2), with EUR respectively being $1.13\times10^8m^3$, $1.05\times10^8m^3$, $1.11\times10^8m^3$. After half a year of production, the area of pressure drop funnel increased by 13.7% compared with that without offset well production, and the production formation pressure level was basically the same. The results show that inter-well interference is inevitable in the development process of small well spacing. For a gas reservoir with a single closed volume, inter-well interference affects 6.8% of the production of adjacent Wells. A sequential approach could reduce the impact to less than 2 per cent.

4 Economic evaluation method of shale gas development benefit in stratified system

When calculating the economic benefit of shale gas development, we should not only consider the cost of shale gas well construction, the law of production decline, but also consider the time value of investment capital. Therefore, net present value method and internal rate of return method are mainly used for economic evaluation. In this paper, the internal rate of return method (IRR) is used to calculate the economic benefits of shale gas development in China. Currently, the benchmark rate of return used by CNPC and Sinopec in shale gas development is generally 8%, and the calculation formula is as follows:

$$\sum_{t=0}^{n}(CI-CO)_t(1+i)^{-t}=0 \quad (1)$$

including CI is the cash inflow, ten thousand yuan; CO is the amount of cash outflow, ten thousand yuan; i is the discount rate; n Is the production life; t is the year of production.

Taking a typical shale gas well in Chongqing as an example, the drilling cost is 35 million yuan, fracturing construction cost is 30 million yuan, other costs are about 5 million yuan, the total cost is about 70 million yuan. According to the above calculation, the cumulative output in 15 years is about 1.2×108 m3. At the same time, due to the different geological characteristics and production technology, the production volume of the area fluctuates, and the production range is 0.6-1.6×108 m3. Therefore, the cost price of shale gas under different production and development costs (yuan /m3) is calculated in the case of 8% benchmark yield.

https://doi.org/10.1051/e3sconf/202132901049

Table 1. Shale gas needs a cost gas price to achieve an 8% benchmark yield.

Investment cost per well / ten thousand yuan	0.6/ hundred million cubic	0.8/ hundred million cubic	1/hundred million cubic	1.2/hundred million cubic	1.4/hundred million cubic	1.6/hundred million cubic
5000	2.18	1.74	1.33	1.06	0.91	0.74
6000	2.63	2.11	1.58	1.25	1.05	0.88
7000	3.08	2.47	1.84	1.44	1.21	1.04
8000	3.53	2.83	2.11	1.64	1.36	1.15
9000	4.04	3.22	2.38	1.85	1.53	1.36
10000	4.52	3.61	2.66	2.06	1.69	1.47

According to the calculated results in Table 1, the cost gas price drops from 3.08 yuan /m^3 to 1.04 yuan /m^3 when the cumulative production of a single well increases from 0.6 million cubic meters to 160 million cubic meters under the cost of conventional investment of 70 million yuan/well. At present, the average single production of shale gas is about 0.8 to 100 million cubic meters, and the current weather price is 2.6 yuan /m^3. If the well construction cost exceeds 80 million yuan/well, then the development benefit of the producer remains at a low level. In this case, a well production of more than 100 million cubic meters would be beneficial to increase revenue and further enhance shale gas development in the area.

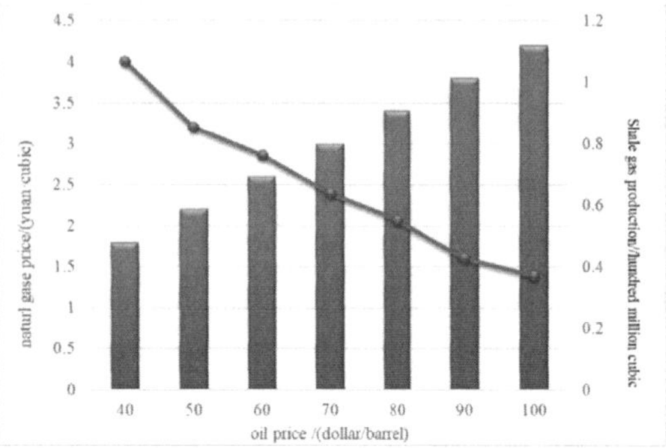

Figure 4. Shale gas production per well at different oil prices.

The required production capacity of a single well is calculated based on the corresponding LNG import gas price under different oil prices, that is to say, the cost price of domestic shale gas to realize profitable development should be lower than the LNG import gas price to achieve productive efficiency. Based on oil price fluctuations in recent years, the cumulative production required for 15 years of shale gas production per well at a production cost of \$40- \$100 per barrel was calculated. As can be seen, when the oil price is \$40 / BBL, the cumulative production of shale gas per well needs to reach 1.07 million cubic meters, which is about the most effective shale gas Wells in China. At an oil price of \$50-60 / BBL, the demand for backflow is reduced to 0.762-0.854 hundred million cubic, which is about the average for most conventional shale gas production wells in the country today. At oil prices of \$80 / BBL and above, the demand for cumulative production is reduced to 0.372-0.553 hundred million cubic for deep shale gas production wells to be productive.

5 Countermeasures for sustainable development of Shale gas in China

5.1 Scientific and technological innovation

The drilling and fracturing technology of shale gas in China mainly learns from the experience of shale gas development in the United States, and has made considerable achievements in the early stage, promoting the production and construction of shale gas in China. However, with the large-scale shale gas exploration and development in China, many technical problems have gradually emerged. The formation conditions of shale reservoirs in the United States are better, and the shale reservoirs in China are more complex, and the physical properties and gas content are relatively poor.

In addition, the country's vast shale gas reserves lie deeper than those in the United States, and existing drilling and production techniques are better applied to reservoirs in the range of 3,000-3,500 meters. For the shale reservoir with a depth of 4000 meters, the gas production effect of most deep shale gas exploration Wells is not as expected, except for some Wells with good geological conditions and high gas production. Therefore, it is necessary to study the formation and enrichment mechanism of shale gas, theory and exploitation technology of deep shale gas in China based on the actual situation in China. On the basis of absorbing and mastering foreign advanced shale gas exploitation technology, the current situation that domestic deep shale gas and atmospheric shale gas are difficult to effectively develop is solved through independent innovation.

5.2 Cost decreasing and benefit increasing

Although China's shale gas production in 2018 increased by 20.96 percent from 2017, it still accounted for less than 8 percent of the country's total natural gas production, far below the level of shale gas development in the United States. One of the main reasons is that the cost of developing shale gas in China is still higher than in the United States. Because domestic shale gas fields such as Fuling shale gas field in Chongqing and Weiyuan shale gas field in Sichuan are located in mountainous areas, the

cost of well site construction, pipeline placement and material transportation has been difficult to reduce.
After several years of development, the construction cost of a single well for shale gas in China has been reduced from about 100 million yuan at the beginning to about 50-70 million yuan. However, compared with conventional natural gas Wells, it has no advantage in terms of output or cost, while the cost of a single well for shale gas in the United States is close to that of conventional gas wells. The high cost of well construction makes it difficult for small and medium-sized enterprises to enter the shale gas development market, and they can only engage in technical services, on-site construction and other work. Therefore, it is necessary to realize the localization of key tools and materials by further digesting and absorbing foreign advanced technology. At the same time, improve shale gas exploration and development technology, through process optimization to minimize the domestic shale gas drilling and fracturing construction costs, such as improving drilling speed, the use of low oil-water ratio oil-based drilling fluid technology, reduce unilateral costs, reduce drilling costs. The use of low-cost quartz sand instead of ceramic, appropriate to reduce the construction scale or reduce the number of sections.

6 Conclusion

This paper studies the economic benefits of shale gas development in Chongqing region from the angle of technology and economy, and puts forward corresponding countermeasures for the sustainable development of shale gas in China.
(1) The internal rate of return (IRR) is used to calculate the impact of different costs and production conditions on shale gas development. The calculation results show that only when the cost per well is controlled at about 70 million yuan and the production is provided at more than 100 million cubic meters, can the shale gas Wells in China have obvious profit effect. Moreover, consider that direct subsidies for shale gas will be phased out. Domestic producers should further reduce the cost of shale gas development, increase the productivity of single well and maximize the benefit of domestic shale gas development by means of integrated innovation and technological optimization.
(2) The countermeasures to improve the economic benefits of shale gas development in China are put forward from the perspectives of technological innovation, cost reduction and efficiency increase, and multi-type policy support, which are helpful to realize the economic and effective development of shale gas resources in China.

References

1. Liu jing, Xu Haitang, Yu Guo, et al. Development indices and potential of shale gas, Sichuan basin[J]. Natural Gas Explora-tion and Development, 2014, 37(2): 45-47.
2. Dong Dazhong, Shi Zhensheng, Guan Quanzhong, et al. Progress, challenges and prospects of shale gas exploration in the Wufeng--Longmaxi reservoirs in the Sichuan Basin[J]. Natural Gas Industry, 2018, 38(4): 67-76.
3. Zhou Shouwei, Jiang Wei, Zhang Chunyang, et al. The enlightenment on shale gas exploration and development in China getting from Eagle Ford in American[J]. Engineering Scinences, 2012,14(6):16-21.
4. Liang Xing, Wang Gaocheng, Zhang Jiehui, et al. High-efficiency integrated shale gas development model of Zhaotong National Demonstration Zone and its practical enlightenment[J]. China petroleum exploration, 2017,22(1):29-37.
5. ZHANG Jiaming, GUO Xiaoxia, HE Wenyuan. Comparative study on shale gas development in southern Sichuan and the ways to reduce costs & increase efficiency[J]. International Petroleum Economics Monthly, 2018, 26 (9) : 31-38.
6. Ma Xinhua & Xie Jun. Advances in shale gas exploration and development in southern Sichuan[J]. Petroleum Exploration and Development, 2018, 45(1): 161-169.
7. Shang Shaofen, Yan Hong, Wu Jian ,et al. Production characteristics and development modes of shale-gas wells, Changning gasfield, Sichuan Basin[J]. Natural Gas Exploration and Development, 2018, 41(4): 69-75.
8. WANG Xiuzhi, SHI Zhiliang, LONG Shengxiang, et al. Study on economic and reasonable well spacing of Changxing Formation reservoir development in Yuanba Gas Field [J].Fault-Block Oil & Gas Field, 2012, 19 (3) : 340-342.
9. GAO Shikui, ZHU Wenli, YIN Cheng. An economic analysis of shale gas resources: A case study of the Marcellus shale play[J]. Natural Gas Industry, 2014, 34(6): 141-148.
10. LI Qinghui, CHEN Mian, JIN Yan. Economic influence of hydraulic fracturing parameters on horizontal wells in shale gas bed[J]. Special Oil & Gas Reservoirs, 2013, 20(1): 146-150.
11. Jiang Xinchun, Sun Qinping, Zhao Qun, etc. Production Characteristics and Economic Benefit Classification Method of Shale Gas Wells in Southern Sichuan[J].SINO-GlOBAL ENERGY,2021,26(8):43-50.
12. Fang Dazhi, Qian Jin, Mei Junwei,etc. Layer series division for development of shale gas of Pingqiao anticline in Nanchuan Block and reasonable well spacing optimization[J].Reservoir evaluation and development, 2021,11(2):78-84.
13. Wen Zhentao, Yuan Dongshan, Chen Lan,etc. Quantitative analysis on the risk parameter interval of ecnomic evaluation of shale gas fields[J].Journal of Yangtez University, 2017,14(15):77-81.

Effect of nano pour point depressant on the flow properties of the waxy crude oil from Changqing Oilfield

Huatao Wen[1], Songzhao Zhang[2], Yubo Lian [1, *], Zhiguang Zhao [3], Weibo Wang [1], Yajun Wei [1], Yu Duan [3], Sanbao Dong [4]

[1] Xi'an Changqing Chemical Group Co. Ltd, Xi'an Shaanxi, 710068, China
[2] The Second Oil Delivery Office of Changqing Oilfield, Xi'an Shaanxi, 710068, China
[3] The First Oil Delivery Office of Changqing Oilfield, Xi'an Shaanxi, 710068, China
[4] Xi'an Key Laboratory of Tight oil (Shale oil) Development, Xi'an Shiyou University, Xi'an, 710 065, China

Abstract. In this paper, graphene oxide (GO) was modified with alkyl amidopropyl diethanolamine to obtain a nano pour point depressant (GO-PPD), which was used to improve the flowability of the waxy oil extracted from Changqing Oilfield, China. Fourier transform infrared (FTIR), differential scanning calorimetry (DSC), polarized optical microscopy (POM) and viscometer were employed to evaluate the performance of the GO-PPD. The results showed that compared with traditional pour point depressant (PPD), the GO-PPD exhibited higher performance in promoting the flowability of waxy crude oil. With the presence of 500 mg/kg GO-PPD in the waxy crude oil, the pour point of which could be reduced by 5.5 °C. Also, with the presence of 500 mg/kg GO-PPD, the viscosity reduction rate of the waxy crude oil can reach up to 52% at 30 °C. Through the observation via polarized microscopy, we have also found that with the introduction of GO-PPD in the crude oil, the formation of the wax crystals can be greatly retarded. This confirmed that the graphene oxide derivates could also be served as PPD, which facilitates the flowability of certain crude oil (e.g., waxy crude oil from Changqing Oilfield).

1. Introduction

Paraffin-wax deposition on the walls of wells and pipelines poses a great challenge to oil & gas production [1,2]. Paraffin wax is a major component in some hydrocarbons extracted from special reservoirs (e.g., shale reservoirs), and also is the main component of diesel and other refined products. The main components of the wax are the linear and branched hydrocarbon molecules which usually have more than 16 carbons and less than 40 carbons [3-4]. n-alkanes with higher molecular sizes would converse into the wax at higher temperatures if no wax inhibitor is injected into the production liquids. The formation of wax would reduce the production rate, and also lead to the blockage of the pipelines [5-7].

The traditional methods to mitigate the risk of pipeline blockage caused by the wax deposition including the injection of chemical additives into the flowlines of the extracted hydrocarbons [1,8]. Usually, the chemical additives can be split into two types: crystal modifiers and wax dispersants [9-11]. The crystal modifiers are usually oil-soluble copolymers, which can interact with the nucleus of the wax crystals and inhibit the deposition of the wax crystals [12]. Also, the wax-like segment in the crystal-modifier would participate in the formation of the wax crystal and herein modify the morphology of the wax [12]. Other segments (e.g., branched hydrocarbon chains, amide groups, etc.) in the molecule of modifier do not cocrystallize with the original paraffin waxes in the crude oil. The non-wax-like segments provide steric hindrance on the wax surface that retard the growth and aggregation of the crystals [7]. There are many publishes claiming that the size of the wax crystals can be significantly reduced by the polyaminoamide additives [2, 12]. However, it should be noted that there are many factors that influence the performance and application of the polymers, for example, the high costs of the preparation process, the low performance at high shearing conditions and the unignorable performance loss at reheating conditions. The wax dispersants are usually surface-active molecules that can adsorb on the surfaces of wax and pipeline walls. Due to the adsorption layers on the pipeline inner surface, the wettability of these surfaces are kept in water-philic conditions which would reduce the adhesion force between crystal and the inner surface of the pipeline [13]. The common crystal modifiers are ethylene vinyl copolymers (EVA), poly (ethylene-butene) copolymers (PEB), poly (maleic anhydride amide co-α-olefin) (MAC) and their derivatives. The content of vinyl acetate (VA) in a molecule (polar group content), the molecular weight, the length of the side hydrocarbon chain and the compositions of the crude oil are the main factors that affect the performance of the crystal modifiers [14-17]. The optimum content of VA in a modifier (e.g., EVA) is ~30% [18-19]. The EVA molecules can alter the wax crystals from plates to spheres. It was reported that the MAC was more effect than PEB in crude oils containing a high amount of asphaltenes [10]. Singhal et al.[20]

* Corresponding author: dicpyblian@163.com

 https://doi.org/10.1051/e3sconf/202132901050

summarized two empirical rules for selecting the crystal modifiers: the carbon numbers of the side chain length of the alkyl ester and the wax component should be equal or close to each other, the melting points of the modifier and the wax should be equal or close to each other. Dispersants are different kinds of surfactants, such as sulfonates surfactants [21], alkyl phenol derivatives, polyamides and naphthalene[13]. The water cut may have a significant effect on the performance of the wax dispersants. The crystal modifiers and wax dispersants have attracted much attention from researchers, but their performance still needs to be further improved.

In recent years, the nano pour point depressant has been researched by many researchers to mitigate the risk of pipeline blockage caused by the wax formation and deposition [22-23]. Compared with traditional crystal modifiers and wax dispersants, the nano pour point depressants exhibited higher performance in reducing the pour point and improving the flowability of the waxy oil. However, the working mechanisms of the nano pour point depressant have not been fully understood. In this paper, graphene oxide (GO) was modified with alkyl amido propyl diethanolamine to obtain a nano pour point depressant (GO-PPD), which was used to reduce the pour point and improve the flowability of the waxy oil extracted from Changqing Oilfield, China. Fourier transform infrared (FTIR), differential scanning calorimetry (DSC), polarized optical microscopy (POM) and viscometer were employed to evaluate the performance of the GO-PPD.

2 Experimental section

2.1 Materials

The waxy crude oil was offered by Chnagqing Oilfield, China. The SARA information of the crude oil is listed in Table 1. The SARA components of the crude oil were separated based on the method described in one of our previous works [2]. The commercial pour point depressant, EVA with a molecular weight of 2000 and vinyl content of 28%, was purchased from Aladdin Biochemical Technology Co., Ltd. (Shanghai). The nano material with a purity of 99% was also purchased from Aladdin Biochemical Technology Co., Ltd. (Shanghai). N, N-bis-(2-aminoethyl) dodecanamide was prepared in our lab.

Table 1 The basical properties of the waxy oil from Jing'an Block (Changqing Oilfield)

Pour point (°C)	P20 (g·cm^{-3})	Saturates (%)	Aromatics (%)	Resins (%)	Asphaltenes (%)
24	0.88	66.83	10.12	17.70	5.35

2.2 preparation of hybrid pour point depressant GO-PPD

The GO-PPD was prepared by the modification of GO by amine (N, N-bis-(2-aminoethyl) dodecanamide). In this paper, the GO was reacted with an amine at vacuum conditions (10-2 Torr) and 180 °C for 2 hours. After the reaction, the product (GO-PPD) was obtained by removing the amine residues at vacuum conditions and 150 °C for at least 1 hour.

2.3 Viscosity measurement

The viscosity properties of the crude oil with and without the introduction of PPD were evaluated using a Viscometer (Brookfield II, USA) at various temperatures. Each sample was measured three times to increase the quality of the results.

2.4 Pour point measurement

The pour point of each waxy crude oil sample with and without the PPD was measure at a certain temperature range of 5-50°C based on the standard of ASTM D5853 (Standard Test Method for Pour Point of Crude Oils).

2.5 Thermogravimetric analysis (TGA) of the crude oil

The TGA test was performed using a Mettler Toledo A851 TGA/SDTA instrument. During each test, 3-6 mg oil sample was placed on the pans of the instrument. Then the sample was heated from 35 °C to 500 °C at a rate of 10 °C/min under the N2 environment (with a flow rate of 20 mL/min).

2.6 DSC test of waxy crude oil samples

The DSC analysis of the oil samples were carried out using a DSC apparatus (Mettler-Toledo DSC822e, Switzerland). In each measurement, 6-8 mg oil sample was placed in the pans of DSC and then the sample was heated (at a rate of 11°C/min) from room temperature to 50°C in the N2 environment (with a flow rate of 20 mL/min). At 50°C, the sample was kept for 5 min to remove the effect of memory effect of wax formation. Then the sample was cooled from 50 °C to -20 °C at a rate of 8 °C/min. During the cooling stage, the change in the heat was recorded.

2.7 Microscope observation of the morphology of wax crystals

The saturates of the waxy crude oil was extracted[2] and placed on the glass slide of a polarized microscopy (BX41-P OLYMPUS, Japan) to observe the process of formation and morphology of the wax crystals. Before each test, the saturates sample was heated to 50 °C to melt the already formed wax crystals. Then the sample was cooled from 50 °C to 10 °C to facilitate the formation of the wax crystals. During the observation, the temperature of the copper stage placed on the microscopy was kept at 10 °C.

https://doi.org/10.1051/e3sconf/202132901050

3. Results and discussion

3.1 Thermogravimetric analysis of crude oil sample

Fig 1 shows the TGA curve of the crude oil sample at the temperature range of 35 °C - 500 °C. It can be found that as the temperature ramps to 350 °C, the mass loss of the sample reaches 86.75%, indicating that the main component of the crude oil sample is light hydrocarbons. When the sample was further heated from 350 °C to 450 °C, the mass of the sample was gradually reduced. At the temperature of 500 °C, the mass loss of the sample turned out to be 96.18%, which indicated that there was a little amount of hydrocarbons with carbon numbers higher than 35 in the original crude oil ample.

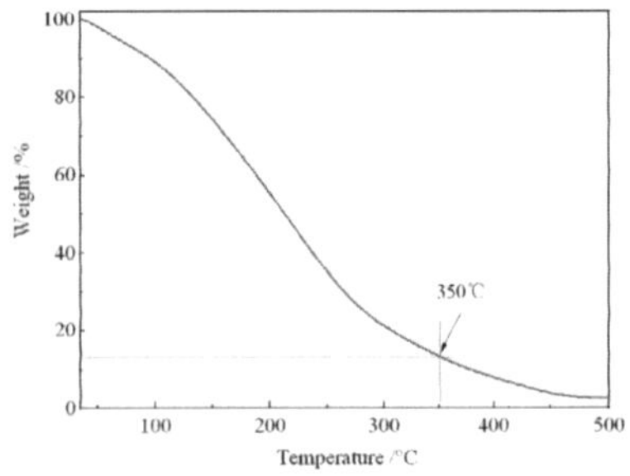

Fig 1. TGA curve of the crude oil sample

3.2 Differential Scanning Calorimetry Analysis of crude oil and saturates samples

The DSC curves of the crude oil and its saturates component were presented in Fig 2. As can be seen from Fig. 2, as crude oil sample (line N0) was cooled from 50 °C to 22.83 °C, the curve encountered an exothermic peak at 22.83 °C, indicating that the wax crystal formed at the temperature of 22.83 °C (wax appearance temperature (WAT)). Based on the area of the exothermic peak, the released heat of wax crystallization was calculated to be 0.53J/g. After the appearance of the peak, the curve gradually decreased till the temperature reduced to -20 °C. The curve of the saturates sample showed a similar trend to that of the crude oil sample. However, the WAT and heat of the wax crystallization were found to be 24.77 °C and 2.90 J/g, respectively. The WAT of the saturates was 2.16 °C lower than that of the crude oil sample, indicating that the polar materials (e.g., resins, asphaltenes, etc.) in the crude oil sample may have inhibition effect on the appearance of the wax crystals.

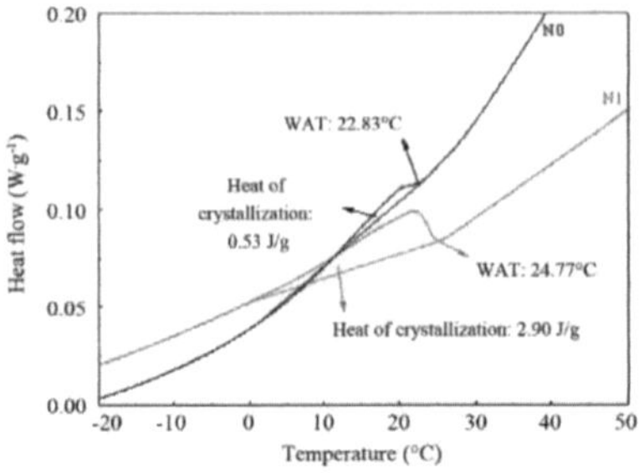

Fig 2. The DSC curves of the crude oil and its saturates component (N0: crude oil sample, N1: saturates sample)

3.3 IR analysis of polar components of crude oil

The component with high polarity was separated using the method described in one of our previous works [2]. The polar component was analyzed using an infrared spectroscopy instrument (Nicolet 5700, USA). The IR results were shown in Fig 3. The peak at the range of 3600-3350 cm^{-1} refers to the hydroxyl stretching vibration. The two peaks at 2600 cm^{-1} and 2550 cm^{-1} may be caused by the stretching vibration absorption of the –SH group. The absorption peaks at 1630 cm^{-1} are exhibited due to the delocalized π bond, the stretching vibration of the C=O bond and amide or bending vibration of the N-N bond. The results show that separated component contains strong polar compounds (e.g., resins, asphaltenes, etc.).

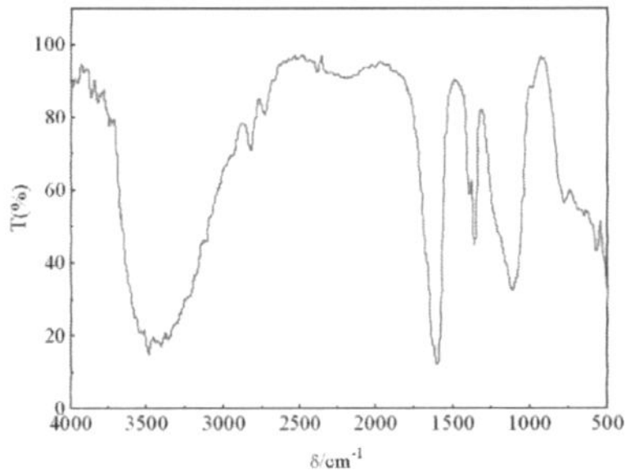

Fig 3. IR spectrum of the adsorption residuals on silica gel

3.4 IR analysis of GO-PPD

The product GO-PPD was analyzed by infrared spectroscopy, the IR curve was shown in Fig 4. The peak at 3340 cm^{-1} refers to the stretching vibration of -OH. The absorption peaks at 1625 cm^{-1} are due to the stretching vibration of -CONH-. The peaks at 3005 cm^{-1}, 2928 cm^{-1} and 2844 cm^{-1} may be caused by the stretching vibrations of $-CH_3$, $-CH_2-$ groups.

 https://doi.org/10.1051/e3sconf/202132901050

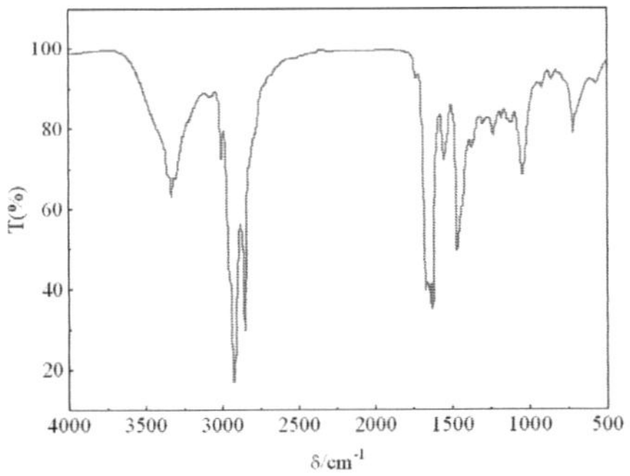

Fig 4. IR analysis of GO-PPD

3.5 Performance of GO-PPD in reducing the viscosity of crude oil

In this section, 500 mg/kg of PPD was introduced into the crude oil sample, and then the viscosity of each mixed oil sample with PPD was measured to see check the PPD's performance. As can be seen in Fig 5, the viscosity of the crude oil sample decreased as the temperature increased. However, at 27 °C, the viscosity of crude oil had a breakpoint. When the sample was further increased from 27 °C, no significant reduction in the viscosity can be found. At 30 °C, the viscosity of the crude oil turned out to be 1855 mPa·s. It should be noted that with the presence of 500 mg/kg GO-PPD (or 500 mg/kg EVA), a similar trend of the viscosity-temperature curve to that of the crude oil sample was observed. With the presence of 500 mg/kg GO-PPD and 500 mg/kg EVA, the viscosity of each mixed oil sample was found to be 146 mPa·s and 126 mPa·s (with a reduction rate of 52%), respectively. Therefore, the performance of GO-PPD in reducing the viscosity of crude oil is higher than EVA.

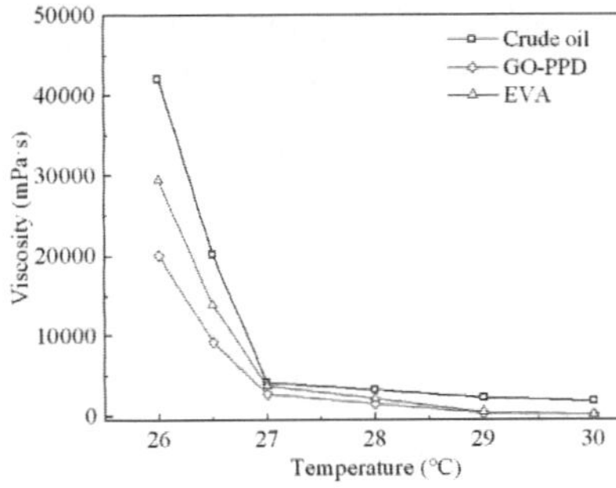

Fig 5 Effect of PPD on the viscosity of the crude oil

3.6 Performance of GO-PPD in reducing the pour point of crude oil

In this section, the effect of GO-PPD on the pour point of the crude oil was measured. As can be seen in Table 2, with the presence of 500 mg/kg GO-PPD in the crude oil sample, the pour point of the oil can be lowered to 18.5 °C, which was lower than the pour point temperature (20.2 °C) of crude oil with 500 mg/kg EVA. This result confirmed the effectiveness of GO-PPD in reducing the pour point temperature of crude oil.

Table 2 The effectiveness of GO-PPD in reducing the pour point temperature of crude oil.

Sample	Crude oil	Crude oil +500 mg/kg EVA	Crude oil +500 mg/kg GO-PPD
Pour point (°C)	24	20.2	18.5

3.7 Wax morphology analysis

In this section, the effects of GO-PPD and EVA on the morphology of the wax crystal were examined, which were present in Fig 6. As can be seen in Fig 6(a), without chemical additive, sword-like wax crystals were observed. As the amount of the crystals increased, a three-dimensional network of wax crystals was observed. However, with the introduction of EVA in the crude oil, the formed wax crystals were effectively dispersed, and also the amount and size of the crystals were significantly reduced (see Fig 6(b)). When GO-PPD was present in the crude oil sample, it was easy to found that the formed wax crystals were further dispersed, and the sizes of each wax crystals were also smaller than that of crystals in the presence of EVA, indicating that the GO-PPD is a promising PPD that can effectively inhibit the formation of wax and promote the flowability of the crude oil (from Changqing Oilfield).

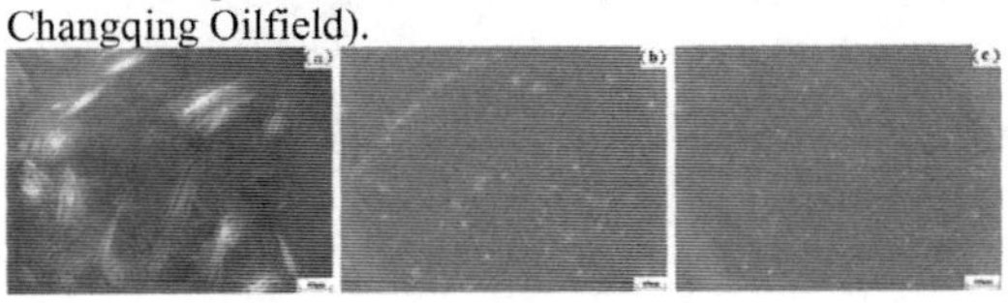

Fig 6. Wax morphologies of sample (a): saturates, sample (b): saturates+500 mg/kg EVA, sample (c): saturates+500 mg/kg GO-PPD

4 Conclusions

In this paper, we have prepared a novel pour point depressant for waxy crude oil in Changqing Oilfield. The performance of GO-PPD was systematically investigated. The results showed that compared with traditional pour point depressant (PPD), the GO-PPD exhibited higher performance in promoting the flowability of waxy crude oil. With the presence of 500 mg/kg GO-PPD in the waxy crude oil, the pour point of which could be reduced by 5.5 °C. Also, with the presence of 500 mg/kg GO-PPD, the viscosity reduction rate of the waxy crude oil can reach 52% at 30 °C. Through the observation via polarized microscopy, we have also found that with the introduction of GO-PPD in the crude oil, the formation of the wax crystals can be greatly retarded. This confirmed that the graphene oxide derivates could also be served as PPD, which facilitates the flowability of certain crude oil (e.g., waxy crude oil from Changqing Oilfield).

Acknowledgments

The work was supported financially by the Open Fund of Shaanxi Key Laboratory of Carbon Dioxide Storage and Enhanced Oil Recovery (YJSYZX20SKF0008), Natural Science Basic Research Program of Shaanxi (Program No.2020JQ-775), Scientific Research Program Funded by Shaanxi Provincial Education Department (Program No. 21JP094) and The Youth Innovation Team of Shaanxi Universities.

References

1. Huang Z., Zheng S., Fogler H.S. Wax Deposition: Experimental Characterizations, Theoretical Modeling, and Field Practices, CRC Press, 2015.
2. Chen G., Lin J., Hu W., Cheng C., Gu X., Du W., Zhang J., Qu C. Characteristics of a crude oil compositions and its in situ waxing inhibition behavior[J]. Fuel, 2018, 218, 213-217.
3. Abdallah D., Weiss G.R., n-alkanes gel n-alkanes (and many other organic liquids)[J]. Langmuir, 2000, 16(2): 352-355.
4. Aiyejina A., Chakrabarti D.P., Pilgrim A., Sastry M.K.S. Wax formation in oil pipelines: a critical review[J]. International Journal of Multiphase Flow, 2011, 37(7): 671-694.
5. Yang J.H., Lu Y.D., Daraboina N., Sarica C. Wax deposition mechanisms: Is the current description sufficient?[J]. Fuel, 2020, 275: 117937.
6. Valinejad R., Nazar A.R.S. An experimental design approach for investigating the effects of operating factors on the wax deposition in pipelines[J]. Fuel, 2013, 106: 843-850.
7. Sun M.W., Naderi K., Firoozabadi A. Effect of crystal modifiers and dispersants on paraffin-wax particles in petroleum fluids[J]. SPE Journal, 2019, 24(01): 32-43.
8. Al-Yaari M. Paraffin wax deposition: mitigation & removal techniques[C]. SPE Saudi Arabia section Young Professionals Technical Symposium, Dhahran, Saudi Arabia, March 2011. SPE-155412-MS.
9. [9] Li L., Guo X.H., Adamson D.H., Pethica B.A., Huang J.S., Prudhomme R.K. Flow improvement of waxy oils by modulating long-chain paraffin crystallization with comb polymers: an observation by X-ray diffraction[J]. Industrial & Engineering Chemistry Research, 2011, 50(1): 316-321.
10. Li L., Tinsley J., Adamson D.H., Pethica B.A., Huang J.S., Prudhomme R.K., Guo X.H. Inprovement of oil flowability by assembly of comb-type copolymers with paraffin and asphaltene[J]. AIChE Journal, 2012, 58(7): 2254-2261.
11. Yi S.Z., Zhang J.J. Relationship between waxy crude oil composition and change in the morphology and structure of wax crystals induced by pour-point-depressant beneficiation[J]. Energy & Fuels, 2011, 25(4): 1686-1696.
12. Radulescu A., Fetters L.J., Richter D. Tailored polymer additives for wax (paraffin) crystal control, Chapter 11 Crude Oil Emulsions: composition Stability and Characterization[M], INTECH Open Access Publisher.
13. Dobbs J.B. A unique method of paraffin control in production operations[C]. SPE Rocky Mountain Regional Meeting, Gillette, Wyoming, May 1999. SPE-55647-MS.
14. Kuzmic A.E., Radosevic M., Bogdanic G., Srica V., Vukovic R. Studies on the influence of long chain acrylic esters polymers with polar monomers as crude oil flow improver additives[J]. Fuel, 2008, 87: 2943-2950.
15. Xu J., Xing S., Qian H., Chen S., Wei X., Zhang R. Effect of polar/nonpolar groups in comb-type copolymers on cold flowability and paraffin crystallization of waxy oils[J]. Fuel, 2013, 103: 600-605.
16. Wu Y., Ni G., Yang F., Li C., Dong G. Modified maleic anhydride co-polymers as pour-point depressants and their effects on waxy crude oil rheology[J]. Energy & Fuels, 2012, 26: 995-1001.
17. Deshmukh S., Bharambe D. Wax control by designing polymeric pour point depressant additives for Indian crude oil[J]. Journal of Dispersion Science and Technology, 2011, 32: 291-297.
18. [18] Machado A.L. Lucas E.F., and González G. Poly (ethylenc-co-vinyl acetate)(EVA) as wax inhibitor of Brazilian crude oil: oil viscosity, pour point and phase behavior of organic solutions[J]. Journal of Petroleum Science and Engineering, 2001, 32(2-4): 159-165.
19. Machado A.L., Lucas E.F. Influence of ethylene-co-vinyl acetate copolymers on the flow properties of wax synthetic systems[J]. Journal of Applied Polymer Science, 2002, 85(6): 1337-1348.
20. Singhal H., Sahai G., Pundeer G., et al., Designing and selecting wax crystal modifier for optimum field performance based on crude oil composition[C]. SPE Annual Technical Conference and Exihibition, Dallas, 6-9 October, 1991. SPE-22784.
21. Chen G., Zhou Z.C., Shi X.D., Zhang X.L., Dong S.B., Zhang J. Synthesis of alkylbenesulfonate and its behavior as flow improver n crude oil [J]. Fuel, 2021, 288: 119644.
22. Wang F., Zhang D.M., Ding Y.F., Zhang L.X., Yang M.S., Jiang B.L. The effect of nanohybrid materials on the pour-point and viscosity depressing of waxy crude oil[J]. Chinese Science Bulletin, 2011, 56: 14-17.
23. Yang F., Paso K., Norrman J., Li C. Oschmann H., Sjöblom J. Hydrophilic nanoparticles facilitate wax inhibition[J]. Energy & Fuels, 2015, 29: 1368-1374.

Effect of a rapeseed oil derived pour point depressant on the flow properties of the waxy oil from Changqing Oilfield

Yongzhen Wang[1], Guibin Liu[2], Xiaoyu Zhou[3], Qingzhu Zhng[2], Wei Jiang[2], Chuanghui Han[2], Sitong Huo[1], Yubo Lian[2,*], Sanbao Dong[4]

[1] The First Oil Delivery Office of Changqing Oilfield, Xi'an Shaanxi, 710068, China
[2] Xi'an Changqing Chemical Group Co. Ltd, Xi'an Shaanxi, 710068, China
[3] The Third Oil Delivery Office of Changqing Oilfield, Xi'an Shaanxi, 710068, China
[4] Xi'an Key Laboratory of Tight oil (Shale oil) Development, Xi'an Shiyou University, Xi'an, 710 065, China

Abstract. In this paper, rapeseed oil was modified with diethylenetriamine to obtain a vegetable oil-derived pour point depressant (PPD), which was used to reduce the pour point and improve the flowability of the crude oil (from Changqing Oilfield), which has a high content of paraffin. Various instruments such as differential scanning calorimetry (DSC), polarized optical microscopy (POM), and a viscometer were used to investigate the performance of the PPD. The results showed that with the introduction of PPD into the waxy oil sample, the pour point of which could be significantly lowered from 20.0 °C to 13.2 °C. At 23 °C, the viscosity reduction rate of the oil sample was 68% in the presence of 500 mg/kg PPD. The wax morphology observation also confirmed that the formation and growth of the wax crystals could be significantly delayed and restrained. The results supported that the PPD synthesized in this paper could effectively improve the flowability of the waxy oil from Changqing Oilfield.

Keywords: Waxy oil, pour point depressant, vegetable oil, wax crystal.

1. Introduction

Flow assurance challenges of petroleum pipelines can be caused by asphaltenes, paraffin wax, scales, and gas hydrate[1]. The blockages of wells and pipelines during the production from shale reservoirs (e.g., Changqing Oilfield, China) are usually attributed to the deposition and buildup of wax crystals in the production liquids.

The wax from crude oil is a mixture of linear paraffins, iso-paraffins and cyclo-paraffins with various carbon numbers (from 18-65)[2]. The deposition of the wax in production liquids occurs below the wax appearance temperature (WAT), causing increased viscosity of the hydrocarbon liquids and consequent interrupt of the production.

Temperature, the composition of hydrocarbon liquids, flowing rate and surface property of the pipeline wall have a great influence on the deposition and buildup of the wax crystals. There are two parameters in paraffin wax deposition, WAT and gel formation temperature (GFT)[3]. During the production, the waxy crude oils are extracted from reservoirs to pipelines, if the temperature of the pipeline wall is lower than the WAT of the waxy oil, the decrease in temperature would facilitate the formation and deposition of wax crystals on the wall surface. The temperature gradient and even the concentration gradient between the bulk liquids and wall surface would certainly promote the buildup of wax layer on the wall surface until the GFT of the production liquids is reached. For this reason, the pressure drop would also be increased, which will significantly raise the risk of pipeline plugging. The temporarily deposited wax crystals would be decomposed by the shearing effect of the production liquids. However, during the shale oil production, the low flow rates (which is more likely) of the production liquids can not delay/prevent the deposition.

The chemical inhibitors used for mitigating wax deposition, which is a promising method, can be split into two categories: crystal modifiers and dispersants[4-6]. The main advantages of dosing these chemicals are low dosage requirement and the simple operation without the need to shut down the production. Crystal modifiers are oil-soluble copolymers containing hydrocarbon branches with certain carbon numbers (needs to be optimized for improving the inhibiting performance) and hydrophilic groups. The hydrocarbon branches can participate in the formation of the wax crystals and herein provide steric hindrance that prevents the aggregation of the wax crystals. Therefore, the growth of the wax crystals can be effectively disrupted and retarded. Wax dispersants are usually inter-surface active agents containing wax-philic groups and water-philic groups. Therefore, the wax dispersants can be adsorbed onto the wax surface and deposition surfaces, altering the wettability of these

* Corresponding author: dicpyblian@163.com

surfaces to water-wet conditions, and preventing further aggregation and buildup of these wax crystals[7-9].

There are many factors affecting the performance of the chemical inhibitors, for example, the type of the hydrocarbon fluid, properties of the wall surfaces, and the crude oil-chemical interactions. The chemicals could reduce the pour point and viscosity of the waxy crude oil, and also alter the morphology of the wax crystals. In this study, various instruments such as differential scanning calorimetry (DSC), polarized optical microscopy (POM), and a viscometer were used to investigate the performance of the rapeseed oil-derived pour point depressants (PPD). The PPD was obtained by the amination reaction between rapeseed oil and diethylenetriamine.

2 Experimental section

2.1 Materials

The waxy crude oil was provided by Chnagqing Oilfield, China. The SARA data of the oil sample is offered in Table 1. The SARA components were separated using the method described in Chen's works [10]. A commercial copolymer, EVA, used as a PPD, and diethylenetriamine (with a purity of 97%) were from Aladdin (Shanghai).

Table 1 The properties of the waxy oil from Nanliang Block (Changqing Oilfield)

Pour point (°C)	P20 (g·cm^{-3})	Saturates (%)	Aromatics (%)	Resins (%)	Asphaltenes (%)
20.0	0.87	51.66	33.12	12.01	3.21

2.2 Preparation of the pour point depressant PPD

The PPD was prepared by the amination of rapeseed oil by diethylenetriamine. In this paper, the rapeseed oil and diethylenetriamine (with a mol ratio of 1:1) were reacted in the toluene phase for 4 hours under flush conditions. After the reaction, the toluene was removed under vacuum conditions, after the separation of the toluene from the mixtures, the products (see Fig. 1) were obtained.

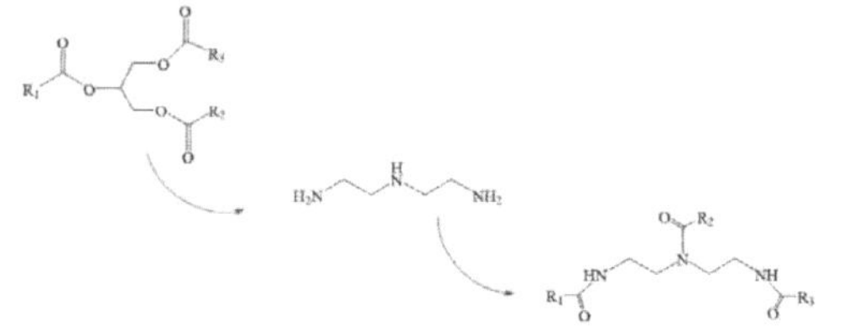

Fig. 1 The preparation of PPD through the amination of rapeseed oil

2.3 Viscosity measurement

The viscosity of each oil sample with and without PPD was measured using a viscometer (Brookfield II, USA). In each measurement, the temperature of the oil sample was increased from room temperature to 30 °C. Each measurement was repeated three times to keep the accuracy of the results.

2.4 Pour point measurement

Based on the method described in the standard of ASTM D5853 (Standard Test Method for Pour Point of Crude Oils), each oil sample's pour point was tested at least three times to find the average value of the pour point.

2.5 Microscope observation of the morphology of wax crystals

After the separation of saturates from the waxy crude oil sample, the saturate sample was placed on a glass slide, which was put on the stage of the polarized microscopy(BX41-P Olympus, Japan). The temperature of the stage was controlled by an externally connected chiller. Before each test, the saturate sample was kept at 50 °C (higher than the melting point of the paraffin wax) for at least 15min to avoid the influence of already-formed crystals. To promote the formation of the wax crystals, the temperature of the sample was reduced from 50 °C to 10 °C at a cooling rate of 4 °C/min. The sample was controlled at 10 °C to keep the growth of the wax crystals. During the whole test, the morphology of the wax crystals was automatically recorded.

3. Results and discussion

3.1 Performance of PPD in reducing the viscosity of crude oil

The wax inhibitors mitigate the risk of pipeline blockage caused by wax deposition via the reduction in viscosity and pour point. In this section, the performance of PPD was analyzed compared with the performance of EVA. In each test, wax inhibitor (EVA or PPD) with a dosage of 500 mg/kg was mixed with the waxy crude oil. Then the viscosity of each mixed oil sample was measured. As shown in Fig 3, without wax inhibitor, the viscosity of the crude oil decreases from 44000 mPa·s to 902 mPa·s in the temperature range of 23 °C - 28 °C. With the presence of 500 mg/kg PPD, the corresponding viscosity of the sample was decreased from 20100 mPa·s to 122 mPa·s in the temperature range of 23 °C - 28 °C, indicating that the PPD can effectively reduce the viscosity of the crude oil. It should be noted that the oil sample with 500mg/kg was higher than that of the oil sample with 500 mg/kg PPD. The performance of EVA was slightly lower than that of PPD as can be seen from Fig 2.

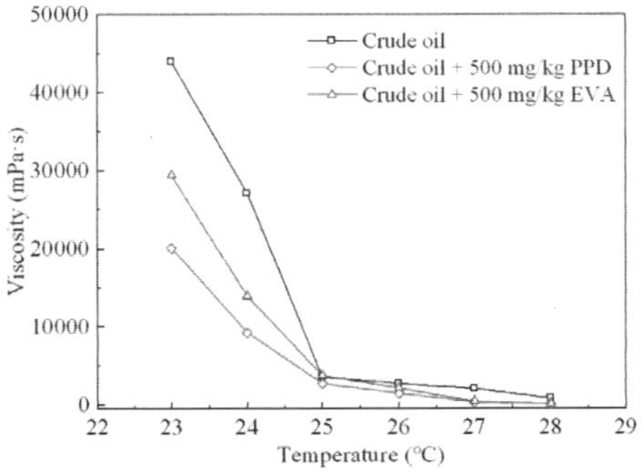

Fig 2 Effect of PPD and EVA on the viscosity of the crude oil

For the effect of PPD's dosage on viscosity reduction, the viscosity of the oil sample gradually decreases with the increase in the PPD's dosage in the waxy crude oil as can be drawn from Fig 3. At the dosage of 500 mg/kg, the viscosity of the crude oil can be lowered to 122 mPa·s at 28 °C.

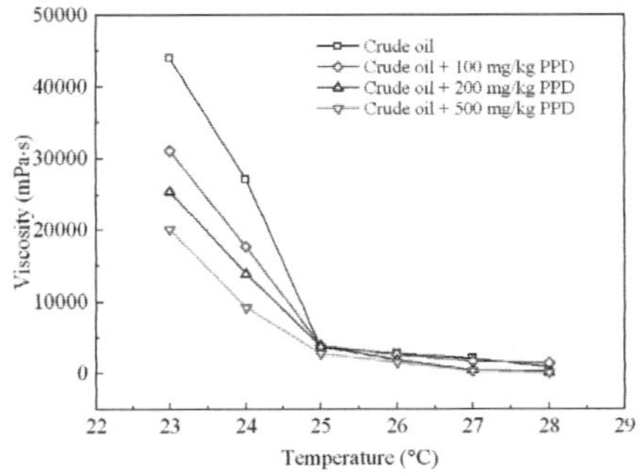

Fig 3 Effect of PPD's dosage on the viscosity of the crude oil

3.2 Performance of PPD in reducing the pour point of crude oil

The effect of PPD on the pour point of the oil sample was listed in Table 2. After the introduction of 500 mg/kg PPD in the crude oil sample, the pour point of which can be reduced to 13.2 °C, which was slightly lower than the pour point (14.4 °C) of the oil sample with EVA at the same dosage. This result verified that the efficiency of PPD was superior to EVA.

Table 2 Efficiency of wax inhibitors on the pour point of waxy oil from Changqing Oilfield

Sample	Crude oil	Crude oil +500 mg/kg EVA	Crude oil +500 mg/kg PPD
Pour point (°C)	20	14.4	13.2

3.3 Wax morphology analysis

The effects of inhibitors on the morphology of wax crystal were present in Fig 4. Without chemical additive, a three-dimensional network of sword-shaped crystals was observed. However, in the samples containing wax inhibitors, the wax crystals with limited sizes were dispersed in the saturates phase, no network of crystals can be found. These results confirmed that the PPD is a promising wax inhibitor that has a similar performance with EVA when tested with the waxy crude oil from Changqing Oilfield.

Fig 4. Physical appearances of (a): saturates, (b): saturates+500 mg/kg EVA and (c): saturates+500 mg/kg PPD

4 Conclusions

A wax inhibitor (PPD) was synthesized through the amination of rapeseed oil with diethylenetriamine. The PPD's efficiency in improving the flowability of the waxy oil sample from Changqing Oilfield was investigated. The pour point of the oil sample can be lowered from 20 °C to 13.2 °C in the presence of 500 mg/kg PPD. At the same dosage of PPD, the viscosity of the oil sample can be lowered to 122 mPa·s at 28 °C. The wax crystals can be well dispersed in the saturates phase.

Competing interests

The authors declare that they have no competing interests.

Acknowledgments

The work was supported financially by the Open Fund of Shaanxi Key Laboratory of Carbon Dioxide Storage and Enhanced Oil Recovery (YJSYZX20SKF0008), Natural Science Basic Research Program of Shaanxi (Program No. 2020JQ-775), Scientific Research Program Funded by Shaanxi Provincial Education Department (Program No. 21JP094) and The Youth Innovation Team of Shaanxi Universities. And we thank the work of the Modern Analysis and Testing Center of Xi'an Shiyou University.

References

1. Kelland M.A. Production Chemicals for the Oil and Gas Industry, 1st Edition[M]. Boca Raton, CRC Press, 2008.
2. Lake L.W. Petroleum Engineering Handbook: Volume 1. General Engineering, SPE, ISBN: 978-1-55563-108-6.
3. Srivastava S.P., Saxena A.K., Tandon R.S., Shekher V. Measurement and prediction of solubility of

https://doi.org/10.1051/e3sconf/202132901051

petroleum waxes in organic solvents[J]. Fuel, 1997, 76(7): 625-630.

4. Li L., Tinsley J., Adamson D.H., Pethica B.A., Huang J.S., Guo X. Improvement of oil flowability by assembly of comb-type copolymers with paraffin and asphaltene[J]. AIChE Journal, 2012, 58(7): 2254-2261.
5. Xu J., Xing S., Qian H., Chen S., Wei X., Zhang R., Guo X. Effect of polar/nonpolar groups in comb-type copolymers on cold flowability and paraffin crystallization of waxy oils[J]. Fuel, 2013, 103: 600-6056.
6. Wei B. Recent advances on mitigating wax problem using polymeric wax crystal modifier[J]. Journal of Petroleum Exploration and Petroleum Technology, 2015, 5(4): 391-401.
7. Sun M., Naderi K., Firoozabadi A. Effect of crystal modifiers and dispersants on paraffin-wax particles in petroleum fluids[J]. SPE Journal, 2018, 24(01): SPE-191365-PA.
8. Dobbs J.B. A unique method of paraffin control in production operations[C]. SPE Rocky Mountain regional meeting, Society of Petroleum Engineers, SPE-55647-MS, 1999.
9. Kar T., Firoozabadi A. Mitigation of paraffin wax deposition and the effect of brine[C]. SPE Annual Technical Conference and Exhibition held in Calgary, Alberta, Canada. Society of Petroleum Engineers, SPE-196032-MS, 2019.
10. Chen G., Lin J., Hu W., Cheng C., Gu X., Du W., Zhang J. Characteristics of a crude oil composition and its in situ waxing inhibition behavior[J]. Fuel, 2018, 218: 213-217.

 https://doi.org/10.1051/e3sconf/202132901052

Characteristics Analysis and Corrosion Inhibition in Typical Produced Water of Jidong Oilfield

Chen Dong [1, 2]

[1] Tangshan Jiyou Ruifeng Chemical Co., Ltd, 063200, Tangshan, China
[2] Hebie Oilfield Chemical Agent Technology Innovation Center, PetroChina Jidong Oil Field, 063200, Tangshan, China

Abstract. The composition of oilfield water is complex and diverse. The development and application of corrosion inhibitors can effectively inhibit the corrosion of pipelines and equipment by oilfield water. The ions in produced water of Jidong Oilfield were determined by instrumental analysis and chemical analysis. Triazine prepared from ethanolamine and formaldehyde as corrosion inhibitor was evaluated by weight loss method. The corrosion inhibition experiments were carried out in ancient water samples, and the effects of concentration and temperature on corrosion inhibition were investigated. Finally, corrosion inhibitors were applied to the corrosion inhibition of produced water with different compositions in Jidong Oilfield, the corrosion inhibition of steel sheet was evaluated under different salinity conditions, and the effects of various ion contents in typical water samples and salinity on corrosion inhibition were analyzed. The experimental results show that the corrosion inhibition effect of triazine is better, and the product can be used in various water samples in oilfield with the inhibition efficiency above 85%.

Keywords: Corrosion inhibition; weight loss method; water analysis.

1 Introduction

In the process of oil and gas field oil production, the produced water is of poor quality and complex composition [1]. It generally has high salt content, high COD value, large polymer content, and large oil content [2,3]. Some also contain small amounts of heavy metals and bacteria [4]. Direct discharge will cause irreversible damage to the surrounding environment and a great waste of water resources. In oil and gas fields, it is generally considered to reinject the produced water after treatment [5]. If the water quality conditions do not meet the injection requirements, scaling, blockage, and corrosion will occur [6,7], which will inevitably cause corrosion on the oil layer [8], water injection pipelines and equipment [9]. In order to ensure that the water quality meets the technical requirements of the industry, oilfield water detection and analysis is required [10]. On the one hand, it can ensure that the water quality requirements for reinjection are met, and scaling and corrosion are reduced [11], and on the other hand, the recovery factor cannot be reduced [12]. In addition to using corrosion-resistant materials, corrosion inhibitors must be selected according to the specific water environment and added to the produced water [13]. However, research on the cost-effective, low-toxicity and environmentally friendly corrosion inhibitors with high corrosion inhibition efficiency for steel is urgently needed [14-16]. As a new corrosion inhibitor, triazine, a synthetic product of ethanolamine and formaldehyde, has excellent corrosion inhibition properties. With the stable spatial structure and inhibitory properties [17,18], triazines are suitable for oilfield produced water containing complex ions.

2 Experimental

2.1 Materials

Ethanolamine was supplied from Tianjin Kermel Chemical Reagent Co., Ltd. Formaldehyde was purchased from Tianjin Tianli Chemical Reagent Co., Ltd. Anhydrous calcium chloride was purchased from Tianjin Shengao Chemical Reagent Co., Ltd. Magnesium chloride hexahydrate was supplied from Chengdu Kelong Chemical Reagent Factory. Sodium bicarbonate was purchased from Tianjin Baishi Chemical Co., Ltd. Anhydrous sodium sulfate was purchased from Tianjin Tianli Chemical Reagent Co., Ltd. Sodium chloride and Potassium chlorid were purchased from Tianjin Shengao Chemical Reagent Co., Ltd. Methyl orange and Phenolphthalei were supplied from Tianjin Kermel Chemical Reagent Co., Ltd.

2.2 Characteristics analysis method

At present, the main methods of water quality analysis are instrumental analysis and chemical analysis. The contents of Na^+, K^+, Ca^{2+}, Mg^{2+}, Ba^{2+}, Sr^{2+}, F^-, Cl^-, NO^{3-}, NO^{2-},

* Corresponding author: 807086230@qq.com

 https://doi.org/10.1051/e3sconf/202132901052

PO_4^{3-} and SO_4^{2-} in typical water samples of Jidong Oilfield were determined by ion chromatography. The flow rate controlled by chromatographic conditions was 1.200mL/min, and the injection volume was 20 L. The content of Fe^{3+} in typical water samples of Jidong Oilfield was determined by spectrophotometry. And the contents of CO_3^{2-} and HCO_3^- in water samples were determined by titration analysis. 0.5% phenolphthalein solution as indicator, CO_3^{2-} in water samples was determined by sulfuric acid standard solution. 0.1% methyl orange solution as indicator, HCO^{3-} in water samples was determined by sulfuric acid standard solution.

2.3 Preparation of simulated oilfield water

According to water type and salinity, four typical water samples from Jidong Oilfield were screened out in this experiment, as shown in Table 1.The simulated water is prepared according to the salinity of oilfield water, and the calculation method is based on the calculation program of simulated water (the volume of simulated water is 1L). The preparation of simulated water for four typical Jidong Oilfield water samples are shown in the table below.

Table 1 The preparation of simulated Jidong Oilfield water

Name	Reagent	Quality (g)
	$CaCl_2$	1.9680
	$MgCl_2{\cdot}6H_2O$	0.4543
ND9	$NaHCO_3$	1.9552
	Na_2SO_4	0.8220
	NaCl	9.4825
	$CaCl_2$	10.4373
	$MgCl_2{\cdot}6H_2O$	6.8758
ND17-1	$NaHCO_3$	2.0158
	Na_2SO_4	0.2029
	NaCl	54.7707
	$CaCl_2$	1.9645
	$MgCl_2{\cdot}6H_2O$	0.6430
ND17-3	$NaHCO_3$	1.4017
	Na_2SO_4	0.6950
	NaCl	4.6031
	$CaCl_2$	0.7417
	$MgCl_2{\cdot}6H_2O$	2.4216
ND38P1	$NaHCO_3$	0.6237
	Na_2SO_4	0.1609
	NaCl	1.1840

2.4 Synthesis of triazine

Ethanolamine and formaldehyde were attained at the molar ratio of 1:1, ethanolamine was put in a 50mL three-necked flask, a small magnet was added in, the three-necked flask was fixed in the magnetic agitator of a constant temperature water bath at 60°C, the mixed liquid was stirred to add formaldehyde within 0.5h. Condensation at reflux for 3h, the synthesized product was cooled to room temperature.

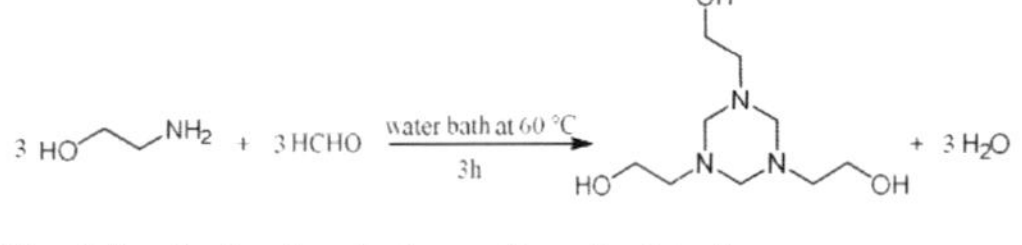

Fig. 1 Synthetic chemical equations for triazine

2.5 Preparation of corrosive solution

The preparation of ancient water samples required firstly, 21.65g $CaCl_2$, 1.4g $MgCl_2{\cdot}6H_2O$, 0.09g $NaHCO_3$, 1.12g Na_2SO_4, 8.24g NaCl, 0.9g KCl were weighed and added in beaker, then dissolved with water and determined volume in 1L volumetric flask. 100mL of ancient water samples were added to 125mL grinding bottle, and the triazine products synthesized by ethanolamine and formaldehyde were added, respectively. Ensure that the concentration in water samples was 0, 0.1%, 0.2%, 0.5%, 1.0%, and 2.0%. Corrosion solutions for other types of water sample corrosion inhibitors are prepared in the same manner as described above.

2.6 Treatment of steel sheets

The tested material is A_3 steel sheets were abraded with a series of emery papers, and the surface area S of the steel sheet was calculated and recorded. The steel sheet was successively soaked in petroleum ether and anhydrous ethanol, and then the surface of the steel sheet was dried with a hair dryer. The mass of each steel sheet is weighed by an analytical balance and recorded as m_1. Every two pieces of steel discs are divided into a group and photographed for recording. Each group of A_3 steel sheet was hung with string into the grinding bottles containing corrosion fluids with different corrosion inhibition agents, and these bottles were put into the thermostatic water bath heated to specific temperature in advance for 48h.The steel sheets were taken out and put on the qualitative filter paper to dry naturally, then photographed for recording. The plates were put in $0.2mol{\cdot}L^{-1}$ dilute hydrochloric acid immersed for 30 s, made the steel rust attached to the surface drop out. The liquid on the surface of the steel sheets was blotted up with filter paper, soaked in the petroleum ether for 2 min, immersed in anhydrous ethanol for 2 min. These steel discs were taken out and placed on qualitative filter paper, dried with hair dryer. The quality of each piece of treated steel sheets were weighed by the analytical balance, noted as m_2. And compared with the m_1, Δm was obtained.

2.7 The method of corrosion inhibition performance evaluation

The weight loss experiments were performed in several kinds of water samples in the absence and presence of various concentration of triazine corrosion inhibitor. The corrosion inhibition performance was mainly evaluated by calculating uniform corrosion rate and corrosion inhibition rate. The mass loss is determined after removing from the corrosion solution. The uniform corrosion rates (W_{corr}) have been obtained as suggested in equation (1):

$$W_{corr} = \frac{\Delta m}{St} \tag{1}$$

Δm—the weight loss, g; *S*—the exposed area of steel sheet, cm^2; *t*—the period of immersion of corrosion, h.

The inhibition efficiency *(E_w%)* was determined by following equation(2):

$$E_w\% = \frac{\Delta m_0 - \Delta m}{\Delta m_0} \times 100 \quad (2)$$

Δm₀—the weight loss of the steel sheet in the absence of triazine in the corrosion solution, g; *Δm*—the weight loss of the steel sheet in the presence of triazine, g.

3 Result and Discussion

3.1 Water samples analysis results

The ions in typical water samples of Jidong Oilfield were determined by ion chromatography and chemical analysis. The results are shown in the Table below.

Table 2 The mass concentration of ions in some water samples of Jidong Oilfield

No.	Name	Cationic concentration (mg/L)					Anion concentration (mg/L)		
		Na^++K^+	Ca^{2+}	Mg^{2+}	Ba^{2+}+Sr^{2+}	Fe^{3+}	Cl^-	SO_4^{2-}	HCO_3^-
1	ND8	5940.5	319.2	32.9	116.9	0	4014.9	1263.8	336
2	ND9	7972.8	710.6	54.4	98.9	0	7167.8	555.8	1420
3	ND17	20633.3	4490.2	1240.6	285.4	1.2	37800.3	116.4	680
4	ND17-1	24942.1	3768.7	823.4	203.1	2.9	42292.3	137.2	1464
5	ND17-3	4655.3	688.4	77.0	3.3	0	4234.6	469.9	1018
6	ND23	4791.3	637.4	38.0	70.5	0.1	3654.3	724.4	1180
7	ND26	7203.9	483.1	33.7	7.4	0	11984.2	0	990
8	ND105	4807.5	225.9	39.1	110.7	0	4528.2	353.8	320
9	ND38P1	1007.7	267.8	2.9	11.9	0.7	1200.4	108.8	453
10	ND38P64	38877.0	8171.1	2481.8	3492.1	0	78483.2	113.8	560

The measured results in Table 2 were processed to calculate the total alkali content, total hardness and salinity of the water sample, and the type of the water sample was determined.

Table 3 Analysis of some water samples in Jidong Oilfield

No.	Name	Total base quantity (mg/L)	Total hardness (mg/L)	Salinity (mg/L)	Water type
1	ND8	336	352.1	12024.2	$NaHCO_3$
2	ND9	1420	765	17980.3	$NaHCO_3$
3	ND17	680	5730.8	65246.1	$CaCl_2$
4	ND17-1	1464	4592.1	73633.7	$CaCl_2$
5	ND17-3	1018	765.4	11146.5	Na_2SO_4
6	ND23	1180	675.4	11096.0	$NaHCO_3$
7	ND26	990	516.8	20702.3	$CaCl_2$
8	ND105	320	265.0	10385.2	Na_2SO_4
9	ND38P1	453	270.7	3053.2	$CaCl_2$
10	ND38P64	560	10652.9	132178.8	$CaCl_2$

It can be seen from Table 3 that the salinity, hardness and chloride ion content of produced water samples in Jidong Oilfield vary greatly, with the salinity span ranging from thousands to tens of thousands, and the salinity and chloride ions have a great influence on the corrosion rate. Therefore, ND38P1, ND17-3, ND17-1 and ND9 were selected to study the corrosion of steel sheet under different salinity conditions.

3.2 Evaluation of corrosion inhibition of triazine

The corrosion inhibition performance of triazine in ancient water quality was evaluated by weight loss method. The results are shown in Table 4. As seen, the uniform corrosion rate decreases with the increase of inhibitor concentration. When the concentration of inhibitor is 2%, the uniform corrosion rate is the lowest 0.0240 $g \cdot cm^{-2} \cdot h^{-1}$. The inhibition efficiency increases to 84.7953%, which has an excellent corrosion inhibition effect on steel sheet.

Table 4 Evaluation of corrosion inhibition of triazine

No.	Inhibitor concentration(%)	W_{corr} ($g \cdot cm^{-2} \cdot h^{-1}$)	$E_w\%$
1002	0.0	0.1751	/
1003		0.1789	
1092	0.1	0.0467	72.92
1093		0.0470	75.00
1097	0.2	0.0432	75.00
1098		0.0434	77.08
1105	0.5	0.0414	76.04
1106		0.0387	78.12
1111	1.0	0.0340	80.21
1112		0.0311	82.29
1114	2.0	0.0269	82.46
1115		0.0240	84.80

It can be seen from Fig. 2 that the corrosion inhibition effect on the steel sheet is the best when the corrosion inhibitor concentration is 2.0% at 343K. The surface of the steel sheet is attached with reddish brown corrosion products, indicating the presence of Fe_3O_4. Black corrosion products are also attached to the surface of the steel sheet, and the specific phase composition needs to be further analyzed by XRD. When the concentration of corrosion inhibitor reached more than 0.5%, spot corrosion was observed, mainly because of the uneven adsorption of corrosion inhibitor, a small anode/large cathode activation - passivation battery system is formed [19,20].

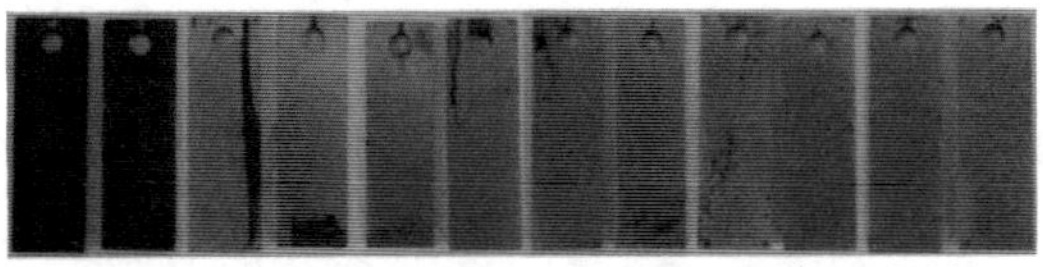

Blank 0.1% 0.2% 0.5% 1.0% 2.0%

Fig. 2 Comparison of corrosion inhibition effect of different concentrations of triazine

3.3 Effect of temperature on corrosion inhibition of triazine

The value of uniform corrosion rate (W_{corr}) and inhibition efficiency *($E_w\%$)* were summarized from the weight loss measurements with and without the addition of various concentration of triazine in ancient water at 323K, 333K, 343K and 353K, respectively. The results were shown in the tables below.

Table 5 Corrosion inhibition evaluation of triazine at 323K

No.	Inhibitor concentration(%)	W_{corr} (g·cm^{-2}·h^{-1})	E_w%
1126	0.0	0.0840	/
1127		0.0850	
1213	0.1	0.0144	82.42
1214		0.0109	86.81
1215	0.2	0.0111	86.81
1216		0.0146	82.42
1217	0.5	0.0091	89.01
1218		0.0129	84.62
1219	1.0	0.0109	86.81
1220		0.0129	84.62

Table 6 Corrosion inhibition evaluation of triazine at 333K

No.	Inhibitor concentration(%)	W_{corr} (g·cm^{-2}·h^{-1})	E_w%
1181	0.0	0.1309	/
1182		0.1397	
1123	0.1	0.0294	78.38
1124		0.0256	81.08
1125	0.2	0.0217	83.78
1126		0.0255	81.08
1127	0.5	0.0239	82.43
1128		0.0201	85.14
1129	1.0	0.0218	83.78
1130		0.0220	83.78

Table 7 Corrosion inhibition evaluation of triazine at 343K

No.	Inhibitor concentration(%)	W_{corr} (g·cm^{-2}·h^{-1})	E_w%
1002	0.0	0.1751	/
1003		0.1789	
1092	0.1	0.0467	72.92
1093		0.0470	75.00
1097	0.2	0.0432	75.00
1098		0.0434	77.08
1105	0.5	0.0414	76.04
1106		0.0387	78.12
1111	1.0	0.0340	80.21
1112		0.0311	82.29

Table 8 Corrosion inhibition evaluation of triazine at 353K

No.	Inhibitor concentration(%)	W_{corr} (g·cm^{-2}·h^{-1})	E_w%
1101	0.0	0.2223	/
1102		0.2259	
1103	0.1	0.1198	46.28
1104		0.1145	47.93
1105	0.2	0.0920	58.68
1106		0.0884	60.33
1107	0.5	0.0747	66.12
1108		0.0724	67.77
1109	1.0	0.0624	71.90
1110		0.0575	73.55

The effect of temperature on the inhibition efficiency of triazine was studied in the temperature range from 323 to 353 K. As seen in Table 6 and 7, the corrosion rate in the brine environment is low under the conditions of 323K and 333K, and the addition of 0. 1% triazine can achieve a good corrosion inhibition effect. As the concentration of corrosion inhibitor continues to increase, the corrosion inhibition rate increases in a gentle trend. However, under the temperature of 343K and 353K, the corrosion rate was relatively high while inhibition efficiency was low. And the inhibition efficiency increased slowly as the concentration of triazine increased. Therefore, under the simulated water environment of this brine, the corrosion inhibition requirement could be met by adding 0.1% corrosion inhibitor at low temperature, while the corrosion inhibition rate could reach 82.29% only when the concentration of triazine was 1.0% at high temperature.

3.4 Effects of ions in produced water on corrosion inhibition performance

The water samples were chosen as ND38P1, ND17-3, ND17-1 and ND9 by weight loss method to study the corrosion of steel sheet under different salinity conditions at 323K.

Table 9 Evaluation of corrosion inhibition of triazine in different typical water samples

Name	No.	Inhibitor concentration(%)	W_{corr} (g·cm^{-2}·h^{-1})	E_w%
ND38P1	14	0	0.1976	/
	1003		0.2202	
	1075	2.0	0.0152	92.89
	1076		0.0114	94.67
ND17-3	1078	0	0.2336	/
	1079		0.2183	
	15	2.0	0.0501	95.59
	1093		0.0420	96.70
ND17-1	16	0	0.1343	/
	1117		0.1535	
	17	2.0	0.0072	94.94
	1121		0.0036	97.47
ND9	18	0	0.1635	/
	1123		0.1511	
	1124	2.0	0.0223	86.05
	1125		0.0205	87.21

It can be seen from Table 9 that the corrosion rate of steel sheet is different in different water quality environments. After comparison in the same environment, it is found that the corrosion rate of steel sheet is lower in water quality of $CaCl_2$ and $NaHCO_3$ than in water quality of $NaSO_4$. The water quality is less than 75000mgL^{-1}, and the corrosion inhibitor can have better corrosion inhibition effect. The inhibition efficiency is all above 85%, which can be applied to the water quality system of oil field to delay the corrosion of steel pipe.

4 Conclusion

The corrosion inhibition effect of triazine on steel sheet was efficient. The inhibition effect was the best when the concentration was 2.0%, and the inhibition efficiency was up to 84.80%. The corrosion rate decreased with the increase of inhibitor concentration and the inhibition efficiency increased with the increase of inhibitor concentration. Some produced water samples in Jidong Oilfield have high salinity and high hardness, which is easy to cause steel pipe scaling and aggravating corrosion. The results showed that when the salinity reached tens of thousands, the corrosion rate increased significantly, and the appropriate addition of triazine corrosion inhibitor had obvious corrosion inhibition effect, with the inhibition efficiency above 85%.

References

1. Igunn E.T., Chen G.Z., Produced water treatment technologies. International Journal of Low-Carbon Technologies, 9(3) (2014) 157-177.
2. Malvis A., Hodaifa G., Halioui M., Seyedsalehi M., Sánchez S., Integrated process for olive oil mill wastewater treatment and its revalorization through the generation of high added value algal biomass. Water research, 151 (2019) 332-342.
3. Yang M., Jing B., Chen W., Li Q., Yin X., Experimental study on COD composition and electrochemical degradation of waste water in offshore oilfields. Journal of the Chinese Chemical Society, 64(1) (2017) 73-79.
4. Zhou Z.F., Wang M.X., Zuo X.H., Yao Y.H., Comparative investigation of bacterial, fungal, and archaeal community structures in soils in a typical oilfield in Jianghan, China. Archives of environmental contamination and toxicology, 72(1) (2017) 65-77.
5. Fischer K., Ferreira F.C., Holzberg B.B., Pastor J.S., Reinli L., Furuie R., Dutra T.A. Integrated modeling of formation damage and multiple induced hydraulic fractures during produced water reinjection. In SPE Latin America and Caribbean Petroleum Engineering Conference. Society of Petroleum Engineers.(2017, May)
6. Azari A., Nazemi S., Kakavandi B., Rastgar A., Survey of scaling and corrosion potential in drinking water resources of Shahrood city by using stability indexes in 2013, (2016).
7. Chen T., Chang F., Liang F., Mukhles A., Corrosion and Scaling Monitoring Under Real Downhole Conditions in Sour Gas Wells. In SPE Kingdom of Saudi Arabia Annual Technical Symposium and Exhibition. Society of Petroleum Engineers, (2018, August).
8. Askari M., Aliofkhazraei M., Ghaffari S., Hajizadeh A., Film former corrosion inhibitors for oil and gas pipelines-A technical review. Journal of Natural Gas Science and Engineering, 58 (2018) 92-114.
9. Andrei M., Ghiban B., Rucai V., Bacterial activity in a deposit from a residual injection water pipeline. REVISTA DE CHIMIE, 70(8) (2019) 2707-2711.
10. Gavrilenko N.A., Volgina T.N., Urazov E.V., Gavrilenko M.A., Transparent polymer sensor for visual and photometrical detection of thiocyanate in oilfield water. Journal of Petroleum Science and Engineering, 172 (2019) 960-963.
11. Liang S.C., A new technique for oil field water analysis by ion chromatography. China petroleum and chemical industry standard and quality, 14 (2013) 25-27.
12. Zou Y.P., Application of water ion chromatography in liaohe oilfield. Logging engineering, 19(1) (2008) 9-13.
13. Liu Y., Zhang Y., Yuan J. Influence of produced water with high salinity and corrosion inhibitors on the corrosion of water injection pipe in Tuha oil field. Engineering Failure Analysis, 45 (2014) 225-233.
14. [14]Li H.J., Zhang W., Wu Y.C., Anti-Corrosive Properties of Alkaloids on Metals. In Alkaloids-Their Importance in Nature and Human Life. IntechOpen, (2018)
15. Dehghani A., Bahlakeh G., Ramezanzadeh B., Ramezanzadeh M., Detailed macro-/micro-scale exploration of the excellent active corrosion inhibition of a novel environmentally friendly green inhibitor for carbon steel in acidic environments. Journal of the Taiwan Institute of Chemical Engineers, 100 (2019) 239-261.
16. Verma C., Kumar A.M., Mazumder M.A.J., Quraishi,M.A., Chitosan-Based Green and Sustainable Corrosion Inhibitors for Carbon Steel. Chitin-Chitosan: Myriad Functionalities in Science and Technology, 143 (2018).
17. Qian P.P., Wang S., Feng K.R., Ren Y.J., Molecular modeling studies of 1, 2, 4-triazine derivatives as novel h-DAAO inhibitors by 3D-QSAR, docking and dynamics simulations. RSC advances, 8(26) (2018) 14311-14327.
18. Obot I.B., Kaya S., Kaya C., Tüzün B., Theoretical evaluation of triazine derivatives as steel corrosion inhibitors: DFT and Monte Carlo simulation approaches. Research on Chemical Intermediates, 42(5) (2016) 4963-4983.
19. Park I.C., Kim S.J., Effect of stabilizer concentration on the cavitation erosion resistance characteristics of the electroless nickel plated gray cast iron in seawater. Surface and Coatings Technology, 376 (2019) 31-37.
20. Du Y., Gao L.Y., Yu D., Liu Z.Q., Comparison and mechanism of electromigration reliability between Cu wire and Au wire bonding in molding state. Journal of Materials Science: Materials in Electronics, (2020) 1-9.

Factors affecting chemical heat washing efficiency of oily sludge and Demulsification Mechanism

Yong Yang [1], Yijun Wang [2]. Zebin Gao [2], Long Zhang [3], Chengtun Qu [3], Bo Yang [4]

[1] Yixing Environmental Protection Industry Development Center Yixing, China
[2] Karamay Sanda testing and analysis Co. LTD Karamay, China
[3] College of Chemistry and Chemical Engineering Xi'an Shiyou University Xi'an, China
[4] Shaanxi Key Laboratory of Environmental Pollution Control Technology and Reservoir Protection of Oilfield, Xi'an, China

Abstract. Oily sludge belongs to the hazardous waste specified in the national hazardous waste list. It will directly or indirectly threaten the ecological environment and human health. This paper introduces the advantages of chemical hot washing of oily sludge. This paper discusses the main factors affecting the treatment effect, Demulsification Mechanism and research progress in the treatment of oily sludge by chemical thermal washing, and puts forward some suggestions on the future research and development direction of oily sludge utilization and disposal.

1 Introduction

Oilfield oily sludge mainly refers to the oil field, the refining of crude oil processing, the process of oil storage, transportation, use, due to the equipment aging, damage, corrosion or operator improper operation and production accidents, create the heavy oil such as crude oil, refined oil, residue "run, run, drip, leak" as the ground oil and mixed with the soil formation of oil, water and mud mixture. The sludge sand discharged from the production system during normal oilfield production can also be classified as oily sludge. It is one of the main non-recyclable pollutants and is listed as "national hazardous waste"[1].China's petroleum and petrochemical industry produces an average of 800,000 tons of oily sludge every year, while the world produces 60 million tons every year. Generally, the oil content of oily sludge is 10% ~ 50%, the moisture content is 40% ~ 90%, and the appearance is brown black viscous. At present, a large amount of oily sludge is often produced in the process of crude oil exploitation at home and abroad. Because of the high water content of oily sludge, it will occupy a large amount of land area. Its composition is complex and changeable, belonging to a number of systems, which are generally composed of water in oil (W/O), oil in water (O/W) and suspended solids, and usually exist in colloidal and semi-solid states[2]. Due to the high emulsification degree of oily sludge, it has high viscosity, poor fluidity and poor heat transfer; moreover, so it is not easy to denaturate, which makes it difficult for the solid phase to settle completely, and the three phases of oil, water and mud are difficult to separate.

2 Classification and Harm of Oily Sludge

According to different sources, oily sludge can be roughly divided into the following three categories[3,4]: landing mud: mainly from drilling operations and crude oil extraction process. During well testing operation, oily sludge containing waste drilling mud and drilling cuttings is produced due to oil spill or blowout due to ground pressure; ground mud is produced during crude oil production due to tubing fracture, blowout, sand spitting from oil well and poor anti-seepage; tank bottom mud mainly comes from oilfield gathering and transportation and oil storage process. It includes the bottom mud of crude oil tanks, settling tanks, sewage tanks and oil separators in transfer stations and combined stations (gathering and transportation stations and sewage stations), as well as oily sludge caused by large amount of crude oil or product oil leakage caused by leakage accidents of gathering and transportation pipelines or storage tanks; three sludge in refinery mainly comes from the oil refining process, commonly known as "three sludge". It includes scum on the surface of flotation cell (air flotation tank) produced by water treatment, bottom sludge of oil-water separation tank (oil separation tank), excess activated sludge from biological aeration tank or sedimentation tank or heat exchanger cleaning sludge, waste oil emulsion, etc. This kind of sludge is generally large in yield, and its composition is complex, difficult to degrade, difficult to settle, difficult to concentrate and difficult to treat.

If the oily sludge produced in the process of oilfield development and production is directly discharged into the environment, it will cause serious pollution and even irrecoverable damage to the surrounding soil, water and air. The residual oil in the oily sludge will directly or indirectly enter the soil, affect the physical and chemical properties and functions of the soil, cause the nutrient imbalance in various parts of the soil, inactivate plant enzymes, penetrate plant structure, and cause plants to grow normally [5]. Oily sludge contains benzene series, phenols, anthracene, pyrene and other odorous toxic substances. The toxic and volatile components will volatilize in the atmosphere, causing serious air pollution. People and animals inhale a large amount of dust will cause lung disease. Cycloalkanes, alkanes, aromatics and olefins in oily sludge can be enriched by fish and shellfish, and accumulate in human and animal bodies through food chain and food web, resulting in various human diseases. Aromatic hydrocarbons also have mutagenic, carcinogen-nic and teratogenic effects [6]. Heavy metals such as chromium, mercury, lead and other heavy metals in oily sludge flow into the farmland through material circulation, which will lead to the reduction of beneficial bacteria in farmland soil, the decline of farmland self-regulation ability, and the change of soil morphology. Toxic and harmful substances such as polychlorinated biphenyls, dioxins and radionuclides in oily sludge may cause coma of human and animals, and even cause death at high concentration. Dioxins also have chemical changes in the environment, producing new substances harmful to the environment and human body. If oily sludge is piled up arbitrarily for a long time, it will produce a large amount of methane and other combustible gases under the action of microorganisms, which will not only pollute the air, but also easily cause fire and explosion.

3 Chemical Hot Washing Technology of Oily Sludge

At present, the treatment technology of oily sludge at home and abroad has been developing rapidly [7,8]. The traditional incineration method consumes a lot of energy and causes serious secondary pollution; the solidification method not only does not completely decompose pollutants, but also wastes oil resources; the landfill method will occupy a large amount of land resources, which is easier to pollute soil and groundwater than other methods. The common pyrolysis method also has the problems of high energy consumption, high cost and easy to pollute the atmosphere; the modulation mechanical separation method is only limited to the small-scale oily sludge treatment, and the treatment process will produce noise; the solvent extraction method can not deal with heavy metals; the biological method generally has a long treatment cycle. The chemical heat washing method [9] not only achieves the purpose of recycling resources, but also improves the environment. It has the advantages of simple process, mature technology, low energy consumption, low cost, wide application range and good treatment effect. It is suitable for sludge with high oil content. The oil content of sludge can be washed to less than 1%. For small-scale oily sludge treatment, centrifuge and air flotation can be used for three-phase separation; for large-scale treatment, static sedimentation method can be used.

Chemical heat washing is the earliest and most commonly used oily sludge treatment technology. Chemical heat washing method is also called thermal desorption, which is to dilute the cleaning agent (such as hot alkaline water, demulsifier, flocculant, cleaning agent, washing powder or surfactant, etc.) with water under auxiliary means such as stirring and direct heating. The oily sludge is fully contacted with it and washed repeatedly to change the force between oil, water and silt, so that oily substances such as oil with high viscosity are resolved and demulsified, and then separated from the solid surface [10].After mixing and stirring, there will be obvious interface between oil, water and mud, and three-phase separation can be carried out; or the mixed oily sludge can be separated by high-speed centrifuge, and then the mixed oil-water can be separated by air flotation. The separated water phase and hot solution can be recycled. The separated oil phase is sent into the storage tank for resource recovery after treatment. The remaining sludge is treated by dehydration and then treated for harmless treatment or resource utilization.

At present, some scholars combine heat washing with other technologies, and the treatment effect has been greatly optimized. Ma Jiang ping [11] and others have combined the chemical hot washing method with the solidification method, which not only effectively improves the recovery rate of the crude oil in the oil sludge, but also greatly reduces the pollutant content in the sludge, so that heavy metal ions can be solidified and the possibility of secondary pollution is reduced. Lei Jiang hui [12] and others combined the thermal washing method with the biological method, which not only improved the traditional chemical hot washing into biological hot washing, but also changed the thermal washing agent from ordinary cleaning agent to biological cleaning agent, and also further composted the sludge after biological hot washing, and the treatment effect was further optimized, and the pollutants in the sludge were basically degraded and utilized by microorganisms, which had good environmental benefits.

4 Factors Affecting the Effect of Chemical Heat Washing

In order to improve the hot washing efficiency of oily sludge and enhance the hot washing effect, it is necessary to study and optimize the key factors in the hot washing process. The factors affecting the treatment effect generally include: liquid-solid ratio of slurry, type of cleaning agent, temperature, stirring intensity, hot washing time and times.

4.1 Liquid Solid Ratio

The hot washing of oily sludge usually needs to dilute the sludge first, so as to facilitate the cleaning agent to wash the petroleum substances, refractory organic substances, toxic and harmful substances and sediment in the solid sludge into the liquid phase. When the liquid-solid ratio is too high, the content of liquid in slurry medium is high, which will lead to the increase of treatment capacity and the cost of sewage treatment. When the liquid-solid ratio is too low, the content of solid components in sludge water medium is high, which will lead to insufficient washing of oily sludge, and some organic and toxic substances remain in the solid, resulting in inadequate degradation and poor treatment effect.

Lei Jiang hui [12] found that when the liquid-solid ratio was 5, the oil content of the residual sludge after cleaning was the lowest, about 6%. When the liquid-solid ratio was increased, the oil content of the remaining sludge had an upward trend. Similarly, Pu Yue qi [13] and others also found that when the liquid-solid ratio is about 4.7, the washing effect is the best. In general, the washing effect is better when the liquid-solid ratio is between 4-7.

4.2 Selection of Cleaning Agent

The key of oily sludge hot washing technology is the selection of cleaning agent. Generally, cleaning agents can be divided into single cleaning agent and compound cleaning agent. Single cleaning agent can also be divided into inorganic salt cleaning agent, organic salt cleaning agent (non-ionic cleaning agent, general anionic cleaning agent) and biological cleaning agent. The selection of cleaning agent should be in line with the principles of high efficiency, environmental protection and economy. According to the different properties of oily sludge, the appropriate cleaning agent should be selected to weaken the force between oil, water and sludge, and promote the separation of oil from sludge.

Li Ying [14] and others found that the same cleaning agent has different treatment effect on different oily sludge. The developed reagent A is suitable for sludge from sewage tank and landing oily sludge, while compound agent C is suitable for biochemical sludge on the surface of sludge pool. Wang Qi [15] and others found that among the non-ionic cleaning agents with long-chain structure, alcohols and phenols SP169, amines ae9901, resin ta1031 and dendrimer SD-2 have better cleaning effect, and the deoiling rate is above 68%, which indicates that long-chain structure has advantages in cleaning oily sludge containing polymer. After the optimized cleaning conditions, the deoiling rate can reach 91.12% with Na2SiO3 and 85.59% with el100. Baihe [16] and others developed a cleaning agent formula (2.0% sodium dodecyl sulfate + 0.25% Triton TX-100 + 0.86% sodium silicate Na2SiO3 · 9h2o + 5.0% n-butanol + 91.89% deionized water) for treating residual oil in oil-based drilling cuttings with high oil content, and the oil removal rate reached 97.3%.

4.3 Hot Washing Temperature

The temperature directly affects the separation of oil, sludge and water. This is because the increase of temperature will intensify the thermal movement of particles, increase the chance of collision between particles, and weaken the surface tension, leading to the occurrence of demulsification, which leads to the weakening of oil film adhesion, that is, the oil viscosity is reduced, the oil quality and sludge adsorption force are weakened, and oil will be desorbed from the surface of sludge; at the same time, The activity of the molecules is enhanced, and it is easy to enter the sludge layer of oil at higher temperature, so that crude oil and sludge can be separated. The oil separated from the sludge also floats on the water surface in the form of oil drops due to the increase of temperature and the density difference from water, and the oil-water interface is clear.

However, when the temperature is too high, the water evaporation will increase, the heat loss will increase, and the cost will increase; at the same time, the cleaning agent will evaporate, which will reduce the growth rate of oil removal rate. Therefore, from the perspective of energy consumption cost, the appropriate temperature should be selected in the washing process.

Wang Yong ping [17] and others found that the removal rate of cleaning agent increased by 28% from 35 °C to 60 °C, but only 7% from 60 °C to 80 °C. When the temperature is over 60 °C, the effect of oil sludge cleaning does not change significantly, so the temperature should be controlled between 55-60 °C to obtain the best treatment effect. Ma Jiang ping [11] and others found that when the temperature exceeds 40 °C, the removal rate of oil quality by the cleaning agent begins to decline, and the removal rate at 60 °C is only about half of that at 20 °C. In general, the washing effect is better when the temperature is between 30-80 °C. For oily sludge with different properties, the optimal treatment temperature is slightly different.

4.4 Hot Washing Stirring Strength

In the process of hot washing, the stirring intensity will also directly affect the hot washing effect. If the stirring intensity is not enough, the separation of oil and sludge will not be complete; if the stirring intensity is too high, the oil quality and water will become w/o emulsion, which can achieve half the result with twice the effort. Zhang Yuanyuan [18] and others found that with the increase of stirring speed, the oil content of sludge gradually decreased. When the stirring speed exceeds 30R / min, the oil content decreases slowly. When the stirring speed reaches 50 R / min, the washing phase emulsifies to become w/o emulsion, and it is difficult to separate the intermediate layer from the washing phase. 30 R / min was selected for hot washing. Li Yichuan [19] and others found that the oil content of sludge first decreased and then increased with the increase of agitator speed, and reached the minimum value at 200R / min, and the oil content of sludge was about 2%. According to the different oily sludge, the best stirring intensity is different, which needs to be analyzed and tested.

4.5 Hot Washing Time

The length of hot washing time will also directly affect the hot washing effect. If the hot washing time is too short, the sludge will not be fully washed, and the oil quality will remain in the sludge; if the hot washing time is too long, it will strengthen the emulsification phenomenon, and is not conducive to the separation of oil quality and sludge.

Luo Bing bing [20] and others found that the removal rate of oil quality by cleaning agent reached 80.2% after hot washing for 30min. After continuous hot washing, the increase of removal rate slowed down, and the hot washing time of sludge in follow-up experiment was set as 30min. Wang Xiao jun [21] and others washed the oily sludge of Liao he Oilfield which oil content is 35.78% for 60min, and the oil content of the original sludge decreased to about 7%. However, with the increase of hot washing time, the emulsification of solution becomes more and more serious. After 120 min, the oil content of the original sludge rises to about 8%. The hot washing time of oily sludge should not be too long, which will not only increase energy consumption, but also weaken the hot washing effect.

4.6 Hot Washing Times

The hot washing times have great influence on the hot washing effect of oily sludge with high oil content. In order to better integrate oily sludge and medicament and achieve more complete deoiling, two or more hot washing is adopted. Yu Lan lan [22] and others found that during the first hot washing, the deoiling rate was 82.14%; in the second hot washing, the deoiling rate was 83.91%; with the increase of hot washing times, the deoiling rate was no longer significantly increased. From the point of view of saving energy and time, two times of hot washing is selected.

4.7 Demulsification Mechanism

Most of crude oil exists in W/O emulsion. There are three conditions for oil and water to form emulsion [23]: there are two immiscible liquids (i.e. crude oil and water); there are emulsifiers (NATURAL emulsifiers in crude oil: asphaltene, gum, resin, silica, clay, etc.); sufficient mixing is required. In recent years, the content of natural emulsifiers in crude oil has increased significantly. They are adsorbed on the oil-water interface to form a layer of viscoelastic membrane, which causes droplet aggregation and kinetic obstacles, resulting in serious emulsification and more stable crude oil emulsion [24]. For different demulsification methods, the demulsifycation mechanism is also different.

Chemical demulsification is to change the interface of emulsion system from stable to unstable by demulsifier, so as to achieve the purpose of demulsification. Demulsifying properties of demulsifier mainly depend on the hydrophilic / lipophilic ability of demulsifier and the ability of breaking the interfacial film. At present, there are four recognized demulsification mechanisms [25]: phase transfer reverse deformation mechanism: after adding demulsifier into emulsion system, surfactant with opposite type to emulsion can be generated as reverse emulsifier (demulsifier), that is, phase transfer occurs. This kind of demulsifier and hydrophobic emulsifier can form complex, which makes emulsifier lose emulsifying property; the mechanism of collision breaking down the boundary film: under heating or stirring conditions, demulsifier will collide with the boundary film, or adsorb on the interface membrane, or replace part of the surface active substance on the interface film, so as to break the boundary film, reduce its stability, and then achieve the demulsification effect through flocculation and coalescence; solubilization mechanism: some molecules of demulsifier form high molecular coil or micelle, which can increase the solubility of emulsifier and cause demulsification of emulsified crude oil; folding deformation mechanism: water in oil emulsion has multi-layer water circles, and there are oil circles between water circles. Under the action of heating, stirring and demulsifier, the water circles in the droplets are connected with each other, which makes the droplets agglomerate and demulsify.

Ultrasonic demulsification [26,27] is a new demulsification method. When a certain intensity of ultrasonic wave propagates in emulsion, a series of mechanical, thermal and optical effects will be produced. The main mechanism of ultrasonic demulsification is as follows. Mechanical action: the ultrasonic wave will cause the suspension water and the crude oil medium to vibrate together. The suspended water of different sizes will collide and bond with each other due to the different relative vibration speed, forming water droplets and settling and separating under the action of gravity; Cavitation: when the ultrasonic wave propagates in the liquid, temporary negative pressure will be formed in some areas, resulting in holes or bubbles. When the ultrasonic pressure is negative half cycle, the liquid is pulled, and the bubble core expands rapidly to several times of the original volume. Then when the ultrasonic pressure is positive half cycle, the bubble suddenly collapses and splits into many small bubbles, forming new cavitation Nuclear. Cavitation can produce local high temperature and local high pressure, and is accompanied by strong shock waves and high-speed micro jets. This huge instantaneous pressure can cause the surface of emulsion suspended in liquid to be damaged rapidly, thus promoting oil and water separation; heat action: when ultrasonic waves propagate in the medium, the vibration energy will be absorbed by the medium into heat energy, which will destroy the stability of the emulsion.

The biological demulsification method made the emulsion demulsifying and dehydrating [28] by adding microbial fermentation medium. It uses the growth and development of microbial cells, metabolism and metabolites to reduce the surface tension of oil-water interface and destroy the stability of emulsion, so as to achieve the purpose of demulsification oil-water separation. Biological demulsification has the advantages of simple process, low cost, low energy consumption and easy commercialization. There are three main ways of demulsification: microbial cell demulsification; biosurfactant demulsification produced in the process of microbial fermentation; and microbial metabolic process demulsification.

Microbial species, age and other factors have an important impact on the demulsification effect, and the screening of bacteria has become one of the important contents of biological demulsification research.

5 Conclusion and Prospect

In order to promote the practical application of chemical heat washing in oily sludge treatment, the following aspects can be strengthened

The research on the interaction mechanism between cleaning agents is less, so it is necessary to study the synergistic effect mechanism between cleaning agents to optimize the detergent formulation;

The results show that there may be synergistic effects among various process intensification methods, and the combination of these enhancement methods is helpful to further improve the oil washing removal effect;

Due to the complexity of oil composition, the properties of oily sludge are complex and changeable, which makes the chemical hot washing process and effect different. Therefore, the influence of oily sludge properties on the chemical hot washing process can be systematically studied to provide basic data for detergent formulation research;

Generally, chemical heat washing process can reduce wastewater discharge by optimizing operation conditions and reusing washing wastewater, but the reuse times of washing wastewater are limited. Therefore, how to effectively treat washing wastewater and reduce the secondary pollution of wastewater is another urgent problem to be solved.

Inorganic pollution in oily sludge only had a single treatment technology which can not meet the requirements of environmental protection, nor can be utilized as resources. Therefore, it is necessary to combine various treatment technologies according to the characteristics of different treatment technologies, so as to expand the scope of use and improve the treatment efficiency. The deep disposal of oily sludge in different grades and stages has become the focus of the development of oily sludge treatment technology.

References

1. Ministry of environmental protection, national develop-ment and Reform Commission, Ministry of public security. National hazardous waste list [EB / OL] (2016-06-14)
2. Chieng S,Wang Y H,Gao N B,et al.Pyrolysis of oil sludge with oil sludge ash additive employing a stirred tank reactor[J].journal of Analytical and Applied Pyrolysis,2016,120:511-520.
3. Wang Yuhua, Chen chuanshuai, Meng Juan, Han Yunying, Li Aimin, Wang Dong. New development and application status of oily sludge disposal technology [J]. Safety and environmental engineering, 2018,25 (03): 103-110
4. Mrayya B,Battikhi M N,Biodegradation of total organic carbons (TOC) in Jordanian petroleum sludge [J].journal of Hazardous Materials, 2005, 120(1/2/3): 127-134
5. HUI K, TANG J, LU H, et al. Status and prospect of oil recovery from oily sludge: A review[J]. Arabian Journal of Chemistry, 2020
6. HU G, FENG H, HE P, et al. Comparative life-cycle assessment of traditional and emerging oily sludge treatment approaches[J]. Journal of Cleaner Production, 2019
7. MOHIT M A, GHAZBAN F, OMIDVAR B. Optimization of influential parameters of hydrocarbon recovery from waste oily sludge by solvent extraction using solvent blend[J]. 2020
8. VARJANI, SUNITA, J. Microbial degradation of petroleum hydrocarbons[J]. Bioresource Technology Biomass Bioenergy Biowastes Conversion Technologies Biotransformations Production Technologies, 2017
9. Bai Yu, Cheng Yuanpeng, Hu Jiujiang, Huang Chunfeng, Tang shanfa. Research status and Prospect of thermal cleaning technology for oily sludge [J]. Applied chemical industry, 2020,49 (06): 1498-1501 + 1507
10. Ying L, Xi C, Xiumin L, et al. Engineering Application of Chemical Thermal Washing for Oily Sludge[J]. Environmental Protection of Oil & Gas Fields, 2018.
11. Ma Jiangping, Zhao Honggang, Wang Feng, et al. Experimental study on oily sludge treatment by chemical hot washing and heavy metal solidification [J]. Environmental protection of oil and gas field, 2015, 025 (003): 18-20
12. Lei Jianghui. Laboratory experimental study on treatment of oily sludge by biological hot washing and microbial degradation technology [J]. Oil and gas field surface engineering, 2020,39 (04): 40-44
13. Pu Yueqi, Yu Chao, Li Zhiliang, et al. Study on rhamnolipid enhanced thermochemical washing of oily sludge [J]. Journal of Zhejiang Ocean University (NATURAL SCIENCE EDITION), 2014 (33): 572-575
14. Li Ying, Chen Xi, Li Xiumin, et al. Engineering application of chemical heat treatment of oily sludge [J]. Environmental protection of oil and gas field, 2018, 028 (001): 14-16
15. Wang Qi, Li Meirong, Zhu Wei, et al. Effects of different types of cleaning agents on the cleaning effect and interfacial properties of oily sludge containing polymer [J]. Journal of environmental engineering, 2012
16. Baihe, Wu Weifeng, Weng Liangyu, et al. Study on chemical cleaning treatment of oil-based drilling cuttings with high oil content [J]. Natural gas and petroleum, 2018, 036 (003): 76-80
17. Wang Yongping, Zong Tinggui, Zhang Lei. Study on ultrasonic enhanced hot washing process of oily sludge [J]. Petrochemical application, 2016, 35 (010): 145-148
18. Zhang Yuanyuan, Yu Yin. Study on harmless treatment technology of oily sludge based on chemical washing [J]. Xinjiang environmental protection, 2018, 40 (03): 32-38

19. Li Yichuan, Wang Dong, Wang Yu, et al. Washing sludge by thermochemical cleaning -- Study on process conditions of oil recovery [J]. Cleaning world, 2008 (11)
20. Luo Bingbing, Huang zuhao. Study on treatment of oily sludge by hot washing process [J]. Yunnan Chemical Engineering, 2018, 45 (08): 75-76
21. Wang Xiaojun, Wu Yue. Screening of washing agents for oily sludge treatment in oil field [J]. Liaoning chemical industry, 2017 (04): 325-328
22. Yu Lanlan, Song Jian, Zheng Kai, et al. Study on hot washing process for oily sludge [J]. Chemical technology, 2014, 22 (001): 29-33
23. .Mu Jianhai. Study on Demulsification Mechanism of crude oil and development of demulsifiers [J]. Chemical technology market, 2002, 25 (004): 26-28
24. Cherney D P, Wu C, Thorman R M, et al. Investigating the Impact of Crude Oil Solubility on Water-in-Oil Emulsion Stability and Its Relation to Molecular Composition of Crude Oil at the Oil–Water Interface[J]. Energy & Fuels, 2015, 29(may-jun.):3616-3625.
25. Feng, Jie, Cao,Effect of Demulsifier Structures on the Interfacial Dilational Properties of Oil-Water Films[J]. Journal of Dispersion Science & Technology, 2016.
26. Li Zhongjie, Tong Zhenzhen. Application status of ultrasonic in petroleum engineering [J]. Contemporary chemical research, 2020 (13): 7-8
27. Wang Yongping, Zong Tinggui, Zhang Lei. Study on ultrasonic enhanced hot washing process of oily sludge [J]. Petrochemical application, 2016,35 (10): 145-148 + 153
28. Cai Q, Zhu Z, Chen B, et al. Oil-in-water emulsion breaking marine bacteria for demulsifying oily wastewater[J]. Water Research, 2019, 149(FEB.1):29.

https://doi.org/10.1051/e3sconf/202132901054

Analysis of logging parameter model of Jurassic complex reservoir in Xinjiang Shenquan Oilfield

Yanjie Li[1,2], Yitong Wang [1,2], Hansheng Ding [1,2], Yongyi Ma [1,2], Yan Shen [1,2], Lilan Chu [1,2], Zhiyang Nie [1,2], Jinxia Du [1,2], Zhaoyi Shao [1,2], Qicai Gao [3] and Shuheng Wang [1,2]

[1] Exploration and Development Research Institute of Daqing Oilfield Co., Ltd., 163712 Daqing Heilongjiang, China
[2] Key Laboratory of tight oil and shale oil accumulation of Heilongjiang province, 163712 Daqing Heilongjiang, China
[3] Heilongjiang Province Key Research Laboratory in Tight Sandstone and Shale Oil Reservoir Accumulation,Daqing, China

Abstract. The study discusses the characteristics of Jurassic reservoir in Shenquan Oilfield in Xinjiang. The reservoir of Qiketai Sanjianfang formation mainly develops delta front facies and shore shallow lake facies sand bodies.The lithological characteristics of the reservoir are mainly fine sandstone and siltstone, and the physical properties of the target layer are mainly medium porosity and medium permeability. The Jurassic reservoir space in Shenquan Oilfield is sandstone pore type, mainly composed of secondary pores, mainly including intergranular dissolved pores, intragranular dissolved pores and intercrystalline pores, with good connectivity between pores. According to the analysis data of well logging, core analysis, production performance and DST in Shenquan Oilfield, porosity, permeability and water saturation models are established respectively:The porosity interpretation model is established by using acoustic wave and shale content; the permeability parameter interpretation model is realized by applying graph based cluster analysis method and neural network algorithm. The oil saturation interpretation model is obtained by Archie formula. According to the logging interpretation model and interpretation results, it can meet the needs of reservoir evaluation, oil-water distribution and original geological reserve parameters. After analyzing the logging parameter model and reservoir parameter calculation method, it is considered that the Jurassic in the Shenquan Oilfield in Xinjiang has better physical properties.

1 Introduction

The aimed area is located in the east of Xinjiang Uygur Autonomous Region, about 8km northeast of Turpan city with about 20km long from east to west and 11km wide from north to south. There are Shenquan oil production plant and some villages in the seismic work area, and the southeast corner is close to Astana ancient tomb group (Fig. 1). The seismic area of Shenquan Oilfield covers 95km2, with a total of 304 wells and more than 40 faults. As a whole, Shenquan Oilfield is dominated by complex fault-block reservoirs controlled by anticline structures, followed by lithologic reservoirs. Due to fault cutting, many fault nose and fault block traps are developed. Oil bearing intervals are principally developed in Tertiary Shanshan Group, Cretaceous Tugulu group and Jurassic Qiketai formation and Sanjianfang formation (Table 1).

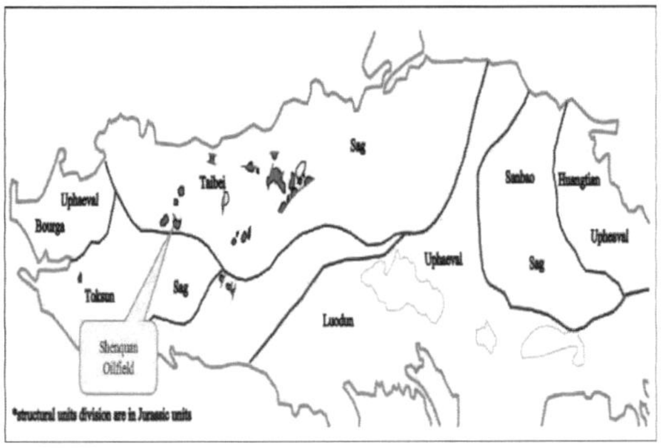

Fig. 1 Three dimensional geographical location of Shenquan in Turpan Hami Basin

---

* Corresponding author: dqlyj@petrochina.com.cn

https://doi.org/10.1051/e3sconf/202132901054

Table 1 Comparison of standard formation profiles in Shenquan Oilfield

Stratigraphy			Lithology section			Pressure coefficient
System	Series	Formation	Thickness (m)	Lithology description	Oil and gas interval	
Quaternary	Holocene-Pleistocene	Q	85	Variegated conglomerate	Oil and gas interval	1.0
Tertiary	Pliocene	N2p	913	Interbedding of variegated conglomerate and brownish yellow mudstone		1.0
	Oligocene-Miocene	N1t	488	Brownish yellow and brownish red mudstone with white gypsum layer		1.0
	Paleocene Esh	Esh	424	Dark purple mudstone is mainly intercalated with gypsum mudstone and silty mudstone, with a layer of gray sandy conglomerate and fine sandstone developed at the bottom	Oil and gas interval	1.0
Cretaceous	Lower	K1tg	155	A large set of dark purple and purplish red mudstone is mainly mixed with siltstone and argillaceous siltstone locally	Oil and gas interval	1.0
Jurassic	Upper	J3q	248	Mainly brownish red and purplish red mudstone		1.0
	Middle	J2q	55	The upper section is gray mudstone with thin layer of gray siltstone, and the lower section is gray argillaceous siltstone, sandstone and gray mudstone interbedding		1.0
		J2s	300	Dark gray mudstone is interbedded with light gray siltstone, fine sandstone, coarse sandstone, gravelly sandstone and glutenite in unequal thickness, intercalated with a thin layer of variegated and grayish yellow mudstone	Oil and gas interval	1.1

2 Reservoir characteristics

The Jurassic Qiketai Sanjianfang formation in Shenquan Oilfield mainly develops delta front facies and shore shallow lake facies sand bodies. The lithological characteristics of the reservoir are dominated by fine sandstone and siltstone, accounting for 37.5% and 25%, respectively, followed by unequal-grained sandstone, accounting for 25% which mostly feldspar lithic sandstone. Sandstone types include gravel-bearing gravel sandstone, medium sandstone, fine sandstone, and siltstone, with various types and low structural maturity. The clastic components are mainly rock debris, feldspar and quartz. The content of feldspar is 10-35%, with an average of 22%. Quartz content accounts for 28.8%, feldspar 21.11% and rock debris 49.18%. Core analysis data describes that the physical properties of the target layer are chiefly medium porosity and medium permeability, followed by medium porosity and low permeability. The highest average porosity of layer S1 is 19.6%, and the highest average permeability is 265.4 mD. The lowest average porosity of layer 2 is 16.6%, besides the lowest average permeability of layer Q is 65.3 mD. The porosity distribution range is basically of 20 ~ 24%; the permeability distribution range is mainly between 100mD and 500mD (Table 2). Depending on the observation and identification of rock flakes, cast flakes, and scanning electron microscopy, the Jurassic reservoir space in Shenquan Oilfield is sandstone pore type, mainly composed of secondary pores, including intergranular dissolved pores, intragranular dissolved pores and intercrystalline pores with diameter of 20μm ~ 90μm, and pores is well connected. Additionally, there are a small amount of residual primary intergranular pores.

Table 2 Analysis of core physical properties in Shenquan Oilfield

Oil deposit	Porosity (%)			Permeability ($10^{-3}\mu m^2$)		
	Max	Min	Ave	Max	Min	Ave
Q	8	26.6	16.9	0.05	323	65.3
S1	4.8	26.9	19.6	0.2	2127	265.4
S2	13	20.7	16.6	2.21	569	131
J2s	4.6	26.5	18	0.2	2127	200

3 Logging parameter interpretation model

3.1 Porosity parameter model

Porosity reflects the ability of fluid storage in the reservoir. Due to the different types of minerals in the target formation in different study areas, the response coefficients of various types of minerals to a certain physical measurement method are also different. If the response equations are established for various types of minerals in the formation rocks, the results will be absolutely uncertain and difficult to operate. Therefore, for sandstone reservoir, it is generally assumed that in the formation rock with volume of "1", its constituent minerals are simplified as argillaceous minerals, skeleton minerals and pores, and the pores contain oil, gas and water fillers, which is commonly referred to as "volumetric physical model".

In the process of establishing the log interpretation porosity model, normally we apply core porosity analysis data to construct a relationship with various logging data reflecting porosity, such as density, sonic and neutron logging curves. Occasionally use multiple regression method, and finally the most reasonable functional relationship is selected as the porosity calculation model.

This is commonly known as "core calibration logging" technology.

$$\Phi=0.1235AC-16.77 \quad (AC\leq305\mu s/m) \quad (N=12,\ R=0.976) \tag{1}$$

$$\Phi=0.1097AC-16.29 \quad (AC\geq305\mu s/m) \quad (N=7,\ R=0.894) \tag{2}$$

This formula is established based on the coring analysis data at different stages. The porosity is divided to two cases with the sonic logging value of 305μs/m as the boundary. The main issue of calculating the porosity is that when the acoustic logging value is at 305μ/m, the calculated porosity will have a discontinuous abrupt value. The reason for this situation is that the establishment of formula (2) is mainly based on the core analysis data of well S1, and the acoustic time difference logging value of the standard formation of well S1 is obviously higher than that of other exploratory wells. Therefore, the sonic logging value of well S1 must be standardized. On the basis of standardized treatment, parameter optimization and correlation comparison show that there is a good correlation between acoustic wave, shale content and porosity (Fig. 2 and Fig. 3).

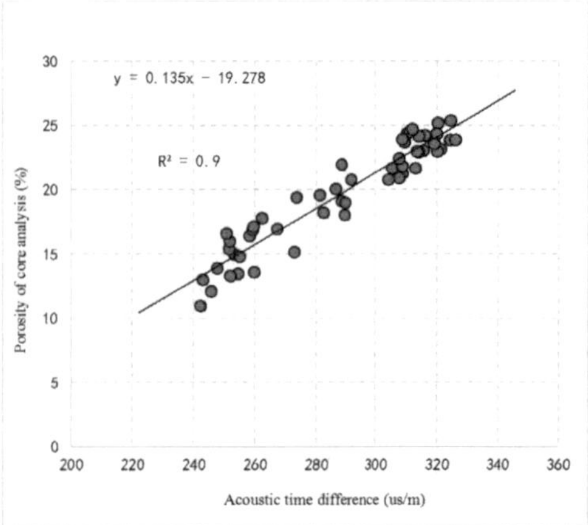

Fig. 2 The relationship between porosity and acoustic time difference

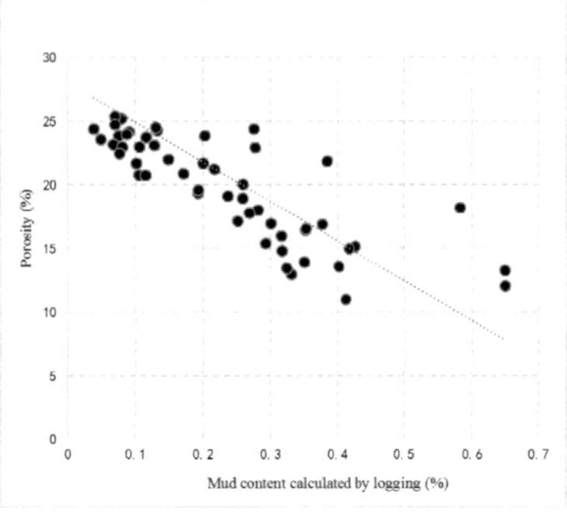

Fig. 3 Relationship between porosity and shale content

In this study, the porosity calculation formula is established by acoustic wave and shale content and 52 sub-layers of 6 wells: Φ= 0.117 * ac-5.616 * vsh-12.686, correlation coefficient 0.925, average relative error 5.1% (Fig. 4).

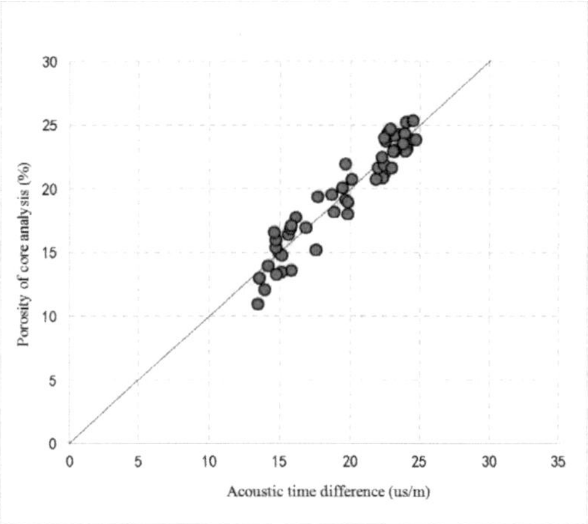

Fig. 4 Comparison of porosity accuracy between logging calculation and core analysis

Figure 5 exhibits an example of porosity calculation. The coring depth of well Shen105 is 2512-2516m, and the acoustic moveout value of the coring section is above 305μs/m,, according to the original calculation method, formula (2) will be practised for calculation. It is obvious that the porosity calculated by logging is too small (the second curve trace on the right in the figure), and the relative error reaches more than 15%, while the calculation by the new regression formula is in good agreement with the core analysis (the first curve trace on the right), and the relative error is less than 5%. The reliability of the porosity method is further verified.

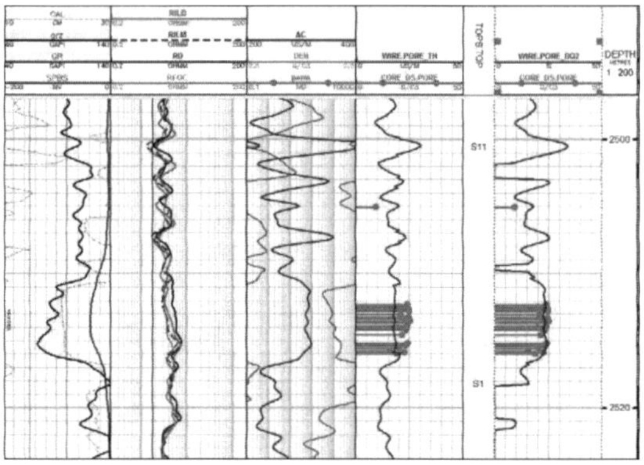

Fig. 5 Porosity calculation effect of coring section of well Shen105

3.2 Permeability parameter model

Reservoir permeability and porosity are important parameters of logging interpretation. Scholars at home and abroad have done a lot of research work on them, and put forward many mature theoretical and empirical calculation formulas based on rock volume physical model. These two parameters reflect the characteristics of reservoir from different aspects. They are both related and different. They are closely related to the pore volume of reservoir, and are also controlled by the geometric size, shape and distribution of pores. In reservoir physics, the well-known Kozeny equation is used to describe and calculate permeability which is only applicable to relatively pure and well consolidated formations with medium and high porosity (> 15%), but not to low porosity and low permeability formations.

Adopting graph based cluster analysis method, this method analyzes and classifies logging curve data considering curve shape, distribution density, and correlation, automatically determines the number of optional clusters, and realizes permeability parameter interpretation via neural network algorithm. Preferably, AC, VSH, DEN, CNL and other curves are interpreted by self-organizing mapping algorithm to achieve permeability parameter interpretation, and the calculation accuracy coincidence rate is 91.2% (Fig. 6).

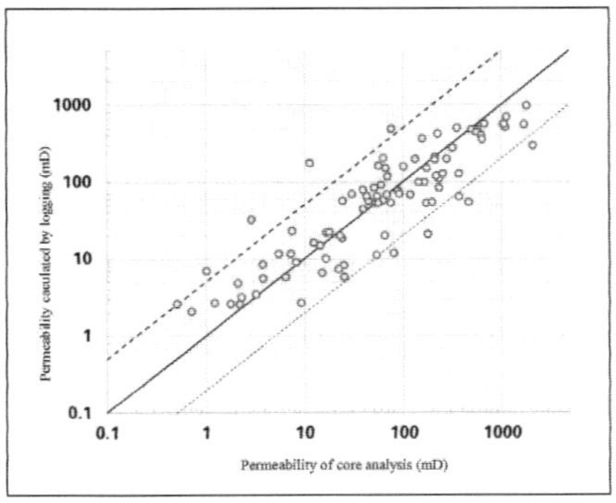

Fig. 6 Comparison of permeability accuracy between logging calculation and core analysis

Take the coring section of well Shen104 as an example (Fig. 7). The permeability calculated by the new method (the second curve channel in the right) obviously better agrees with the core analysis data and the trend is significantly better than the permeability calculated by the original method(the first curve channel in the right).

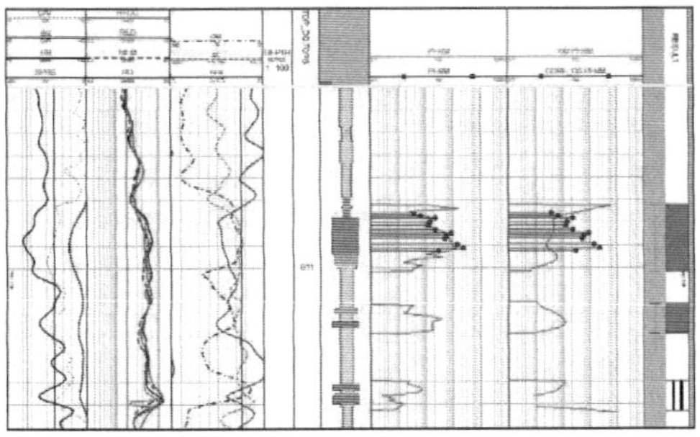

Fig. 7. Comparison of permeability effect of well Shen104 logging calculation

3.3 Water saturation model

As an important index to evaluate the oil bearing property of reservoir, saturation is one of the essential parameters in logging quantitative interpretation. Over the years, logging academic circles at home and abroad have carried out in-depth research around the oil saturation model of argillaceous sandstone reservoir. Each model indicates different argillaceous sandstone reservoir characteristics from a certain aspect, or develops different understanding of argillaceous conductivity to some extent.

For a long time, the interpretation model based on resistivity has been dominant. For example, the W-S model and two water model based on the cation exchange properties of clay minerals. Ran et al. (1990) proposed the extended Archie formula (EAE), Iivar et al. (1991) suggested the particle conductivity method of argillaceous sandstone, Charles (1995, 1996) constructed the effective medium model (EMM), and Givens (1987, 1989) proposed the rock skeleton conductivity model (crmm), Xu Dong Jing et al. generalized empirical relationship model (GEM) between resistivity increase coefficient and saturation. These conductivity models have certain application value for evaluating oil and gas reservoirs using logging data in China. In the process of selecting the saturation model, the rock composition, mineral structure and related influencing factors of the reservoir in the study area is necessary to be analyzed. According to the application conditions of different models, Archie formula, Indonesian equation, W-S model and double water model should be extensively applied to calculate the reservoir saturation parameters. Although some other logging techniques such as carbon-oxygen ratio and nuclear magnetic resonance can also provide relevant information on reservoir saturation, in the current logging interpretation, the most used and most widely studied is still based on resistivity saturation model.

The distribution form of shale in the study area is closer to dispersed shale, and the formation water salinity is between 160,000 and 250,000 mg/L, which is a high-salinity reservoir with low shale content.Therefore, Archie formula is selected. The formula is as follows:

$$S_w = \left(\frac{abR_w}{\Phi^m R_t} \right)^{1/n} \tag{3}$$

$$S_o = 1 - S_w \tag{4}$$

S_o, S_w - oil saturation, water saturation, f; m, n - rock cementation index and saturation index;

R_w - formation water resistivity,Ωm; R_t - reservoir resistivity, Ωm; Φ - Reservoir porosity,f.

The critical parameters of Archie saturation model is the determination of a、b、m、n, as well as the formation water resistivity R_w and the formation resistivity R_t, in the meantime the accuracy that directly determines the oil saturation precision. Water saturation is obtained by Archie formula, and according to the reservoir characteristics and the petroelectric test results of well Shen2, well Shengnan 2 and well Shengnan 203 (Fig. 8 and Fig. 9). The model parameters are decided as follows: a = 1.0, B = 1.3, m = 1.83, n = 1.71

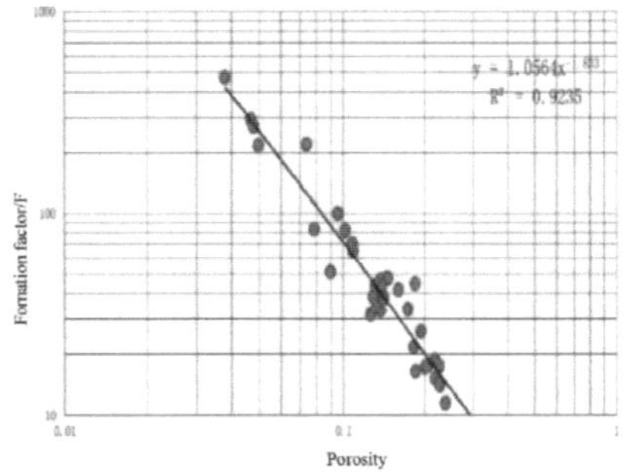

Fig. 8 Relationship between resistivity increase rate and saturation

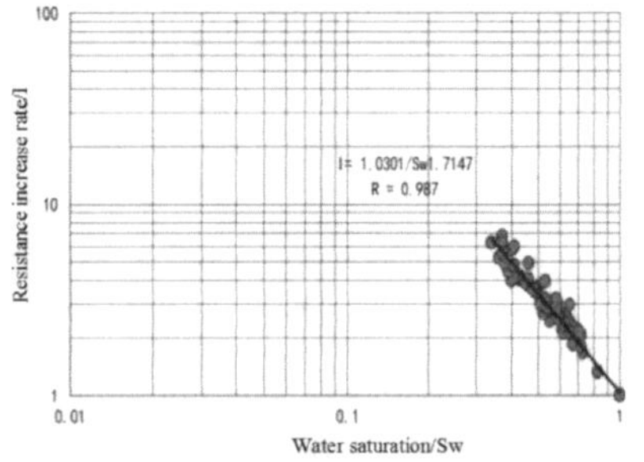

Fig. 9 Relationship between resistivity increase rate and saturation

Due to the large amount of perforation production data and water analysis data in the study area, in the process of saturation calculation, the well logging intersection Pickt map is established by the well logging data of the production well (well Shen103) at the edge of the structure, and the petroelectric parameters of the saturation model are modified through the "water line" of the intersection map (Fig. 10).

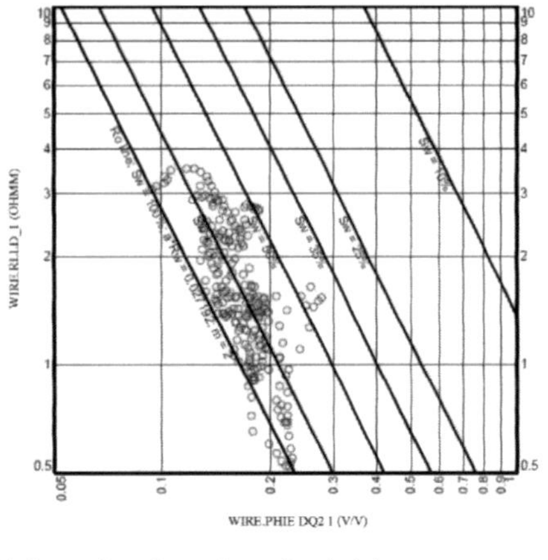

Fig. 10 Cross plot of porosity and resistivity

As the oil-water relationship of Shenquan Oilfield is complex, fault blocks are developed, reservoirs are developed with multiple and thin layers, and the original oil-water interface has been destroyed under the influence of later structure, it is difficult to accurately determine the original oil column height, which affects the calculation of original oil saturation.

4 Conclusion

(1) Delta front facies and shore shallow lake facies sand bodies are primarily developed in Jurassic Qiketai - Sanjianfang formation of Shenquan Oilfield in Xinjiang. The lithological characteristics of the reservoir are mainly fine sandstone and siltstone, and the physical properties of the target layer are mostly medium porosity and medium permeability.

(2) After analyzing the logging parameter model and the calculation method of reservoir parameters, it is concluded that the Jurassic of Xinjiang Shenquan Oilfield has better physical properties. The average porosity is 18.0%, and the average permeability is 230.6mD, belonging to medium porosity and medium permeability reservoir.

(3) The appropriate porosity, permeability, and water saturation models of the study area were established through logging, and parameters such as porosity, permeability, and water saturation were calculated, which supplies a good fundamental value for reservoir identification accuracy and reserve parameter logging evaluation method, meanwhile effectively supports the exploration and evaluation project.

References

1. Zeng W.C., Well logging evaluation technology of oil and gas reservoirs. Well logging evaluation technology of oil and gas reservoirs, (1947).
2. He X.P., Reservoir logging evaluation technology of oil and gas reservoirs. Petroleum geology and engineering, 20, 5 (2006).
3. Zhai Y.M., Cai B.J., Zhang Q.M., Analysis of oil and gas accumulation conditions in the south slope of Turpan Depression . Xinjie oil and gas, 9, 1 (2013)
4. Ren J.W., La Q.L., Study on logging parameters of oil reserves in an area . Petrochemical technology, 2, (2017)
5. Luo S.Q., Wu Y.F., Chen X., Application of model inversion technology in shallow reservoir research. Xinjiang Petroleum Geology, 21, 4 (2000).
6. Qiao T.T., Qi X.L., He T.H.,Research on logging interpretation technology of low porosity and permeability reservoir in Shan 23 member of Su 77 block . Foreign logging technology, 6 (2013) .
7. Yin H.R., Zhou H., He W.J., Interpretation model of reservoir logging parameters and lower limit standard of reservoir evaluation in Zhuang 9 Chang 8_ (2) in Heshui Oilfield .Journal of Changjiang University: Natural Science Edition, 8, 2 (2021).
8. Liu Z.D., Shi Y.J., Zhou J.Y., ummary and applicability analysis of methods for determining the lower limit of effective reservoir physical properties. progress in Geoscience, 33, 3 (2018).

 https://doi.org/10.1051/e3sconf/202132901055

Cambrian Ordovician sedimentary facies and its evolution in Tadong area

Ding Hansheng [1,*]

[1]Exploration and Development Science Research Institute of Daqing Oilfiled Company Ltd., Daqing 163712, China

Abstract. In order to clarify the sedimentary development law under the Cambrian Ordovician regional stratigraphic framework in Tadong area, and lay a theoretical foundation for further oil and gas exploration in the study area. The distribution, characteristics and evolution of main sedimentary facies belts of Cambrian Ordovician are studied by means of drilling core observation, cast thin section identification, logging curve feature analysis, seismic profile and well connection profile. The results show that the Cambrian middle lower Ordovician in Tadong area is equivalent to a second-order sequence and can be further divided into 12 third-order sequences. Each third-order sequence is mainly composed of transgressive and highstand tracts. Carbonate platform margin beach facies and Reef (mound) beach complex facies are favorable reservoir development facies belts in this area; Under the regional stratigraphic framework, three types of sedimentary facies can be identified in Cambrian Ordovician, and a total of 10 subfacies are developed; The evolution of sedimentary facies is mainly controlled by the rise and fall of sea level, which is characterized by the migration of platform margin facies and the change of platform facies.

1 Introduction

Carbonate strata are developed in Cambrian-Ordovician in Tarim Basin. Through the research work done by predecessors on sedimentary facies of Cambrian-Ordovician in Tarim Basin [1-14] provided guidance on the discovery of many large-scale oil and gas fields of reef-beach complex, such as Tazhong slope break zone and Shunnan slope, which indicates that Tarim Basin has a good exploration prospect. After Daqing Oilfield took over the Tadong block at the end of 2012, TJ13, TJ13-2, TJ13-3 and TJ13-4 wells were drilled in Gucheng platform one after another, and high-yield industrial gas flow was obtained, showing the exploration potential of platform facies in Tadong area. However, due to the lack of drilling in Tadong area in the early stage, little study has been done on the development law of sedimentary facies belt. In such case, in order to further accelerate the exploration efforts in Tadong area and effectively guide the exploration deployment, on the basis of previous studies, this paper makes full use of a large number of outcrops, drilling and seismic data, and systematically compiles the Cambrian-Ordovician sedimentary facies under the second-order sequence framework in Tadong area, and clarifies the development characteristics and evolution law of sedimentary facies.

The study area is located in the eastern region of Tarim Basin, part of the Tarim plate, bounded by Kuruktag in the north, southeast bulge in the south, TJ16 well in the west and Laokaiping in the east, with an area of about 12×104km2. The north and south of the study area are controlled by Cherchen fault and Kongquehe fault respectively, and can be divided into six secondary structural units, such as Gucheng low uplift, Tadong low uplift, Lop Nur uplift, part of Manjiaer sag, Yingjisu sag and Kongquehe slope (Figure 1). TJ13 well and TJ13-2 well have obtained industrial high-yield gas flow in dolomite reservoir of Middle and Lower Ordovician, while well TJ15 well and TJ8 well have seen better oil and gas in Cambrian, indicating that carbonate reservoir of Cambrian-Middle and Lower Ordovician is one of the main oil and gas exploration reservoirs in this area at present.

During Cambrian-Early Ordovician, stable carbonate platform-basin sedimentary system was developed in Tadong area, and its main body was located in deep water basin area, where undercompensated semi-deep sea-deep sea basin facies was developed, and the deposited dark mudstone was one of the most important source rocks in the basin. Two carbonate platforms were developed on east and west sides of the basin respectively, platform slopes and platform margins were developed between the platforms and the basin, and reef-beach deposits developed in the platform margins and platforms are favorable reservoirs. At the end of Early Ordovician, due to the subduction and collision of the mid-Kunlun island arc in the southern margin of the basin, the basin began to be in a compressional tectonic environment, and Tazhong, Tabei, Bachu and Tadong uplifts began to take shape, while the amplitude of Tadong uplift was relatively small. Turbid flow basins are developed in Tadong area, and a set of dark mudstone is deposited, which can be used as potential source rocks and favorable caprocks in Tadong area.

* Corresponding author: dinghansheng@petrochina.com.cn

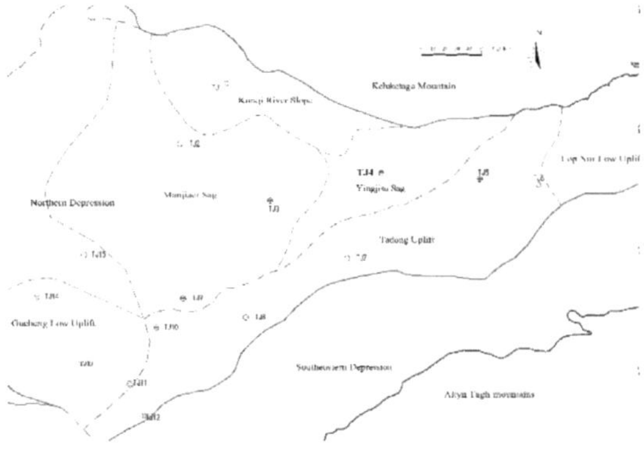

Fig. 1. Geographical location and structural map of the study area.

2 Sequence stratigraphic framework

The rising and falling rate of sea level obviously controls the carbonate platform structure. When the sea level rises rapidly, accretion and retrogradation platform structures are usually formed and the platforms may be even submerged, and the platform edges remain stationary or migrate towards the platform. When the sea level drops rapidly, platform structures which constantly migrates to the sea and with a foreset-and-decline stacking pattern are often formed. Therefore, on the basis of fully combining the previous basin research results, well-seismic analysis and systematic tracking and interpretation of seismic profiles, the sedimentary sequence division scheme of Cambrian-Middle and Lower Ordovician was determined, and the sequence stratigraphic framework and system tract composition style were established. The Middle-Lower Ordovician in the region is equivalent to the second-order sequence (Figure 2), which is generally composed of a large transgression-regression cycle. The middle and lower Ordovician in eastern Tarim area can be divided into seven third-order sequences, namely SQ_O1p-1, SQ_O1p-2, SQ_O1y-1, SQ_O1y-2, SQ_O1y-3, SQ_O1y-4 and SQ_O2y. Each sequence is mainly composed of transgressive and high-level domains, and condensed segments are developed during the maximum transgressive period. Carbonate platform beaches are widely developed in high-level domains of each sequence, which is a favorable reservoir development period. SQ_O1y-4 is generally characterized by relative transgression. The CAMBRIAN in Tadong area is equivalent to the second-order sequence, which is generally composed of a large transgression-regression cycle, and can be divided into five third-order sequences, namely SQ_C1, SQ_C2, SQ_C3-1, SQ_C3-2 and SQ _ C3-3. Each sequence is mainly composed of transgressive and high-level domains. Carbonate platform beaches in SQ_C3 sequences are widely developed, which is a favorable reservoir development period, and is generally characterized by retrogression sedimentation. The sequence division of Middle and Lower Ordovician in Tadong area can be compared with the previous research results in Tazhong and Tabei.

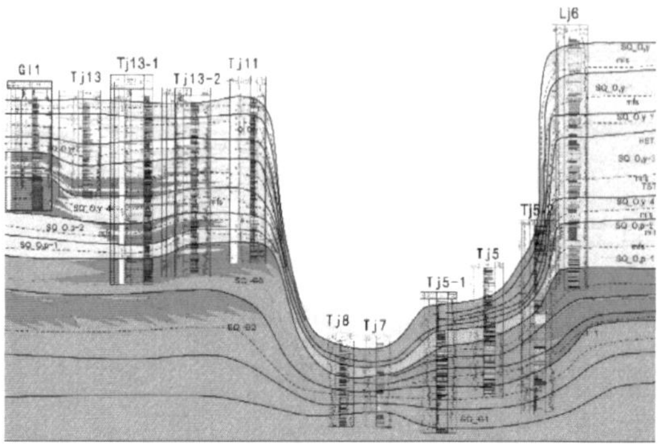

Fig. 2. Sequence framework and system tract composition of Gucheng-Luoxi in comparison of Ordovician series wells in Tadong area.

3 Distribution and characteristics of sedimentary facies belt

Guided by sequence framework, according to drilling cores, microscopic observation, logging curve characteristic analysis and macroscopic seismic facies interpretation, three types of sedimentary facies are identified, which are basin facies, platform margin facies and platform facies. Ten subfacies are developed, including open platform, limited-semi-limited platform, platform margin reef-beach complex and mud mound, platform margin front slope, shallow sea shelf and deep water basin. The sedimentary facies evolution model is comprehensively established, and the distribution and characteristics of main sedimentary facies belt in Cambrian-Ordovician Gucheng platform- Luoxi platform of Tadong is systematically summarized.

3.1. Basin facies

The basin facies is located at the front end of the platform edge slope facies, and distributed in a sheet shape. The reflection is characterized by medium-weak amplitude, good continuity and parallel structure, reflecting mudstone and shale deposits distributed stably in low-energy environment (Figure 3). The lithology is dominated by marl and dark mudstone (Figure 4a), with horizontal bedding, with radiolaria, sponge spicule and other fossils observed. TJ7 well, TJ8 well, TJ5-1 well and TJ5-2 1 well all reveal carbonate basin facies, in which the lithology of TJ5-1 Well is black mudstone mixed with argillaceous dolomite, with grain bedding.

3.2 Platform margin facies

The platform margin facies lies between the deep water shelf and the shallow carbonate platform facies, and the sedimentary boundary is above the wave strike surface. With strong wave action and high water energy, the platform margin facies is a subtidal high-energy facies belt, which is distributed in Cambrian-Ordovician in

Tadong area. The platform margin facies can be further divided into platform front slope, platform margin reef-beach complex and mud mound. On the seismic profile, it shows a hillock reflection seismic phase, with weak amplitude, weak continuous chaotic or disordered internal reflection structure and continuous top reflection with medium-strong amplitude (Figure 3).

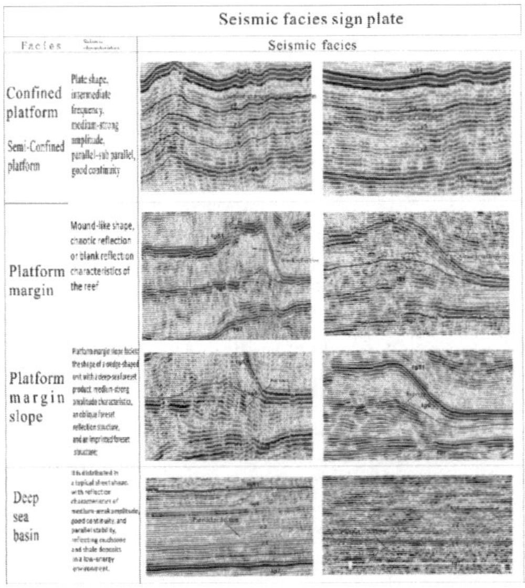

Fig. 3. Seismic facies characteristic chart of Cambrian-Ordovician main sedimentary facies in Tadong area.

3.2.1 Platform front slope facies

The platform front slope facies is located in the slope zone of the platform edge near the shallow sea basin, and the sedimentary interface is above the wave strike surface, with a water depth ranging from zero to tens of meters, and the facies belt is narrow, with an average width of more than ten kilometers, but it extends far in the longitudinal direction. Its lithology is dominated by dark gray argillaceous limestone and argillaceous cloud limestone mixed with black mudstone (Figure 4b). A large number of collapsed breccias are developed at the foot of the slope, which is an obvious sedimentary feature of this facies belt, and is represented by Turshaktag Formation of TJ5-2 well and TJ5-1 well. On the seismic profile, it mainly shows the continuous reflection seismic facies with medium amplitude (Figure 3).

3.2.2 Reef-beach complex and mud mound at platform margin

The reef-beach complex and mud mound at the platform margin are located in the slope of the platform margin. The sedimentary interface of the reef-beach complex is above the average wave strike surface, more than ten meters above or below sea level. The wave action is strong, and a narrow medium-high energy facies belt is formed. The lithology of platform margin beach is mainly granular limestone (especially sparkly granular limestone). All sequences of Penglaiba Formation, Yingshan Formation to Yijianfang Formation of Ordovician in Gucheng area (TJ11, TJ13, TJ13-1 and TJ13-2 wells) and Luoxi area (TJ6 1 well) have typical platform margin facies characteristics. High-energy sand debris and bioclastic beach are widely developed at platform margin (Figure 4c,d), which are mostly composed of granular debris and are of granular beach. Grain cuttings are mainly raw cuttings, sand cuttings and oolites, followed by pellets. Most organisms are echinoderms, brachiopods, ostracods and gastropods. There are small trough-shaped cross bedding and vertical wormholes, and the phenomenon of layer scouring is widespread. When the platform margin beach was formed, the hydrodynamic energy was strong, and the beach was usually distributed around the platform margin in a belt shape. The reef-beach complex has obvious outline on the seismic profile, especially in the areas where reef-building organisms are active and reef uplift is obvious, which are often shown as the external mound shape, while the internal reflection is often chaotic or non-reflective due to energy shielding. For example, in the front edge of Cambrian reef-beach complex on the edge of Gucheng platform where TJ11 well is located, multiple sets of biological mound-beach bodies are laterally deposited and migrated to form a composite S-shaped or imbricate-shaped prestack structure with low angle. The Luoxi platform margin also shows obvious uplift characteristics in seismic profile, and the platform margin beaches developed in different Cambrian-Ordovician periods are superimposed in the form of progradation and accretion.

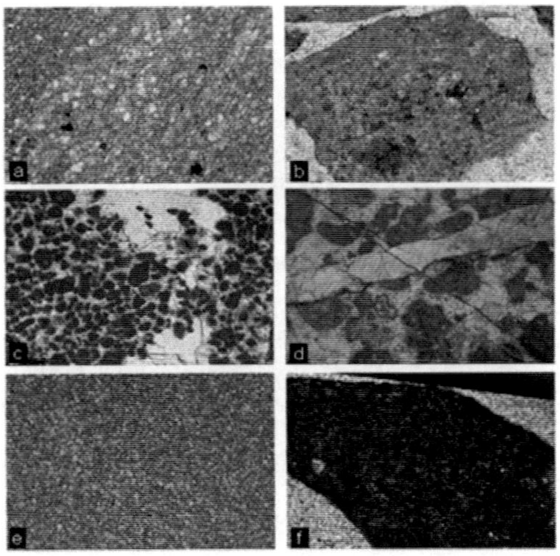

Fig. 4. Microscopic photos of different sedimentary facies in Tadong area. (Note: a. Silty limestone of basin facies, TJ8 well, 4569m,; B. Muddy powder crystalline dolomitic limestone of platform front slope facies, TJ13-2 well, 5598m, Tumuxiuke Formation; C. Reef-beach complex and mud mound at platform margin, GJ13-1 well 6192.3m, silty-mesomorphic dolomite; D. Platform margin sparkly sandy limestone, LJ6 well, 4994.96m, p; E. Semi-confined platform silty dolomite, TJ13-1 well, 6499m, p; F. Microcrystalline limestone of limited platform, TJ13-1 well, 6767m,).

3.3 Platform facies

Tadong carbonate platform facies includes Gucheng platform and Luoxi platform, which are mainly distributed in the west of Manjiaer Sag and the south of Kuruktag. Gucheng platform extends from west to east to basin area, while Luoxi platform extends from east to west to basin area. On seismic profile, the platform facies shows parallel and continuous reflection facies with moderate and strong amplitude (Figure 3).

3.3.1 Open platform facies

Open platform develops within the platform, which refers to shallow sea with good seawater circulation and normal salinity, and its water energy is low and its organisms are developed. Open platform facies developed in various periods from Cambrian to Ordovician, mainly distributed in the east and west ends of Tadong area. Except for the Middle and Late Ordovician, the eastern strata were denuded in other periods, and open platform facies was not developed. The sedimentary environment of this sedimentary facies is shallow, ranging from several meters to tens of meters, with vast sea area, medium water circulation and salinity close to normal or slightly higher. This sedimentary environment is suitable for the growth of all kinds of organisms except those with narrow salinity. Rock types are mainly granular limestone and marl, with occasional terrigenous clastic deposits such as mudstone and siltstone. The species and quantity of organisms are abundant, and most of them are normal salinity organisms. It can be further divided into intra-platform granular beach, inter-beach sea, intra-platform algae reef (mound) and inter-mound depression. The lithology of the inner beach (inner beach) of the platform is mainly granular limestone, and there are brilliant granular limestone and plaster granular limestone according to the different hydrodynamic strength. The inner beach is represented by Yijianfang Formation of Gl 1 well, and the scale distribution of the beach is limited.

3.3.2 Semi-confined platform facies

Semi-confined platform facies mainly developed intra-platform beach and inter-beach sea, which developed in the Early Cambrian and Late Cambrian, and distributed in the eastern side of the Gucheng nose in Tadong area. This facies belt is located inside the open platform facies belt, and the seawater circulation is not smooth. During the Early Cambrian, a set of rocks was mainly deposited, which was mainly composed of dark gray thin-layer silty dolomite and muddy dolomite, mixed with sand cuttings and gravelly dolomite. Late CAMBRIAN is dominated by dark gray silty limestone and dark gray algal microcrystal dolomite, with argillaceous laminae. Texture develops, often with mud cracks, stromatolites and other structures. Syngenetic silty-fine dolomite is developed in the upper part of Penglaiba Formation in TJ13-1 well, while high-energy sparkly sandy limestone, oolitic limestone and low-energy mud sparkly limestone are developed in the lower part (Figure 4e), showing a semi-confined platform sedimentary environment with unsmooth water flow and coexistence of high and low-energy lithofacies.

3.3.3 Limited platform facies

The confined platform mainly developed tidal flat, lagoon and barrier beach bar, while the Upper Cambrian was mainly confined platform facies, which evolved from the semi-confined platform in the Middle Cambrian. The sedimentary environment of the sedimentary facies belt is closed, low energy, inactive seawater circulation and high salinity. The rock types are mainly dark gray, brown gray dolomite, gypsum dolomite mixed with black mudstone and purplish red argillaceous dolomite. Local gypsum-salt rock lumps and dry cracks are found. The content of biological debris is low, mainly blue-green algae and ostracods with wide salinity. Tidal flat develops at the local high terrain, which is especially beneficial to the development of tidal flat during the period of sea level decline or relatively stable period. It mainly includes the supratidal zone and intertidal zone. In this area, the tidal flat is deposited with thin layers of light gray silty dolomite and plaster limestone, and horizontal bedding and stromatolites are developed. The upper CAMBRIAN developed quasi-contemporaneous dolomite formed in early diagenetic stage, recrystallized dolomite formed by later diagenetic transformation, and microcrystalline limestone in low-energy environment, showing limited platform deposits with limited water flow and low energy (Figure. 4f).

4 Evolution of sedimentary system

4.1. Early Cambrian

Early Cambrian Tadong area was dominated by deep-water basin facies, with a large basin area, and its sedimentary center was near Md 1 well, and its lithology was mainly black mudstone and siliceous rock. The platform margin slope facies was developed on both the east and west sides of the basin, and the Lunnan-Gucheng slope facies was narrow and banded in the north-south direction, with an average width of about 28 km. The slope of the Luoxi slope belt was relatively gentle, with a width of about 15 km. The shallow sea shelf facies developed locally (Figure 5).

4.2. Middle Cambrian

In Tadong area of Middle Cambrian, large-scale regression began, and the development area of deep-sea basin facies decreased, but Manjiaer Sag was still the sedimentary center in this period. Shallow shelf facies and slope facies are widely developed between platform and deep-water basin, and platform margin facies is distributed in narrow bands on both sides of the basin. Large-scale regression led to the successive migration of various facies belts in Tadong area to the basin. In the Gucheng slope facies area, platform margin reef-beach deposit (Figure 6) appears, and its lithology is mainly clotted lime mud mound, which is the identification mark of early platform margin facies belt.

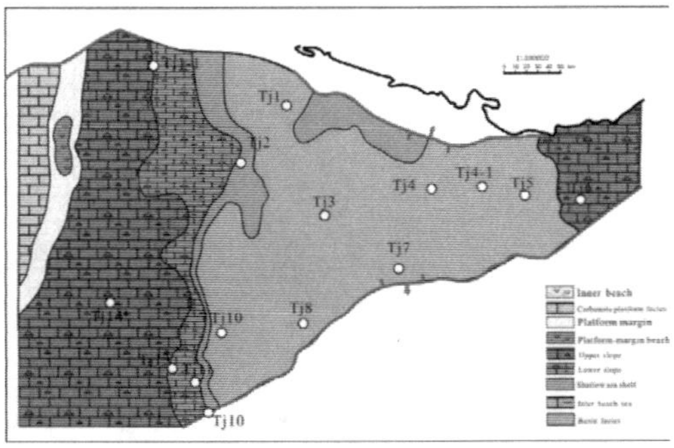

Fig. 5. Sedimentary facies map of Early Cambrian in Tadong area.

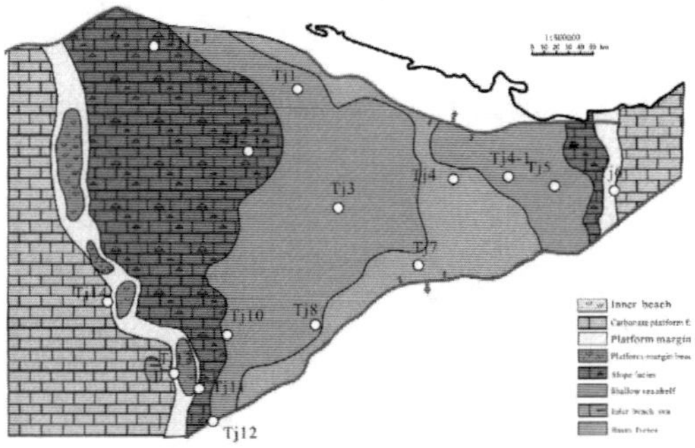

Fig. 6. Sedimentary facies map of Middle Cambrian in Tadong area.

4.3. Late Cambrian

The area of trough and basin in Late Cambrian continued to decrease, and shallow sea shelf facies developed mainly between platform and deep-water basin, with basin facies seen locally, and platform marginal facies belt and reef-beach complex continued to migrate towards the basin (Figure 7). The platform margin and platform margin beach in Gucheng area were developed, and the shoal facies deposits were the main ones. Platform front slope facies was developed around TJ5-1 well and TJ5-2 well, which is mainly composed of collapse sedimentation and carbonate turbidity current sedimentation, and TJ6 well is characterized by typical platform edge shallow beach sedimentation.

4.4. Early and Middle Ordovician

In the early Ordovician, the sea level began to rise slowly, the scope of basin facies expanded, the scope of shallow sea shelf facies decreased, and the sedimentary center and boundary of each facies belt moved westward (Figure 8). The turbidite, marl, carbonaceous mudstone and radiolarian silicalite of semi-deep-deep sea facies were deposited in the Early-Middle Ordovician of TJ7 well and TJ8 well. The carbonate gentle slope in Gucheng-Lundong area gradually developed into platform margin with obvious slope break, and reef-beach complex with high-energy facies belt developed at the platform margin. At this time, shelf-building organisms began to appear in platform margin mound-beach facies, and mud mound deposition in the middle and late Cambrian evolved into bio-mound, bio-beach, algae reef and other deposits. The Middle-Lower Ordovician in TJ11 well is mainly composed of sparkly sandy limestone, algal sandy limestone and oolitic limestone, and trilobite fragments, spines and a large number of shrubby algae fragments with algae reef function are found in rock slices, reflecting the strong hydrodynamic conditions and the sedimentary characteristics of platform margin beach facies. In the early-middle Ordovician, the Luoxi slope break belt basically did not migrate, and the reef beach was also developed. The Middle-Lower Ordovician in TJ6 well is mainly composed of glutenite limestone, oolitic limestone and bioclastic limestone, with algal fragments, which is a typical platform margin high-energy granular beach deposit.

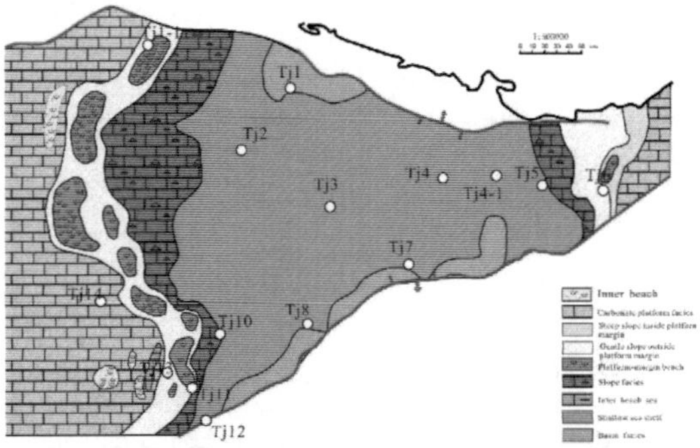

Fig. 7. Sedimentary facies map of Late Cambrian in Tadong area.

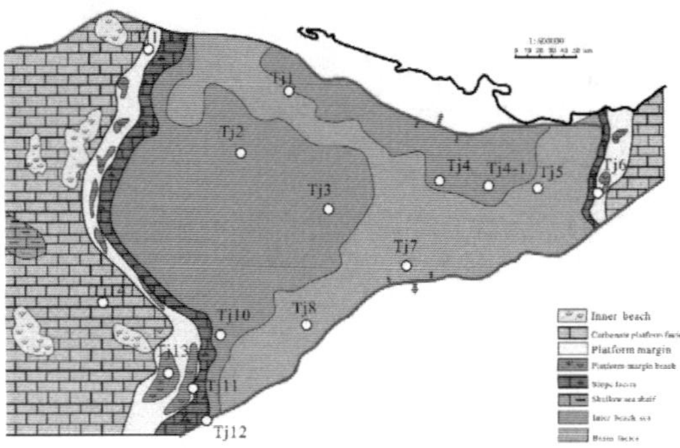

Fig. 8. Sedimentary facies map of Early Ordovician in Tadong area.

 https://doi.org/10.1051/e3sconf/202132901055

5 Conclusion

(1) Cambrian and Middle-Lower Ordovician in Tadong area are equivalent to second-order sequences, which are generally composed of a large transgressive-regressive cycle, and are divided into 12 third-order sequences. Each third-order sequence is mainly composed of transgressive and high-level domains, and argillaceous limestone/dolomite or mudstone deposits are developed during the maximum transgressive period. The carbonate platform (platform margin) beaches in the high-level domain of the third-order sequences of Penglaiba Formation and Yingshan Formation in the Middle and Lower Ordovician are widely developed, which is a favorable reservoir development period. The Cambrian third-order sequences are mainly composed of transgressive and high-level domains, and the carbonate platform beach or reef (mound) beach complexes in SQ_C2, SQ_C3-1, SQ_C3-2 and SQ_C3 sequences are developed, which can form favorable reservoirs.

(2) In Tadong area (Gucheng platform and Luoxi platform), the early Cambrian mainly consists of platform margin slope facies-deep water basin sediments, while the middle and late Cambrian consists of (semi) limited platform facies-open platform-platform margin hill (reef) beach facies-platform margin slope facies-basin facies. Compared with the middle Cambrian, the sea level in the late Cambrian decreased further and the platform margin moved into the basin. Transgression began in Early Ordovician, and the sedimentary pattern was open platform-platform margin beach facies-platform margin slope facies-basin facies from west to east.

References

1. Li Wenhou, Chen Qiang, Li Zhichao, et al. Lithofacies palaeogeography of early Paleozoic in Ordos region [J]. Journal of palaeogeography, 14 (1): 85-100 (2012).
2. Feng Zengzhao, Bao Zhidong, Wu Maobing, et al. Cambrian lithofacies palaeogeography in Tarim area [J]. Journal of palaeogeography, 8 (4): 427-439 (2006).
3. Feng Zengzhao, Bao Zhidong, Wu Maobing, et al. Ordovician lithofacies palaeogeography in Tarim area [J]. Journal of palaeogeography, 9 (5): 447-460 (2007).
4. Yang Yongjian, Liu Jiaduo, Tian Jingchun, et al. Congratulations on the lithofacies palaeogeography of Cambrian sequence in Tarim Basin [J]. Natural Gas Geoscience, 22 (3): 450-459 (2011).
5. Yu Kuanhong, Zhenkui jin, Li Peng, et al. Analysis of Cambrian-Ordovician lithofacies palaeogeography in Tadong area [J]. Northwest Geology, 43 (4): 113-118 (2010).
6. Liu Yang. Study on sedimentary facies characteristics of Cambrian reservoirs in Tadong area [J]. Journal of Yangtze University (Natural Edition), 11 (31): 17-20 (2014).
7. Xu Kang, Liu Guangxiang, Hu Wenxuan, et al. Distribution and sedimentary facies of reefs and beaches in the Upper Ordovician sequence framework in Tazhong area [J]. Petroleum Geology and Development of Daqing, 37 (2): 27-30 (2018).
8. He feng, Lin changsong, Liu jingyan et al. sedimentary characteristics and main controlling factors of carbonate rocks in Gucheng area [J]. Special oil and gas reservoirs, 23 (5): 17-19 (2016).
9. Zhang Tao, Gui Xiangbin, Wang Shuyi, et al. Characteristics and genesis of reef-beach dissolved pore reservoirs in Yijianfang Formation of Ordovician in Tahe Oilfield [J]. Petroleum and Natural Gas Geology, 25 (4): 462-466 (2004).
10. Yang Haijun, Li Yong, Liu Sheng, et al. Main understanding of division and correlation between middle and upper Ordovician in Tazhong area [J]. Xinjiang Petroleum Geology, 21 (3): 209-212 (2000).
11. Kang Zhijiang and Li Hongkai. Characteristics of Ordovician carbonate reservoirs in Tahe Oilfield [J]. Petroleum Geology and Development of Daqing, 33 (1): 21-26 (2014).
12. Lv xiuxiang. Characteristics of Ordovician carbonate reservoirs in Tazhong area and their hydrocarbon significance [J]. Journal of Xi' an Shiyou university (natural science edition), 19 (4): 72-76 (2004).
13. Yan Xiangbin, Li Tiejun, Zhang Tao, et al. Differences in formation conditions of Ordovician karst reservoirs in Tazhong and Tahe areas [J]. Petroleum and Natural Gas Geology, 26 (2): 202-207 (2005).
14. Wu guanghui, Li qiming, Zhang baoshou, et al. Structural characteristics and exploration fields of slope break zone of Tazhong No.1 fault [J]. Acta petrolei sinica, 26 (1): 27-30 (2005).

https://doi.org/10.1051/e3sconf/202132901056

Fine evaluation and favorable area prediction of deep source rocks in northern Songliao Basin

Fan Zhang[1,2], Yanjie Li[1,2], Xiaoshan Ji[1,2], Qiuli Huo[1,2], Yuming Wu[1,2], Yitong Wang[1,2], Zhigao Ren[1,2], and Junchi Bao[1,2]

[1]Exploration and Development Research Institute of Daqing Oilfield Co., Ltd., 163712 Daqing Heilongjiang, China
[2]Key Laboratory of tight oil and shale oil accumulation of Heilongjiang province, 163712 Daqing Heilongjiang, China

Abstract. Focusing on Xujiaweizi fault depression, the geological conditions and geochemical characteristics of deep natural gas formation in the north of Songliao basin are evaluated, the natural gas resources are estimated, and the favorable areas are optimized. Shahezi Formation shale is a set of coal bearing sediments with high organic matter abundance (TOC is 1% ~ 12%), high over maturity (R_o is 1% ~ 4%) and shore shallow lake facies, which are mainly distributed in Xujiaweizi fault depression, Gulong-Lindian fault depression and Yingshan fault depression. The thickness, TOC, R_o and hydrocarbon generation of four third-order sequences with different lithology (mudstone and coal) are carefully evaluated for the Shahezi Formation shale in the deep layer of Songbei. The comprehensive evaluation shows that the mudstone thickness of Es4 member in Anda and Xuzhong areas of Xujiaweizi fault depression is large (150 ~ 525m), TOC is high (1% ~ 4%), thermal evolution degree is high (R_o is 1.2% ~ 3.4%), and gas generation intensity is high (20 ~ 815) × $108m^3$ / t), moderate buried depth (3000~4500m) and overlapping area of $756km^2$. It is a favorable exploration area for natural gas and shale gas in Daqing Oilfield.

1 Deep geological features in Songbei

The deep strata in the north of Songliao Basin mainly include lower Cretaceous Huoshiling formation, Shahezi formation, Yingcheng formation and Denglouku formation from bottom to top. The basement is Carboniferous Permian (Fig. 1), belonging to volcanic eruption and clastic rock coal bearing deposition in fault depression period. Among them, the coal mudstone of Shahezi formation dominated by shore shallow lake facies is thick, with good hydrocarbon generation conditions and great resource potential. Among them, the coal mudstone of Shahezi Formation dominated by shore shallow lake facies is thick, with good hydrocarbon generation conditions and great resource potential. Among them, Xujiaweizi has discovered the Xushen gas field in the volcanic rocks of the Yingcheng formation, and has obtained industrial gasflow in several sets of formations such as Shahezi and Denglouku formations. There are also many exploratory wells that have seen oil and gasd in Gulong-Lindian fault depression in the west and the Yingshan fault depression in the east [1-3].

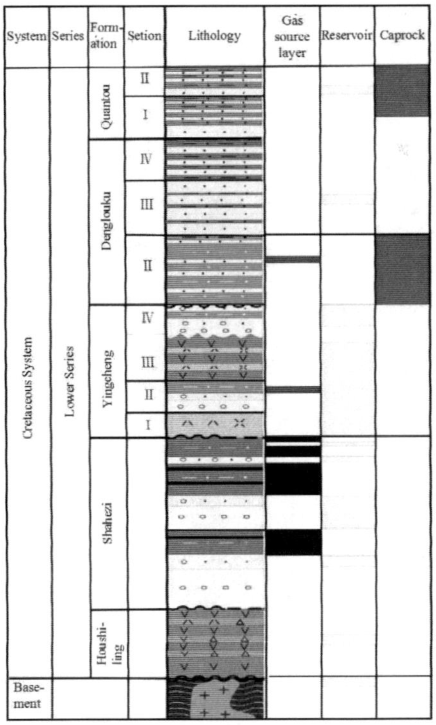

Fig. 1. Comprehensive column section of deep layer in northern Songliao Basin.

* Corresponding author: zhangfan1@Petrochina.com.cn

2 Geochemical characteristics of deep source rocks in Songbei

The abundance of organic matter in gas-bearing shale not only determines the amount of shale gas generation, but also affects the occurrence and enrichment of shale gas, which in turn influences the resource abundance of shale gas [4]. For thermogenic shale gas reservoirs, with the increase of shale R_o, when R_o>1.1%, shale oil begins to crack and generate gas; when R_o>3%, the volume of gas generation is significantly reduced [5], moreover, the large reduction of organic matter is not conducive to the formation of adsorbed gas. Therefore, the range of R_o value of 1.1% ~ 3% is a favorable distribution area for thermally derived natural gas reservoirs. In this study, four sets of source rocks of deep Huoshiling formation, Shahezi formation, Yingcheng formation and Denglouku formation were systematically analyzed and evaluated from three aspects of abundance, type and maturity in the north of Songliao Basin. In terms of region, the organic abundance index of Xujiaweizi fault depression located in the northern and central part of Songliao basin is higher than that of gulong, Yingshan and other peripheral fault depressions (Fig. 1). From the horizon, the geochemical index of Shahezi formation shale is better than other horizons as a whole (Table 1).

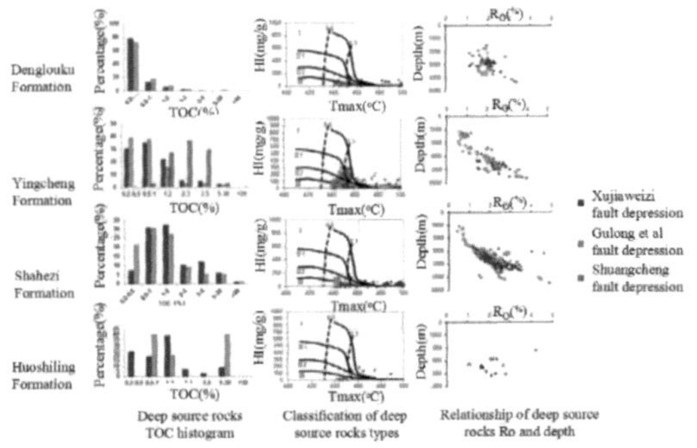

Fig. 2. Geochemical comprehensive evaluation map of deep source rocks in northern Songliao Basin.

Table 1. Evaluation table of geochemical characteristics of deep source rocks in northern Songliao Basin.

Formation	TOC	Type	R_o
Denglouku	0.20% ~ 3.70% (0.48%)	III	1.13% ~ 2.62% (1.94%)
Yingcheng	0.21% ~ 14.25% (1.14%)	II ~ III	1.50% ~ 3.14% (2.25%)
Shahezi	0.2% ~ 76.6% (2.62%)	II ~ III	1.07% ~ 3.56% (2.22%)
Huoshiling	0.20% ~ 26.19% (2.11%)	III	1.62% ~ 2.39% (2.01%)

*Minimum~Maxmux (Average)

3 Fine geochemical evaluation of deep source rocks in Songbei

As the main component of natural gas generation and occurrence, a certain thickness of gas-bearing shale is the basic condition for the formation of natural gas enrichment areas, and it is also one of the important factors influencing the abundance of natural gas resources. The greater the thickness of shale rich in organic matter, the larger the scale of gas-bearing shale. At the same time, it can meet the needs of large-scale fracturing transformation and large-scale development of natural gas. When the thickness of gas bearing shale and the abundance of organic matter are determined, the area of gas bearing shale determines the amount of natural gas resources [6].

Due to the limited analysis data of core samples and the continuous formation information provided by logging data, organic matter has specific logging response because of its inherent physical and chemical properties. In recent years, a variety of logging combined evaluation methods have emerged: Based on the response characteristics of organic matter to sound wave, resistivity, density and natural gamma, an interpretation model is established to calculate the organic carbon value [7]. In this study, the Δ LOG method was used to accurately evaluate the total TOC of 137 exploratory wells in five fault depressions in the northern Songliao Basin, and effectively predict the total organic matter of source rocks in this area. Taking Well DS17 as an example, the average TOC of the source rocks of the four members of the Shahezi Formation is 4.93%, and the thickness of "mud-coated sand" interval is greater than 30m, which has the geological conditions for forming of self-generating and self-storing natural gas [8] (Figure 3). On this basis, in order to meet the needs of fine research and source rock feature evaluation, this study made a fine comparison of the source rocks of the Shahezi Formation in Xujiaweizi. The results show that three cycles of high-quality source rocks can be developed in each sequence set of Shahezi Formation, in which the TOC value of Shahezi Formation source rocks in the north of Xujiaweizi fault depression is mainly 1% ~ 3%, and the TOC value of some wells is greater than 3%. It is describedthat the source rocks of Shahezi Formation in the north of Xujiaweizi fault depression are getting better and better in abundance and thickness from north to south from the TOC well connection profile of source rocks of well DS3-SS6 (Fig. 4). High quality source rocks are principally distributed in Anda area, which is generally characterized by high abundance and good continuity of Sha4 member.

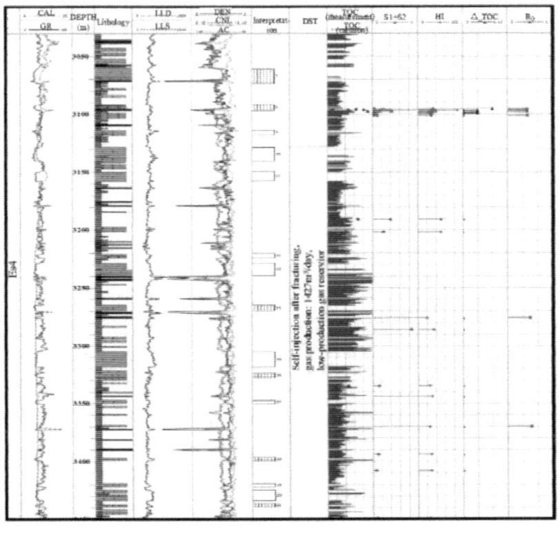

Fig. 3. Comprehensive histogram of Shahezi Formation in well DS17.

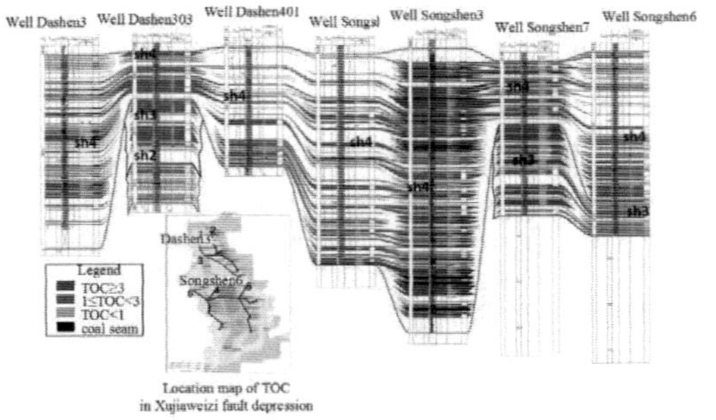

Fig. 4. TOC profile of Shahezi Formation source rocks in well DS3—SS6.

Based on the fine evaluation of single well and connected well of source rock, we calculate TOC results (removing oil-base mud pollution) and R_o according to all exploratory wells, mud logging, lithology, and well logging, besides draw the mudstone thickness map, TOC plane (Fig. 6) and R_o plane (Fig. 7) of the four sequences (sh1, sh2, sh3, sh4) of the source rocks of the Shahezi Formation in the Xujiaweizi fault depression, referring to the formation thickness and sedimentary facies trend (Fig. 5).

It can be seen from the figure that the distribution trend and characteristics of the thickness, abundance and maturity of the source rocks of Shahezi Formation in the four vertical sequences are not different, and the plane distribution characteristics are also similar, indicating that the sedimentary center of this area is stable, multi cycle high-quality source rocks are developed, and the overall performance is relatively high abundance coal bearing source rocks, of which SH4 layer is the best. Abundance of organic matter in the source rocks of the Shahezi Formation in Xujiaweizi is mainly distributed between 0.5% and 5.0% of the TOC classification thickness map. Among them, the TOC value is mainly 1.0% to 2.0%, accounting for about 50% of the mudstone thickness of the whole Shahezi Formation.

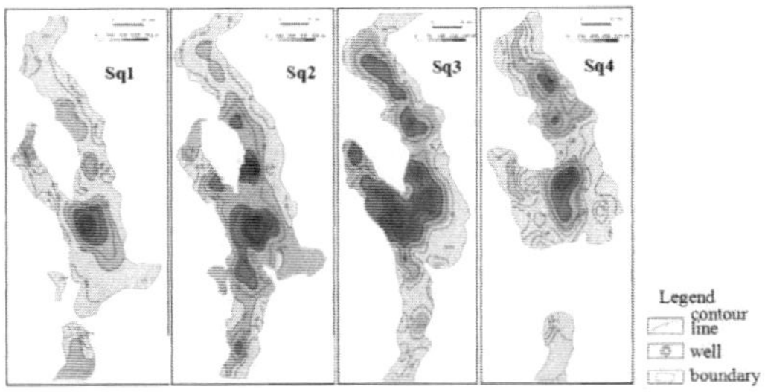

Fig. 5. Comparison map of mudstone thickness of four sequences of sourcerocks in Shahezi Formation of Xujiaweizi fault depression.

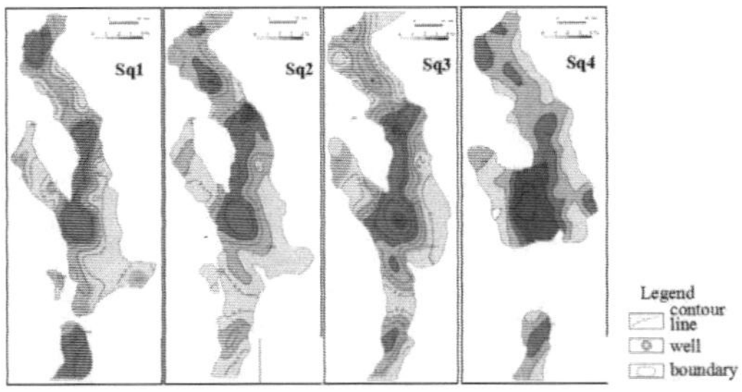

Fig. 6. TOC distribution comparison map of four sequence mudstone of Shahezi Formation in Xujiaweizi fault depression

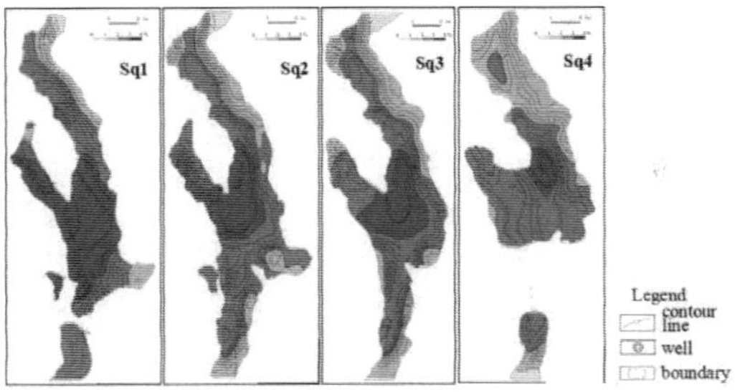

Fig. 7. Ro comparison map of sourcerocks of four sequences of Shahezi Formation in Xujiaweizi fault depression

4 Deep natural gas resource potential and favorable area optimization in Songbei

According to the latest drilling and seismic data, the mudstone thickness map is predicted, and the TOC value is calculated by logging curves, the measured R_o value is corrected for the thermal history, and the gas generation volume of the subdivision layer is calculated by the genetic method (Table 2), The calculation results show that the proportion of gas generation of four sets of source rocks is Shahezi Formation > Yingcheng Formation > Huoshiling formation > Denglouku formation. The gas generation intensity of source rocks in Shahezi Formation is high, more than 20 in most areas × 108 / km^2, which

 https://doi.org/10.1051/e3sconf/202132901056

can form large and medium-sized gas fields [9 ~ 12] (Fig. 8).

Table 2. Calculation table of gas generation proportion of subdivision layers in Xujiaweizi fault depression.

Formation	Thickness of source rocks	TOC (%)	Gas generation (10^2m^2)	Percentage (%)
Denglouku	0 ~ 200	0.36	1805	0.61
Yingcheng	0 ~ 300	0.78	9361	3.16
Es4 mudstone	0 ~ 600	1.63	52626	17.78
Es4 coal seam	0 ~ 80	60	19004	6.42
Es3 mudstone	0 ~ 300	1.39	44531	15.05
Es3 coal seam	0 ~ 60	55	14522	4.91
Es2 mudstone	0 ~ 500	1.53	61097	20.64
Es2 coal seam	0 ~ 45	53	7959	2.69
Es1 mudstone	0 ~ 800	1.19	72056	24.35
Es1 coal seam	0 ~ 60	57	9596	3.24
Huoshiling	0 ~ 100	0.72	3399	1.15
In total			295956	100

Fig. 8. Shale gas generation intensity plan of Shahezi Formation in Xujiaweizi fault depression

Since the main part of shale gas is adsorbed gas, according to the principle of hydrocarbon generation and expulsion, part of the natural gas that not discharged but adsorbed in the source rock occupies the main source of shale gas. According to Pepper (1995), the gas adsorption coefficient is 0.02, and in the middle of the fault depression, the hydrocarbon generation intensity is more than 100 × 10^8 / km^2 which has shale gas resource potential.

The favorable natural gas area with TOC > 2% of mudstone in n the fourth section of Shahe in Xujiaweizi fault depression is $756km^2$ and R_o is 1.1% ~ 4.0%, which has good natural gas resource potential; Shahezi Formation mudstone has high gas generation intensity, ranging from 100 to 815 × $108/km^2$ in the Anda and Xuzhong areas (Fig. 9). The measured porosity of 9 shale samples is 1.2% ~ 5.3%, and the buried depth is 3000 ~ 4500m, which is a favorable exploration area for natural gas and shale gas in the next step [13 ~ 16].

https://doi.org/10.1051/e3sconf/202132901056

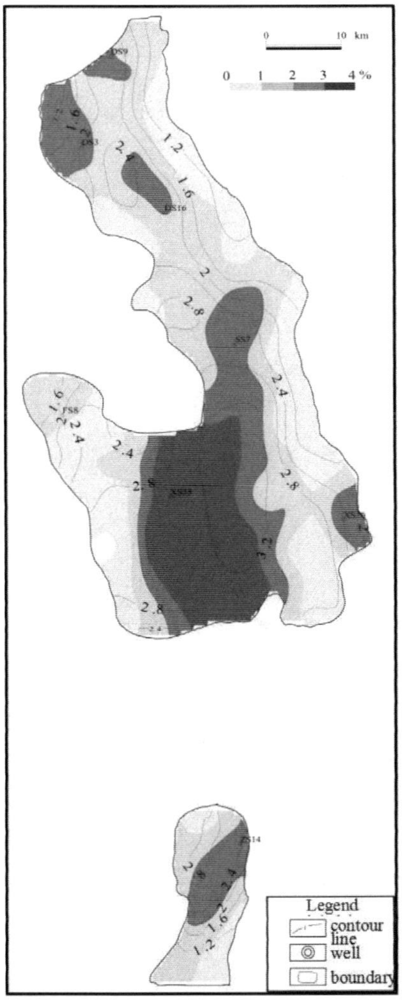

Fig. 9. Superposition of Ro and TOC of mudstone in the fourth member of Shahejie Formation inXujiaweizi fault depression.

5 Conclusion

1)Shahezi Formation develops coal-bearing sedimentary shale dominated by shore-shallow lacustrine facies, with high abundance and over-mature organic matter. It is the main source rock layer in the deep northern Songliao Basin, mainly distributed in the Xujiaweizi fault depression;
2)The gas generation of mudstone in Xujiaweizi fault depression is dramatically higher than coal gas generation. The mudstone gas generation volume of the first, second, third and fourth members of the Shahezi Formation is similar, and the mudstone of the fourth member of Shahezi Formation accounts for the largest proportion due to the highest maturity;
3) The fourth member of Shahezi Formation in Xujiaweizi fault depression has large mudstone thickness, high abundance and continuous distribution. Anda and Xuzhong, e.g. areas have moderate burial depths, which are favorable exploration areas for natural gas and shale gas in the next step of the Daqing Oilfield.

References

1. Feng Z. Q., Exploration potential of large Qingshen gas field in the SongliaoBasin [J]. Natural Gas Industry, 26(6):1-5 (2006) .
2. Zhang F. et al, Study on geochemical characteristics andResource potential of coal measures source rocks in Daqing exploration area [R]. Daqing Oilfield Co., Ltd., Daqing, Heilongjiang, 1-51 (2014).
3. Wang W. G., et al, Gas generation characteristics and resource potential of the deepsourcerock in Xujiaweizi fault depression, northern Songliao Basin[J]. Natural Gas Geoscience, 25(7):1011-1018 (2014).
4. Chen S. Y., Liu H. J., Carboniferous Permian lithofacies and paleogeography in the eastern part of the north china platform [J]. Regional Geology of China,16(4):379-386 (1997).
5. Xie D. N., Zhou L. F., Discussion on hydrocarbon generation potential and secondary hydrocarbon generation of organic matter from Permo-Carboniferous coal measures in the southern North China [J]. Coal Geology & Exploration, 34(1):30-34 (2006).
6. Ren Z. L., Tectonic and thermal evolution history of sedimentary basins in northern China[M]. Beijing:Petroleum Industry Press, 6-42 (1999).
7. Yan J.P., Liang Q., Geng B., et al, Log calculation method of geochemical parameters of lacustrine shale and its application: a case of lower Es3 in Bonan subsag, ZhanhuaSag. Lithologic Reservoirs, 29(4):108-116 (2017).
8. Fu W., Hou H.S., Zhang J. D., Liu C., Feng X., Lu.W., Structural characteristics of Shahezi Formation beneath the Well SK-2 and its periphery: An analysis of logging and seismic data. Geology in China, 46(5): 1052-1060 (2019). (in Chinese with English abstract).
9. Song Yan, Dai Jinxing, Li Xianqi, etal. Main geochemical and geological characteristics of large and medium-sized gas fields in China [J]. Acta petrolei sinica, 1998,19(1) :1 ~ 5.
10. Fu G., Zang F. Z., Favorable geological conditions for Formation of Xushen large gas field. [J]. Journal of Jilin University (Earth Science Edition), ,41(1):12-20 (2011).
11. Nie H. K., Zhang J.C., Li Y.X., Accumulation conditions of the Lower Cambrian shale gas in the Sichuan Basin and its periphery [J]. Acta Petrolei Sinica, 32(6):959-967 (2011).
12. Zou C.N., Tao S. Z., Zhu R. K., etal, Formation and distribution of "continuous" gas reservoirs and their giant gas province:A case from the Upper Triassic Xujiahe Formation giant gas province, Sichuan Basin

[J].Petroleum Exploration and Development, 36(3):307-319 (2009).

13. Yang S.Y., Zhang J. C., Tang X., Three-dimensional basin modeling study on Zhangjiatan shale gas area in Ordos Basin [J]. Natural Gas Geoscience, 27(5):932-942 (2016).
14. Wang S. J., Wang L.S., Huang J. L., etal, Accumulation conditions of shale gas reservoirs in Silurian of the Upper Yangtze Region [J]. Natural Gas Industry,29(5):45-50 (2009).
15. Guan D. S., Niu J. Y., Guo L., Unconventional Petroleum geology in China [M]. Beijing:Petroleum Industry Press (1995).
16. Zhang J. C., Jiang S.L., Tang X., etal, Accumulation types and resources characteristics of shale gas in China [J]. Natural Gas Industry, 29(12):109-114 (2009).

https://doi.org/10.1051/e3sconf/202132901057

Reservoir forming conditions and exploration potential of Lower Paleozoic carbonate rocks in Gucheng area, Tarim Basin

Liu Zheng[1,*],Yan Liping[1] ,Sun Guoxin[1],Ding Hansheng[1], Li Yanjie[1],

[1] Exploration and Development Research Institute of Daqing Oilfield Co Ltd., Daqing 163712, China

Abstract. The Gucheng region is oil and gas accumulation area of Tarim Basin, dominated by beach controlled lithologic oil and gas reservoirs, which has superior oil and gas accumulation conditions and broad exploration prospects. The lower Paleozoic strata in the Gucheng region are the Carboniferous, Ordovician, and Cambrian systems from top to bottom. The Silurian, Devonian, and Permian systems are missing. The lower Ordovician – Cambrian are the main exploration target layers in the area. There are four reservoirs: the Ordovician group, the Yingshan group, the Penglaiba group and the Cambrian group in The Gucheng region, The upper part of the Yijianfang formation and Yingshan formation are limestone reservoir, which is a set of potential reservoir. The lower part of the Yingshan formation, the Penglaiba formation, and the Cambrian formation developed three sets of dolomite reservoirs. The distribution of oil and gas in Gucheng region was mainly found in Yijianfang formation, Yingshan formation, Penglaiba formation and Cambrian of the Ordovician. The Yingshan group and the Hanwu group were the main exploration target layers. Yingshan formation and Cambrian are the main exploration targets, and Yijianfang formation and Penglaiba formation of Ordovician can be used as the concurrent exploration targets. Gucheng region can form two major realistic exploration areas: dolomite reservoir of Ordovician and reef beach of Cambrian. The Ordovician dolomite reservoirs are mainly searched for the inner shoal of the platform edge high-energy facies belt, and the Cambrian system is mainly search for the high-energy reef beach far away from the platform edge facies belt in the north of cherchen fault.

1 Introduction

The marine sedimentary basins in China were mainly developed in the Paleozoic Era and contained abundant oil and gas resources [1-4]. However, the formation of marine carbonate reservoirs in China is early and the stratigraphic age is old [5-6]. The exploration target layer is deeply buried and multi-phase tectonic movements are superimposed. The rules of hydrocarbon accumulation and enrichment are very complicated, which affect the breakthrough in the oil and gas exploration of China's marine carbonate reservoirs. Tarim Basin is a superimposed basin [7]. With the continuous discovery of oil and gas in the Yingshan Formation and Cambrian dolomite in Tazhong and Tabei, it has become one of the largest oil and gas basins in China. The Gucheng Low Uplift is located in the central-southern part of the northern depression area of the Tarim Basin, adjacent to the East Uplift in the east, and adjacent to the Du Uplift in the west. (Figure 1). It is a large, wide, gentle nose-like structure dipping from north to west [8-9], with an area of about 6100km2. At present, TJ13 well, TJ13-2 well, TJ13-3 well, and TJ13-4 well, which have been drilled in the three-dimensional area, have respectively obtained 260,000 cubic meters, 478,000 cubic meters, 1.0789 thousand cubic meters and 227,300

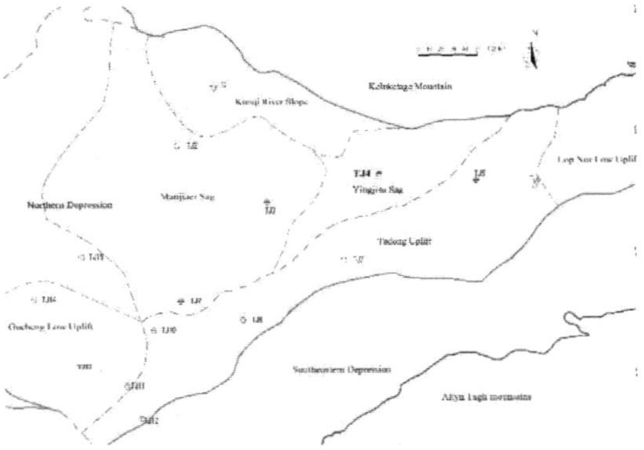

Fig. 1. Structural location map of the Gucheng area.

cubic meters of high-yield airflow in the Ordovician Eagle III. The Gucheng area has a favorable structural location, well-developed carbonate reservoirs and favorable accumulation conditions. It has the petroleum geological conditions for the formation of large-medium-sized oil and gas fields. It has large exploration potential and abundant resources. A number of research results have been obtained by researchers on the oil and gas accumulation conditions in the adjacent Tazhong area [10-14], but the current exploration level in the Gucheng

* Corresponding author: liuzheng3@petrochina.com

 https://doi.org/10.1051/e3sconf/202132901057

area is very low, and there is no clear understanding of the oil and gas accumulation conditions and rules, which restricts the understanding of the oil and gas resource potential of the Gucheng area, and some problems remain to be resolved.

2 Geological characteristics

2.1 Stratigraphic characteristics

According to seismic and drilling data, the Paleozoic strata in the Gucheng area developed Cambrian, Ordovician and Carboniferous from bottom to top, and Devonian and Permian were absent. The thickness of the Lower Paleozoic is more than 5000m. The Silurian is only distributed at the northern tip of the rhinoplasty of Gucheng. The Cambrian, Ordovician and Carboniferous are basically distributed throughout the region. The Lower Ordovician-Cambrian is the main exploration target layer in this area.

2.2 Structural features

From the perspective of the tectonic evolution history of the Gucheng area, the Gucheng area has experienced Caledonian, Hercynian, Indosinian-Yanshanian, Himalayan and other tectonic movements. With slight adjustments, the late-active faults are mainly distributed near the Cherchen fault in the south. Therefore, since the Late Caledonian, the Gucheng area has been a long-term inherited large-scale nose structure with early formation and weak late tectonic activity and is well preserved (Figure 2). In addition, the Gucheng area is close to slopes and basins—favorable source rock development areas, which have favorable conditions, and are long-term oil and gas migration points. The barrier blocks and local structural traps in the Gucheng area, which are located in relatively high structural positions, are favorable areas for late natural gas accumulation.

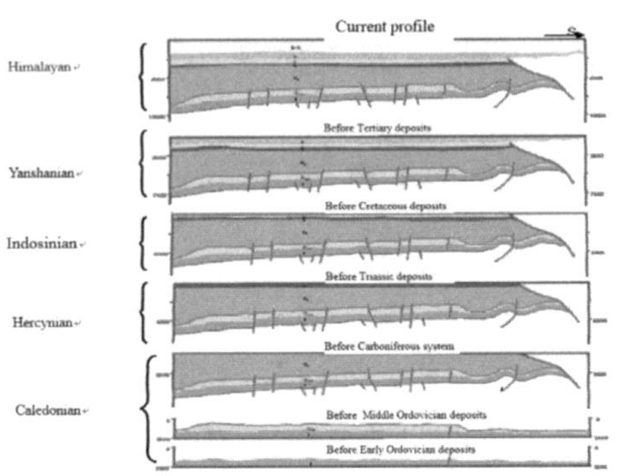

Fig. 2. Sectional diagram of structural development in Gucheng area

2.3 Structural features

Sedimentary facies control the nature of the reservoir and affect the enrichment of oil and gas [14]. Controlled by the rise and fall of sea level, the Lower Paleozoic Cambrian to Middle-Lower Ordovician water bodies in the Gucheng area experienced a process from the Cambrian water body becoming shallower to the Ordovician gradually becoming deeper, and it experienced a gentle slope → rim The evolution process of type platform→submerged platform, Cambrian-Early Ordovician, Gucheng area is a confined platform-semi-restricted platform deposition, and the quasi-contemporaneous dolomite formed under the control of the Cambrian confined sedimentary environment is relatively stable., And low-energy environment microcrystalline limestone, showing a restricted platform sedimentary environment with limited water flow and low energy. The Penglaiba Formation is a semi-limited platform, mainly composed of sprite sandstone limestone and oolitic limestone. Early Ordovician in the late-early Ordovician, the Gucheng platform was uplifted strongly due to the collision between the West Kunlun Island and the Tarim board [15]. The Yingshan Formation~Yijianfang Formation developed platform marginal sandy shoal deposits, mainly brightly developed. Crystalline sand clastic limestone, microcrystalline limestone, medium-fine crystalline dolomite, the Upper Ordovician Tumuxuke Formation mainly develops bioclastic (including mud) limestone, micrite, microcrystalline limestone, and nodules Limestone, etc., reflect the low-energy environmental deposits of the submerged platform in relatively deep water.

2.4 Reservoir characteristics

Lower Ordovician-Cambrian reservoir lithology is mainly fine-medium crystalline dolomite. The core is parallel and the oblique bedding is developed (cracking along the bedding). The residual grain structure and phantom can be seen under the microscope, reflecting the high-energy grain beach facies which are formed by cloudification (Figure 4). The reservoir spaces are mainly of intercrystalline (dissolved) pores and caves, and the main type is porosity (Figure 3). Fractures are partially improved to form fracture-pore (cavity) reservoirs. The overall reservoir is characterized by "low porosity and ultra-low permeability", and the porosity and permeability have good correlation. The measured porosity of the core is mainly distributed between 1.8% and 5.0%, and the permeability is between 0.01 and 10mD.

Fig. 3. Core slice in Gucheng area.

2.5 S Source-reservoir-seal combination

The Gucheng area is bordered by the Manjiaer Cambrian-Ordovician Lower Paleozoic oil-generating depression in the north, which has relatively rich oil and gas sources [16-17]. The source rock is mainly composed of two sets of argillaceous rocks (mudstone and shale) and argillaceous limestone from the Cambrian to the Middle-Lower Ordovician. The Lower Paleozoic source rock in the Manjiaer Sag of the Late Caledonian entered the peak of hydrocarbon generation and expulsion, and it was a condensate source kitchen [18-19]. The analysis of cuttings in TJ13-1 well and TJ13-2 well showed that the TOC of Ordovician hydrocarbon source is 0.03 to 0.81%, averaging at 0.22%, and the abundance of organic matter is low. The Cambrian-Ordovician source rocks in the Gucheng area are not developed, and the oil source comes from the Manjiaer sag on the east side.

The Gucheng area developed four sets of reservoirs, namely, Ordovician Yijianfang, Yingshan, Penglaiba and Cambrian. The upper part of the Yijianfang and Yingshan formations are limestone reservoirs, with micrite limestone and sandy limestone. There are three sets of dolomite reservoirs developed in the lower part of Yingshan Formation, Penglaiba Formation and Cambrian, with powdery, fine and medium-coarse crystal dolomites, among which fine-medium-coarse crystal dolomites form favorable reservoirs.

Three sets of reservoir-cap combinations developed in the Gucheng area. The thickness of the upper tight limestone caprock of the Yingsan Formation dolomite is 400-500m, the thickness of the upper dolomite caprock of the Penglaiba Formation is 100-180m, and the thickness of the caprock developed on the top of the Cambrian dolomite is 100-120m. Affected by the thickness and fracture of the caprock, the first set of reservoirs and caps has good oil and gas bearing capacity. The Upper Ordovician Queerchak Formation, which is several kilometers thick, is dominated by mudstone deposits, which can be used as a regional caprock, conducive to the preservation of oil and gas generated in the Lower Paleozoic (Figure 4).

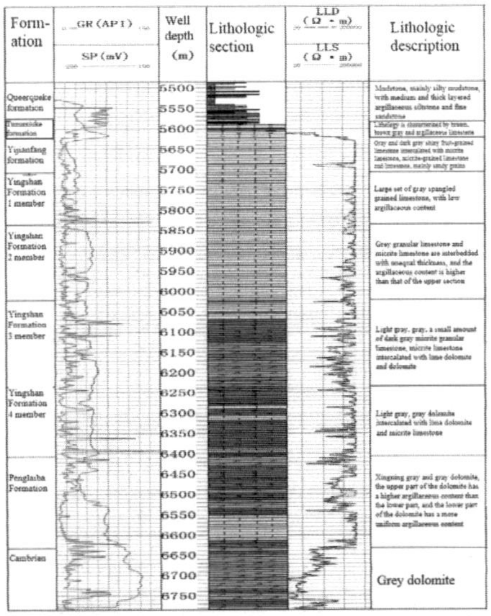

Fig. 4. Distribution map of reservoir-cap combinations in Gucheng area.

3 Oil and gas distribution characteristics and prediction of favorable areas

Three sets of reservoir-cap combinations developed in the Gucheng area. The thickness of the upper tight limestone caprock of the Yingsan Formation dolomite is 400-500m, the thickness of the upper dolomite caprock of the Penglaiba Formation is 100-180m, and the thickness of the caprock developed on the top of the Cambrian dolomite is 100-120m. Affected by the thickness and fracture of the caprock, the first set of reservoirs and caps has good oil and gas bearing capacity. The Upper Ordovician Queerchak Formation, which is several kilometers thick, is dominated by mudstone deposits, which can be used as a regional caprock, conducive to the preservation of oil and gas generated in the Lower Paleozoic (Figure 4). The Gucheng area has a geological development history of early uplift, mid-subsidence stereotypes, and late stability. It is more favorable for later preservation of oil and gas and is a favorable accumulation area for natural gas. At present, the oil and gas distribution in the Gucheng area is mainly found in the Yingshan Formation and Cambrian of the Ordovician. Among them, the Yingshan Formation has obtained industrial oil and gas discoveries, and the upper Cambrian reefs and banks have also seen good oil and gas shows. The Gucheng area has good oil and gas accumulation conditions. The effective reservoir-cap configuration relationship controls the distribution of oil and gas. The high-quality shoal facies reservoirs control the degree of oil and gas enrichment. Studies have shown that oil and gas are distributed in a quasi-layered large area in the horizontal direction, and by multiple stages of later reservoir transformation in the vertical direction, and are not controlled by structural heights and lows. The distribution and heterogeneity are obvious. The reservoir type should be a quasi-layered lithologic gas reservoir. On this basis, combined with the current drilling data, two realistic exploration areas can be identified in the Lower Paleozoic in Gucheng area: Ordovician dolomite reservoirs and Cambrian platform margin reef banks. Ordovician dolomite reservoirs are mainly used to search for intra-platform beaches in the high-energy facies zone dolomitized at the edge of the platform. The middle and lower parts of the Ordovician Yingshan Formation are favorable exploration targets. TJ13 well and TJ13-2 well in the dolomite reservoirs in the third member of the Ordovician Ying platform have obtained industrial oil and gas flow. The Penglaiba Formation in TJ13-1 well encountered dolomite, with good reservoir properties and well-developed fractures and cavities. TJ13-1 well obtained 3810 cubic meters of gas. Penglaiba formation can be used as the alternative target for exploration and exploration. The Cambrian platform margin belt runs through the north-south Gucheng-Lunnan of the Tarim Basin, with a length of 560km, a width of 15-90km, and an area of 23380km2. The platform margin belt in the Gucheng area has a favorable exploration area of 1,400km2, with large-scale reef-shoal traps developed. The biological reef shoals in the high-energy facies belt in the platform margin can form more favorable reservoirs, which are close to the oil source. This is conducive to the formation of large-scale lithologic oil and gas reservoirs. The Cambrian platform margin zone encountered in TJ11 well is a collapsed front facies, with fracture-cavity reservoirs and breccia-like asphalt. The main reason for the failure was that it was close to the southern fault, and the later preservation conditions were not good, and the oil reservoir was destroyed; to the south, it was severely affected by the Cherchen fault. The Cambrian platform margin belt in the north developed four stages of reef-banks as shown on the seismic profile, extending from the platform margin to the east, away from the Cherchen fault on the south margin [20], the conditions for oil and gas accumulation are favorable. The Cambrian dolomitized high-energy reef shoal reservoir of Ct 1 well has resulted in good oil and gas show, and the highest total hydrocarbon gas surveyed is 83.7%. The northern platform margin belt in the Gucheng area is the main target for future Cambrian platform margin exploration.

4 Conclusion

(1) The Lower Paleozoic strata in the Gucheng area are composed of Carboniferous, Ordovician and Cambrian from top to bottom, and Silurian, Devonian and Permian are missing. The Lower Ordovician-Cambrian is the main exploration target of this area.
(2) There are four sets of reservoirs in Gucheng Low Uplift, that is, Ordovician Yijianfang Formation, Yingshan Formation, Penglaiba Formation and Cambrian. The upper part of Yijianfang Formation and Yingshan Formation are limestone reservoirs, which are a set of potential reservoirs. There are three sets of dolomite reservoirs developed in the lower part of Yingshan Formation, Penglaiba Formation and Cambrian.
(3) The oil and gas distribution in the Gucheng area is mainly found in the Ordovician Yijianfang Formation, Yingshan Formation, Penglaiba Formation and Cambrian. Yingshan Formation and Cambrian are the main exploration targets, and Ordovician Yijianfang Formation The Penglaiba Formation can be used as an alternative exploration target.
(4) The Lower Paleozoic in the Gucheng area can form two realistic exploration areas: Ordovician dolomite reservoirs and Cambrian reef banks. The Ordovician dolomite reservoirs are mainly looking for intra-platform beaches in the platform margin high-energy facies belt, and the Cambrian is mainly looking for high-energy reefs and beach bodies away from the platform margin facies belt in the northern part of the Cherchen fault.

References

1. Zhu Guangyou, Yang Haijun, Zhu Yongfeng, et al. Study on hydrocarbon geological characteristics and accumulation of carbonate rocks in Hatang area of Tarim Basin [J]. Acta petrologica sinica, 27 (3): 827-842 (2011).
2. Zhang Shuichang, Zhang Baomin, Li Benliang, et al. History of oil and gas accumulation in marine basins across major tectonic periods in China: taking Tarim Basin as an example [J]. Petroleum Exploration and Development, 38 (1): 1-15 (2011).
3. Luo chunshu, Yang Haijun, Li jianghai, et al. Characteristics of Ordovician high-quality reservoirs and fault control in central Tarim basin [J]. Petroleum exploration and development, 38 (6): 716-724 (2011).
4. Zhao Wenzhi, Shen Anjiang, Hu Suyun, et al. Geological conditions and distribution characteristics of large-scale development of carbonate reservoirs in China [J]. Petroleum Exploration and Development, 39 (1): 1-12 (2012).
5. Zhu Guangyou, Zhang Shuichang. Conditions of deep oil and gas accumulation and exploration potential in China [J]. Acta petrolei sinica, 30 (6): 793-800 (2012).
6. Zheng Herong, Wu Mao, Wu Xingwei, et al. Oil and gas exploration prospect of lower Paleozoic dolomite reservoir in Tarim Basin [J]. Acta petrolei sinica, 28 (2): 1-8 (2007).
7. Tian Lei, Cui Haifeng, Liu Jun, et al. [J]. Paleogeography and Sedimentary Evolution of Early and Middle Cambrian in Tarim Basin [J]. Petroleum and Natural Gas Geology, 39 (5): 1011-1021 (2018).
8. Wang Zhaoming, Yang Haijun, Qi Yingmin, et al. Major breakthroughs in Ordovician natural gas exploration in Gucheng area of Tarim Basin and its enlightenment [J]. Natural Gas Industry, 34 (1): 1-9 (2014).
9. Liu Yanting, Fu Heng, Kelvin Chen, et al. Sequence stratigraphic characteristics of Cambrian in Bachu-Tazhong area of Tarim Basin [J]. Lithologic Reservoir, 22 (2): 48-53 (2002).
10. Yang Haijun, Han Jianfa, Chen Lixin, et al. Characteristics and models of hydrocarbon accumulation of marine carbonate rocks in Tazhong Middle Uplift. Petroleum and Natural Gas Geology, 28(6):784-790 (2007).
11. Xiao Zhongyao, Lu Yuhong, Sang Hong, et al. A typical CAMBRIAN reservoir: genetic analysis of well Tazhong 62 in Tarim Basin [J]. Geochemistry, 34 (2): 155-160 (2005).
12. Hua Xiaoli, Lu Xiuxiang, Huaqiao. Distribution law and reservoir forming conditions of fractured-vuggy carbonate rocks: a case study of Yingshan Formation in the east of Tazhong North Slope [J]. Journal of Modern Geology, 27 (2): 346-355 (2013).
13. Zhou Xinyuan, Wang Zhaoming, Yang Haijun, et al. The fifth exploration example of marine oil and gas fields in China-Exploration and discovery of Ordovician large condensate gas fields in Tazhong [J]. Marine Oil and Gas Geology, 11 (1): 45-51 (2006).
14. Yang jun, Jiang zhenxue, Xiang caifu, et al. hydrocarbon characteristics and genetic mechanism

 https://doi.org/10.1051/e3sconf/202132901057

of carbonate rocks in Tazhong 83- Tazhong 16 well area of Tarim basin [j]. petroleum and natural gas geology, 33 (1): 101-109 (2012).

15. Li Minggang, Pang Xiongting, Ma Zhongzhen, et al. Analysis of main controlling factors of hydrocarbon accumulation in Wuerxun-Beier Sag [J]. Daqing Petroleum Geology and Development, 26 (5): 1-4 (2007).
16. Han Jianfa, Zhang Haizu, Yu Hongfeng, et al. Reservoir-forming characteristics and exploration of marine carbonate large condensate gas field in Tazhong uplift [J]. Acta Petrologica Sinica, 28 (3): 769-780 (2012).
17. Liu keqi, Jin Zhijun, Lu xiuxiang, et al. hydrocarbon accumulation of Ordovician carbonate rocks in Tazhong low uplift of Tarim basin [J]. Petroleum Experimental Geology, 26 (6): 531-536 (2004).
18. Yi Shiwei, Du Jinhu, Yang Haijun, et al. Controlling Factors of Lower Paleozoic Reservoir Formation in Tarim Basin and Exploration Ideas [J]. China Petroleum Exploration, 3 (1): 1-7 (2012).
19. Qian Ling, Li Jianghai, Lu Yuhong, et al. Difference analysis of oil and gas accumulation between Gucheng 4 well and Luoxi 1 well in Tadong area [J]. Natural Gas Geoscience, 22 (4): 628-634 (2012).

Study of seismic attributes recognition method of high quality igneous rock reservoir

Wang Yitong[1,2], Li Jingsheng[1,2], Sam Zandong Sun[3], Qiao Wei[1,2], Li Yanjie[1,2], Ma Yongyi[1,2], Shen Yan[1,2] and Wang Shuheng[1,2]

[1]Exploration and Development Research Institute of Daqing Oilfield Co., Ltd., 163712 Daqing Heilongjiang, China
[2]Key Laboratory of tight oil and shale oil accumulation of Heilongjiang province, 163712 Daqing Heilongjiang, China
[3]BGP, 052751 Zhuozhou Hebei, China

Abstract. Aimed target area is deeply buried, complex lithology, dual media, reservoir development degree is controlled by a variety of factors, meanwhile, lateral thickness and lithofacies change rapidly, and strata formation is poor. Therefore, igneous rock reservoir has difficulty in predicting, since seismic is complicated to track trace, reservoir attribute analysis is hard to determine the time window, and inversion modeling requires sophisticated. By analyzing, the basalt in the target research area accounts for the principal component of the igneous rock, however, the igneous rocks with relatively developed reservoirs are mostly distributed in the trachyte breccia which has good productivity. The results of petrophysical study indicate that frequency-dependent AVO inversion method is an important means to identify fluid and reservoir prediction, notwithstanding it is difficult to distinguish high-quality reservoirs barely by P-wave impedance. Consequently, AVOF inversion method is appropriately proposed to identify igneous rock reservoir. Foremost, eliminating the effects of algorithm,frequency, spectrum balancing and other factors, then put the improved three-term Aki&Richards frequency-dependent AVO inversion method applying to distinguish igneous reservoir fluid and lithology, for the purpose of carrying out the identification of high-quality reservoirs.

1 Introduction

Mason and Mcskinin first pointed out the relationship between wavelength attenuation and frequency by measuring the acoustic attenuation of some metals and glasses and the relationship between frequency and particle size [1]. Toulis and Shumway measured the acoustic velocity and attenuation of sedimentary rocks by resonance method, which was an earlier experimental measurement to study the relationship between frequency, velocity and attenuation when waves propagate in sedimentary rocks [2]. Chapman proposed that the seismic wave reflection coefficient changes with frequency and has a great relationship with the type of AVO [3]. Wilson established a practical method for calculating the dispersion properties [4]. Wu Xiaoyang calculated seismic dispersion values from actual data and achieved good results [5]. An improved algorithm by employing Aki &Richards approximation to obtain more accurate inverted P-wave dispersion was presented and analyzed multiple technical steps affecting dispersion, such as amplitude-preserved processing, high-resolution time-frequency analysis technology, frequency components and reference frequency [6,7]. Both physical model test and real seismic example indicating that the magnitude of P-wave dispersion among three typical fluids (i.e. gas, oil and water) are obviously different, concretely, strong, moderate, weak for gas, oil and water respectively [8].

Conventional AVO inversion, ignoring the dispersion phenomenon of fluid, cannot be accurately distinguished from fluid type and lithology. In spite of post-stack methods such as spectral decomposition can be used to point hydrocarbon in seismic data, these type of methods still have many limitations. For instance, incorrect superposition may produce false "low frequency anomalies" and suppress high frequency information; or introducing large offset information that is stretched by dynamic correction during the superposition process, thereby weakening useful information and making the superposition profile, which will hide low frequency information. Additionally, due to the presence of seismic resolution processed by post-stack methods occasionally cannot meet the requirements of reservoir prediction. For this reason, exploring more effective fluid identification and detection methods is indispensable and significant. Considering that the fluid in the reservoir causes the seismic wave dispersion effect, we propose a pre-stack frequency-dependent AVO inversion method based on the more accurate three-term formula Aki & Richards algorithm with variational Vs/Vp obtained from pre-stack AVO inversion to substitute traditional Smith & Gidlow two-term inversion approximate formula with constant Vs/Vp(0.5). The modified method expands the

* Corresponding author: wangyitong_sj@foxmail.com

approximate formula of the Zoeppritz equation based on the pre-stack AVO inversion, introduces the frequency term, and solves the dispersion value. The experiments and field data massively demonstrate that the improved frequency-dependent AVO inversion method provides more reliable P-wave dispersion gradient, can more effectively indicate fluid types and cumulative production and suppress non-reservoir energy, paramountly, it can achieve the role of identifying high-quality reservoirs.

2 Workflow

The proposed workflow is based on the intergration of different angle stack seismic data and is composed of four processing stages under the supervision of Chapman rock physic theory.

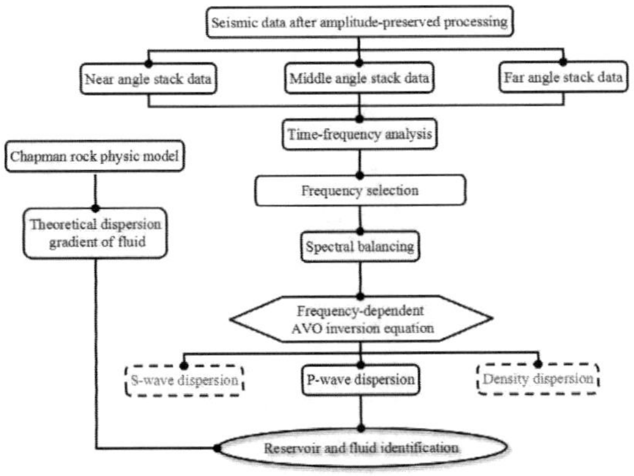

Fig. 1. Workflow of frequency-dependent AVO inversion.

3 Affecting factors

3.1 Factor of model

Three types of frequency-dependent AVO inversion algorithms are selected to further explore the influence of frequency dispersion results. Well 19 and Well 16 are oil wells with medium yield, and Well 3 and Well 1 are water wells. Compared figure 2(a) with figure 2(b) and figure 2(c), it is obvious that P-wave dispersion gradient values at the two oil wells is higher, and the dispersion response is much stronger, which cannot meet the actual production, as well as poorer suppression of non-reservoir, especially near the water well. The non-reservoir in figure 2(c) has a better suppression effect and fluid dispersion indicator can describe the wells' production more elaborately by comparison with figuir 2(b).

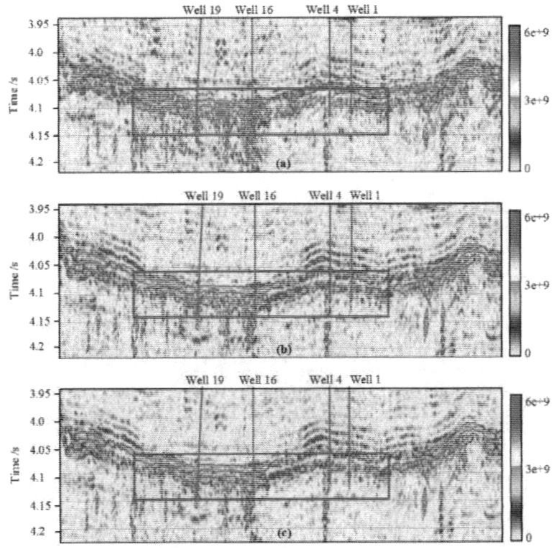

frequency-dependent AVO inversion algorithms
((a) Smith & Gidlow approximation with constant-Vs/Vp;
(b) Aki & Richards approximation with constant-Vs/Vp;
(c) Aki & Richards approximation with variant-Vs/Vp)

Fig. 2 P-wave dispersion gradient sections comparison by different

Figure 3 shows the RMS P-wave dispersion values of three frequency-dependent AVO inversion algorithms. Wells of circles represent P-wave dispersion gradient is inconsistent with oil yield. Table 1 shows the daily oil production of 30 wells,which comprised 25 oil wells including 17 high-yield oil wells, besides 5 water wells. Well 1 and well 24 are water wells, well 15, well 16 and well 20 are low-yield oil wells, as well as well 28 is high-yield oil wells. During the detection of fluid types, through different types of inversion algorithms, an image of P-wave dispersion is obtained from various approximate Zoeppritz formulas. It is illustrated that whether water or oil wells display distinct discrepancy between dispersion response and fluid type in figure3(a) comparison of the three obtained results. Despite P-wave dispersion gradient in figure 3(c) does not efficaciously detect Well 20 and Well 24 influenced by faults, in contrast with figure 3(b), the fluid characterization conformance is more accurate about low-yield well 15 and well 16.

 https://doi.org/10.1051/e3sconf/202132901058

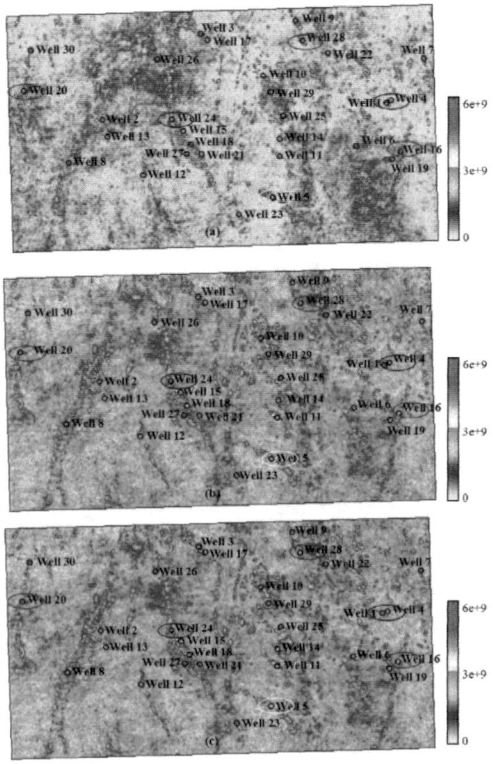

frequency-dependent AVO inversion algorithms
((a) Smith & Gidlow 2-iterm approximation with constant-Vs/Vp;
(b) Aki & Richards 3-iterm approximation with constant-Vs/Vp;
(c) Aki & Richards3-iterm approximation with variant-Vs/Vp).

Fig. 3. P-wave dispersion gradient attributes comparison by different

Table 1. Well production statistics.

Well name	Production (t/day)	Well name	Production (t/day)
Well 1	0.91	Well 16	10.58
Well 2	1.17	Well 17	10.54
Well 3	0.72	Well 18	10.85
Well 4	0.51	Well 19	11.09
Well 5	1.13	Well 20	7.13
Well 6	1.75	Well 21	6.38
Well 7	1.66	Well 22	5.40
Well 8	4.91	Well 23	17.03
Well 9	5.09	Well 24	19.42
Well 10	5.99	Well 25	28.74
Well 11	7.15	Well 26	22.94
Well 12	8.52	Well 27	24.27
Well 13	8.98	Well 28	28.43
Well 14	9.79	Well 29	28.31
Well 15	10.03	Well 30	34.14

*production of 1mm nozzle

Analyzing the coincidence rate statistically between P-wave dispersion gradient and the well production, the agreement applying Smith & Gidlow approximate formula with constant value Vs/Vp is 66.7%, the accordance introducing Aki & Richards approximate formula with constant value Vs/Vp is 76.7%, moreover, the rate of Aki & Richards approximate formula with variable value Vs/Vp is 86.7%, which prominently demonstrate that frequency-dependent AVO inversion model will affect the P-wave dispersion gradient results. Generally speaking, the P-wave dispersion gradient calculated based on the Aki & Richards three-term approximation formula with varying Vs/Vp can reflect the fluid type more veritably and effectively.

3.2 Spectral balancing

For a certain seismic data volume s(t,n), there are t sampling points on each seismic trace, and time-frequency transform method is performed to obtain a series of single-frequency data volume S(t,n,f)

$$s(t,n) \leftrightarrow S(t,n,f) \qquad (1)$$

For the purpose of eliminating the energy difference of the seismic amplitude spectrum corresponding to the different frequency, it is fundamentally to allow energy "compensation" for each single-frequency amplitude spectrum. This compensation process is called spectrum equalization.

$$S_b(t,n,f) = S(t,n,f)\omega(n,f) \qquad (2)$$

In the above formula, $\omega(n,f)$is weighting factor responding to relative single-frequency amplitude spectrum data volumes:

$$\omega(n,f) = \frac{\max(A_{ref}(n))}{\max(A_f(n))} \qquad (3)$$

At the position of channel n, $\max(A_{ref}(n))$ refers to the maximum amplitude value of reference frequency f_0, and $\max(A_f(n))$ refers to the maximum amplitude value of the single frequency f.

The weight function is acquired through an appropriate elastic model. According to Gassmann's theory, the reflection coefficient obtained from the elastic interface is independent of frequency in the seismic frequency band. For dispersion model, there is no such assumption. (The weight function under the elastic model is used to spectraLly equalize the amplitude spectrum of the dispersion model.) That is, in field case, the elastic interface can be regard as a reference to equalize the amplitude spectrum of the dispersive medium. In practical applications, selecting a continuous strong lithological interface (theoretically assumed to be an elastic interface) as the reference interface to perform spectral equalization in the frequency-dependent AVO inversion, and the sampling points of each channel of the frequency division data volume by time-frequency transform method are weighed to obviate the energy discrepancy of the frequency-divided data itself, thereby highlighting the true difference of geological frequency on the seismic signal.

Figure 4 shows the seismic profiles and the P-wave dispersion gradient profiles without and with spectral equalization of the connected wells. The blue circle represents water well and the remaining three are oil wells. The P-wave dispersion gradient profile processed by spectrum equalization in figure 4(c) is more consistent with the well type. The obtained results show more details that seismic amplitude and P-wave dispersion gradient without spectral balancing process neither can distinguish oil wells from water wells, and has poorer suppression effect on non-reservoir, which is not conducive to subsequent fluid prediction. Moreover, this is consistent with theoretical analysis that it is prerequisite to perform spectrum equalization in the frequency-dependent AVO inversion process.

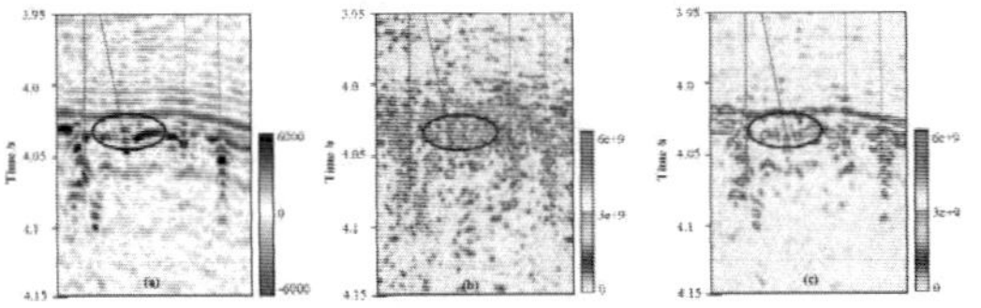

((a) seismic amplitude ; (b) P-wave dispersion gradient with spectral balancing; (c) P-wave dispersion gradient without spectral balancing).

Fig. 4. Comparison of sections

The results show that the diversity of P-wave dispersion gradient indication area in the two plans is significantly (Fig. 5). On the basis of the well production information, the high P-wave dispersion in figure 5(b) has better consistency of the oil and gas production such as Well 18, Well 21, Well 28 and Well 27, additionally, it has better recognition effect for water wells such as well 2. Not only the suppression effect of non-reservoir is better, but the trend of the high frequency dispersion is similar to that of the fault. It demonstrates that the spectrum equalization processing is validated to eliminate the interference of the frequency-divided data itself to a certain extent, thus highlighting the real seismic frequency response characteristics under geological conditions. Practical applications show that spectrum equalization processing is performed has a great influence on the frequency-dependent AVO inversion result. The P-wave dispersion gradient value executed by spectral equalization process can more accurately identify the fluid type.

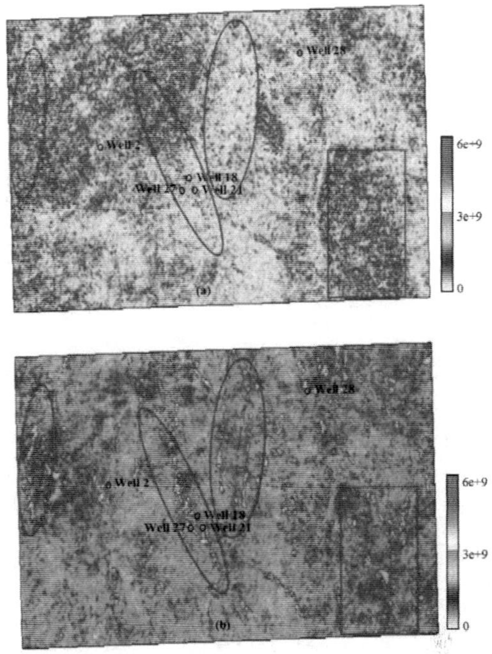

((a) spectral balancing; (b) without spectral balancing)

Fig. 5. Comparison of P-wave dispersion RMS attributes extracted 50ms down along Yijianfang group (To3t)

4 Application

4.1 Fluid identification

Figure 6 shows P-wave dispersion gradient of six wells and circles are for DST. It describes that P-wave dispersion gradient is comparatively stronger at Well 33 (gas) contained less water, whereas the strength at Well 35 (water) is weaker. Well 31 and Well 34 are both oil well,with similar porosity and saturation, manifesting similar degree of P-wave dispersion gradient, which is slightly stronger than Well 32 (oil and water). Application effects bring into correspondence with theoretical theory of fluid dispersion response. Furthermore, the frequency-dependent AVO inversion algorithm has a markedly higher degree of fluid recognition and significantly better application effect in igneous rock reservoirs.

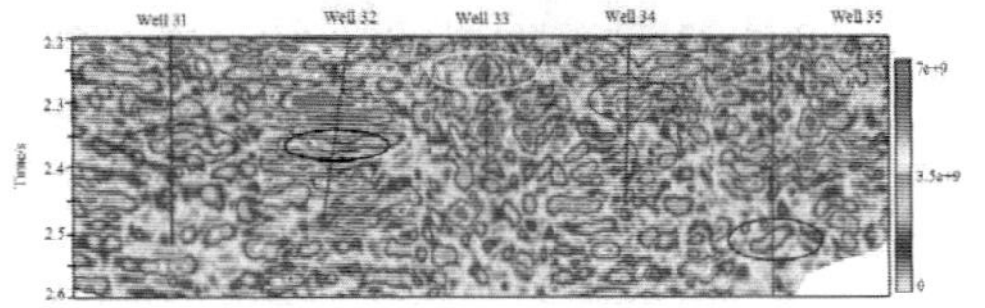

Fig. 6. Sections of inverted P-wave dispersion.

https://doi.org/10.1051/e3sconf/202132901058

4.2 Lithology identification

The lithology in the red box of figure 7 is explosive facies trachyte and trachyte breccia lava, which is the high-quality lithology sought in the study area. It is obvious that higher P-wave dispersion gradient value is displayed in well 36 and well 37, while well 38 and well 39 in figure 8 correspond to lower dispersion gradient value where the lithology is overflow facies basalt and explosive facies basalt, which is consistent with the display of root mean square attribute values on the plane (Fig. 9).

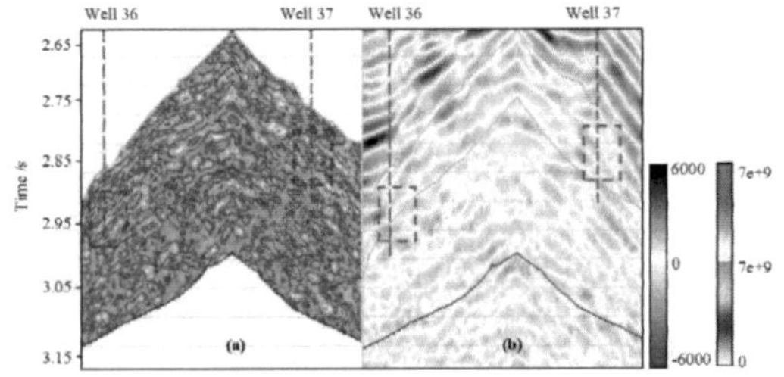

((a) P-wave dispersion gradient; (b) seismic amplitude)

Fig. 7. Comparison of sections

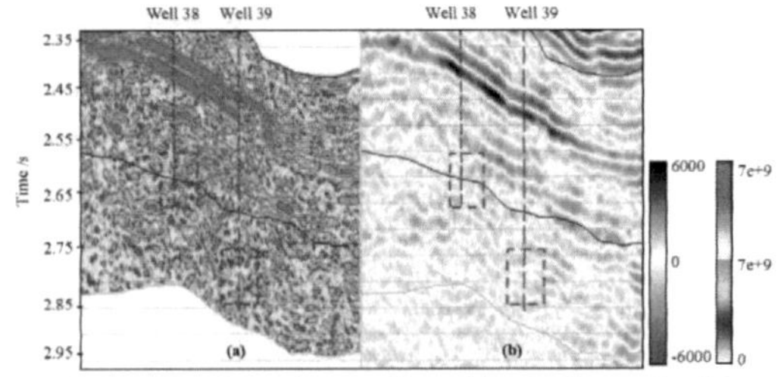

((a) P-wave dispersion gradient; (b) seismic amplitude)

Fig. 8. Comparison of sections

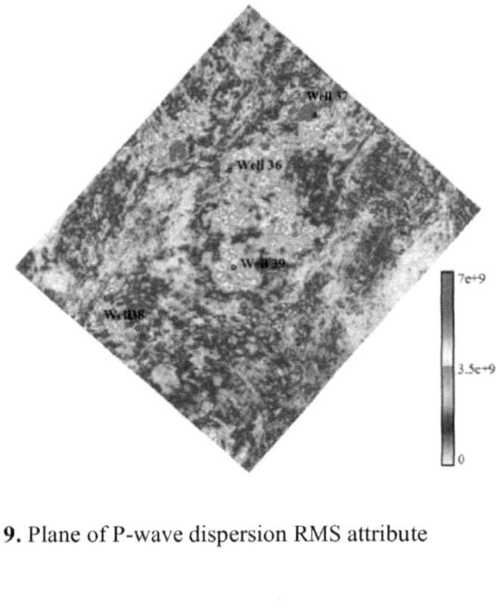

Fig. 9. Plane of P-wave dispersion RMS attribute

5 Conclusion

Based on the research of the frequency-dependent AVO inversion method, we primarily analyze the factors model and spectra balancing, hoping that the method can truly and effectively reflect the fluid type and identify high-quality reservoirs. Then applied the improved method to igneous reservoirs. The result describes that the method can distinguish fluid types and the inverted P-wave velocity dispersion gradient can have a better recognition effect with high-quality reservoir lithology. In conclusion, the high recognition effect indicates that the frequency-dependent AVO inversion method has high accuracy for fluid detection and high-quality reservoir identification, and improves the understanding of fluid detection and identification of complex igneous reservoirs.

References

1. W.P. Mason, H.J. McSkimin, Attenuation and scattering of high frequency sound waves in metals and glasses. The Journal of the Acoustical Society of America, 19, 3 (1947).
2. W.J. Toulis, Theory of a resonance method to measure the acoustic properties of sediments. Geophysics, 21, 2 (1956).
3. M. Chapman, Liu E., Li X.Y., The influence of fluid-sensitive dispersion and attenuation on AVO analysis. Geophysical Journal International, 167, 1 (2006)
4. A. Wilson, M. Chapman, Li X.Y., Frequency-dependent AVO inversion. SEG Annual Meeting, 79 (2009)
5. Wu X.Y., Reaserch on dispersion AVO inversion based on spectrum analysis technology. (Doctoral Dissertation, China University of Geosciences (Wuhan), Wuhan, China, 2010)
6. Sun S.Z., Yue H.Y., Hu L.G., Zhang Y.Y., Du. T., An Improved Frequency-dependent AVO Inversion Algorithm for Fluid Detection. SEG Annual Meeting, 84 (2014).
7. Du T., Sun S.Z., Zhang Y., Yue H.Y., Li C., Liu Z., A Study on Influence of Frequency selection to Frequency-dependent AVO Inversion. EAGE Conference and Exhibition, 76 (2014) .
8. Sun S.Z., Yue X.T., Liu L.F., Du T., Karst Carbonate Reservoir Identification Using Frequency-dependent AVO inversion in Tarim Basin. EAGE Conference and Exhibition, 78 (2016).

https://doi.org/10.1051/e3sconf/202132901059

Study on the boundary of development adjustment policy of B Oilfield

Simian Sun

Qing xin oil development limited liability company of Daqing Oilfield Company Ltd, Daqing 163000, China

Abstract. After more than ten years of water injection development, the water cut in the main block of oilfield B has increased rapidly and the production has decreased significantly. In view of the rapid rise of comprehensive water cut in the current oilfield, it is necessary to carry out the research on the development law of oilfield B and the reasonable boundary of various development adjustment indexes, so as to clarify the different production stages The development status and development index change law of the oilfield under different oil production rates provide an accurate and reliable theoretical basis for the rational development and adjustment of the oilfield. This paper provides theoretical basis and effective technical support for the adjustment in the middle and later stage of B oilfield development, which is of great significance to the stable production and long-term sustainable development of the oilfield.

1. Study on reasonable limits of oilfield development indexes

1.1 Reasonable ratio of oil-water wells

The production injection index ratio method is used to calculate the reasonable oil-water well number ratio. The reasonable oil-water well number ratio mainly depends on the ratio of water absorption index and liquid production index.

$$R_h = \sqrt{\frac{1}{B_o(1-f_w)+f_w}\frac{I_w}{J_L}} \quad (1)$$

In the formula:
B_o—Crude oil volume factor, m^3/m^3,1.075
f_w— water cut
I_w,J_L—water absorption index and liquid production index respectively ,t/d.MPa
The water absorption index and liquid production index of the oilfield over the years are counted, and the relationship between water absorption index, liquid production index and water cut is regressed. The water absorption index and liquid production index under different water cut are calculated by using the regressed equation, and then the reasonable oil-water well number ratio under different water cut conditions is obtained See (Figure 1 , Figure 2and Table 1).

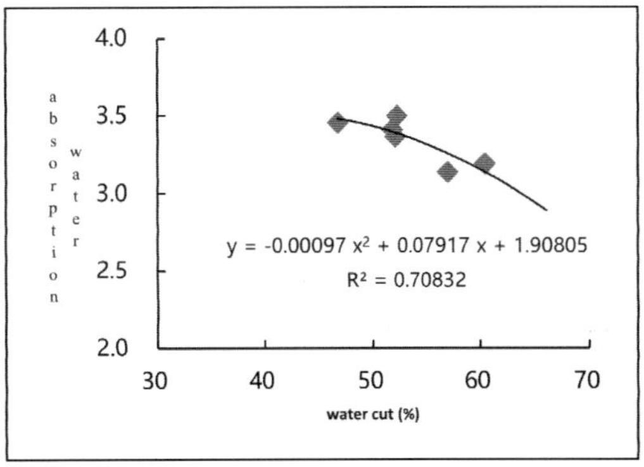

Fig. 1. Relationship between water absorption index and water content

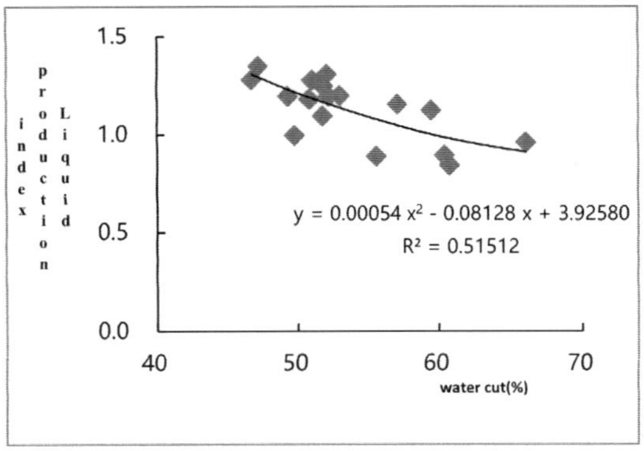

Fig. 2 Relationship between liquid production index and water cut

Table 1. Ratio of oil-water wells under different water cut conditions

Water cut (%)	Water absorption index (m^3/MPa.d)	Liquid production index(t/MPa.d)	Oil water well number ratio
60	3.166	0.993	1.8
70	2.697	0.882	1.7
80	2.034	0.879	1.5
90	1.176	0.985	1.1

According to the calculation, under the current water cut conditions, the reasonable oil-water well number ratio is 1.7, and the actual oil-water well number ratio of B oilfield is 2.33, so the injection production system still needs to be further optimized[1-5].

1.2Reasonable injection production ratio

Whether the injection production is balanced can be reflected in the change of formation pressure. According to the change relationship between injection production ratio and formation pressure, a reasonable injection production ratio can be determined.

$$\frac{dP_R}{dt} = a + b * IPR_h \tag{2}$$

In the formula:

dP_R/dt—Formation pressure variation,Mpa ;

a、b—Regression coefficient of equation.

Based on the actual data, the linear relationship between injection production ratio and total differential pressure change can be established, and the reasonable injection production ratio limit under different pressure changes can be obtained (Table 2). At present, the formation pressure of oilfield B is relatively low. According to the annual recovery of 0.3MPa, the injection production ratio is 1.6, which is basically consistent with the actual situation (Fig. 3).

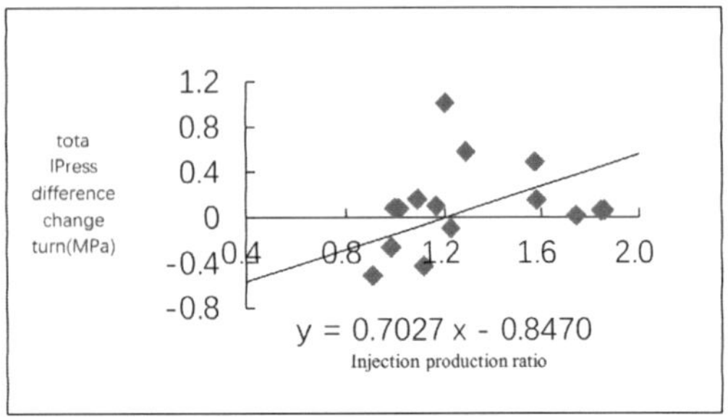

Fig. 3 Relationship between injection production ratio and total differential pressure

Table 2 Injection production ratio limit under different pressure recovery

Annual pressure change(MPa)	-0.2	-0.1	0	0.1	0.2	0.3	0.4	0.5
Reasonable injection production ratio	0.9	1.1	1.2	1.3	1.5	1.6	1.8	1.9

1.3 Reasonable production rate

Oil production rate has an important impact on oilfield development effect, especially for bottom water or conglomerate reservoirs, too high oil production rate will cause rapid rise of water cut and sudden water flooding. Based on the relative permeability curve and water cut diversion equation, the relationship between oil production rate and water cut and water cut rise rate can be established. The expression of underground volume moisture content is:

$$f_w = \frac{1}{1+\frac{\mu_w k_{ro}}{\mu_o k_{rw}}} \tag{3}$$

In the formula:

μ_o , μ_w oil phase and water phase viscosity respectively,mPa.s ;

k_{ro} , k_{rw} Relative permeability of oil phase and water phase ;

According to relevant studies:

$$\frac{k_{ro}}{k_{rw}} = a \cdot e^{-b \cdot S_w} \tag{4}$$

The recovery degree can be expressed as:

$$R = \frac{S_w - S_{wi}}{1 - S_{wi}} \tag{5}$$

From the above formulas:

$$R = A \cdot \ln\frac{1-f_w}{f_w} + B \tag{6}$$

$$A = -\frac{1}{b(1-S_{wi})} \tag{7}$$

$$B = -\frac{1}{b(1-S_{wi})} \ln\frac{\mu_o}{\mu_w a} - \frac{S_{wi}}{1-S_{wi}} \tag{8}$$

Take the derivative of time t at both ends of the above formula and take DT = 1a to obtain the calculation formula of oil production rate :

$$v_o = \frac{1}{b(1-S_{wi})} \cdot \frac{1-f_w}{f_w} \cdot F_w \tag{9}$$

In the formula:

v_o is the production rate, decimal

F_w is annual appreciation,%。

In order to obtain the relationship curve between theoretical oil production rate and water cut of oilfield B, first fit and regress the relative permeability curve of the block to obtain B = 28.68578, and then obtain the relationship curve between oil production rate and water cut when the water cut of oilfield B rises in different years. It can be seen that when the water cut is the same, the higher the annual water cut appreciation, the greater the oil production rate; The effect of water cut on oil production rate is negative correlation when the water cut is less than 50%, and then positive correlation. According to the statistics of the actual production data of oilfield B, the annual water cut appreciation in 2015 was 4.5, the

 https://doi.org/10.1051/e3sconf/202132901059

comprehensive water cut of the block was 66.7%, the corresponding theoretical oil production rate was 1.02, and the actual oil production rate of the oilfield was 1.04, slightly higher than the theoretical value.

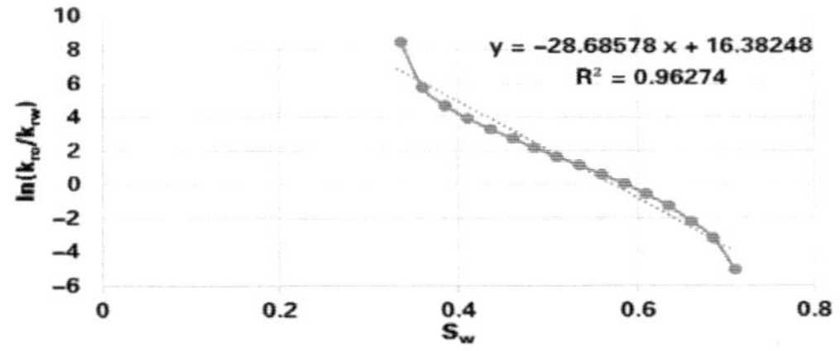

Fig.4 Relationship between natural logarithm of oil-water relative permeability ratio and water saturation

2 Other indicators

2.1 Reasonable water injection intensity

According to the principle of injection production balance, considering the reasonable injection production ratio when maintaining formation pressure, IPR:

$$Q_i = IPR * Q_L \tag{10}$$

In the formula:

Qi is the annual cumulative water injection of the block,m^3;

Q_L the annual liquid production of the block,m^3;

IPRis the reasonable injection production ratio of the block。

According to the definition of water injection intensity, there are:

$$Q_i = 365\alpha q_i H_i \tag{11}$$

In the formula:

α Is the production hour rate of water injection well, decimal;

q_i Is the average water injection intensity of the block,m^3/d.m ;

H_i is the total thickness of water injection wells in the block,m

Production rate in block VO and annual average moisture content f_wWhen w is certain, the annual liquid output can be expressed as:

$$Q_L = \mathrm{N_p}\frac{B_o V_o}{\rho_o(1-f_w)} \tag{12}$$

In the formula:

$\mathrm{N_p}$ Is the original geological reserves,t ;

B_o Is the volume factor of crude oil, decimal ;

V_oIs the production rate, decimal;

ρ_o Is the density of crude oil,kg/m^3

According to the above formulas, the water injection intensity at a given reasonable oil production rate is:

$$q_i = IPR * \mathrm{N_p}\frac{B_o V_o}{365\alpha\rho_o H_i(1-f_w)} \tag{13}$$

According to the above research, the reasonable injection production ratio of B oilfield is 1.6, the reasonable oil production rate is 1.02, and the original geological reserves are 1913.8 × 10^{4}t, crude oil volume coefficient is 1.075, crude oil density is 0.868, block water cut is 66.7%, the total perforation thickness of water injection well is 718.1m, and the production rate of water injection well is 0.733. By substituting the above formula, the reasonable water injection intensity is 3.94m^3 / D.M. Statistics show that the average water injection intensity of oilfield B in 2015 is 4.38m^3 / D.M, slightly higher than the reasonable value. The water injection intensity should be appropriately reduced according to the reasonable oil production rate and reasonable injection production ratio.

2.2 Reasonable decline rate

Natural decline rate and comprehensive decline rate are important indicators to reflect oilfield development. Natural decline refers to the production decline of old wells without measures in the previous year; Comprehensive decline refers to the production decline after the measures of old wells last year.

Namely:

$$D_n = \frac{Q_{i-1} - (Q_i - Q_{xj} - Q_{cs})}{Q_{i-1}} = 1 - \frac{Q_i - Q_{xj} - Q_{cs}}{Q_{i-1}} \tag{14}$$

$$D_R = \frac{Q_{i-1} - (Q_i - Q_{xj})}{Q_{i-1}} = 1 - \frac{Q_i - Q_{xj}}{Q_{i-1}} \tag{15}$$

In the formula:

D_n—Natural decline rate of annual production of Oilfield Block,% ;

D_R—Comprehensive decline rate of annual production of Oilfield Block,% ;

Q_{i-1}, Q_i Annual oil production in year i-1 and year i respectively,10^{4}t/a ;

Q_{xj}—Production of new wells in the I year of the Oilfield Block,10^{4}t ;

Q_{cs}—Production measures of old wells in the I year of Oilfield Block ,10^{4}t;

In a certain development stage, when the proportion of new well production and old well measure oil production is basically unchanged every year, the natural decline rate and comprehensive decline rate have a linear relationship.

$$D_n = aD_R + b \tag{16}$$

Where a and b are constants.

Using the above formula, the natural decline and comprehensive decline of B oilfield since 2006 are calculated respectively (table 3). Considering that oilfield B has been in the stable production stage since the tenth five year plan, the water cut rises slowly and the development situation is in a good state, therefore, the comprehensive decline rate and natural decline rate under different water cut conditions are characterized by establishing the relationship between comprehensive water cut and comprehensive decline rate and the relationship between comprehensive decline rate and natural decline rate. It can be seen that the water content is between 40% - 60%, the comprehensive decline is 19-27%, and the natural decline is 21-28%; The water content is between 60-70%, the comprehensive decline is 14-19%, and the natural decline is 16-21%; The water content is between 70-80%, the comprehensive decline is 7-14%, and the natural decline is 10-16%.

Table 3 Natural decline and comprehensive decline of B oilfield over the years

time (year)	2006	2007	2008	2009	2010	2011	2012	2013	2014	2015
Natural decline rate(%)	30.2	24.0	29.6	24.9	17.9	20.0	19.5	26.6	19.6	18.2
Comprehensive decline rate (%)	29.7	22.4	26.7	24.5	16.4	18.5	17.7	24.4	16.2	15.9

2.3 Water content rise

Water cut rising rate is the rising value of water cut after 1% geological reserves are produced. It is an important index to evaluate the development effect of oil field. The relative permeability curve is used to draw the relationship curve between water cut and water cut rise under different permeability, which can intuitively reflect the water cut change law in different water cut stages. From the curve, when the water content is 50%, the water content rise rate is the largest; The water content increases from 0 to 50%, and the water content increases with the increase of water content; The water content increases from 50% to 98%, and the water content decreases with the increase of water content. Under the same water cut conditions, the water cut increases with the increase of permeability. Therefore, the values of water content rise under different water content conditions are given (Table 4). The water content is between 40% - 60%, and the water content rise rate is 5.4% - 5.6%; The water content is between 60% - 70%, and the water content rise rate is 4.7% - 5.4%; The water content is 70% - 80%, and the water content rise rate is 3.6% - 4.7%. At present, the water cut of B oilfield is 66.7%, and the water cut increase rate is 4.6%, which is relatively reasonable.

Table 4 Variation of water cut rise rate in different water cut stages (theoretical)

classification	Water content rise under different water content(%)									
	10%	20%	30%	40%	50%	60%	70%	80%	90%	98%
K=20MD	1.2	2.2	2.9	3.4	3.5	3.4	2.9	2.2	1.2	0.3
K=120MD	1.5	2.7	3.6	4.1	4.3	4.1	3.6	2.7	1.5	0.3
K=250MD	2.4	4.4	5.8	6.6	6.8	6.6	5.8	4.4	2.4	0.6
K=500MD	2.6	4.6	6.1	7	7.3	7	6.1	4.6	2.6	0.6
normalization	2	3.6	4.7	5.4	5.6	5.4	4.7	3.6	2	0.5

3. Conclusions and recommendations

Through calculation, the policy limits of 7 important development indicators such as reasonable oil-water well number ratio, reasonable injection production ratio and reasonable flow pressure in B oilfield are obtained, which points out the direction for the subsequent optimization of water injection well adjustment and pressure system adjustment, improving the degree of reservoir production, and controlling the rate of production decline and water cut rise.

References

1. Du Qinglong, Zhu Lihong, Guo Junhui. A Quick-and-Novel Method in Evaluating the Producing Performance of Single-Layer in the Multilayered Sandstone Oilfield. SPE131482, 2010.
2. M.D. Dunn. Simulation Based Dimension less Type Curves for Predjcting waterfloo Reeovery. SPE68839, 2001.
3. Liu Jin-feng, Ma Li-jun, Mu Bo-Zhong, et al. The field pilot of microbial enhanced oil recovery in a high temperature petroleum reservoir, Journal of Petroleum Science and Engineering, 2005, 48:265-271.
4. Jang Long-Kuan, Chang P W, John E. Findley. Selection of bacteria with favorable transport properties through porous rock for the application of microbial-enhanced oil recovery[J]. Applied and Environmental Microbiology, 1983:1066-1072.
5. Cunningham A B, Sharp R R, Caccavo F, et al. Effects of starvation on bacterial transport through porous media[J]. Advances in Water Resources,2006.
6. Jiang Feng guang,Wang Xiao lin. Research on influence factors and theoretical value on recovery rate [J]. Journal of Southwest Petroleum University, 2013,6(5):104-108.
7. Xue Wei. SPSS application and analysis statistic [M]. Beijing: Publishing House of Electronics Industry,2004:242-243.
8. Tang Zi chang, Cheng Lin song, Li Chun lan. Application of partial correlation analysis to oil recovery rate influencing factor analysis [J]. Xinjiang Petroleum Geology,2010,31(4):423-425.
9. Jiang Han qiao, Yao Jun, Jiang Rui zhong. The fundament and practice of reservoir engineering [M]. .Dong ying, Shandong:China Petroleum University Press,2006:236.
10. Wang Peng. Research on reasonable production rate of low permeability injection reservoir [J]. Inner Mongolia Petrochemi-cal Industry,2012,6(3):139-141.
11. Yu Qi tai, Li Wen xing. Production stabilizing pattern for waterflolding reservoir [J]. Petroleum Geology & Oilfield Devel-opment in Daqing,1998,17(1):20-25.
12. 1CHEN Yuan qian, ZOU Cun you. Model's typical curve and its application for forecasting production and recoverable reserves of oilfields
13. WANG Xiao lin,CHANG Yu wen,DOU Hong en. The experiences from east Texas oilfield being of great significance for the large oilfield development in China[J]. Foreign Oilfield Engineering, 2010, 26(9): 15-19.

14. SUN Long de,WU Xiao lin,ZHOU Wan fu,et al. Technologies of enhancing oil recovery by chemical flooding in Daqing oilfield,NE China[J]. Petroleum Exploration and Development, 2018, 45 (4) :636-645.
15. ZHOU Yun, ZHOU Fu jian, FENG Lian yong. A new model for predicting oil and gas production [J]. Petroleum Geol-ogy and Oilfield Development in Daqing,2018,37(5):76-80.

Improvement and Application of Key Tools for Hydraulic Jet Fracturing

Yibo Zhang

Daqing Oilfield Production Engineering Research Institute, Daqing 163453, China

Abstract. Aiming at the problems of poor reliability, low construction efficiency, and short service life of fracturing string in the application of hydraulic jet technology in Daqing Oilfield, targeted research has been carried out on key tools such as rubber cylinder, slip, and nozzle.A unidirectionally-sealing rubber cylinder adopting composite sealing method was designed. The rubber cylinder could bear the pressure of 80 MPa and temperature resistance of 150℃ under the setting force of 20 kN, with tensile strength increased by 23% and residual deformation rate less than 3%, showing good overall sealing performance and recovery. The finite element method was used to optimize the structural parameters, materials and process of the slip to increase the ability of slips to withstand alternating loads by 1.5 times. The arc-shaped nozzle was improved to reduce the pressure drop and increase the jet density. By improving the performance indicators of key tools, the optimized fracturing string could meet the construction requirements of low setting force, high sealing pressure difference, and repeated setting and fracturing. The improved technical string was applied to section 3459 of well 218 well in the field, with the success rate reaching 100%. The large-scale application of the improved fracturing string satisfied the construction requirements for large-scale volume fracturing of the for the 5 1/2" inch casing new well completion in the low-permeability and tight reservoirs of Daqing Oilfield.

Keywords: Hydraulic jet; fracturing string; packer; nozzle; perforation.

1 Introduction

Hydraulic jet fracturing technology with coiled tubing is a new and efficient reservoir reconstruction technology for horizontal wells. It uses high-speed sand-carrying liquid jets to penetrate through the casing and cement ring to achieve combined operation of perforation and fracturing, so as to achieve the large-scale volume reconstruction of reservoirs [1-3]. This process has the advantages of short operation cycle, less times of lifting and lowering the string, and low construction cost. It is gradually becoming one of the main fracturing processes for horizontal well development wells in the periphery of Daqing Oilfield. However, there are still some problems to be solved in the process of field application. Specifically, it is difficult to control the setting and releasing of the packer, deformation and damage of the rubber cylinder, long perforation time of the spray gun, short perforation distance, and serious damage to the outer surface of the spray gun by erosion and backsplash abrasion [4]. Aiming at the above problems, this paper aims to optimize the hydraulic jet fracturing tools and improves the technical indicators and reliability of the tools.

2 Improvement of the structure of key tools for hydraulic jet fracturing

2.1 Improvement of the structure of the bottom seal rubber cylinder of the packer

Aiming at the problems that the traditional three-rubber cylinder and the compression-type single-rubber cylinder with a V-shaped angle structure at the end have the disadvantages of low pressure capacity, poor temperature resistance, releasing after repeated setting, and easy damage, an unidirectionally-sealing pressure-bearing rubber cylinder with pressure cap and tapered shoulder is designed. The pressure cap limits the pressure-bearing deformation of the rubber cylinder and improves the releasing performance of the rubber cylinder. The tapered shoulder reduces the setting force and improves the pressure-bearing performance. At the same time, we improve the high tensile strength rubber formula and its compatible additives, and research the secondary molding and vulcanization process of the rubber cylinder. Through multiple rounds of 150°C oil immersion experiments on the structure, the new structure has achieved the pressure bearing of 80 MPa, with tensile strength increased by 23%,

* Corresponding author: yibozhang@petrochina.com.cn

the residual deformation rate less than 3% and the fatigue repeated 80 times, under 20 kN setting force. The overall sealing performance and recovery of the rubber cylinder are good, and meet the needs of on-site volume fracturing construction.

Fig 1. Structure schematic diagram of packer

Fig 2. 3D diagram of rubber cylinder

2.2 Bottom seal slip structure of the packer and material optimization

Horizontal well completions mostly use high-hardness P110 steel-grade casing, which has higher requirements for the rationality of slip structure, material toughness and hardness. For the cases where slip's tooth tip is flattened or the slip body is broken, leading to the failure of covering the casing, the ANSYS finite element software is used to optimize the design of the slip structure. It can be seen from the axial stress distribution law of slip teeth that there is obvious stress concentration near the side of the slip. Combined with the results of numerical analysis, it is determined that the number of slip teeth is 7, the tooth width is 8 mm, the arc angle at the contact with the hoop is 2°, and the arc angle of the tooth bottom is 1°, which improves the stress state, reduces stress concentration, and ensures that the slip can fully anchor the casing. The ultra-high-strength and corrosion-resistant A-100 steel is adopted, the carburizing and laser treatment processes are used, so that the slip has high tensile strength, hardness and fatigue resistance, as well as high fracture toughness and ductility, with the ability to withstand alternating loads being increased by 1.5 times.

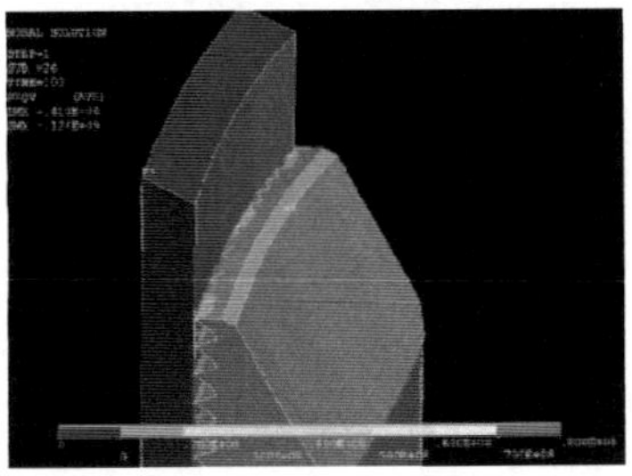

Fig 3. Mises stress distribution of 3D contact

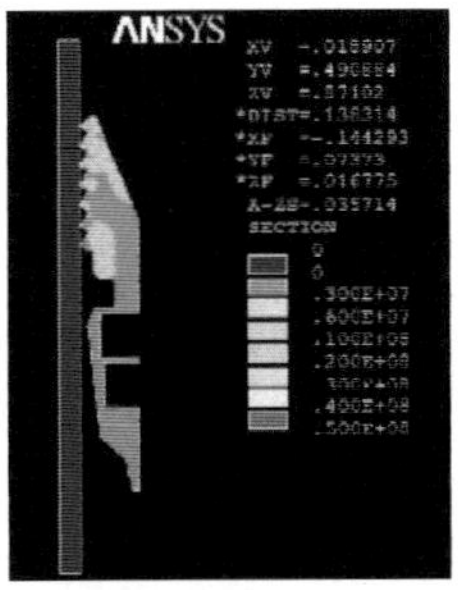

Fig 4. Mises stress distribution of axial contact

2.3 Improvement of the nozzle structure of built-in spray gun

As a key component of jet formation, the structural characteristics of the nozzle directly affect the quality of the jet. Aiming at the problems of long jet time, poor concentration, and large pressure loss under the conditions of high displacement and large sand fracturing, the single-phase flow model and multi-phase flow model are used to simulate the jet performance of the cone-shaped and curve-shaped nozzles, respectively. The single-phase flow model adopts the k-epsilon turbulence model and the simple algorithm, and uses the second-order upwind style to improve the calculation accuracy [5-6]. The multi-phase flow model adopts Euler model and couple coupling algorithm.

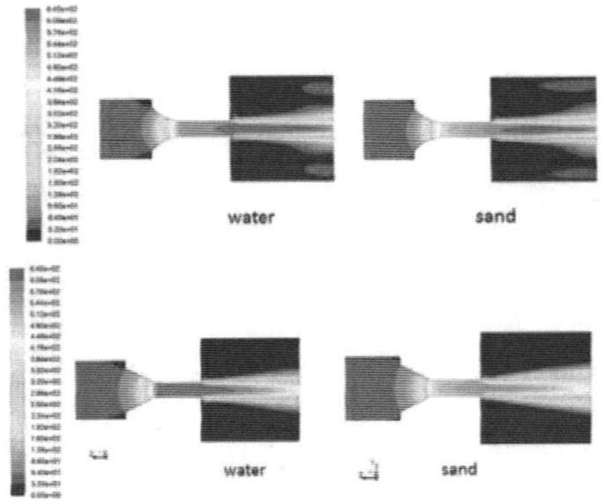

Fig 5. Contours of sand phase and water phase velocity

The simulation results of the two-phase flow model show that the jet density of the arc-shaped nozzle is significantly better than that of the cone-shaped nozzle. The orthogonal analysis of the outlet diameter, the length of the straight column and the radius of the inlet curve shows that the jet performance (jet impact force and density) of the arc-shaped nozzle is mainly affected by the outlet diameter, followed by the length of the straight column.

 https://doi.org/10.1051/e3sconf/202132901060

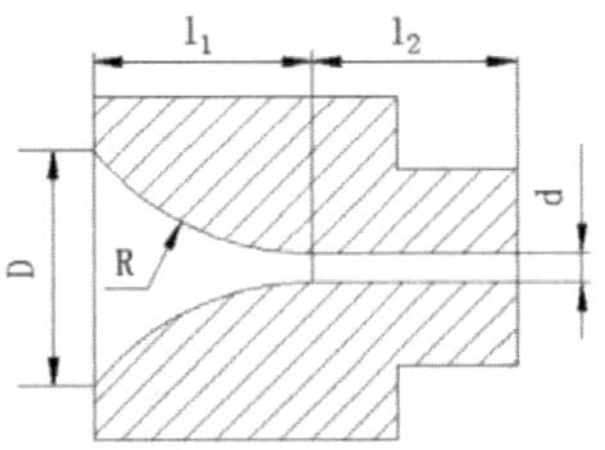

Fig 6. The structure of curved-shape nozzle

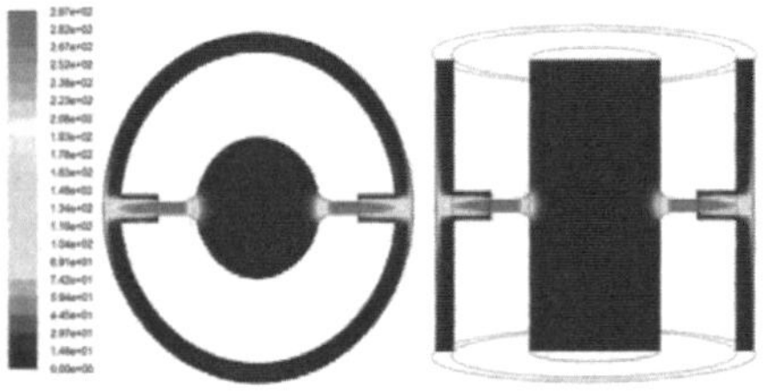

Fig 7. Jet velocity distribution at the nozzle

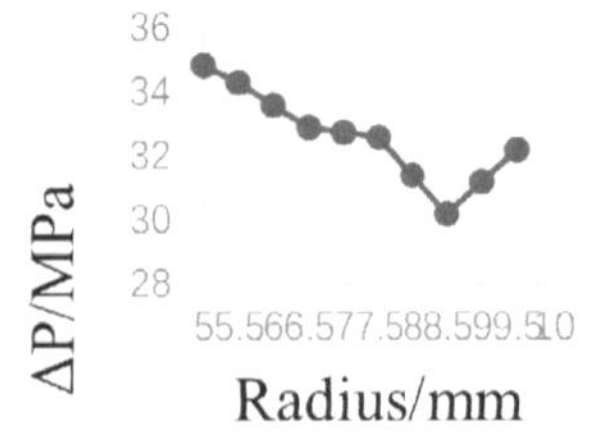

Fig 8. The influence of the inlet curve radius on the differential pressure

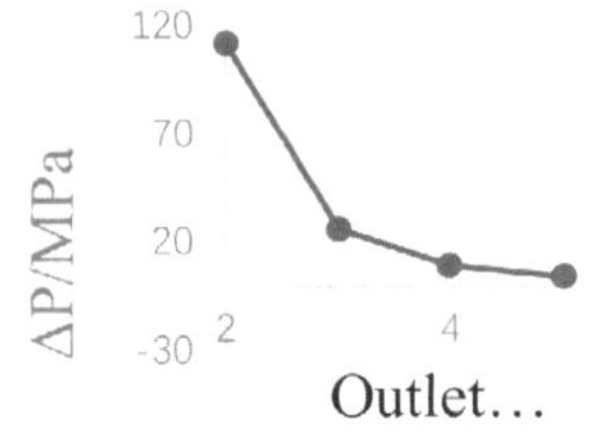

Fig 9. The influence of the outlet diameter on the differential pressure

We select the 8.5 mm inlet curve radius corresponding to the minimum pressure difference between the nozzle inlet and the outlet, and increase the nozzle outlet diameter as much as possible under the premise of meeting the jet velocity. Then, we choose the outlet diameter *d* in the range of 4.5-5 mm. In order to control the jet pressure difference while ensuring a certain jet density, the length l_2 of the straight column is selected to be 8-11 mm. In order to reduce the difficulty of tool processing, the inlet diameter D is selected as 2d. The results of digital simulation and field tests have proved that we can effectively reduce the pump pressure, ensure sufficient jet impact force, and increase the jet density of the nozzle by applying arc-shaped nozzle and selecting reasonable range of structural parameters.

Table 1. Three Scheme comparing.

Optimal parameters	Outlet diameter (mm)	Straight column (mm)	Entrance curve Radius (mm)
Optimal pressure Loss	$d \geq 4.5$	$l_2 \leq 11$	8.5
Optimal jet velocity	$d \leq 5.0$	/	/
Optimal jet density	$d \geq 4$	$l_2 \geq 8$	/
Optimal jet quality	$4.5 \leq d \leq 5.0$	$8 \leq l_2 \leq 11$	8.5

3. Field application

Well Gp 1 is a horizontal well located in the Putaohua structure of the Daqing Placanticline in the central depression area of the Songliao Basin. The field test was carried out in Well Gaoping 1 using the coiled tubing hydraulic jet fracturing technology . The technical string was composed of safety joint, centralized spray gun, packer, positioner, and guide head. Within construction time of 42 hours, we completed 19-section fracturing construction in one procedure, with high construction pressure of 62 MPa, maximum displacement of 6.7 m^3/min, liquid volume of 8814.5 m^3, sand volume of 830.5 m^3, the cumulative number of string setting and releasing reaching 52 times,100% success rate . This has created a new record for the number of fracturing sections, liquid volume, sand volume and construction efficiency of one procedure in Daqing Oilfield. In 2018-2020, the coiled tubing hydraulic jet annulus sanding fracturing process was promoted and applied to 3459 sections of 218 wells. The technical string showed temperature resistance of 120°C, pressure resistance of 70 MPa, tubing displacement of 1 m^3/min, and annulus displacement 10 m^3/min, and 20 sections fracturing within one procedure in average, which solved the problems of high environmental pressure and low construction efficiency of conventional tubing fracturing technology in peripheral low-permeability tight reservoirs, and improved single well productivity and potential reservoir production.

4.Conclusion

1. An unidirectionally-sealing rubber cylinder with a composite sealing method and a tapered shoulder protection structure are designed to achieve the technical indicators of 80 MPa pressure and temperature resistance of up to 150°C under 20 kN setting force. The overall sealing performance and recoverability of the rubber cylinder are good, so that the packer has good setting and releasing performance.

2. The finite element method is used to optimize the slip structure parameters, ultra-high-strength materials are selected and carburizing and laser processing techniques are used to improve the stress state, reduce stress concentration, and improve the slip's ability to withstand alternating loads.
3. The arc-shaped nozzle with better jet performance is selected and reasonable structural parameters are selected, which can effectively reduce the pump pressure, ensure sufficient jet impact force, and improve the jet density of the nozzle.
4. Through tool improvement and matching, the coiled tubing hydraulic jet fracturing technology has been improved. From 2018 to 2020, the improved fracturing technology has been promoted and applied to 3459 sections of 218 wells. The technical string showed temperature resistance of 120°C, pressure resistance of 70 MPa, and 20 sections fracturing within one procedure in average, which could meet the construction requirements of new wells for high-efficiency volumetric fracturing of 5 1/2" casing cementing completions in low-permeability tight reservoirs.

Acknowledgments

We greatly appreciate the assistance in terms of the theory, technology and experiment of this paper provided by the Oil Production Engineering Research Institute of Daqing Oilfield.

References

1. HUANG Zhongwei, LI Gensheng, WANG Yongzhang, et al. Hydra-jet fracturing applied in a well with three-layercasings [J]. Oil Drilling & Production Technology, 2012, 34(5): 122-124.
2. WANG Jinyou, XU Guowen, LI Lin, et al. Technology of coiled tubing abrasive perforating and annulus fracturing [J]. Oil Field Equipment, 2016, 45(5): 69-72.
3. WANG Fengshan, ZHANG Shujin, WANG Wenjun, et al. New development of horizontal well fracturing stimulation technologies for low-permeability reservoirs in Daqing [J]. Petroleum Geology & Oilfield Development in Daqing, 2009, 28(5): 234-238.
4. XU Guowen. Numerical simulation research of coiled tubing hydraulic jet erosion [J]. Oil Field Equipment, 2016, 45(6): 12-15.
5. FU Dangang, LI Xianwen, REN Yong, et al. Application of hydraulic jet perforation in oilfield development [J]. Special Oil & Gas Reservoirs, 2011, 18(3): 97-99.
6. SURJAATMADJA J B, BEZANS0N J, LINDSAY S D. New hydra-jet tool demonstrates improved life for perforating and fracturing applications [R]. SPE113722, 2008.

https://doi.org/10.1051/e3sconf/202132901061

Analysis on carbon emission reduction path and expected effect of inland river ships in China

Kun Li, Wendi Liu, Fangfang Jiao , Yongbo Ji

China Waterborne Transport Research Institute, Beijing 100088, China

Abstract. In order to systematically study the carbon emission reduction path suitable for China's inland river ships, this paper analyzes the actual situation of China's inland river ships ' operation scale, fleet structure, fuel consumption, etc., and combines the development of domestic and foreign carbon emission reduction technologies and the ship's own operating characteristics. Carbon emission reduction technology paths for China's inland river ships have been proposed in this paper, which includes different tonnage classes, different ship ages, and different types of river ships. Also, the expected application effects of each technology path are analyzed in depth in this paper. The research results can give a certain support for the implementation of carbon emission reduction in China's shipping industry.

1 Introduction

In recent years, China has continuously promoted the adjustment of the transportation structure, vigorously promoting the transfer of bulk cargo and medium and long-distance cargo to more environmentally friendly and green water transportation. While contributing to the realization of China's dual carbon goals by reducing the overall carbon emissions of the transportation industry, it is imperative to develop green and low-carbon technologies that adapt to China's specific national conditions(Hou et al.,2013) and future shipping emission requirements to promote the green and low-carbon transition of shipping(Li et al.,2013).

2 Current situation of inland river ships in China

According to incomplete statistics, As shown in Figure 1 and 2,China has 118,305 million inland river ships, including 105,225 freight ships, and 13,080 passenger ships, The tonnage and age structures of ships, are shown in Table 1 and Table 2. According to survey estimates, the annual fuel consumption of inland river ships is about 7.0168 million tons, of which the annual fuel consumption of freight ships is 6,814,200 tons, and the annual fuel consumption of passenger ships is 132,600 tons.

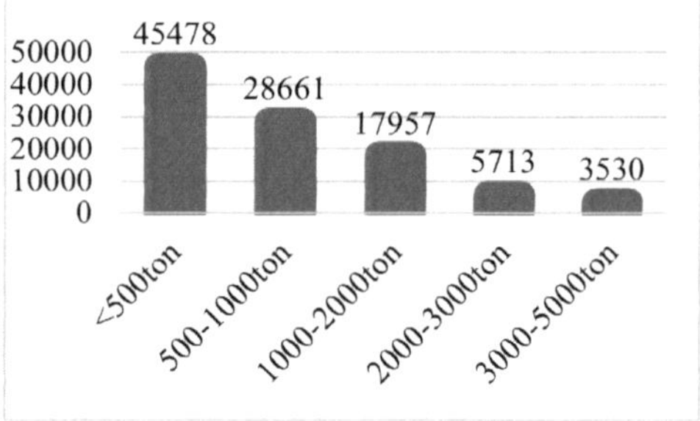

Figure 1. Statistics of the number of inland river freight ships of different tonnage classes

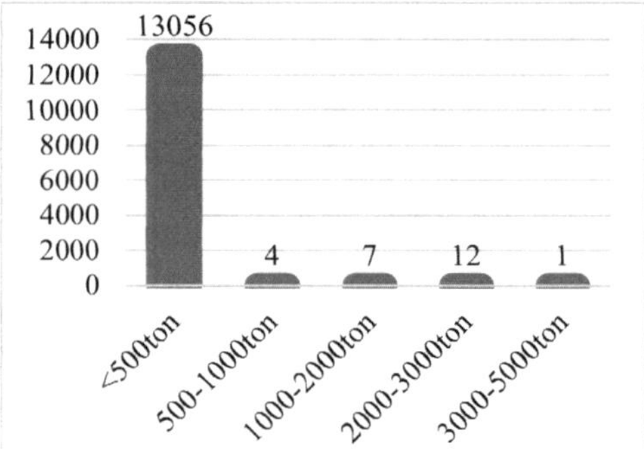

Figure 2. Statistics on the number of inland river passenger ships of different tonnage classes

Table 1. Estimated fuel consumption of different tonnages of inland river freight ships

DWT	Estimated annual fuel consumption per vessel(tons/year)	Estimated total fuel consumption (tons/year)
<500	15	682170
500-1000	35	1003135
1000-2000	70	1256990
2000-3000	120	685560
3000-5000	200	706000
5000-8000	400	971600
8000-10000	800	479200
>10000	1200	1029600
total		**6814255**

Table 2. Estimated fuel consumption of different tonnage classes of inland river passenger ships

DWT	Estimated annual fuel consumption per vessel (tons/year)	Estimated total fuel consumption (tons/year)
<500	10	130560
500-1000	30	120
1000-2000	80	560
2000-3000	100	1200
3000-5000	200	200
>5000	500	0
Total		**132640**

3 Carbon emission reduction path and expected effect of inland river ships

Facing the requirements of the waterway transportation industry to achieve the carbon peaking target, the shipping industry is actively seeking various feasible technological paths to achieve the ultimate carbon peaking target and carbon neutral vision. At present, in addition to the carbon trading market mechanism and carbon capture and storage (CCS), the main measures that are relatively mature and operable that can be used to reduce carbon in the waterway transportation industry include: application of low-carbon, zero-carbon new energy or clean Energy, accelerate the elimination of dismantling old ships, continue to promote large-scale ships; innovate and optimize the transportation organization model, etc.

3.1. Promote the application of new energy and clean energy to accelerate the adjustment of energy structure

Reducing the application of traditional fuels for ships and increasing the proportion of low-carbon clean energy such as LNG, methanol, batteries, fuel cells (including hydrogen, etc.) on ships is the most fundamental and effective solution to reduce carbon dioxide emissions from ships. LNG is currently the most mature low-carbon alternative energy source for ships, and battery is a relatively mature zero carbon alternative energy. It is expected that LNG and battery will be the main contributors to inland water ships to achieve " peaking carbon dioxide emissions " in 2030.

According to "General Principles of Comprehensive Energy Consumption Calculation" (GB/T2589-2020) and "Provincial Greenhouse Gas Inventory Compilation Guide" (NDRC Climate (2011) No. 1041), the calculation of carbon dioxide emission coefficients for different energy sources is shown in Table 3.

Table 3. Reference Coefficients of Diesel and LNG (Liquefied Natural Gas) Carbon Emissions

Energy name	Average low calorific value	Converted into standard coal factor	Carbon content per calorific value (ton carbon/TJ)	Carbon oxidation rate	carbon dioxide emission factor
Diesel	42 705 kJ/kg	1.457 1 kgce/kg	20.2	0.98	3.0998 kg-co2/kg
LNG	32238 kJ/m3 ~38979 kJ/m3	1.100 0 kgce/m3 ~ 1.330 0 kgce/m3	15.3	0.99	1.7905 ~ 2.1649 kg-co2/kg

According to calculations, when LNG is used as a ship fuel, theoretically, natural gas with the same calorific value can reduce carbon dioxide emissions by about 30% compared with diesel. Considering that the existing technology of natural gas engines is not very mature, there are still CH4 fugitive emissions. The ability to reduce greenhouse gas (GHG) emissions has been weakened to a certain extent, and the overall effect of reducing emissions is currently about 10%-20%.

Electric ships use batteries as their power source, which has the characteristics of "zero emission and zero pollution". Compared with the use of diesel, it can achieve zero carbon dioxide emission and achieve up to 100% GHG benefits.

Considering the technical maturity, endurance, ship space demand, cost and other factors of LNG and battery, combined with the characteristics of Chinese inland ships, it is recommended that cargo ships less than 1000 dwt and all passenger ships be transformed into electric ships powered by battery through elimination, renewal, dismantling, new construction and reconstruction of existing ships; LNG power is adopted for the elimination and renewal of cargo ships of 1000 DWT and above, dismantling and new construction, and reconstruction of existing ships.

Cargo ship. 50% of ships under 1,000 dwt are upgraded to electric ships. According to Table 1, the amount of fuel can be replaced by 842,700 tons; according to Table 3, carbon dioxide emission reductions of 2,612,200 tons can be achieved. 50% of ships above 1,000 dwt are upgraded to LNG-powered ships. According to Table 1, the amount of fuel can be replaced by 2.5645 million tons. According to Table 3 and SGMF's forecast, the 22% GHG benefit of replacing fuel with natural gas can be achieved, which can reduce carbon emissions by 1,768,800 tons.

Passenger ships. 50% of passenger ships are updated to electric ships. According to Table 2, the amount of fuel can be replaced by 66,300 tons. According to Table 3, carbon emissions can be reduced by 205,500 tons.

By 2030, 50% of the existing inland river cargo ships and passenger ships will be upgraded to LNG-powered ships and electric ships, and a total of 4,586,500 tons of carbon emissions can be reduced.

3.2 Implement the elimination and dismantling of old ships to build new ones, and promote the optimization of the fleet structure

Old ships have the characteristics of high energy consumption, low energy efficiency, and relatively high pollutant emissions. In addition to scrapping a part of the ships according to the prescribed natural life limit, through economic encouragement of old transport ships to be scrapped in advance and renewed into new energy clean energy ships, can achieve dual carbon reduction capabilities: Firstly, with the continuous development of China's shipbuilding industry, ship products have achieved rapid upgrades, especially the energy efficiency of China's ship engines during the "13th Five-Year Plan" period, and the unit carbon emission intensity has decreased; secondly, Second, the old ships will be demolished and newly built into new energy and clean energy ships, with low or even zero carbon emission intensity compared with traditional diesel.

According to the survey, during the 13th Five Year Plan period, the energy efficiency level of ship engines continued to improve, and the unit carbon emission intensity gradually decreased. After the old ships were eliminated and disassembled, the energy efficiency level of engines increased by about 20% compared with the old ships, which means that the carbon emission intensity decreased by 20%.According to statistics, the total number of ships of different tonnages built before 2015 is 117,739. According to Table 1 and Table 2, the total fuel consumption of ships is 6,873,900 tons. After the elimination, dismantling and new construction, the amount of fuel saved is 1,374,700 tons. Table 3 estimates that the carbon emission reduction can be achieved by 4.2615 million tons.

3.3. Promote the large-scale development of ships and continuously improve the energy efficiency of ships

The speed of large-scale inland river ships depends on the two major factors of the large-scale new and updated ships(Jia et al.,2007). The large-scale addition of new ships is mainly due to the improvement of infrastructure, market demand and market competition, and the adjustment of the transportation capacity structure will accelerate the update(Jia et al.,2015). Promoted the historical process of large-scale. Since the beginning of the new century, China's inland river ships have continued to grow in size. In 2020, the average dwt of inland river motor cargo ships will reach 1443 tons, As shown in Figure 3,which is 18.4 times that in 2000 and 2.7 times that in 2010(Jia et al.,2019).

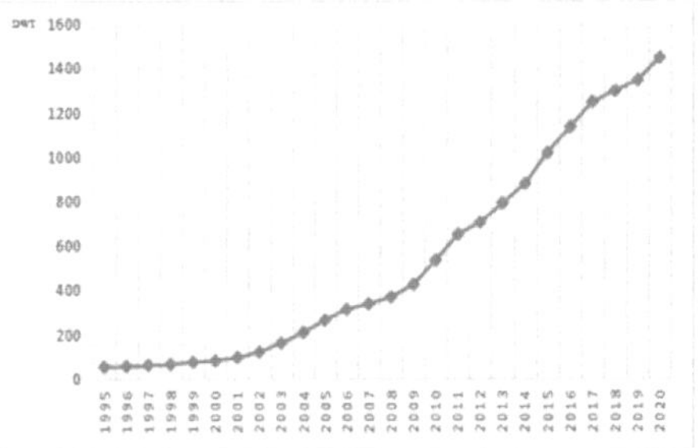

Figure 3 Average deadweight of inland water motorized cargo ships

The increase in the size of inland watercrafts has made the increase in total power slower than the rate of increase in total deadweight. For example, the power of a dry bulk carrier of about 1,000 dwt is about 300 kilowatts, and that of a dry bulk carrier of about 2,000 dwt is about 400 kilowatts. In 2019, the total deadweight of inland water-driven cargo ships was 10.17 times that of 2000 and 1.95 times that of 2010, while the total power was only 4.29 times that of 2000 and 1.33 times that of 2010. It can be seen from Figure 4 that the average power per deadweight tons of inland water-driven cargo ships has been declining year by year due to the large-scale ships. There is a linear relationship between ship fuel consumption and power under other conditions unchanged, so it can be explained that under the development trend of large-scale ships, if the ship's total deadweight remains the same, the ship's total fuel consumption will decrease.

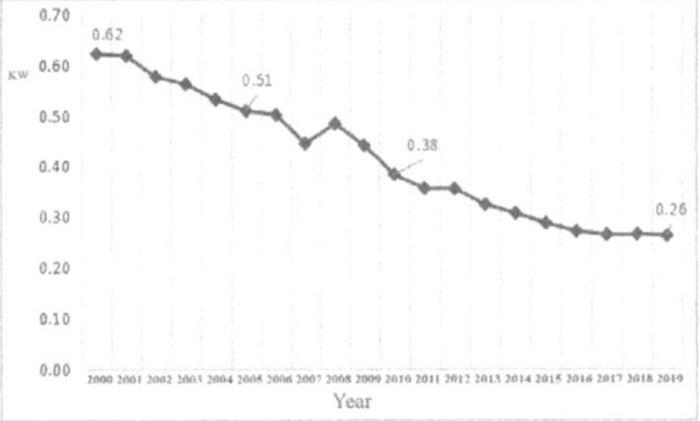

Figure 4 Average power change per deadweight ton of inland water-driven cargo ships

At present, China's inland river infrastructure construction continues to increase, infrastructure improvement and long-distance transportation demand growth, there is still a large potential for large-scale ships. According to the current large-scale rate of newly built cargo ships, China's inland river motor cargo ships will be estimated by 2030. The average tonnage reaches about 2000t. According to the calculations of Data for 2019, every increase of 100 tons in average deadweight of inland river motor cargo ships can achieve a power drop of about 0.005 kilowatts per deadweight ton. It is estimated that the power per deadweight ton of inland river motor cargo ships in 2030 will decrease by 0.025 kilowatts compared with 2020, a decrease of approximately Is 9.5%. Due to the linear relationship between fuel consumption and power, the carbon dioxide emission reduction per deadweight ton of ship capacity that can be brought about by the large-scale ship by 2030 is about 9.5%. According to Table 1 and Table 2, the fuel consumption of ships and the calculation in Table 3 can achieve a carbon emission reduction of 2.0663 million tons.

3.4. Optimize the organization mode of ship transportation and improve the level of transportation efficiency

The improvement of ship transportation organization efficiency with emission reduction contribution is mainly reflected in reducing empty driving, berthing and speed optimization. The organizational efficiency of ship transportation can be reflected by the annual cargo turnover that can be completed per DWT. In 2019, as shown in Figure 5,the average cargo turnover of inland ships per DWT will be 12500 ton-kilometer, which is 1.57 times that in 2010 and 2.07 times that in 2003. (Affected by the epidemic situation and the adjustment of statistical data, the turnover of inland river goods decreased in 2020.)

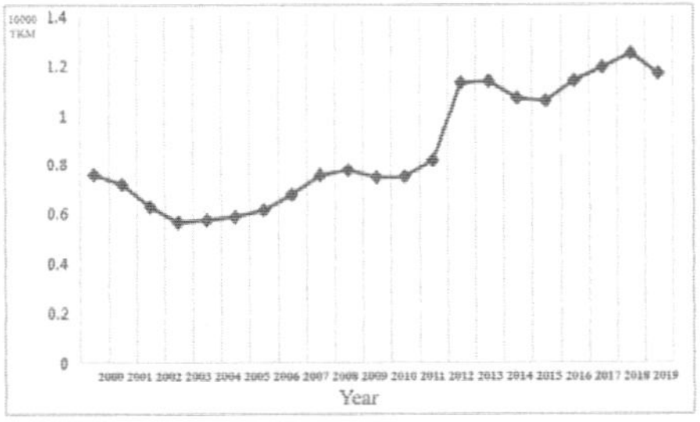

Figure 5 The average capacity of inland ships per deadweight ton completed and the change in cargo turnover

However, according to market research, the current empty rate of inland water ships is still relatively high. In the future, with the improvement of logistics technology such as freight platform economy, the efficiency of transportation organization can be effectively improved. With the development of port automation, the improvement of port efficiency will also reduce the number of ships waiting for berths. Comprehensive consideration of reducing the operating turnover efficiency of ships such as empty ships and awaiting berths. It is estimated that the completed cargo turnover per deadweight ton in 2030 can reach 13,500 tons. The sailing time of ships carrying cargo is about 80 days per year, accounting for about 27% of the ship's operating time. In this scenario, the improvement of transportation organization efficiency can bring about 8% of carbon dioxide emission reduction per deadweight ton ship capacity. According to the fuel consumption of ships in Table 1 and Table 2, and according to Table 3, the carbon emission reduction can be achieved by 1.74 million tons. Speed optimization refers to the use of assisted driving technology and intelligent ship technology to rationally optimize the speed of each period according to the turnaround time, so as to achieve fuel saving and emission reduction. However, when the supply of goods is sufficient, it may be possible to increase the speed to speed up the turnover of the ship. The scenario is more difficult to predict, so here No quantitative analysis, but the overall intelligent technology will promote ship emissions reduction in terms of speed optimization and intelligent energy efficiency.

Conclusion

The structure of China's inland watercraft is complex. The path selection of ship carbon emission reduction should comprehensively consider factors such as economic development, infrastructure construction, ship type, tonnage, route, ship age, etc., combined with shipping carbon reduction targets and carbon reduction technology maturity and economy, etc., this paper proposes four Different paths include "promoting the application of new energy and clean energy", "accelerating the adjustment of energy structure, implementing the elimination of old ships and building new ones", "promoting the optimization of fleet structure and promoting the development of large-scale ships", and "continuously improving ship energy efficiency Level, optimize the ship transportation organization model, and improve the level of transportation efficiency. "These four different technological paths have their own characteristics. The selection of carbon emission reduction technology paths that are suitable for ships in the region must be tailored to local conditions and lay a solid foundation for China to achieve its dual-carbon goals.

Acknowledgments

Fund Project: This research is supported by the research project on the safe operation guidelines and technical standards of the shore power supply of international sailing ships, the special self-help project for the innovation of green and intelligent inland ships and the special project of green ecological and environmental protection ship engineering.

References

1. Li Jing and Li Ruiyu. (2013). Research on the impact of carbon tax on China's shipping industry and Its Countermeasures. Navigation technology, 000(005), 67-69.
2. Hou guiguang, Xu Yi ,Jia Jielin and Lu jing. (2013). Quantitative study on the correlation mechanism between energy structure adjustment and carbon dioxide emission. Annual meeting of Chinese society of Environmental Sciences

3. Jia Dashan.(2007). Exploration on the development strategy of water transportation in China. Dalian Maritime University Press
4. Jia Dashan and Ji Yongbo.(2015). Inland advantage strategy. People's Communications Press
5. Jia Dashan ,Ji Yongbo and Jiao Fangfang.(2019). Vigorously promote ship standardization and promote the high-quality development of inland ships. China water transportation (7).

Study on the application of natural gas to inland and coastal ships in China

Kun Li [1], Xiang Li [1], Jianghe Zhang [2]

[1] China Waterborne Transport Research Institute, Beijing, 100088, China
[2] CNOOC Gas & Power Group, Beijing, 100028, China

Abstract: In order to promote the green and low-carbon development of ships and promote shipping energy conservation and emission reduction. This paper systematically combed the development of liquefied natural gas (LNG) powered ships in China's inland ships, deeply studied and analyzed the development characteristics of China's new LNG powered ships and reconstructed LNG powered ships, so as to provided reference experience and lay a foundation for China to continue to promote the application of LNG in the water transportation industry in the future.

Key words: Ship; liquified natural gas; newly build; reconstruction; LNG.

1 Introduction

Strengthening the construction of ecological civilization is an important content for the CPC Central Committee to comprehensively promote the overall layout of "five in one" and the strategic layout of "four comprehensives" in the new era. The nineteen Party Congress made clear arrangements for speeding up the construction of ecological civilization and building transport power. In April 2018, general secretary Xi Jinping made an important speech when he inspected the Yangtze River. He stressed that the development of the Yangtze River Economic Belt should be placed in an overwhelming position to restore the Yangtze River's ecological environment. LNG is a clean energy that the state gives priority to promoting. The opinions of the CPC Central Committee and the State Council on Comprehensively Strengthening Ecological and environmental protection and resolutely fighting a tough battle of pollution prevention and control issued by the CPC Central Committee and the State Council and the three-year action plan for winning the blue sky defense war put forward the promotion of natural gas clean energy ships as a key task. The promotion and application of LNG in water transportation industry is the need of developing green transportation and building ecological civilization. It is an effective measure to realize energy conservation and emission reduction, transformation and upgrading and optimize energy consumption structure in water transportation industry. In order to thoroughly implement the spirit of the 18th CPC National Congress, take the scientific outlook on development as the guidance, adhere to the basic national policies of saving resources and protecting the environment, and focus on optimizing the energy consumption structure and reducing emissions in the water transportation industry, the Ministry of transport issued the guiding opinions on promoting the application of liquefied natural gas in the water transportation industry in 2013, Take the promotion and application of LNG as an important starting point for energy conservation, emission reduction and ecological civilization construction of water transportation industry, accelerate the pace of green water transportation construction, and promote the comprehensive, coordinated and sustainable development of water transportation[1]. In 2018, the CPC Central Committee and the State Council issued the opinions on Comprehensively Strengthening Ecological and environmental protection and resolutely fighting the battle of pollution prevention and control, which clearly proposed to "encourage the promotion and use of clean energy ships". The Ministry of transport also issued the implementation opinions on Comprehensively Strengthening Ecological and environmental protection and resolutely fighting the battle of pollution prevention and control, which proposed to "promote the construction and transformation of LNG powered ships and electric ships"[2]. In 2019, the Ministry of transport issued the guiding opinions on accelerating the use of shore power and promoting the application of LNG ships on the Yangtze River trunk line, and proposed to accelerate the promotion of LNG ship application on the Yangtze River trunk line[3-5]. In September 2020, President Xi Jinping announced at the seventy-fifth general debate of the UN General Assembly that China's carbon dioxide emissions will reach a peak by 2030, and strive to achieve carbon neutralization by 2060. Since then, local governments along the river have also issued a series of policies and

 https://doi.org/10.1051/e3sconf/202132901062

measures to support the application of LNG powered ships, and the promotion of LNG in the water transportation industry has been gradually launched.

2 Analysis on the development status of LNG power ship market

As of December 2019, according to the incomplete statistics of our institute, there are 284 LNG powered ships in China, including dry bulk carriers, container ships, port tugs, LNG carriers, official ships, tourist passenger ships, etc.[6]. Among them, 157 LNG powered transport ships (1 LNG electric propulsion), 4 port tugs and 2 inland river LNG transport ships are newly built; One official ship, one tourist ship and one teaching ship. A total of 118 LNG powered ships were reconstructed, 72 were reconstructed by means of overall renewal of power system, and 46 LNG powered ships were reconstructed directly on the original ship engine.

3 New LNG powered ship in inland river

Since April 2014, the Ministry of Finance and the Ministry of transport issued the measures for the management of inland ship type standardization subsidy funds to give financial subsidies to new LNG powered ships. According to incomplete statistics of the Water Transportation Research Institute of the Ministry of transport, up to now, there are 157 new LNG powered ships in China, including 154 LNG powered demonstration ships and 3 non demonstration ships.

1) Ship type

The newly-built LNG power demonstration ships in China are mainly ordinary dry bulk carriers, including 146 general dry cargo ships, bulk cargo ships and multi-purpose ships, accounting for 94.8% of the newly-built LNG power demonstration ships in China, and the rest are container ships, distribution ships and self unloading sand ships.

Table 1 Types of new LNG power demonstration ships

Ship type	Quantity
General dry cargo ship	69
Bulk cargo ship	43
Multipurpose ship	34
Container ship	6
Distributed dual-purpose ship	1
Self unloading sand ship	1
Total	154

2) Tonnage structure

The measures for the administration of subsidies for inland river ship type Standardization (CJ No. 2014-61) issued by the Ministry of Finance and the Ministry of communications stipulates that the tonnage of newly-built LNG power demonstration ships shall not be less than 400 gross tons. According to statistics, the tonnage (deadweight tons) of newly-built LNG power demonstration ships are mainly 500-1500 tons. See the following table and figure for details.

Table 2 Tonnage (DWT) distribution of new LNG power demonstration ships

Tonnage	Quantity	Tonnage	Quantity
500-1000	99	2500-3000	6
1000-1500	29	3000-4000	7
1500-2000	11	6000 以上	2

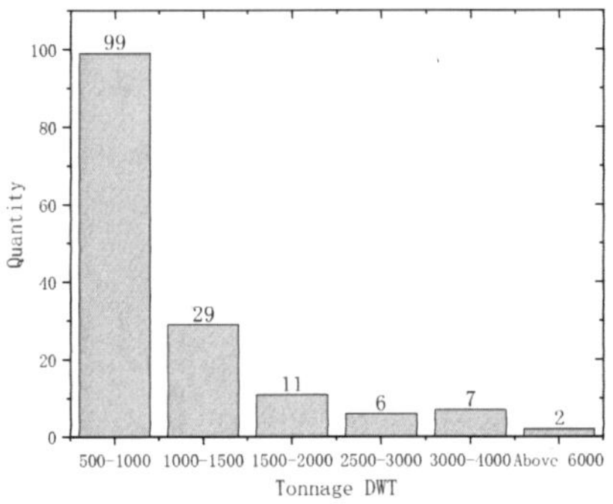

Fig. 1 Tonnage (DWT) of new LNG power demonstration ship

3) Regional distribution

According to statistics, China's new LNG power demonstration ships are mainly distributed in Shanghai, Jiangsu and Guangxi, including 90 in Shanghai, 24 in Jiangsu and 26 in Guangxi. The specific distribution is shown in the following table and figure.

Table 3 Regional distribution of new LNG power demonstration ships

Region	Quantity	Region	Quantity
Shanghai	90	Chongqing	2
Guangxi	26	Hubei	2
Jiangasu	24	Guizhou	2
Guangdong	4	Anhui	2
		Shandong	2

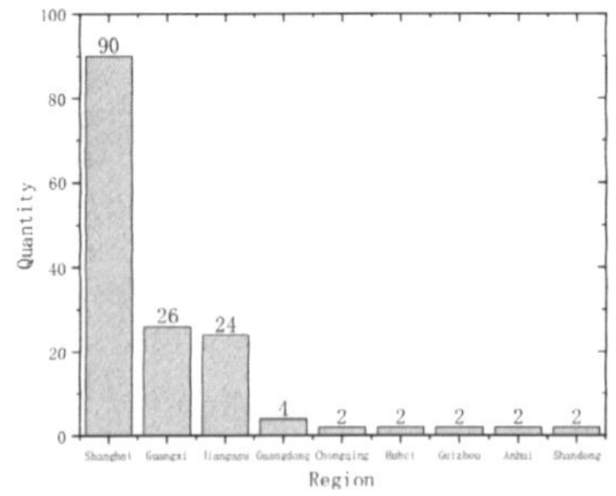

Fig. 2 Regional distribution of new LNG power demonstration ship

 https://doi.org/10.1051/e3sconf/202132901062

3 Power system update LNG powered ship

In order to encourage existing ships to use clean energy and promote the supply side structural reform of the shipbuilding industry, in 2016, the Ministry of Finance issued a supplementary notice on the measures for the administration of subsidies for ship scrapping, dismantling and ship type Standardization (finance [2016] No. 418), which adjusted the subsidy policy for LNG powered ships, The original subsidy for newly-built LNG powered demonstration ship is adjusted to the subsidy for the reconstruction of LNG powered demonstration ship by overall renewal of power system[7] [8]. Up to now, according to the incomplete statistics of the Water Transportation Research Institute of the Ministry of transport, 72 LNG powered ships have been reconstructed by the overall renewal of the power system, including 71 demonstration ships.

1) Ship type

According to statistics, the LNG power demonstration ships reconstructed by the overall renewal of power system mainly include dry cargo ships, dry bulk cargo ships, cement tank ships and container ships, of which dry cargo ships are the main, dry bulk cargo ships and cement tank ships are the second, and container ships are the least. See the figure below for details.

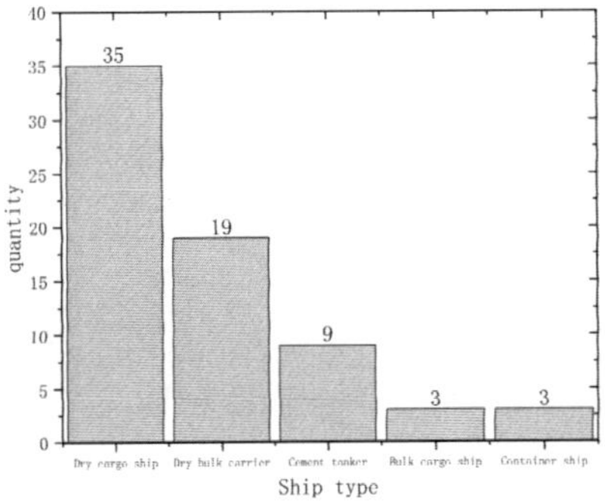

Fig. 3 Type of LNG power demonstration ship reconstructed by overall renewal of power system

2) Regional distribution

Jiangsu Province is the most active Province in responding to the national development of green water transportation. Among the 71 LNG powered demonstration ships that have been reconstructed by adopting the overall renewal of the power system of the existing ships in China, there are 68 in Jiangsu, accounting for 95.77%; The remaining three ships are two in Chongqing and one in Wuhan, Hubei. See the table below for details.

Table 4 Regional distribution of reconstructed LNG power demonstration ships in the overall renewal mode of power system

Ship type	Quantity
Jiangsu	68
Chongqing	2
Hubei	1

3) **Tonnage structure**

In the supplementary notice of the Ministry of Finance on the measures for the administration of subsidies for ship scrapping, dismantling and ship type Standardization (finance [2016] No. 418), it is stipulated that the existing ships shall be reconstructed into LNG powered demonstration ships by means of overall power system renewal, and the ship tonnage shall not be less than 400 gross tons. According to statistics, the tonnage (deadweight tons) of the reconstructed LNG powered demonstration ships are concentrated in 500-2000 tons. See the following table and figure for details.

Table 5 Tonnage (DWT) of LNG power demonstration ship for overall renewal and reconstruction of power system

Tonnage	Quantity	Tonnage	Quantity
500-1000	26	2000-2500	6
1000-1500	17	2500-3000	3
1500-2000	16	4000-6000	3

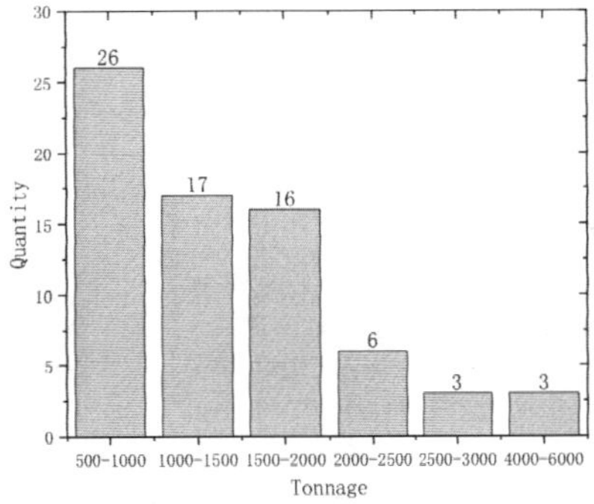

Fig. 4 Tonnage (DWT) of LNG power demonstration ship after overall renewal and reconstruction of power system

4) Ship age distribution

In the supplementary notice of the Ministry of Finance on the measures for the administration of subsidy funds for ship scrapping, dismantling and ship type Standardization (finance [2016] No. 418), it is stipulated that the existing ships shall be reconstructed into LNG powered demonstration ships by means of overall power system renewal, and the ship age must be within 10 years (inclusive). Therefore, the ship age of the reconstructed LNG powered demonstration ships is generally small, concentrated in 3-5 years and 7-8 years. See the following table and figure for details.

Table 6 Age distribution of LNG power demonstration ship reconstructed by overall renewal of power system

Ship age / year	Quantity	Ship age / year	Quantity
1	3	6	6
2	10	7	11
3	10	8	11
4	11	9	1
5	7	10	1

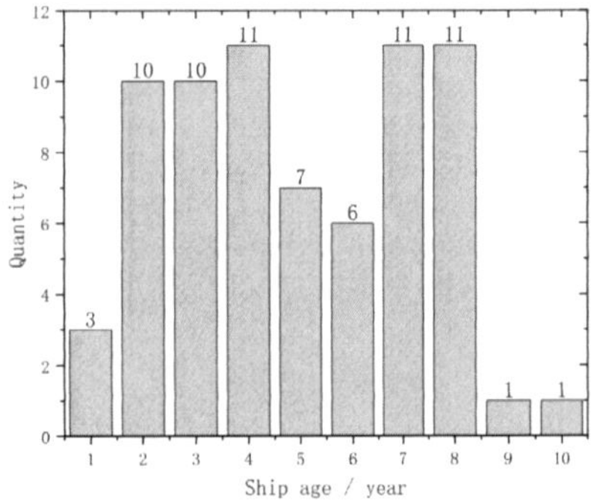

Fig. 5 Age of LNG power demonstration ship reconstructed by overall renewal of power system

4 Summary

According to the above analysis, it can be seen that the development of LNG powered ships in China is dominated by new construction and the combination of reconstruction and reconstruction; Ship tonnage is relatively small; The ship age is relatively small; The economically developed areas have the characteristics of relatively rapid development, and the overall development is relatively slow. In order to further implement China's policies on energy conservation and emission reduction, the development of new energy and clean energy, and accelerate the realization of the dual carbon goal of China's shipping industry, it is suggested that China give relevant support policies in promoting the application of LNG in the water transportation industry:
1) Introduce the continuation policy of financial subsidies, give certain financial subsidies to new LNG and reconstructed LNG powered ships, and improve the enthusiasm of ship owners to use LNG;
2) Improve the network of LNG filling facilities to ensure the operation safety of LNG powered ships;
3) Strengthen the supervision on the use of fuel oil by ships and establish a good LNG market application environment.

Acknowledgments

Fund Project: this research is supported by the research project of safety operation guidelines and technical standards for shore power of international navigation ships, the special project of green intelligent inland river ship innovation and the special project of green ecological and environmental protection ship project.

References

1. Wu Qiwei. Comparative analysis of LNG as alternative fuel for gasoline and diesel [J]. Shanghai gas, 2013, (5): 32-78.
2. Li Kun, Liu Chang, Zhu Lin. development of LNG fueled ships in China [J]. File, 2016, (7): 727-728.
3. Sun Hui, Wang Zhanli, Tian Ying. Opinions on coordinated and stable development are issued, and the construction of terminal has stepped into the fast lane - the development of China's natural gas industry in 2018 and the outlook for 2019 [J]. International petroleum economy, 2019, (6): 33-40.
4. Gao Zhenyu, Gao Peng, Liu Qian, et al. Current situation analysis and development suggestions of China's LNG industry [J]. Natural gas technology and economy, 2019, (6): 14-19.
5. Bi Shanshan, Liu Changjian. Adapting to the new demand and planning the Yangtze River LNG transportation in advance [J]. China water transportation, 2018, (10): 14-15.
6. Wang Sijia. Formal "ventilation" of the Three Gorges ship lock of the Yangtze River [J]. China ship inspection, 2019, (6): 74-78.
7. Zhao Bo. Gradual progress of LNG powered ship [J]. China ship inspection, 2018, (10): 21-24.
8. Gong Linlin, Qu Hongling. Development status and Countermeasures of green channel in Jiangsu Province [J]. China water transportation, 2019, (1): 19-20.

https://doi.org/10.1051/e3sconf/202132901063

Life Cycle Carbon Emission Analyzing of Rural Residential Energy Efficiency Retrofit-A Case Study of Gansu province

Qiangnian Li[1], Tongze Han[1, *], Changlin Niu[1,2] and Ping Liu[1]

[1]School of Civil Engineering, Lanzhou University of Technology, Lanzhou, China
[2]GCIGC Technology Development CO., LTD., Lanzhou, China

Abstract. Objective To study and analyze the life-cycle carbon emissions of existing rural residential energy retrofit projects to provide theoretical and data support for local rural green development and sustainable construction. Methods Life cycle analysis (LCA) was used to analyze and compare the life cycle carbon emissions (LCE) of a rural residential envelope energy efficiency retrofitting project in central Gansu. Results It was found that rural dwellings have a very high potential for energy efficiency retrofitting, and the contribution of retrofitted homes to CO_2 emissions reduction can reach more than 30% over the whole life cycle. Secondly, during the retrofitting process, neglected in previous studies, carbon emissions account for about a quarter of the LCE. It is concluded that introducing LCA into evaluating rural residential energy retrofit projects' energy-saving and emission reduction benefits is more scientific, reasonable, and necessary.

1 Introduction

Due to a large amount of greenhouse gas emissions, global climate system warming accelerated in just past 2019, and many historical records such as early weathering period, glacier melting, and sea-level rise were set, and climate extremes increased significantly. In response to the global warming trend, China proposed at the United Nations General Assembly in September 2020 that "carbon dioxide emissions should strive to peak by 2030 and work towards carbon neutrality by 2060" [1]. China's construction industry currently accounts for about 40% of total energy consumption and 20% of greenhouse gas emissions [2]. Some statistics show that the total floor area of existing residential buildings in rural areas of China is about 24 billion square meters, most of which do not adopt energy-efficient technologies and have backward and inefficient heating methods, leading to a yearly increase in energy consumption in rural residential buildings, resulting in a large amount of unnecessary CO2 and other harmful gas emissions [3, 4]. In response to the above problems, the Ministry of Housing and Urban-Rural Development and the Ministry of Industry and Information Technology attached great importance to this issue and issued the Notice on the Construction of Green Rural Houses. Thus, it can be seen that a large number of existing dwellings in rural areas of China have great potential for energy-saving renovation. In addition, the energy-saving retrofit and life cycle emissions (LCE) analysis of such houses is of great significance for China to achieve the goal of "double carbon."

Many scholars have studied the whole life cycle carbon emissions of buildings. Xie Baichen et al. introduced the emissions of CH_4 and N_2O based on LCE and evaluated a research building [5]; Cai Weiguang used the Monte Carlo model to simulate the uncertainties in the life cycle of a building, to measure the LCE of a building more accurately [6]; Qin Ardian et al. who also studied the LCE of subway stations from a life-cycle perspective [7].

For the study of carbon emission of rural houses, both domestic and international have focused on calculating and evaluating energy consumption and carbon emission in its operation phase, especially in the winter heating phase. For example, Guo Jianhong et al. conducted a comparative analysis of the operational energy consumption and carbon emissions of existing rural dwellings in Hebei Province after different retrofitting schemes [8]; J Y Dong et al. used a BIM platform to simulate and compare the energy consumption of rural dwellings in cold regions before and after retrofitting [9]; Xu Juan used a questionnaire method to systematically investigate and calculate the housing energy consumption of 660 rural households in 11 provinces in northern China over two years [10]. Xu Juan conducted a systematic study and calculation of the energy consumption of 660 rural households in 11 provinces in northern China using a questionnaire method over two years [10].

According to the current status of related studies, it can be found that there are few LCE studies for energy efficiency retrofitting of existing rural houses. Since rural homes have a large inventory from a macroscopic point of view and generate a large amount of energy consumption at all life cycle stages, carbon emission measurement only from the operation stage is not comprehensive enough. This study is expected to provide theoretical and data support for local rural green development and sustainable construction.

* Corresponding author: hantzacademic@163.com

https://doi.org/10.1051/e3sconf/202132901063

2 Methodology

With the increasing awareness of environmental protection and the better understanding of the impacts that may accompany the production and consumption of products, there is a desire to establish methods to better understand and account for these impacts. Life cycle assessment (LCA) is one of the techniques developed with this purpose in mind [11]. Life cycle refers to the entire process of a product (or service) from the acquisition of raw materials, production, use until disposal, i.e., from the cradle to the grave. As defined by ISO 14040, life cycle assessment is a method used to assess the environmental factors and potential impacts associated with a product (or service) by compiling inventory records of the relevant inputs and outputs of a system, evaluate the potential environmental impacts associated with these inputs and outputs, and interpreting the results of the analysis of inventory records and ecological impacts according to the objectives of the life cycle assessment study.

2.1. Determination of the calculation boundary

At present, most of the existing building carbon accounting models divide the life cycle of buildings into the physical phase and the operational phase. The physical step includes carbon emissions from the production and transportation of building materials, construction, maintenance, and demolition of buildings. The active phase includes carbon emissions from heating, air conditioning systems, ventilation, lighting, electrical equipment, domestic hot water, and other service facilities [12].

However, rural existing residential energy retrofit projects are different from a new build. The energy retrofit projects for existing buildings need to be adjusted according to the existing carbon emission measurement model to redefine the time boundary of the project, as shown in Figure 1 [12].

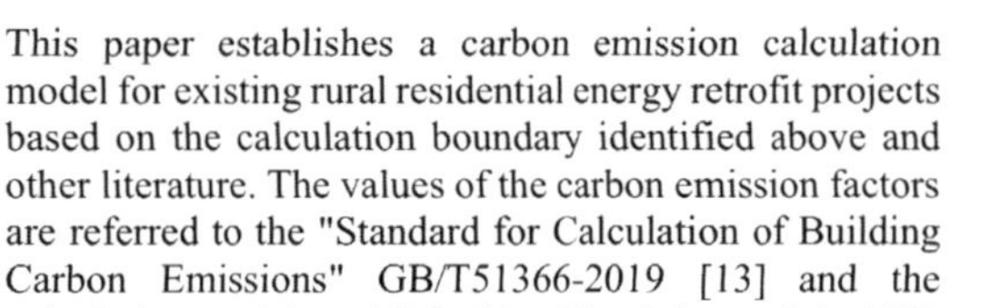

Fig.1. The life cycle of rural existing residential energy retrofit projects

2.2 Carbon emission calculation model

This paper establishes a carbon emission calculation model for existing rural residential energy retrofit projects based on the calculation boundary identified above and other literature. The values of the carbon emission factors are referred to the "Standard for Calculation of Building Carbon Emissions" GB/T51366-2019 [13] and the calculation model established by Qian Prim et al. in 2018 for the whole life cycle carbon emissions of energy efficiency retrofitting of existing buildings in Shenzhen [14].

2.2.1 Building materials production stage

According to the bill of quantities of the renovation project for the types and amounts of building materials required, the carbon emissions generated in the process are calculated according to the "Standard for Calculation of Carbon Emissions from Buildings" GB/T51366-2019, with the following formula. $C_{JC} = \frac{C_{sc}+C_{ys}}{A}$ (1). Where: CJC is the carbon emission per unit building area in the production and transportation stage of building materials ($kgCO_2e/m^2$); Csc is the carbon emission in the production stage of building materials ($kgCO_2e$); Cys is the carbon emission in the transportation process of building materials ($kgCO_2e$); A is the building area (m^2). The calculation of the production stage of building materials mainly considers the type of building materials required, the consumption of building materials, and the carbon emission factor of producing such materials. The specific calculation formula is as follows: $C_{SC} = \sum_{i=1}^{n} M_i F_i$ (2) Where: Csc is the carbon emission of the production stage of building materials ($kgCO_2$); Mi is the consumption of the ith means building material; Fi is the carbon emission factor of the ith means building material ($kgCO_2$/unit quantity of building materials).

2.2.2 Building materials transportation stage

The primary considerations of the building materials transportation stage are the consumption of building materials required for the renovation project, the transportation distance, and the carbon emission factor of the transportation method. The specific calculation formula is as follows: $C_{ys} = \sum_{i=1}^{n} M_i D_i T_i$ (3) Where: Cys is the carbon emission of the transportation process of building materials ($kgCO_2$); Mi is the consumption of the ith means building material (t); Di is the average transportation distance of the ith building material (km); Ti is the carbon emission factor of the transportation distance per unit weight under the transportation mode of the ith building material [$kgCO_2$/(t-km)].

2.2.3 Renovation construction stage

The carbon emission accounting in the renovation of existing residential energy-saving projects should include the carbon emissions generated by each sub-project and each construction measure. The calculation time should be calculated from the beginning of construction to the end of completion. The carbon emissions from various types of machinery, small temporary equipment, on-site mixing of concrete, mortar, paint, and other materials during construction should be included in the total emissions of the renovation construction phase. Specific calculation public as follows. $C_{JZ} = \sum_{i=1}^{n} E_{jz_i} EF_i$ (4). Where: CJZ is the carbon emission of the building construction stage ($kgCO_2$); Ejz, i is the total amount of the ith type of energy used in the building construction stage (kWh or kg); EFi is the carbon emission factor of the ith type of energy ($kgCO_2/kWh$ or $kgCO_2/kg$).

2.2.4 Operation stage

The primary sources of carbon emissions in the use phase of the energy-efficient renovated rural houses are related to summer cooling, winter heating, domestic water, and electricity consumption. According to the Energy Conservation Design Standard for Rural Residential Buildings GBT 50824-2013 [15] and the Notice on the Construction of Green Agricultural Houses issued by the Ministry of Housing and Construction, this study defines the carbon emission calculation period of the operation phase as 30 years. The specific calculation formula is as follows. $C_M = [\sum_{i=1}^{n}(E_i EF_i) - C_p]y$ (5). Where: CM is the carbon emission in the operation phase of the building ($kgC0_2$); Ei is the annual consumption of the i-th type of energy of the building (unit/a); EFi is the carbon emission factor of the i-th type of energy, i is the type of terminal energy consumed by the building, including electricity, gas, oil, municipal heat, etc.; j is the type of energy system of the building, including heating and air conditioning, lighting, domestic hot water system, etc.; Cp is the carbon sink of the green space of the building system annual carbon reduction ($kgC0_2/a$); and y is the building design life (a). In this study, all types of energy consumption in the post-retrofit use phase were simulated using eQUEST 3.65 software.

2.2.5 Dismantling stage

Since there are few studies on carbon emission measurement for the demolition stage, this study draws on the existing literature to argue that the main factors that should be considered to measure the carbon emission of the demolition stage of a building are its engineering volume, the process and the carbon emission factor of the process [12]. The specific calculation formula is as follows. $C_{cc} = \sum_{i=1}^{n} C_{iFi}$ (6). Where: Ccc is the carbon emission of the building demolition stage ($kgC0_2$); Ci is the engineering volume of the i-th demolition process; Fi is the carbon emission factor of the i-th process.

2.2.6 Waste disposal stage

The sources of carbon emissions in the post-demolition waste disposal phase are mainly in the transportation and landfill of waste. If there are building materials that can be recycled, they can moderately offset some of the carbon emissions generated during the production of building materials. The specific calculation formula is as follows [11]. $C_w = C_t + C_d + C_g$ (7). $C_t = \sum_{i=1}^{n} m_i l_i k_{tj}$ (8). $C_d = \sum_{i=1}^{n} Dm_i(1-R)$ (9). $C_g = \sum_{i=1}^{n} m_i R k_{mi} R_i$ (10). Where: Cw is the carbon emission of the waste treatment process ($kgC0_2$); li is the transportation distance of waste (km); D is the carbon emission factor per unit of construction waste landfill ($kgCO_2$); Ri is the recycling rate (%); R is the ratio of construction waste recycling to save carbon emission of building materials production (%). Among them, the recycling rate of steel, aluminum, and other metal materials is 100%, the recycling rate of waste such as thermal insulation wool, asphalt waterproofing membrane, and polystyrene sheet is 0, and the remaining part of the list is all landfilled.

3 Case Study

3.1 Project Introduction

The rural residence is located in Yongdeng County, Lanzhou City, Gansu Province, with a residential base area of 200 m^2, a land area of 112.7 m^2, a building area of 221.8 m^2, usable space of 185.2 m^2, a functional area coefficient of 0.835, a building density of 56.4%, and a floor area ratio of 1.1. The resident population is four people. The structure was modified according to the specific needs, living habits, and winter resident population of the occupants [16-18], the Energy Conservation Design Standard for Rural Residential Buildings GBT-50824-2013 [15], the Green Retrofit Assessment Standard for Existing Buildings JGJ26-2018 [19] and the Technical Specification for External Wall Internal Insulation IGJT261-2011 [20]. A retrofitting scheme was proposed for the building envelope, and a BIM model was established (Fig 2). Furthermore, the specific practices of the envelope are shown in Table 1.

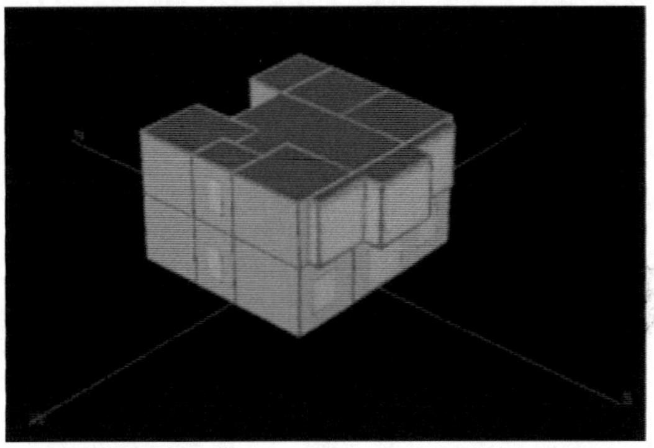

Fig.2. Retrofit Case Model

Table 1 Specific approach to energy efficiency retrofits

	Structure	Structure practice/mm	Heat transfer coefficient/ (W-(m²-K)⁻¹)	Amount of retrofitting work/m²
Existing Buildings	Wall	20 white plaster mortar + 240 brick wall + 20 cement mortar + exterior decoration (from inside to outside)	1.312	/
	Roofing	20 cement mortar + 80 slag + 200 reinforced concrete (from top to bottom)	1.607	/
	Ground	Floor tiles + 20 cement mortar + 40 gravel or pebbles (from top to bottom)	12.400	/
	Windows	Single-glazed windows	5.700	/
	Door	Single-layer wooden doors	5.930	/
Retrofit Program	Wall	20 white mortar + 240 brick wall + 20 cement mortar leveling + bonding + 70 expanded polystyrene board + 5 anti-cracking mortar + alkali-resistant glass fiber mesh cloth + exterior wall decoration (from inside to outside)	0.434	228.0
	Roofing	20 cement mortar + protective layer + 1.2 PVC waterproof layer + 90 expanded polystyrene board + 20 cement mortar for slope + 80 slag + 200 reinforced concrete + gypsum board ceiling (from top to bottom)	0.091	127.2
	Ground	Floor tile + 20 cement mortar + 5 anti-cracking slurry anti-alkali glass fiber mesh + 40 polystyrene board + 20 cement mortar leveling + 100 concrete bedding (from top to bottom)	0.720	127.2
	Windows	Aluminum ordinary insulating glass window(6+12+6)	2.900	56.1
	Door	Aluminium double sliding doors	3.120	22.06

3.2 Life cycle carbon emissions accounting

Based on the above research methodology, this study uses the corresponding accounts provided by the householder and Table 1 to prepare a bill of quantities to calculate and compare the life-cycle carbon emissions of the existing house with those of the retrofitted house, to analyze the carbon benefits of the retrofit project. eQUEST software was used to simulate the energy consumption during its use phase, with the indoor temperature set at 26°C in summer and 18°C in winter (heating from October 15 to March 15), 5W/m^2 for lights, 10W/m^2 for equipment, and four people. The specific results are shown in Table 2.

Table 2 Calculation results of life cycle carbon emissions

Stage	Exciting	Retrofitting
Building materials production ($kgC0_2$)	/	46611.07
Building materials transportation ($kgC0_2$)	/	6397.1
Retrofit construction ($kgC0_2$)	/	1993.8
Operation stage ($kgC0_2$)	333128.1	174458.3
Dismantling ($kgC0_2$)	558.9	558.9
Waste disposal ($kgC0_2$)	3320.4	-1553.4
Aggregate ($tC0_2$)	337.1	228.5

3.3 Life cycle carbon emission analyzing

3.3.1 Carbon Emission Analysis in the Retrofit Phase

The retrofit phase in this study includes the production of building materials for the retrofit project, the transportation of building materials, and the construction of the retrofit. As shown in Figure3, the energy efficiency retrofit project produced a total of 55 tons of CO2 during this phase, with the production and transportation of building materials accounting for about a quarter of the total life-cycle emissions of 53 tons. As shown in Figure 4, the three building materials with the highest carbon emissions during the production phase are aluminum doors and windows (19.87 tons), silicate cement (11.20 tons), and polystyrene insulation board (7.43 tons). The rest of the primary carbon emissions of building materials production stage is minor, such as PVC waterproofing membrane is about 5.84 tons, lime is about 1.43 tons, while carbon emissions are generated by sand, stone, water, etc. are negligible. According to the default transportation distance of building materials in the "Construction Carbon Emission Calculation Standard," the carbon emission at this stage accounts for 3% of the total, and the carbon emission is relatively small, accounting for 1% of the whole because the construction of rural residential renovation is mostly self-built and dominated by manual work. It can be seen that the carbon emissions in the stage of the energy-saving renovation project have a non-negligible contribution to the whole life cycle carbon emissions, and it is more scientific and reasonable to use the entire life cycle analysis method rather than just the use stage after the renovation to evaluate the carbon emission benefits of the energy-saving renovation of existing buildings.

 https://doi.org/10.1051/e3sconf/202132901063

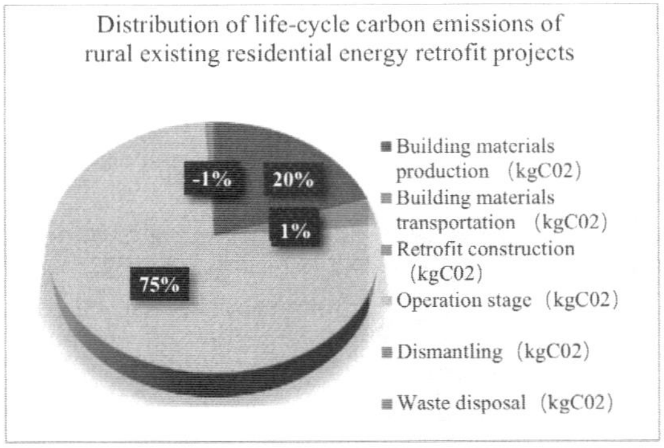

Fig.3. Distribution of life-cycle carbon emissions of rural existing residential energy retrofit projects

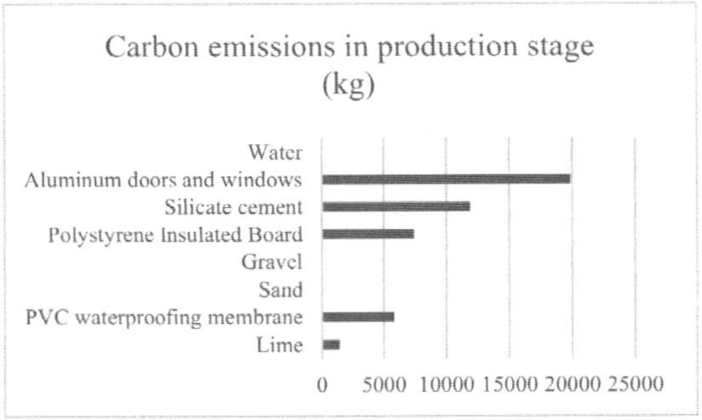

Fig.4. Carbon emissions by building materials production stage

3.3.2 Carbon emission analysis of the use phase after retrofitting

In the 30-year life cycle, the carbon emissions from the use phase of the retrofitted house account for about 75% of the total life-cycle carbon emissions, with the preprimary carbon emissions coming from energy consumption for winter heating and electricity for daily use. According to the simulation results of eQUEST software, the carbon emission in the use phase of the renovated house is reduced from 50.25 $kgCO_2{\cdot}m^2{\cdot}a^{-1}$ before the renovation to 26.22 $kgCO_2{\cdot}m^2{\cdot}a^{-1}$ per unit ar47.82%, which proves that the energy-saving renovation has a significant emission reduction effect. However, the reality is that domestic energy consumption in the area is mainly fuelwood, straw, non-renewable and inefficient [21].one of the main factors contributing to the unsatisfactory carbon emission benefits of the energy efficiency retrofit.

3.3.3 Analysis of carbon emissions in the closing phase

The end-stage is the demolition and waste disposal stage of the project. As shown in Figure 3, the carbon emissions from the demolition and waste disposal phases account for a tiny portion of the total life-cycle emissions, about 1%. It is noteworthy that the carbon emissions from the waste disposal phase of the retrofitted rural houses are harmful due to the use of bridged aluminum windows and doors, which have a high recycling value and can offset some of the carbon emissions from the production of the building materials. This phenomenon also proves that using environmentally friendly recyclable building materials can contribute to energy saving and emission reduction goals in the building industry.

In summary, although the retrofitting project generated an additional 55 tons of carbon emissions during the retrofitting process, the energy efficiency of the original dwelling was significantly improved during the use phase after retrofitting, and the emission reduction contribution per unit area was as high as more than 47%. As a whole, the total CO_2 emission of existing rural houses is about 337.1 tons in the whole life cycle, while the total CO_2 emission of the places after energy-saving renovation is about 228.5 tons. The carbon emission reduction benefit reaches 32.2%, which proves that rural houses have a high potential for energy saving and emission reduction and the need for energy-saving renovation.

4 Conclusion

With the increasing concern about environmental problems, China, as the world's largest developing country, has put the goal of achieving "peak carbon and carbon neutrality" on its agenda. This study analyzed the life-cycle carbon emissions of a rural housing retrofit project in Gansu Province using a life-cycle analysis method. The following conclusions were drawn to reference the early achievement of the "double carbon" goal and the construction of an ecologically livable countryside.

(1) Through calculation, it is found that the retrofitting of existing residential buildings also generates a large number of carbon emissions in the pre-use retrofitting phase accounting for about 1/4 of the whole life cycle emissions, and this part of carbon emissions is often neglected in previous studies. Therefore, it is more scientific and reasonable to apply the entire life cycle analysis method to evaluating energy-saving and emission reduction benefits of existing buildings' energy-saving renovation.

(2) Existing rural dwellings have great potential for energy saving and emission reduction retrofitting. By analyzing the carbon emissions of the cases, it can be seen from the analysis that the energy-efficient retrofitted dwellings can reduce CO_2 emissions by more than 30%. The use phase of carbon emissions accounts for about 75% of the LCE, most caused by inefficient fossil fuels. It will make a significant contribution to achieving the "double carbon" goal.

(3) From the results of this study, it can be seen that the recycling of waste can offset part of the carbon emissions in the waste disposal stage after building demolition. However, in reality, the disposal methods for building debris are still mainly landfill-based. Strategies to improve waste recycling rate include secondary use of waste concrete to make aggregates [22]. It can also effectively reduce the LCE of buildings.

This study analyzes the LCE of a rural residential energy efficiency retrofit project in Gansu from a microscopic perspective. Other methods can be used in the future to analyze existing rural buildings in different regions at a macroscopic level. Other GHG emission analyses should also be introduced [5], providing a more informative data basis for local governments when formulating emission reduction planning policies.

Acknowledgments:

This work was financially supported by the National Natural Science Foundation of China (72061023) and the Natural Science Foundation of Gansu Province (20JR10RA173) fund.

References

1. Xi Jinping, speech at the general debate of the 75th session of the UN General Assembly [R]. Beijing, September 30, 2020
2. China Building Energy Conservation Association Energy Consumption Special Committee, China Building Energy Consumption Research Report (2020) [R].2021.
3. Zheng, G., and W. Bu. "Review of Heating Methods for Rural Houses in China." Energies 11.12(2018).
4. He, B. J., et al. "Overview of rural building energy efficiency in China." Energy Policy 69. Jun. (2014):385-396.
5. Xie, B. C., et al. "Assessment of energy and emission performance of a green scientific research building in Beijing, China." Energy and Buildings 224(2020):110248.
6. Ma, M., et al. "Low carbon roadmap of the residential building sector in China: Historical mitigation and prospective peak." Applied Energy 273 (2020): 115247.
7. Qin Ardian, Yuan Yanping, Jiang Fujian. " A study on the whole life cycle carbon emission of subway station buildings: an example of a station on Line 3 in Chengdu. " Construction Economics, 2020(S1).
8. Guo Jianhong, Zhang Shun, Ma Kunru. "Analysis of the benefits of on-demand energy-saving renovation of existing buildings in rural areas in cold northern regions: an example of a farmhouse in Fuping County, Baoding City, Hebei Province. " Hebei Industrial Science and Technology, 37(5):8.
9. Dong, J. Y., et al. "Thermal environment analysis and energy conservation research of rural residence in cold regions of China based on BIM platform." IOP Conference Series Earth and Environmental science 69.1(2017):012009.
10. Xu, J., W. Gao, and Huoxiaoping. "Analysis on energy consumption of rural building based on a survey in northern China." Energy for Sustainable Development 47.DEC.(2018):34-38.
11. "Environmental Management Life Cycle Assessment Principles and Framework." GB/T 24040-2008
12. Chen Sha, Cui Dongge, Zhang Huijuan. "Carbon emission calculation method and case study of buildings. "Journal of Beijing University of Technology, v.42.04(2016):120-126.
13. The standard for Calculating Carbon Emissions from Buildings.GB/T 51366-2019
14. Qian Snap, Feng Tao, Ma Yaoyao, et al. "Carbon emission calculation method for retrofitting existing public buildings in Shenzhen." Construction Technology, 47.S3(2018):98-101.
15. Energy Conservation Design Standards for Rural Residential Buildings.GBT 50824-2013
16. Ma, L., et al. "Farmers' Rural-To-Urban Migration, Influencing Factors and Development Framework: A Case Study of Sihe Village of Gansu, China." International Journal of Environmental Research and Public Health 16.5(2019).
17. Chen, N. C., et al. "Environmental Satisfaction, Residential Satisfaction, and Place Attachment: The Cases of Long-Term Residents in Rural and Urban Areas in China." Sustainability 11(2019).
18. Liu, Y., et al. "Field study of different thermal requirements based on the indoor activities patterns of rural residents in winter in Northwest China." Science and Technology for the Built Environment (2018):0-0.
19. Evaluation Standards for Green Retrofitting of Existing Buildings.JGJ26-2018.
20. Technical Specification for Thermal Insulation in External Walls.IGJT261-2011.
21. Wang, R., and Z. Jiang. "Energy consumption in China's rural areas: A study based on the village energy survey." Journal of Cleaner Production 143.FEB.1(2017):452-461.
22. Yang M. B., Su Y., Li Z. J., Chen C. P., Chen B. X., Guo W. E., Wang H. C. "Pilot process research on the production of fully recycled fine aggregate from waste concrete. "Concrete No.372.10(2020):85-89.

 https://doi.org/10.1051/e3sconf/202132901064

Analysis of Petroleum Geological Exploration and Reservoir Evaluation Methods

Xin Guan

Liaohe Oilfield Exploration and Development Research Institute, Panjin, Liaoning, 124000. China

Abstract. As one of the important natural resources, oil resources play an irreplaceable role in China's social and economic development. At present, the social demand for oil is increasing gradually, and China's oil production is in the forefront of the world, which is both an opportunity and a challenge for the development of the oil industry. The research on petroleum geological exploration and storage in China is relatively lacking. As a non-renewable resource, excessive exploitation of oil will lead to the shortage of natural resources in China, which is not only extremely unfavorable to the development of the oil industry, but also leads to the collapse of China's social economy. Therefore, the relevant staff must do a good job in the corresponding reservoir work, conduct in-depth research on the exploration technology, and fully implement the reservoir evaluation scheme system, so as to improve the overall comprehensive level of China's petroleum geological exploration, carry out protective development on the utilization of petroleum resources on the basis of meeting the needs of China's social development, and improve the utilization rate of petroleum resources.

Keywords: Oil, Geological exploration, Reservoir evaluation method, Explore.

1 Introduction

With the continuous development of China's social economy, China's comprehensive national strength has been greatly improved. As one of the important strategic resources, oil resources have an excellent role in promoting China's comprehensive strength. However, the national dependence on oil has reached more than 70%. At present, most of China's oil fields are in the middle and late stage of exploration, which will not only seriously affect the stable development of China's society, but also hinder the improvement of China's comprehensive strength, resulting in the limitation of China's survival and development. Therefore, when carrying out daily exploration, relevant staff must do a good job in geological exploration technology, conduct in-depth research on its technology, and fully implement the storage evaluation system, so as to improve the quality and efficiency of oil exploitation, improve the oil utilization rate, alleviate the current situation that China's oil supply exceeds demand, and provide power for the stable and long-term development of China's economy, Comprehensively improve China's comprehensive strength.

2 Summary of petroleum geological exploration

Petroleum geological exploration technology is mainly to ensure that petroleum enterprises can obtain high-quality petroleum resources in the mining process and improve the mining safety and reliability of petroleum enterprises. Therefore, petroleum geological exploration technology plays an irreplaceable role in the development of petroleum industry, and its quality and efficiency will directly determine the output of Chinese petroleum enterprises. Although China has abundant oil reserves, compared with other developed countries, China has less exploitable oil reserves and low production. Most of the oil resources are still in the undeveloped stage, mainly because most of the oil is located in several kilometers underground, and the surface environment will be damaged during the exploitation of stone oil, It is easy to cause collapse of geological structure. In the process of oil exploitation, relevant staff need to do a good job in corresponding exploitation, and conduct in-depth analysis on the specific structure, reservoir conditions and surface environment of the oil area, so as to formulate a scientific and reasonable exploitation scheme, so as to avoid potential safety problems in the exploitation process and affect the overall quality and efficiency of oil exploitation. China's petroleum geological exploration is mainly divided into sedimentary structure exploration, permeability rhythm exploration, horizontal and vertical permeability and particle size rhythm. Among them, sedimentary structure exploration is mainly to explore the underground bedding. In the process of oil exploration, the underground bedding is different, and its permeability will also be different. For example, horizontal bedding, massive bedding, etc., both of which have different penetration directions. The staff can judge the specific situation of the oil resource stratum by exploring the sedimentary structure, so as to judge the oil resource area. The horizontal permeability and vertical permeability are mainly used to judge the formation permeability during reservoir water injection. Generally, the horizontal permeability and pure plant permeability are located in a high position, so the water washing range is large and the overall fluid permeability is ideal. The particle size rhythm is mainly based on the law of debris particle change in sand dust to explore the oil underground. The physical shape of sand layer will be different with different particle size rhythm. The staff can clearly identify the specific conditions of the formation by using the particle size rhythm exploration technology. Finally, the permeability rhythm exploration contains more contents. Such as positive rhythm transformation, anti rhythm transformation and consistent rhythm transformation. In the process of petroleum geological exploration, the rhythmic exploration of permeability can distinguish the oil permeability of different rhythms, so as to master the actual situation of oil formation, formulate corresponding production measures according to the actual situation, and do a good job in safety while ensuring the quality and efficiency of oil production.

3 Current situation of petroleum exploration in China

3.1 Higher degree of practicality

With the decrease of China's oil exploitation, the available oil resources are gradually shrinking. Therefore, the relevant staff need to focus on the practicality of oil geological exploration technology research, improve the practicality of oil geological exploration work in an all-round way, and re-integrate and adjust oil exploitation projects, so that they can adapt to the development of the times and meet the needs of the public. Relevant staff can change the focus of traditional petroleum geological exploration, change the research object, and conduct petroleum geological exploration technology research based on medium-term and short-term needs, so as to ensure the long-term development of petroleum exploration technology and solve the situation that China's oil resources are in short supply.

3.2 Gradually introduce innovative ideas

With the changing structure of China's national economy, China's oil industry is also facing many challenges, especially the shortage of oil resources. The shortage of oil resources will not only affect the harmonious development of Chinese society, but also hinder the overall comprehensive strength of the country, resulting in the decline of the overall strength of the country and the inability to gain the right to speak in the world system. Therefore, relevant staff gradually began to pay attention to petroleum geological exploration work, actively introduced new concepts, and reformed the traditional concepts according to the actual situation of China's petroleum exploitation work, so as to make the concept of petroleum exploration work meet the national development needs.

At present, China has paid more and more attention to the study of petroleum geological exploration. During the research, the researchers analyzed various plate structures and geology of oil-bearing basins, such as oil-bearing basin dynamics, oil-bearing system, reservoir lateral prediction, natural gas geology and so on. Through the research, we can master the specific topographic features and relationships of different petroliferous basins in an all-round way, thus improving the overall quality and efficiency of geological exploration. In addition, China's information technology is now developing well, and most oil companies gradually apply information technology in the process of petroleum geological exploration, optimizing and innovating traditional oil exploitation concepts and technologies, and forming new exploration concepts and advanced technical guidance, which is an excellent opportunity for petroleum geological exploration. Therefore, the relevant staff must seize this opportunity to optimize the petroleum geological exploration work, so as to provide power support for the harmonious development of Chinese society.

3.3 Improvement of scientific and technological level of geological exploration

Due to the influence of market economy, oil companies, as China's energy industry, are facing increasingly fierce competition pressure. China's science and technology have developed vigorously in recent years, which has a certain impact on oil exploration. Most oil companies in China gradually use science and technology to innovate exploration work, and increase the proportion of investment in science and technology. The purpose is to independently develop first-class exploration technology, thereby improving the comprehensive strength of enterprises and enabling enterprises to gain certain competitive advantages in the market economy. Therefore, the scientific and technological work of geological exploration in China plays a great role in the process of oil exploitation, which can improve the quality and efficiency of exploration work, reduce the exploration cost of enterprises and maximize the economic benefits of enterprises.

3.4 Research on realizing cooperation

Talent team is one of the important factors that determine the development process of oil enterprises, and it is also the key to determine the quality and efficiency of geological survey in oil enterprises. As one of the important energy sources in China, in order to improve the overall output of oil enterprises, the national government has issued many policies, the purpose of which is to combine enterprises and institutions, strengthen the cooperation between enterprises and institutions, and enable enterprises and institutions to cooperate with each other to study exploration technology. Therefore, the cooperative research mode has been widely used in petroleum geological exploration. The cooperative research mode can not only improve the utilization rate of resources to a certain extent, but also improve the exploration technology and rationally allocate the exploration capacity of oil enterprises.

4 Classification of petroleum exploration technology in China

4.1 Logging technology

This technology mainly measures the physical parameters of petroleum geology, and measures the physical properties of rock formations, such as acoustic characteristics and electrochemical characteristics. Nowadays, the development of information technology in China is ideal. Under this background, various logging technologies of Chinese petroleum enterprises have been greatly improved, such as imaging technology, magnetic resonance, casing and so on. Imaging logging technology mainly uses imaging instruments to collect and transmit its data in the exploration process, so as to provide accurate values for oil exploitation, thus improving the quality and efficiency of oil exploitation in all directions. Imaging instruments can also effectively improve the exploration depth, greatly improve the sampling rate of oil resources and improve the accuracy and reliability of its data. In addition, due to the rapid development of information technology in China, the data processing capability of Chinese oil companies in applying logging technology has also been greatly improved, such as data acquisition, data processing and data analysis, which provides technical support for the development of logging technology industry.

4.2 Drilling technology

Drilling technology is not widely used in petroleum exploration, mainly because drilling technology has higher construction cost compared with other exploration technologies. Therefore, the relevant staff need to innovate the drilling technology and reform its technology on the basis of reducing the cost of oil exploration. Workers can optimize traditional drilling technology and use information technology to establish visual drilling technology, and can also study deep well drilling technology and special process drilling technology according to the characteristics of oil wells, so as to ensure the quality and efficiency of drilling technology and reduce the production cost of enterprises.

5 Content of petroleum reservoir evaluation system

5.1 Comprehensive classification evaluation

The comprehensive classification evaluation system is mainly to clarify some parameters in the reservoir, and make an in-depth study on the factors that may affect the quality and efficiency of the reservoir based on these parameters, and make a scientific and reasonable evaluation of the storage, so as to formulate corresponding solutions. Because of the difference of reservoir types, geological exploration workers will have different ways in the process of exploration. Only by doing a good job of reservoir evaluation and ensuring that the staff can fully understand the actual situation of underground reservoirs, can the risks in the process of oil exploration be reduced to a certain extent, and the safety production work can be done while ensuring the quality and efficiency. Attention should be paid to the need for staff to make clear the reservoir parameters when evaluating the reservoir, so as to ensure the scientific and effective results and provide data support for oil exploration and development.

5.2 Underbalanced drilling

This method is one of the most widely used methods in reservoir evaluation system of petroleum enterprises at present. It mainly evaluates reservoirs when drilling with corresponding detection tools and equipment, which can not only improve the quality and efficiency of reservoir evaluation work to a certain extent, but also reduce the evaluation cost and improve the economic benefits of petroleum enterprises. Scientific and rational use of underbalanced drilling technology by workers can keep the bottom hole under pressure for a long time, thus introducing the fluids in the formation into the well. The staff can analyze and collect all the information of oil reservoir and dynamically evaluate the reservoir when obtaining the specific data of oil production, so as to ensure the quality and efficiency of evaluation. Compared with other evaluation techniques, underbalanced drilling evaluation technique has lower cost and higher work efficiency. Professionals comprehensively collect data from all aspects of the reservoir, and make scientific and reasonable analysis based on the data, so as to ensure the scientificity and rationality of its mining work, make its data more reliable and professional, and avoid potential safety hazards in the later mining process, endangering the personal safety of workers.

5.3 Comprehensive quantitative

The geological environment of oil geological reservoir is special, which is easily affected by many factors, resulting in changes in geological environment, such as diagenetic characteristics, reservoir characteristics, sedimentary characteristics and so on. Therefore, the relevant staff need to comprehensively consider the corresponding factors when evaluating the reservoir, so as to effectively improve the quality and efficiency of reservoir evaluation. The comprehensive quantitative evaluation system of storage is to comprehensively evaluate it according to its storage characteristics, which can comprehensively control the oil quality of different oil reservoirs in different regions and different stages. This method mainly combines various evaluation methods, such as grey correlation method and fuzzy comprehensive evaluation. The staff can master the corresponding information in all directions, so as to ensure the quality and safety of oil exploitation in the later stage. Because there are many aspects involved in reservoir evaluation, when applying comprehensive quantitative evaluation method, relevant staff need to select corresponding evaluation methods according to different reservoir areas and characteristics, and determine the number of related rights and interests, so as to improve the scientificity and accuracy of evaluation results in an all-round way, and reasonably distinguish reservoir characteristics of different intervals and regions. When the related staff evaluate the reservoir, they also need to comprehensively evaluate the related rock strata, so as to comprehensively collect all the influencing factors, provide accurate and reliable information for the mining personnel, and comprehensively improve the safety in the process of oil exploitation.

6 Conclusions

To sum up, although China's science and technology are developing well, and the development of new energy is strong, as a basic resource, the demand for oil, whether it is the development of China's industrial industry or the daily life of the public, is only increasing. At present, the petroleum resources developed by China can no longer meet the needs of the market. Therefore, relevant staff must actively introduce advanced scientific mining technologies, conduct in-depth research on petroleum geological exploration, improve the utilization rate of resources, and give full play to the role of petroleum resources. At the same time, the relevant staff also need to improve the storage evaluation system, so as to ensure the scientific and reasonable mining work, improve the output of oil enterprises in all directions, and provide power for the harmonious development of Chinese society.

References

1. Wan Peng, Wang Xinyi. Analysis of petroleum geological exploration and reservoir evaluation methods[J]. China Petroleum and Chemical Standards and Quality, 2021, 41(02): 4-6.
2. Lin Ziwei. Analysis of petroleum geological exploration and reservoir evaluation methods[J]. China Petroleum and Chemical Standards and Quality, 2020, 40(15): 7-8.
3. Ning Guolan. Innovating petroleum geological exploration and reservoir evaluation methods with information technology[J]. Information System Engineering, 2019(07): 156.
4. Wang Xuansu. Discussion on petroleum geological exploration and reservoir evaluation methods[J]. China Petroleum and Chemical Standards and Quality, 2019, 39(13): 3-4.

Study on Reasonable Injection-Production Parameters of Water-Flooding Reservoir: A Case Study of Daqing Oilfield

Cong Nie

The Sixth Oil Production Plant, Daqing Oilfield Co., Ltd., Daqing 163114, Heilongjiang, China

Abstract. This paper, by conducting comprehensive analysis on water-flooding reservoir, intensively studied and discussed the characteristics of geological development, determined the evaluation indexes of development effect, carried out research from five steps, the first is the introduction part, it conducted in-depth study and discussion from research background, research significance, literature review, and research methods four aspects. The second outlined the related theory from the optimization principle and mechanism of injection-production parameters, and the distribution mechanism of remaining oil two aspects. The third analyzed the current situation of reasonable injection-production of water-flooding reservoirs in Daqing Oilfield from reservoir characteristics and reservoir deposition. The fourth evaluated the water-flooding development effect from combined water cut, changing conditions of production and predicted recovery ratio. The fifth intensively studied the reasonable injection-production parameters of water- flooding reservoirs in Daqing oilfield from design of separated layer water injection plan, maintaining level reservoir formation pressure, and reasonable plan of injection-production ratio and oil recovery rate three aspects, and provides theoretical guidance for later targeted excavation of water-flooding reservoirs.

Keywords: Water-flooding reservoir; injection-production parameters; Daqing oil field; structural characteristics.

1 Analysis of the Current Situation of Reasonable Injection-Production of Water-Flooding Reservoirs in Daqing Oilfield

1.1 Survey of Daqing Oilfield

Daqing Oilfield was discovered in 1959 and started to develop in 1960, Daqing Oilfield is the largest oilfield in China, and it is one of the largest continental sandstone oilfields in the world. Daqing Oilfield is located in the northern part of the Songnen Plain and the middle and western part of Heilongjiang, the oil and gas in this oil field consists of oil and gas in Lamadian, Hailar, Chaoyanggou, Saertu, Xingshugang and so on. There are more than 90000 oil and gas wells in Daqing Oilfield. Generally speaking, after fracture, the faults in Daqing Oilfield mainly contain oil and gas layers, they extend upward and cross the cap rock and oil layer of group I and group II of Nenjiang River, and some even extend to the area of Mingshui group. Due to the channel nature of the faulted water layer, oil and gas can move upward, they can be stored in the top of the section II of Nengjiang, the section III of Nengjiang and the section V of Nengjiang area, and even reach the Mingshui group, thus forming the secondary oil and gas reservoir. They are still controlled by the structure. The reservoir types in the Daqing Oilfield are relatively simple, due to the influence of anticlinal structural factors, it plays an important role in the formation of oil and gas reservoirs in this area, due to the influence of faults, lithology and other factors, it also has a certain impact in some positions, then structural oil-gas reservoirs and different types of composite oil-gas reservoirs are formed.

1.2 Characteristics of reservoir deposit

The production oil layers of Daqing oil field have the following layers: the first is the Putaohua oil layer. This layer belongs to the important containing layer in the southern of Daqing oil field; in addition, there are mainly fine sands in the southern of Aobaota tip. The average porosity is 17% to 20%, the thickness of fine sands is generally 9 to 18m, the permeability is 10-3 m^2×(100-240). The second is the Salto oil layer. This layer is the main production layer of the oil field, this layer is mostly composed of medium sandstone and fine sandstone, this layer mainly includes four storage oil groups, they are mainly S0, S1, S2, and S3. The northern S2 group of the oil field belongs to the medium prototype river deposition; the shape of the single sand body is geometrical, presents

https://doi.org/10.1051/e3sconf/202132901065

the belt distribution to the south of Changyuan, presents the shape of sandstone lens body, and it is deposited in the front facies of the delta. The type of sand layer in the area, its thickness is 5-10m, the average porosity of this layer is 18%-20%, and its permeability is 10-m^2×100, this area belongs to permeable oil reservoir. The third is Gaotaizi oil layer. This layer is mainly distributed in the northern area of Daqing oil field, its porosity is 18% to 26%, the thickness of its single layer is only about 1m, and the permeability is 10-3 m^2×(50~350).

2 Effect Evaluation of Water-Flooding Reservoir

This paper studies the side energy water of Daqing Oilfield, the amount of water is weak, natural energy is weak, stratum pressure and saturation pressure is relatively small, self-spurt capacity of oil wells, if we want to obtain better collection effect, we should supplement the energy of stratum by artificial way. Since it is more convenient to obtain water resources on water, and it has the advantages of water-flooding development technology and economic maturity, therefore, water injection development must be used to collect. However, the water-flooding process is complicated due to the influence of geological factors and development factors. The following conducts comprehensive evaluation for the existing geological development data and development effect of oil field, its evaluation is also to provide a certain theoretical basis for the appropriate adjustment of oil field development later. There are many approaches to evaluate the oilfield development, which mainly include the multi-index evaluation comprehensive evaluation approach and the single-index evaluation approach, and they contain the integrated water content, production change situation, predicted recovery rate and other aspects of contents.

2.1 Comprehensive Water Content

Comprehensive water content evaluation is a comprehensive concept, which can reflect the water content of crude oil in oilfield and the development effect of the oilfield. When a field which is developed with water-flooding and reaches a high water content stage, the comprehensive water content evaluation approach plays an important role in adjusting the field development plan and studying the field development dynamics.

2.1.1 Longitudinal Distribution of Water Content

Due to the shallow burial depth, many stratums, and thick oil layer in test area, and the longitudinal inhomogeneity of the reservoir is more serious. In addition, there is a certain difference in the watered-out degree in the longitudinal direction of the reservoir. Therefore, there are large differences in the development effect of water flooding. The author used eclipse numerical simulation software for the longitudinal distribution evaluation of water content in the study area, divide the reservoir water flooding longitudinally in accordance with sections, the vertical distribution of water flooding in each main oil layer can be obtained, as shown in Fig.1. It can be concluded that the water content in the oil layer group is generally high via specific analysis and research. The water content of each oil layer group is relatively low, and the layered water injection amount is adjusted appropriately based on the water content of different levels, so achieve the purpose of improving the development water-flooding effect.

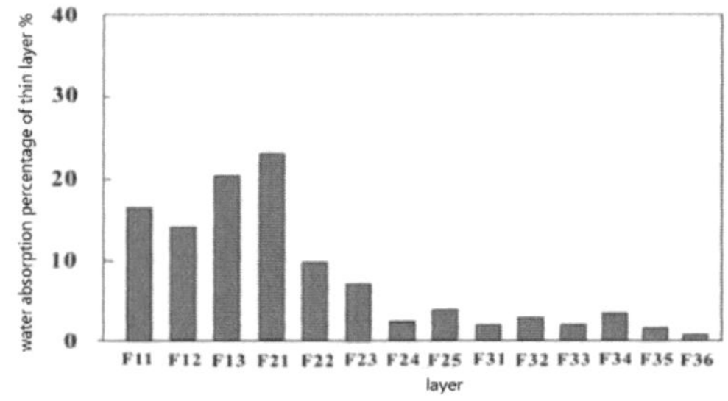

Fig.1 Water content distribution in the test area

2.1.2 Relationship between Recovery Layer and Water Content

At present, the commonly used methods to analyze the relationship between the recovery layer and water content have flow pipe method, phase permeation curve method and Tong's calibration curve method. Relatively speaking, the principle of phase permeation curve method is relatively simple and quick to calculate, and it is a common method in reservoir engineering. According to the curve, it can be seen visually that different adjustment methods can be used for different conditions at different time points, and the different adjustment effect can be obtained. The development effect at different stages can be correctly evaluated by effectively comparing the theoretical curve with the actual curve. In addition, it is also possible to predict the condition of recovery ratio.

The prediction for the test area can obtain the water content, by comparing the change curve between standard chart and recovery degree, as shown in Fig.2. There are mainly five curves within the standard chart, the reservoir replacement characteristics and oil storage properties of each curve are somewhat different, the relationship among the gap structure, wettability, and fluid properties of reservoir rock are comprehensively analyzed. The formula relationship:

Formula 1: convex curve: $R=B(\ln-f\_w)$

Formula 2: the transition curve between S-type and curved: $A+B(\ln-f\_w)=\ln(1-R)$

Formula 3: the transition curve between concave line S type: $A+B(\ln 1—f\_w)=\ln R$

Formula 4: S-type curve is R: $R=B\{\ln 1-f\_w/(1-f\_w)\}+A$

Formula 5: concave curve: $\ln R=B\ln f\_w+A$

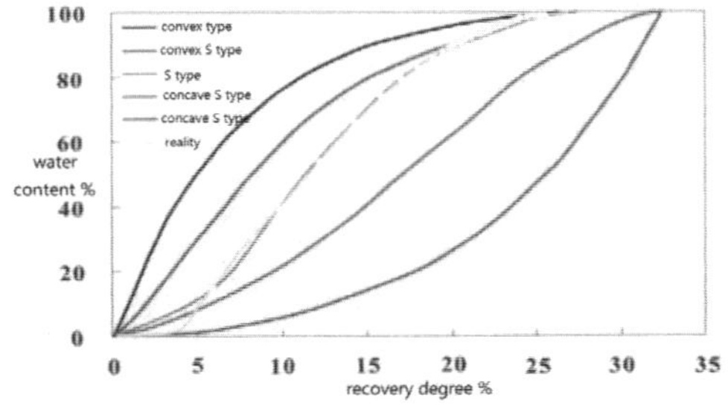

Fig.2 Relationship curve between the recovery degree and water content

At present, by drawing recovery degree and standard charts of water content, and comparing them with relationship curves, we can conclude that the curve is consistent with the concave shape in the early stage, and roughly keep with the S type in the later stage, the water content rises quickly in the early stage. This is because the reservoir type in the test area is a highly porous and highly permeable oil layer. When water is injected into the oil wells, the concentration of oil and water is relatively large and the water content rises fast, so we can conclude that most of the reservoir reserves are recovered from the high water content period.

2.2 Changing Conditions of Output

The evaluation of the dimensionless fluid recovery and oil recovery index can be used to obtain the changing conditions of oil amount. First, the actual fluid parameters, reservoir physical properties and phase permeability curves of the test area are applied to the mathematical model of oil recovery index, and the relationship among dimensionless fluid recovery, oil recovery index and water content in the test area is drawn, as shown in Fig.3. The dimensionless fluid recovery index in the test area continuously increases as water content rises; in addition, the dimensionless recovery index continuously increases as water content rise, its variation is relatively large. This is because the study area actually belongs to the reservoir with high oil-water viscosity and high permeability. When the flow pressure and stratum pressure are stable, the index of dimensionless fluid recovery will be influenced by the oil-water viscosity and permeability, it will and more severe. In general, the permeability will affect the changing trend of the dimensionless fluid recovery index, but the oil-water viscosity will affect the dimensionless fluid recovery index, so there will be a large upward trend in the high water-bearing period.

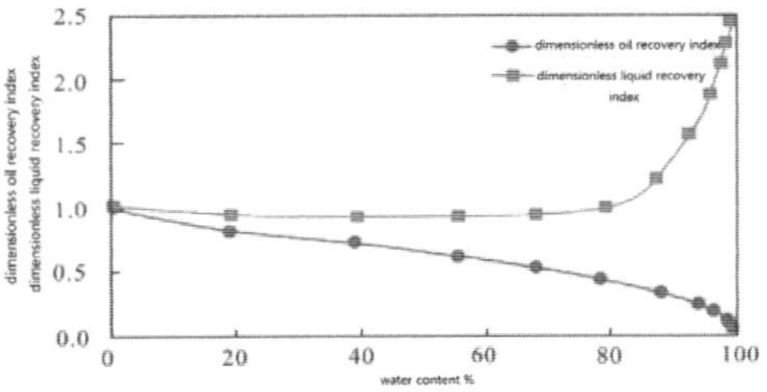

Fig.3 Relationship curve among water content, dimensionless oil recovery and liquid recovery

2.3 Predicted Recovery Ratio

The development level of the oil field can be measured by collecting the pre-recovery ratio, which is influenced by the reservoir management level, structural characteristics, extraction mode, extraction technology and other relevant factors. According to the reservoir characteristics of the oil field, more prediction methods of recovery ratio will be generated. At present, the more commonly used methods are water-flooding law curve method, empirical formula method, water-flooding oil experiment method, core analysis method, etc. In the following evaluation of oil collection condition, the water-flooding law curve method is mainly used, and theoretical derivation is used to predict the recovery ratio of the test area.

The water-flooding law curve method is mainly based on the oil-water two-phase permeation mechanism, creates the mathematical relationship based on the production law, and the quantitative relationship of production data is built by using regression curves in the median coordinate system, the recovery ratio conditions during the limit water cut period can be obtained. The test area of this study belongs to layered reservoir with high crude oil viscosity. According to the curve selection criteria of water flooding, the regular tendency of water flooding of type B can be collected. Predecessors obtained the half pair curve relationship between cumulative oil production and cumulative liquid production via the specific form of empirical formula, later, it was uniformly named type B water flooding. The following is the specific expression formula.

$$log L_P = \alpha + bN_P$$

In the formula. α、b are regression constants and are dimensionless. L_P is the cumulative fluid production 10^4t. The relation between water content and cumulative oil production: $N_P = \frac{log\left(\frac{1}{1/w}\right) - c}{b}$

In the formula: $c = \alpha + \log\ (2.303.b)$

If in the state of extreme water content, therefore, the formula of the amount of reserves that can be recovered can be obtained:

$$NN_{lim} = \frac{log\left(\frac{1}{1/wlim}\right) - [\alpha + \log\ (2.303b)\ ]}{b}$$

Taking the specific experimental area as an example, the water drive law curve of type B is effectively fitted, and the fitted law which we come to is shown in Fig.4.

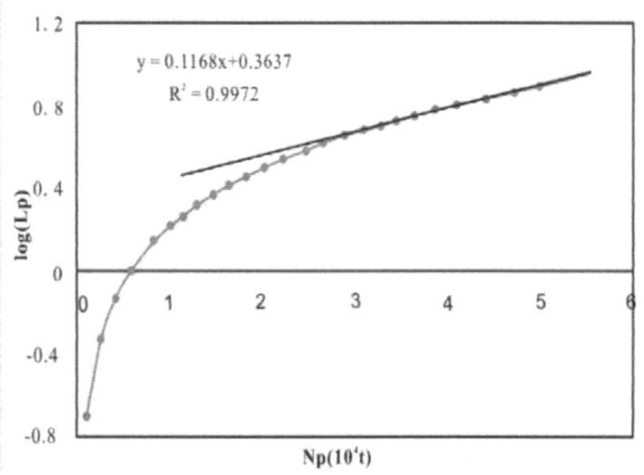

Fig.4 Curve of B-type water-flooding law in the test area

Table 1. Gets the calculation results of predicted recovery ratio of water-flooding law curve in the test area.

parameter	a	b	recovery degree %	
			water content 95%	water content 90%
numerical value	0.3635	0.1166	28.84	26.99

According to the above table, when the extreme water content reaches 90%, the predicted recovery ratio is 28.84%, and when the extreme water content reaches 95%, its predicted recovery ratio is 28.84%.

3 Conclusion

In conclusion, the development of oilfield projects is the key contents of national project, and its research and construction is an important content in promoting the sustainable development of China's petroleum cause. Moreover, the development of water-flooding reservoirs in our country has entered into the middle and late stage, many problems and contradictions have gradually emerged, this causes the relevant departments to deepen their exploration of the development law of water-flooding reservoirs, and continuously study water-flooding reservoirs as well as water-flooding reservoirs with good water content in depth, and then help people better solve the problem of the conflict between supply and demand.

References

1. Lv Duanchuan, Wu Wei. Injection-Production Parameter Adjustment of Water Flooding Development: Taking Two Fault-Blocks on Both Sides of F Fault of Daqing Oilfield Bei'Erxi as Examples[J]. Fault-Block Oil & Gas Field, 2017(06):83-86.
2. Huang Bin, Xu Rui, Song Kaoping, Fu Cheng, Zhang Wei, Shi Zhenzhong. Identification of Thief Iones Between Injection Wells and Production Wells of Standstone Reservoirs Flooding by Fuzzy Comprehensive Analysis Method[J]. Mathematics in Practice and Theory, 2017(13), 127-135.
3. Long Ming, Xu Yanan, Yu Dengfei, et al. Quantitative Optimization of Water Injection for Bohai Heavy Oil Reservoir with Bottom Water: A Case Study of the West Block of QHD32-6 Oilfield [J]. China Offshore Oil and Gas, 2017(4), 91-97.
4. Li Shujing, Zhang Tianjie, Yang Jian, et al. Study on Prediction Method of Reasonable Injection-to-Production Ratio in High Water-cut Stage of Yan'an Formation Reservoir: Taking A Reservoir in Ansai Oilfield as an Example[J]. Technology Supervision in Petroleum Industry, 2017 (10), 12-14.
5. Wang Langbo, Gao Xiangrui, Dai Bo, et al. Study on Single Sand Body in Ultra-Low Permeability Reservoir and Adjustment to Injection-Production: An Example from Wangyao Area of Ansai Oilfield[J]. Journal of Petrochemical University, 2018, 31 (2): 82-88.
6. Zhang Jingtao. Optimization of Injection-Production Parameters and Development Index of Mixed Well Pattern in Daqing M15-1 Block[D]. Yanshan University, 2016.
7. Shi Pengtao, Chen Penggang. Study on Reasonable Injection-Production Ratio of Wangjiawan Chang 2 Reservoir [J]. Petrochemical Industry Application, 2017(04): 82-83,93.

E3S Web of Conferences **329**, 01066 (2021) https://doi.org/10.1051/e3sconf/202132901066
GESD 2021

Reservoir Development by Waterflooding and Reservoir Performance Analysis

Jiali Li

Geological Brigade of the Sixth Oil Production Plant of Daqing Oilfield Co., Ltd., Daqing, Heilongjiang, 163453, China.

Abstract. With the rapid development of China's society and economy, relevant industries get good development opportunities, and the demand for oil is increasing. The oil production in the oilfield is increasing day by day. At present, most oilfields gradually enter the middle and late stage of water injection exploitation, which increases the exploitation difficulty and reduces the exploitation amount. In this paper, a comprehensive analysis of waterflooding reservoirs and reservoir performance is made to promote better and faster development of petroleum industry.

Keywords: Water injection development, Reservoir, Reservoir performance.

1 Introduction

In the process of oil exploitation, water injection can effectively improve the overall quality and efficiency of oil exploitation, and deal with the water injection development technology accordingly. So as to more comprehensively grasp the key methods of reservoir performance analysis, improve the scientific rationality of water injection production technology, and effectively improve the overall level of oilfield production.

2 Reservoir analysis of water injection development

According to the analysis of actual oilfield production, it is necessary to ensure the accuracy of experimental information and data, and at the same time have sufficient information, so as to make corresponding accurate judgments on reservoir dynamic analysis. In addition, the relevant analysts need to process the data for the geological model, ensure the information accuracy of the data model, give full play to the advantages and functions of computer technology, start the simulation test on the data model, and then process the oilfield production effect well. In the development process of waterflooding reservoirs, some oil layers are flooded, resulting in poor fluidity, so it is necessary for the mining personnel to quickly drain the internal water and promote the oil wells to resume normal and smooth production [1]. The staff can use the cement test to judge the detailed position of the water layer, timely formulate targeted technical scheme and carry out water injection design, so as to ensure the consistency of water injection effect and water flow direction, and prevent the adverse impact of water outlet on the normal and smooth flow of oil and gas flow. In addition, in the process of water injection reservoir development, it is necessary to carry out reservoir management (as shown in Figure 1), fine description and understanding of oil and gas reservoirs, and adopt perfect and reasonable engineering design and implementation to realize efficient development and production of oil and gas fields. The main feature of reservoir management is multidisciplinary cooperation.

Figure 1 Reservoir operation and management

3 Dynamic analysis of waterflooding reservoir

3.1 Analysis content

The main content of reservoir dynamic analysis of waterflooding development is the dynamic change of reservoir during waterflooding, understanding and mastering the actual transformation of reservoir during the change process, and grasping the actual situation of reservoir treatment in time. The reservoir development content in water injection system mainly includes the related connection aspects of various reservoirs. Real-time monitoring of the situation and mastering the changes of bathing places, such as the internal reservoir and associated water, plays a major role in contacting the related water bodies and triggering the eruption of water bodies in the process of reservoir development. The water body in the reservoir plays an important role in the actual situation of the reservoir. In addition, monitoring the distance between fault and interlayer is conducive to relevant staff to timely control the actual situation in the process of oil reservoir exploitation. The distance between fault and interlayer is very important for the distance of reservoir production. In the actual work, the production personnel need to ensure that they will not touch the block of the interlayer and there will be no landmark collapse and fracture. At the same time, for the development of the reservoir, they need to monitor the relevant information data in real time.

During the exploitation of the reservoir, the production personnel need to timely and accurately grasp the changing data of oil, gas, water and rock. Changing data play a vital role in the whole reservoir production [2]. Staff need to timely, comprehensively and accurately grasp the first-hand information in reservoir development, timely control the oil and gas on the surface, and analyze the connection between disciplines by controlling the problems of various disciplines involved in reservoir development. Reservoir dynamic analysis in waterflooding belongs to a detailed and rigorous attention process, which requires relevant staff to control various information in time. The reservoir development content and related aspects in water injection system involve a lot of information and data in reservoir development. Dynamic reservoir development needs comprehensive and timely control of various information data, so as to avoid a variety of bad problems in reservoir development during water injection development.

3.2 Basic situation analysis

In China, the oil field has a relatively complex geology. The chemical and physical properties of the crude oil produced are different from those in the past, and the oil production rate of the oil field is obviously reduced. Therefore, China needs to continuously increase production reserves and research and apply related technologies to enhance oil recovery. In the process of oil production in China, some oil companies are affected by oilfield geology of some multi-layered sandstone reservoirs, resulting in high permeability of crude oil among layers, layers and planes, which has a negative impact on normal and smooth oil production. At the same time, under the condition of serious heterogeneity of the reservoir, there are certain differences in the water absorption of each layer, resulting in the continuous advancement of the water injected into the reservoir under the influence of the high-permeability formation, resulting in great changes in the advancement speed in the vertical and plane, so as to promote the staggered distribution of crude oil and water, resulting in the scattered distribution and mutual accumulation of the remaining crude oil, It further increases the difficulty of oilfield production. In the actual production process of oil field, affected by many factors, injecting a large amount of water into the formation has a great impact on its properties. At the same time, under the condition of long water injection production time, the heterogeneity of reservoir is more serious.

3.3 Characteristic analysis

①Changes of reservoir fluid properties in waterflooding reservoirs. When oilfield water injection development is carried out for a long time, it is most feared that the salinity of formation water, formation temperature, original composition, pressure in formation and other related information data will change dynamically. For example, during the research of China A Oilfield Company, the water saturation of each pore in the reservoir will be continuously improved by the continuous water injection in the oilfield. At the initial stage of waterflooding development, the pore water saturation showed a significant increase, and with the implementation of waterflooding development in oil fields, its increasing trend gradually decreased in the middle and late stage of exploitation. In addition, the water injection production project of the oilfield causes certain dynamic changes in the salinity of formation water to a certain extent. When fresh water is injected into the formation of the reservoir, the salinity of formation water is continuously reduced with the progress of water injection. When bottom water and edge water are injected into the formation of the reservoir, the substances contained in the water body are similar to those in the oilfield, and the overall mineralization in the formation will not change significantly. When crude oil separated water is injected into the formation, the salinity of water in the formation shows an upward trend.

②Dynamic changes of reservoir properties. When people analyze the reservoir properties, they use the model theory to show the reservoir properties to researchers more intuitively. Researchers can use dual media model to show the change of reservoir properties [3]. The water injection pressure of oil field increases gradually with the water injection collection. When the water injection pressure reaches the upper limit of the fracture pressure of the ground layer around the oil well, cracks will occur on the ground surface. After the cracks are restored, the model will still change to a certain extent, which obviously represents the traces of cracks produced by artificial hydraulics. According to the change analysis of well test curve (as shown in Figure 2), the researchers prove that with the continuous water injection, the reservoir will gradually change from homogeneous to composite. At the same time, during the continuous water injection development, the fluid properties of the reservoir will change to a certain extent. The relevant changes in the reservoir are mainly due to the application of water injection production and other methods, which directly leads to the change of reservoir properties, and then leads to a series of conditions.

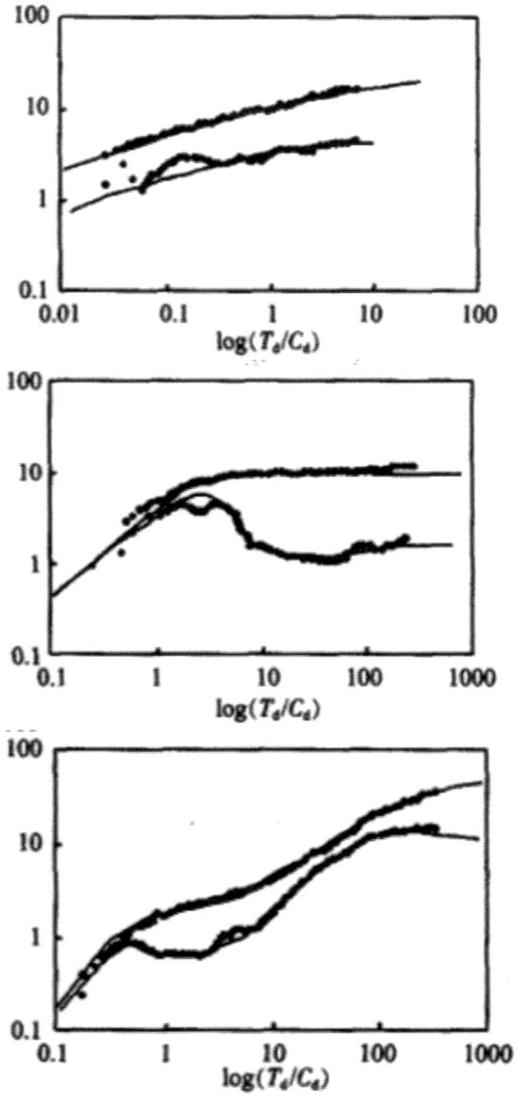

Figure 2 Comparative analysis of the characteristics of the three well test curves of a certain well

③Analysis of reservoir parameter variation. In the process of oilfield exploitation by related workers using water injection development technology, the crude oil and mud between formation pores are constantly washed out by the influence of water pressure of water injection, which increases the porosity of oilfield formation as a whole, and at the same time continuously expands the diameter of pore throat, which greatly increases the connectivity between formations and improves the pore throat diameter ratio of formation pores. In addition, under the influence of water injection flow, the mud between pores will be washed away. When accumulation occurs in the formation of the oilfield, it will affect the permeability of the formation of the oilfield, which is not conducive to the smooth exploitation of oil. The rock surface in the reservoir stratum is affected by the water flow, resulting in the continuous reduction of the oil film attached to the surface, increasing the hydrophilicity of the formation rock and the wettability of the oilfield production formation.

④Analysis of remaining oil distribution. The distribution of remaining oil in oilfield reservoirs will change correspondingly with the continuous development of oilfield water injection [4]. We analyze according to the distribution area, shape and formation reasons. The remaining oil outside the water injection development area is not washed out by the water flow in the process of exploitation, and most of the remaining oil in the bottom layer presents a flaky distribution. In the process of cluster residual oil exploitation, the permeability channel of reservoir stratum is relatively complicated by geological structure, and the periphery of larger permeability channel mostly contains network-like small permeability channel. In the process of water injection exploitation, part of water flow can not wash out the original small permeability channel, resulting in the original surplus being mostly distributed in the bottom channels. The columnar residual oil is mainly distributed in more than two channels, mainly because the water flow does not pass through the channels during water injection production, or the washed crude oil caused by capillary action in the channels cannot be taken away by the water flow. Angular residual oil is mainly in the corners of the bottom layer and areas where water flow is difficult to contact. Island like residual oil is mainly distributed in the macropores of the reservoir. In water injection production, the crude oil in the center of the pore is difficult to be efficiently exploited, resulting in oil surplus. The film like residual oil is mainly attached to the rock surface, which is mainly due to the insufficient scouring force of water flow in water injection production, which can not effectively overcome the adsorption force between rock and oil film to produce crude oil surplus.

3.4 Analytical method

①Numerical simulation analysis method. With the continuous progress and development of China's scientific informationization, numerical simulation analysis method is widely used in the dynamic analysis of oil reservoirs, mainly relying on computer software to effectively and dynamically simulate the development process of oilfield water injection, which is more convenient for researchers to formulate scientific and reasonable oil exploitation plans according to the actual situation of the model [5]. In the process of numerical simulation, researchers need to comprehensively collect a number of information data of the oilfield, create a targeted geological model with data, and divide the model into cells. The whole production process is simulated by the operation of computer software, and the dynamic changes of multiple data of oil reservoirs during exploitation are comprehensively analyzed.

②Well test analysis method. The staff used this method to analyze the performance of waterflooding reservoirs, and effectively analyzed the dynamic changes of corresponding data by relying on the relevant data such as oilfield pressure, temperature and production measured by well testing operations. At the same time, the staff scientifically and reasonably applied the seepage mechanics theory, analyzed and collected information data, effectively obtained the production status, actual productivity and physical property data of oilfield water injection development, and comprehensively analyzed the oil fluid communication in sandstone layer.

③Empirical formula method for plate analysis. This analysis method is mainly used by staff to analyze the changes of oil and gas properties in strata and predict the changes of many parameters, which belongs to the dynamic analysis method of reservoir parameters. Workers need to analyze relevant data in practical production. In the process of measuring and extracting oilfield oil and gas parameters, practical operation is simple, which improves the practical effectiveness of this method. In the process of applying the empirical formula method of chart analysis, the staff provide a large amount of information data collected by the chart method as the basis of data analysis, and then comprehensively analyze and summarize the actual data, determine a set of empirical formula, and use the formula to calculate the relevant data. During the dynamic analysis of water injection reservoir, the staff collected and sorted out various data, and selected appropriate analysis methods according to the characteristics of data change, so as to improve the actual effect of water injection development.

4 Conclusions

In the process of continuous water injection development of the reservoir, many data of the reservoir change to varying degrees, which has a certain impact on the follow-up oil production. Therefore, relevant personnel need to adopt appropriate reservoir dynamic analysis methods for waterflooding development, comprehensively analyze the distribution of remaining oil, understand its change characteristics, comprehensively improve the quality and efficiency of waterflooding, and obtain more comprehensive benefits.

References

1. Liu Cong. Dynamic analysis of water injection development reservoirs[J]. Chemical Engineering and Equipment, 2020(1):141-141.
2. Jiang Xueying. Analysis of dynamic analysis of waterflooding development reservoir[J]. Chemical Engineering and Equipment, 2019, 000(003):171-171.
3. Xu Maohe. Discussion on dynamic analysis of waterflooding development reservoir[J]. Chemical Industry Management, 2019(21):224-224.
4. Pu Menglei. Discussion on dynamic analysis of waterflooding development reservoir[J]. Chemical Management, 2019, 000(026):213-214.
5. Wang Rui. Research on the dynamic analysis method of waterflooding development reservoir[J]. China Petroleum & Petrochemical, 2019, 09(No.368):91-92.

Evaluation of the influence of geographical factors on the difficulty of grid operation cost

Geli Zhang[1*], *Heng Zhang*[1], *Yan Zheng*[1], *Yaqiong Liu*[1], *Tong Li* [1], *Meibao Wang and Xiaohui Wang*[1]

[1]State Grid Economic and Technological Research Institute Co. Ltd., Beijing, China

Abstract. The cost of power network operation and maintenance is affected by many factors, which will affect the cost. Among the many influencing factors, the geographical environment is one of the more important factors. The influence of geographical factors on cost is realized by geographical conditions including distance, topography, climate, etc. . When the distance is longer, the topography is more complex, and the climate change is more severe, for the operation and maintenance costs will have a higher increase in the effect.This paper analyzes the process that the internal and external composing mechanism of Power Grid Enterprise's operation cost is affected by geographical factors, and analyzes the influence way and result of geographical factors on power grid's operation cost.

1 Introduction

The study of power grid cost is always an important research hotspot. For the enterprise management, the cost determines the final business results of the important output indicators: The level of profits. When the profits of enterprises are high, more resources will be put into reproduction, and the enterprises will be more capable of stepping onto the track of sustainable development.

Guided by China's domestic power reform policies, the power industry sector, especially grid enterprises, needs to deal with the regulatory authorities: The Energy Bureau's Electricity Price Regulation, in order to limit the monopoly profit of power grid, which is a natural monopoly, the supervision of transmission and distribution cost is an important part of the price supervision. To some extent, the price of electricity directly depends on the transmission and distribution cost of power grid enterprises, especially the cost of operation and maintenance.

But at present, the closed-loop Cost Management of power grid enterprises has not formed a perfect management system, and the current management level still can not meet the high requirements of accurate management and control costs. In order to push the management level of power grid enterprises to a higher level and increase the grasp of cost management, it is of great significance to carry out the research on transmission and distribution cost of power grid enterprises.

Among them, operation and maintenance cost is the main cost of power grid operation, its cost management was extensive in the past, in 27 provincial companies management differences, did not form a unified and efficient company-level management system. In order to achieve the objectives of management system such as life-cycle management, cost budget management, cost externalization and multi-dimensional lean management, it is necessary to link the labor, material, outsourcing, mechanical cost of operation and maintenance equipment with operation and maintenance activities, to realize the quantitative control of operation and maintenance cost by operation and maintenance standardization cost.

In order to study the cost of power grid, we first need to determine the factors that affect the cost. For the operation and maintenance of substations and lines located in the mountains, hills and plains outside the city, geographical factors will be a major factor affecting the cost of operation and maintenance, so it should be explored from the inside and outside of the impact of geographical factors on the type of power grid cost.

In the related research field of power system, the research of comprehensive evaluation method focuses on the combination of subjective and objective weight and the comprehensive utilization of several methods. The objectivity of the evaluation method is often considered in the traditional evaluation, and the combination weight method is used to make the weight calculation process objective, delphi, AHP, FAHP, principal component method, entropy weight method, grey correlation method, sequence synthesis-g1 method, matter-element extension method and other methods are used to improve the objectivity of weight, but the consistency of different methods is ignored.

* Corresponding author: fredistheone@126.com

2 Geographical factors assessment of grid operation cost

2.1 Evaluation Index

In order to analyze the internal and external influence factors, we should first establish the influence factors index and construct the index system. The source of influencing factor index is the expert experience of Transportation and inspection management. By summarizing the key index in the past, the key index under the geographical influence of cost is formed and the index system is constituted.

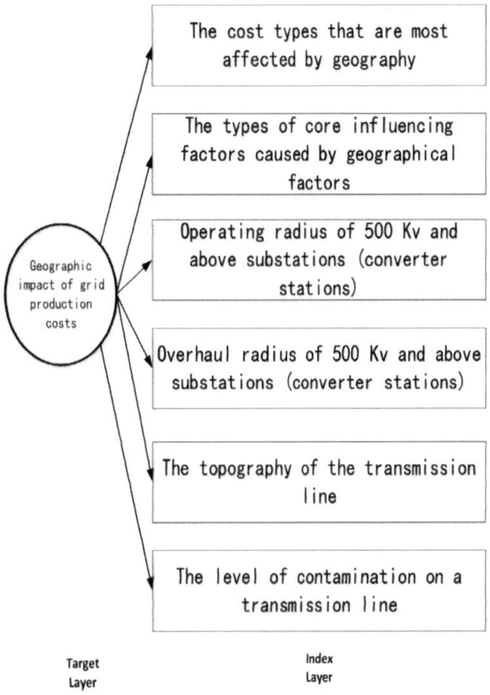

Fig. 1. Hierarchy of indicator system

(1) Among the cost types which are most affected by geographical factors: The value types include labor cost, material cost, travel cost, outsourcing cost[1-10], storage cost, and other costs.

(2) The value of the core factors caused by geographical factors: The value of the type of operation and maintenance radius, climate factors, channel environment, terrain factor, pollution level, three-span quantity, other factors.

(3) The radius of operation and maintenance of 500kV and above substations (converter stations), the range of which is less than 50km, 50-100km, 100-200km and more than 200km. The design corresponding 4 grade values are 1 under 50km, 2 under 50-100km, 3 under 100-200km and 4 over 200km.

(4)The maintenance radius of 500kV and above substations (converter stations) is less than 50km, 50-100km, 100-200km and more than 200km. The design corresponding 4 grade values are 1 under 50km, 2 under 50-100km, 3 under 100-200km and 4 over 200km.

(5)The terrain in which the transmission line is located shall be defined as plains, hills, mountainous areas, and others. According to the measurement index of operation, maintenance and repair difficulty, the design corresponds to 4 levels of values, 1 for the plain, 2 for the hills, 3 for the mountains, and 4 for the others.

(6)The pollution level of the transmission line is below grade C, above grade C and above Grade D. The corresponding value of grade 3 is 1 below grade C, 2 above grade C and 3 above Grade D.The title is set in bold 16-point Arial, justified. The first letter of the title should be capitalised with the rest in lower case. You should leave 35 mm of space above the title and 6 mm after the title.

2.2 Evaluation Index System

When using AHP to do the system analysis, we need to divide the index system into several levels and form the index system of several levels. Hierarchical structure is divided into objectives, elements and indicators of the three layers. The target layer[11-13] is a comprehensive index formed by the Final Assembly, and it is the basis of ranking. The indicator layer represents the various measures, policies and programs to be chosen to solve the problem.

According to the evaluation content, this paper puts forward the two-layer structure of the evaluation system of operation and maintenance cost geographic impact as follows: Index name, index type, index direction, cost type whose value interval is most affected by geographical factors, the radius of operation and maintenance of 500 kv and above substations (converter stations) , the interval index, the greater the impact on the results, the greater the maintenance radius, the interval index of 500 kv and above substations (converter stations) , the greater the impact on the result, the greater the impact on the result, the greater the impact on the result, the greater the impact on the terrain, the pollution level and interval index of the transmission line.

The calculation method of the index is formed by the historical cost data, so the index data has practical significance and can guide the future cost management through the cost data that happened in the past. Taking the operation and maintenance cost of 500 kv substation as an example, this paper illustrates that the cost is affected by geographical factors.

2.3 Comprehensive evaluation method

After the formation of the index system, it is necessary to combine the comprehensive evaluation method to form the final analysis results.In order to determine the degree of analyzing the geographical influence cost, it is necessary to use the index Quantitative analysis to determine the influence of the overall geographical factors on the cost management, and to adopt the comprehensive evaluation method to form the quantitative conclusion of the comprehensive evaluation.

https://doi.org/10.1051/e3sconf/202132901067

The method of comprehensive evaluation can choose analytic hierarchy process or other evaluation methods. Through the index value of the set index system, the weight is given to realize the aggregation of evaluation conclusions. The weight can be calculated synthetically by the entropy weight method.
In the analysis example of this paper, the cost of materials, labor, outsourcing costs can be considered as the basic cost of the cost composition, referred to as the three maintenance costs, through the above three costs as the basis, can form the basic framework of cost analysis, combined with the geographical attributes of the cost, we can analyze the changes of the above three kinds of expenses under the factors of terrain, climate, distance, etc.

3 Analysis on the influence of geographical factors on the three-fee of maintenance

3.1 Content of the three-fee of maintenance

According to the foregoing, the three maintenance fees include material costs, labor costs, outsourcing costs. These three types of costs are respectively the labor costs resulting from the wages of the maintenance personnel and the wages of other personnel involved in the maintenance activities. These types of costs are related to the types of maintenance operations that occur, and the more complex the maintenance operations, the more labor costs, generally speaking, labor costs should be standardized quota data, easier to control, but also by the national wage standards with changes in economic development, generally shows an upward trend. With the increase of manpower, the maintenance cost, that is, the maintenance cost will also rise, which brings pressure to control the maintenance cost.
Material cost is the cost of purchasing materials used in maintenance, which is usually related to the price and quantity of materials used in maintenance. When the maintenance work is more mature, the use of materials is smaller, so the cost of materials is smaller, when the price of materials is lower, the cost of materials is smaller. Therefore, in the management of material costs, the first priority should be to detect changes in material prices in a timely manner, to increase the volume of purchases as prices fall, to decrease as prices rise, and to predict future changes in material prices and propose solutions. Constantly promote the progress of maintenance technology, with more cost-effective materials to complete higher-quality maintenance, innovation, maintenance business to achieve high-quality development.
The three maintenance fees are the comprehensive result of the deep development of the financial dimension of the cost management of power grid enterprises, that is to say, to control the maintenance cost of power grid is to control the three maintenance fees, from the expenses of Labor, materials and outsourcing, lean control of the cost of maintenance expenditure. In the digital background, to determine the core of the maintenance cost types, the cost of a good path to enhance the purpose of data collection, and the follow-up use of data assets analysis of cost and expenditure characteristics and weak links have greater significance.

3.2 Major geographical factors

The power transmission and transformation facilities of the company cover a wide area and vary greatly. Through investigation and research, the first is to understand the influence factors and priority of transmission line and Substation Geographic Environment on operation, maintenance and repair cost; the second is to understand the topographic distribution and pollution grade distribution of transmission lines 35 kv and above, the distribution of operation and maintenance radius and maintenance radius of 500 kv and above substations can provide a basis for setting the Adjustment Coefficient of transmission and substation operation and maintenance cost budget.
It is found that geographical environment mainly affects labor cost, travel cost and outsourcing cost. According to the factors, it is found that the main factors affecting the cost of power transmission inspection are the radius of operation and maintenance, channel environment and terrain. The operation and maintenance radius, maintenance radius and pollution level are the main factors that affect the cost of substation operation and inspection. Specifically, 84% of the company's 500 Kv and above substations have an operating radius of less than 100 km, while the maintenance radius is evenly distributed between 1-200 km companies.
The company's 35 kv and above transmission lines are located mainly in the plain and hilly Hoe San Estate, with the length of transmission lines accounting for 93%. According to the pollution level, the pollution level of transmission lines is mainly distributed below grade C, Grade C and Grade D, 89% of the line length.

4 Operation maintenance management system

This part of research aims to understand the provincial power grid assets and cross-regional power grid assets operation and maintenance management system. Survey found that the provincial power grid assets operation and maintenance management of provincial inspection and maintenance companies, transmission and distribution companies and power supply companies. The operating voltage range of maintenance assets varies among maintenance companies. Generally speaking, provincial maintenance companies mainly operate and maintain power facilities of 500 KV and above, among them, the maintenance companies of Beijing, Shanghai, Chongqing and Heilongjiang Operate and maintain 500 kv and above and some 220 kv power facilities, and the maintenance companies of Tianjin operate and maintain 220 kv and above power facilities, operation and maintenance of 330 kv and above power facilities in Qinghai maintenance company.

In addition, at the level of prefectural and municipal companies, there are cases of co-working or complete separation between the transportation and Inspection Department (Equipment Department) and the inspection and repair branch, of which 9 are co-working units, accounting for 33.33%, and 13 are completely separated, accounting for 48.15%. As the transmission and Distribution Company is a subsidiary of the provincial company, the operation and maintenance of the assets of the provincial company and maintenance of assets is mainly based on the provincial inspection and repair companies signed contracts.

As for the operation and maintenance of trans-regional assets, the main operation and maintenance organizations of trans-regional assets are provincial inspection and maintenance companies and transmission and distribution companies, and the main operation and maintenance organizations of trans-regional assets are provincial inspection and maintenance companies, and maintenance cost management is mainly incorporated into the provincial company after the unified budget management.

The provincial power grid assets operation and maintenance units are mainly provincial maintenance companies, power transmission and transformation companies and local power supply companies. The situation involves 15 units (accounting for 55.56%) , the operation and maintenance organizations of 5 provincial grid assets are provincial inspection and Maintenance Company (DC transportation and Inspection Company) and local power supply company, accounting for 18.52% . In addition, the assets of Shanxi Province also includes the provincial transmission maintenance company, management and other units are different.

The transportation and Inspection Department (Equipment Department) of the prefectural and municipal company is completely separated from the inspection and repair branch and coexists with the joint office. Among them, 13 units transportation and Inspection Department and Inspection and repair branch completely separate management, accounting for 48.15%, 9 companies transportation and Inspection Department and Inspection and repair branch co-office, accounting for 33.33%. In addition, Hebei and Sichuan companies have some professional co-working situations, the co-working ratio is 204% and 43.48% respectively.

Provincial Maintenance Company (DC transportation and Inspection Company) mainly operate and maintain 500 kv and above power facilities, some units exist maintenance company operation and maintenance 220 kv and above provincial assets. The maintenance companies of 18 units operated and maintained 500 KV and some 220 kv power facilities, accounting for 66.67% ; the maintenance companies of 4 units in Beijing, Shanghai, Chongqing and Heilongjiang operated and maintained 500 kv and above and some 220 kv power facilities, accounting for 14.81% ; Tianjin Maintenance Company operation and maintenance of 220 kv and above power facilities; Qinghai Maintenance Company mainly operation and maintenance of 330 kv and above power facilities.

As the transmission and distribution companies are subsidiaries of provincial companies, there are problems of compliance and legality of the assets of the operation and maintenance companies, the basis for the operation and maintenance of the assets of the provincial company and the assets of the maintenance agent of the power transmission and distribution company is mainly through the signing of contracts between the power transmission and distribution company and the provincial maintenance and repair company, involving 18 units, accounting for 66.67% of the total, of which 12 units sign contracts through decision-making procedures, six units based on the framework form of bidding. In addition, Shanxi Company set up a separate power transmission and maintenance company, operation and maintenance of 500 kv and above transmission lines and their power facilities, Shanghai, Shandong and Zhejiang Company transmission and Distribution Company is not operating and maintenance province assets, Beijing company no transmission and Distribution Company.

5 Conclusion

First, the preparation of production cost project management standards. According to the activity types of production cost, different project types, different project management processes, responsibilities of each management link, process control depth and settlement filing requirements are defined.

Second, to promote the entire process of production cost information management. Based on PMS, ERP, or self-built system, it realizes the information management of requirement, plan, implementation process, settlement and filing of cost project.

Third, carding and improving the standard item library. According to the nature of different projects, the requirements of project management, and other factors such as the special focus of each specialty, the standard project database of production cost is revised on a rolling basis, through the standard project library clear project specification naming, research preparation and evaluation, project process control and data archiving project management norms.

Fourth, further standardize the production cost project WBS structure. materials, services and other, respectively used under the project material detail, service list, and can research, design, supervision, review cost classification management.

On the basis of the application of the research results in Article 11, combined with the actual management situation, equipment status, geographical environment and other factors of each unit's operation and maintenance business, work closely with the finance department to carry out the standard cost localization transformation. Based on the "operation cost standard of power grid production and operation" mentioned in Article no. 11 of State Grid Finance [2020] , and in the light of the actual situation of this province, the localization transformation is carried out, including local operation addition and calculation of Adjustment Coefficient, power grid production and operation activity-based cost standards" related content can consult the Ministry of Finance.

One is to carry out the localization transformation of the Standard Operation Library. For the types of equipment and activities not covered in the standard for operating costs of power grid production and operation, the additional fields are basically consistent with the fields covered in the standard for operating costs of power grid production and operation, including operation name, voltage grade, capacity, equipment unit, personnel name, unit, unit price, quantity, amount. The other is to calculate the regional adjustment coefficient. According to the actual situation of operation, maintenance and repair work in this province, the adjustment Coefficient of the equipment layer of this unit is calculated, taking into account the factors such as operation frequency, unit price of materials, geographical environment, etc., it makes the product of the original equipment cost and the adjustment Coefficient in the standard of power grid production and operation cost accord with the situation of this province basically. The municipal units of each locality may measure and determine the adjustment Coefficient of each locality according to the actual situation.

Acknowledgments

The fund support comes from: The research of reasonable configuration of operation and maintenance cost difference based on equipment status (ZZKJ-2021-10)

References

1. Katzenstein W, Apt J. The cost of wind power variability[J]. Energy Policy, 2012, 51(DEC.):233-243.
2. Murr W C. Remote power cost display system: US, US5635895 A[P]. 1997.
3. Lamedica R, Esposito G, Tironi E, et al. A survey on power quality cost in industrial customers[C]// Power Engineering Society Winter Meeting. IEEE, 2001.
4. Morris R P. Methods, Systems, And Computer Program Products For Controlling Data Transmission Based On Power Cost[J]. US, 2007.
5. Chung C Y, Chung T S, Lin C W Y X J. Cost-based reactive power pricing with voltage security consideration in restructured power systems[J]. Electric Power Systems Research, 2004, 70(2):85-91.
6. Cole F A, Deben W A, Swartz R C. Power-cost efficiency of eight macrobenthic sampling schemes in Puget Sound[J]. Canadian Journal of Fisheries and Aquatic Sciences, 1989, 46(12):2157-2165.
7. Batista J, Alfonso J L, Martins J S. Low-cost power quality monitor based on a PC[C]// IEEE International Symposium on Industrial Electronics. IEEE, 2003.
8. Chen S L, Wang J G. VLSI implementation of low-power cost-efficient lossless ECG encoder design for wireless healthcare monitoring application[J]. Electronics Letters, 2013, 49(2):91-92.
9. Sanyal S K. Cost of Geothermal Power and Factors that Affect it[J]. proceedings world geothermal congress, 2005.
10. Yan D, Yixin N, Fushuan W, et al. REACTIVE POWER PRICING BASED ON COST ALLOCATION THROUGH POWER FLOW TRACING[J]. AUTOMATION OF ELECTRIC POWER SYSTEMS, 2000.
11. Weng Z, Shi L, Xu Z, et al. Power system dynamic economic dispatch incorporating wind power cost[J]. Proceedings of the Csee, 2014, 34(4):514-523.
12. Achayuthakan C, Ongsakul W. TVAC-PSO based optimal reactive power dispatch for reactive power cost allocation under deregulated environment[J]. IEEE, 2009.
13. Chang H C, Chang H K, Jin S K. Power Cost Analysis of Go-ri Nuclear Power Plant Units 1 and 2[J]. Physics Letters B, 1976, 215(4):617-623.
14. Hicks M, Egan C, Christianson B, et al. Reducing the Branch Power Cost in Embedded Processors Through Static Scheduling, Profiling and SuperBlock Formation[C]// Advances in Computer Systems Architecture, 11th Asia-Pacific Conference, ACSAC 2006, Shanghai, China, September 6-8, 2006, Proceedings. Springer-Verlag, 2006.
15. Schwieger, Bob. Northside's CFB repowering halves power cost, reduces emissions. [J]. Power, 2002.

E3S Web of Conferences **329**, 01068 (2021) https://doi.org/10.1051/e3sconf/202132901068
GESD 2021

Research on power grid operation and maintenance cost based on analysis of internal and external influence factors

Geli Zhang[1*], *Heng Zhang*[1], *Yan Zheng*[1], *Yaqiong Liu*[1], *Tong Li* [1], *Meibao Wang and Xiaohui Wang*[1]

[1]State Grid Economic and Technological Research Institute Co. Ltd., Beijing, China

Abstract. In the context of the company's cost-reduction reform, lean production operation and maintenance cost management as the goal, production and maintenance cost analysis as the carrier, the existing power grid production and maintenance business and financial accounting costs to achieve "industry-finance integration" as the goal of management projects. Under the strategic background of reducing cost and increasing benefit, the implementation of this project is of great significance to optimize the main part of the cost of the company to optimize the configuration of production cost, to control reasonably, to enhance management.

1 Introduction

The reasonable arrangement of production cost can ensure the normal operation of equipment is the basic condition of the company's production and operation. The objective of this project is to achieve a reasonable allocation of production costs, control production costs. The production cost of the equipment is the main component of the company's operating cost. The analysis of the production cost of the equipment is of great significance, cost analysis is carried out according to the operation cost standard of power grid maintenance, pre-schedule and quota of power grid maintenance, PMS equipment account, business data and financial accounting cost data. Statistical basic data, determine analysis indicators, put forward the current management status and future-oriented lean management indicators and methods.

Guide the company operation and maintenance cost management, operation and Inspection Strategy Optimization, power grid asset management overall upgrade. In the company's system to carry out work in stages, through the main grasp of typical units in-depth mining of the analysis of production costs, the formation of a general analysis of the outline, at the same time, the relevant units based on practical experience and theory with practice, taking equipment management as the core, combining finance, radiation development, material and so on, sum up the configuration experience of production cost. Through the promotion of the headquarters, fully promote the company's advanced management experience in the company-wide units to point-to-point, in the company's production cost management-related professional, play the role of this lean management promotion model. This project is complementary to the standard cost and multi-dimensional lean management and control of finance, and is consistent with the company's strategic direction of cost reduction. Be able to check the capital management of the company's transportation and inspection costs by means of the actual completion of the cost budget analysis, cost input and output effectiveness analysis, and feed back to the formulation of the company's financial professional cost budget, make the cost budget more reasonable, through mining the cost allocation method of advanced units, management experience of continuous summary and enhance communication, for the rational use of production costs of the company has a model, enhance the role.

The direction of the project development is scientific and reasonable: through the "financial integration" management perspective, fully carry out the analysis of operation and maintenance costs, in line with the company's internal and external business situation to the production cost management requirements and expectations, it is beneficial to the scientific management of the company to carry out basic analysis by lean analysis and reduce operating cost reasonably by means of forecasting and problem mining. 2. Methods Scientific and reasonable: to establish the normal analysis mechanism of equipment transportation and inspection cost of headquarters, provincial companies and prefectural and municipal companies.

Design production cost analysis using statistical analysis, correlation analysis and other analysis ideas, design the system of cost analysis indicators and methods [1-5], based on the historical operating data analysis, multi-dimensional display of the use of production costs, after checking and checking the analysis data, the basic reliability of the analysis conclusion is high, which can guide the company to manage the production cost. 3. The division of production cost is scientific: taking the operation and maintenance cost of a single station and a single line as the analysis object, fully considering the factors of self-operation and outsourcing, and reasonably calculating the cost level.

* Corresponding author: fredistheone@126.com

2 Accurate input of power grid cost

To guide the accurate input of the operation and inspection cost into the substation-level cost object [5-10], and to rationalize the cost input and control the unreasonable cost by analyzing the reasonable allocation of the cost. By forming the final analysis report, the curing analysis outline makes it have the normal development condition, the analysis result guides the equipment selection, enhances the equipment economy under the life cycle cost angle of view, promotes the equipment own operation cost ratio. By analyzing the problems of high operation and low operation and maintenance, supporting the related decision, considering the cost factor, analyzing the reasonable time of technical renovation and overhaul, improving the overall economic benefit of power grid. The political and economic conditions in the new normal have changed profoundly, which brings great challenges to the development of power grid enterprises. General Industrial and commercial electricity price decline, transmission and distribution price regulation, economic changes at home and abroad are increasingly complex, economic downturn is expected to increase and so on.
In order to maintain the normal operation of the Power Grid Enterprise, the power grid enterprise needs to reduce the operating cost, control the cost input reasonably and reduce the cost burden. With the increasing difficulty of operation, if power grid enterprises want to continue to improve quality and efficiency, they need to raise the level of input and output, raise the level of cost management and control, and improve the profitability of enterprises by reducing costs and increasing efficiency, ensure the healthy development of power grid industry. The strengthening of the supervision of power transmission and distribution price is reflected in the reform of power system that the power transmission and distribution price should be approved according to the principle of "permitted costplus reasonable income". The transmission and distribution price of the power grid shall first verify the permitted income of the transmission and distribution business of the power grid enterprise, and then verify the transmission and distribution price on the basis of the permitted income. The permitted income of power grid is composed of permitted cost, permitted income and tax.
Therefore, in the case of a certain amount of electricity, the transmission and distribution price is directly proportional to the permitted income of the power grid company, and the verification of the income and taxes is relatively clear in the process of price verification, which can be based on the identification principle of effective assets, as well as the income rate, tax rate and other parameters accurate calculation, so the approval of the permitted cost becomes the key to straighten out the transmission and distribution price. In order to adapt to the supervision and examination of power transmission and distribution cost, the State Grid Co., Ltd., through the reform and optimization of the operation and Management Strategy Work Program, proposed to study the reasonable level of operation and maintenance cost of each province and each kind of operation and management strategy in the key work task of optimizing the operation and management strategy, to provide support basis for cost supervision policy. It is helpful to realize the standardization and standardization of operation and maintenance cost, to adapt to the supervision and examination of transmission and distribution cost, to improve the cost management level of power grid enterprises, and to push forward the reform of power system. First, the research results of the project provide a standard basis for the supervision and examination of transmission and distribution costs, effectively enhance the transparency of transmission and distribution costs, and provide effective guidance for the apportionment of operation and maintenance costs among different voltage levels, it can effectively check and reflect the cost difference between different regions and different voltage levels, and lay a foundation for the proper treatment of cross-subsidy of transmission and distribution price.
The other is to standardize the pricing standard of the company's operation and maintenance cost, to improve the standardization level of the cost management of the power grid company, to help guide the scientific and lean management and control of the enterprise's cost, and to further promote the reasonable optimization of the transmission and distribution cost. The third is to provide cost calculation standards for project management personnel, assist in guiding transportation and inspection management personnel to carry out maintenance work, improve the accuracy of capital investment, and enhance the level of project management.

3 Factor Analysis

In order to adapt to the company's new business situation, to meet the company's strategic cost reduction and efficiency needs [16,17], cost management lean is imperative. Combined with the current pilot analysis and research work, combing the practice, summarized the current cost lean management of specific measures and objectives are as follows.
(1) carrying out the measures to realize the budget of production cost of "standard operating cost" equipment in operation and inspection: comprehensively promoting the trial use of "standard operating cost" in the companies in all provinces of China, provincial companies began to be familiar with the "standard operating costs" of the application method and the scope of the trial. Promote the localization of "standard activity cost", supplement the local characteristic activity, calculate the standard cost level of the characteristic activity, calculate the local adjustment coefficient, optimize the application of the standard cost in the local, use in the cost budget estimation and formulation. Compare the differences between the standard activity-based cost and the actual cost to further understand whether the cost level is reasonable. Continuously improve the existing "standard operating cost" system to promote the application of standard costs.
(2) collecting single-station, single-line, asset group actual cost realization measures: using the existing PMS production system equipment information and the financial management system ERP project actual cost information, the actual cost incurred by a single station, a

single line and an asset group. Through data background export and manual processing, the actual production cost is collected to station, line level and asset group level. For the company's equipment management leading financial management departments to jointly promote the co-dominant cost management to provide a preliminary approach.

(3) promoting project management of Production Cost: Implementation Measures: constructing the basic system of project management of production cost, including: sorting out the typical types of local operation and maintenance cost and designing cost project prototype. Classify and build the basic production cost items, build the local production cost standard items to manage the provincial reserve production cost items. Project base of local cost items according to the dimensions of specialty, equipment and operation. Formulate the cost project management method, according to the cost project driving type, categorize the whole process of project management rights and responsibilities, requirements and depth. Establish the norm standard of typical cost item.

(4) strengthening data integration to realize industry financial integration: MEASURES TO REALIZE: strengthening effective integration of PMS work task list, overhaul work list and ERP work list, strengthening data management and filling quality control; The PMS2.0 establishes the cost project management module in the PMS, strengthens the informationization support. To make the cost information collected directly in the system complete and perfect to reflect the real situation, through the system, strengthen the linkage of cost management between equipment management and Financial Accounting Management, build an interactive bridge, promote cost management.

(5) PROMOTION OF EXPLICIT MANAGEMENT OF PRODUCTION COSTS: Implementation Measures: Establishment of a management system in an all-round way, sorting out of regulations to form an enterprise standard, standardization of the operating cost management system, based on digital means, make cost explicit management efficient, accurate, flexible, interactive. Through the guiding role of the Enterprise Standard, optimize the existing cost management system in each province, make the cost management gradually transparent and open, gain the trust of the society and the supervisor. At the grass-roots level, standardize the cost information input system, digital management, avoid manual bills, through the construction of information systems to promote cost management.

(6) strengthening the measures to realize the precise input of production cost: based on the individual factors such as equipment status, age structure, terrain conditions and regional conditions, a differentiated cost input model of many influencing factors is established and constructed, assist the decision-making of capital investment of provincial company. This paper constructs the differential input model of maintenance cost, studies the Adjustment Coefficient of the differential influence factors, calculates the differential maintenance cost of different substations and overhead transmission lines, and implements the differential precise input of maintenance cost, guide the rolling correction of standard operating cost.

(7) measures to optimize the life cycle management of assets: In the stage of Type Selection and procurement, the quality of source equipment is investigated, the technical standards of equipment selection and procurement are improved, the quality of network equipment is improved, and initial investment is increased, pay attention to the life cycle cost of the equipment in the running period, form the effective assets and reduce the later maintenance cost. In the operation and maintenance stage, we should study and improve the strategy of assets operation and maintenance, raise the technical standard, strictly implement the strategy of maintenance and repair, carry out the work of standardized maintenance, improve the efficiency of operation and maintenance, strengthen the management and control of the whole process of on-site operation, and ensure the quality and safety of operation, promote the performance of maintenance operation.

4 LCC research

One is to continue to promote the "each device" value lean management research results. Each unit, in combination with its own business management practice, selects three kinds of operation and maintenance workers' single link connecting modes, namely "Hubei", "Jiangsu" and "Hebei (Beijing) " , to link up with the PMS system's single link, carry out a single equipment initial investment cost, operation and maintenance cost, decommissioning and disposal cost collection (or allocation) , to achieve the operation and maintenance cost to each equipment accurate collection. Second, develop analysis scenario application based on LCC. Such as supplier analysis and evaluation and technical renovation decision-making. In the aspect of supplier evaluation, the average level of key equipment operation and maintenance cost is analyzed through the data collection of LCC cost, and the supplier evaluation is carried out with the indexes of equipment defect frequency, Operation State, availability coefficient, etc., to provide decision-making basis for the optimization of the company's purchasing strategy. On the basis of establishing LCC model of single equipment, calculate the operation and maintenance cost of different voltage levels, different regions, different models, different suppliers and other equipment for each year of operation and maintenance (distinguish operation and maintenance, daily maintenance, overhaul and rush repair costs) , according to the results of calculation, the cost curve of operation, maintenance and repair of the main equipment is drawn, including annual input and cumulative input. The annual average operation and maintenance cost input of the equipment after technical renovation or overhaul is estimated by integrating the actual cost, and the optimal cost input is taken as the objective to assist the decision-making of technical renovation.

5 Information support situation

ERP system is still the main platform for the cost management of the operation and maintenance project, with the application unit accounting for 88.89% , 40.74% of the units realize the project construction through the

Integrated Planning and Management Platform; A, B for materials and outsourcing services overall Erp service order collection, internal labor cost (Man-hour) collection carrier is not unified standards and norms; 85.19% of the implementation process data are managed by the traditional paper-based Archive (including electronic version and scanning part) , and the application of physical ID can cover all specialties and assist the project implementation process management The PMS2.0 system has not been fully differentiated, only 29.63% of the units have been differentiated; the production cost management system has different functions, the overall coverage function is not comprehensive, and the data interaction with the financial budget management system is insufficient, the correlation granularity and the correlation maintenance object are insufficient.

At present, companies in all provinces have adopted the Information System to manage the operation, maintenance and repair items, the cost management of operation and maintenance project mostly adopts PMS2.0(9 units) , ERP (24 units) , financial budget management system (12 units) , integrated planning management platform (11 units) and other production cost management system (4 units) , most of the provincial companies use multiple systems for operation and maintenance project management.

The companies of all provinces have more than 2.0 integration with ERP system, less integration with financial budget management system, integrated planning management platform and production cost management system, production cost management system and financial budget management system, the development of Integrated Planning Management Platform is rarely integrated; The data interface, Web Service and big data are the main integration methods among the systems, the PM work order, material/Service purchase order and maintenance object are the data source of ERP system In the production cost information, the project basic information, the work task list, the overhaul work list, the overhaul object mainly take PMS system data as the source data, most of the provincial companies do not take the production cost management system data as the source data.

6 Conclusion

Aiming at the management of lean production operation and maintenance cost and relying on the analysis of production operation and maintenance cost, we can realize the reasonable allocation of production cost, control production cost, and realize the "industry-finance integration" of power network operation and maintenance cost and financial accounting cost. It is an innovative practice of cost management of municipal company to carry out the analysis of equipment production cost. Innovative production cost analysis of statistical analysis, correlation analysis, special cost analysis, special analysis is the analysis of different professional, voltage levels and other dimensions, to grasp the relationship between operation and maintenance cost including assets, safety, benefit, maintenance efficiency and so on, so as to excavate the investment effect of production cost.

Ensuring input into grid operation, maintenance and repair costs. First, it is suggested that provincial-level companies should take practical measures to guarantee the cost of maintenance and repair of power grid equipment, keep an appropriate match with the growth trend of their own power grid equipment scale, and strengthen the management of power transmission and distribution costs, scientific and rational allocation of professional cost quotas, reverse the situation of low or decreasing input level year by year, effectively ensure the safety of power grids and better serve users; second, provincial-level companies should actively integrate with the existing transmission and distribution tariff reform policies, focusing on the fact that the proportion of the newly added maintenance costs to the original value of the newly added assets is gradually approaching 2.5% , with the goal of power grid security and the company's long-term development as its goal, to ensure the input of its own unit's maintenance costs, for the next transmission and distribution nuclear price cycle to lay a good foundation.

Improving Midwestern Sectional Figure Skating Championships operation and maintenance standards. First, the mechanism of coordinated development should be studied. The Midwestern Sectional Figure Skating Championships economy is relatively backward, and the profit level is generally low, so it is suggested that the company should study and try to put in place the compensation and pricing mechanism for power supply services in remote areas at the national level, so as to improve the profitability of the companies in the relevant provinces. Strive for "east-west help" and other internal cooperative development mechanism to ensure that it has the necessary operation and maintenance cost input capacity. Second, optimize the assessment system. Comprehensive optimization, overall coordination and adjustment of the company's profit, debt ratio, cost input and other evaluation and assessment system, combined with the research results of transmission and distribution price verification, to find a reasonable balance point and guide the maintenance cost of power grid operation to maintain a reasonable level, to ensure the company's long-term healthy and stable development.

Third, we should face up to the differences between the East and the West. The differential operation and maintenance standards are adopted to reasonably reduce the demand for high power supply reliability in the western region, and to improve the operation and maintenance quality and reduce the pressure on the operation and maintenance cost in the less developed areas by paying close attention to the basic, basic and basic skills. The cost of operation, maintenance and repair is the necessary guarantee for the safety production of power grid. Through this investigation, it is found that there is indeed a gap in the operation and maintenance cost of some units, which needs to be solved through measures such as studying policy, perfecting mechanism, carrying out differential operation and maintenance, and optimizing management mode, so as to ensure the safe and stable operation of the power grid, for the company to win a comprehensive "one strong three excellent" modern company, accelerate the building of a world-class

https://doi.org/10.1051/e3sconf/202132901068

enterprises with global competitiveness to lay a solid foundation.

Acknowledgments

The fund support comes from: The research of reasonable configuration of operation and maintenance cost difference based on equipment status (ZZKJ-2021-10)

References

1. Niknam T, Azizipanah-Abarghooee R, Narimani M R. An efficient scenario-based stochastic programming framework for multi-objective optimal micro-grid operation[J]. Applied Energy, 2012, 99:455-470.
2. Mohamed A, Radwan A A. Hierarchical Control System for Robust Microgrid Operation and Seamless Mode Transfer in Active Distribution Systems[J]. IEEE Transactions on Smart Grid, 2011, 2(2):352-362.
3. Zeineldin H H, El-Saadany E F, Salama M. Distributed Generation Micro-Grid Operation: Control and Protection[C]// Power Systems Conference: Advanced Metering. IEEE, 2006.
4. Al-Saedi W, Lachowicz S W, D Habibi, et al. Power quality enhancement in autonomous microgrid operation using Particle Swarm Optimization[J]. International Journal of Electrical Power & Energy Systems, 2012, 42(1):139-149.
5. Rui G E, Dong Y, Yue-Chun L. Analysis of Large-Scale Blackout in UCTE Power Grid and Lessons to be Drawn to Power Grid Operation in China[J]. power system technology, 2007, 31(3):1-6.
6. Kim H M, Lim Y, Kinoshita T. An Intelligent Multiagent System for Autonomous Microgrid Operation[J]. Energies, 2012, 5(12).
7. Kim H M, Kinoshita T. A Multiagent System for Microgrid Operation in the Grid-interconnected Mode[J]. Journal of Electrical Engineering and Technology, 2010, 5(2):246-254.
8. Kim, HakMan, Lim, et al. Energies, Vol. 5, Pages 3347-3362: An Intelligent Multiagent System for Autonomous Microgrid Operation[J]. 2012.
9. Pei W, Li S, Li H, et al. Key technology and testbed for microgrid operation control[J]. Automation of Electric Power Systems, 2010, 34(1):94-96.
10. Daryoush, Habibi, Octavian, et al. Power flow control in grid-connected microgrid operation using Particle Swarm Optimization under variable load conditions[J]. International journal of electrical power and energy systems, 2013.
11. Tang X, Deng W, Li N, et al. Control technologies of micro-grid operation based on energy storage[J]. Electric Power Automation Equipment, 2012, 32(3):99-103+108.
12. Gupta, R. A, Kishor N. A robust optimization based approach for microgrid operation in deregulated environment. [J]. Energy Conversion & Management, 2015.
13. Kim H M, Kinoshita T, Lim Y, et al. A Bankruptcy Problem Approach to Load-shedding in Multiagent-based Microgrid Operation[J]. Sensors, 2010, 10(10):8888-8898.
14. Sachs J, Sawodny O. A Two-Stage Model Predictive Control Strategy for Economic Diesel-PV-Battery Island Microgrid Operation in Rural Areas[J]. IEEE Transactions on Sustainable Energy, 2016, 7(3):1-11.
15. Zhang C, Chen M Y, Wang Z C. Study on control scheme for smooth transition of micro-grid operation modes[J]. Dianli Xitong Baohu yu Kongzhi/Power System Protection and Control, 2011, 39(20):1-5+10.
16. Kim, HakMan, Kinoshita, et al. Sensors, Vol. 10, Pages 8888-8898: A Bankruptcy Problem Approach to Load-shedding in Multiagent-based Microgrid Operation. 2010.
17. Huang Z, Ning Z, Tuffner F K, et al. MANGO – Modal Analysis for Grid Operation: A Method for Damping Improvement through Operating Point Adjustment[J]. power transmission & distribution, 2010.

A well selection method of CO_2 huff and puff in heavy oil reservoir with edge and bottom water based on water cut

Jia Wang[1], Tongjing Liu[1,], Hengyu Shi[1], Pengxiang Diwu[1], Jian Zhou[1], Lekun Zhao[2], Juan Ni[1] and Yan Shi[1]*
[1]The Unconventional Oil and Gas Institute, China University of Petroleum, 18 Fuxue Road, Changping, Beijing 102249, China
[2]The School of Petroleum Engineering, China University of Petroleum, 18 Fuxue Road, Changping, Beijing 102249, China

Abstract. The water cut of heavy oil reservoir with edge and bottom water rises rapidly and the recovery degree of crude oil is low. CO_2 huff and puff is an effective measure to improve the recovery of this kind of reservoir, and scientific well selection is the premise of the measure effect. Because the existing well selection methods of CO_2 huff and puff in heavy oil reservoir with edge and bottom water mostly take the oil increase of oil well as the evaluation index, ignoring the characterization of well water cut after huff and puff, it is unable to accurately screen all potential wells. Therefore, a quantitative well selection method of CO_2 huff and puff in heavy oil reservoir with edge and bottom water based on water cut is proposed. The method is based on fuzzy comprehensive evaluation theory and analytic hierarchy process, and takes the water cut after CO_2 huff and puff as the evaluation index. Several groups of typical models are designed to screen the sensitive factors and laws of CO_2 huff and puff well selection in heavy oil edge and bottom water reservoir from three aspects of geology, water production law and technology. Than the judgment matrix is established. There is little interference from human factors in the process of well selection, and the rationality of the method has been verified by the effect of field actual well measures. This method is helpful to improve the well selection method system of CO_2 huff and puff in this kind of reservoir, and is a reasonable supplement to the existing well selection method which takes the oil increment as the only evaluation index.

1 Introduction

Heavy oil reservoir with edge and bottom water has the characteristics of high oil-water viscosity ratio and low water displacement efficiency[1]. In the process of long-term depressurization production, the invasion of water leads to the rapid rise of oil well water cut, and the recovery ratio of the reservoir is poor[2-4]. A large number of studies and field practice show that CO_2 huff and puff can not only reduce viscosity and supplement formation energy, but also control water and increase oil production in heavy oil reservoirs. It is an effective measure to improve the recovery of heavy oil reservoir with edge and bottom water[5-7]. Optimizing measure wells through scientific well selection method of CO_2 huff and puff is the premise to ensure the effect of CO_2 huff and puff measures.

In recent years, experts have carried out a lot of research on CO_2 huff and puff and put forward a variety of well selection methods and principles: in 2003, Zhao Junsheng and others[8] put forward qualitative well selection conditions in terms of reservoir structure, reservoir characteristics, crude oil characteristics and construction technology through the analysis of 12 CO_2 huff and puff test wells in low-permeability reservoir; In 2009, Wang Xiande and others[9] summarized the examples of CO_2 single well huff and puff at home and abroad, according to the mechanism of CO_2 oil increase, combined with the characteristics of reservoir, crude oil and reservoir rock, they summarized and quantified the well selection criteria by using cluster analysis method; in 2013, Bi Yongbin and others[10] collected the evaluation factors of CO_2 huff and puff construction in Nanpu oilfield, they used the statistical principle to calculate the possibility of successful huff and puff when the value of a single factor changed, and applied the fuzzy evaluation method to establish an evaluation method suitable for huff and puff well selection potential of horizontal wells in complex fault blocks; in 2019, Jin Chunyu and others[11] established a set of quantitative well selection methods for CO_2 huff and puff in heavy oil reservoirs with edge and bottom water using the fuzzy comprehensive evaluation method based on the oil increase index after measures. To sum up, the quantitative well selection method for CO_2 huff and puff in heavy oil reservoir with edge and bottom water has been proposed, however, the oil increment is the only evaluation index of these methods. In fact, although some CO_2 huff and puff wells can not achieve significant oil increase, they can effectively inhibit water production, which also reflects the effective mitigation of interlayer contradictions. Therefore, the "water content" index is indispensable for the accurate characterization of the effect of CO_2 huff and puff measures.

Based on this, a new quantitative well selection method of CO_2 huff and puff in heavy oil reservoir with edge and bottom water is established. The method is based

* Corresponding author: ltjcup@cup.edu.cn

on fuzzy comprehensive evaluation theory[12] and analytic hierarchy process, and takes the water cut after CO_2 huff and puff as the evaluation index. We have designed 175 typical models to screen the sensitive factors and laws of CO_2 huff and puff well selection in heavy oil reservoir with edge and bottom water from three aspects: reservoir geological conditions, water production law and construction technology. We establish the judgment matrix according to the obtained factors and laws The scientific well selection process of this method can reduce the interference of human factors and meet the needs of quantitative characterization of multi factor well selection. The rationality of the method is verified by the effect of field practical measure wells. Moreover, this method is helpful to improve the well selection method system of CO_2 huff and puff in this kind of reservoir. It is a reasonable supplement to the existing well selection method which takes the oil increment as the only evaluation index.

2 well selection method

2.1 Establishment process

The method is based on fuzzy comprehensive evaluation theory and analytic hierarchy process. Since the water cut after CO_2 huff and puff has a good correlation with the final oil increment, the water cut after CO_2 huff and puff is taken as the evaluation object.

The first step: determination of factor set and evaluation set

Firstly, a set of n factors that affect the water cut after CO_2 huff and puff is established, which is called factor set.

$$U = (u_1, u_2, \cdots, u_n) \quad (1)$$

In this method, 18 sensitive factors (n = 18) obtained from 175 typical models of heavy oil reservoir in three rounds of CO_2 huff and puff are taken as target factors. These sensitive factors have different degrees of influence on the water cut after CO_2 huff and puff.

Then, the evaluation set is determined according to the influence of various factors on the water cut after CO_2 huff and puff.

$$V = (v_1, v_2, \cdots, v_m) \quad (2)$$

The second step: single factor evaluation

The single factor evaluation of water cut after CO_2 huff and puff is carried out by using the single ranking method for continuous variables. The established factor set and evaluation set are evaluation tools.

The third step: determination of initial weight vector

According to the influence degree of each sensitive factor on water cut after CO_2 huff and puff, the weight of each factor is calculated by AHP. Then, the weight set is established, which is expressed as weight vecto.

$$\overset{\circ}{A} = (a_1, a_2, \cdots, a_n) \quad (3)$$

In this formula, a_i is the weighted value of the i-th factor. general provisions:

$$\sum_{i=1}^{n} a_i = 1 \quad (4)$$

The fourth step: comprehensive evaluation calculation

The single factor fuzzy evaluation of the i-th factor is a fuzzy subset of Evaluation set V:

$$\overset{\circ}{R}_i = (r_{i1}, r_{i2}, \cdots, r_{im}) \quad (5)$$

Then the evaluation matrix on Evaluation set V is:

$$\widetilde{R} = \begin{bmatrix} r_{11} & r_{12} & \text{K} & r_{1m} \\ r_{21} & r_{22} & \text{K} & r_{2m} \\ \text{M} & \text{M} & \Lambda & \text{M} \\ r_{n1} & r_{n2} & \Lambda & r_{nm} \end{bmatrix} \quad (6)$$

So the fuzzy comprehensive evaluation subset $\overset{\circ}{B}$ for the evaluation object is a fuzzy subset on Evaluation set V:

$$\overset{\circ}{B} = \overset{\circ}{A} \otimes \overset{\circ}{R} \quad (7)$$

In the fuzzy transformation, $\overset{\circ}{A}$ is the fuzzy subset (weight vector) on U, $\widetilde{R}$ is the evaluation matrix, then $\overset{\circ}{B}$ is the evaluation result matrix of the evaluation object. For $\overset{\circ}{B} = (b_1, b_2, \cdots, b_m)$, in practical application, the multiplication of $\overset{\circ}{A}$ and $\overset{\circ}{B}$ is usually performed:

$$b_j = \sum_{i=1}^{n} a_i \times r_{ij}, \quad i = 1, 2, \text{L} \;, n;\; j = 1, 2, \text{L} \;, m \quad (8)$$

In the formula above, b_j is the evaluation result of the j-th evaluation object (It is the decision factor corresponding to the well selection method).

The fifth step: correction and adjustment of well selection method

The decision factors corresponding to 525 wells-times (175 wells × three rounds of huff and puff) of typical models are obtained by comprehensive evaluation and calculation. After ranking the decision factors in descending order, the correlation between the ranking and the water cut after CO_2 huff and puff in the typical model is tested. Adjust the initial weight vector and repeat the third, fourth and fifth steps to modify and adjust the well selection method. When the ranking of decision factors is well positively correlated with the water cut after CO_2 huff and puff (that is to say, the smaller the sort number, the lower the water cut after huff and puff), the final weight vector is determined.

2.2 Determination of factor set and evaluation set

According to 175 typical models of heavy oil reservoir, 18 sensitive factors were selected to establish target factor set (Table 1). These sensitive factors have different degrees of influence on the water cut after CO_2 huff and puff. Take the first round of CO_2 huff and puff data as an example: the larger the difference and ratio between the maximum and minimum water cut after huff and puff, the more sensitive the factor is.

To determine the evaluation set, it is necessary to determine the corresponding values or ranges of the 18 influencing factors. For non-continuous variable factors, the conventional factor level is directly used to determine

Table 1. List of well selection factors.

Factors	Sensibility / first round	Sensitivity law	Difference between maximum and minimum water cut / first round	Ratio of maximum to minimum water cut / first round
Stratigraphic dip /°	Relatively weak	Positive correlation	0.002	1.05
Location of interlayer	Relatively weak	It's the highest in the middle	0.020	1.87
pay-zone thickness /m	Relatively strong	Positive correlation	0.047	11.74
Sedimentary rhythm	Relatively strong	Anti rhythm is better	0.051	13.75
Heterogeneity (lorentz coefficient)	Strong	Positive correlation	0.192	Infinity
Water cut after commissioning /f	Relatively strong	Positive correlation	0.006	1.22
Length of water producing section	Relatively weak	Negative correlation	0.026	2.63
Location of water producing section	Relatively weak	Relatively weak	0.030	31.00
Heterogeneity degree of water producing section	Relatively weak	Negative correlation	0.025	1.93
Position of horizontal section	Relatively strong	The upper part of positive rhythm is good, the rule of anti prosody is opposite	0.054	55.00
Water cut before huff and puff /f	Relatively weak	Positive correlation	0.013	Infinity
cyclic steam injection volume /sm^3	Relatively strong	Negative correlation	0.052	3.30
Gas injection rate /sm^3/d	Relatively weak	Positive correlation	0.009	1.40
soak time /d	Relatively weak	Negative correlation	0.029	2.75
Liquid recovery rate after well opening/rm^3/d	Strong	Positive correlation	0.242	10.44
Distance between parallel wells /m	Strong	Negative correlation	0.734	26.75
Liquid production of parallel wells /rm^3/d	Strong	Positive correlation	0.867	31.42
Turn of huff and puff	Strong	Negative correlation	0.292	6.31

the factor value. Taking the dip angle as an example: three conventional values 3°、6°、10° were used as factors value. For continuous variable factors, it is necessary to determine the value range of factors according to the conventional value range. Taking the pay-zone thickness as an example: the conventional value range of 0m ~ 14m is taken as the factor value interval.

For a certain factor, according to the definite factor value range, the influence law of all factors on water cut after huff and puff is compared. Here, the value range of 1 ~ 9 is used to assign values to different factors. The lower the water cut after huff and puff, the larger the value of the evaluation parameter (Table 2).

Table 2. Parameter value of evaluation set

Parameter value	Meaning
1	The numerical value of this factor makes the water cut reach the highest level after huff and puff
…	The numerical value of this factor makes the water cut reach the middle value after huff and puff
9	The numerical value of this factor makes the water cut reach the lowest level after huff and puff

Table 3. Evaluation of stratigraphic dip factors.

Stratigraphic dip /°	Water cut in the first round of well opening	score	Water cut in the second round of well opening	score	Water cut in the third round of well opening	score
3	0.030	9	0.158	9	0.261	9
6	0.030	6	0.182	5	0.269	7
10	0.032	1	0.204	1	0.290	1

2.3 Single factor evaluation

For discontinuous variables, the values of each factor can be directly assigned according to the correlation between the value of the factor and the water cut after huff and puff. Taking the stratigraphic dip as an example, it has three levels: 3°、6°、10°. According to the conclusion of typical model, the value of this factor is positively correlated with the water cut after huff and puff. The assignment is shown in Table 3.

For continuous variables, it will lead to the inconsistency of the judgment matrix if the judgment matrix is established based on subjective experience. Here, a single sort method for continuous variables is established. For a certain factor, according to the correlation law between the value of the factor and the water cut of the well after huff and puff, the inflection point and extreme point values are assigned and then fitted by sections. In this way, the correlation formula between the factor value and the evaluation parameter value is obtained. The evaluation parameters of any other factor values are calculated by the correlation formula. Taking cyclic steam injection volume as an example, the overall assignment rule is shown in Figure 1.

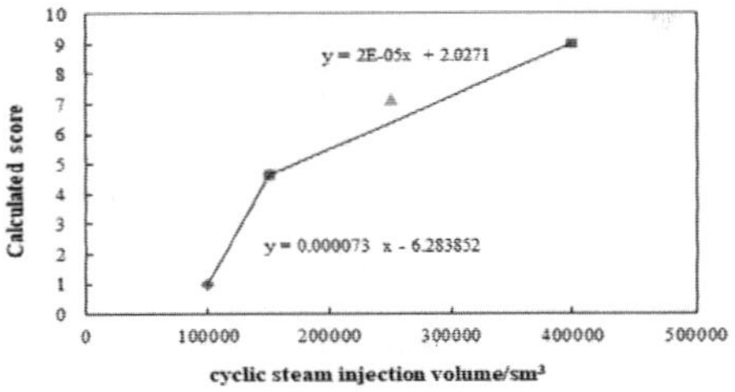

Figure 1. Regular chart of cyclic steam injection volume and Calculated score.

2.4 Determination of initial weight vector

According to the influence degree of each sensitive factor on water cut after CO_2 huff and puff, the weight of each factor is calculated by AHP. In this paper, the difference between the maximum and minimum water cut after huff and puff corresponding to different parameter values of each factor is used to characterize the influence of each sensitive factor on the water cut after CO_2 huff and puff. The initial weight of each sensitive factor is obtained by normalizing the difference between the maximum and minimum water cut after huff and puff (Table 4).

Table 4. Initial weight table.

Factors	Initial weight
Stratigraphic dip /°	0.00074
Location of interlayer	0.00743
pay-zone thickness /m	0.01747
Sedimentary rhythm	0.01895
Heterogeneity（lorentz coefficient）	0.07135
Water cut after commissioning /f	0.00223
Length of water producing section	0.00966
Location of water producing section	0.01115
Heterogeneity degree of water producing section	0.00929
Position of horizontal section	0.02007
Water cut before huff and puff /f	0.00483
cyclic steam injection volume /sm^3	0.01932
Gas injection rate /sm^3/d	0.00334
soak time /d	0.01078
Liquid recovery rate after well opening/rm^3/d	0.08993
Distance between parallel wells /m	0.27276
Liquid production of parallel wells /rm^3/d	0.32219
Turn of huff and puff	0.10851

2.5 Correction and adjustment of well selection method

For each factor, the decision factors corresponding to 525 wells-times (175 wells × three rounds of huff and puff) in the typical model are obtained by using the evaluation matrix. After ranking the corresponding decision factors of 525 wells-times in descending order, the correlation between the ranking and the water cut after CO_2 huff and puff in typical model is tested. Adjust the initial weight vector and repeat steps three, four and five to modify and adjust the well selection method. Until the ranking of decision factors has a good positive correlation with water cut after CO_2 huff and puff (Figure 2). At this point, the lower the score of the well, the higher the water cut after CO_2 huff and puff, reflecting the worse the effect of measures (Figure 3).

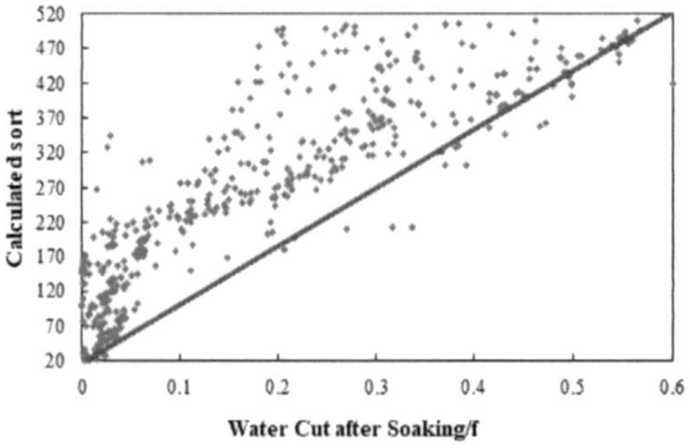

Figure 2. Correlation between water cut after soaking and decision factor ranking of 525 well-times in typical model after huff and puff

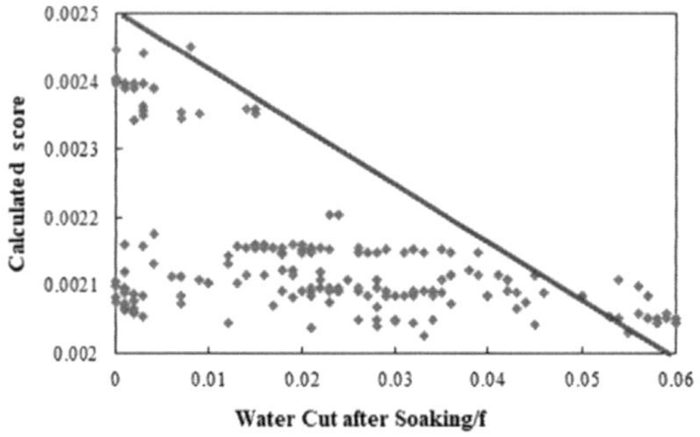

Figure 3. Correlation between water cut and theoretical score of 525 well-times in typical model

At this time, the determined weight vector is the final weight vector, as shown in the Table 5.

Table 5. Final weight of factors.

Factors	Final weight
Stratigraphic dip /°	0.00282
Location of interlayer	0.0169
pay-zone thickness /m	0.08263
Sedimentary rhythm	0.0169
Heterogeneity（lorentz coefficient）	0.0939
Water cut after commissioning /f	0.01502
Length of water producing section	0.01972
Location of water producing section	0.02441
Heterogeneity degree of water producing section	0.01878
Position of horizontal section	0.02066
Water cut before huff and puff /f	0.03099
cyclic steam injection volume /sm^3	0.07042
Gas injection rate /sm^3/d	0.01315
soak time /d	0.0385
Liquid recovery rate after well opening/rm^3/d	0.14178
Distance between parallel wells /m	0.14178
Liquid production of parallel wells /rm^3/d	0.15775
Turn of huff and puff	0.0939

 https://doi.org/10.1051/e3sconf/202132901069

3 Field application of well selection method

Taking 37 horizontal wells of multi round CO_2 huff and puff in A oilfield as an example. The well selection method was applied to evaluate the 37 wells. The final sorting results are shown in the Table 6.

The correlation between water cut and well selection decision factors of 37 wells after multiple rounds of CO_2 huff and puff is shown in Figure 4. The correlation between water cut and well selection decision factor ranking of 37 wells after multiple rounds of CO_2 huff and puff is shown in Figure 5.

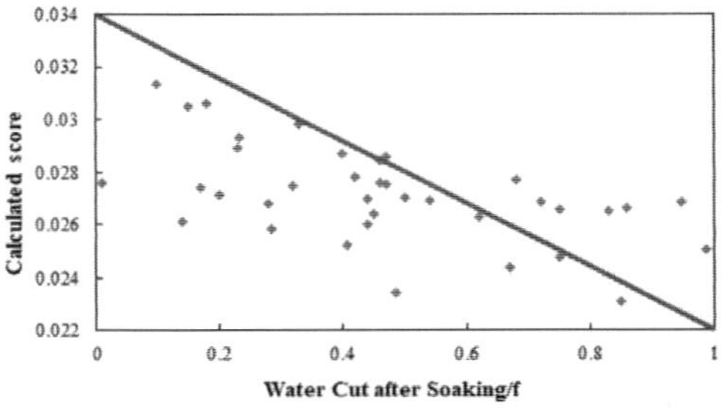

Figure 4. Correlation between water cut and theoretical score of actual wells after multiple cycles of huff and puff

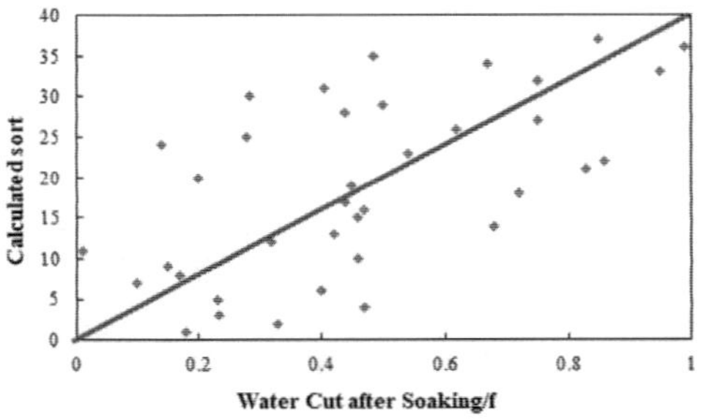

Figure 5. Correlation between water cut and calculation sort of actual wells after multiple cycles of huff and puff

Table 6. Well selection results of actual horizontal wells for multiple rounds of CO_2 huff and puff

Well Name	Water cut after huff and puff /f	Calculation score	Actual sort	Calculation sort
1stG120-P1CP1	0.01	0.0276006	1	12
1stG104-5P125	0.1	0.0313372	2	1
2ndG104-5P105	0.14	0.0261232	3	29
1stG104-5P106CP1	0.15	0.030537	4	3
1stG104-5P105	0.17	0.0274394	5	16
1stG104-5P78CP1	0.18	0.0306097	6	2
1stG104-5P115	0.2	0.0271678	7	17
1stG24-P4	0.23	0.0289228	8	6
1stG24-P2	0.233	0.0293259	9	5
1stG104-5P10	0.28	0.0267854	10	23
1stG104-5P13	0.285	0.0258526	11	31
1stG104-5P28	0.32	0.0274509	12	15
1stG104-5P44	0.33	0.0298705	13	4
1stG104-5P112	0.4	0.0287163	14	7
1stG106-5CP1	0.407	0.0252184	15	32
1stG104-5P100	0.42	0.0278219	16	10
1stG24-P3	0.44	0.0269984	17	19
1stG104-5P97	0.44	0.0260188	18	30
1stG104-5P117	0.45	0.0264135	19	27
1stG104-5P93	0.46	0.0284233	20	9
1stG104-5P91CP1	0.46	0.02758	21	13
1stG104-5P1	0.47	0.0286056	22	8
1stG104-5P79	0.47	0.0275332	23	14
2ndG104-5P115	0.484	0.0234217	24	36
1stG104-5P101	0.5	0.0270513	25	18
1stG24-P1	0.54	0.026902	26	20
1stG104-5P32	0.62	0.026285	27	28
1stG104-5P102	0.67	0.024408	28	35
1stG104-5P12	0.68	0.0277	29	11
1stG104-5P34	0.72	0.0268442	30	22
1stG104-5P82	0.75	0.0265636	31	25
1stG104-5P96	0.75	0.0247866	32	34
1stG104-5P85	0.83	0.0265052	33	26
2ndG104-5P97	0.85	0.023106	34	37
1stG104-5P21	0.86	0.0266206	35	24
3rdG104-5P97	0.95	0.0268761	36	21
3rdG104-5P115	0.99	0.0250627	37	33

According to the comparison between the evaluation results of well selection and the actual data: there is an obvious inverse correlation between water cut and well selection decision factors in 37 wells after multiple rounds of CO_2 huff and puff; There is a positive correlation between the water cut and the order of well selection decision factors in 37 wells after multi round CO_2 huff and puff. It is proved that the well selection method can optimize the target well based on the water cut after huff and puff.

4 Conclusions

(1) A new quantitative well selection method of CO_2 huff and puff in heavy oil reservoir with edge and bottom water is established. The method is based on fuzzy comprehensive evaluation theory and AHP, and takes the water cut after CO_2 huff and puff as the evaluation index. According to the sensitive factors and their influence rules obtained from 175 typical models of heavy oil during three rounds of CO_2 huff and puff, the judgment matrix is established. This method is helpful to improve the well

selection method system of CO_2 huff and puff in this kind of reservoir. It is a reasonable supplement to the existing well selection method which takes the oil increment as the only evaluation index.

(2)Two efforts have been made to improve the accuracy and applicability of the judgment matrix of the method: a set of single ranking method for continuous variables in evaluation factors is established to solve the problem of inconsistent judgment matrix of continuous variables; The 525 well-times production data in typical model is used to verify and adjust the weight of well selection factors, so that the well selection factor ranking has a good positive correlation with water cut after CO_2 huff and puff.

(3) The well selection method is used to evaluate the well selection of 37 horizontal wells in A oilfield, which is a typical heavy oil reservoir with edge and bottom water. By comparing the evaluation results of well selection with the actual production data, it can be seen that: the water cut of 37 wells after multiple rounds of CO_2 huff and puff has obvious inverse correlation with the well selection decision factors, while the well selection sequence has good positive correlation with the water cut after huff and puff, which indicates that the better the well selection sequence is, the better the measure effect will be. It also proves that the well selection method can optimize the target well based on the water cut after huff and puff.

References

1. Jiang Weidong, Ren Yanbin, Zhang Yunlai, et al. Study on Oil and Water Relative Permeability and Water Flooding Efficiency in Heavy Oil Reservoirs [J]. Petroleum Geology and Oilfield Development in Daqing, 2008, 27(4), 50-53.
2. Li Kuiying, Chen Hui. Study on Development Law of Heavy Oil Reservoir with Edge and Bottom Water [J]. Journal of Southwest Petroleum University (Natural Science Edition), 2008, 30(3), 93-96.
3. Zhang Jichang, Zhang Ying, Jiang Youwei,et al. Research on edge and bottom water control technology of heavy oil reservoirs in Liaohe Oil Region[J]. Special Oil & Gas Reservoirs, 2007, 14(2):73-75.
4. Haihua Pei, Guicai Zhang, Jijiang Ge, Ping Jiang, Jian Zhang, Yuxin Zhong. Study of polymer-enhanced emulsion flooding to improve viscous oil recovery in waterflooded heavy oil reservoirs[J]. Colloids and Surfaces A: Physicochemical and Engineering Aspects, 2017, 529.
5. Sun Lei, Pang Hui, Sun Yang,et al. Mechanism study on water control and enhanced oil recovery by CO_2 huff and puff for shallow heavy oil reservoir[J]. Journal of Southwest Petroleum University(Science & Technology Edition),2014,36(6):88-94.
6. Xiang Zhou, Qingwang Yuan, Zhenhua Rui, Hanyi Wang, Jianwei Feng, Liehui Zhang, Fanhua Zeng. Feasibility study of CO_2 huff 'n' puff process to enhance heavy oil recovery via long core experiments [J]. Applied Energy, 2019, 236.
7. Luo Ruilan, Cheng Linsong, Li Chunlan, et al. Research on the Adaptability of Cyclic CO_2 Injection for Heavy Oil Reservoir [J]. Journal of Xi'an Shiyou University (Natural Science Edition), 2005, 20(1), 43-46.
8. Zhao Junsheng, Qian weiming, Lang Chunyan. Field test of CO2 huff 'n' puff in Subei low permeability reservoir[J]. Fault-block Oil & Gas Field,2003, 10(01):73-75.
9. Wang Xiande, Han Zhenzhe, Zhao Tongqing,et al. Well selection principle of CO_2 injection single well huff and puff[J]. Xinjiang Geology,2009,27(02):188-190.
10. Bi Yongbin, Zhang Mei, Liu Zhenlin, et al. The Evaluation Method for Selection of Horizontal Well for CO_2 Huff and Puff in Complex Fault Block Reservoir [J]. Offshore Oil, 2013,33(01), 64-68.
11. Jin Chunyu, Xu Meixiang, Hu Chaoyang, et al. The Study of Fuzzy Comprehensive Evaluation Method for Heavy Oil CO_2 Huff and Puff Well Selection in Nanpu Oilfield [J]. Mathematics in Practice and Theory, 2019, 49(15), 139-145.
12. Yin Daiyin, He Rui, Wang Yong. Economic profile-control well optimization with fuzzy comprehensive evaluation in Yushulin Oilfield[J]. Special Oil & Gas Reservoirs,2020,27(3):82-87.

https://doi.org/10.1051/e3sconf/202132901070

Adapting sustainability and energy efficiency principles to architectural education: A conceptual model proposal for the design studio sequence

Salih Ceylan[1,*]

[1]Faculty of architecture and design, Bahçeşehir University, Istanbul, Turkey

Abstract. Architectural education is the first step into the professional career of an architect. It has strong connections with the profession itself regarding the technological trends and the needs of the society. Therefore, emerging challenges and developments in the world of architecture like environmental problems and sustainability issues need to be addressed by the educational programs of architecture. In addition to this, sustainable Development Goals (SDG) Program of United Nations puts an important responsibility on the shoulders of architectural educators. SDG includes both architectural and educational goals such as quality education, affordable and clean energy, sustainable cities and communities, and climate action. Accordingly, architectural education needs to be formed in a way to respond to the requirements of the contemporary global society. Design studio is the heart and core of architectural education. It is the place where all the theoretical and technical knowledge and skills gained in other courses become useful for the students to come up with design ideas and products. Additionally, design studios are not isolated environments. They form a series of courses by the consecutive repetition throughout the continuum of architectural education. Thus, they need to be treated as a developing sequence. Therefore, it is very important and valuable that the structure of the design studio sequence is improved and updated with suitable revisions towards emerging needs of the profession and society like the adaptation of sustainability principles, to reflect the dynamic character of architectural education itself. This paper presents a conceptual model proposal for the design studio sequence for the adaptation of sustainability principles to architectural education.

1 Introduction

Since ancient times, the practice of architecture and its products are affected from the environment they exist in. These effects can be taken physically in terms of conditions such as the accessibility of the material, topographic and geographical situation, climate, as well as in the sense of psychological, social and cultural influence. Rapoport [1] mentions that the form transformed into a structure is determined by cultural factors, and natural conditions have only a limited effect in this process. As a result of the effect of all these factors and conditions, the concept of architectural identity emerges. According to Hacıhasanoğlu [2], architectural identity, as a sub-expansion of cultural and urban identity systems, interacts with city and conservation plans, architectural styles, architectural languages, building and environmental policies, materials and technology, behaviors and attitudes towards the environment. All architectural products that emerged throughout the history of humanity have revealed an identity belonging to the time, geography and culture they are in, and on the other hand, contributed to the formation of culture. This cycle has continued in this way from the first ages until the twentieth century without exceptions. In the early 1900s, changes began to emerge in the world order as a result of the industrialization process. Mass production takes the place of handwork, and gigantic factories take the place of small workshops. The rapid development in technology and communication facilities not only allowed events and developments to spread around the world in a short time, but also contributed to the replacement of traditional architectural movements by innovative trends like modern architecture which are based on technology and industrialization. Modern architecture has changed the course of architectural history both through its intellectual infrastructure and its expression in form, and has led to the first step to be taken in reconsidering the concept of "architectural identity".

Architecture is a profession that benefits from technology, follows the economy, and is affected by political events. Therefore, the profession and education of architecture are affected by emerging developments. Architecture has been developing with the globalization of technology, reaching everywhere in a short time, and obtaining information more easily wherever it is, and architectural products started to be similar to each other [3]. After the 1950s, architecture in the world has started to be applied on a basis of technology and globalism. An important result emerged from this development is the increasing

* Corresponding author: salih.ceylan@arc.bau.edu.tr

amount of resources and energy required in architectural and construction activities. The energy consuming structures required by the increased populations and living comfort standards, together with technological systems used in the buildings have exceeded the sustainable limits, especially since the late 20th century. According to the World Energy Council, global energy demand will increase until 2030 [18].
Today, the construction sector and its components, which is the most important application area of architecture, is the sector that has the largest share in the world's total energy use. 50% of the energy consumed worldwide, 50% of greenhouse gases that cause global warming, 24% of air pollution, and 50% of CFC and HCFC emissions are due to activities related to buildings [4]. Therefore, any action within the architectural practice is expected to have a major impact on energy and environmental issues. Consequently, serious approaches to mitigate energy and resource use have been developed within the practice of architecture especially in the last 40 years.
Sustainability is an urgent theme for humanity. The responsibility of architecture about sustainability and energy efficiency problems is emphasized in the Sustainable Development Goals Program of the United Nations as well, considering that many of the goals are directly or indirectly related with architecture [8]. Especially goal number 7, affordable and clean energy, and goal number 11, sustainable cities and communities shall be considered under the responsibility of architecture profession. To contribute to the achievement of these goals, and to provide more sustainable built environments, architecture profession needs to fulfill its responsibilities towards the society. But firstly, these responsibilities and strategies to overcome the energy and resource problems must be clarified, and architects need to be made aware of the situation. Considering the strong connection between the profession and education of architecture and that both are affecting and being affected from each other, the process of awareness most effectively starts in the education period. Consequently, increasing environmental problems and sustainability issues need to be initial elements of architectural education so that it can contribute to the profession to stay up-to-date.

2 Sustainability and architectural education

Education is a powerful agent that raises awareness for emerging developments and contributes to social change [12]. The profession of architecture is a special field that differs from other professions in terms of its prominent features at the education stage. Unlike other fields of science, architectural education requires the students to work with all their senses and emotions. In architectural education, the ability of the mind, eye and hand to work together should be developed [5]. According to Cook, the most enjoyable and at the same time most disturbing aspect of architecture is its open-endedness, which is a mixture of measurable and immeasurable features [6]. The transfer of knowledge and experience in architecture can only be carried out in a healthy way with a mutual interaction between the educator and the student. The rapid development and change seen in today's world of architecture requires a rapid change in architectural education as well. Therefore, it is an important step to be taken by the institutions providing architectural education to design their programs and contents in such a way that they can be changed and updated frequently, to ensure the continuity of education and its competence in meeting professional needs.

It is assumed that environmental and energy problems, and therefore energy efficiency and sustainability concepts in design will continue to have the importance they have today, and will even become more critical in the upcoming years. Therefore, energy efficient design is not a temporary architectural trend, but a design approach that should be permanently embedded in architectural practice. It is necessary that mandatory requirements of sustainability is a core issue for the formation of professional competence [13]. The training of architects who will implement this approach is only possible if they recognize and apply this approach during their education. Architectural education has great responsibility in the training of young architects who are aware and capable of solving environmental problems [11]. Architectural education is failing to respond to the rapid changes in the world [9]. Architects should have more knowledge, competence and awareness about energy efficiency and sustainability than anyone else. UIA [14] states that architectural education is expected to prepare architects to bear the responsibility for the health, safety, welfare and cultural interests of the public and for the sustainability of the built environment. Accordingly, sustainability principles need to be implemented into the architectural education curriculum as fundamental elements. The concerns about sustainability of the built environment and the sustainability of the society are important aspects of architectural education [15]. However, this is a complex process, including theory and practice-based interdisciplinary knowledge [10]. The issue about which phases and which sections of the curriculum these principles are supposed to be inserted into is a question of curriculum design. The curriculum needs to be revised and developed by the adaptation of sustainability principles to the relevant and suitable points.

Contemporary architectural education can be divided into 4 main categories: In the first category, there are courses that form the theoretical and conceptual foundations of architecture as in the second category, the concepts that constitute the technical infrastructure of architecture are taught; in the third category courses that develop presentation and expression techniques are; and in the fourth category there are studio lessons in which the lessons learned in the other three categories are applied by bringing them together [7]. Therefore, the implementation of sustainability principles may happen at various courses or course groups, but as the heart and core of architectural education, the design studio is the most suitable place to create connections between design issues and sustainability.

2.1. The importance of the design studio

The common aspect of architectural education in all institutions is that it is a branch that advocates learning by practicing. Studios are the places where the practice in architectural education happens most effectively. Design studio is the environment where the students learn various ways of design and nourish their creativity [16]. In the design studio, students learn different aspects of design education such as visualization and architectural thinking [17]. While theoretical, technical or practical information is given to students in all courses, studios are the place where all this information is brought together and the acquired skill is transformed into a product. Studio courses have the greatest weight in the curriculum of all architectural education institutions. The studio lessons, which continue regularly from the first semester of architectural education to the end, are the most important factors that shape the student's view and style of architecture throughout his/her education life. In this context, the content of architectural design or studio courses should be examined and prepared very well, both as a single course and as a series of courses. While planning the curriculum as a whole, the general structure of the architectural project series should be established, and the scope and subjects of the successive studios should be fictionalized within a certain context. It is important that the topics, plots and systems in question are not repetitive, but complement each other. The individual contents of the project courses should also be created in accordance with the general framework, including the information and practices that the student will need according to the level she/he is in. Since the project course environment is an environment where the lecturer and the students are in direct contact, it is also important that these courses are designed in a way that makes it possible to be personalized. In terms of energy efficient design approach, studios draw attention as the most suitable application areas. It should be among the objectives of the design courses that many different topics, from the basic principles of energy efficient design to the latest technological applications and technical details, are handled in the studios and that various applications that provide energy efficiency are integrated into the design by the student. Which energy efficient application will be used in which periods and in what kind of projects and how it will be adapted to the design are also important issues to be decided during the design of the studio sequence and the preparation of the curriculum.

Design studio courses are the most suitable parts of architectural education regarding the adaptation of sustainability principles. However, the studio courses need to be considered as a complete sequence from first to the last year of education, instead of producing independent solutions for each studio separately. It should be taken into consideration that the coordination of the studios is important to hinder repetitive, irrelevant, and incomplete attempts.

3 A conceptual model proposal for the design studio sequence

Adapting sustainability and energy efficiency principles to architectural education is an important issue for the future of architecture profession. Accordingly, the implementation needs to be executed firmly. The design studios are appropriate courses to be the first steps of a revision of this kind because they are flexible and open for customization. Each design studio is unique and has its own character and requirements. However, within the frame of the curriculum, all design studios must be connected with each other within a context. This study evaluates the design studio sequence as a continuous structure, and proposes a holistic conceptual model for the whole sequence. Therefore, the definitions for each design studio are open to interpretation. Following chapter presents the specifications for each design studio from first to the last year of education.

3.1. Details of the proposed model

Adapting sustainability and energy efficiency principles to architectural education is an important issue for the future of architecture profession. Accordingly, the implementation needs to be executed firmly. The design studios are appropriate courses to be the first steps of a revision of this kind because they are flexible and open for customization. Each design studio is unique and has its own character and requirements. However, within the frame of the curriculum, all design studios must be connected with each other within a context. This study evaluates the design studio sequence as a continuous structure, and proposes a holistic conceptual model for the whole sequence. Therefore, the definitions for each design studio are open to interpretation. Following chapter presents the specifications for each design studio from first to the last year of education.

3.1.1 First semester design studio

The design studio in the first semester is mostly considered as an introduction to architectural practice. It is the first design experience of the students, so that the structure of the studio needs to be kept simple but effective. The architectural program of the design problem shall be very simple, consisting of a single space or volume with only a natural environment in the surroundings. The size of the construction area might be up to 50 square meters, and the emphasized sustainability element shall be orientation and relationships with the land and topography.

3.1.2 Second semester design studio

In the second semester of their educations, students are still in their freshman year, and their introduction to architecture continues. In the design studio, the basic definitions of design shall be emphasized. The architectural program shall be a simple one with a singular function, and the context shall be natural or rural environment with low density. The structure can be between 50 and 200 square meters, and the sustainability applications can focus on the relations between indoor and outdoor spaces, and passive systems like natural ventilation and daylight use.

3.1.3 Third semester design studio

In the third semester, the design studio shall become relatively complicated. Architectural program is still low in complexity with a basic function like residential or commercial purposes. Environmental context shall be an urban area with low density, and the area of construction can be between 200 and 500 square meters. The sustainability applications emphasized in this studio are simple insulation and indoor environmental quality issues.

3.1.4 Fourth semester design studio

Architectural program gets more complicated in the fourth semester, considering two interconnected actions like working and residing together. The building size shall be between 500 and 1000 square meters, and the surrounding area shall be an urban space with average density. The sustainability applications proposed for this studio are artificial lighting, artificial ventilation and recycled materials.

3.1.5 Fifth semester design studio

Fifth semester design studio deals with a more complicated design problem like a local public building in an urban area with high density. The total construction area in this design studio can be between 1000 and 2000 square meters, and the related energy efficiency concerns are focused on green roofs and rainwater management.

3.1.6 Sixth semester design studio

In the sixth semester, the architectural program gets complex through a mixed-use building that most possibly consists of multiple floors. The environment around the building may be a special urban area, probably one with a historic value. The total construction area is 2000 to 3000 square meters, as the sustainability applications are focused on acoustic solutions, wastewater management, and solid waste management.

3.1.7 Seventh semester design studio

Seventh semester design studio deals with a complicated design problem with a unique function that requires specialization like education or healthcare. The context is urban with high density, and the total construction area is supposed to be between 3000 and 6000 square meters. Related energy efficiency applications in this design studio are focused on alternative and sustainable energy production, and energy efficient façade design.

3.1.8 Eighth semester design studio

Eighth semester is the last one of the proposed architectural curriculum, and the design studio shall reflect the advanced level the students have reached at the end of their education. The architectural program in this design studio represents a public building with multiple functions. The context of this design studio consists of an urban environment with high density and special features, and the total required construction area is between 6000 and 10000 square meters. Sustainability applications in this design studio are smart building systems, life-cycle cost analysis and optimization, and sustainable construction methods.

Table 2. The proposed design studio sequence.

Studio Course #	Architectural Program	Environmental Context	Area	Sustainability applications
Design Studio 1	Single volume with a sole function	Simple natural environment	Up to 50 m²	Orientation, land and topography relations
Design Studio 2	Simple program with a single function	Natural or rural environment with low density	50 to 200 m²	Natural ventilation, daylight use, indoor-outdoor relations
Design Studio 3	Single program with a basic function	Urban environment with low density	200 to 500 m²	Basic insulation, indoor environmental quality
Design Studio 4	Combination of two interconnected functions	Urban environment with medium density	500 to 1000 m²	Artificial lighting, artificial ventliation, recycled materials
Design Studio 5	Moderately complex program with a public function	Urban environment with high density	1000 to 2000 m²	Green roofs, rainwater management
Design Studio 6	Complex program for mixed-use functions	Urban environment with historical value	2000 to 3000 m²	Acoustics, wastewater management, solid waste management
Design Studio 7	Complex program with unique function	Urban environment with high density	3000 to 6000 m²	Alternative energy production, energy efficient façades
Design Studio 8	Complex program for a public building with various functions	Urban environment with high density and special features	6000 to 10000 m²	Smart building systems, life-cycle cost analysis and optimization, and sustainable construction methods

The proposed design studio sequence is continuous, and each studio builds upon and develops the one that comes before (See Table 1). The architectural program starts with a single volume and gets more complicated in the upcoming studios, the environmental context gets denser and complex, and constructed area gets gradually bigger each semester. Accordingly, applications about sustainability and energy efficiency also start from simple and basic ones in the first semesters, and they get more technical, complicated, and technology-based in the advanced semesters. Consequently, the proposed design studio sequence covers a wide perspective of sustainability and energy efficiency approaches. However, it must be not forgotten that the design studio courses are not alone in the curriculum, and must be supported with theoretical and technical courses that answer to the questions revealed in the relevant design studios.

4 Conclusion

Architectural education is a multidimensional and multi-component process. Energy efficient design has recently started to be included in these training programs as an important component and it is predicted that this situation will not change in the future. In this context, it is an undeniable fact that architectural education should be redesigned to include an energy efficient design approach. It is important to ensure the continuity of education that this redesign is carried out not as an action that can be done in a single step, but as a planned process that spans a certain time. In addition, taking into account the factors depending on the place and time will ensure that this process works in a healthy way. Existing architectural, educational and social conditions have led to the process of adapting energy efficient design principles to architectural education to progress more slowly and without determination. However, it is seen that it is possible to accelerate this process and make it healthy and holistic with the measures to be taken and the practices to be made.

The approach of using the design studio sequence as the focal point for the implementation sustainability principles into the architectural education is an effective one. However, this operation requires expertise and strong collaboration among education curriculum designers. Design studio educators, students, and all other stakeholders of the design studio need to be a part of this process, to eventually come up with a structure that responds to the needs of students and educators, and the requirements of the emerging professional medium at the same time.

The proposed model is conceptual and creates a general frame for the design studio sequence. For more concrete implementations, the local factors for the architecture and education of the region or country must be considered. Accordingly, revisions and improvements for better adaptation might be required as well.

This study is purely focused on the design studio sequence as the revisions in other technical and theoretical courses are not considered. Even though this study provides a lot of information and knowledge about the possibilities for the improvement of the architectural education curriculum, for better and more holistic results, all courses must be considered as a whole. The interaction between design studios and other courses shall ensure a more effective result in terms of the desired objectives regarding the implementation of sustainability principle into architectural education.

This paper and similar studies that aim to provide connections between architectural education and sustainability are important and beneficial for creating a framework of the problem. However, to be more effective and sufficient, these studies need to increase in numbers and become more realistic through experience-based case studies, together with other empirical and statistical implementations concerned about the same or similar research subjects.

References

1. A. Rapoport, *House Form and Culture* (Prentice-Hall inc., New Jersey, 1964)
2. O. Hacıhasanoğlu, Multi-Functional Space Organization: A Comparison Of Traditional And Squatter Houses In Turkey, *Open House International*, 22, 87-93 (1997).
3. H. Nalçakan, Ç. Polatoğlu, Türkiye'deki ve Dünyadaki Mimarlık Eğitiminin Karşılaştırmalı Analizi ile Küreselleşmenin Mimarlık Eğitimine Etkisinin İrdelenmesi (Education Of Architectural In World And Turkey With Comparison Analyses And The Effect Of The Globalisation To The Architectural Education), *Megaron*, 3 79-103 (2008)
4. D. Eryıldız, Sürdürülebilirlik ve Mimarlık Dosyasında Ekolojik Mimarlık (Ecological architecture in the folder of sustainability and architecture), *Arradamento Architecture*, 154, 71-75 (2003)
5. J.N. Erzen, Eğitimin Estetik Süreç Olarak Yorumu ve Mimarlık Eğitimi (Interpretation of education as a process and architectural education), *Journal of METU Faculty of Architecture*, 2, 175-185 (1976)
6. P. Cook, *Primer* (Academy Editions, Londra, 1996)
7. B. Uluoğlu, Mimari Tasarım Eğitimi: Tasarım Bilgisi Bağlamında Stüdyo Eleştirileri, Ph.D. thesis, İTÜ, İstanbul, Turkey (1990)
8. United Nations, Sustainable Development Goals, available at: https://sdgs.un.org/goals (2015)
9. X. Xiang, X. Yang, J. Chen, R: Tang, L. Hu, A Comprehensive Model of Teaching Digital Design in Architecture that Incorporates Sustainability, *Sustainability* 12, 1-29 (2020)
10. J. Jablonska and S. Ceylan, Sustainable architecture in education, *World Transactions on Engineering and Technology education* 19, 96-101 (2021)
11. S. Ceylan, Students' approach to recycling and sustainability: A design study on affordable container houses, in: J. Charytonowicz (ed.), *Advances in Human Factors in Architecture, Sustainable Urban Planning and Infrastructure*, 175-181, Springer, New York (2020)
12. M. Taleghani, H.R. Ansari, P. Jennings, 2011, Sustainability in architectural education: A comparison of Iran and Australia. *Renewable Energy*, 36, 2021–2025 (2010)
13. S. Altomonte, Environmental education for sustainable architecture, *Review of European Studies*, 1, 12-21 (2009)
14. UIA, Architectural Education Reflections and Recommendations, *International Union of Architects*, Paris (2002)
15. S. Ceylan and M. Soygeniş, design studio experience: impacts of social sustainability, *Archnet-IJAR*

International Journal of Architectural Research, 13, 368-385 (2019)

16. S. Ceylan, P. Şahin, S. Seçmen, M.E. Somer, K.H. Süher, An evaluation of online architectural design studios during COVID-19 outbreak, *Archnet-IJAR International Journal of Architectural Research*, 15, 203-218 (2021)
17. S. Ledewitz, Models of design in studio teaching, *Journal of Architectural Education*, 38, 2-8 (1985)
18. R. Davies, Global demand for energy will peak in 2030 says World Energy Council, *The Guardian*, available at :

    https://www.theguardian.com/business/2016/oct/10/global-demand-for-energy-will-peak-in-2030-says-world-energy-council (2016)

https://doi.org/10.1051/e3sconf/202132901071

Analysis of Leakage Cause of L415M Spiral Submerged Arc Welding Pipe

Yan Xu *, *Yinglai Liu, Xianghui Nie, Zhenjun Feng, Liang Li, Fengping Yang*

Tubular Goods Research Center of CNPC, Xi'an, Shaanxi, 710077, China

Abstract. The causes of longitudinal crack in gas pipeline during pressure were analyzed by means of mechanical property test, microstructure analysis, sem analysis and energy spectrum analysis. The results show that the leakage of L415M spiral submerged arc welded pipe is caused by the contact between the base material of steel pipe and low-melting copper alloy in the process of high temperature, resulting in the brittleness of liquid metal. Specifically, when the leakage site of the steel pipe is heated at an abnormal high temperature, copper penetration along the grain boundary causes embrittlement, strength reduction and crack along the grain boundary, finally leading to the crack and leak of the steel pipe.

1 Introduction

The failure of product parts caused by liquid metal is a common problem in the current engineering production process. This brittle fracture often occurs in an instant, and the consequences are also very serious [1–2]. As the carrier of oil and gas transportation, steel pipes need to have excellent mechanical properties, because once accidents occur, the consequences will be disastrous [3–6]. Under certain temperature and tensile stress, the low melting point metal penetrates into the material from the surface of the part along the grain boundary, causing the embrittlement of the metal material and leads to the failure of the part. It is called the low melting point metal contact embrittlement crack, which is referred to as the metal erosion embrittlement crack or the low melting point metal thermal pollution crack. The design pressure of a gas pipeline is 6.3MPa. When the pressure rises to 1.25 times the design pressure during the hydraulic strength test of the process pipeline system in the construction unit, it is found that the pressure drops slowly at 0.1MPa per hour. The operator organizes personnel to patrol and find that the leakage point is 18.4km from the upstream station, which is located in the straight section and the buried depth of the pipeline is about 2.3 meters. The leakage pipe is Φ508mm × 7.1mm L415M spiral submerged arc welded pipe.

2 Macroscopic examination

The sample is the pipe section at the leakage site cut by a flame, and the leakage point is about one meter from the nearest pipe end. The morphology of the steel pipe after stripping the external coating is shown in Fig. 1. There are obvious cracks in the inner and outer walls of the steel pipe at the leakage point, and the cracks are distributed along the longitudinal direction of the pipe body. The crack center is 36 mm away from the center of the spiral weld, as shown in Fig. 1(b).

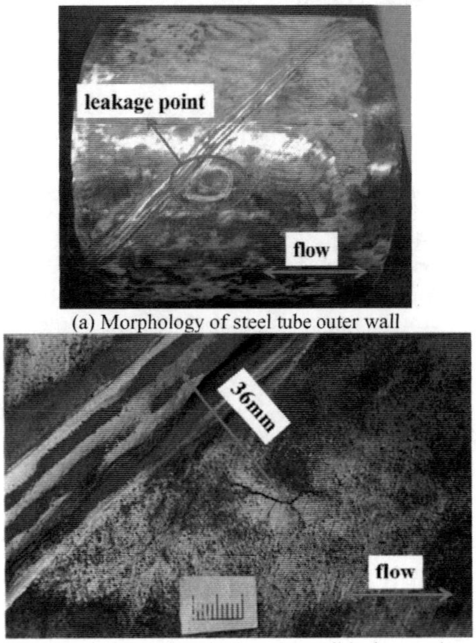

(a) Morphology of steel tube outer wall

(b) Morphology of steel tube outer wall leakage point

Fig 1. Macroscopic appearance of leakage steel pipe

The Smartzoom5 super-depth digital microscope was used to observe the cracks at the leakage, and the results are shown in Fig 2. The length of the outer wall crack is 21 mm, and the length of the inner wall crack is 31 mm. Fig. 2 (b) shows that there is a pit in the inner wall of the

* Corresponding author: xuyanxjtu@163.com.cn

 https://doi.org/10.1051/e3sconf/202132901071

steel pipe, 6 mm from the crack, and the maximum diameter is 2.4 mm. The wall thickness of the steel pipe near the crack was measured by ultrasonic thickness gauge. The wall thickness of the pipe at the pit was 5.9 mm, and that at other locations was about 7.1 mm.

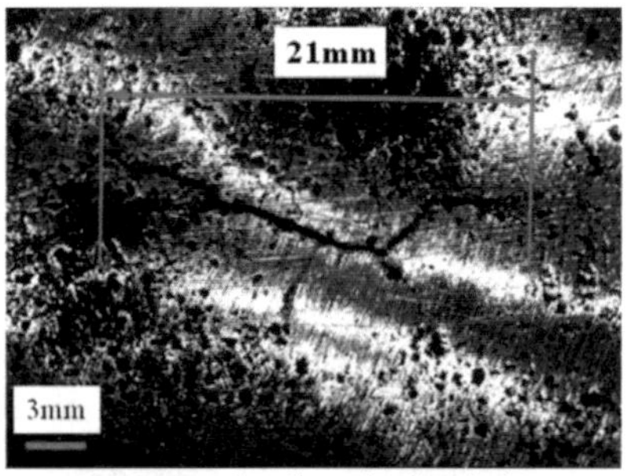

(a) External wall crack of steel pipe

(b) Internal wall crack of steel pipe

Fig 2. Macromorphology of cracks

3 Physical and chemical properties test

According to the standard requirements, chemical composition analysis, tensile properties, Charpy impact test, Vickers hardness test and metallographic examination were carried out on the pipe body of the steel pipe. All physical and chemical test samples were sampled at the position far from the defect.

3.1 Chemical constituents analysis

The chemical composition analysis samples were taken from the tube far from the defect position, and the chemical composition was analyzed by using ARL 4460 direct-reading spectrometer according to ASTM A751-14a standard. The results are shown in Table 1.

Table 1. Chemical constituents (Wt× 100%)

element	C	Si	Mn	P	S	Cr	Mo	Ni	Cu
content	0.085	0.23	1.38	0.011	0.026	0.029	0.022	0.016	0.04

3.2 Mechanical properties test

Tensile properties, Charpy impact test and Vickers hardness test were carried out on the steel tube. The tensile strength is 563MPa and the yield strength is 453MPa. The Vickers hardness is 174~194HV10. When the test temperature was −20°C, the minimum absorption energy of the tube Charpy impact test was 232J, and the average was 266J. The above results meet the standard requirements.

3.3 Metallographic analysis

The microstructure of the tube far from the defect position is shown in Fig. 3, which is ferrite and a small amount of pearlite. The grain size is 11.0, and there is no obvious abnormality.

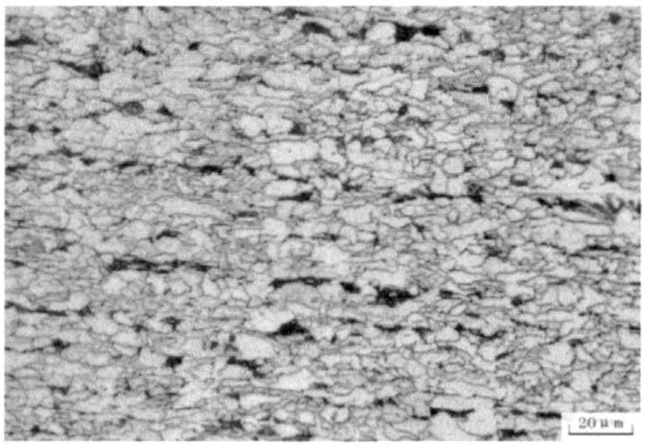

Fig 3. Tube structure away from leakage point

In order to understand the causes of cracks at the leakage of steel pipe, the T1 and T2 sections are intercepted along the direction perpendicular to the crack, as shown in Fig. 4. The full-thickness metallographic specimens are processed and observed by Smartzoom5 ultra-depth digital microscope and OLS 4100 laser confocal microscope.

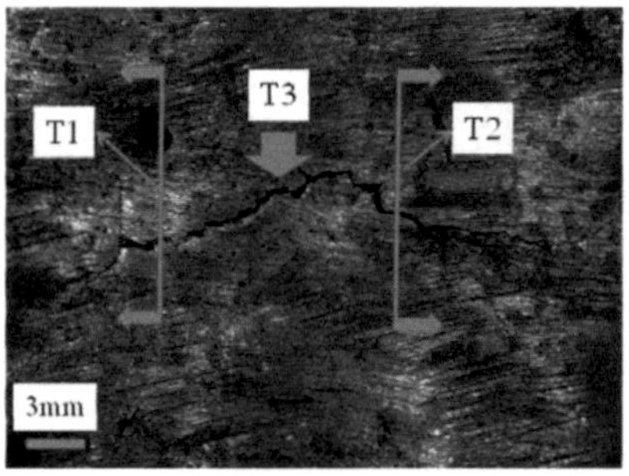

Fig 4. Sampling position of steel pipe leakage point

Fig 5 (a) is a photograph of the T1 section. The pit in the lower left corner corresponds to the inner wall pit found by the macroscopic inspection. The depth of the pit is about 1.2 mm, and the color of the pit and its adjacent cross section is abnormal, and the local color is purple yellow. The abnormal area is semi-elliptical, and the deepest part is close to the central position of the steel tube

 https://doi.org/10.1051/e3sconf/202132901071

wall thickness. There are holes at the interface between the color abnormal area and the tube matrix. The crack originates from the color abnormal area and extends to the outer wall. The microstructure at 1–1 is shown in Fig. 5 (b). The grains are abnormally coarse, and the color is different from the normal microstructure of the steel pipe, especially at the grain boundary. Fig. 5 (c) is 1-2 metallographic photographs. The grains near the main crack are coarse, the fine cracks propagate along the grain boundaries, and some purple inclusions exist in the grain boundaries.

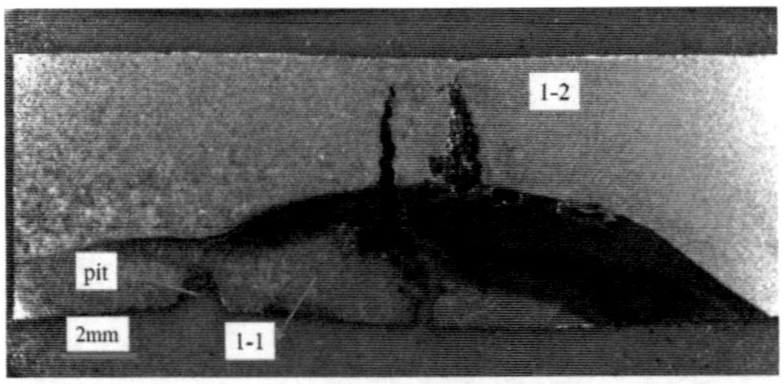

(a) T1 section

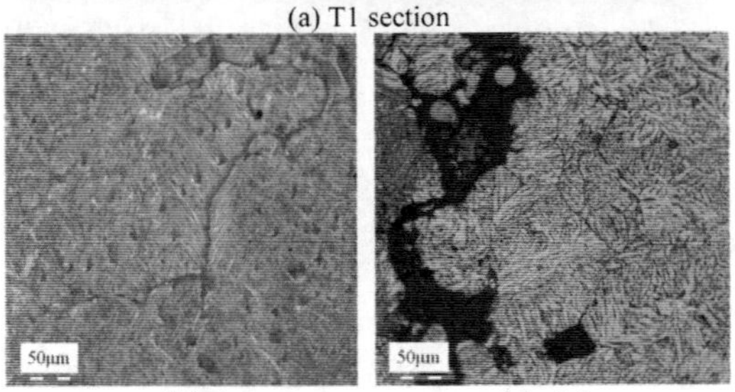

(b) 1-1 (c) 1-2

Fig 5. T1 section metallographic examination results

Fig. 6 (a) shows the photograph of T2 section, and there is still an abnormal area in the cross section of steel pipe. Fig 6 (b) is 2-1 metallographic photographs of the tube, which is located at the interface between the abnormal area and the normal tissue of the tube. The interface gap is obvious, and there are holes near the inner wall. The microstructure near the crack of the tube near the outer wall is overburned. Fig. 6 (c) in 2-2 metallographic photos, visible part of the grain boundary cracking, there are purple foreign bodies between grain boundaries, microstructure is overheating.

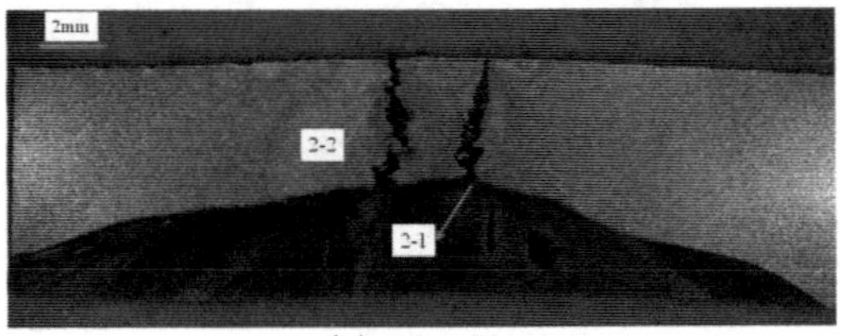

(a) T2 section

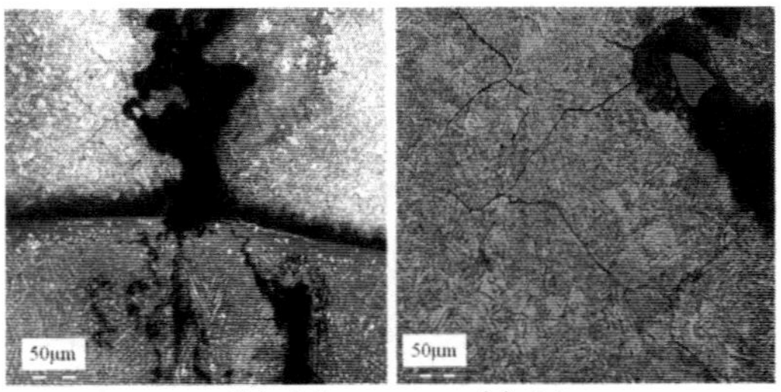

(b) 2-1 (c) 2-2

Fig 6. T2 section metallographic examination results

3.4 SEM micrograph and EDS analysis

The INCA 350 spectrometer was used to analyze the energy spectrum at different positions of the T1 section. The results are shown in Fig. 7. The main elements in the outer region are Fe, C and Mn, while the main elements in the inner color anomaly region are Fe, Cu and Mn. Among them, the copper element is abnormally high.

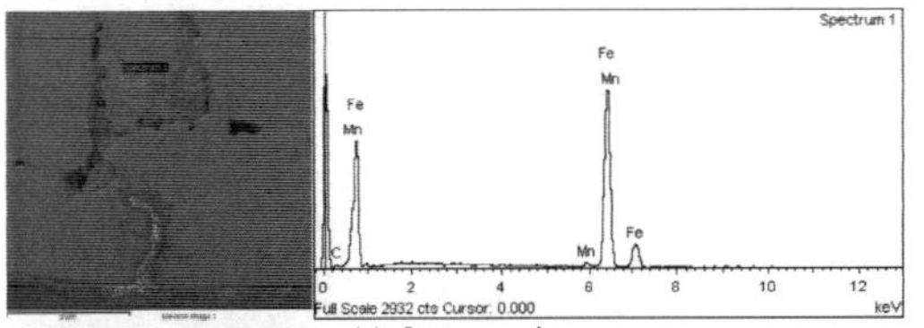

(a) Outer region

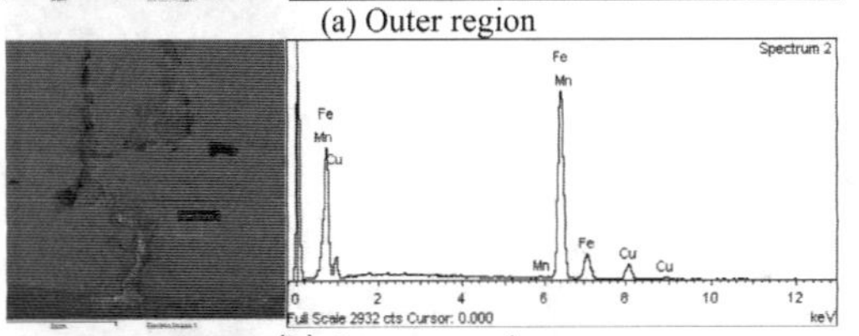

(b) Inner anomaly area

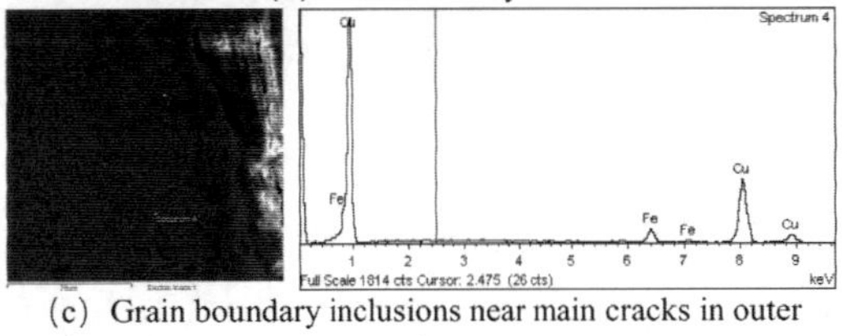

(c) Grain boundary inclusions near main cracks in outer region

Fig 7. T1 section morphology and energy spectrum analysis

In order to better understand the distribution of Cu element, the T1 section and T2 section were analyzed by backscattered electron imaging. The results are shown in Fig 8. The contrast of backscattered electron imaging is caused by the difference in atomic number, so the distribution area of Cu element can be clearly distinguished. The area is consistent with the cross-sectional abnormal area found in metallographic observation, and it is semi-elliptical. In backscattered electron imaging, the contrast between the inner abnormal area and the outer tube interface area is higher, reflecting the higher content of Cu element.

 https://doi.org/10.1051/e3sconf/202132901071

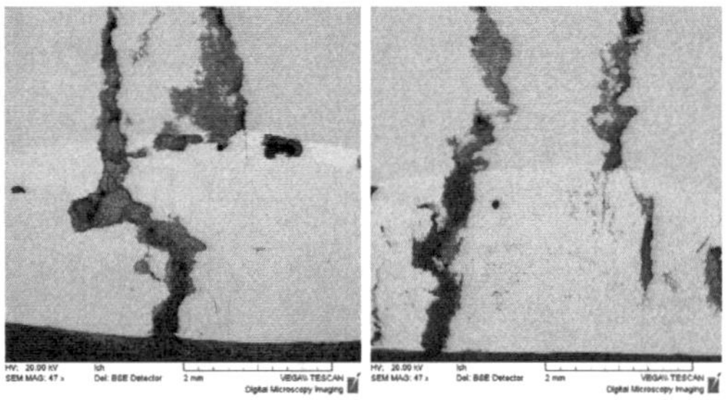

(a) Section T1 (b) Section T2

Fig 8. Backscattered electron imaging morphology

The cracks between T1 and T2 sections in Fig 7 were broken off, and the morphology of T3 fracture surface was observed, as shown in Fig 9. The fracture surface is obviously oxidized at high temperature. Fig 9 (a) is the fracture surface morphology of the inner layer, and Fig 9 (b) is the fracture surface morphology of the outer layer.

(a) Fracture surface in the inner anomaly area

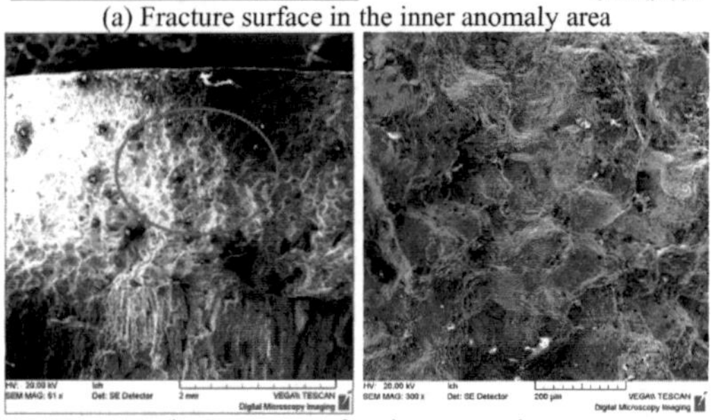

(b) Fracture surface in outer region

Fig 9. T3 Fracture Surface Morphology

The energy spectrum analysis of the inner and outer layers of the T3 fracture surface is carried out. The results are shown in Fig. 10. There are Cu elements in the abnormal area of the inner layer and the outer layer of the fracture surface, indicating that Cu elements are enriched in the crack area.

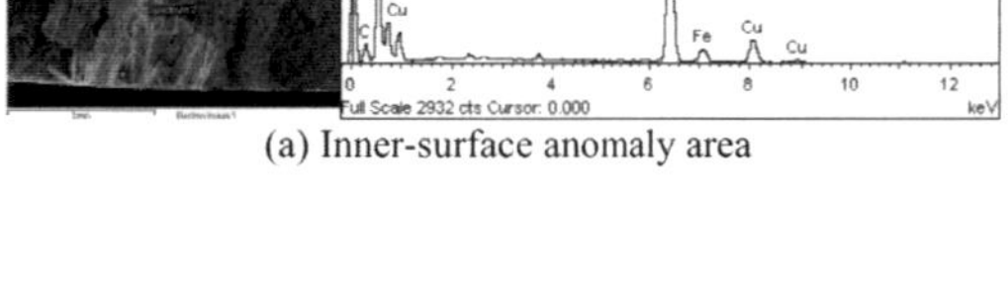

(a) Inner-surface anomaly area

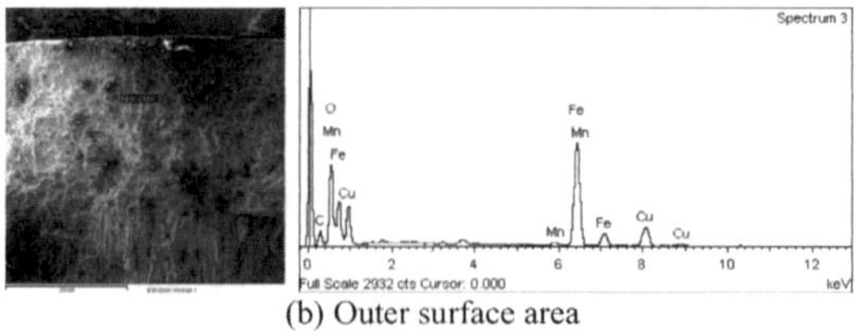

(b) Outer surface area

Fig 10. T3 fracture surface energy spectrum analysis

4 Analysis and discussion

According to the product standard, the chemical composition, tensile test, Charpy impact test and Vickers hardness test results of the steel pipe body far from the leakage point are in line with the standard requirements. The metallographic structure is polygonal ferrite and a small amount of pearlite, and there is no abnormality. The macroscopic inspection results show that there are obvious cracks in the inner wall and outer wall of the steel pipe at the leakage, and the cracks are distributed along the longitudinal direction of the pipe body. The length of the inner wall crack is larger than that of the outer wall, and there is no obvious deformation of the pipe body at the leakage. The metallographic examination results show that the microstructure of the inner layer of the steel pipe at the leakage is abnormal, there are holes and pits, the grain is abnormally coarse, and the color is different from that of the normal pipe. The area is semi-elliptic and the deepest part is close to the center of the wall thickness, which is characterized by high temperature burn. The crack originated from the abnormal area in the inner layer of the steel pipe, expanded to the outer wall, and the fracture surface was obviously oxidized at high temperature. The microstructure near the main crack was coarse, and the fine cracks near the main crack extended along the grain boundary. There was Cu element enrichment in the crack gap.

The above results show that the main reason for the leakage of the steel pipe is that the leakage site experienced an abnormal high temperature process. When the working environment or heating temperature is abnormally increased, the low melting point metal Cu is melted into liquid, and part of the metal on the inner wall of the steel pipe is melted. The liquid Cu diffuses into the molten steel pipe matrix, resulting in abnormal Cu content in the inner layer of the pipe wall, and the color of the burning area is different from that of the normal part. However, part of the grain boundary of the steel pipe near the burnt part melts, and Cu diffuses along the grain boundary. The Cu element is enriched at the normal grain boundary, that is, the molten metal at the burnt part, the molten grain boundary and the crack jointly constitute the diffusion channel of Cu element, which further deteriorates the material strength and reduces the effective bearing capacity of the steel pipe. When the hydraulic test is carried out on the site, the crack propagates along the weak grain boundary, and finally leads to the leakage of the steel pipe.

It is generally believed [7,8] that when the actual working temperature of the part reaches about 2/3 or even

1/2 of the melting point temperature of the low melting point metal, under the action of tensile stress, the low melting point metal will infiltrate into the metal along the grain boundary to make it brittle, and gradually form cracks. When the low melting point metal is heated and liquefied, if it is directly contacted with the solid metal surface, the solid metal is often wet and brittle. Under the action of tensile stress, the crack starts from the surface, and the crack tip adsorbs low melting point liquid metal atoms, which further reduces the crystal bond strength of the solid metal and leads to the brittle propagation of the crack. In the absence of tensile stress and a certain temperature under the joint action will not produce low melting point metal contact embrittlement failure, tensile stress can be external tensile stress, can also be the residual stress formed in the process of parts.

According to the characteristics of high temperature burn and Cu pollution at the leakage site, it is speculated that the cause should be due to improper operation and contact between the grounding Cu electrode and the pipe body in the process of pipemaking or field pipeline laying. When the contact between the two is not good, the resistance becomes larger and the arc is generated. The high temperature arc leads to the burning of the pipe body and the melting of the Cu electrode, which leads to Cu pollution and cracking in the part.

5 Conclusions

The leakage of the steel pipe is mainly caused by the crack of the steel pipe caused by high temperature burning of the pipe wall and Cu pollution.

References

1. Ning Mei, Sun Meihong, Zhong Shuming, et al. Analysis on Crack of High Pressure Air Bottle [J]. Tianjin Metallurgy, 2005(06):27-29+83.
2. Zhao Jun, Chen Yi, Chen Rong, et al. Failure Analysis on Copper Brittleness of Gas Cylinder [J]. Hot Working Technology, 2014,43(04):220-222.
3. Wang Gaofeng,Wu Jing,Nie Xianghui, et al. Cause Analysis of L415M LSAW Pipe Cracking [J]. Pressure Vessel Technology, 2016,33(07):58-62.
4. Xu Yan, Liu Yinglai, Nie Xianghui, et al. Research on Microstructures and Properties in Transition Zone of L415M Steel Grade Induction Heating Bending Pipe [J]. Hot Working Technology, 2016,45(16):175-178.
5. Wang Bin, Zhou Cui,Yang Li, et al. Cracking analysis of induction heating bended L415M pipeline steel [J]. Heat Treatment of Metals, 2013,38(08):123-127.
6. Xu Yan, Liu Yinglai, Nie Xianghui, et al. Prediction on the burst pressure of small diameter and thin wall steel pipe [J]. Oil & Gas Storage and Transportation, 2017,36(11):1265-1269.
7. Zhang Quanming, Chi Chun, Zhang Yong, et al. Study on Failure Cases of Liquid Metals Embrittlement [J]. Physics Examination and Testing, 2008(06):51-53.
8. Cao Xiaoyan,Li Tianlei, Luo Guangwen. Analysis on the Reasons Causing Cracks in Hot Bends [J]. Natural Gas and Oil, 2011,29(1):59-62.

Atmospheric environmental monitoring and evaluation based on experimental statistical methods

Lihong Sun

The Basic Science College, Harbin University of Commerce, Harbin 150028, China

Abstract: With the gradual improvement of social economy, the ecological environment is gradually deteriorating. In order to obtain higher economic benefits, enterprises often sacrifice the natural environment. After entering the new era, the public pays more and more attention to the ecological environment, and the relevant state departments also begin to use a variety of scientific and technological means to monitor the environment in order to obtain accurate information, find pollution problems in time, and take effective measures to control them. This is of great significance to improve the problem of environmental pollution. In this paper, combined with the actual situation, the application and layout methods of atmospheric environmental monitoring are analyzed in detail, in order to provide reference and reference for the related work in the future. In this paper, it is pointed out that atmospheric environmental monitoring is the process of measuring the concentration, observation and analysis of the change of pollutants in the atmosphere and the shadow response of the environment. The experimental statistical method is the product of the digital age, and it has its own advantages for monitoring the atmospheric environment, such as climate assessment, environmental assessment, atmospheric disaster prevention and early warning, etc., all based on the application of the experimental statistical method. Taking the digital measurement as the cut-in point, this paper describes the object and appearance of atmospheric monitoring, and focuses on the application strategy of atmospheric environmental monitoring test statistics, in order to clarify the importance of digital measurement and provide reference.

Keywords: Atmospheric environmental monitoring, experimental statistical methods, monitoring and evaluation

1 Introduction

Digital measurement as an effective means of atmospheric environment monitoring, the use of digital instruments to measure the atmospheric environment, has a good human-computer interaction function, especially conducive to the measurement of some complex environments. How to apply digital measurement in atmospheric environment monitoring is the focus of atmospheric environment monitoring [1]. Therefore, it is of great practical significance to explore the digital measurement strategy of atmospheric environment monitoring. In modern society, green environmental protection has become the theme of the times, and people begin to pay more attention to environmental issues. However, the gradual expansion of various production industries has had a certain impact on atmospheric environmental protection. This makes atmospheric environmental monitoring become a monitoring content that the relevant departments must pay attention to [2]. It is necessary to ensure its reasonable application, select appropriate layout methods, and ensure the accuracy of atmospheric environmental monitoring data [3].

Atmospheric environmental monitoring is a process of measuring the concentration of pollutants in the atmospheric environment, observing and analyzing their changes and their impact on the environment. Air pollution monitoring is to determine the types and concentrations of pollutants in the atmosphere and observe their temporal and spatial distribution and changes [4]. Usually, the pollutants in the monitoring area mainly include sulfur oxides, nitrogen oxides, carbon monoxide and so on, and the granular pollutants include falling dust, total suspended particulates, floating dust and acid deposition. With the development of the times, now the atmospheric environment is deteriorating, atmospheric change is easily affected by various factors, especially industrial production [5]. Through the rational application of atmospheric environmental monitoring, on the one hand, we can timely understand the concentration, type and range of various pollutants in the atmospheric environment, so as to obtain more basic data and compile atmospheric environmental quality reports [6]. On the

* Corresponding author: honghong03050305@163.com

https://doi.org/10.1051/e3sconf/202132901072

other hand, it is conducive to early warning, once large-scale changes in pollutants in the atmospheric environment are found, an alarm can be made as soon as possible to avoid the possibility of air pollution to the greatest extent and provide adequate protection for people's health [7]. Combined with the actual situation of atmospheric environmental monitoring at the present stage, the specific application is mainly reflected in three aspects as shown in Figure 1.

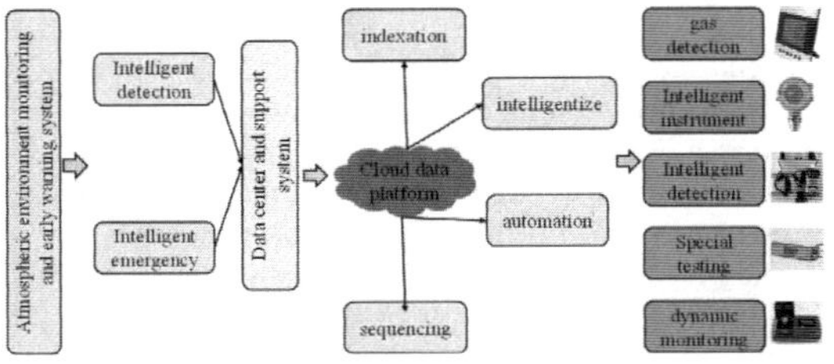

Figure 1 The method of Atmospheric Environment Detection

Based on the above analysis, it is not difficult to understand that atmospheric environmental monitoring is usually used in the monitoring of atmospheric pollutants, and provides necessary data reference for atmospheric research. The monitoring department must choose a reasonable monitoring method according to the characteristics of the environment to ensure the accuracy of the data to the maximum extent. In this paper, combined with the actual situation, the application and layout methods of atmospheric environmental monitoring are analyzed in detail, in order to provide reference and reference for the related work in the future. In this paper, it is pointed out that atmospheric environmental monitoring is the process of measuring the concentration, observation and analysis of the change of pollutants in the atmosphere and the shadow response of the environment. Digital monitoring technology is the product of the digital era, which has its own advantages in monitoring the atmospheric environment, such as climate assessment, environmental assessment, atmospheric disaster prevention and early warning and so on. Taking the digital measurement as the cut-in point, this paper describes the object and appearance of atmospheric monitoring, and focuses on the application strategy of digital measurement of atmospheric environment monitoring, in order to clarify the importance of digital measurement and provide reference.

2 Test statistical method

2.1 Grid layout method

Grid distribution is to divide the monitoring area into multiple grids, follow specific requirements and principles, and divide the grid evenly on the basis of ensuring reasonable division. In general, the setting of the sampling point should be at the center of the grid. If the situation is special, the sampling point will be set at the corner and position of the two straight lines of the grid, so as to ensure the representativeness of the sampling point. At the same time, in order to ensure the accuracy of atmospheric environmental monitoring, it is also necessary to determine the grid size, combined with the number of population, manpower and material conditions, the degree of pollution to make a comprehensive determination, a comprehensive analysis of atmospheric information, to determine the final sampling point. In general, the grid layout method is mainly applied to the rest of the serious pollution, or where the degree of pollution is uniform in the area, only in this case, can it meet the principle that "the location of the site should ensure that the pollution concentration is consistent with the surrounding concentration", and reflect the spatial information of pollution more truly and comprehensively.

2.2 Fan layout method

This method is similar to the concentric circle distribution method, in which a fan-shaped area is drawn in the downwind direction allowed by the regional condition and range, and the angle is set to 45 °. According to the different conditions in the area, the angle can also be set to 60 degrees. No more than 90 degrees as far as possible. After the drawing is completed, the sampling points will be designed in fan-shaped arcs, with a number of 3-4. It should be noted that the angles of the two adjacent sampling points are controlled between 10 and 20 degrees, and the maximum cannot exceed 20 degrees. This layout method is mainly used in some areas where the dominant wind direction is more obvious, specific particularity, and the existence of relatively isolated high point sources, it can be said to be the most obvious difference among many methods. It is also the layout method with the most requirements and conditions, and the scope of application is too limited, so it needs specific analysis of specific problems to determine whether the sector layout method can be used.

2.3 Functional area layout method

Similar to the general statistical test method, as shown in Figure 2, this layout method is mainly based on the division of functional areas for layout and monitoring. Therefore, in the process of practical application, it is necessary to achieve accurate division of business, residential areas, industrial areas and so on, and this method can be chosen only when the conditions are available. However, affected by various factors, the division of these regional functions is often very difficult to achieve "fine". In order to change this situation, the relevant staff should have a detailed understanding of the regional population and pollution in the process of monitoring. In order to dynamically adjust the atmospheric environment monitoring work. If there are no special circumstances, in the process of monitoring, regular points can also be selected for monitoring, spoiler routine characteristics, the accuracy will be slightly reduced. Staff need to be fully aware of this and make a full study of each point when arranging points, so that a more representative location can be selected to ensure that each set of data is more accurate. In the process of practical application, this method requires that the pollution center point in the monitoring area must be determined, the concentric circle with different radius should be drawn around the center store, and the radiation should be done from different angles from the center of the circle. The position of the sampling point can be set at the focus of the ray. Affected by the monitoring environment, the number of sampling sites is not very stable, and the distribution is uneven. Generally, more sampling sites are arranged in the downwind direction of the dominant risk, and the upwind direction is relatively less.

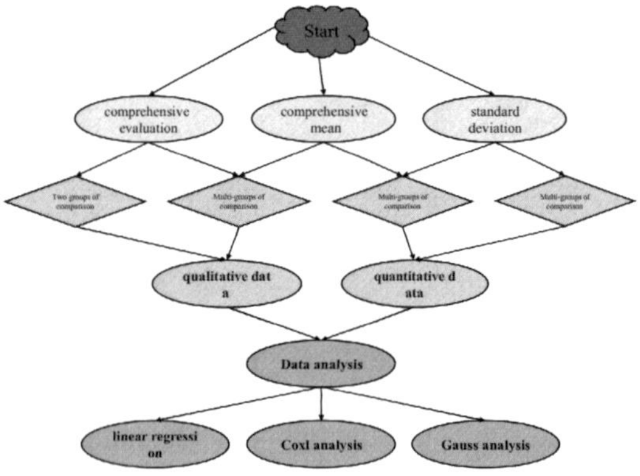

Figure 2 Statistical test method

3 Effective way to optimize the layout of Atmospheric Environmental Monitoring

3.1 Reduce the scope of influence

In the process of actual monitoring, the sampling points are likely to be affected by many factors and lead to judgment errors, which need to be further optimized to reduce the scope of influence as much as possible. At present, the commonly used optimization methods are as follows: first, the correlation coefficient method, in which the number of grids is set to m, the number of monitoring points is n, and the plot points of Shangfeng line are unlimited. When using the grid layout method, it is necessary to calculate the correlation coefficient R between m and n, and then calculate the coefficient of variation and the variance of each point according to the regional pollution, the law of change and the average concentration, and then determine the location of the points after comprehensive consideration. Second, matter-element analysis requires that in the process of detection, the maximum and minimum values should be found, the average value should be calculated, and the average value should be combined with the highest value and the lowest value to form a standard matter-element matrix, from which the correlation function can be calculated. Draw the distribution point aggregation diagram to determine the best location of the layout point. Third, the testing method, after using the above method, we can divide and calculate the high, middle and low pollution areas according to the average concentration of pollutants in the region, determine the layout scheme according to the evaluation scheme, and choose the optimal layout method to ensure the rationality of the layout. Fourth, the optimal index method is a way and method based on the formation of TOPSIS. In the process of practical application, we should first establish the initial monitoring data matrix M, get the optimal index box girder matrix N, unify M to establish the final most decision-making matrix Z, calculate the difference between the monitoring point index and the most level, follow the order from big to small, and select the optimal location. Fifth, the characteristic analysis method, mainly according to the pollution concentration to achieve the division of the telephone, in which each category of beans should set up a representative distribution point, to meet the monitoring principle of the representative requirements of the distribution point, establish the correlation matrix and calculate, get the best location.

3.2 Virtual measurement

The digital measurement of atmospheric environment monitoring is inseparable from the application of virtual instrument and technology, the virtual instrument and computer technology are combined with each other, the high-performance modular hardware is used to measure the atmospheric environment, and the measured data can be used as the basis for atmospheric environment monitoring. It is more simple and convenient, and the accuracy is higher. Specifically, the panel of the virtual instrument is virtual, the main thing is to integrate the network communication technology and electronic technology, and the measurement function is mainly completed by software to realize the sharing of monitoring information. Among them, the panel of variability, multi-layer and self-creation in the virtual instrument can better configure the atmospheric data resources, calculate, analyze and judge the atmospheric environment monitoring data flexibly, and realize the interconnection between the virtual instrument and the peripheral equipment. The automatic test system with virtual instrument technology as the trend is also conducive to the realization of real-time monitoring of atmospheric environment and has more practical application value. For example, the application of excited light radar is very effective in detecting atmospheric components and their pollutants. Lidar can detect the specific distribution of air-dissolved limbs and cloud particles, and at the same time, it can detect large gas components and pollute environmental gases, so as to obtain high-precision data and improve the quality of atmospheric data.

As the application front end of digital measurement, the application of digital server function just realizes the human-computer interaction. The function of the data server is a comprehensive system support. According to the sampling, storage and analysis of the atmospheric environment monitoring data, and the sampling server recording the atmospheric environment monitoring data, it should be noted that the sampling work seems simple, but in fact it is not, although it is the most basic work of atmospheric environment monitoring. However, it is also the most critical part, and whether the sampling data is accurate or not directly affects the results of atmospheric environment monitoring. The selection of sampling points and the time of sampling are very important. First of all, the sampling site must be representative, that is to say, the sampling site should represent the general pollution level of the area and reflect the pollution situation of the atmospheric environment in the region. Secondly, the sampling time should be marked and marked at the right time to ensure that the number of samples collected is true and useful. After the networking, the data is supported and finally stored in the data database for use. Here, it involves the storage of atmospheric environmental monitoring data, and the storage function can store the data tested in the past. The data of atmospheric environmental monitoring are more intuitive and clear, which lays a foundation for the analysis of atmospheric environmental monitoring data. Not only that, but in the application of the data server, manual operation can also be carried out. When necessary, according to the needs of the atmospheric monitoring and measurement work, the manual input modifies the parameter setting, the management of the data is more convenient, and the man-machine interaction is coordinated and cooperated with each other. The amount of work in data processing is reduced. So that the atmospheric environment control methods and control hands are more optimized. For example, in the process of investigating air pollution, a large number of real-time monitored data can be collected by carrying the gas sensor on the UAV.

4 Atmospheric environment assessment

4.1 Collect accurate environmental data

Ensuring the accuracy of basic data is the basic requirement of atmospheric environmental assessment. In addition to conventional data, such as meteorological data, topographic data, pollution data, etc., we should also focus on the detection data of the current regional situation, changes in the distribution of regional pollution sources, project plans and environmental management materials. The detection data of the current situation of the area can directly reflect the change degree of air quality in this area as shown in Figure 3, and the distribution change materials of regional pollution sources can provide reference for the change of the number of pollution sources in each stage of the implementation of the planning project. The feasibility of the planning project can be effectively judged by analyzing the project proposal through environmental management materials.

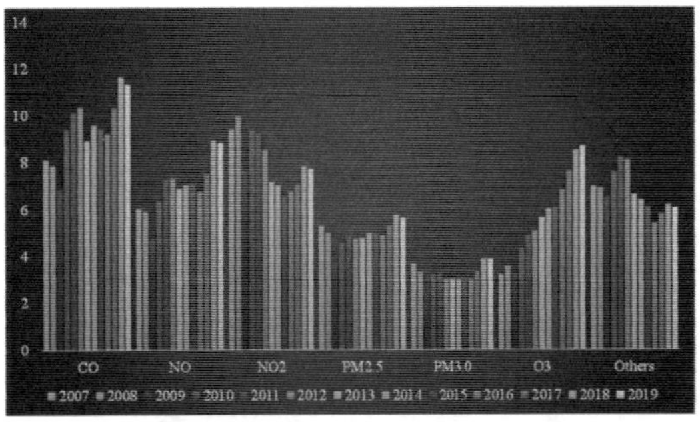

Figure 3 Collect and collate environmental data

4.2 Develop a prediction scheme for environmental impact

The environmental impact prediction scheme is carried out through the selection of pollution sources and emission locations in different planning periods, including simulating environmental quality changes, short-term environmental impact prediction, long-term environmental impact prediction and so on. In order to reduce the workload and improve work efficiency. Among them, it is necessary to accurately analyze the air pollution sources and understand the type, discharge and distribution of the pollution sources. In the process of construction, the emissions, discharge volume and location are uncertain, so it is impossible to carry out accurate quantitative analysis. Therefore, it is necessary to use effective methods to carry out analysis, such as analogy analysis, to compare the contents involved in the planning project with the relevant data in the industry, and to analyze the comparison of scale, process and output. As a result, the quantity, category and nature of pollutants can be predicted as shown in Figure 4. Coefficient method, this method is based on the database constructed by data analysis in the industry, and the prediction results can be obtained according to the emission coefficient. Finally, the prediction scheme also needs to determine the evaluation factors, calculation points and prediction areas, so that the workload can be reduced. Detailed analysis of the planning plan, with reference to the local meteorological characteristics, local leading industries and combined with the latest environmental management policy to analyze the content of the planning scheme to judge the maneuverability of the plan. If a certain area is planned, it is necessary to select the atmospheric index and formulate the planning plan according to the regional characteristics. China's environmental control quality standards have put forward clear indicators for various types of pollutants, which also need to be referred to when implementing the planning plan. In addition, it is necessary to compare and select the planning schemes according to the results of the prediction and evaluation, according to the regional concentration and key points mentioned in the planning scheme, and synthesize the factors such as location, scale, layout and industrial structure, and finally get the adjustment opinion. According to the characteristics of different planning projects, optimization suggestions are put forward mainly

 https://doi.org/10.1051/e3sconf/202132901072

from three aspects: site layout, scale and industry, and analysis results. The optimization of the scheme is carried out according to the analysis results. If the deviation between the project analysis results and the reality is too large, the planning scheme should be cancelled to avoid economic losses.

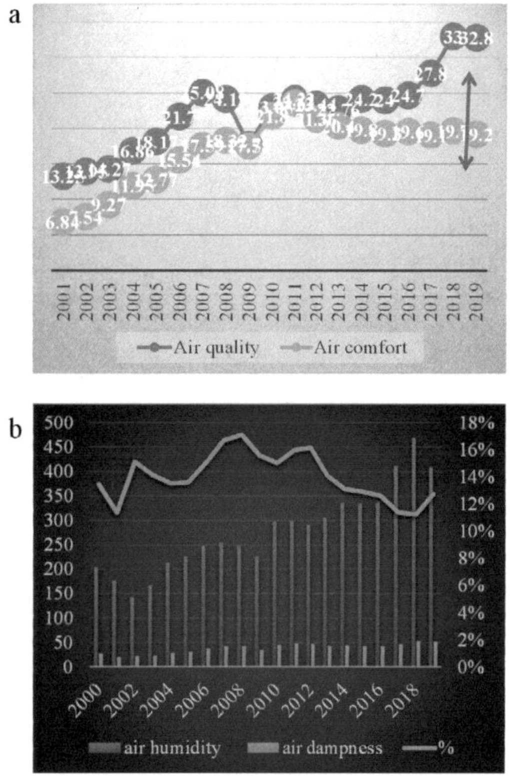

Figure 4 Effect of polymerization degree

5 Conclusion

Atmospheric environmental monitoring is usually used in the monitoring of atmospheric pollutants and provides necessary data reference for atmospheric research. The monitoring department must choose a reasonable monitoring method according to the characteristics of the environment to ensure the accuracy of the data to the maximum extent. In this paper, combined with the actual situation, the application and layout method of atmospheric environment monitoring will be analyzed in detail by using the method of experimental statistics, in order to provide reference and reference for the related work in the future. In this paper, it is pointed out that atmospheric environmental monitoring is the process of measuring the concentration, observation and analysis of the change of pollutants in the atmosphere and the shadow response of the environment. Digital monitoring technology is the product of the digital era, which has its own advantages in monitoring the atmospheric environment, such as climate assessment, environmental assessment, atmospheric disaster prevention and early warning and so on. Taking the digital measurement as the cut-in point, this paper describes the object and appearance of atmospheric monitoring, and focuses on the application strategy of digital measurement of atmospheric environment monitoring, in order to clarify the importance of digital measurement and provide reference.

Acknowledgements

If any, should be placed before the references section without numbering.

References

1. Hunter, James F., et al. "Comprehensive Characterization of Atmospheric Organic Carbon at a Forested Site." Nature Geoscience, vol. 10, no. 10, pp. 748–753, 2017.
2. Ochoa-Hueso, Raúl, et al. "Ecological Impacts of Atmospheric Pollution and Interactions with Climate Change in Terrestrial Ecosystems of the Mediterranean Basin: Current Research and Future Directions." Environmental Pollution, vol. 227, pp. 194–206, 2017.
3. Yu, Haili, et al. "Development of Atmospheric Acid Deposition in China from the 1990s to the 2010s." Environmental Pollution, vol. 231, pp. 182–190, 2017.
4. Ramsay, David B., et al. "Making Air (Quality) Visible: Exploiting New Technology to Dramatically Improve Atmospheric Monitoring." IEEE Pervasive Computing, vol. 17, no. 3, pp. 90–94, 2018.
5. Hofman, Jelle, et al. "Biomagnetic Monitoring of Atmospheric Pollution: A Review of Magnetic Signatures from Biological Sensors." Environmental Science & Technology, vol. 51, no. 12, pp. 6648–6664, 2017.
6. Chen, Long, et al. "Atmospheric Mercury Outflow from China and Interprovincial Trade." Environmental Science & Technology, vol. 52, no. 23, pp. 13792–13800, 2018.
7. Bazarov, A. V., et al. "Mobile Measurement System for the Coupled Monitoring of Atmospheric and Soil Parameters." Russian Meteorology and Hydrology, vol. 43, no. 4, pp. 271–275, 2018.

Analysis and Research on prefabricated building technology based on green low carbon environmental protection

Wenjie Ni [a]

Gannan University of Science and Technology, Jiangxi Ganzhou, 341000, China

Abstract. With the continuous progress of socialist construction, urban development is changing with each passing day, and various construction projects are continuously completed. At present, the State advocates low-carbon environmental protection and pays attention to ecological stability. Therefore, urbanization construction needs to be carried out under the concept of low-carbon environmental protection. The past construction technology has exposed many deficiencies in the continuous development and construction, such as serious waste of resources, neglect of environmental protection and low production efficiency, which has seriously hindered the transformation and development of the construction field, Prefabricated building technology has the characteristics of resource intensive, green environmental protection and high technical content, which perfectly meets the requirements of the development of the times. In the development of the times, prefabricated buildings gradually appear in people's sight, and are widely recognized by the public because of their green environmental protection, resource intensive and high technical content. Based on this, this paper mainly studies the prefabricated building technology under the current background of Green Low Carbon Environmental Protection.

Keywords: Green low carbon environmental protection; Prefabricated building; Green building.

1 Introduction

Since the 21st century, with the rapid development of China's construction industry, the scale of the construction industry has continued to expand, and a large number of laborers transferred from rural areas have joined the construction industry. The rapid development of the construction industry has contributed to the development of society and the construction of the economy. And the improvement of people's livelihood has made important contributions [1]. However, it should be noted that the rapid development of the construction industry consumes huge natural resources and has also caused certain damage to China's environment. The rapid development of the construction industry has caused the air quality to continue to decline, affecting people's daily life and health. Based on the current situation, the concept of realizing the green and sustainable development of the construction industry has gradually become popular [2]. The development of prefabricated buildings is still in its infancy. As a representative of building industrialization, we can see from the policies of the State Council and local governments that it is the national development goal to make "house components" flow on the assembly line and form the process of "building blocks" to build houses. As an efficient construction mode, The production mode of prefabricated buildings represents the development direction of construction industry in the future [3].

Assembly-type building is different from the previous building forms. Instead of completing the components in the construction site, it adopts prefabricated form, which is designed and constructed in advance, and then transported to the construction site after completion. The construction site only needs to complete the relevant assembly work. This way realizes the streamlined production of building construction, greatly improves the work efficiency of construction personnel, and at the same time reduces their workload and effectively reduces the construction cost [4]. The application of low-carbon concept makes prefabricated building technology more energy-saving and environment-friendly, and becomes the main development direction of modern building construction [5]. At present, China's prefabricated construction technology still needs further development, and the development of related research work still needs to be strengthened. It is necessary for related workers to carry forward the spirit of hard work and stand hard work and continue to study, so as to improve the quality of China's construction projects [6]. Traditional building forms are all the components needed to complete the construction project at the construction site, but the prefabricated building is different. It refers to the use of the form of appointment to produce the components. After the production is completed, it is transported to the construction site and then assembled according to the

[a] ni007sd@163.com

procedure. Components, and then realize the construction of the entire building [7]. In terms of advantages, prefabricated buildings prefer the construction form of flow production, which is not only conducive to the improvement of the work efficiency of constructors, but also effectively reduce the resource cost of construction [8]. In terms of environmental protection, it also minimizes the damage of construction projects to the environment, thus avoiding the problem of pollution to the environment, effectively implements the concept of green low carbon environmental protection, and further promotes the development of prefabricated buildings in a more green and environmental protection direction [9].

2 Application of Low Carbon and Environmental Protection in Assembly Building Design

2.1 Advantages and disadvantages of prefabricated buildings

Advantages:

Prefabricated buildings are formed by mechanized production of prefabricated components in factories and installation on site. Because standardized production is adopted, prefabricated components can be processed by flow process, which can not only ensure the production speed, but also effectively ensure the continuity of supply. Moreover, prefabricated members only need to command the construction of structural nodes when they faint, which reduces the construction difficulty and workload to a certain extent, and reduces the waste of components generated on site [10]. Prefabricated building construction adopts factory centralized production and unified transportation, so that the concrete mix ratio, cement mortar mix ratio and the number of steel bars are controlled more accurately, and the pollution caused by concrete is reduced. As the prefabricated building uses integrated, integrated and integrated precast concrete components, the wet operation of the formwork on site can reduce various pollution phenomena that often occur in traditional building construction to a certain extent, and can reduce To a certain extent, the cement mortar operation is saved, and the cleanliness of the construction site is effectively improved [11]. Prefabricated buildings can effectively reduce material stacking, loading and unloading, so as to reduce the impact of dust on the surrounding environment. Prefabricated buildings convert most of the on-site construction to factory processing, which can help reduce the discharge of solid waste and sewage, reduce dust and noise pollution, and be conducive to environmental protection [12].

Disadvantages:

At present, there is a lack of some targeted laws and regulations at the national level, while some binding policies and legal norms have little impact on promoting the development of prefabricated buildings. Because the building industrialization is still in the primary stage, the government has not issued specific incentive policies, which restricts the development of large enterprises and the progress of building industrialization [13]. In order to improve the construction efficiency of prefabricated components, it is necessary to have stronger substitutability and universality, and government departments and research institutes at all levels also need to increase the research on standardization, modularization and serialization of components. Due to the large one-time investment in the construction of the prefabricated component factory, and the small market demand at this stage, the component amortization cost is high [14]. The current construction industry management system and talent training mechanism cannot meet the needs of this aspect. Basically, universities and colleges across the country have not set up "prefabricated component" majors, nor have they opened up channels for training skilled workers.

Compared with traditional buildings, the production mode of prefabricated buildings has various obvious advantages, such as building safety and quality, personnel, construction period, environmental protection, etc. see Table 1 for details.

Table 1 Comparison of production methods of prefabricated buildings and traditional building production methods

Compare content	Traditional building	Prefabricated building
Engineering quality and construction safety	The open-air construction environment has greatly restricted the improvement of the quality of engineering projects, and the characteristics of high-altitude operations will increase the incidence of safety accidents	The parts and components are uniformly mass-produced in advance in the factory, which can more conveniently and effectively control quality problems, reduce high-altitude operations, greatly reduce safety benefits, and ensure construction safety
Construction period	The construction period is long, and it is greatly affected by the external environment such as natural environment, and cross-operations cannot be realized between various majors	A large number of parts and components are prefabricated in the factory, which is not affected by weather and temperature; various professional constructions can achieve three-dimensional crossover operations, which greatly reduces the construction period
Construction staff	The number of workers required is large, the stability of personnel is poor, and the professional level is different, which increases the difficulty of management	The high degree of mechanized operation has greatly reduced the number of workers required, but the requirements for the professionalization of workers have increased
Labor productivity	Low labor productivity: using traditional wet methods	High production efficiency: parts and components are prefabricated by the factory in batches, and on-site mechanized construction
Energy saving and environmental protection		The construction site is clean and tidy, produces less construction waste, and has a higher recovery rate. The semi-closed production environment of the factory can effectively reduce noise, dust and other pollution phenomena
Building quality	There is a lot of construction waste at the construction site, and water resources and noise pollution are serious. Material consumption is high, and it is easy to produce waste	The prefabricated production of parts and components mainly relies on the mechanized numerical control equipment in the factory, which can ensure the accuracy of the components, thereby greatly improving the quality of the building. And the parts produced are better in terms of impermeability, etc.

https://doi.org/10.1051/e3sconf/202132901073

2.2 Embodiment of low carbon environmental protection in prefabricated building design

The prefabricated building technology based on the concept of low-carbon and environmental protection requires comprehensive consideration at the beginning of the design. First of all, through scientific design, the building takes away indoor heat through the flow of natural wind, instead of artificial means such as air conditioners and electric fans, so as to create a comfortable indoor environment. Figure 1 shows the design of green, ecological and environmentally friendly buildings.

Figure 1 Green ecological and environmental protection building design

Secondly, water resources in construction projects can also be reduced by reasonable water supply and drainage design, and solar energy can be used for water heating and power generation. Finally, the energy utilization in the construction project can be replaced by new clean energy, so as to effectively control greenhouse gas emissions [15]. Unreasonable design of prefabricated buildings will lead to poor indoor air circulation, resulting in indoor air pollution, and long-term living will be harmful to the body. Ideally, a permeable north-south indoor layout will make the indoor air unblocked. However, the current city has a small amount of land and it is difficult to achieve such an indoor layout. In this case, the indoor walls and spaces can be rationally laid out. Enhance the circulation of indoor air and the entry of light. For example, designing a ceiling in the interior can appropriately reduce the indoor space, thereby effectively reducing the consumption of resources, and at the same time increasing the indoor thermal insulation efficiency.

Different scholars have different views on the theory of technological innovation process. Many scholars at home and abroad have put forward many models about technological innovation process and technology promotion model from different angles. This model points out that technological innovation mainly covers several stages, such as research and development, production and marketing. Technological innovation begins with scientific R & D and technological creation, and finally the market through the continuous development of enterprise technology. The realization of technological innovation is inseparable from the application of innovative achievements such as new technologies and new products developed by enterprises in the laboratory. In other words, the market receives the enterprise's new technology, and the customer is the passive recipient. Only by investing enough factor resources in the innovation process, the enterprise will have more opportunities to carry out technological innovation, and the possibility of innovation success will be greater. The model is shown in Fig. 2.

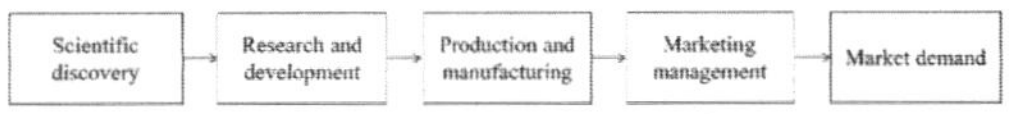

Fig. 2 Technology promotion model

3 Prefabricated building technology based on Green low carbon environmental protection

3.1 Dynamic scheduling in prefabricated buildings

In the process of assembly construction, if the useful information needed by construction scheduling can not be fed back to the scheduling system in time, it will lead to the scheduling system failing to make timely and accurate response according to the actual construction state and produce errors. A complete assembly construction process should involve three links, namely, construction preparation, construction scheduling and construction execution, in which construction data are continuously generated and exchanged. As shown in Fig. 3.

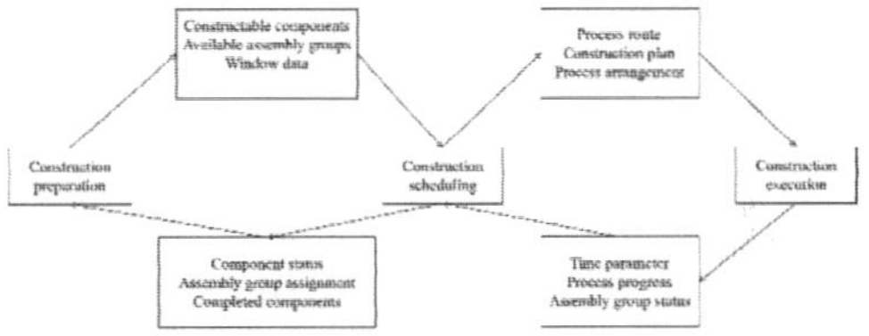

Fig. 3. Data flow chart

The construction process of prefabricated buildings involves multiple building components and multiple processes, and multiple assembly groups are required to complete the construction at the same time. Therefore, it is necessary to coordinate the building components, assembly processes, and assembly groups, and optimize the optimization goals by rationally arranging the resources between each component. The mathematical model of prefabricated building schedule optimization can be described as: suppose there are a total of n building components, and each component has k assembly procedures. These k procedures are constructed by an assembly group of m different components composed of several people. There are x types of groups, and each type contains different assembly groups. The objective function of the schedule optimization problem is to find the shortest time required to complete all the processes of all building components.

According to the construction characteristics of prefabricated buildings, the main constraints of the schedule optimization of prefabricated buildings are shown as follows:

$$P_{ij} + T^a_{ij} \le C_{ij} \quad (1)$$

$$C_{ij} \le P_{i(j+1)} \; C_{ij} \le P_{i(j+1)} \quad (2)$$

$$P_{ij} + T^a_{ij} \times W^a_{ijfg} \le P_{fg} + T^a_{fg} \times \left(1 - W^a_{ijfg}\right) \quad (3)$$

Among them, formula (1) and formula (2) represent the front and rear constraints of the process in the same component. The start time and the construction time of a process shall not be greater than its construction completion time, and the construction completion time of any process of the same component shall not be greater than the start time of its next process. Formula (3) indicates the sequence constraint between different component processes in the same assembly group, and the completion time of a component process in the assembly group should not be greater than the start time of another component process in the assembly group. The purpose of this paper is to use scheduling strategies to optimize the construction schedule of the project and minimize the maximum completion time. Therefore, for the evaluation of the pros and cons of the scheduling strategy of prefabricated buildings, this paper will take the minimization of the maximum completion time as the goal and establish an objective function. The function expression is:

$$C = \min\left(\max_{l \le i \le n, l \le j \le k} C_{ij} \right) C = \min\left(\max_{l \le i \le n, l \le j \le k} C_{ij} \right) (4)$$

Among them, $\max_{l \le i \le n, l \le j \le k} C_{ij}$ represents the largest one of the time when all components complete the last process.

3.2 The main features of Green low carbon environmental protection in prefabricated buildings

In terms of prefabricated buildings, the opinions on promoting the sustainable and healthy development of the construction industry issued by the general office of the State Council on February 17, 2017 clearly proposed to promote prefabricated buildings, adhere to standardized design, industrialized production, prefabricated construction, information management and intelligent application, promote the innovation of construction methods and vigorously develop prefabricated concrete buildings, Strive to increase the proportion of prefabricated buildings in new buildings to 30% within 10 years.

According to the statistics of the Ministry of Housing and Urban-Rural Development, as shown in Table 2, in 2015, the newly-built prefabricated building area in China was 73.7 million square meters, accounting for 2.9% of the newly-built urban building area. In 2016, the newly built prefabricated building area was 125 million square meters, accounting for 5.2% of the newly built urban building area, an increase of 57% compared with 2015. From January to October, 2017, about 130 million square meters of prefabricated buildings have been built nationwide.

Table 2 Analysis of the development of prefabricated buildings from 2015 to October 2017

Time	Newly built prefabricated building area (ten thousand square meters)	Growth rate of newly built prefabricated building area (%)	Percentage of newly built building area (%)
2015	7370	—	2.9
2016	12500	62	5.2
January-October 2017	13000	—	

In November 2017, the Ministry of Housing and Urban-Rural Development announced the first batch of prefabricated building demonstration cities and industrial bases for prefabricated buildings. As shown in Table 3, 30 cities including Beijing and Hangzhou were identified as the first batch of prefabricated building demonstration cities. 195 companies including Beijing Zhuzong Group Co., Ltd. and Hangxiao Steel Structure Co., Ltd. are the first batch of prefabricated building demonstration industrial bases.

Table 3 Fabricated building industry development analysis table

Time	Newly built assembly-type demonstratio n cities (one)	Assembled constructio n industrial base (one)	Demonstratio n project of prefabricated buildings (one)
November 2017	40	195	—
2020	70	59	510
Completio n rate (%)	60	98.2	—

Different from previous buildings, the exterior wall design of prefabricated buildings can better meet the requirements of residents. According to different changes in region, climate and temperature, the exterior wall of prefabricated buildings can not only retain personalization, but also have energy-saving effect better than traditional buildings. In the roof structure design of prefabricated buildings, the commonly used structural types are waterproof panel, thermal insulation panel, etc. load materials such as thermal insulation roof board and cement composite board are the main construction materials of roof structure.

Use information software for spatial modeling to improve the fineness of interior design. As shown in Figure 4.

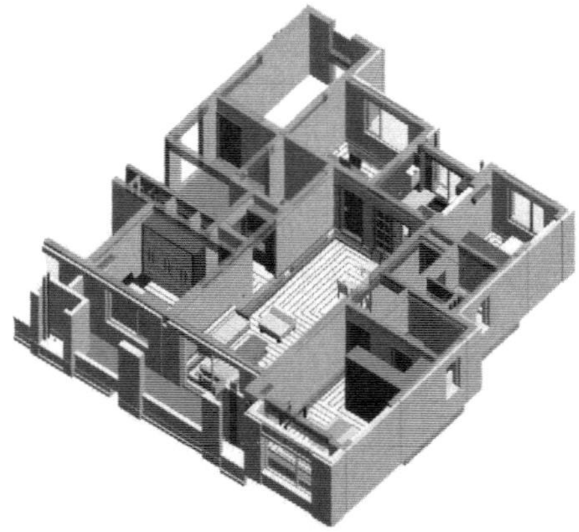

Figure 4. Prefabricated building design

In addition to the wall surface, the exterior windows of prefabricated buildings have also changed. In the construction of exterior windows, we can no longer rely solely on the traditional single-layer glass, and we need to choose double-layer hollow plastic-steel window glass to satisfy the use of exterior windows of prefabricated buildings and achieve better thermal insulation effect. At the same time, the floor treatment of prefabricated buildings is quite different from that of previous buildings, and it is necessary to consider the suitable materials needed for different floor thicknesses in the selection of materials.

The operation and management process of green buildings touches on many aspects of building envelope, water supply and drainage, HVAC, electrical, renewable energy, environmental greening, garbage disposal, traffic safety, etc. The following two points are mainly used for renewable energy and measurement and testing equipment. There are many differences with traditional buildings, and more standardized requirements are put forward for operation and management. Therefore, these two requirements are listed as special items. The technical content includes the operation and maintenance of renewable energy and measurement and testing equipment, optimization, and data collection and analysis of testing equipment.

4 Conclusions

In the current social development, the advancement of urbanization requires people to have more and more demand for buildings. They not only need buildings to have more functions, but also pay more attention to green energy conservation and environmental protection. Prefabricated building is an important product that more conforms to the development of the concept of green, low-carbon and environmental protection in the current era. The building method based on prefabricated components and assembly can not only effectively reduce the loss of energy and the waste of construction cost, but also reduce the generation of construction waste. The use of green, low-carbon and energy-saving building materials further realizes the protection of the environment. Assembly-type green building is a relatively environmentally friendly, systematic and integrated way. In order to protect the environmental resources we have now, we must adopt green, reasonable and feasible construction methods. At the same time, in order to meet the standards of building environmental protection, green construction technical measures should also be taken to ensure the quality, quantity and timely completion of the project.

References

1. Guo Wenlong. Construction technology advantages of concrete prefabricated residential buildings[J]. Building Materials and Decoration, 2019, 000(034): 42-43.
2. Niu Xinchun. Exploring the key technology analysis and quality supervision measures of house building fabricated structure[J]. China Construction, 2020, No.203(02):124-125.
3. Zhang Mengjie. Research on prefabricated steel structure building system based on BIM technology[J]. Building Materials and Decoration, 2019, 567(06):130-131.
4. Zhang Ying. Design and construction of prefabricated concrete structure[J]. China Science and Technology Information, 2018, 000(021):38-40.
5. Chen Liru. On prefabricated buildings-the main force of green buildings[J]. Building Development, 2020, 4(5):97-98.
6. Lv Yizhe, Ding Jie. The application and breakthrough of green building technology for steel structure prefabricated houses-new strategies for healthy, low-carbon, and smart houses[J]. Architectural Techniques, 2019, No.289(10):120-123 .
7. Hai Rihan, Zhou Jiaxu, Jia Xiaohu. The expression of the prefabricated architectural concept in the modern new-style yurt design[J]. Industrial Construction, 2020(3): 58-63.
8. Wang Zhenyu. Talking about the construction technology of prefabricated concrete structure of house building[J]. Architecture and Budget, 2020, No.292(08):70-72.
9. Fu Lei. Application of prefabricated building technology under the background of green building[J]. Green Building Materials, 2020, No.155(01):55+57.
10. Zhao Xiaoni. Application of BIM technology in green construction of prefabricated buildings[J]. Building Materials and Decoration, 2018, 558(49):25-26.
11. Zhang Guoyin. Low-carbon energy-saving prefabricated building technology research[J]. China Housing Facilities, 2020, No.209(10):81-82.
12. Ye Guoren. The application of BIM technology in prefabricated structure[J]. Gansu Science and Technology, 2017, 33(014):85-86.

13. Ma Xinyi. Application of prefabricated houses in green buildings[J]. Construction Technology Development, 2018, v.45; No.391(13):8-9.
14. Xue Jingdong. Prefabricated buildings: the integration of green and innovative technologies [J]. China Economic Information, 2017, 000(023): 36-37.
15. Wang Jian. Research on low-carbon energy-saving prefabricated building technology[J]. China Construction, 2017, No.151(12):138-139.

https://doi.org/10.1051/e3sconf/202132901074

Analysis and Discussion on Burning Low Calorific Value Fuel of Circulating Fluidized Bed Boiler

Yi Yan

School of Energy and Power Engineering, Huazhong University of science and technology, Wuhan, China.

Abstract. China's coal-dominated energy structure will not change in a short time. In order to implement the relevant national policies on energy conservation and emission reduction, it is a good response to adopt circulating fluidized bed boilers in coal-fired power stations. Circulating fluidized bed boiler adopts fluidized combustion, which has high reliability and high utilization rate because of its special combustion mode and particle recycling device. For the low calorific value inferior fuel which is difficult to burn in common boiler, CFB boiler also has certain adaptability and compatibility. In this background, this article with coal gangue as the representative of the inferior coal are discussed its application in the circulating fluidized bed boiler, the combustion process, boiler equipment and special fly ash produced by combustion is proposed that often appear in the three aspects of problems, and accordingly put forward the solution and optimization measures, provides reference for the actual process.

Key words: Circulating fluidized bed boiler; Low calorific value fuel; Operation problem; Retrofit scheme.

1 Introduction

Circulating fluidized bed combustion technology, a new combustion technology which has been widely built and running in modern thermal power stations, is relatively mature and can achieve clean combustion. When the air flow passes through a bed where solid particles are accumulated, if its flow rate reaches such a level that the frictional resistance between the fluid and the particles is just equal to its net gravity, solids will be suspended to have certain fluid appearance characteristics, which is called fluidization. Fluidized bed combustion shows this phenomenon. Continue to increase the air flow rate, when it exceeds a certain speed, the particles will just be blown out of the combustion chamber, but if a separator is installed at the outlet end to capture solid particles according to their size, the unburned particles can be captured from the flue gas and sent back to the combustion chamber for circular combustion through the return feeder, so that the fuel can be recycled, which is called circulating fluidized bed combustion technology. Therefore, compared with other types of boilers, it is easier to deconstruct inferior fuels and further decompose the nonflammable parts. The fly ash obtained from combustion can also be returned to the furnace to improve its burnout rate. The bed materials for fluidized bed combustion are not limited to traditional fossil fuels, but also include solid wastes and various biomass fuels. Because of its advantages of high combustion efficiency, wide adaptability to coal types, uniform temperature distribution in combustion chamber and large load regulation interval, through research, the basic principles of heating, mechanics, materials and other engineering problems such as overtemperature of heating surface, expansion of furnace body and wear of materials have been well solved. Circulating fluidized bed technology has been in a priority position in the field of clean coal power generation in China [1-3].

Schematic Diagram of Circulating Fluidized Bed Boiler

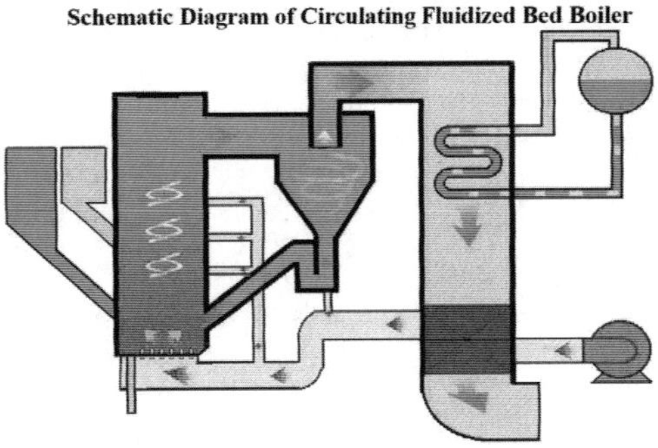

Fig 1. Schematic Diagram of Circulating Fluidized Bed Boiler

Coal gangue is a black rock that is mixed with organic and inorganic compounds co-deposited with coal during the coal formation process. It is a solid waste discharged during the process of excavation, mining and coal washing. It has low carbon content and hard texture. Coal occupies an absolutely dominant position in Chinese energy utilization, and a large amount of coal gangue has

*Corresponding author: u201812078@hust.edu.com

been accumulated during the long-term mining and utilization of coal. The coal gangue pile not only covers a large area, but also after a long period of sun and rain, it may cause spontaneous combustion and cause an explosion, releasing a large amount of dust and harmful gases, which seriously affects the air quality of the surrounding environment [4-5]. At the same time, as a resource, coal gangue is a kind of inferior fuel with utilization value, and it is of practical significance to mix and burn with coal.

Circulating fluidized bed boilers designed according to a specific fuel or coal are not suitable for other fuels with different properties. However, based on the economical operation and the extensive utilization of CFB technology, some coal-fired power stations choose to mix and burn coal with different properties from the design coal, so the combustion conditions in the furnace will deviate from the design value during actual operation. As a result, the fuel consumption of the whole plant increased and the economy decreased. Based on this, taking coal gangue as an example, this paper will discuss the possible problems when burning low calorific value fuel in circulating fluidized bed boiler from three aspects and put forward corresponding solutions.

2 Combustion Conditions

Coal gangue has complex chemical composition, low fixed carbon content and low calorific value, and its calorific value is generally 3000~11000kJ/kg. The ash content is generally over 60%, up to 90%, and the main components in the ash are aluminum oxide and silicon dioxide. Some coal gangue also has the characteristics of high sulfur content and high hardness. All these will have an adverse effect on combustion [6].

2.1 Difficulties in Ignition and Extinguishment

First of all, due to the uneven distribution of fuel particle size, the part composed of finer particles is easy to catch fire, while the large particles are difficult to decompose and ignite, which leads to the difficulty of boiler operation. In the initial stage of power plant application, deflagration occurred due to large input. In the later application, although deflagration no longer occurs, the volatile matter of coal is too low. It is very difficult for boiler to catch fire, volatile matter cannot be precipitated stably, which leads to unstable combustion state and slow rise of bed temperature. After the hearth is in a burning state, the bed layer is an area where coal gangue and air are preliminarily mixed, and the burning is intense, so it is impossible to control the bed temperature to drop in a short time, which makes it difficult for the boiler to extinguish the fire. In addition, the high temperature of the primary air chamber adjacent to the bed easily causes the burner under the bed to burn out due to overtemperature.

After analyzing the characteristics of this kind of refractory coal and certain tests, the following methods can be adopted to solve the problem of ignition and fire extinguishing difficulties:

(1) Choose higher combustion share in dense phase zone. At the initial stage of ignition, more fine particles are put in, which makes the combustion share of dense phase zone in the lower bed of furnace larger. Fine particles are easy to burn and release more heat, which makes the bed temperature rise rapidly and reach a stable combustion state. After the bed temperature is normal, large particles are put into combustion step by step.

(2) Adopt higher primary air rate. Under the condition of (1), high primary air rate must be used at the same time to bring the released heat out of dense phase zone, avoid local high temperature caused by excessive heat accumulation, and send the heat to the dilute phase zone where the upper combustion temperature is still low, so as to improve the upper combustion conditions and make the whole furnace heat load uniform. However, when a higher fluidization speed is adopted, under the condition of the same cycle ratio, the residence time of fuel particles in the combustion area may be short, which may reduce the combustion efficiency; When the fluidization velocity is lower, besides the above-mentioned overtemperature phenomenon, it may also cause the deposition of large particles and coke on the nearby water wall. Therefore, it is necessary to ensure that the fluidization speed is in a proper range.

(3) Because the volatile matter content and calorific value of coal gangue are low, and the heat release in dense phase area is less, in order to prevent the bed temperature from being too low at the initial ignition stage, the amount of circulating materials should be reduced to reduce the heat carried by the burnt particles transported back by the return feeder, and the lower circulation ratio can be adopted, which can be realized by appropriately reducing the fluidization speed.[7]

2.2 Local High Temperature in Dense Phase Area of Mixed Coal Burning Stone.

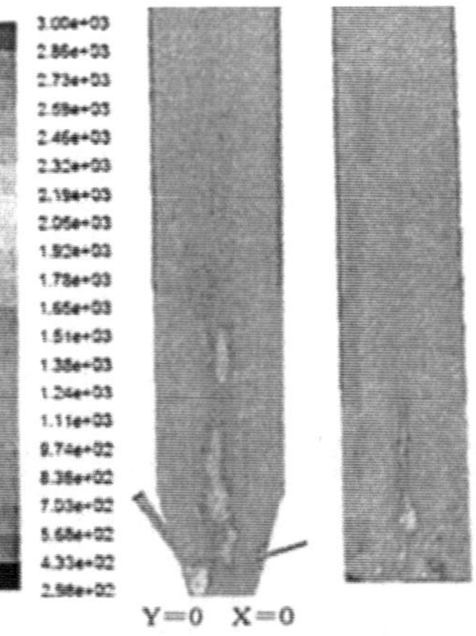

Longitudinal section temperature distribution

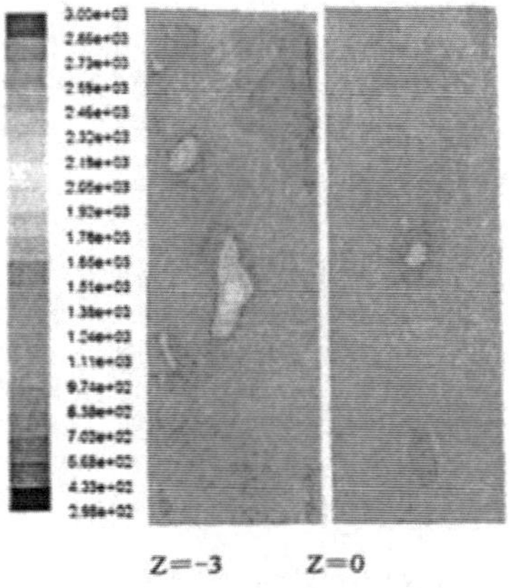

Cross section temperature distribution

Fig. 2. Temperature distribution in furnace

When burning coal gangue, a high temperature zone appears in the dense phase area above the coal feeding port close to the front wall, and the temperature could reach more than 1000K K. This is because when the mixed particles of coal gangue and pulverized coal are blown up and burned, the airflow near the side wall appears backflow phenomenon, which is mixed with the primary air coming out from the bottom, and the coal gangue particles with larger density and volume are in the bed. Continuous combustion on the surface and unable to bring out the heat generated by the combustion in time. The occurrence of this phenomenon on the one hand causes the high temperature area to stick to the front wall, causing the heating surface to coke, and it is easy to cause over-temperature damage to the air distribution plate under the bed; on the other hand, it blocks the upward flow of primary air and affects the fluidization state in the combustion chamber. This causes uneven heat absorption of the water wall on the radiant heating surface, and reduces the combustion efficiency in the dilute phase zone.

Existing boiler operation cases show that the above phenomenon is particularly obvious when coal is mixed with 20% coal gangue, which can be solved by adjusting the secondary air volume, especially by increasing the air intake of the secondary tuyere of the side wall of dense phase area to realize large-scale interference to the return air flow, so that it can be realized in the mixed flow in three-dimensional space. Prevent hot air from staying near the bed for too long, which can optimize the fluidization state. In addition, pre-throwing coal gangue with finer particles mentioned in the above-mentioned ignition and fire extinguishing difficulties can also effectively alleviate this problem, so it is necessary to increase the grinding intensity of coal in the coal pulverizing system and the separation accuracy of the separator. When 10% coal gangue is blended, due to the small proportion of coal gangue, it can't have much influence on the whole combustion situation, there is no obvious flame sticking to the wall in dense phase area, and the temperature distribution in the furnace is relatively uniform as a whole, so the damage phenomenon of heating surface in boiler operation is obviously reduced.[8]

3 Boiler and Its Auxiliary Equipment

3.1 Coal Transportation and Slagging

Since the calorific value of coal gangue is generally low, when the boiler load is constant, the coal consumption of the boiler increases, the ash produced in the furnace increases, and the slag discharge increases, thus increasing the use frequency of slag cooler and slag conveyor, and shortening the service life when overloaded. When the boiler ash is large, it is easy to block the slag discharge pipe, but it cannot discharge slag normally for a long time. Ash will accumulate on the bed to block the feeding and block the passage of primary air from the bottom. In severe cases, it will lead to coking of the bed and aggravate the blocking phenomenon. However, most coal gangue is hard and hard to grind, so it puts forward higher requirements for the mechanical strength of coal mills, which increases the use failures of coal mills, separators and other equipment suitable for ordinary coal quality. Operation and maintenance costs of the pulverizing system increase.

Low calorific value and large ash content increase the operation burden and failure frequency of the coal conveying and slag removal system. If it is planned to burn coal gangue with poor quality for a long time, the traditional wind-water combined slag cooler unit often has the problem of insufficient slag discharge capacity when the slag discharge amount is large, and if unburned combustible materials are discharged into the wind-water slag cooler, its continuous combustion will lead to coking and burning of slag cooler. After a large number of large fluidized bed units have been put into operation in China, the practical results show that the slag discharge capacity can be improved and the problem of slag blockage can be solved by changing to the drum slag cooler. Because the slag is always driven to rotate by the roller in the roller slag cooler, there is no dead angle which is easy to deposit

https://doi.org/10.1051/e3sconf/202132901074

in the structure. And is easy to be fully cooled, so coking phenomenon is not easy to occur; In addition, the roller slag cooler makes use of the pressure difference between the hearth and its interior to make the slag automatically enter the slag cooler through the slag dropping pipe, so its structure is simpler and its volume is smaller than that of the wind-water combined slag cooler.

3.2 Sulfur Content

Coal gangue generally contains a large amount of sulfur, and its sulfur substances burn to form composite sulfate, which condenses on the outer wall of water wall at high temperature, destroys the oxide film on it, gradually reduces the thickness of pipe wall, and finally leads to pipe explosion; While in the economizer and air preheater where the wall temperature is lower than the dew point of flue gas, Sulphuric acid steam produced by a series of reactions such as combustion will corrode at low temperature, destroy materials and affect the normal operation of two kinds of equipment. And when the sulfur content of fuel is large, the deformation temperature DT of ash decreases, which makes the return feeder and bed more prone to coking under the condition of keeping the same combustion temperature in the circulation loop.

When the sulfur content is high, it is necessary to combine desulfurizer with flue gas desulfurization to meet the environmental protection requirements. In order to achieve desulfurization efficiency, the desulfurizer is mixed in the fuel, and a higher Ca/S ratio is adopted. The desulfurizer captures the generated sulfur oxides in time, reducing the content of the latter in the flue gas, and the products will be discharged together with ash and slag without pollution.[9]at the same time, the flue gas desulfurization device is added, which requires higher desulfurization efficiency than the traditional wet desulfurization technology.

4 About Fly Ash

Strictly speaking, a series of problems caused by fly ash from combustion belong to the combustion conditions and the loss of machinery and equipment easily caused by burning low calorific value fuel (coal gangue) in circulating fluidized bed boiler discussed before, but according to statistical data, the problems generated in the operation process account for a large proportion of the whole. This kind of problem becomes particularly obvious when burning coal gangue, so it is discussed separately.

4.1 Fly Ash Erosion

4.1.1 Abrasion of Heating Surface

In a circulating fluidized bed boiler burning coal gangue and other fuels with low calorific value and high ash content, due to the large amount of fly ash in flue gas and high flue gas flow rate, in which the content of solid particles is much larger than that of ordinary pulverized coal boilers, the wear caused by scouring the furnace wall and heating surfaces at all levels is much more serious than that of traditional boilers. The wear of solid surface caused by a large number of particles carried by flue gas can be divided into two types: scouring wear and impact wear.

Erosion wear refers to the impact of solid particles on the material surface at a small angle relative to the solid surface. The partial velocity parallel to the normal of the solid surface makes the particle wedge into the object surface, while the partial velocity parallel to the tangent of the solid surface makes it still slide along its surface after wedging. The combined effect of these two partial velocities plays a "cutting" role. When the strength of the surface material is not enough to bear the impact, a tiny tissue structure will be cut off. This kind of phenomenon occurs constantly, and local materials on the surface are continuously lost, which will lead to obvious wear.

Impact wear refers to the larger impact angle of particles relative to the solid surface. This kind of impact makes the surface structure produce tiny plastic deformation or microcrack along the depth direction, which reduces the strength or fatigue of the material; Under the repeated impact of a large number of particles, the whole plastic deformation layer gradually falls off to form wear [10].

4.1.2 Anti-Abrasion Measures

The data show that the wear of the heating surface and furnace wall in the furnace is mainly scouring wear. The reason of wear is that the kinetic energy of ash particles will destroy it when it hits the wall. The analysis shows that the wear problem of heating surface exists in two aspects: particle and worn structure. Therefore, the concrete measures can be divided into two aspects: improvement of boiler operating conditions and equipment management and transformation. It is complicated and costly to replace all the materials that are easy to wear, so it can be realized by installing anti-wear structures at the parts that are easy to wear [11].

In terms of operation optimization, the following measures can be taken:

(1) Reasonable air distribution. Specifically, it can be realized by reducing the air volume of primary air at the bottom and increasing the air volume of transverse secondary air, so as to reduce the flue gas flow rate while ensuring the combustion efficiency.

(2) Reduce the ash concentration. Specifically, it can be realized by appropriately reducing the primary air volume and excess air coefficient, strengthening the grinding strength and reasonably controlling the particle size of the incoming particles below a certain level.

In terms of equipment management and transformation, the following measures can also be taken to reduce wear and tear:

(1) metal surface grinding and spraying.

(2) Thick-walled pipes are used locally. In places that may be seriously worn, a section of thick-walled pipe should be properly used to prolong the service life.

(3) Reasonable construction of bosses. Control the height of anti-wear bosses, and repair the anti-wear bosses of

https://doi.org/10.1051/e3sconf/202132901074

water walls at four corners of the furnace to areas with low particle concentration.

(4) Install anti-wear beams. The characteristics of hydrodynamic field in circulating fluidized bed boiler determine that the root of fly ash wear lies in the wall flow in the boiler. Anti-wear beam can reduce the ash concentration and velocity of wall-attached flow by setting multiple obstacles on the outer surface of water wall to ensure reasonable structure and heat transfer performance, thus reducing the fly ash wear of heating surface.

4.2 Large Loss of Carbon Residue in Fly Ash

When coal gangue is burned, the carbon residue in fly ash can be as high as 20-40% or more, which greatly increases the loss of incomplete combustion of solid. But this is also a problem that can be solved, is being solved or has been partially solved. In the past, measures such as increasing excess air coefficient, increasing furnace temperature or air temperature, and feeding with positive pressure were taken, but the improvement effect was not significant or even decreased. It is effective to recycle the fly ash into the combustion chamber, but the particle size of the fly ash is small and the activity of carbon residue has decreased, so the complete combustion of carbon residue still cannot be realized. At present, the most effective measure is to send the fly ash to an external special furnace to gasify and burn it, which can further burn the residual carbon and remove more than 95%. The discharged ash can be directly used as building materials.[12]

5 Conclusion

This paper first introduces the working principle of circulating fluidized bed boiler, and points out its main characteristics as a clean combustion technology, especially highlights its wide adaptability to various fuels. Evidently, as most of low-calorific fuels are characterized by their poor quality, various problems will inevitably arise in the combustion process. In this paper, coal gangue is taken as an example. This paper expounds from the following three aspects, and puts forward feasible solutions:

(1) Due to the difference between coal gangue and pulverized coal, some problems that will not occur in the combustion of the latter frequently occur in the combustion conditions of the former, such as difficult ignition, unstable combustion and local high temperature in dense phase zone, etc. Production practice has proved that adopting lower circulation ratio and properly mixing pure pulverized coal to drive coal gangue to burn can appropriately reduce the above problems.

(2) The combustion of inferior fuel is easy to increase coal consumption and produce a large amount of slagging, which will have adverse effects on coal mills, slag coolers and other equipment. High-strength pulverizing system can be adopted and corresponding equipment can be modified to adapt to working conditions; The problem of high sulfur content in inferior fuel can be solved by desulfurization agent in furnace and desulfurization treatment of flue gas.

(3) The problems caused by fly ash are various, among which the problem of high residual carbon can be improved by using the special structure of CFB boiler. However, the serious wear problem of heating surface is closely related to the physical and chemical properties of inferior fuel, so the loss can only be avoided by pretreating fuel and strengthening the structure of boiler related equipment.

References

1. Yue G., Lu J., et al. The Up-To-Date Development and Future of Circulating Fluidized Bed Combustion Technology [J]. Electric Power, 2016,49(1):1-13.
2. Feng C., Yue G., et al. Circulating fluidized bed combustion boiler [M]. Beijing: China Electric Power Press, 2003.
3. Wang Z., Pan L., et al. Review of present situation and prospect for clean development of coal-fired power in China [J]. Electric Power Technology and Environmental Protection, 2018,34(1):1-8.
4. Liu R., Wang Q., & Liu J. Hazards and Countermeasures of the Gangue in Northeast of China [J]. Journal of Geographical Sciences, 2002, 22(1):110-113.
5. Jiang H. Harm of coal gangue to environment and its comprehensive treatment and utilization [J]. Coal Processing & Comprehensive Utilization, 2003, (3):43-46.
6. Lou B., Shi J. The Problem Analysis of Co-combustion of Coal and Coal Gangue in CFB Boiler [J]. Power System Engineering, 2009,25(05):21-22.
7. Sin L., Pan J., & Wu H. The Commissioning and Operation of Coal-gangue-fired CFB Boilers. 2006, 027(003):36-40.
8. Wang F. Numerical Simulation and Combustion Optimization of Circulating Fluidized Bed Boiler Mixed with Coal Gangue [D]. South China University of Technology, 2011.
9. Zheng G. Influence of coal quality change on combustion of circulating fluidized bed boiler and countermeasures [J]. Shandong Coal Science and Technology, 2007(04):21+23.
10. Zhang H. Study on Wear Mechanism and Anti-wear Measures of CFB Boiler Water Wall [D]. North China Electric Power University (Beijing), 2011.
11. Wang X. Equipment management and operation optimization of 300MW circulating fluidized bed boiler [D]. north China electric power, 2012.
12. Zhao Z. Progress in industrial application of fluidization technology-on the development of fluidized bed boilers with inferior fuel [J]. Chemical Engineering, 1974(01)

Research on the Output Characteristics of Emerging Market Entities Serving the Development of New Energy

Shuo Yin [1], Xing Chen [1], Zhe Chai [1], Yao Lu [1], Danni Zhang [2,*]

[1] State Grid Henan Electric Power Company Economic and Technological Research Institute, Zhengzhou, China
[2] School of Electrical Engineering, Zhengzhou University, Zhengzhou, China

Abstract. The power emerging market entities connected to the grid in a decentralized manner can increase the local new energy consumption rate. Under the background of accelerating the development of new energy, it is urgent to clarify the output characteristics of emerging market entities such as wind power, photovoltaics, energy storage, and electric vehicles to adapt to the development of distributed energy and the progress of power market reform. Major emerging entities contribute to mathematical analysis, promote the transaction design of emerging entities, and promote the rapid and healthy development of new energy.

Keywords: New energy; Emerging market; Electricity trading; Output characteristics

1 Introduction

The realization of the carbon peak, carbon neutral target and the goal of building a new power system with new energy as the main body requires the number of new energy installed capacity and the development of power market to guarantee. At the same time, as China's power market reform enters the deep water area, the specific circumstances in the transaction process after the power sale side is liberalized are very complicated, and will also face a variety of risks. The output characteristics of emerging entities such as wind power, photovoltaics, and energy storage are clarified. It is the key to the penetration of the new power system into the traditional power system.

2 Wind power output characteristics

Performance characteristics of wind power output curve:
(1) Randomness. From the fitting curve of the annual output curve of wind power, it can be clearly observed that for the hourly output of wind power, the output at each moment shows obvious uncertainty.
(2) Intermittently. The wind speed is obviously intermittent, and it is also affected by the cutting-in speed and the cutting-out speed. Therefore, there are some points in the curve where the output of the fan is zero. The point where the output of these fans is zero may be due to the fact that the cut-in wind speed has not been reached, or the cut-out wind speed has been reached, and the fan has been cut off, so the fan output is not continuous.
(3) Seasonal characteristics. The output of wind power has certain seasonal variation characteristics. The output of wind power is the largest in winter, and the output of wind power in autumn is the smallest. The difference is relatively obvious. At the same time, the output curve of each season is quite different from the typical sunrise curve, and the typical sunrise curve cannot well reflect the characteristics of the output change in each season.
To study the characteristics of wind power output, we must first establish a mathematical model of wind speed. A large amount of literature shows that the mathematical model of wind speed (that is, the distribution of wind speed) fits well with the two-parameter Weibull distribution. The mathematical function expression of the two-parameter Weibull distribution is as follows:

$$H(V)=\left(\frac{\beta}{\alpha}\right)\left(\frac{v}{\alpha}\right)^{\beta-1}\exp\left[-\left(\frac{v}{\alpha}\right)^{\beta}\right]$$

Among them, v represents the wind speed; β represents the shape parameter; α represents the size parameter.
Similarly, the probability distribution function of the two-parameter Weibull is as follows:

$$G(v)=\int_{0}^{+\infty}H(v)dv=1-\exp\left[-\left(\frac{v}{\alpha}\right)^{\beta}\right]$$

In a more ideal state, each group of wind speed data has a unique power corresponding to it, but the relationship between wind speed and power is affected by the characteristics of the wind turbine itself in engineering. The wind speed-power characteristic curve is shown in Figure 1(left):

* Corresponding author: yinshuo1111@163.com

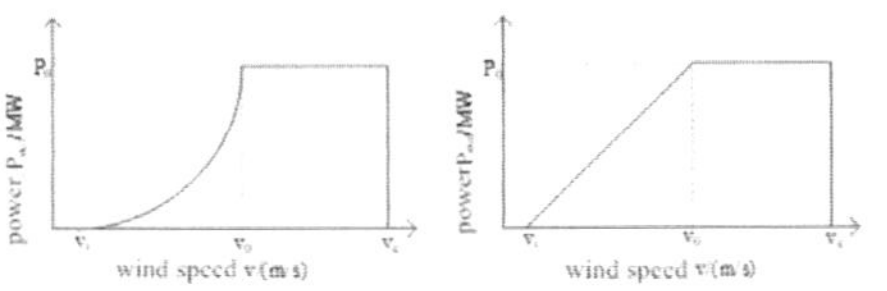

Figure 1 Wind speed-power curve

In engineering applications, the linear curve in Figure 1 (right) can be used to replace the non-linear curve, which makes it easier to obtain the required data. When using the wind speed-power linear curve, the mathematical expression of the fan output power section function is as follows:

$$P_0 = \pi R^2 \rho v^3 \frac{\left(1+\frac{v_0}{v}\right)^2\left(1-\frac{v_0}{v}\right)}{4}$$

$$P_{out} = \begin{cases} 0, & 0 \le v \le v_i \\ \frac{v-v_i}{v_0-v_i} P_0, & v_i < v \le v_0 \\ P_0, & v_0 < v \le v_c \\ 0, & v > v_c \end{cases}$$

Among them, P_0 is the rated output of the wind turbine; v_i is the cut-in wind speed of the fan; v_0 is the rated wind speed; v_c is the cut-out wind speed; v is the wind speed at the height of the wind turbine hub; R is the fan radius; ρ is the air density where the fan is located.

3 Photovoltaic output characteristics

Distributed photovoltaic power generation has the characteristics of strong intermittent and strong randomness, and will be affected by environmental conditions and other factors, such as: season, geographic area, time, etc. It will also form a power load restriction with the connected neighboring photovoltaic. The output level of photovoltaic power generation is closely related to the changes in the solar radiation received by the ground. The output level of the photovoltaic power generation system increases first and then decreases with the change of solar radiation, reaching the peak output at noon, showing a "semi-envelope" shape overall. The width of the half envelope is related to the sunshine time, and the height of the half envelope is related to the season and meteorological conditions. For this reason, the output characteristics of photovoltaic power generation can be discussed from the following two aspects.

(1) The influence of seasonal attributes

The time period of photovoltaic power output is limited by the time of sunrise and sunset, so it is greatly affected by the characteristics of the season. The shape of the photovoltaic output curve is quite different. The initial power generation time of the photovoltaic power station in summer and winter differs by up to 2 hours, and the difference in the duration of the full-day output is close to 4 hours.

Generally speaking, "sunrise time plus 0.5h" can be used to characterize the start time of photovoltaic power generation on the day, and "sunset time minus 0.5h" can be used to characterize the end time of power generation. It can generally be considered that in a certain season, the time of sunrise and sunset of a certain place are basically unchanged. Therefore, for the photovoltaic output curve, the corresponding seasonal attributes are very important. If the season attributes are known, the start time and end time of photovoltaic power generation can be roughly determined according to the time of sunrise and sunset, and the duration of photovoltaic power generation can be further calculated.

(2) The influence of meteorological conditions

Because photovoltaic power generation is directly affected by solar radiation, different weather conditions will also affect the output level of the photovoltaic power generation system. For example, under rainy or cloudy weather conditions, the output of the photovoltaic power generation system may drop sharply. This brings greater uncertainty to the output of photovoltaic power generation systems. Weather conditions have a great influence on the output of photovoltaic systems and are an important constraint on the output of photovoltaic systems.

There are many methods to study the output characteristics of photovoltaic power generation in engineering, but the main method is to use the multivariable nuclear density estimation method to establish a mathematical model of photovoltaic power generation output for characteristic analysis and the project uses this method to believe that this method conforms to the Bata distribution.

The output of photovoltaic power generation is mainly affected by the light intensity. The standardized light intensity is as follows:

$$H_t = \frac{I_t}{I_{max}}$$

Among them, I_t is the actual light intensity; I_{max} is the maximum light intensity; according to the principle of Bata distribution, its probability density function is

$$F_{H_t}(H_t,\gamma,\theta) = \frac{\Gamma(\gamma+\theta)}{\Gamma(\gamma)\Gamma(\theta)} H_t^{\gamma-1}(1-H_t)^{\theta-1}$$

Among them, γ,θ are the shape parameters of Bata, $\Gamma(.)$ is the gamma function, $\Gamma(\gamma) = \int^{\infty} x^{\gamma-1}e^{-x}dx$.

Photovoltaic power plant receives solar radiation and converts it into usable electric energy, then the light energy-power conversion can be calculated by the following formula:

$$P_{out} = \lambda SI\left[1-0.005(t_0+25)\right]$$

Among them: λ is the conversion efficiency of light energy and electric energy; S is the area of the photovoltaic panel; I is the solar irradiance; t_0 is the ambient temperature where the photovoltaic device is installed.

4 Energy storage output characteristics

This section takes the static energy storage device lithium battery as an example to analyze. In the discharge process of the lithium battery, it uses constant current discharge. The change of the working voltage in this process can be divided into 3 stages. In the initial stage of discharge, the battery working voltage drops rapidly, and then enters the linear drop zone. When the discharge is close to the end, the battery working voltage begins to drop again. When the voltage is lower than a certain value, the discharge stops. Combining with the charging and discharging characteristics of lithium batteries, it can be seen that during the charging and discharging stage, the charging and discharging power of lithium batteries is constantly changing. Combining with the charging and discharging characteristics of lithium batteries, it can be seen that during the charging and discharging stage, the charging and discharging power of lithium batteries is constantly changing, in order to play the role of peak shaving and valley filling after being connected to the large power grid, it is necessary to rationally arrange the electrochemical energy storage capacity in combination with the peak load of the power grid.

In order to explore the distribution characteristics of the battery during charging and discharging more in-depth, this study uses a variety of distribution functions to fit the data of the battery during charging and discharging, and determine the best fitting distribution. This study uses the distribution in the @risk software The fitting function is used to specifically analyze the quantitative characteristics of the battery during charging and discharging, and the probability distribution of the battery during charging and discharging under certain conditions can be obtained. At the same time, the Anderson-Darling (A-D) test was performed on the fitting results. The test results showed that among the many distribution functions, the A-D test value of the Logistic distribution is the smallest, that is, the logistic distribution has the highest degree of fit, so it can be judged that the battery charge and discharge roughly obey Logistic distribution, the specific probability density function expression is shown in the following formula:

$$F(x)=\frac{e^{\frac{x-\alpha}{\beta}}}{\beta\left(1+e^{\frac{x-\alpha}{\beta}}\right)^2}$$

Where, $F(x)$ represents the probability that the sample data is near a certain value point, α represents the average value of the data (location parameter), β represents the variance (scale parameter). In the process of sample fitting, α,β need to be estimated based on different sample data. According to the collected data, the Monte Carlo simulation (5000 simulations) is performed, and the logistic probability density function of the charge and discharge output of the battery is fitted. The mean parameter is 0.21 and the variance parameter is 0.89. There may be differences in the probability distribution functions or related parameters obtained by fitting different types of chemical energy storage.

5 Electric vehicle output characteristics

Conventional electric vehicle charging and discharging control uses each electric vehicle and its battery status as research variables, and the simulation process is relatively complicated, while the electric vehicle switching mode can focus on the modeling of the switching station itself, which is equivalent to an energy storage device. Modeling the energy storage state in adjacent periods. On the one hand, the power exchange station interacts with the power grid through the charging pile, and on the other hand supplies electric energy to the electric vehicle in the form of battery pack replacement. The total remaining power in the station at time t can be expressed as:

$$Q_t=Q_{t-1}+\left(P_t^c\eta_c-\frac{P_t^d}{\eta_d}\right)\Delta t-D_{t-1}\left(w-L_{t-1}\right)$$

Among them: Q_t represents the remaining electric energy of the switching station at time t; P_t^d represents the power injected into the switching station from the grid at time t; η_c represents the power fed into the grid by the switching station at time t; η_d represents the charging efficiency of the switching station receiving power from the grid; D_t represents the number of electric vehicle battery packs that need to be replaced in the t time period; w represents the rated capacity of the battery pack; L_t Represents the average remaining power of the battery packs replaced in the t time period.

Considering the charging and discharging status of the entire swap station, assuming that the swap station will reserve a charging position for the battery pack replaced by an electric vehicle at time t, according to the power swap demand, the charge and discharge power constraints of the swap station at time t can be obtained as:

$$0\le P_t^c\le c_t\left(M-D_t\right)P_b^c$$

$$0\le P_t^d\le d_t\left(M-D_t\right)P_b^d$$

Among them: P_b^c represents the charging power of the battery pack; P_b^d represents the discharge power of the battery pack; M represents the total number of charging positions in the switching station; c_t,d_t represent the charging and discharging control variables (0/1) at time t, and the two variables are mutually exclusive.

Assuming that both the battery swap station and the electric vehicle use battery packs with a uniform standard capacity, in order to ensure the charging demand in a certain period of time in the future, set the lower limit of the remaining power at time t in the station as, and the power storage limit and reserve demand of the swap station can be expressed as:

$$\left(1+\eta_{ev}\right)Q_{\min,t} \le Q_t \le wM_B$$

Among them: η_{ev} stands for the power reserve rate in the swap station; M_B stands for the number of backup batteries in the swap station.

6 Conclusion

The power emerging market entities connected to the grid in a decentralized manner can increase the local new energy consumption rate. Distributed energy not only improves the utilization level of new energy, but also brings huge economic and environmental benefits, and helps to achieve sustainable economic and social development. With the expansion of power emerging market entities and the implementation of market-oriented trading trials, it is necessary to accurately evaluate and analyse the economic and environmental benefits of power emerging market entities. Therefore, accurate evaluation models need to be established to guide grid companies and government departments to better develop and utilize the main technology of the power emerging market. From an overall point of view, clarifying the output characteristics of the power emerging market entities will help to connect to the large power grid and help save investment, reduce energy losses, and give full play to the respective advantages of distributed power sources and power grids. From the perspective of the main projects in the power emerging market, it will also help investors to build costs, make reasonable investments, and provide accurate quotations.

Acknowledgments

This work was financially supported by the State Grid Henan Electric Power Company Science and Technology Project (5217L021000F).

References

1. Yang Bai, Ang Li, Qing Xia, Electricity marketing model and new electricity price system under the new situation, J. Power System Protection and Control.2016,44(05):10-16.
2. Yang Bai, Le Xie, Qing Xia, Qixin Chen, Haiwang Zhong, China's system design and suggestions for promoting the marketization of the power sales side, J. Automation of Electric Power Systems, 2015,39(14):1-7.
3. Chengshan Wang, Zhen Wu, Peng Li, Application prospects and challenges of distributed electric energy storage technology, J. Automation of Electric Power Systems, 2014,38(16):1-8+73.
4. Shiming Tian, Beibei Wang, Jing Zhang, Key technologies of demand response under smart grid conditions, J. Proceedings of the Chinese Society for Electrical Engineering, 2014,34(22):3576-3589.
5. Li Ma, Menghua Fan, Lei Guo, Song Xue, Kun Li, The latest development trend of foreign power market and its enlightenment, J. Automation of Electric Power Systems, 2014,38(13):1-9.
6. Jiajia Yang, Junhua Zhao, Fushuan Wen, Yusheng Xue, Liang Li, Haohua Lv, Bidding strategy of virtual power plant including electric vehicles and wind turbines, J. Automation of Electric Power Systems, 2014,38(13):92-102.

Research on Prediction Methods of Wellbore Hydrate Formation of Ultra-Deep Gas Wells in Northwest Sichuan

Qilin Liu *, Jian Yang, Lang Du, Jianxun Jiang, Dan Ni, Rong Zheng, Zhaoqian Luo, Peng Liu

Northwest Sichuan Gas Mine of Petro China Southwest Oil & Gasfield Company, Sichuan, China;

Abstract. According to the formation and handling situation of hydrate in ultra-deep high-pressure sulfur-containing gas wells in northwest Sichuan, the formation conditions of natural gas hydrate was studied based on previous studies on hydrate, the molecular dynamics of natural gas hydrate and the multiphase flow law of high-temperature high-pressure high-sulfur-containing gas wellbore were combined, and the pressure prediction model with high-temperature high-pressure sulfur-containing gas wells as the target was built. The chemical and physical control methods of wellbore hydrate plugging were discussed to provide the scientific theoretical basis for the prediction and control of hydrate in high-temperature high-pressure high-sulfur-containing gas wells.

Keywords: high-temperature and high-pressure, sulfur-containing gas well, wellbore, hydrate, prediction method.

1 Introduction

There are many oil and gas-containing reservoirs in the northwest Sichuan, at present, exploration and development targets are mainly the Qixia, Maokou, Changxing and Feixianguan group, gas wells show ultra-deep, high-pressure and sulfur-containing characteristics, stratum pressure ranges from 91.13 to 131.09 MPa, stratum temperature ranges from 121.70 to 164.11°C, hydrogen sulfide content range from 0.02 to 88 g/m3, and total depth of four stratums all exceed 5500m, at present, the total depth in the Shuangyushi block is average 7746m, which is the record of the deepest wells of land average total depth of CNPC. The actual drilling confirmed that most of the new wells put into production in the above four stratums belong to ultra-deep high-pressure sulfur-containing gas wells [1].

The hydrate formation temperature at the wellhead of ultra-deep high-pressure sulfur-containing gas wells in northern Sichuan is as high as about 35℃, which is much higher than that of normal-pressure gas wells without hydrogen sulfide, and it seriously affects safe production. Recently, the wellbore hydrate blockage was formed during the production process, and after the measures of pumping glycol and wellhead insulation were ineffective, finally, the self-heating unblocking process was adopted to remove the blockage, the operation from start to finish is a total of 5 months, affecting the output $1812\times10^4 m^3$; it is easy to form hydrate and freeze plugging in the production process; wellbore hydrate blockage is formed in the process of oil test, which affects the project progress; after opening the safety valve in the pit and preparing to pump ethylene glycol, hydrate blockage was rapidly formed in the wellbore (within 2 hours), affecting the normal production.

There are still gaps between the current theoretical research and field practice in hydrate formation mechanism and law, the degree of influence of pressure, hydrogen sulfide content, gas-liquid ratio, wellbore working conditions and other factors on hydrate formation and accurate prediction of hydrate formation process and time, which need further research, and provide theoretical basis and guiding basis for the injection system of injecting hydrate inhibitor in the wellbore [2]. In wellbore hydrate control, in addition to the existing self-heating unblocking process that needs further optimization; exploration and research on other ultra-deep high-pressure sulfur-containing gas wellbore unblocking processes need to be carried out.

2 Present Situation of Hydrate Prediction at Home and Abroad

The long-term research has been carried out at home and abroad in the prediction of hydrate formation conditions, for effective hydrate control schemes, several thermodynamic prediction models of hydrate formation conditions have been proposed and applied in natural gas development at home and abroad, including van der Waals-Platteeuw model, improved version of vdW-P model (Parrish-Prausnitz, Ng- Robinson model), and Chen-Guo model, etc. On the basis of CSMHyK hydrate

* Corresponding author: liuql10@petrochina.com.cn

kinetic model and OLGA multiphase flow equation, Jing-Yi Shi and Jian-Chun Fan et al. built the numerical simulation method of hydrate formation in deep water drilling, and predicted the formation scope and degree of wellbore annulus hydrate by introducing the degree of supercooling density to characterize the hydrate formation possibility [3]. Antonin Chapoy et al. Heriot-Watt University, UK, achieved good results in describing the phase equilibrium of natural gas hydrates by modifying the vdW-P model. Peter Jorgensen Herslund et al. Technical University of Denmark built the model that can describe the phase equilibrium of natural gas hydrates containing CO_2 based on the CPA equation of state and the vdW-P model, and he obtained high computational accuracy by adsorbing this model with modified binary interaction parameters and Langmuir. Ioannis Tsivintzelis, Technical University of Denmark, M. Naveed Khan, Colorado School of Mines, and other scholars have studied the association effect containing H_2S and other polar components, which provides the references for the modification of the association parameters of H_2O, H_2S, CO_2, and alcohol materials. Therefore, on the basis of CPA equation of state and vdW-P model, predicting the formation conditions of sour gas hydrates has a good application in sour gas fields, modifying binary interaction parameters or association parameters in polar molecules such as H_2S, this is expected to solve the difficult problem of predicting gas hydrate formation conditions in ultra-deep high-pressure sulfur-containing gas wells in northwest Sichuan [4].

3 Formation Conditions of Natural Gas Hydrate

The formation of hydrate is not only related to the component, composition and free water content of natural gas, but also needs certain thermodynamic conditions, namely a certain temperature and pressure. Russian scholar Rozeboom et al. believed that only when the pressure of gas components in the system is greater than its hydrate decomposition pressure, the gas containing saturated water vapor can spontaneously generate hydrate[5], which can be expressed by fugacity as follows:

$$f_{decomposition}^{hydrate} < f_M^{system} \le f_M^{saturation}$$

The relative density of natural gas composed of different components is different, natural gas with relatively high density is more likely to form hydrate [6]. The main conditions of forming hydrate are: partial free water in the natural gas fluid and fluid temperature is less than or equal to the dew point of free water. The fluid is prone to generate hydrate when it passes through the pipeline, throttling components or ground equipment with intense disturbance.

4 Prediction of High-Temperature High-Pressure Wellbore Hydrate

4.1. Research on flow law of sour gas in wellbore

According to the formation conditions of hydrate, when predicting where hydrates are likely to form and how much they will crystallize during production, it is necessary to calculate the temperature and pressure of the fluid at every moment in the wellbore; therefore, the study of wellbore multiphase flow law of sulfur-containing gas wells is the prerequisite for predicting hydrate formation. The flow pressure and temperature change of fluid in the wellbore vary greatly, the parameter calculation methods of Hagedorn brown and other multiphase flow correlation formulas do not consider the change of wellbore phase state, amd the parameters are obtained based on the method of crude oil, in order to make Hagedorn Brown model and other correlation formulas more suitable for multiphase flow calculation of gas wells containing water and H_2S, the parameters in the correlation formula should consider the characteristics of wellbore phase change[7], and the correction of phase change should be considered under the influence of temperature, pressure and other factors. This problem is still explained based on the Hagedorn Brown correlation formula.

$$\frac{\Delta p}{\Delta h} = g\rho_m \sin\theta + \frac{f_m q_L^{\ 2} M_t^{\ 2}}{9.21\times10^9 d^5 \rho_m}$$

According to previous studies, there are two dissolution mechanisms of sulfur in natural gas with high H_2S content, namely physical dissolution and chemical dissolution. In the following, relevant theories of association models in supercritical fluids are first used to explain these two dissolution mechanisms, and then the association model containing two dissolution actions is built [8]-[10].

$$\frac{y}{(1-y)^n} = \left[\frac{k\varphi_1^n \varphi_s^s P_s^s \exp\left(\int_{P_s^s}^{P} \frac{V_s^s}{RT} dp\right)}{\varphi_3}\right] P^{n-1}$$

In the formula:

V_s^s ------------Molar volume of solid solute sulfur;

P_s^s -------------Saturated vapor pressure of solid solute sulfur;

φ_s^s -----------Fugacity coefficient of pure solid solute sulfur;

$\exp\left(\int_{P_s^s}^{P} \frac{V_s^s}{RT} dp\right)$ -----------Poynting correction factor

φ_1 ---------Fugacity of $H_2S_{(f)}$;

φ_3 ---------Fugacity of $H_2S_{x+1(f)}$;

P ----------- System pressure

 https://doi.org/10.1051/e3sconf/202132901076

The fugacity of solid is calculated by the following formula:

$$f_s^s = \varphi_s^s P_s^s \exp\left(\int_{P_s^s}^{P} \frac{V_s^s}{RT} dp \right)$$

4.2. Prediction research on hydrate formation

4.2.1. Diagram method

In mine practice, the commonly used empirical diagram method is based on hydrate formation temperature and pressure curve calculated in accordance with relative density of natural gas, combine the calculation results of natural gas with different relative densities, thus, the relationship curve between temperature and pressure of forming gas hydrate is drawn. The empirical diagram method can be used to predict the approximate formation area of hydrate. The empirical diagram method can be used as the judgment basis and plays a key role in rough prediction of the formation area of gas hydrate[11]. The following figure can be used as the basis for judging hydrate formation.

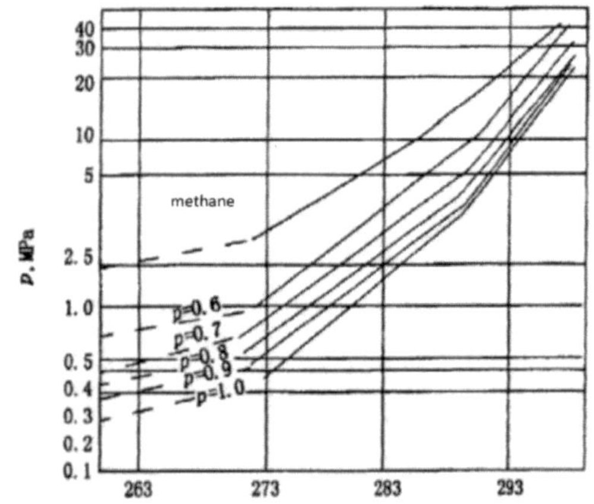

Fig.1 Predicted P-T phase diagram of relative density and hydrate formation

The figure gives the temperature and pressure curves of methane and five natural gas fluids with 0.6, 0.7, 0.8, 0.9 and 1.0 relative density, respectively. If the relative density of natural gas is known to fall in the curves in the figure, the formation conditions of hydrate can be determined by linear interpolation.

Given R and T, when P is generated by hydrate, the formula is:

$$P = P_1 + \left[(P_1 - P_2) \frac{r_1 - r}{r_1 - r_2} \right]$$

Given R and P, when T is generated by hydrate, the formula is:

$$T = T_1 + \left[(T_2 - T_1) \frac{r_1 - r}{r_1 - r_2} \right]$$

In the formula, r-- the relative density of natural gas, $r_1 < r < r_2$

P_1, P_2 formation pressure of gas hydrate with relative density r_1, r_2;

T_1, T_2 – formation temperature of gas hydrate with relative density r_1, r_2.

4.2.2 Statistical thermodynamics method

The prediction algorithm of hydrate formation is derived by statistical thermodynamics method via Newton's method[12], The general form of the equation is:

$$\ln Z - Y = 0$$

For natural gas without H_2S, when P>6.865MPa:

$$\lg Z = 8.9751 - 0.03304T$$

When P < 6.865 MPa when:

$$\lg Z = 3.51517 - 0.01436T$$

For natural gas with H_2S:

$$\lg Z = -5.40964 - 0.02133T$$

For Y:

$$Y = \gamma_1 \ln\left(1 - \sum_1^n \theta_{A1}\right) + \gamma_2 \ln\left(1 - \sum_1^n \theta_{A2}\right)$$

θ_{A1} and θ_{A2} reflect the filling degree of gas in the large and small cavity of hydrate, respectively, which can be expressed as:

$$\theta_{ij} = \frac{C_{ij} P_i}{1 + C_{1j} P_j + C_{2j} P_j}$$

In the formula, Cij can be calculated as follows:

$$C_{ij} = \exp\left(A_{ij} - B_{ij} T\right)$$

In the formula, the values of A_{ij} and B_{ij} coefficients are shown in Table 1.

Table.1 Relationship between A_{ij} and B_{ij} and natural gas components

natural gas component j	small hole		big hole	
	A_{1j}	B_{1j}	A_{2j}	B_{2j}
CH_4	6.0499	0.02844	6.2957	0.02845
C_2H_6	9.4892	0.04058	11.9410	0.04180
C_3H_8	-43.670	0.00000	18.2760	0.04661
C_4H_{10}	-43.670	0.00000	13.6942	0.02773
N_2	3.2485	0.0262	7.5990	0.02447
CO_2	23.0350	0.0904	25.2710	0.09781
H_2S	4.9285	0.0093	2.4030	0.00633

The hydrate formation temperature under a given pressure is calculated and solved by Newton's method, the iterative format is:

$$T^{n+1} = T^n - \frac{f(T^n)}{f'(T^n)}$$

4.2.3. Van der Waals-Platteeuw model

In 1959, van der Waals and Platteeuw et al. [13] proposed the thermodynamic model based on the phase equilibrium theory. That is:

$$\mu_W^H = \mu_W^\alpha$$

In the Formula: μ_W^H -- chemical potential of phase H;

μ_W^α -- chemical potential in the water-rich phase.

The chemical potential of hydrate lattice (β phase) as the reference state, the equilibrium condition can be expressed as:

$$\Delta\mu_W^{H-\beta} = \Delta\mu_W^{\alpha-\beta}$$

The thermodynamic model of water-rich phase is as follows:

(1) Hydrate phase

Van der Waals and Platteeuw derived the expression of chemical potential difference $\Delta\mu^{H}$ based on statistical thermodynamics and isothermal adsorption theory:

$$\Delta\mu_{W}^{H} = \mu_{W}^{H} - \mu_{W}^{\beta} = -RT\sum_{m=1}^{2}\gamma_m \ln(1-\sum_{j} Y_{jm})$$

In the formula: -- chemical potential of the fully empty hydrate, J/ mol;

μ^{H} -- Chemical potential of fully filled hydrate, J/ mol;

γ_m-- Characteristic constant of hydrate structure, for structure I: γ_1 take 1/23, γ_2 take 3/23; For structure II: γ_1 take 2/17, γ_2 take 1/17;

R -- Gas constant, 8.31434J/ (mol•K);

T -- temperature, K;

Y_{jm} -- The fraction of m type hole occupied by j group.

The fraction Y_{jm} of m type hole occupied by j group is calculated based on the following formula:

$$Y_{jm} = \frac{C_{jm} f_j}{1+\sum_{j} C_{jm} f_j}$$

In the formula: f_j -- Fugacity of j component, MPa;

C_{jm} -- Langmuir constant of the j component in the m type hole, 1/ MPa.

In the above formula, the expression of Langmuir constant C_{jm} was calculated by using the empirical relation proposed by Parrish and Prausnitz[14], thus simplifying the cumbersome steps of vDW-P model in solving C_{jm}, The expression of C_{jm} is shown as follows:

$$C_{jm} = \frac{A_{jm}}{T}\exp\left(\frac{B_{jm}}{T}\right)$$

In the formula, A_{jm} , B_{jm} - experimental fitting parameters. Most of the fitting parameters of structure I and II hydrates in this model come from studies by Munck[15], Rasmussen and Pedersen et al. [25].

(2) Water-rich phase

Holder et al. believed that the chemical potential deviation between the fully empty hydrate lattice in the water-rich phase and the water phase was only related to temperature, and proposed the following expression $\Delta\mu_{W}^{\beta\text{-}\alpha}$:

$$\frac{\Delta\mu_{W}^{\beta\text{-}\alpha}}{RT} = \frac{\Delta\mu_{W}^{0}}{RT_0} - \int_{T_0}^{T}\left(\Delta h_{W}^{0} + \int_{T_0}^{T}\Delta C_{PW}^{0}\,dT\right)\frac{\Delta h_{W}}{RT^2}dT + \int_{0}^{P}\frac{\left(\Delta V_{W}^{0} + \Delta V_{W}^{1} P\right)}{RT}dP - \ln x_{W}$$

In the formula: Δh_{W}^{0} and ΔC_{PW}^{0} --represent the enthalpy difference (J/mol) and heat capacity difference [J/(mol•K)] of phase β and phase αat T_0 (T_0=273.15K), respectively;

ΔV_{W}^{0} and ΔV_{W}^{1} -- represent the molar volume difference (m^3/mol) under the reference condition (T_0=273.15K) and the actual condition, respectively, m^3/mol;

x_{w} -- represents the molar fraction of water in the solution [13].

5 Control Hydrate Blockage in Wellbore

In the control of hydrate blockage in wellbore, the common ways are chemical unblocking and physical unblocking. Chemical unblocking is in hydrate inhibitors and unblocking solution research, methanol and ethylene glycol are commonly used at home in the early stage, but because the methanol has toxicity, and unblocking effect of ethylene glycol is not good, therefore, the inhibitors in the follow-up study mainly consider the needs of environmental protection and safety based on having the function of inhibiting hydrate, a new non-toxic hydrate inhibitor Z-6 has been developed and applied in Dongsheng Gas field, the field application is carried out, the results show that this new inhibitor can replace methanol in hydrate prevention and unblocking effect[16].

The coiled tubing operation is usually used in physical unblocking, however, The coiled tubing operation of ultra-deep high-pressure sulfur-containing gas wells has high requirements for coiled tubing strength, sulfur resistance, injection head capacity, coiled tubing performance, control of jacking force at dredging moment, well control device, operation tools in the pit, performance of sand washing fluid, friction pump pressure and risk identification and control during operation [17]. In the early stage of Tarim Oilfield, in allusion to the problems such as deep buried reservoir (6500~ 8038m), high temperature (150~190 ℃), high stratum pressure (105~136mpa) and difficult coiled tubing operation in Kucheshan structure, the pressurized operation method was successfully used to solve the wellbore wax blockage, but the cost was high [18]. In order to solve the difficulties of coiled tubing operation, the coiled tubing unblocking process and supporting equipment suitable for Kucheshan had been successfully applied in Tarim Oilfield through software simulation calculation and process argument, and they were improved based on the liquid and process problems in the operation process, had been successfully applied to the follow-up oil tube blocking in high-temperature, high-pressure and high sulfur-containing wells and achieved good results.

6 Conclusions and Suggestions

The formation mechanism of gas hydrate and the research on wellbore flow law of high-temperature, high-pressure and high sulfur-containing gas wells were combined, the prediction method of wellbore hydrate formation in high-temperature, high-pressure and high sulfur-containing gas wells was built.

Three prediction methods of hydrate formation, diagram method, statistical potential method and phase equilibrium calculation method, were discussed, it is concluded that diagram method is convenient to determine the hydrate formation condition in the case of knowing temperature and pressure.

Statistical thermodynamics method and Van der Waals-Plateaus model are conducive and calculate the temperature and pressure of the whole wellbore section in real time to predict the formation of natural gas hydrate.

The usual approaches are chemical unblocking and physical unblocking in the control of wellbore hydrate blockage. The new non-toxic inhibitor Z-6 is suggested for chemical unblocking.

Acknowledgments

Scientific Project of Petro China Southwest Oil & Gasfield Company "Research on Wellbore Risk Evaluation and Control Technology of Ultra-Deep and High-Pressure Sulfur-Containing Gas Wells in Northwest Sichuan", No. 20200303-24.

References

1. Deng Ke. Research on Hydrate Control Technology in Gas Field of West Sichuan[D]. Southwest Petroleum University, 2007.
2. Tan Linlin. Research on Effusion Law and Hydrate Control Technology of Surface Gathering and Transmission Pipeline in Dongsheng Gas Field[D]. Southwest Petroleum University, 2013.
3. Tang Daobing. Quantitative Research on Wellhead Temperature of Gas Wells to Prevent Hydrate Formation[J]. China Petroleum and Chemical Standard and Quality, 2018, 38 (07): 91-92 + 94.
4. Zhou Lianlian, Chen Lin, Lan Qiong, Xie Xulong, Wang Jia. Prevention and Application of Natural Gas Hydrate in Surface Gathering and Transportation of Longgang Gas Field[J] Natural Gas Technology and Economy, 2018, 12(01): 41-44+82-83.
5. Xu Changhui. Prediction and Control System Technology of Wellbore Hydrate in Gas Field Production System[D]. Yangtze University, 2016.
6. Liu Shixin. Experimental Study And Prediction of Gas Hydrate Control in Gas Field Production[D]. Southwest Petroleum Institute, 2005.
7. Huang Taihong, Wang Yang. Research on Prevention and Control Measures of Hydrate Ice Blockage in Gas Pipeline[J]. China Petroleum and Chemical Standard and Quality, 2018, 38 (07): 141-142.
8. Wang Li, Zhou Keming, et al. Experimental Investigation of Hydrate Forming Mechanism in Gas Reservoirs with High Sulfur Conten[J]. Natural Gas Industry, 2003.5, 23(3), 97-100.
9. Kennedy, H.T. and Wieland, D.R. Equilibrium in the Methane-Carbon Dioxide-Hydrogen Sulfide-Sulfur System, Traits, A1ME(1960}21.
10. erdgas-zeitschrift, 99.jg., Heft 9, Septembet 1983, 297300.
11. Kunal Karan,Robert A.Heidemann and Leo A.Behie,Sulfur Solubility itt Sour Gas Predictions with an Equation of State Model,Ind.Eng.Chem.Res. 1998, 37, 1679-1684.
12. Yang Xuefeng, A Study On Special Phase Behavior State Of Multiphase Fluid In Gas Reservoirs With High Sulfur Content And Formation Damage Caused By The Sulfur Deposition[D]. Southwest Petroleum University.
13. Platteeuw J C, Waals J H V D. Thermodynamic properties of gas hydrates[J]. Molecular Physics, 1958, 1(1) :91-96
14. Parrish W R, Prausnitz J M. Dissociation Pressures of Gas Hydrates Formed by Gas Mixtures[J]. Ind. Eng.chem. process Design Devlop, 1972, 22(3) :26-35.
15. Munck J, Skjold-J Rgensen S, Rasmussen P. Computations of the formation of gas hydrates[J]. Chemical Engineering Science, 1988, 43(10) :2661-2672.
16. Wu Zhiliang. Prediction Of Gas Hydrate Formation Conditions In Wellbore[J]. Petroleum Geology and Experiment, 2008(03):315-320.
17. Chen Yuchuan, Shi Bohui, Li Wenqing, Liu Yang, Song Shangfei, Ding Lin, Gong Jing. Progress Of Influence Mechanism Of Kinetic Hydrate Inhibitors[J]. Chemical Industry and Engineering Progress, 2018, 37(05): 1726-1743.
18. Miao Chunhui, Zhang Qinghu, Qiu Shujuan. Summarize the Formation of Gas Hydrates and its Prevention Technique in Long -Distance Transportation Pipeline[J]. Shanghai Gas, 2007(02): 23-25.

Research on intelligent operation and maintenance management technology based on Microgrid group

Wu Min [a]

Tibet Autonomous Region Energy Research Demonstration Center, Lasa 850000, China

Abstract. In this paper, the research and platform development of cloud based microgrid cluster operation and maintenance management technology are carried out in Tibet, the solar thermal power combined microgrid system is built at decentralized load points, the data of each independent new energy system is integrated and shared by using Internet technology, and the unified operation, maintenance, monitoring and management of microgrid cluster are realized through the cloud platform, It can greatly improve the use efficiency of clean energy, improve the power supply quality and reliability of users, reduce the operation and maintenance cost, improve the work efficiency and system operation reliability, realize the intelligent monitoring, operation and maintenance and management of photovoltaic power generation and micro electric network group, and provide new operation and maintenance management methods and development ideas for the multi energy complementary and comprehensive utilization of Tibet Autonomous Region.

Keywords: Tibet region; Solar energy utilization; Microgrid group; Intelligent operation and maintenance

1 Introduction

Although Tibet has made outstanding achievements in the development and utilization of solar energy, through investigation, investigation and consultation, it is found that there is still a lack of real-time monitoring, management, analysis and statistics on the many photovoltaic power stations built and distributed in the whole region in Tibet. In addition, due to the lack of timely and effective power station operation and maintenance due to the sparsely populated area, extreme climate and traffic interruption, the operation status of the power station has been changed "Small problem" becomes "big fault" , the maintenance of the failed photovoltaic power station was delayed due to the failure of timely and effective solution, and finally the function of the power station was continuously weakened to complete loss. The lack of energy supply made the already difficult living conditions of residents in agricultural and pastoral areas even worse and extremely difficult. In addition, the people's increased energy consumption and practical needs such as pleasant and livable, environmental protection and energy conservation brought about by the improvement of living conditions made them unnecessary or reduced The solar photothermal photoelectric technology and system with the characteristics of less external power input, promoting the coupling development of new energy use and economy, giving full play to Tibet's advantageous solar resources and integrating power supply + heating is an effective way to solve the power consumption problems of people in areas where the power grid cannot cover and areas with unstable power supply, meet the power demand, and take into account the future development trend Good layout[1].

The future of Tibet's new energy development must be inseparable from innovative measures in improving quality and efficiency, and later management and monitoring is an effective means. Therefore, the energy research and demonstration center of Tibet Autonomous Region has carried out relevant research work many years ago[2]. In 2014, the energy research and demonstration center of Tibet Autonomous region was launched in cooperation with Chongqing Institute of green intelligent technology, Chinese Academy of Sciences "Distributed solar power plant monitoring system" In the project, by building a remote monitoring and acquisition platform for photovoltaic power station data, remote monitoring is realized for the three photovoltaic power stations selected in Lhasa City, Naqu area and Yangbajing. By monitoring the output and input current, voltage and actual power of photovoltaic power stations, as well as the operation of batteries, inverters and photovoltaic modules, the daily power generation and input power of photovoltaic power stations are collected And monitoring of power consumption at the user's side. Although certain achievements have been made in the monitoring and management of photovoltaic power plants through continuous exploration and attempts, due to the technical limitations and actual conditions at that time, we still failed to realize data sharing and intelligent unified dispatching management among new energy power

[a] 853969982@qq.com

 https://doi.org/10.1051/e3sconf/202132901077

generation systems, let alone carry out tasks such as load forecasting and dispatching based on big data analysis technology Exploration of advanced applications such as power generation prediction and Microgrid energy optimal dispatching[3].

To sum up: in order to adapt to the natural conditions of sparse population and harsh environment in Tibet, and in view of the current situation that various photovoltaic power stations are located in remote and remote areas, the local people's technical and cultural level is relatively low, and it is difficult to conduct on-site operation, maintenance and management, the project plans to develop and build a set of intelligent operation and management system of distributed generation and Microgrid system based on Cloud Architecture application. After completion, the system can support multiple distributed generation systems (such as distributed photovoltaic power generation system) and Microgrid systems to access wirelessly or wirelessly, realize data sharing and intelligent unified dispatching management between distributed generation systems and Microgrid, reduce operation and maintenance costs and reduce operation risks; Solar distributed power generation and heating system have the advantages of clean and environmental protection, but also have the disadvantages of volatility and uncontrollability. Based on big data analysis technology, the project carries out advanced applications such as load forecasting, power generation forecasting and Microgrid energy optimal dispatching, which can realize one site, one microgrid, multi-point layout, multi-point interconnection and energy exchange. In addition, in order to test the effectiveness of the intelligent operation and maintenance management platform of the microgrid group and explore a new mode of green heating and power supply in Tibet, the project will also make full use of solar energy supply (solar heating and heating, solar power generation), energy storage technology, thermoelectric microgrid energy interconnection management and other technologies through technological integration and innovation to develop a thermoelectric microgrid system in line with the energy consumption characteristics of Tibet, Provide new operation and maintenance management methods and development ideas for the comprehensive utilization of solar energy in the region, and become a strong scientific and technological support for practicing the concept of green and low-carbon economic development, promoting the improvement of people's livelihood, protecting ecology, strengthening "stabilizing Tibet by science and technology, enriching the people and strengthening the county by science and technology, targeted poverty alleviation by science and technology", and building a beautiful Tibet.

2 Content of the project

2.1 Project overview

In this paper, the project has carried out research on Intelligent interconnection management technology of unified information monitoring cloud platform based on big data analysis, and developed an intelligent operation and maintenance management platform of microgrid group based on cloud platform application, so that it can support multiple new energy systems (cogeneration microgrid system, wind solar complementary power station, photovoltaic power station, photothermal system and air source heat pump system) And Microgrid system are connected in wireless or wired mode, so as to achieve the purpose of data sharing and intelligent unified dispatching management between various photovoltaic power generation systems and Microgrid, and realize big data analysis technology to carry out advanced applications such as load forecasting, power generation forecasting, microgrid energy optimal dispatching and so on. In addition, in order to test the effectiveness of the intelligent operation and maintenance management platform of microgrid group and explore a new mode of green heating and power supply in Tibet, the project will also make full use of technologies such as solar energy supply (solar heating, solar power generation) energy storage technology and Microgrid energy interconnection management for combined heat and power through technological integration and innovation, A thermoelectric microgrid system in line with the energy consumption characteristics of Tibet has been developed to verify the performance of the platform together with the existing new energy facilities (wind solar complementary power station, photovoltaic power station, photothermal system and air source heat pump system), so as to provide new operation and maintenance management methods and development ideas for the comprehensive utilization of multi energy in the whole region. As shown in Figures 1 and 2, the project in this paper is located at No. 22 anjuyuan, Lhasa City, and the total charging capacity and discharge capacity from September 1 to September 8.

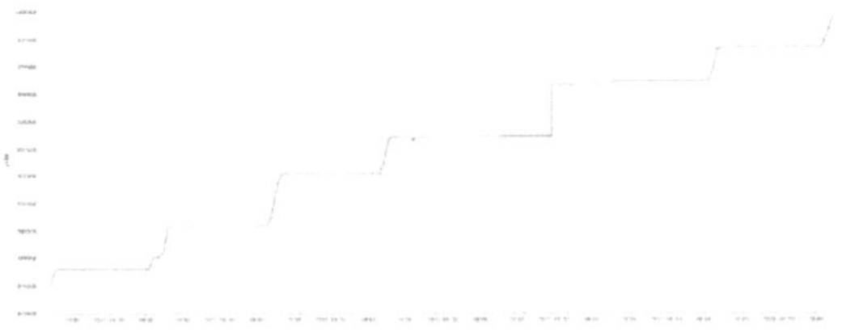

Fig.1 Total charge from September 1 to September 8

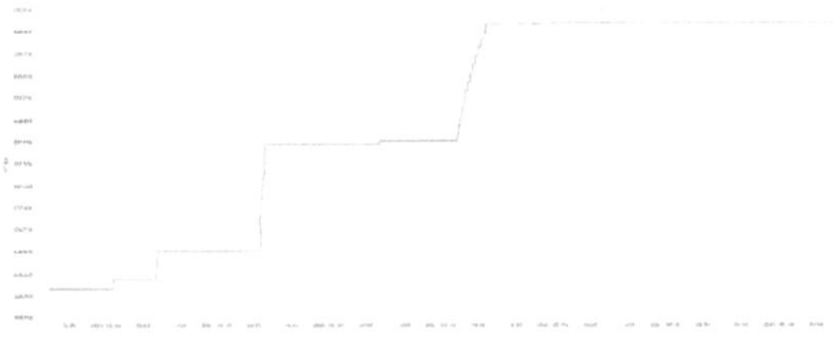

Fig.2 Total discharge from September 1 to September 8

2.2 Specific construction contents

In March 2021, four representative project implementation sites with different energy demand and climate were located in maker space of the testing center of the energy research demonstration center of Tibet Autonomous Region, anjuyuan No. 22 of Lhasa Tibetan Medical College, the duty room of the township government of wumatang Township, Dangxiong County, Lhasa City, and the village Committee of Jirong village, zaxigang Township, Razi County, Shigatse City, A distributed cogeneration microgrid system with solar energy guarantee rate of no less than 50% has been built (including 1 set of 4kWp photovoltaic power generation system, 1 set of 12kWh energy storage system, 1 set of optical storage integrated machine with output power of 5kW and air energy heat pump with total power of 5.5kW, 4.96kW, 4.54kW and 3.3kW respectively). Through the above equipment, combined with the existing projects in Naqu and Ali demonstration sites, it is verified that the micro network group intelligent operation and maintenance management platform developed by the project carries out the research and demonstration of micro network energy consumption management, prediction, monitoring, operation and maintenance analysis and intelligent regulation of photoelectric, photothermal, storage and load integration. Figure 3-6 shows the real-time data of the projects involved in this paper.

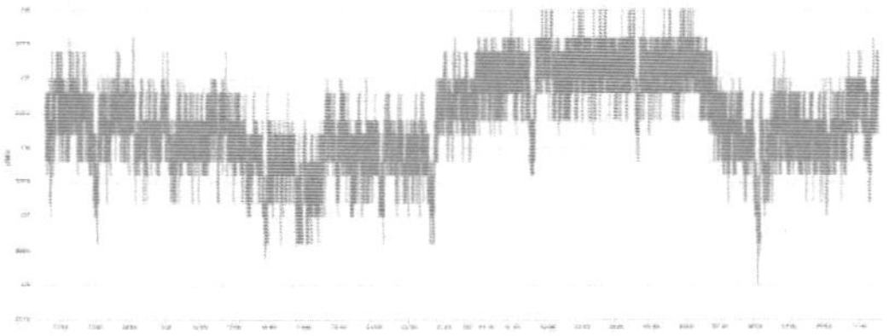

Fig. 3 Real time input voltage

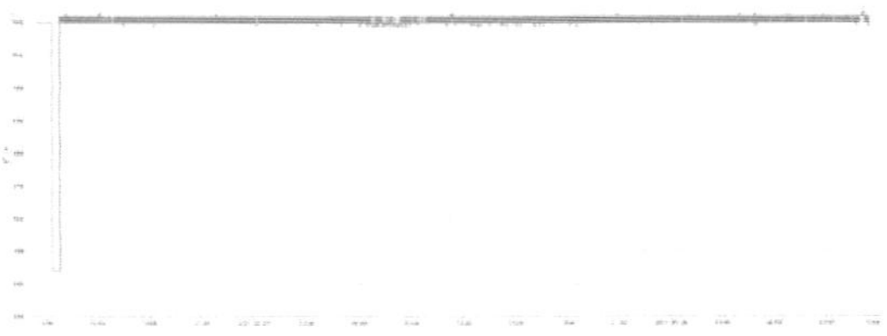

Fig.4 Real time output voltage

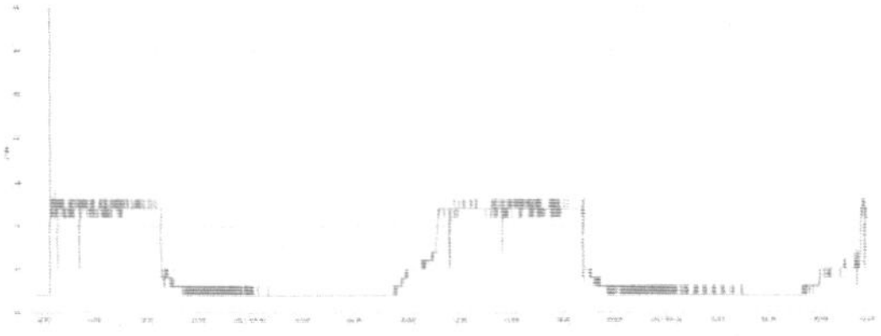

Fig.5 Real time battery voltage

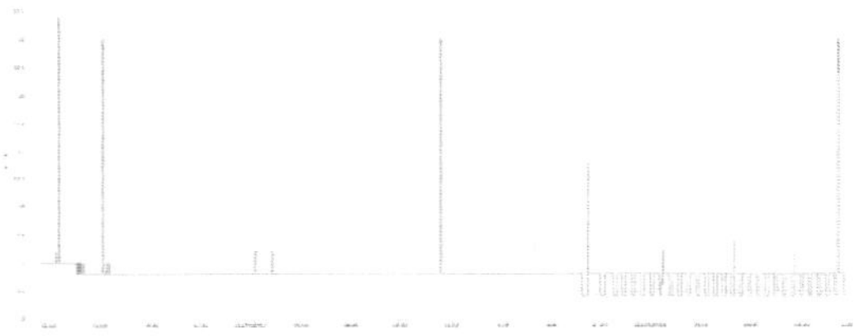

Fig.6 Load percentage from September 1 to September 8

3 Benefit analysis

Through the research of the project technology and the construction of the system in this paper, intelligent monitoring, operation and maintenance can be carried out for all kinds of photovoltaic power stations in Tibet and the power generation systems such as newly-built distributed power stations, microgrid systems and independent photovoltaic power stations, so as to improve the operation reliability of the connected power generation system, reduce the operation and maintenance cost, reduce the operation risk and reduce the maintenance cycle, Considerable economic benefits will be achieved. In addition, the existing photovoltaic power stations used to solve the power consumption of the masses and government organs are mainly invested by the state. The implementation of the project in this paper can avoid the idleness and waste of state-owned assets to a certain extent. In addition, according to the calculation of the average solar radiation in Lhasa, the cumulative installed capacity of the distributed photovoltaic microgrid system is not less than 16kW. The average daily power generation of the distributed photovoltaic microgrid system is about 85.3 kWh and the annual average power generation is about 31134.5 kWh. According to the calculation of 0.7 yuan per kilowatt hour, the direct economic benefit is more than 21000 yuan per year. Coupled with the regulation of energy storage system, combined with air energy heat pump and energy consumption regulation, it not only solves the problem of power consumption, but also solves the problem of hot water heating.

The ecological status of Tibet Plateau is very important and it is an important ecological security barrier in China and even Asia. After the implementation of the project in this paper, it can make all kinds of photovoltaic power stations lacking management and maintenance operate normally for a long time, improve the reliability of power stations and reduce the use of conventional energy, which will certainly make greater contributions to reducing pollutant emissions and protecting the ecological environment of Tibet. In addition, the direct environmental benefits brought by the implementation of the project in this paper are equivalent to saving 11.2 tons of standard coal per year, reducing 8.45 tons of pollution dust, 31.01 tons of carbon dioxide, 0.93 tons of sulfur dioxide and 0.47 tons of nitrogen oxides.

4 Conclusion

At present, after the overall completion of the project in March 2021, after four months of trial operation, the project in this paper runs in good condition, and all functions meet the user's requirements. Give full play to the intelligent monitoring, operation and maintenance and management of photovoltaic power generation and micro grid group, and play the established goal of improving the operation reliability of micro grid system, reducing operation and maintenance cost and reducing operation risk. Through the implementation of the project, the problems of power and heating in the duty room of wumatang township government in Dangxiong county and the village committee of Jirong village, zhaxigang Township, Shigatse have been effectively solved. According to the characteristics of high altitude and cold climate in Tibet, a number of scientific research projects in the field of solar energy that adapt to the plateau environment, can be replicated, have good effects and can be popularized and applied in the whole region have been built, Provide strong scientific and technological support for the high-quality economic and social development of the region, help industrial development and promote rural revitalization.

Acknowledgments

Research and platform development of micro network group intelligent operation and maintenance management technology based on Cloud Architecture (XZ202001YD0009C).

References

1. Zhang Feng, Zhang Jianhua. Research status and application prospect of smart microgrid [J]. Science and technology and enterprises. 2013 (14): 116.
2. Peng C H, Xie P, Pan L, et al. Flexible robust optimization dispatch for hybrid wind/photovoltaic/hydro/thermal power .
3. Kaufmann M, Tüysüz A, Kolar J W. New optimum modulation of three-phase ZVS triangular current mode GaN inverter ensuring limited switching frequency variation[A]. 8th IET International Conference on Power Electronics, Machines and Drives (PEMD 2016)[C]. Glasgow, 2016.

Distributed solar panel heating scheme with auxiliary energy in Tibet

Liqing Zhou [a]

Tibet Autonomous Region Energy Research Demonstration Center, Lasa 850000, China

Abstract. Tibet is located in the high cold and high altitude, the ecological environment is fragile, and the clean utilization of energy has attracted much attention. On the basis of clean energy and reliable power supply, winter heating in Tibet is an urgent livelihood problem to be solved. Based on the current situation of solar heating in Tibet, a distributed solar flat heating scheme with auxiliary energy is proposed, and the heating dispatching mode is analyzed. The research results not only provide a solution for energy efficient and clean heating in Tibet, but also have a certain reference value for the construction of clean energy base in Tibet.

Keywords: Tibet; Multi energy complementarity; Distributed heating; Large flat plate collector

1 Introduction

At present, in addition to cogeneration central heating in cities and towns in Tibet, there are still a large number of small boiler rooms for heating. In rural winter, the main heating is the burning of cow dung by a single household small stove, which has not realized central heating. Low energy efficiency and destruction of ecological environment [1]. Due to the fragile ecological environment in Tibet, the clean utilization of energy has attracted much attention. Tibet is an area of solar energy resources, and solar heating has been preliminarily popularized and applied [2]. As a recognized clean energy, solar energy has gradually become the focus of human development and utilization because of its clean, sustainable, rich source and inexhaustible advantages. Strengthening the application and demonstration of solar heat utilization technology is of far-reaching significance to Tibet in China. However, solar energy has unstable factors such as dispersion, randomness and intermittence. In case of seasonal changes or long-term cloudy and rainy weather, the operation of the heating system is unstable, which limits the development process of solar heat utilization [3]. If solar energy is integrated and coupled with conventional heating technology, develop solar distributed heating system with auxiliary energy, optimize efficient heating equipment, and protect the ecological environment while ensuring heating quality. The analysis shows that using multi energy complementary heating mode to form the heating source in Tibet will effectively improve the comfortable quality of life of green public buildings and residential buildings.

2 Current situation of solar heating

Because most renewable energy sources are unstable and can not provide continuous heating, many scholars and research institutions are discussing how to make better use of renewable energy. The realization of multi energy complementary distributed heating will be one of the effective ways to solve this problem. At present, the complementary utilization of energy mainly includes solar energy and heat pump, solar energy and biomass energy, solar energy and electric cooker [4,5].

2.1 Solar + Heat pump

Solar heat pump is an effective combination of solar heat collection technology and heat pump technology. Solar energy provides the main heat source for the evaporator, and takes the energy improved by heat pump technology as the auxiliary heat source. Cao Wenpu et al.[6] preliminarily analyzed the energy-saving effect of solar hybrid heat pump and obtained that the combination of solar heat pump system and other heat pumps can not only improve the average temperature of heat pump heat source, but also effectively improve the energy utilization rate. However, due to the operation and maintenance of heat pump equipment and other reasons, the operation effect of heating systems such as Tibetan hospitals that have been put into operation is poor.

[a] 853969982@qq.com

2.2 Solar + Biomass energy

Biomass solid briquette refers to a kind of solid granular fuel that can be burned directly after crushing agricultural and forestry wastes under a certain temperature and pressure. Its energy flow density is equivalent to the median bituminous coal. There is no CO_2 emission during combustion, and the emissions of NOx and SO_2 are much lower than that of coal. Biomass has been put into operation as a fuel to provide heating source for residents in Ali Region of Tibet, but it can not be sustained because of the cost. Wang Zelong et al. [7] established the economic analysis model of biomass solar energy system, and used linear programming to optimize the system design parameters, which is suitable for small and medium-sized users; Yuan Shijie et al. [8] established a solar biomass complementary heating system from the perspective of realizing complementary advantages, easy control and ensuring heating continuity and stability, which can operate according to the set control mode through the automatic control unit.

2.3 Solar + Electric boiler

Electric boiler is a kind of boiler that converts electric energy into heat energy and transmits steam, high-temperature water or organic heat carrier to the outside in the form of boiler. Liu Xinyu [9] studied the optimization design of the performance of the combined solar and electric boiler heating system. By establishing the dynamic mathematical model of each component and using MATLAB programming software to simulate and analyze, the operation characteristics in the heating season are obtained; Feng Junzhi [10] analyzed the complementary system of solar heat pump and electric boiler in Lanzhou, and discussed the heating capacity of each part of the complementary heating system in the heating season under different guarantee rates of solar collectors. This heating method is suitable for areas with reliable power supply.

3 Solar collector and auxiliary heat source

Combined with the advantages of abundant solar energy resources in Lhasa, taking natural gas as the auxiliary heat source to better meet the needs of residents for heating and domestic hot water, this paper puts forward a distributed multi energy complementary coalless heating system scheme with auxiliary energy, and discusses the solar heat utilization technology and the process of natural gas integration.

3.1 Large flat plate collector

At present, vacuum tube collectors are usually used in many areas. Although this scheme saves operation cost and protects the environment, the vacuum tube collectors used have some disadvantages, such as fragile, easy to damage, high maintenance cost and difficult to transport and install; At the same time, it has no overheating protection function and poor frost resistance. It cannot be used normally below - 20 °C; The glass tube gap is large, the unit effective heat collection area is small, the heat gain is low, and it needs to occupy more building space. Therefore, the high-efficiency large flat plate collector is selected for the heating system in this paper, as shown in Figure 1. The basic size is 5920mm×2224mm×130mm. The antifreeze heat transfer medium is used to collect heat. Under the irradiation of sunlight, the light energy is converted into heat energy, and then the heat is transferred to the cold water through the plate heat exchanger to gradually increase the temperature of the cold water.

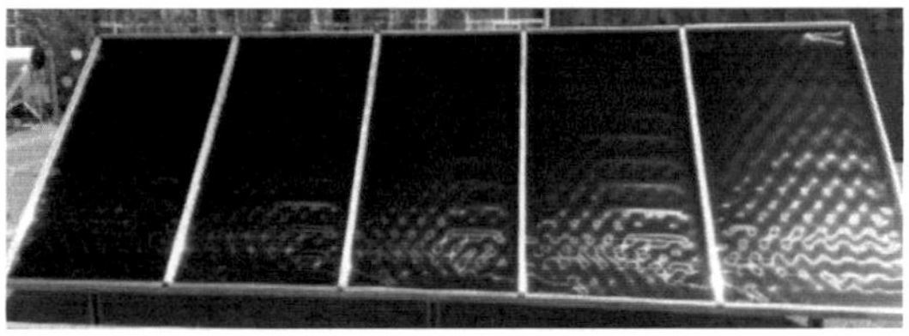

Fig.1 High efficiency large flat plate collector

The high-efficiency large flat plate collector of Tianpu New Energy Technology Co., Ltd. is selected for the system, which not only makes up for the deficiency of vacuum collector tube, but also has the following main advantages compared with ordinary small flat plate collector:

a) The flat plate collector designed in cooperation with Senzo company of Spain adopts imported blue film, with an absorption ratio of 0.94 (national standard requirement $\geq$ 0.92) and an emissivity of 0.033 (national standard requirement $\leq$ 0.2);

b) High permeability and low iron cloth tempered glass is adopted, with light transmittance $\geq$ 95%; The instantaneous efficiency can reach 0.80, the heat absorption is fast, the thermal efficiency is increased by more than 20%, and the thermal performance is higher;

c) Ultra thick insulation is adopted, with small heat loss, which is better than the national standard by 25%, meeting the requirements of efficient heat collection under different working conditions. The heat lasts for a long time, and the air drying temperature can reach 208 °C;

d) It adopts natural color ultra-high strength aluminum alloy frame, which is corrosion-resistant;

e) The heat collecting area of a single block is greater than 12m2, which reduces the relative surface area per unit area and reduces the heat loss;

f) High system safety: the single collector forms an array, removes the connection between small collectors, has reliable sealing and high safety performance;

g) Integrated design of heat absorption channel, less welding, corrosion resistance and high safety performance.

3.2 Auxiliary heat source

Common auxiliary heat sources of solar heating system include coal-fired boiler, oil fired boiler, gas wall mounted boiler, electric boiler, air or water source heat pump, biomass boiler, etc. As a traditional heating method, coal-fired boiler and oil-fired boiler have a long development time and high technical level, but they are unfavorable to the ecological environment. Although biomass boilers can effectively reduce energy consumption and environmental pollution, their use is relatively limited due to short development time, relatively low technical level of equipment and poor adjustability. Combined with the geographical location, resource distribution and environmental protection of Tibet, the gas wall mounted furnace is adopted in the heating scheme. Due to the low energy flow density of solar energy and inconvenient storage, it can only be stored and utilized in other forms. The complementary application of gas wall mounted furnace and solar heat collection technology can alleviate the shortage of natural gas resources.

4 Integration of solar distributed heating system with auxiliary energy

Take the office building of the energy research and demonstration center of Tibet Autonomous Region as an example. The office building is a three storey building facing north and south, with a heating area of 570m^2. Usually, it only needs heating during the day. At present, the solar distributed heating system with auxiliary energy has been applied in office buildings, and the preliminary heating relying on solar energy as a single heat source has been realized. The heating effect is ideal, and the room temperature of each room in the office building can reach the design temperature.

4.1 Heating system

The solar distributed heating system with auxiliary energy takes solar energy as the main heat source and gas as the auxiliary heat source. Natural gas can provide heat source support for the system when solar energy does not meet the demand. It is the best coupling mode of solar energy and gas. At the same time, the heat storage tank can not only overcome the shortcomings of solar energy, but also reduce conventional energy consumption and protect the environment. The working principle of the system is shown in Figure 2.

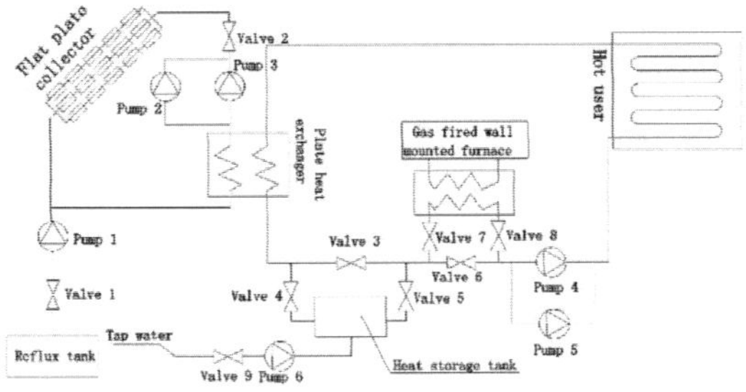

Fig.2 Solar distributed heating system with supplementary combustion

4.2 Multimodal analysis

The solar distributed heating system with auxiliary energy can be controlled by full-automatic heating mode and manual heating mode. Among them, the manual heating mode includes the following four types:

a) Full solar heating mode. Using the temperature difference control mode, when the solar energy is sufficient, the solar system provides heat for heat users for direct heating; When the room temperature is higher than the maximum limit, too much heat is collected. At this time, the excess heat will be stored in the heat storage tank. The gas wall mounted furnace does not work and there is no fossil energy consumption;

b) Dual energy coupled heating mode. When the solar energy is reduced and the room temperature is lower than the set value, the solar energy cannot meet the needs of residents, and the heat storage tank will provide heat. If it still cannot meet the needs, start the gas wall mounted furnace to increase the heating water temperature to the set value;

c) Full wall mounted boiler heating mode. When the room temperature is lower than the minimum limit, it will no longer exchange heat with the solar heat collection system, and the heat of the system will be provided by the gas wall mounted furnace;

d) Anti freezing mode of solar collector. When heating is not required at night or in low-temperature weather, in order to prevent the flat plate collector from freezing and cracking, only the heat transfer medium in the solar heat collection system circulates to achieve the purpose of protecting the heat collection system.

For the solar distributed heating system with auxiliary energy, multi-mode joint control is adopted, and remote monitoring can be carried out through mobile app; The heat collection system adopts high-efficiency large flat plate collector, and the installation angle can be adjusted to maximize the use of solar energy and improve the heat collection efficiency. Under the condition of ensuring the heating quality, make full use of clean energy, optimize its control and operation methods, and increase the economy of the heating system.

5 Conclusion

a) The solar large flat plate heating system with auxiliary energy has simple structure and convenient operation. It adopts multi-mode joint control and remote monitoring through mobile App. The heat collection system adopts high-efficiency large flat plate collector to make maximum use of solar energy and improve the solar energy guarantee rate of the heating system. It can adapt to the heating environment under a variety of complex working conditions and realize clean heating in Tibet.

b) Improving the quality of urban heating is one of the key issues to improve people's livelihood. Through the research on a solar distributed heating system with auxiliary energy and its optimal control method, under the condition of ensuring the heating quality, make full use of clean energy, reduce the energy consumption such as natural gas in the heating period, reduce the cost and protect the ecological environment.

References

1. Wang Xiaofeng, CIREN pingcuo, Wang junle. Some thoughts on the utilization of renewable energy in Tibet [J]. Tibet Science and technology, 2006 (8): 13-16.
2. Jin Hongguang, Sui Jun. heat utilization and comprehensive utilization of renewable energy [J]. Journal of the Chinese Academy of Sciences, 2016 (2): 208-215.
3. Li Zhongyi. Analysis on the development trend of solar thermal utilization technology [J]. Resource conservation and environmental protection, 2014 (8): 32-33.
4. Liu lifeI, Luo Huilong, Tian Panyu, et al. Heat utilization analysis based on solar cogeneration [J]. Low temperature building technology, 2016 (11): 134-136.
5. Chen Chen. Research and application of optimal design of solar heating system [D]. Shijiazhuang: Shijiazhuang Railway University, 2013.
6. Cao Wenpu, Li Haiying, Zhang Yan, et al. Energy saving effect analysis of solar hybrid heat pump [J]. Applied energy technology, 2016 (6): 41-44.
7. Wang Zelong, Tian Yishui, Zhao Lixin, et al. Optimal design of biomass solar complementary heating system [J]. Journal of agricultural engineering, 2012,28 (19): 178-184.
8. Yuan Shijie, Wu yunyun, Lu Shihua. Research on heating system based on combined application of solar hot water and biomass energy [J]. Urban development research, 2015,22 (1): 1-3.
9. Liu Xinyu. Study on the performance of solar energy and electric boiler combined heating system [D]. Beijing: North China Electric Power University, 2016.
10. Feng Junzhi. Analysis of combined operation system of solar heat pump and electric boiler in Lanzhou [D]. Harbin: Harbin Engineering University, 2010.

https://doi.org/10.1051/e3sconf/202132901079

Operation analysis of primary and secondary networks of distributed solar heating system in Tibet

Chengpeng Liu [a]

Tibet Autonomous Region Energy Research Demonstration Center, Lasa 850000, China

Abstract: Tibet is located in a high cold and high altitude area. Heating in winter is an important problem to be solved urgently, and the distributed solar heating system provides a solution for the efficient and clean utilization of energy in this area. Through the test and commissioning of the distributed solar heating system with supplementary combustion in Tibet, the operation characteristics of the heating system are analyzed from the temperature changes of the primary network and the secondary network. The research results can provide technical support for the application and promotion of the solar distributed heating system with supplementary combustion in Tibet.

Keywords: Solar energy; Heating; Supplementary combustion; Distributed energy

1 Introduction

Limited by the lack of fossil fuel resources and the fragile ecological environment, it is impossible to establish a large-scale coal-fired power station with cogeneration for heating, which restricts the heating mode of residents in Tibet [1-3]. Tibet is extremely rich in solar energy, with an average annual radiation of 6000 ~ 8000 MJ / m2, an average annual sunshine of about 3000 h and high stability [4,5]. As an important part of the future energy structure, solar energy is mainly used in photovoltaic power generation and solar heat utilization. Solar heating is a green heating method that uses solar radiant energy as a heat source, which not only saves resources but also meets the requirements of environmental protection [6,7]. However, solar energy has the characteristics of instability and intermittence, and can not meet the heating needs of users in special weather [8,9]. Therefore, it is the best choice to adopt multi energy complementary heating mode to form the heating source of Tibet Autonomous Region from the technical and economic aspects [10].

The solar heating system in Tibet is selected in this research. Distributed solar heating can effectively improve the solar assurance rate, and auxiliary heat sources such as gas wall mounted furnace can overcome the disadvantage of discontinuous solar energy and realize the continuous and stable heating of Tibetan residents in winter [11,12]. In this paper, combined with the supplementary combustion solar distributed heating system in Tibet, the heating commissioning test is carried out, and the operation of the system is analyzed from the daily temperature changes of primary network and secondary network, which provides technical support for the popularization and application of the system in Tibet.

2 Integration of heating system

2.1 Introduction to the system

Take the heating of the office building of the energy research and demonstration center of the Tibet Autonomous Region as an example, build a solar heating system with supplementary combustion with a heating area of no less than 570m2, and the solar energy utilization rate is no less than 70%. The solar distributed heating system with supplementary combustion takes solar energy as the main heat source and gas as the auxiliary heat source. Natural gas can provide heat source support for the system when solar energy does not meet the demand. It is the best coupling mode of solar energy and gas [13,14]. At the same time, the hot water storage tank can not only solve the problem of energy peak and valley, but also reduce conventional energy consumption and protect the environment [15,16]. The working principle of the system is shown in Figure 1.

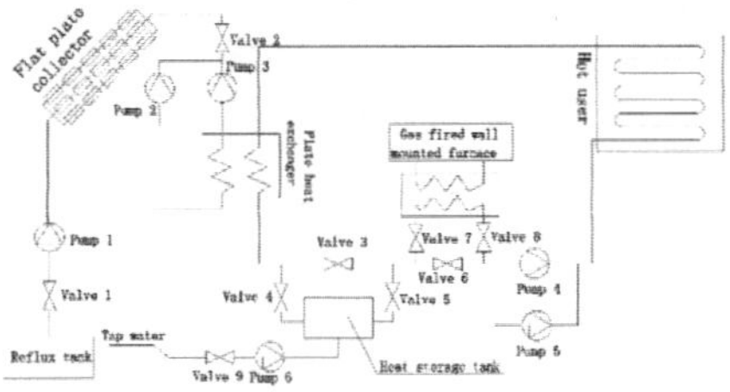

Fig. 1 Working principle of solar distributed heating system with supplementary combustion

[a] 1426382426@qq.com

 https://doi.org/10.1051/e3sconf/202132901079

2.2 Main equipment

a) Flat plate collector

In the solar heating system, the flat plate collector is placed at a certain angle, and the sunlight passes through the cover plate and incident on the heat absorbing plate coated with selective coating, in which the heat absorbing plate collects most of the radiant energy and converts it into heat energy. 15 large flat plates and 3 small flat plates of two thicknesses are selected for the system. The main technical parameters of large and small flat plate collectors are shown in Tables 1 and 2.

Tab.1 Technical parameters of large flat plate collector

Parameter / Unit	Numerical value
Total area / m^2	13.17
Daylighting area / m^2	12.35
Heat absorber area / m^2	12.37
Medium content / L	11.35
Maximum pressure / bar	10
Air drying temperature / °C	218
Geometric dimension / mm	5920*2224*135
Recommended working flow / ($L \cdot m^{-2} \cdot h^{-1}$)	10-25

Tab.2 Technical parameters of small flat plate collector

Parameter / Unit	Numerical value	
National standard model	P-G/0.6L/LM-1.84	P-G/0.6L/LM-1.76
Total area / m^2	2	2
Daylighting area / m^2	1.84	1.76
Core volume / L	1.5	1.5
Maximum pressure / MPa	0.6	0.6
Air drying temperature / °C	218	218
Geometric dimension / mm	2000*1000*80	2000*1000*120

b) Heat storage tank

In the process of solar energy utilization, heat storage is mainly to make up for the impact of solar energy discontinuity and convert the absorbed solar radiation energy into heat energy for storage. A vertical cylindrical pressure heat storage tank is selected for the system, and its technical parameters are shown in Table 3.

Tab.3 Technical parameters of heat storage tank

Parameter / Unit	Numerical value
Design pressure / MPa	0.33
Pressure of pressure test / MPa	0.42
Design temperature / °C	90
Volume / m^3	8.5
Working medium	Water
Main material	Q245R
Product standard	GB150

c) Wall mounted gas boiler

Two LN1GBQ50-NCN52HD condensing gas wall mounted furnaces are selected as auxiliary heat sources for the system, and their main technical parameters are shown in Table 4.

Tab.4 Technical parameters of gas wall mounted furnace

Parameter / Unit	Numerical value
Rated voltage / V	220
Frequency / Hz	50
Rated gas pressure / Pa	2000
Rated heat output / kW	47.7
Rated condensation heat output / kW	52.3
Minimum heat output / kW	5.1
Minimum condensation heat output / kW	5.6
Rated heating heat input / kW	49.8
Minimum heating heat input / kW	5.2
Maximum heating pressure / MPa	0.3

3 Test data analysis

As the system is in the trial operation stage, the gas wall mounted boiler has been installed but has not been debugged, and the heat storage tank cannot be foamed and insulated due to low ambient temperature, so both are not put into operation. The commissioning started on November 1, 2017, and the preliminary heating of the system relying on the solar flat collector as a single heat source was realized on December 6. The heating effect is ideal. The radiators in each room of the office building can reach the ideal temperature during the solar energy work, and the indoor temperature of the office building is maintained at 20 °C for about 7h in the daytime (the data on December 30, 2017 are selected for analysis).

3.1 Solar radiation

The variation of solar radiation with time is shown in Figure 2. As can be seen from Figure 2, from 9:00 to 11:00, the solar radiation intensity increases sharply from 209 W / m^2 to 863 W / m^2, and the average daily solar radiation intensity lasts more than 800 W / m^2 for nearly 6 hours.

https://doi.org/10.1051/e3sconf/202132901079

Therefore, the unique solar radiation characteristics in Tibet provide conditions for the promotion of solar heating system with supplementary combustion in this area.

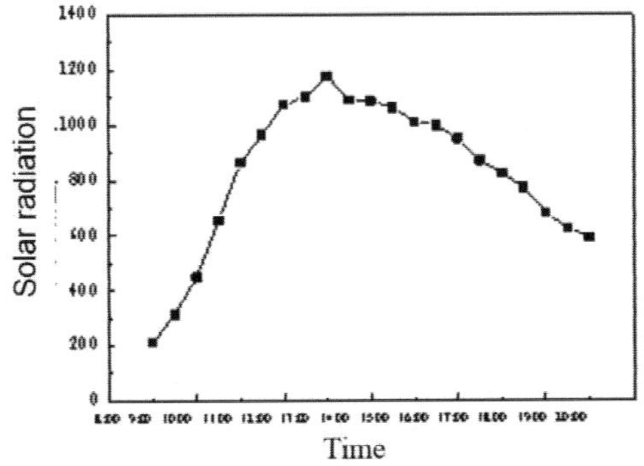

Fig. 2 Variation curve of solar radiation with time

3.2 Indoor and outdoor temperature

The ambient temperature has a great influence on the heat collection and heat loss of solar heating system. The changes of outdoor temperature and indoor temperature in the range of 9:00-19:00 are shown in Figure 3. Because the heat storage tank is not put into use, the change trend of indoor temperature and outdoor temperature is basically the same, and it is obviously affected by the change of outdoor temperature. If excess heat is stored, the indoor temperature change will be more stable and basically maintained in a narrow temperature range.

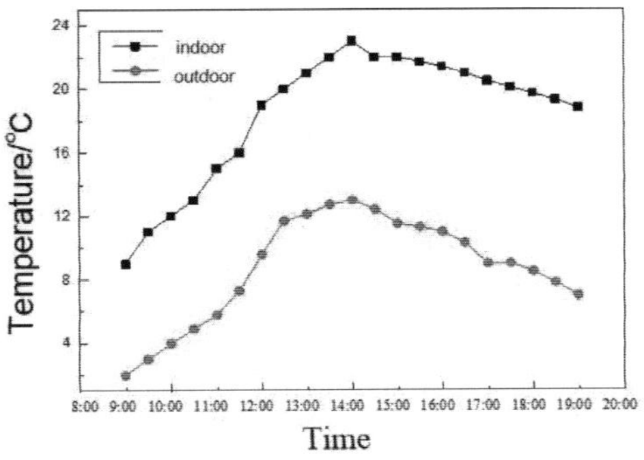

Fig. 3 Variation curve of indoor and outdoor temperature with time

3.3 Inlet and outlet temperature of secondary network

The variation curve of inlet and outlet temperature of secondary network with time is shown in Figure 4. During the test, the system only relies on solar energy as a single heat source for heating, so the inlet temperature of the secondary network rises slowly from 9:00 to 10:00, the temperature rises rapidly from 10:00 to 14:00 and reaches the highest value, and the temperature decreases slowly after 14:00. After 18:30, the secondary network will be used as a heat source to heat the primary network until the secondary network is used as the end again after 9:00 the next day to absorb heat from the primary network. Compared with Fig. 3, the change curve of the inlet and outlet temperature of the secondary network with time is basically consistent with the change trend of the solar radiation with time, indicating that the magnitude of the solar radiation directly affects the inlet and outlet temperature of the secondary network.

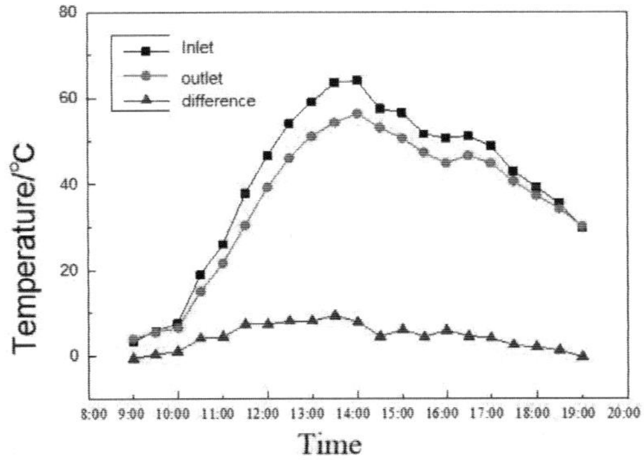

Fig. 4 Variation curve of inlet and outlet temperature of secondary network with time

3.4 Inlet and outlet temperature of primary network

The changes of inlet and outlet temperatures of primary network collector are shown in Figure 5. The inlet and outlet temperature of the collector is relatively high. The minimum temperature at the inlet of the collector is 3.7 ℃ and the maximum temperature is 60 ℃; The minimum temperature at the collector outlet is 3.1 ℃, and the maximum temperature is 65.6 ℃. The daily average values of the inlet and outlet temperatures of the collector are higher than the outdoor ambient temperature, indicating that the collector operates at high temperature and has great heat dissipation loss, which is not beneficial to increasing the heat collection efficiency. The temperature difference between the inlet and outlet of the collector is basically maintained above 4 ℃, and the maximum temperature difference is about 6.6 ℃ at noon. The outlet temperature of the collector is higher than the inlet temperature after 9:00, indicating that the collector absorbs sunlight radiation in the daytime and the heat collection is positive. The temperature change trend of the medium at the inlet and outlet of the collector is the same. With time, it increases first and then decreases, reaching the maximum at about 14:00. At the same time, the change trend is that it increases rapidly before 14:00, and the temperature change is stable after 14:00. The temperature difference between the inlet and outlet of the collector remains basically unchanged during operation, and its operation is in good condition.

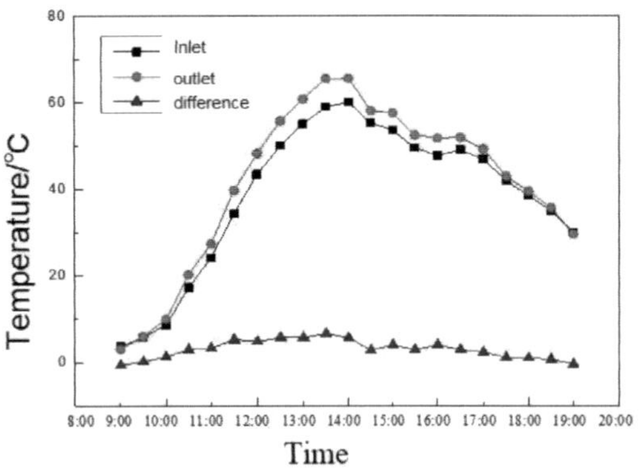

Fig. 5 Variation curve of inlet and outlet temperature of collector with time

4 Conclusion

a) According to the geographical location and solar irradiation characteristics of Tibet, the radiation absorbed by flat plate collector is measured by solar irradiator, and its variation trend with time is obtained. The results show that the solar radiation increases sharply from 9:00 to 11:00 every day, and reaches the maximum from 13:00 to 14:00. However, because the heat storage tank is not put into use, the indoor and outdoor temperature changes have the same trend, which is obviously affected by the outdoor temperature change.

b) Because the solar distributed heating system with supplementary combustion relies on solar energy as a single heat source for heating during the test, the secondary network is used as the heat source for heating the primary network after about 18:30; Until about 9:00 the next day, the secondary net is used as the end to absorb heat from the primary net. However, the system only provides heat for office buildings rather than residential buildings. If the weather is sunny, solar energy as a single heat source can meet the heating needs of users for most of the day.

References

1. Wang Dengjia, Liu Yanfeng, Wang Lijuan. Study on flow law of active solar heating water supply pipe [J]. Journal of solar energy, 2011 (9): 1387-1390.
2. CARPANETO E, LAZZERONI P, REPETTO M. Optimal integration of solar energy in a district heating network[J]. Renewable Energy, 2015 (75) : 714-721.
3. Liu lifeI, Luo Huilong, Tian Panyu, et al. Heat utilization analysis based on solar cogeneration [J]. Low temperature building technology, 2016 (11): 134-136.
4. Qu Shilin, sun Zhifeng, Ji Yubao, et al. Analysis and Experimental Research on the influence of heat storage tank on the performance characteristics of solar water source heat pump system [J]. Journal of solar energy, 2012, 33 (8): 1405-1410.
5. Meng Erlin, sun Zhigao, Liu Chenggang, et al. Experimental study on system performance of solar floor radiation system for night heating [J]. Thermal science and technology, 2016,15 (2): 149-153.
6. LIU T, LI Z, HE G. Experiments of a Heat Pump Water Heating System Using Stored Solar Energy to Defrost [J]. Energy Procedia,2017 (105):1130 -1135.
7. Ling Deli, Li Ming, Luo Xi, et al. Study on drying characteristics of cut tobacco based on trough concentrating solar heating [J]. Journal of solar energy, 2015 (2): 460-466.
8. Guo song, Zheng Xin, Qiu Zejing, et al. Study on the performance of steam generation system using solar medium temperature heat collection [J]. Thermal science and technology, 2017,16 (2): 159-165.
9. ZHANG H, HONG H, GAO J, et al. Thermodynamic performance of a mid-temperature solar fuel system for cooling, heating and power generation[J]. Applied Thermal Engineering, 2016,106:1268-1281.
10. Chen Hongbing, Li Qiang, Chen Xilin, et al. Performance study of centralized solar water heating system [J]. Renewable energy, 2017, 35 (4): 613-620.
11. Xiangdong. Water saving, energy saving and environmental protection transformation of gas water heater [J]. Coal technology, 2011, 30 (10): 235-236.
12. Wang Xiaofeng, CIREN pingcuo, Wang junle. Some thoughts on the utilization of renewable energy in Tibet [J]. Tibet Science and technology, 2006 (8): 13-16.
13. Wang Xinming. Solar heat pump central hot water system [J]. Tibet Science and technology, 2007 (9): 20-22.
14. KICSINY R. Transfer functions of solar heating systems with pipes for dynamic analysis and control design[J]. Solar Energy,2017 (150): 596–607.
15. Li Xinguo, Hu Xiaochen, Wang Jian. Energy analysis and experiment of solar energy, heat storage and ground source heat pump combined system [J]. Journal of solar energy, 2012 (4): 640-646.
16. Yang Wansheng, Li Lirong. Experimental study on sunshade solar phase change collector [J]. Renewable energy, 2015 (7): 965-970.

https://doi.org/10.1051/e3sconf/202132901080

Application of distributed photovoltaic power generation technology in electric vehicles

Min Wu [a]

Tibet Autonomous Region Energy Research Demonstration Center, Lasa 850000, China

Abstract: In this paper, a photovoltaic grid connected electric vehicle charging station suitable for plateau application is developed by using two-way energy storage photovoltaic grid connected power generation technology; Introduce a medium-sized electric bus to carry out long-term operation test in Tibet; The detection experimental device and evaluation method suitable for electric vehicle power battery in plateau area are constructed; Carry out strategic research on the impact evaluation of the development of electric vehicles on the power grid side in the plateau weak current network area.

Keywords: Tibet region; Distributed photovoltaic; Electric vehicle; Research application

1 Introduction

At present, when the global automobile industry is facing the great challenges of financial crisis and energy and environmental problems, a broad consensus has been formed in the world to develop electric vehicles, realize the electrification of automobile energy power system and promote the strategic transformation of traditional automobile industry. Especially in recent years, with the significant progress of high energy storage density and low-cost battery technology, the driving technology has become more and more mature and perfect, which promotes the performance of electric vehicles closer to fuel vehicles. At present, an upsurge of electric vehicle development has formed in the world[1].

All or part of the vehicles driven by motors and equipped with high-capacity electric energy storage devices are collectively referred to as electric vehicles. The components of electric vehicles include: electric drive and control system, driving force transmission and other mechanical systems, working devices to complete established tasks, etc. At present, there are three types of electric vehicles: pure electric vehicle, hybrid electric vehicle and fuel cell electric vehicle. Pure electric vehicle is the main development goal of all countries in the world. It is a vehicle powered by secondary batteries (such as lead-acid battery, nickel cadmium battery, nickel hydrogen battery or lithium-ion battery).

At present, China has formulated a unified standard for charging facilities, which compulsorily regulates the specification and size of charging interface and power supply. The power supply adopts 50Hz 380V and 220V AC to obtain DC through rectification to charge the battery. However, the battery type, parameters and corresponding DC charging parameters have not been standardized. After consulting the literature, there is no report on photovoltaic charging products[2].

Electric vehicle is a vehicle with broad development prospects in China in the future. Electric vehicle charging station provides energy supply for the operation of electric vehicles, which is an important supporting infrastructure for the development of electric vehicles. At the same time, the development of electric vehicle charging station has also become a key problem restricting the wide promotion and use of electric vehicles. The electric vehicle charging station based on the large power grid should not only solve the impact of the instantaneous load on the power grid when the electric vehicle is rapidly charged, but also reduce the impact of harmonic pollution on power quality during charging. Therefore, the conventional power grid is difficult to meet and must be targeted[3]. This involves the transformation of the entire national power grid, which costs a lot. It can not be realized overnight from discussion, project approval to network completion. Therefore, photovoltaic power generation with the advantages of safety, energy conservation and environmental protection, simple maintenance and inexhaustible resources is used as the charging power supply of electric vehicles. The design and research of two-way energy storage photovoltaic grid connected electric vehicle charging system is carried out to meet the charging demand of electric vehicles without transforming the conventional power grid and affecting the power quality.

[a] 853969982@qq.com

2 Objectives of the project

2.1 Project requirements analysis

Electric vehicles replace oil with electricity, which can achieve "zero emission" and low noise. It is an important means to solve energy and environmental problems. With the shortage of oil resources and the development of battery technology, electric vehicles have been close to or even better than traditional fuel vehicles in performance and economy, and began to be gradually popularized and applied all over the world. The new generation of energy-saving and environment-friendly vehicles represented by electric vehicles is the inevitable trend of the development of automobile industry, which has become a general consensus. To promote electric vehicles, perfect supporting infrastructure is indispensable, and the construction of electric vehicle charging station is the key. The project will make use of the abundant solar energy resources in Lhasa and select the photovoltaic power generation system and energy storage system to form a two-way energy storage photovoltaic grid connected electric vehicle charging system. On the premise of not adding additional conventional power grid capacity, the impact of temporary rapid charging of electric vehicles on power grid load and the occurrence of load unevenness will be avoided through the buffer effect of energy storage unit, Eliminate the influence of harmonic pollution on power grid during charging[4].

The implementation of the project in this paper can not only solve the charging problem of electric vehicles in trial operation in Lhasa by using Lhasa's rich solar energy resources without increasing power grid capacity, investment and quantities, broaden the development and utilization channels of clean energy represented by solar energy resources, and achieve the purpose of energy conservation and emission reduction, Explore electric vehicle charging technology and electric vehicle detection and evaluation methods suitable for the actual situation of Tibet, make electric vehicles acceptable to the masses, accelerate the application and popularization of electric vehicles in our region, improve the professional skills of project participants through professional and technical personnel training, and reserve talents for the implementation of similar projects in the future.

2.2 Difficulty analysis of the project

(a) Complete the design and research of two-way energy storage photovoltaic grid connected electric vehicle charging system which can be connected to the grid and operate independently, and master the integration technology. Conventional photovoltaic power stations use low current and long-time discharge, while energy storage photovoltaic power stations used for electric vehicle charging are high current and short-time discharge. Special research needs to be carried out on the control system, the ratio of inverter device and storage bottle, and the battery management system (CMS) of such power stations.

(b) Carry out the research and development of bidirectional converter and key technology of lead-acid battery energy storage system. At present, electric vehicle charging facilities are only applicable to AC, and there is no DC charging technology and equipment, which can be obtained only through special research and development.

(c) Design and development of two DC / DC electric vehicle charging piles with fast charging, slow charging, multiple power levels and multiple output modes to meet the charging needs of different electric vehicles;

(d) Carry out power battery performance detection, evaluation and test research, and actually verify and test the actual adaptability of electric vehicles in plateau areas, so as to provide experimental data for electric vehicle manufacturers to improve and improve existing models to meet plateau applications;

(e) Explore the operation and management mode of electric vehicles and photovoltaic charging stations to provide reference for large-scale promotion and application in the future.

3 Technology roadmap

3.1 Distributed photovoltaic system

Electric vehicles are different from fuel vehicles. Because the voltage of power supply system and motor power system on the vehicle, whether AC or DC, is greater than the safety voltage, or even more than 380V, the detection of safety performance of electric vehicles is particularly important. It is before other performance tests and is the premise for the safe operation of electric vehicles. The test of electric energy consumption rate is an important special project of electric vehicles. How to test various electric parameters of electric vehicles, especially to obtain accurate real-time parameters under operation state, is the accurate basis for the study of electric energy consumption of electric vehicles. The energy consumption of pure electric vehicles is affected by many factors, including charger efficiency, control system energy consumption, motor efficiency and other energy consumption, which are the total energy consumption. It mainly tests the maximum speed, average speed, energy consumption, energy consumption per 100 km, as well as the battery terminal voltage and output current, power consumption, feedback power, etc. during the operation of electric vehicles.

As shown in Figure 1, the principle of the two-way energy storage photovoltaic grid connected charging station is to store the electric energy generated by the photovoltaic power generation system and the electric energy obtained from the low point of the power grid with the battery, and charge the electric vehicle when needed. When there is no electric vehicle to be charged, when its own battery is full, the excess power is sent to the power grid to obtain additional income. The two-way energy storage photovoltaic grid connected charging station stores the electric energy obtained during its own power generation and the trough of the power grid, so it is not affected by power failure and cloudy and rainy weather; Through the buffer function of the energy storage system, the adverse impact on the power grid is avoided; Two way energy storage design enables the system to be connected to the grid or run independently.

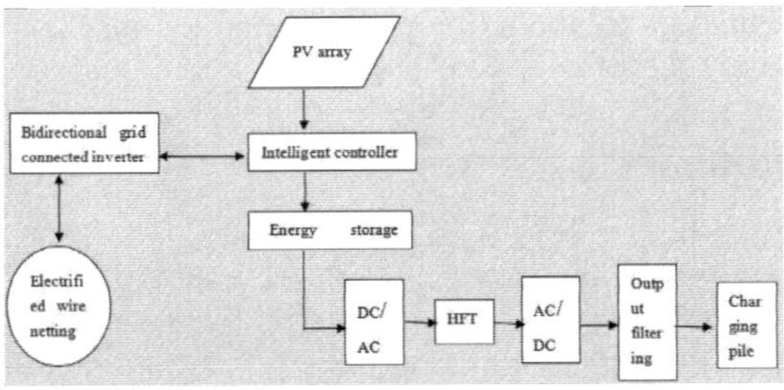

Fig. 1 Schematic diagram of bidirectional energy storage photovoltaic grid connected charging station

3.2 Energy control system

As shown in Figure 2, the power station judges and completes the automatic switching of power supply from PV array or power supply from power grid through intelligent controller, coordinates the output power of charging, battery energy storage and solar grid connected system, realizes the control and island detection of distributed power supply system, ensures safety and verifies the possibility of off grid operation. In case of power loss of the energy storage system, it is preferred to obtain electric energy from the whole PV train. When the PV array has no power at night or in cloudy and rainy weather, the power station obtains electric energy from the power grid through the bidirectional grid connected inverter according to the setting instruction, and then supplies power to the energy storage system through the intelligent controller. When the power of the energy storage system is sufficient, the PV array passes through the intelligent controller and supplies power to the power grid through the bidirectional grid connected inverter. The energy storage system is equivalent to UPS for the whole system. After passing through the energy storage system, all output electric energy of the charging station is boosted and filtered, and then transmitted to the charging pile. Energy flow of charging station circuit: Loop 1 realizes photovoltaic energy storage function: the electric energy emitted by solar cell modules is stored in the energy storage system, which is DC / DC conversion. As we all know, the electric energy emitted by the sun is unstable and changes with time, illumination and temperature, which requires a wide input range of the front stage DC, so as to make the most efficient use of solar energy. Loop 2 realizes the photovoltaic power generation function: the sunlight is sufficient, and the electric energy generated by the photovoltaic system fully meets the power supply demand of the charging station. At the same time, there is surplus. It can be considered to sell electricity to the power grid through this circuit to pursue economic benefits. Loop 3 realizes the power collection function of the energy storage system from the power grid: when the power generation of the PV array is insufficient, the system is controlled by the intelligent controller through the preset command, and takes power from the power grid through the bidirectional grid connected inverter to charge the energy storage system. Loop 4 realizes the power supply function of the charging station to the charging pile. When the charging pile works, the intelligent controller controls the energy storage system to output the DC power after boosting and filtering.

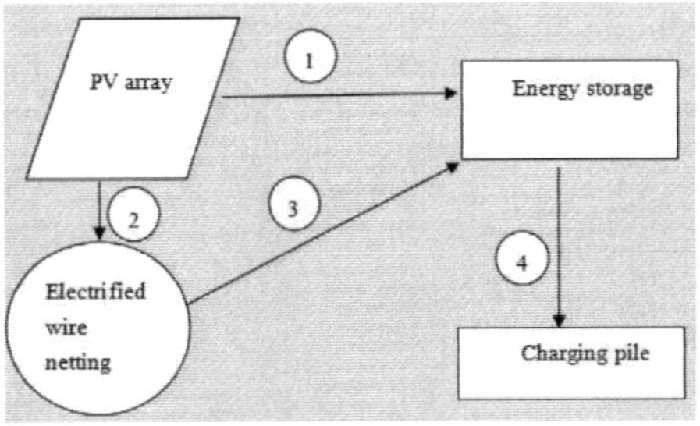

Fig.2 Block diagram of energy flow

For the bidirectional grid connected inverter, in order to realize the power exchange between the energy storage photovoltaic charging system and the grid connected power grid, the whole energy storage photovoltaic grid connected charging device can operate safely, stably and reliably. The bi-directional grid connected inverter will complete the research, design and development of the grid connected inverter based on the mode smooth switching technology based on droop characteristics, seamless switching of charging mode and grid connected and off grid charging control technology, and produce a 380V / 20kW bi-directional grid connected inverter prototype.

For the intelligent control system, in order to realize the management, monitoring and regulation of the whole charging station, the project team will design and develop an intelligent control system. The system will be responsible for monitoring the operation of the whole charging station system, completing the automatic switching between power supply of photovoltaic power generation system and power supply from the power grid, and coordinating the output power of charging, battery energy storage and solar grid connected system, It can realize the control and island detection of distributed power supply system, ensure safety and verify the possibility of off grid operation. A man-machine interface is established to control the switching management system of photovoltaic power generation and power grid power supply and the data acquisition and storage system of electric vehicle power battery charge according to the commands input in the man-machine interface.

For the charging pile, due to the use of the battery provided by the photovoltaic power station for charging, the inverter link and the resulting power loss are reduced, and the system is simpler than the AC charging pile. However, the electrical insulation strength and equipment heat dissipation level need to be increased. At the same time, in order to meet the different needs of various electric vehicles for charging parameters, the charging line is more inclusive, that is, the same line and its control unit can meet the requirements when the charging parameters are adjustable.

The charging pile will be composed of man-machine exchange interface, monitoring system, display terminal, actuator, reserved terminal and terminal at the output end, or special charging gun and safety protection device connected to electric vehicle. The charging pile is equipped with a specific human-computer interaction interface to log in. Through the interface, corresponding charging operations such as charging mode and charging time setting can be carried out. The charging pile display screen can display data information such as charging amount and charging time, which can provide users with more power options and more output modes to meet the needs of users for fast charging and slow charging.

4 Benefit analysis

Through the implementation of the project in this paper, the demonstration and promotion of 18kW charging station is calculated according to the average sunshine of 5 hours per day, the PV array power of the power station is 18kW, and the annual power generation is $18W \times 5$ hours / day $\times 300$ days / year $\times 0.8$ (conversion efficiency) = 27000kW·h / year. Calculated at 1 yuan per kilowatt hour, the direct economic benefit each year exceeds 27000 yuan; The completed charging station is expected to generate 27000 kWh per year, which can also alleviate the local scarce power resources. Through the centralized application and promotion of photovoltaic grid connected power generation technology, the project will effectively promote the development of photovoltaic power generation technology and product market.
After the completion of the charging station, only the electric energy generated by the PV array can fully meet the driving needs of two small pure electric vehicles. The average daily driving mileage of each small electric vehicle is calculated as 50km. Compared with similar fuel powered vehicles, the fuel cost for fuel car driving 100km is calculated as 80 yuan, The cost of pure electric vehicles per kilometer is only 1 / 8 of that of fuel powered vehicles, and each electric vehicle only saves about 10000 yuan of fuel cost per year.

5 Conclusion

Through the implementation of the project in this paper, it can be realized: (a) there is no need to expand the capacity of the existing power grid, which reduces the capital investment and engineering quantity; (b) When the electric energy generated by the photovoltaic power generation system is not enough, it can take electricity from the power grid at the low ebb of the power grid, which plays the role of peak cutting and valley filling, so as to make effective use of the excess electric energy at the low ebb of the power grid. (c) the electric energy stored in the battery can quickly charge the electric vehicle, which reduces the charging time, does not affect the power quality of the power grid and poses a threat to the power grid. (d) The energy storage photovoltaic grid connected charging system adopts modular design and simple capacity expansion. It only needs to increase the battery pack without reconstruction. (e) The two-way energy storage design enables the charging station to complete both grid connection and independent operation. After the power grid outage, the charging station can also serve as a backup power supply system to temporarily supply power to local buildings.

References

1. Zahedi A. Maximizing solar PV energy penetration using energy storage technology[J]. Renewable and Sustainable Energy Reviews, 2011, 15 (1) : 866-870.
2. García-Villalobos J, Zamora I, San Martín J I, et al. Plug-in electric vehicles in electric distribution networks: A review of smart charging ap- proaches[J . Renewable and Sustainable Energy Reviews, 2014, 38: 717-731.
3. Zhao J, Wang C, Zhao B, et al. A Review of Active Management for Distri- bution Networks: Current Status and Future Development Trends[J]. Elec- tric Power Components and Systems, 2014, 42(3-4) :280-293.
4. Cheng L, Chang Y, Huang R. Mitigating Voltage Problem in Distribution System With Distributed Solar Generation Using Electric Vehicles[J]. IEEE Transactions on Sustainable Energy, 2017, 6(4) : 1475-1484.

The effect of global warming to California water cycle

Yi Yang*

Department of Civil and Environmental Engineering, University of California-Davis, California, United State

Abstract. As global warming, climate variability becomes more and more extreme, and the frequency and intensity of extreme weather events are also increasing. Climate change has changed the hydrological cycle, and its change trend has put forward a severe test for the management of world water resources. Climate change has effects on several components of the water cycle, including precipitation, evapotranspiration, runoff, snow and snow cover, glaciers, and frozen soil. This essay analyses the vulnerability of California's various systems to climate change. California's water cycle model is particularly vulnerable to global climate change. In the past hundred years of California, the average precipitation has decreased with fluctuation; the average temperature has increased; extreme weather phenomena have occurred frequently; the evaporation of the water surface has decreased gradually; the temporal and spatial distribution of runoff has changed greatly; and glaciers have accelerated retreat. California's hydrological cycle is consistent with global water cycle change but also exhibits more complicated regional characteristics, with large spatial differences among watersheds.

1 Introduction

Global temperatures have risen, and this trend will continue. The water cycle is driven by climate variability in terms of its temporal role and spatiality. Energy and water, as two active factors in the climate system, are controlled by the water cycle at the same time. The water cycle is a vital component of the earth's material cycle. It transfers water, energy, and agrochemical substances throughout the earth system, thereby forming a global continuous and orderly dynamic system [1]. The water cycle connects the sea and land, shaping the surface morphology and having a profound effect on the formation, evolution, and development of the earth's surface structure. The core of the water cycle is transferring the material and energy which maintains the balance and coordination of the ecosystem by continuously delivering renewable freshwater resources [2]. As a result, the water cycle plays a major role in the evolution of the earth's surface structure and human sustainability.

At present, the world is facing more and more challenges from water problems, which include floods, droughts, and lack of access to clean water supply. There is an urgent need to monitor and manage the water cycle system according to the trend of global climate change, to deal with the problems of water resources in excess, scarcity or excessive pollution aspects [3]. The average temperature, average precipitation, glacier number and runoff in California have changed greatly since the middle and late 1990s. Based on the analysis of meteorological change data, this essay discussed the change trend and causes of the water cycle in California under the trend of global warming.

2 California's climate

2.1 California meteorological conditions

California is a state on the west coast of the United States, surrounded by the Pacific Ocean and encompassing 424,000 square kilometers. In 2020, California has about 40.12 million people and is the most populous state in the United States. Its gross social product (GSP) is $2003 billion, accounting for 12.9% of the US gross domestic product (GDP). If it participates in the world ranking as a separate economic group, it is the ninth-largest economy in the world. The huge population and climate change also put great pressure on the environment of California. California's climate types are relatively diverse, according to geographic factors. To be more precise, the southern Sierra Nevada has a subtropical desert climate, the western coastal region has a Mediterranean climate, and the eastern Sierra Nevada has an alpine climate [4]. In recent summers, the Colorado Desert in the southeast has reached temperatures of over 54 °C, which is comparable to the world's highest temperature; in winter, the Nevada highlands are as frigid as the Arctic. However, the temperature near the Pacific coast is typically between 32 and 0 degrees Celsius. Los Angeles and San Francisco have average temperatures of 18 °C and 14 °C, accordingly. The northwest receives approximately 4420 mm of precipitation, the Colorado Desert receives between 50 and 75 mm, and the Central Valley receives between 200 and 500 mm. Therefore, climate change will have different effects on different regions of California. The complex terrain makes the fluctuation trend of

* Corresponding author: wxyyang@ucdavis.edu

https://doi.org/10.1051/e3sconf/202132901081

temperatures in California relatively dramatic, and the precipitation distribution becomes uneven.

2.2 Climate change in California

Climate change is already evident in some parts of California: the summer is getting hotter, the number of high-temperature days is increasing, the winter has become relatively warmer, the snow on the mountains melts earlier, and the sea level is rising. Climate change in California over the last century has been primarily reflected in temperature, precipitation, and extreme weather. In general, the surface temperature in California continues to rise, but there are obvious regional and seasonal variations; precipitation is declining significantly, and winter snowfall is significantly reduced. Additionally, severe precipitation and drought events are frequent. Extreme weather events such as high temperatures, heatwaves, flooding, and storms have increased in frequency and intensity.

2.2.1 Temperature change

According to the data from 1895 to 2020 released by the national climate data center of the National Oceanic and Atmospheric Administration, the yearly average temperature in California has increased by about 1.4 Fahrenheit in the past 126 years [5]. The data fluctuation is shown in Figure 1.

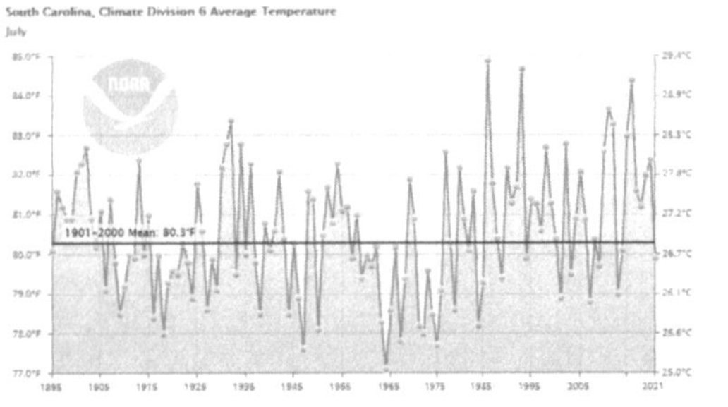

Figure 1. Annual average temperature and mean value in California from 1985 to 2020 (Department of Water Resources, 2020)

The annual hot weather may persist all summer due to the rise in temperatures. In 2100, for example, the number of hot days in Los Angeles might rise from 15 to 96 days. In recent years, Sacramento has experienced temperatures of 101 degrees Fahrenheit or greater four times per year on average, with up to 25 times possible. California's water supplies will be severely impacted by rising temperatures. As the largest reservoir in California, Sierra snow mountain may reduce its water storage by one third by 2060 and a half by 2090 [6].
The increase in temperature will make the snow melt faster and more difficult to store and use. The picture taken in 2005 showed the appearance of the Sierra snow mountain (Figure 2.). Early snow melting combined with heavy rainfall may exacerbate problems between water supply, flood control, and hydropower, leading to a longer and more serious drought, and that is a potential hazard in California.

Figure 2. Photograph of Sierra Nevada glaciers in 2005 (MacDonald, 2007)

2.2.2 Precipitation change.

Climate change has accelerated the evaporation of water vapor on the earth's and ocean's surface, accelerating the water cycle and altering global precipitation patterns. Figure 3 shows the annual average precipitation of California from 1895 to 2020 [5]. As the precipitation curve reveals, the average annual precipitation in California varies considerably but generally follows a steady declining trend. Rainfall, rather than snow, is predicted to increase in the future, as will short-term high precipitation, putting a strain on California's water supplies. Between 2007 and 2009, the drought exacerbated California's already-scarce water supply and wreaked havoc on the lives of Californians. Almost all areas of California were classified as extreme drought and normal drought in the autumn of 2009. Some areas were classified as severe drought in January 2009. On March 30, 2011, Governor Jerry Brown announced the end of the drought, but the drought continued until December 2011. According to the data from the California Department of water resources, the drought from 2011 to 2013 is the twelfth consecutive drought period in the state's measurement records. The drought began in 2010 and lasted until the autumn of 2013. In early 2015, climate monitoring showed that most areas of California's Central Valley were moderately dry from January 2015 to April 2015 [7]. Drought monitoring in August 2020 showed that more and more serious droughts occurred in eastern California.

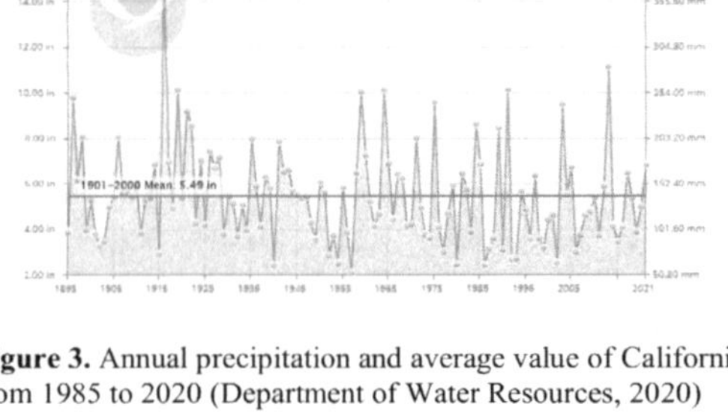

Figure 3. Annual precipitation and average value of California from 1985 to 2020 (Department of Water Resources, 2020)

2.2.3 Sea-level rise

In the past century, the sea level in the coastal area of California has increased by about 7 inches. It is expected that the sea level will increase by 8-12 inches in 2050, compared with 2020. If the sea level rises by 16 inches, the San Francisco Bay area will face the risk of flooding. Simultaneously, the likelihood of saltwater invading coastal groundwater and entering the Sacramento San Joaquin Delta grows as a result of sea level rise. Many islands in the Sacramento San Joaquin sink 0.2 inches per year due to factors such as compaction, dehydration, wind corrosion, and oxidation. If sea level rise continues at its current rate, this region will soon be entirely submerged.

2.2.4 Glacier melting

Sierra Nevada glaciers rank sixth among the world's top ten melting glaciers. Sierra Nevada glaciers include parts of the Liput glacier, which is 16 miles north of Sequoia National Park in Tulare County. Liput glacier is the southernmost glacier in the United States. For more than a century, most glaciers in the Sierra Nevada have been monitored and photographed. Although many of these glaciers were expanding in the 1980s, now all glaciers are moving towards a steady retreat. Most of the smaller glaciers in the Sierra Nevada are located in the north or on the slope facing northeast, and the bright California sunlight can't directly contact them. However, rising temperatures will eventually ring the death knell of these southernmost glaciers in the United States.

3. Causes of water cycle change in California

Generally speaking, the vulnerability of regional systems is generally divided into natural ecosystem vulnerability and human social system vulnerability. The complex terrain structure and diverse climate types make the ecosystem of California more sensitive to adverse effects such as increased temperature, abnormal precipitation and sea-level rise caused by climate change. This situation contributes to a vicious cycle of California's water cycle, Thus, the natural ecosystem and social life system are more fragile.

3.1. Evaporation acceleration

The hydro-logical cycle includes evaporation, water vapor transport, precipitation, infiltration, runoff and other links, in which evaporation is a very important part. Intuitively, global warming makes surface water evaporate rapidly [8]. When evaporation accelerates, precipitation will increase and the warming Earth will become wetter. However, regional climate differences will lead to regional differences in temperature and precipitation response. According to the principle of water balance, the increase in evaporation will reduce runoff. Therefore, unnecessary evaporation should be reduced as much as possible in areas with relatively scarce water resources. However, the evaporation of arid and humid areas in California increases at the same time. Due to the special terrain and the obstruction of some mountain peaks, the water vapor formed by evaporation can not realize a large-scale virtuous cycle.

3.2. Insufficient water supply

3.2.1. Precipitation reduction.

The water cycle is a dynamic and well-organized system. According to system analysis, each connection in the water cycle is both a component and a subsystem of the system. Each subsystem is connected to the others by a network of inputs and outputs. Precipitation is the output of the atmospheric subsystem. However, California has been in a drought state for a long time in recent years, and the precipitation has decreased year by year, resulting in insufficient water cycle output and supply.

3.2.2 Glacier recharge reduction.

California's surface runoff is highly dependent on glaciers (snow). With the warming of the climate and the intensification of extreme climate and hydrological events, the change of glacier water resources will be more complex. Rivers supplied by glacier melt water with relatively developed glaciers may have high runoff fluctuations for a long time. For those watersheds with small areas and numbers of glaciers, under the increase in temperature, glaciers will retreat and glacier water resources will decrease. There will be an inflection point in glacier melting. The amount of glacier melt will decrease sharply, the glacier regulation function will decline, and the variation rate of river runoff will increase due to the influence of abnormal precipitation.

It is worth noting that for those watersheds with small areas and quantities of glacier, due to the limited share of glacier melt water, the impact of glacier water resources change on runoff is limited. Besides, the increase of runoff in the process of temperature rise is not obvious. But the impact of the glacier melting inflection point on glacier shrinkage or glacier disappearance can not be ignored.

3.3 Reduction of vegetation coverage

California has more trouble adapting to the loss of forests because of its current state of drought and tree loss. When the plant life living on the Earth's surface is taken away, the Lithosphere has a difficult time absorbing and capturing rainfall and groundwater, creating runoff and erosion in the area. The removal of forests also drastically increases the amount of carbon dioxide released into the atmosphere and affects multiple of its components, such as the hydrosphere [9]. Due to the influence of surface cover, especially vegetation, the water cycle also shows latitude or seasonal differences. Global warming may also have a negative impact on the ecological effects of arid areas. Many recent research results show that the reduction in vegetation coverage should be related to the changes in temperature, precipitation, and evaporation in arid areas. Since 1965, the precipitation in the arid area of California has decreased, while the temperature and evaporation potential have increased continuously, resulting in the loss of a large amount of soil water. This will result in the death of some shallow-rooted desert plants with weak drought resistance that rely on groundwater and soil water. Species diversity and vegetation coverage will also be reduced [10]. The decrease in vegetation coverage will change the runoff, destroy the vegetation, and remove the cover on the soil surface. Weakening the interception of plants to precipitation, more precipitation will form surface runoff, and the underground runoff will be reduced. And vegetation transpiration is also an important source of water vapour in the atmosphere. Therefore, due to the reduction of vegetation coverage in arid areas, the water cycle efficiency is reduced. When the time comes, the devastation will spread throughout the region and if these problems are still apart from California's resume, then the resilience after the natural disaster will be more difficult unless significant change and progress are made [11].

4. Conclusion

Under the weak action of vapor sinks in the arid areas of California, the change of climate and hydro logical elements has an important impact on the formation and change of temporal and spatial water resources. Global warming not only increases the frequency and intensity of extreme climate hydrological events, but also intensifies the hydrological fluctuation and uncertainty of water resources in California. Each part of the water cycle is significantly affected by the climate. The changes in precipitation reduction and temperature rise caused by greenhouse effects will cause significant changes in runoff, which will have a significant impact on the ecological and hydrological processes of California. The change of climate intensifies the instability of the system elements of the mountain water cycle. Under the influence of global warming, the water resource system based on mountain precipitation and glacier melt water supply in the majority of California's dry regions has become more vulnerable, significantly altering the state's natural water cycle process.

References

1. Hayhoe, K., Cayan, D., Field, C. B., Frumhoff, P. C., Maurer, E. P., Miller, N. L., ... & Verville, J. H. (2004). Emissions pathways, climate change, and impacts on California. Proceedings of the national academy of sciences, 101(34), 12422-12427.
2. Vicuna, S., Maurer, E. P., Joyce, B., Dracup, J. A., & Purkey, D. (2007). The sensitivity of California water resources to climate change scenarios 1. JAWRA Journal of the American Water Resources Association, 43(2), 482-498.
3. Dettinger, M. D. (2005). From climate-change spaghetti to climate-change distributions for 21st-century California. San Francisco Estuary and Watershed Science, 3(1).
4. Stine, S. (1994), Extreme and persistent drought in California and Patagonia during mediaeval time, Nature, 369(6481), 546– 549.
5. Department of Water Resources (2020), The 1985-2020 California Drought: A review, California Department of Water Resources, September 2020.
6. MacDonald, G. M. (2007), Severe and sustained drought in southern California and the West: Present conditions and insights from the past on causes and impacts, Quat. Int., 173, 87– 100.
7. Engineers take on drought in California. (2015, June 09). Retrieved March 13,2021,from https://magazine.engineer jobs.com/2015/engineers take on drought in California. Htm
8. Anderson, B. T. (2003), Tropical Pacific sea-surface temperatures and preceding sea level pressure anomalies in the subtropical North Pacific, J. Geophys. Res., 108(D23), 4732, doi:10.1029/2003JD003805.
9. Hertogh, T. (2020, November 22). Photo by Tomas Hertogh on Unsplash. Retrieved March 13, 2021, from https://unsplash.com/photos/eS6rqvLx7o0
10. Fried, J. S., Torn, M. S., & Mills, E. (2004). The impact of climate change on wildfire severity: a regional forecast for northern California. Climatic change, 64(1), 169-191.
11. Mann, M. E., & Gleick, P. H. (2015). Climate change and California drought in the 21st century. Proceedings of the National Academy of Sciences, 112(13), 3858-3859.

https://doi.org/10.1051/e3sconf/202132901082

On the utilization and development of solar energy in Tibet

Wenya Wang [a]

Tibet Autonomous Region Energy Research Demonstration Center, Lasa 850000 China

Abstract: In order to better promote the application demonstration of clean energy such as solar energy, wind energy and geothermal energy in Tibet, scientifically develop and make rational use of new energy such as photovoltaic, photothermal, geothermal and wind power, optimize the energy supply structure, improve the guarantee capacity of energy use, and form a comprehensive energy system dominated by clean energy and complemented by other energy. This paper mainly discusses the importance, application and development of solar energy in Tibetan people's life by discussing the utilization of photoelectric and light and heat.

Keywords: Solar energy utilization; Photoelectric utilization; Photothermal utilization

1 Introduction

In 2021, the Arctic witnessed a high temperature of 48 °C, and the coldest Antarctic on earth measured an extreme high temperature of nearly 20 °C. In the past 50 years, the average temperature on the Antarctic Peninsula has increased by 3 °C, glaciers have melted and sea level has risen, A series of chain reactions caused by climate change cry to us that the greenhouse effect caused by a large amount of carbon dioxide emitted by the combustion of fossil fuels is gradually destroying the earth's ecological environment on which people depend. However, people's demand for energy is still growing, and a large number of fossil energy is about to be consumed. Renewable energy dominated by solar energy, wind energy, hydropower, biomass energy and tidal energy accounts for an increasing proportion in the energy consumption structure. In the utilization of clean energy, due to the inexhaustible and wide distribution of solar energy, its development and utilization are more and more favored by people [1]. In order to cope with the surge of greenhouse gas emissions, China has also put forward the goal of "achieving carbon peak and carbon neutralization", saving energy and reducing carbon, reducing pollution and reducing carbon. As the "third pole of the world", the ecological environment of the Qinghai Tibet Plateau has attracted much national attention. Tibet is located on the Qinghai Tibet Plateau, with an average altitude of more than 4000 meters. The atmosphere is thin and clean, with high transparency, low latitude, strong ultraviolet rays and long sunshine time. Nature has endowed these characteristics with convenient conditions for the development and utilization of solar energy resources.

According to relevant data, the main characteristics of the distribution of solar energy resources in China are: the high-value center and low-value center of solar energy are located in the area of 22 ~ 35 ° north latitude. The Qinghai Tibet Plateau is the middle and west of high-value. Therefore, it can be said that Tibet is the most abundant area of solar energy in China. At the same time, according to the data provided by China natural resources database (China Institute of Natural Geographical Sciences and resources), the solar energy resources in Lhasa, Tibet: the annual total solar radiation value is as high as 745561J / cm^2 (1998 data), which is very close to the theoretical solar radiation (solar constant = 1353kW / m^2) derived from solar energy, and the sunshine time is as long as 3007.8 hours, The sunshine rate over the years is as high as 69%, the sky is clear and less cloudy, the solar radiation acceptance is good, and it is stable all year round.

With the continuous improvement of people's living standards, the requirements for living and living environment are becoming higher and higher. To create a comfortable environment, it is inevitable to consume energy. With the development of society, the depletion of non renewable resources and the destruction of ecological environment have been caused, which makes the contradiction between global social development and environmental protection more and more prominent. In this context, a new energy revolution with the concept of energy conservation, environmental protection and sustainable development is rising all over the world. As a renewable and clean energy, solar energy has attracted more and more attention all over the world in the process of seeking the sustainable development of human society. China is rich in solar energy resources. According to statistics, the total amount of solar radiation received by China's land every year is equivalent to 2.4 × 1012 t standard coal, the annual average daily radiation in most areas is more than 4 kWh / (m^2 · d), and the annual radiation in more than 2 / 3 areas in China is more than 5.02 × 106 kJ / m2, the annual sunshine hours are more

[a] 853969982@qq.com

than 2000 h, and even more than 3000 h in some areas in Northwest China. Therefore, the development and utilization of solar energy in China will create good social, environmental and economic benefits.

2 Distribution of solar energy resources in Tibet

Tibet is located in the southwest border of China. Its land area is second only to Xinjiang. The climate is generally cold and dry in the northwest and warm and humid in the southeast. In the distribution of solar energy resources in China, the total solar radiation in Western Tibet belongs to class I area [2], and the solar energy resources are the most abundant. The annual sunshine time is about 2600 ~ 3400h, and the total annual solar radiation is about 6680 ~ 8400MJ / m^2. The total solar radiation in southeastern Tibet belongs to class II area, with abundant solar energy resources, and the annual sunshine time is about 3000 ~ 3200h, The total annual solar radiation is about 5852 ~ 6680MJ / m^2. Lhasa, known as "sunshine city", has an annual average sunshine duration of about 3005h and an annual total solar radiation of about 8160MJ / m^2. On the whole, Tibet is extremely rich in solar energy resources.

3 Utilization of solar energy resources in Tibet

At first, Tibet used solar energy in a more traditional way, mostly burning cow dung, new firewood and other fires for cooking and heating, as well as building houses facing the north and the south. Since the 1950s, China began to develop and utilize solar energy and renewable energy. After the 1990s, the solar energy industry began to enter the stage of rapid development [3]. With the national laws and regulations on renewable energy and the encouragement and support policies for solar energy development, Tibet introduced advanced technology, talent training and assistance projects at home and abroad, Gradually improve the utilization of solar energy, which is mainly divided into photoelectric utilization and photothermal utilization.

3.1 Photoelectric utilization

Because of its special geographical location and climate environment, Tibet has unique advantages in power generation. Among them, the annual irradiance of the best inclined plane for photovoltaic power generation in the Middle East of Xigaze City, the northwest of Shannan City, the south central part of Lhasa city and the west of Ali is as high as 2500 kWh / m2, which is 1.13 times that of the best inclined plane for photovoltaic power generation in Tibet, and the amount of technology development is about 2 billion kW. The utilization of solar energy in Tibet is similar to that in most parts of China. It is mainly divided into photovoltaic and photothermal power. Farmers and herdsmen in remote villages still use biomass energy such as cow dung as the main combustion energy. The Institute of Geophysics of Tibet University has tested and prepared a dye-sensitized solar cell with a mixed natural pigment of walnut and Arnebia, which has an efficiency of 8.2% [4]. At present, it has not been put into production.

(a) Photovoltaic power generation

Tibet's photovoltaic power generation mainly uses silicon-based polysilicon photovoltaic modules as power generation units. Under the irradiation of sunlight, the internal semiconductor materials generate potential difference through photovoltaic effect, which converts light energy into electric energy and directly uses solar energy. According to the power demand, under the distribution of the controller, part of it is transmitted to the inverter, and the DC generated by the photovoltaic module is converted into AC under the action of the inverter for use by users; The excess power is stored in the battery. After the battery is full, the remaining power will be connected to the Internet. This often refers to the photovoltaic grid connected power generation system, which can be equipped with or without the battery. Independent photovoltaic systems are often used in remote villages in Naqu and Ali. Due to the high average altitude compared with other cities, the early construction is extremely sparsely populated and pastoral areas, and the power supply cost of the State Grid is very high. The independent photovoltaic system is generally installed on the user's side, and the generated electric energy is not connected to the Internet. Part of it is used by the user, and the rest is stored in the battery. However, households in pastoral areas often like to configure the user's system, and the power generation basically ensures the daily power consumption of the user's family. It is small and easy to carry and install.

Some enterprises and institutions have built and installed monocrystalline silicon solar panels and amorphous silicon thin film solar cells for research and demonstration. Due to the relatively poor cost performance and other factors, they have not been promoted on a large scale in Tibet. Individual machines have built photovoltaic smart microgrid and Microgrid platform for research and analysis.

In 1990, the first photovoltaic power station in Tibet (with an installed capacity of 10kWp and an independent photovoltaic power station, which has been expanded, at least 60kWp at present) was built in Ali Geji county. Up to now, it still supplies power to the office building of the county government and the dormitory of the county Party committee, and the operation is in good condition. In 2011, Longyuan Tibet Yangbajing photovoltaic power station (installed capacity 10MWp), the first large-scale grid connected photovoltaic power station in Tibet, was successfully connected to the grid. In 2021, a 50MW "photovoltaic + energy storage" photovoltaic power station was built in Xigaze, Tibet, and successfully connected to the grid for power generation, which is the largest "photovoltaic + energy storage" investment project in Tibet.

(b) Photothermal power generation

Photothermal power generation is an indirect solar power generation method that converts light energy into heat energy and then into electric energy [5]. It mainly includes slot linear focus thermal power generation system,

Stirling butterfly thermal power generation system, tower solar thermal power generation system and Fresnel thermal power generation system. Disadvantages: low efficiency and high cost. At present, it can only be used in special occasions on a small scale, and large-scale utilization is economically uneconomical compared with other ways of power generation.

As the first solar thermal power generation project to be started in Tibet, the 50MW trough type solar thermal power station in Gongga County, Tibet is still under preparation as of 2018.

3.2 Photothermal utilization

The use of light and heat in Tibet is mostly used for daily heating and domestic water. The main utilization methods include the construction of solar collector, solar stove and solar house.

The solar heating system has the following characteristics: a) the solar collector can operate under pressure, the heat pipe is specially treated, and the normal operation of the system will not be affected when the glass pipe is damaged. The design of collector and support is reasonable, and can play the role of roof insulation layer. b) The constant temperature and pressure device is adopted to ensure the temperature and pressure of water at the water terminal, and the automatic circulating insulation device realizes the instant opening and heating of hot water. c) It is used together with the auxiliary heat source low-temperature heat pump, so that the heating system is not affected by season, day and night and weather. When the system cannot meet the heat energy needs of users in cloudy and rainy days and at night, the insufficient part is provided by auxiliary energy to ensure the normal operation of the heating system in the absence of solar radiation. d) Energy saving and consumption reduction. Building energy consumption accounts for nearly 1 / 3 of the total social terminal energy consumption in China. The perfect combination of solar heating system and buildings can greatly reduce conventional energy consumption and promote the development of society towards "low input, low energy consumption, low pollution and high output". e) Green environmental protection. As a clean renewable energy, solar energy has no environmental pollution.

(a) Solar collector is a device that absorbs solar radiant energy and transfers the generated heat energy to heat transfer working medium [6]. It is the core component of water heater, mainly including vacuum tube, flat plate type, trough type collector and so on. Vacuum tube collectors are mostly distributed in domestic water of household users, and most of them use vacuum tube collectors for collective heating of urban households and institutions (companies, etc.); The second is flat plate collector, which is mostly used for government collective housing or scientific research analysis; Trough collectors exist on the research platforms of individual enterprises and institutions. Individual families use solar air conditioning.

(b) Solar cooker is a device for cooking food. It has the functions of boiling water, cooking, frying, frying and frying. It is easy to operate. It mostly exists in farmers and herdsmen's families, and some urban users are also using it to facilitate them to cook and boil water.

(c) Sunshine houses have basically spread to all regions and families in Tibet. Solar houses are divided into active and passive. Solar houses in Tibet are mostly corridors and courtyards with glass structure. Individual government enterprises and institutions use solar collectors to obtain solar heat and send it to indoor heating through heat distribution system for pilot research.

(d) The sun in Tibet is often directly used for drying dried meat and preserved fruits, and solar dryers are rarely used.

(e) The working principle of solar greenhouse is greenhouse effect, also known as artificial greenhouse, commonly known as "Greenhouse". It is distributed in Lhasa, Shigatse and Ali, and is mostly used to grow vegetables, melons and fruits.

4 Prospect of solar energy resources in Tibet

As a green and pollution-free renewable energy, although the energy density of solar energy is low and intermittent and random, it does not affect the development and utilization of solar energy in Tibet. After years of development and accumulation, the solar energy utilization technology tends to be more mature and perfect, the utilization methods are more diverse, and the application fields are more and more broad. According to the development and utilization of clean energy and the realization of the zero carbon goal during the 14th Five Year Plan period, Tibet will continue to increase investment in the solar energy industry. Energy storage and heat storage are important parts of the development of solar photovoltaic and photothermal respectively. With the technological innovation, the service life will gradually grow, the storage cost will continue to decline, and the phenomenon of solar light and heat abandonment will be slowly removed.

5 Conclusion

In the development and utilization of renewable energy in Tibet, the development and utilization of solar energy resources is only one of them. Relevant utilization technologies need to be improved and upgraded, and the development and utilization of other new energy resources is also developing rapidly. Promote the integration of energy resources, coupling multiple energy resources with complementary advantages, and optimize the structure of energy supply and consumption.

Acknowledgments

Study on the change of power consumption structure in xumai Township, Nimu County (XZ202101ZR0037G)

References

1. Yan Yunfei, Zhang Zhien, Zhang Zhang Zhang, Dai Changlin. Solar energy utilization technology and its application [J]. Journal of solar energy, 2012,33 (S1)
2. Luo Yunjun, he Zinian, Wang Changgui. Solar energy utilization technology (Second Edition) [M]. Beijing: Chemical Industry Press, 2013
3. Wu Juanxiu. Research on laws and regulations of solar energy development and utilization [J]. Chinese and foreign entrepreneurs, 2016, (23)
4. Nusan, tsogawangm, et al. Preparation of natural pigment dye-sensitized solar cells [J]. Materials guide, 2013,27 (S2)
5. Cai Shijie. Research status and development prospect of solar energy utilization technology [J]. China high tech, 2018, (21)
6. Chen Deming, Xu Gang. Overview of solar thermal utilization technology [J]. Physics, 2007, (11)

https://doi.org/10.1051/e3sconf/202132901083

Cause analysis of pipeline vibration in natural gas station

Jia Wu [1, 2], Shuai Kong [1, 2, *], Li Xia [1, 2] and Ping Tang [1, 2]

[1] Zhejiang Academy of Special Equipment Science, Hangzhou, China
[2] Key Laboratory of Special Equipment Safety Testing Technology of Zhejiang Province, Hangzhou, China

Abstract. Natural gas station is an important part of the whole natural gas pipeline transmission system, in which the pipeline vibration is a hidden danger that cannot be ignored for the safe operation of the station. In view of the severe vibration of the pipeline in a natural gas station of an enterprise, the vibration acceleration sensors are used to test the vibration of each measuring point under different working conditions, and then the three-dimensional finite element model of the pipeline is established by using numerical simulation software to study its structural characteristics and gas column characteristics. Through the analysis of field test results and simulation results, it is found that there are two main vibration frequencies in the pipeline: 116.75 Hz and 123.0 Hz. The former is caused by the resonance between the gas column frequency and the natural frequency in pipe B; The latter is due to the impact of the gas column force in pipe A on pipe B.

1 Introduction

In the process of natural gas transmission, the abnormal vibration of the pipeline in the natural gas station will loosen and rupture the connection parts of the pipeline accessories, ranging from leakage to explosion, resulting in serious accidents. Therefore, it is necessary to analyze the vibration of the pipeline system, find out the causes of the vibration, and take effective measures to reduce the vibration.

The research on pipeline vibration has been favored by technicians of scientific research institutes and major engineering companies. Its main research directions focus on two aspects: the analysis of dynamic characteristics of pipeline structure and the analysis of aerodynamic characteristics in pipe.

In the analysis of dynamic characteristics and dynamic response of pipeline structure, the characteristics of natural frequency, vibration amplitude and stress of pipeline structure have always been the focus of theoretical research by many scholars. Housner [1] replaced the pipe model with the beam model, and assumed to ignore the structural damping, gravity, fluid pressure and external tensile pressure, calculated the vibration equation of long straight pipe with equal wall thickness. However, this method is not applicable to calculate the natural frequency of complex pipeline system under space conditions. In this case, the methods often used to calculate the natural frequency are finite element method and transfer matrix method. Jia et al. [2] established the finite element model of variable cross-section liquid filled pipeline, analyzed the vibration frequency of the pipeline system, and judged whether the resonance frequency may occur through the analysis results. Sutar et al. [3] simulated the model of fluid transmission pipeline through finite element analysis, and studied the effects of end conditions on the natural frequency, static and fluid flow conditions of empty pipe. Irie et al. [4] referred to the Timshenko beam model and deduced the vibration and stability of the flow transmission pipeline by using transfer matrix method. Lesmez et al. [5] made a further innovation on the vibration transfer matrix method of spatial pipeline system, and gave priority to the variable separation method when deriving the characteristic matrix of pipeline. Li et al. [6] studied the vibration analysis of liquid pipeline system, and extended the frequency domain analysis of fluid-structure interaction from single pipe to multi-pipe system by using the transfer matrix method. Liu et al. [7] derived the transfer formulas of deflection, angle, bending and shear between two adjacent spans of multi span pipeline by using the matrix transfer method, and gave the natural frequency equation and vibration mode function between adjacent spans of multi span pipeline. Li et al. [8-10] established a 14 equation model to describe the fluid-structure coupling behavior of the pipeline, and then analyzed the effects of support, structural performance and fluid parameters on the dynamic response and natural frequency of the pipeline by using the transfer matrix method. Based on the transfer matrix method, Batura et al. [11] proposed an effective method for calculating the static and dynamic stress and residual strength of complex space pipeline considering the internal vibration of fluid and Poisson coupling. Guidara et al. [12] used the transfer matrix method and Laplace transform to calculate the natural frequencies of the system under different boundary conditions considering fluid-structure coupling.

In studying the gas dynamic characteristics of pipeline, the main means of scientific researchers is to calculate the natural frequency of gas column by analyzing the gas column parameters (such as gas pressure, gas density and

* Corresponding author: wujiachn@126.com

https://doi.org/10.1051/e3sconf/202132901083

gas flow velocity), and establish and improve the mathematical model. Kinsler et al. [13] deduced the plane wave theory through the study of gas in pipe and a series of specific assumptions. Mohri and Hayama [14] appropriately modified the plane wave theory by assuming that the damping is directly proportional to the square of the gas velocity, and expanded the application range of the plane wave theory to apply it to the resonance region. MacLaren et al. [15] considered the influencing factors of the resonance region, such as the average velocity and the heat exchange of the gas, and obtained the unsteady equations. With the help of computer technology, the equations can solve the results of gas pulsation with high precision. The experimental verification is relatively consistent, and the mathematical model of gas pulsation is further improved. Kriesels et al. [16] studied the flow characteristics in a pipe system with two opposite closed side branches of equal length. Tonon et al. [17] discussed the flow-induced pulsation in a resonant pipe network with closed branches, and proposed a single-mode model for predicting self-sustained oscillation. Okuyama et al. [18] conducted several air flow experiments at room temperature and pressure to study the effects of geometric parameters and flow parameters on general acoustic resonance. Xiao et al. [19] experimentally studied the effects of branch length and edge geometry on the acoustic resonance characteristics of closed side branches. With the in-depth research on the basic theory and experimental methods of pipeline vibration by scientific researchers and technicians, more and more pipeline vibration problems have been effectively solved.

In this paper, aiming at the problem of severe pipeline vibration in a natural gas station of an enterprise, the vibration acceleration sensor is used to test the variable condition vibration of each measuring point of the pipeline, and then combined with numerical simulation, the vibration data are analyzed to find out the causes of pipeline vibration.

2 Model and Method

2.1 Distribution of measuring points

The tested pipeline in the natural gas station are shown in Figure 1. Natural gas is transmitted from pipe A to pipe B through tee, and then transmitted to the downstream through buried pipeline. The nominal diameter of pipe A is 800 mm and the wall thickness is 16 mm. The nominal diameter of pipe B is 1000 mm and the wall thickness is 18.4 mm.

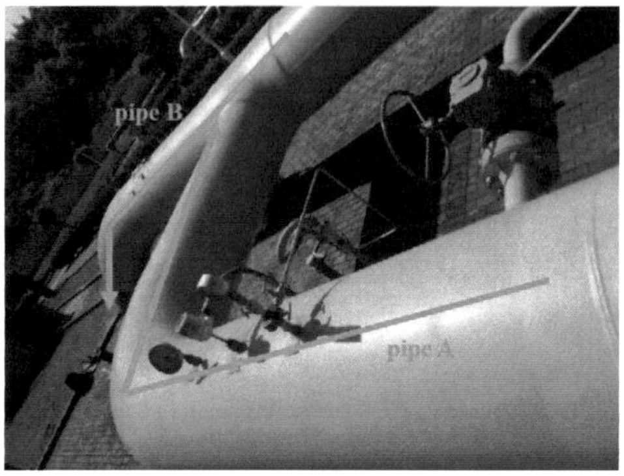

Figure 1. Tested pipeline in natural gas station

The vibration acceleration sensors are used to test the vibration of the pipeline. The measuring points are distributed on pipe A and pipe B. The measuring points on pipe A are measuring point 4 (MP4) and measuring point 5 (MP5) respectively, and the measuring points on pipe B are measuring point 1 (MP1), measuring point 2 (MP2) and measuring point 3 (MP3) respectively, as shown in Figure 2.

(a) MP1 and MP2

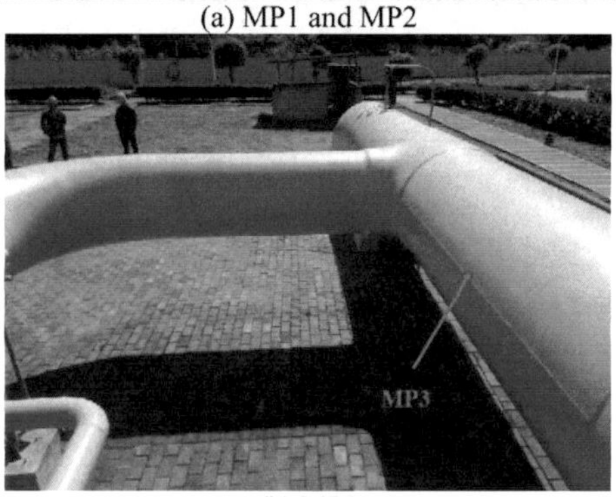

(b) MP3

(c) MP4

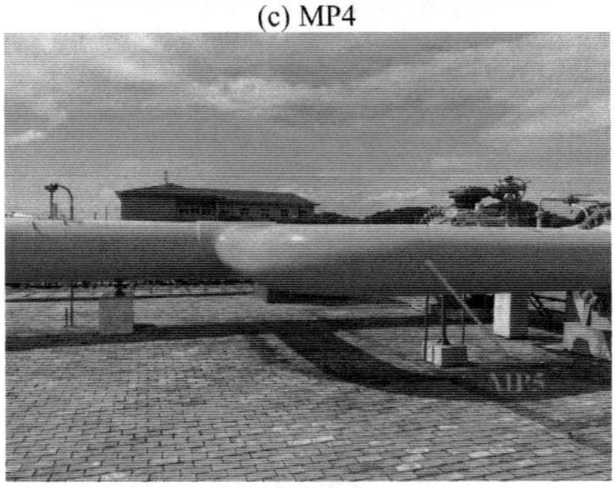

(d) MP5

Figure 2. Distribution of measuring points

2.2 Simulation model

The three-dimensional model of the pipeline is established by using ANSYS software, as shown in Figure 3. The unit type adopts three-dimensional sound unit, the medium in the pipe is natural gas, the density is 44.12 kg/m^3, and the sound velocity is 430.1 m/s. At the closed valve, the boundary condition is set as acoustic closure. At the pipe inlet, the boundary condition is set as acoustic opening.

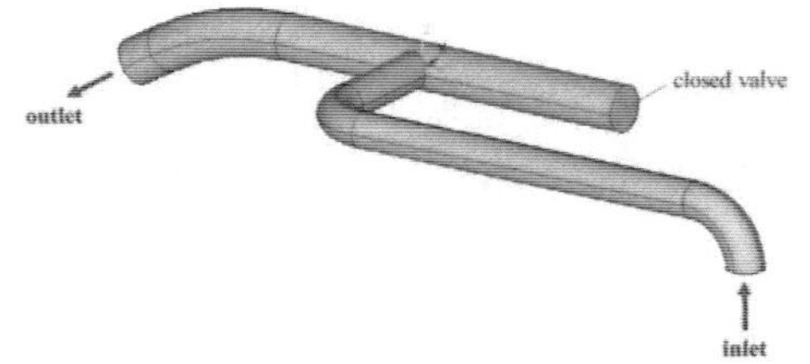

Figure 3. Three-dimensional model of pipeline

2.3 Simulation method

It is assumed that the gas in the pipeline is compressible and there is no average flow. According to the continuity equation and momentum equation, the acoustic wave equation [20] can be obtained:

$$\nabla\cdot\left(\frac{1}{\rho_0}\nabla p\right)-\frac{1}{\rho_0 c^2}\frac{\partial^2 p}{\partial t^2}+\nabla\cdot\left[\frac{4\mu}{3\rho_0}\nabla\left(\frac{1}{\rho_0 c^2}\frac{\partial p}{\partial t}\right)\right] = -\frac{\partial}{\partial t}\left(\frac{Q}{\rho_0}\right)+\nabla\cdot\left[\frac{4\mu}{3\rho_0}\nabla\left(\frac{Q}{\rho_0}\right)\right] \quad (1)$$

where, ρ_0 is the average fluid density, p is the acoustic pressure, c is the sound velocity in the fluid medium, t is the time, μ is the dynamic viscosity and Q is the mass source.

The finite element expression of acoustic wave equation is obtained by using Galerkin method and finite element approximate shape function:

$$\iiint_{\Omega_F}\frac{1}{\rho_0 c^2}\{N\}\{N\}^T dV\{\ddot{p}_e\}+\iiint_{\Omega_F}\frac{4\mu}{3\rho_0^2 c^2}[\nabla N]^T[\nabla N]dV\{\dot{p}_e\} +\iiint_{\Omega_F}\frac{1}{\rho_0}[\nabla N]^T[\nabla N]dV\{p_e\}+\oint_{\Gamma_F}\{N\}\{n\}^T\{N'\}^T ds\{\ddot{u}_{e,F}\} = \iiint_{\Omega_F}\frac{1}{\rho_0}\{N\}\{N\}^T dV\{\dot{q}\}+\iiint_{\Omega_F}\frac{4\mu}{3\rho_0^2}[\nabla N]^T[\nabla N]dV\{q\} \quad (2)$$

Where, $\{N\}$ is the element shape function of pressure, dV is the volume differential of sound domain Ω_F, $\{P_e\}$ is the node pressure vector, and ds is the surface differential of sound field boundary Γ_F, $\{n\}$ is the normal vector outside the fluid boundary, $\{N'\}$ is the element shape function of displacement, $\{u_e\}$ is the node displacement vector, $\{q\}$ is the node mass source vector, and $\{\dot{q}\}$ is the first derivative of the node mass source vector.

The above equation can be expressed by matrix to obtain the following discrete wave equation:

$$[M_F]\{\ddot{p}_e\}+[C_F]\{\dot{p}_e\}+[K_F]\{p_e\}+\bar{\rho}_0[R]^T\{\ddot{u}_{e,F}\}=\{f_F\} \quad (3)$$

were,

$[M_F]=\bar{\rho}_0\iiint_{\Omega_F}\frac{1}{\rho_0 c^2}\{N\}\{N\}^T dV$, is the mass matrix of acoustic fluid;

$[C_F]=\bar{\rho}_0\iiint_{\Omega_F}\frac{4\mu}{3\rho_0^2 c^2}[\nabla N]^T[\nabla N]dV$, is the damping matrix of acoustic fluid;

$[K_F]=\bar{\rho}_0\iiint_{\Omega_F}\frac{1}{\rho_0}[\nabla N]^T[\nabla N]dV$, is the stiffness matrix of acoustic fluid;

$[R]^T=\oint_{\Gamma_F}\{N\}\{n\}^T\{N'\}^T ds$, is the boundary matrix of acoustic fluid;

$\{f_F\}=\bar{\rho}_0\iiint_{\Omega_F}\frac{1}{\rho_0}\{N\}\{N\}^T dV\{\dot{q}\}+\bar{\rho}_0\iiint_{\Omega_F}\frac{4\mu}{3\rho_0^2}[\nabla N]^T[\nabla N]dV\{q\}$, is the load vector of acoustic fluid;

$\bar{\rho}_0$ is the mass density constant of acoustic fluid.

3 Measurement results

3.1 Natural frequency measurement

The natural frequency measurement is carried out by pulse excitation method. By analyzing the frequency spectrum of pulse response, the natural frequency of each pipe section can be obtained.

Figure 4 shows the vibration spectra of MP1 in the vertical direction without excitation and excitation, and Figure 5 shows the vibration spectra of MP1 in the horizontal direction without excitation and excitation. As can be seen from Figure 4, when not excited, the frequency of MP1 is mainly 116.0 Hz, while when excited, the frequency still exists, but the amplitude increases sharply. This shows that the natural frequency of MP1 in the vertical direction is 116.0 Hz. Similarly, it can be seen from Figure 5 that the natural frequency of MP1 in the horizontal direction is also 116.0 Hz.

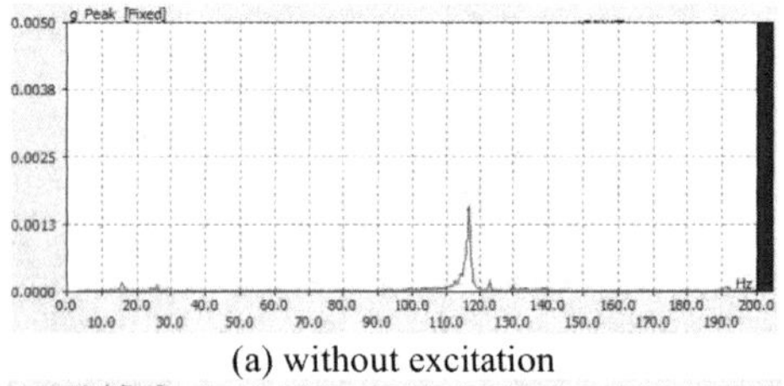

(a) without excitation

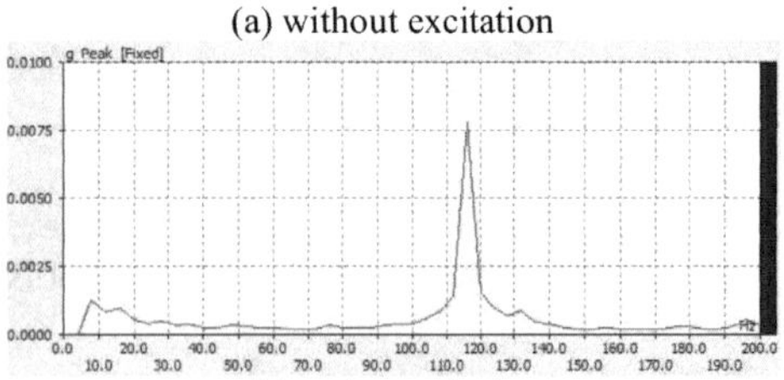

(b) under excitation

Figure 4. Vibration spectra of MP1 in vertical direction without excitation and excitation

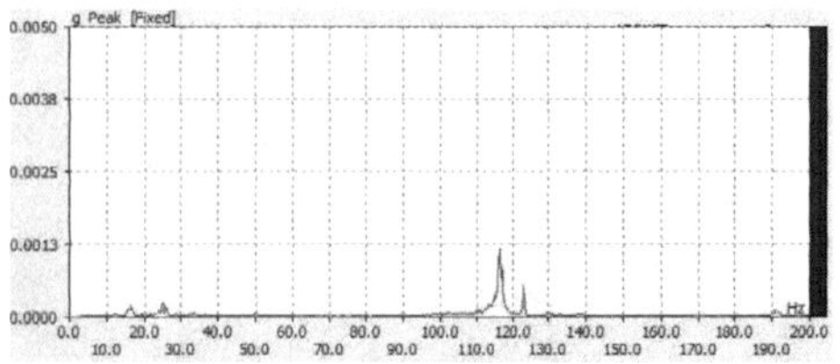

(a) without excitation

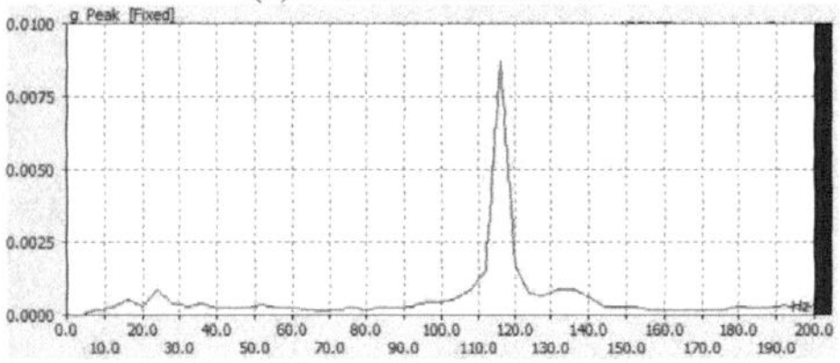

(b) under excitation

Figure 5. Vibration spectra of MP1 in horizontal direction without excitation and excitation

The discussion of other measuring points is similar to the above. In order to reduce redundancy, the vibration spectrum figures of other measuring points are no longer placed in this paper. Instead, Table 1 lists the natural frequencies of all measuring points in both vertical and horizontal directions. It should be noted that there is a vertical support near MP2 (see Figure 2(a)), so the natural frequency measured by pulse excitation method is not obvious.

Table 1. Natural frequencies of all measuring points in vertical and horizontal directions

	MP1	MP2	MP3	MP4	MP5
Vertical direction	116.0 Hz	—	198.0 Hz	284.0 Hz	272.0 Hz
Horizontal direction	116.0 Hz	200.0 Hz	198.0 Hz	284.0 Hz	288.0 Hz

3.2 Comparison between normal and vibration conditions

Figure 6 and 7 show the time domain waveforms and spectra of MP4 under normal and vibration conditions. It can be seen from the figures that under normal condition, the vibration is very small, while under vibration condition, the vibration increases sharply, and the vibration frequencies are 116.75 Hz and 123.0 Hz.

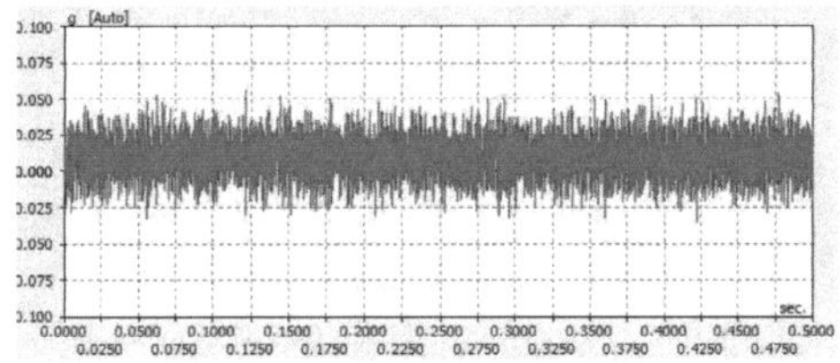

(a) time domain waveform

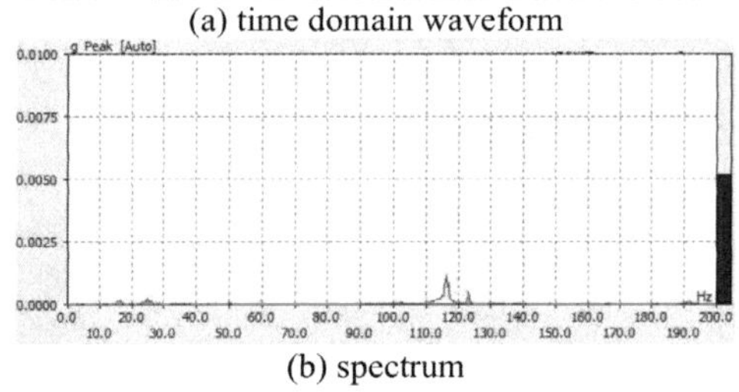

(b) spectrum

Figure 6. Time domain waveform and spectrum of MP4 under normal condition

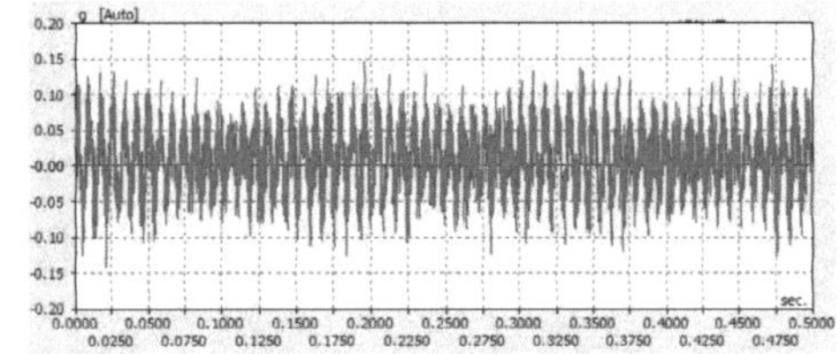

(a) time domain waveform

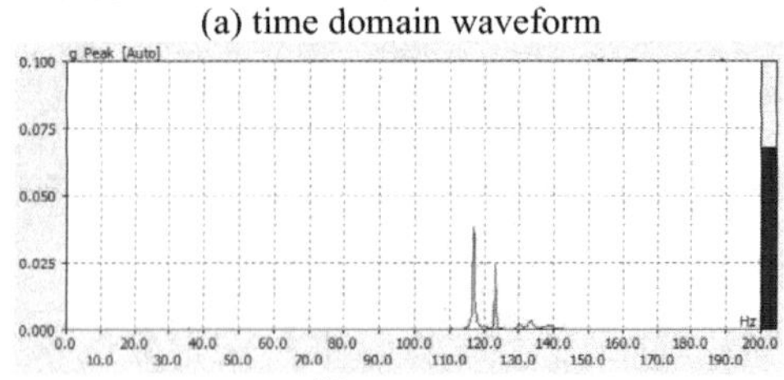

(b) spectrum

Figure 7. Time domain waveform and spectrum of MP4 under vibration condition

https://doi.org/10.1051/e3sconf/202132901083

The discussion of other measuring points is similar to the above. In order to avoid redundancy, the time domain waveforms and spectra of other measuring points under two working conditions are not placed in this paper. Instead, Table 2 lists the main vibration frequencies and peak-peak values of all measuring points under normal and vibration conditions. It can be seen from the table that the main vibration frequencies of each measuring point are about 116.75 Hz and 123.0 Hz under both normal and vibration conditions. Under the vibration condition, the vibration of each measuring point increases sharply (more than ten times larger than that under the normal condition). The maximum vibration is located at MP1, because the natural frequency of MP1 is 116.0 Hz, which may form resonance and cause severe vibration. In addition, the vibration component of 123.0 Hz is also large. In order to understand this phenomenon, it is necessary to conduct numerical simulation analysis of the pipeline.

Table 2. Main vibration frequencies and peak-peak values of all measuring points under two working conditions

Working condition	Measuring point	Direction	Main vibration frequencies	Peak-peak value
normal condition	MP1	vertical	116.75Hz, 123.0Hz	0.025g
		horizontal	116.5Hz, 123.75Hz	0.025g
	MP2	vertical	116.5Hz, 123.0Hz	0.021g
		horizontal	116.5Hz, 123.25Hz	0.025g
	MP3	vertical	116.5Hz, 123.5Hz	0.025g
		horizontal	116.5Hz, 123.25Hz	0.025g
	MP4	vertical	116.75Hz, 123.0Hz	0.031g
		horizontal	116.75Hz, 123.0Hz	0.031g
	MP5	vertical	116.5Hz, 123.5Hz	0.031g
		horizontal	116.5Hz, 123.25Hz	0.030g
vibration condition	MP1	vertical	116.75Hz, 123.25Hz	0.50g
		horizontal	117.0Hz, 123.25Hz	0.57g
	MP2	vertical	116.75Hz, 123.25Hz	0.27g
		horizontal	116.75Hz, 123.25Hz	0.38g
	MP3	vertical	116.75Hz, 123.0Hz	0.35g
		horizontal	116.75Hz, 123.25Hz	0.48g
	MP4	vertical	116.75Hz, 123.0Hz	0.37g
		horizontal	116.75Hz, 123.0Hz	0.17g
	MP5	vertical	116.75Hz, 123.0Hz	0.32g
		horizontal	116.75Hz, 123.0Hz	0.36g

4 Simulation results

Aiming at the problems of unknown vibration frequency and unknown vibration cause in the measurement results, the finite element model of the pipeline is established by using ANSYS software, the variation laws of different order vibration modes are obtained, the gas column frequency of the pipeline is calculated and compared with the excitation frequency, so as to effectively determine the cause of pipeline vibration.

Figure 8 shows the vibration modes of pipe A and pipe B obtained by simulation calculation. As can be seen from the figures, pipe B has both vertical and horizontal vibration, while pipe A is mainly horizontal vibration.

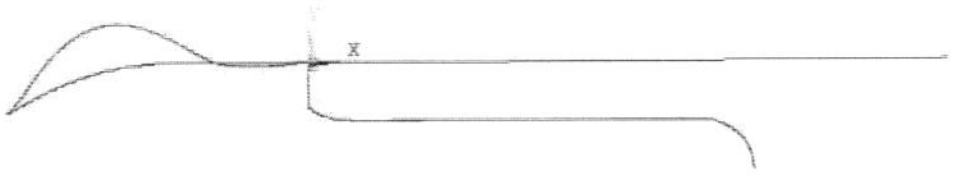

(a) vertical vibration

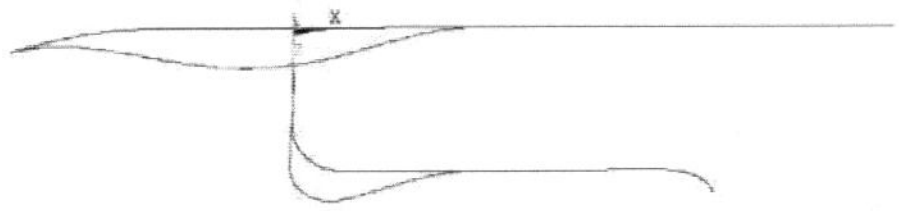

(b) horizontal vibration

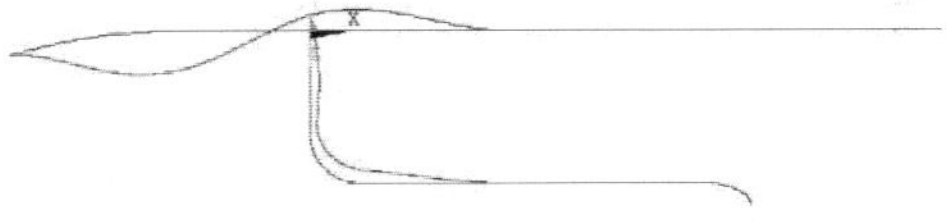

(c) horizontal vibration

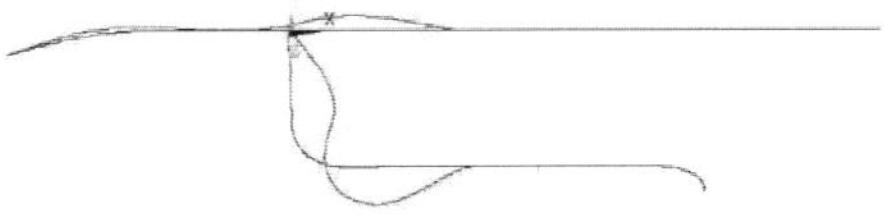

(d) horizontal vibration

Figure 8. Vibration modes of pipe A and pipe B

The contour plots of gas column are obtained through simulation calculation, as shown in Figure 9. In Figure 9(a), the gas column is mainly located on pipe B. It is generated due to the acoustic closure of pipe B. Its frequency is estimated to be 117.8Hz by simulation, which is close to the vibration frequency of 116.75 Hz. In Figure 9(b), the gas column is mainly located on pipe A, which will impact pipe B. The simulation estimates that the gas column frequency is 128.0 Hz, which is close to 123.0 Hz obtained from field measurement.

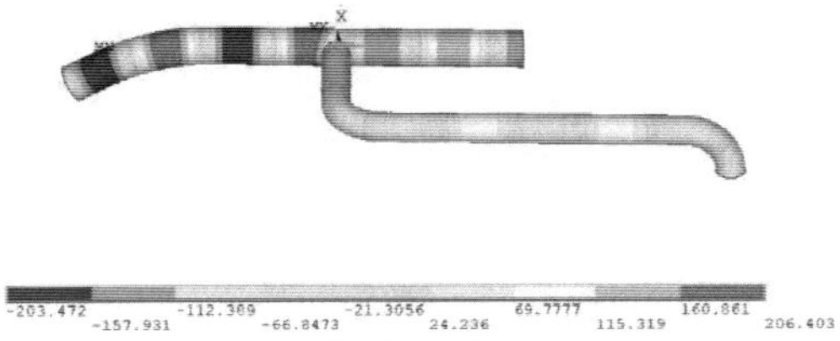

(a) 117.8 Hz

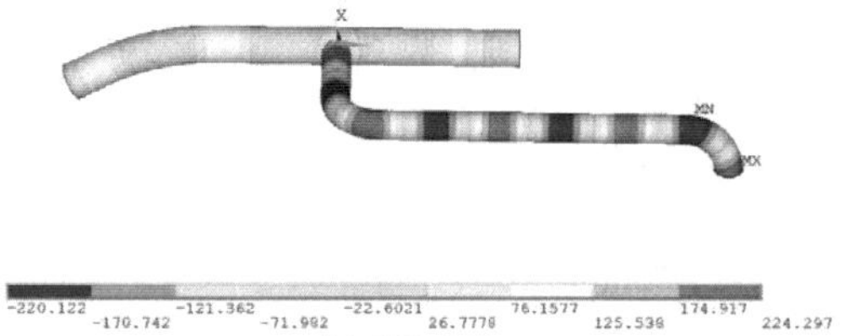

(b) 128.0 Hz

Figure 9. Contour plots of gas column at different frequencies

5 Conclusion

The field vibration measurement and simulation calculation show that there is a gas column excitation source in the pipeline. The gas column propagates outward along the pipeline and will arouse strong vibration when encountering the unreasonable pipeline. There are two main vibration frequencies: 116.75 Hz and 123.0 Hz. The former is caused by the resonance between the gas column frequency and the natural frequency in pipe B; The latter is due to the impact of the gas column force in pipe A on pipe B. In order to reduce the vibration of the pipeline, it is recommended to increase the stiffness of the buried pipe section in pipe B and change the length of the straight pipe section between the tee and the elbow in pipe A.

Acknowledgments

This work was financially supported by the science and technology project of Zhejiang Provincial Administration for Market Regulation (NO. 20210116).

References

1. Housner, G. W. (1952). Bending vibrations of a pipe line containing flowing fluid. Journal of Applied Mechanics, 19(2), 205-208.
2. Jia, Y., Madeira, R. E., & Just-Agosto, F. (2005). Finite element formulation and vibration frequency analysis of a fluid filled pipe. In ASME International Mechanical Engineering Congress and Exposition (Vol. 42282, pp. 127-132).
3. Sutar, S., Madabhushi, R., & Poosa, R. B. (2016). Finite element analysis of piping vibration with guided supports. Int. J. Mech. Eng. Autom, 3(3), 96-106.
4. Irie, T., Yamada, G., & Takahashi, I. (1980). Vibration and stability of a non-uniform Timoshenko beam subjected to a follower force. Journal of Sound and Vibration, 70(4), 503-512.
5. Lesmez, M. W., Wiggert, D. C., & Hatfield, F. J. (1990). Modal analysis of vibrations in liquid-filled piping systems. Journal of Fluids Engineering, 112(3), 311-318.
6. Li, Q. S., Yang, K., Zhang, L., & Zhang, N. (2002). Frequency domain analysis of fluid–structure interaction in liquid-filled pipe systems by transfer matrix method. International Journal of Mechanical Sciences, 44(10), 2067-2087.
7. Liu, J., He, X., Liu, Q., Naibin, J., & Chen, H. (2014). Vibration-modal analysis model for multi-span pipeline with different support conditions. Computer Modelling & New Technologies, 18(5), 14-18.
8. Li, S. J., Liu, G. M., & Kong, W. T. (2014). Vibration analysis of pipes conveying fluid by transfer matrix method. Nuclear Engineering and Design, 266, 78-88.
9. Li, S., Karney, B. W., & Liu, G. (2014). Application of transfer matrix method to dynamic analysis of pipes with FSI. In Pressure Vessels and Piping Conference (Vol. 46025, p. V005T11A005). American Society of Mechanical Engineers.
10. Wu, X. D., Liu, G. M., & Chen, H. (2012). The analysis of natural characteristics of pipeline structure systems based on frequency-domain transfer matrix method. Advanced Materials Research, 383-390, 4541-4545.
11. Batura, A., Novikov, A., Pashchenko, A., & Dubyk, Y. (2019). An application of the transfer matrix approach for a dynamic analysis of complex spatial pipelines. Nuclear Engineering and Design, 349, 174-182.
12. Guidara, M. A., Taieb, L. H., & Taïeb, E. H. (2015). Determination of Natural Frequencies in Piping Systems Using Transfer Matrix Method. In Design and Modeling of Mechanical Systems-II (pp. 765-774). Springer, Cham.
13. Kinsler, L. E., Frey, A. R., Coppens, A. B., & Sanders, J. V. (2000). Fundamentals of acoustics. John wiley & sons.
14. Mohri, Y., & Hayama, S. (1987). Resonant Amplitudes of Pressure Pulsation in Pipelines: Calculations by the Equivalent Linearization Method of| q| q): Vibration, Control Engineering, Engineering for Industry. JSME International Journal, 30(262), 602-607.
15. MacLaren, J. F. T., Tramschek, A. B., Sanjines, A., & Pastrana, O. F. (1975). A comparison of numerical solutions of the unsteady flow equations applied to reciprocating compressor systems. Journal of mechanical engineering science, 17(5), 271-279.
16. Kriesels, P. C., Peters, M. C. A. M., Hirschberg, A., Wijnands, A. P. J., Iafrati, A., Riccardi, G., ... & Bruggeman, J. C. (1995). High amplitude vortex-induced pulsations in a gas transport system. Journal of Sound and Vibration, 184(2), 343-368.
17. Tonon, D., Hirschberg, A., Golliard, J., & Ziada, S. (2011). Aeroacoustics of pipe systems with closed branches. International Journal of Aeroacoustics, 10(2-3), 201-275.
18. Okuyama, K., Tamura, A., Takahashi, S., Ohtsuka, M., & Tsubaki, M. (2012). Flow-induced acoustic resonance at the mouth of one or two side branches. Nuclear engineering and design, 249, 154-158.
19. Xiao, Y., Zhao, W., Gu, H., & Gao, X. (2018). Effects of branch length and chamfer on flow-induced

acoustic resonance in closed side branches. Annals of Nuclear Energy, 121, 186-193.

20. ANSYS Inc. (2013). "ANSYS® Mechanical APDL, Release 15.0, Help System, Acoustics," ANSYS, Canonsburg, PA.

E3S Web of Conferences **329**, 01084 (2021) https://doi.org/10.1051/e3sconf/202132901084
GESD 2021

Study on a collecting structure for impact energy and its dynamic characteristics

Kai Zhang *, Yongyong Zhu and Qizheng Zhou

Dept. of Weaponry Engineering, Naval Univ. of Engineering, Wuhan, China

Abstract. The impact energy harvesting technology is studied in this paper, a piezoelectric harvesting structure with additional resonant structure including spring mass components is proposed, and the system dynamics model of the structure is studied. The influence of the structural parameters of spring mass block on the system output voltage is studied by using the model. The research results have certain application value for impact energy harvesting.

1 Introduction

The piezoelectric harvesting for impact energy is a hot research direction in the field of energy collection technology[1,2,3]. Because the impact character of short-time and instantaneous, which is different from the collection of continuous vibration energy, the energy conversion and storage time for the impact energy harvesting device is limited by the impact pulse width[4]. The intermittent pulse nature of impact response means that it contains rich frequencies. In contrast, The vibration is continuous, and the continuous supply of vibration energy makes the transducer and collection circuit have enough time for energy conversion, collection and storage. Therefore, collecting impact energy is more difficult than collecting vibration energy.

In the collection of single impact energy, the adverse effects of impact time and excitation frequency should be considered. Apparently, the process of energy conversion can be maintained by prolonging the action time of excitation force and changing the natural frequency of the whole vibration system, so as to optimize the system parameters to improve the energy collection efficiency. Therefore, a buffer device is designed in this paper, which can store the transient and impact energy in the form of displacement energy, and then release it slowly. The working principle of this structure is similar to that of the vibration absorber. It is mainly composed of hard springs and mass blocks. During vibration, it can be regarded as an additional resonator to change the resonance frequency of the energy collector.

Under the impact action, the structure converts the impact energy into the kinetic energy of the mass and the elastic potential energy of the spring, and uses the reciprocating motion of the spring to transform the single pulse impact load into multi pulse excitation. In addition, by changing the design parameters of the spring mass block, the resonator frequency is adjusted so that the resonance frequency of the resonator is close to that of the transducer, so as to achieve the purpose of multi-modal energy collection. Compared with the simple piezoelectric structure, the additional resonant structure also brings great flexibility to the design of energy collector.

2 Structural design

In order to realize the collection of impact energy, by fixing the spring-mass block assembly on the end cap of the piezoelectric transducer, in this way, the mass will displace when impacted, the impact energy will be transformed into the kinetic energy of the mass and the potential energy of the spring. In the reciprocating motion of the mass, the piezoelectric transducer intermittently outputs electric energy. The piezoelectric transducer adopts cymbal structure to improve the electromechanical conversion efficiency[5,6]. The structural diagram is shown in the figure1. In addition, changing the design parameters of spring-mass block can adjust the resonance state of energy harvester, which can realize multi-modal energy harvesting.

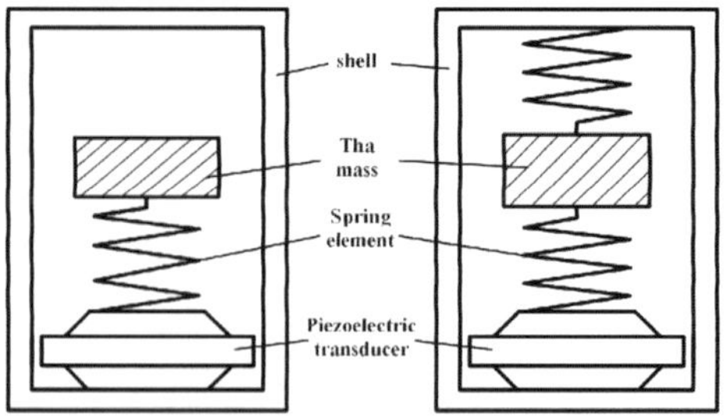

Figure 1. Design of impact energy harvester

* Corresponding author: zys88488@163.com

3 Dynamic model

The dynamic model of the impact energy harvester is shown in Figure 2. m_1、c_1、k_1 are the mass of mass block, spring stiffness and damping in the additional resonator respectively, m_2、c_2、k_2 are the equivalent mass, equivalent damping and piezoelectric structure stiffness of the piezoelectric transducer respectively, $x_1(t)$, $x_2(t)$ are the displacement of m_1 and m_2. Considering that the load is pure resistance R, during the working process, External excitation to the piezoelectric transducer under the action of the impact force $F(t)$, and the generated voltage is V. The piezoelectric plate of the piezoelectric transducer is axially polarized. Assuming that the axial stress is T_3, the piezoelectric equation is adopted

$$S_3 = s_{33}^E T_3 + d_{33} E_3 \tag{1}$$

$$D_3 = d_{33} T_3 + \varepsilon_{33}^T E_3 \tag{2}$$

Where, S_3 and T_3 are axial strain and axial stress respectively, E_3 and D_3 are axial electric field strength and electric displacement, ε_{33}^T is the dielectric constant in free state, s_{33}^E is the elastic compliance constant under constant electric field.

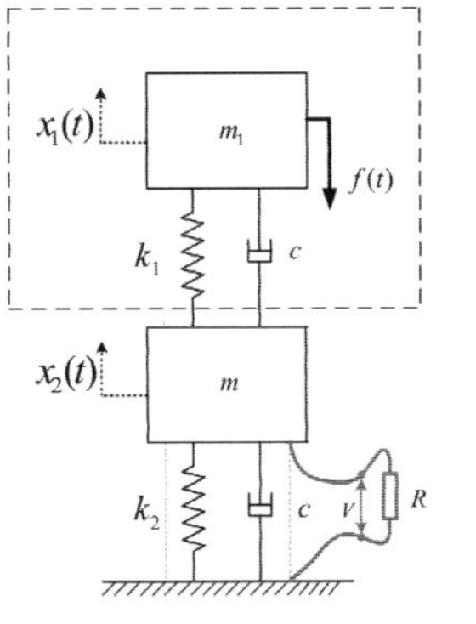

Figure 2. Model of piezoelectric impact energy harvester

Combined the equation (1) and equation (2), the electric displacement of piezoelectric vibrator is

$$D_3 = \frac{q}{A} = d_{33} c_{33}^E S_3 + \frac{\left(1-k_{33}^2\right)\varepsilon_{33}^T}{h_p} v$$

Further, the charge expression at both ends of the piezoelectric wafer is

$$q = A d_{33} c_{33}^E S_3 + C_p v$$

Where, A is the area of the piezoelectric disc, V is the voltage at both ends of the electrode, $c_{33}^E = 1/s_{33}^E$ is the Young's modulus, $C_p = (1-k_{33}^2)c_p$ is the clamping capacitance, $c_p = A\varepsilon_{33}^T/h_p$ is the non-clamping capacitance and k_{33} is the electromechanical coupling coefficient, h_p is the thickness of the piezoelectric ceramic sheet.

According to the dynamic analysis, the differential equation of motion of the system can be expressed as

$$\begin{bmatrix} m_1 & 0 & 0 \\ 0 & m_2 & 0 \\ 0 & 0 & 0 \end{bmatrix}\begin{bmatrix} \ddot{x}_1(t) \\ \ddot{x}_2(t) \\ \ddot{v}(t) \end{bmatrix} + \begin{bmatrix} c_1 & -c_1 & 0 \\ -c_1 & c_1+c_2 & 0 \\ 0 & d_{33}k_2 & C_p(1-k_{33}^2) \end{bmatrix}\begin{bmatrix} \dot{x}_1(t) \\ \dot{x}_2(t) \\ \dot{v}(t) \end{bmatrix} + \begin{bmatrix} k_1 & -k_1 & 0 \\ -k_1 & k_1+k_2 & d_{33}k_2 \\ & & 1/R \end{bmatrix}\begin{bmatrix} x_1(t) \\ x_2(t) \\ v(t) \end{bmatrix} = \begin{bmatrix} f(t) \\ 0 \\ 0 \end{bmatrix} \tag{3}$$

Assuming that the sinusoidal half wave pulse acts on the mass m_1, the expression of the pulse force is

$$f(t) = \begin{cases} f_0 \sin\omega_0 t & 0 \le t \le t_1 \\ 0 & t > t_1 \end{cases}$$

Where f_0 is the amplitude of the force and ω_0 is the angular frequency of the exciting force. Laplace transform equation (3) to obtain

$$(s^2 + 2\zeta_1\omega_1 s + \omega_1^2)X_1 - (2\zeta_1\omega_1 s + \omega_1^2)X_2 = L\left(a\sin(\omega_0 t)\right) \tag{4}$$

$$X_2 = -\frac{(C_p Rs + 1)}{Rk_2 d_{33} s} V \tag{5}$$

$$\left[s^2 + 2\zeta_2\omega_2 s + \omega_2^2 + \mu\left(2\zeta_1\omega_1 s + \omega_1^2\right)\right]X_2 - \mu\left(2\zeta_1\omega_1 s + \omega_1^2\right)X_1 + d_{33}\omega_2^2 V = 0 \tag{6}$$

Then, combined the equation(4), (5) and (6), the follow expression can be obtained.

$$V = -\left(2\zeta_1\omega_1 s + \omega_1^2\right)Rak_2 d_{33} s^2 \Big/ \left[\left(s^2+\omega_0^2\right)\left(a_5 s^5 + a_4 s^4 + a_3 s^3 + a_2 s^2 + a_1 s + a_0\right)\right] \tag{7}$$

Where,

$$a_5 = \mu C_p R$$

$$a_4 = \mu + C_p R\left(2\mu\zeta_2\omega_2 + 2(\mu+1)\zeta_1\omega_1\right)$$

$$= \mu R\left(C_p - k_2 d_{33}^2\right)\omega_2^2 + (\mu+1)C_p R\omega_1^2 + 4\mu RC_p\zeta_1\zeta_2\omega_1\omega_2$$

$$a_2 = 2\mu R\left[\left(C_p - k_2 d_{33}^2\right)\zeta_1\omega_1\omega_2^2 + C_p\zeta_2\omega_2\omega_1^2\right] + \mu\omega_2^2 + 4\mu RC_p\zeta_1\zeta_2\omega_1\omega_2 + (\mu+1)\omega_1^2$$

$$a_1 = \left(C_p - k_2 d_{33}^2\right)\mu R\omega_1^2\omega_2^2 + 2\mu\zeta_1\omega_1\omega_2^2 + 2\mu\zeta_2\omega_2\omega_1^2$$

$$a_0 = \mu\omega_1^2\omega_2^2$$

$\mu = m_2/m_1$, $\omega_1^2 = k_1/m_1$, $\omega_2^2 = k_2/m_2$, $\zeta_1 = c_1/2\sqrt{k_1 m_1}$, $\zeta_2 = c_2/2\sqrt{k_2 m_2}$.

When $t > t_1$, $F(t) = 0$. At that time, the system was affected by the initial displacement and system damping, and the system began to have damped free vibration. The system output voltage in the frequency domain can be calculated by equation (7). The influence of additional spring mass block components on the output voltage can be investigated by changing the spring stiffness, mass block mass, damping coefficient and other parameters.

4 Numerical calculation and analysis

According to the established dynamic model, the effects of the masses m_1, k_1 and c_1 in the additional resonator on the system output voltage are investigated. The main parameters in the simulation are selected as follows: PZT-5A（d_{33}=374 × 10^{-12}m/V, s_{33}^E =18.8 × 10^{-12} m^2/N,

ε_{33}^{T}=1700) , a=3000g, A=4.5×10^{-4} m^2, t_p=0.01m, R=1M Ω.

Firstly, the influence of spring stiffness on the output voltage is investigated. On the basis of not changing the structural parameters and material parameters of piezoelectric transducer, m_1 = 0.02 kg, c_2=0.05, R=1MΩ are selected, and k_1 are seted on 3500N/ m, 5000 N/ m, 10000 N/ m and 20000 N/ m respectively. The output voltage curve is obtained as shown in Figure.3. It can be seen from the figure that when the exciting force frequency $\omega_0 = 100 rad/s$, the first-order frequency does not change with the increase of spring stiffness, the second-order frequency gradually shifts to high frequency, and the second-order vibration peak gradually decreases. When $\omega_0 = 1000 rad/s$, the first-order frequency gradually shifts to high frequency, the second-order frequency does not change.

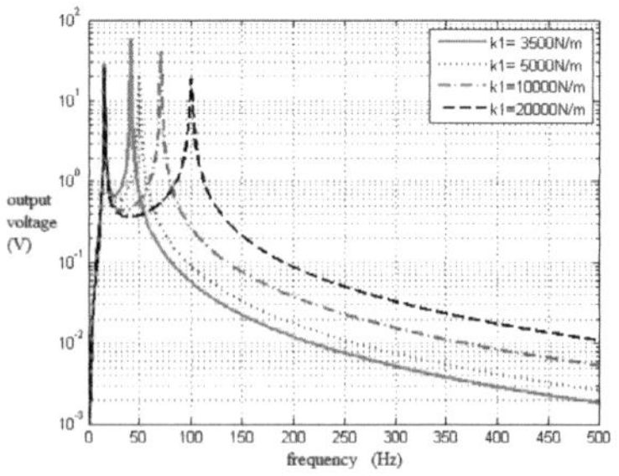

(a) $\omega_0 = 100 rad/s$

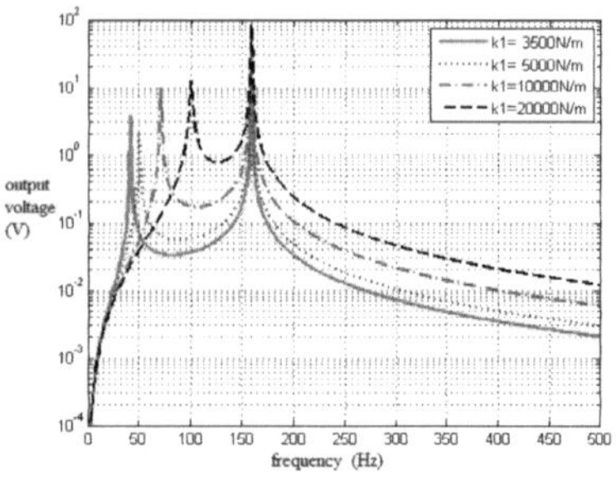

(b) $\omega_0 = 1000 rad/s$

Figure 3. Output voltage curve of different spring stiffness k1

The influence of mass on the system output voltage is investigated. Considering k_1 = 10000N/m, and 0.02kg, 0.05kg, 0.08kg and 0.15kg are selected for m_1 respectively. The corresponding output voltage curve is shown in Figure.4. when $\omega_0 = 100 rad/s$, the first-order resonance, frequency is 20Hz, and the second-order common vibration frequency decreases with the increase of m_1. When $\omega_0 = 1000 rad/s$, the second-order resonance frequency was about 165Hz, and the first-order resonance frequency increased with the increase of m_1.

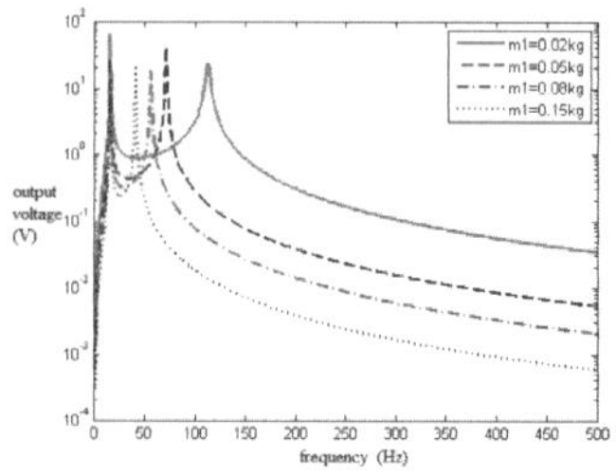

(a) $\omega_0 = 100 rad/s$

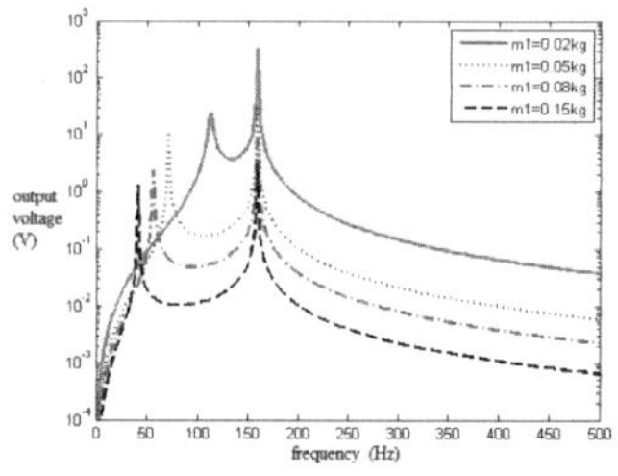

(b) $\omega_0 = 1000 rad/s$

Figure 4. Output voltage curve of different mass m_1

Similarly, considering k_1=10000N/m, and the damping coefficient c_1 are 0.1, 0.3, 0.5 and 0.8 respectively. As shown in Figure.5, with the increase of damping coefficient, the second-order resonance peak decreases gradually. When $\omega_0 = 100 rad/s$, the second-order peak voltage of the system corresponding to c_1 = 0.1 and c_1 = 0.5 is 50.73Vand 9.79V respectively. At this time, the second-order resonance frequency is 71Hz. When $\omega_0 = 1000 rad/s$, c_1 only affected the first-order resonance peak.

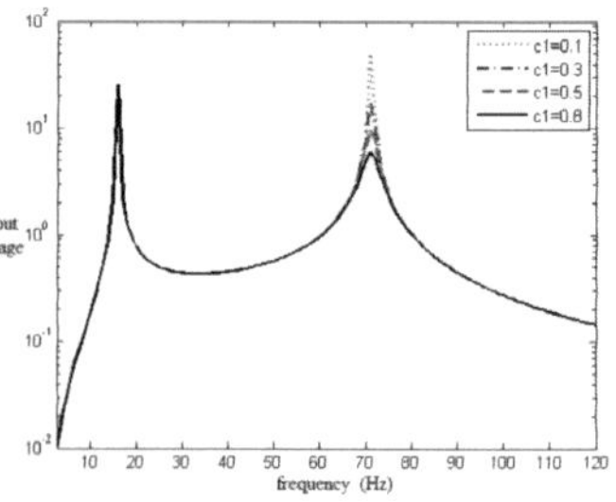

(a) $\omega_0 = 100 rad/s$

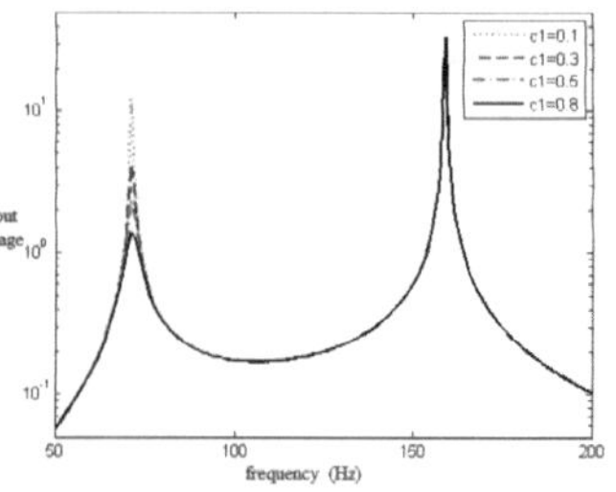

(b) $\omega_0 = 1000 rad/s$

Figure 5. Output voltage curve of different damper c_1

5 Conclusion

Based on the design of the impact energy harvester, the electromechanical coupling dynamic model of piezoelectric impact energy harvester with additional resonator is derived, the frequency domain expression of system output voltage is given, and the effects of the resonator structural parameters, excitation frequency and load resistance on the output piezoelectric performance of the system are discussed, The conclusions are as follows: the additional spring mass module changes the vibration characteristics of the piezoelectric energy capture system and makes the piezoelectric energy capture appear second-order resonance; Increasing the excitation frequency can shift the first-order resonant frequency to the second-order resonant frequency, and the first-order resonant peak also increases.

References

1. Ali Mohammed Abdal-Kadhim, Kok Swee Leong, Impact-driven energy harvesting:effect of stress on piezoelectric bender. J. Tel. Ele and Comp.Eng, 10(2018)87-92.
2. S. Ju, C.H. Ji, R.A. Lupton, Impact-based piezoelectric vibration energy harvester, Applied Energy. 214(2018) 139-151.
3. J. Caban, G. Litak. Impact-based piezoelectric vibration energy harvesting system excited from diesel engine suspension. Applied computer science, 16(2020)16-29.
4. E. Köhlera, L.G.H. Staafa. Impact of designed asymmetries on the effective bandwidth of a backfolded piezoelectric energy harvester.Sensors and Actuators A, 292 (2019) 77-89.
5. Yashwardhan Sahi and Swet Chandan, Finite element analysis of lead-free ceramic based cymbal transducer for energy harvesting, Integrated Ferroelectrics, 212(2020)81-88.
6. Z. Hongduo, Y. Jian, Finite element analysis of cymbal piezoelectric transducers for harvesting energy from asphalt pavement, J. Ceram. Soc. Jpn., 909(2010).118.

https://doi.org/10.1051/e3sconf/202132901085

Study on the law of gas desorption under the condition of water flooding of the abandoned block coal in goaf

Ran Wang [1,2,*], Wenbin Wu [1,2]

[1] State Key Laboratory of the Gas Disaster Detecting, Preventing and Emergency Controlling, Chongqing, 400037, China
[2] Chongqing Research Institute of China Coal Technology and Engineering Group Crop., Chongqing, 400037, China

Abstract. China's coal production has formed a large number of goaf, which contains a large amount of coal and coalbed methane resources. If these resources are stored in the goaf, it will not only cause a waste of resources, but also cause environmental pollution if the coalbed methane in the goaf leaks into the atmosphere. Therefore, the utilization of goaf resources is imminent. In order to achieve the purpose of efficient extraction, it is necessary to evaluate the amount of goaf resources. As an important part of goaf resources, the desorption law of abandoned coal is affected by many factors, among which water is an important factor affecting the desorption law of abandoned coal. At present, there are few studies on the desorption law of block coal under different watered-out volumes.

1 Introduction

In recent years, scholars at home and abroad have done a lot of experimental studies on the law of gas desorption of pulverized coal and lump coal. Zeng Shejiao, Nie Baisheng et al. [1-2] studied the law of gas desorption of pulverized coal with different particle sizes at different desorption temperatures. Wang Yibo et al. [3] conducted a corresponding study on the desorption law of coal samples at ultra-low temperature, and the desorption rate of coal samples at ultra-low temperature significantly decreased. Chen Xiangjun et al. [4] conducted gas desorption experiment for coal with 1mm-3mm particle size under water content, and found that with the gradual increase of water content, coal gas desorption speed gradually decreased. The studies by Yang Qiluan and Wang Zhaofeng et al. [5-6] show that there is a certain limit in the relationship between coal desorption and particle size. When the particle size of coal sample is smaller than the limit particle size, the attenuation coefficient of gas desorption strength increases with the decrease of particle size. Jiang Yongdong et al. [7] studied the desorption law of coal samples under different acoustic fields, and the gas desorption rate accelerated under the action of acoustic fields. Nie Baisheng et al. [8] found that electromagnetic field would promote coal sample gas desorption by adding electromagnetic field influencing factors in coal sample desorption. Wang Ran [9] studied the law of gas desorption of lump coal with different particle sizes under different temperatures and negative pressures.

Through the analysis of existing studies, it is found that the existing gas desorption experimental methods have the following shortcomings in the case of water flooding for the abandoned block coal in goaf:

(1) There are few experimental studies on the desorption law of lump coal, most of which are coal powder with small particle size. There are even fewer studies on the desorption law of lump coal under different volumes of water flooding, which cannot reflect the real situation of gas desorption of coal abandoned in goaf.

(2) Most of the existing desorption experimental devices are filled with coal powder for desorption. In the case of flooding of large coal samples, there are some problems such as inconvenient to inject water into the coal sample tank, imprecise control of adsorption equilibrium pressure and watered-out volume in the coal sample tank, which cannot meet the existing experimental requirements.

(3) The existing empirical formula of gas desorption cannot correctly reflect the calculation of the desorption amount of the left block coal in the case of goaf, which affects the accuracy of goaf resource assessment.

To sum up, it is necessary to trial-produce more suitable experimental devices and methods and deduce a more accurate empirical model of coal gas desorption rate on the basis of existing research and aiming at the desorption situation of coal abandoned in goaf in humid environment.

2 Trial manufacture of experimental equipment

The main purpose of the experimental device is to accurately control the adsorption equilibrium pressure and water flooded volume of block coal. So on the basis of existing coal sample tank, the tank increased the tray can lift device, the dry coal samples placed on the tray, smoking inside the tank to the negative state, open the water valve, under the influence of negative pressure, through measuring cylinder injection of water, water injection coal sample at this time not contact with water,

* Corresponding author: 2015082@cqccteg.com

after water injection, by adding mouth filling the gas, Make the coal sample reach the set adsorption equilibrium state, then lower the coal sample tray to soak the lump coal in water, and the soaking volume is controlled according to the experimental requirements. The desorption device is shown in Figure 1.

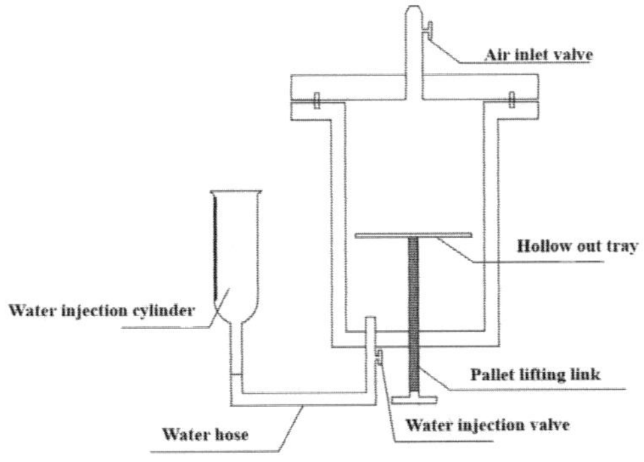

Figure. 1 water flooding gas desorption device of lump coal

3 Coal sample trial production and experimental parameter setting

3.1 Physical parameters of coal samples and trial production of experimental coal samples

The physical characteristics of experimental coal samples are as follows: true density is $1.4t/m^3$, porosity is 4.01%, solidity coefficient F value is 0.96, initial velocity of gas emission ΔP is 9.

The test coal sample is the original coal sample, which is cut into a cube of about 4cm4cm4cm. The trial-produced coal sample is shown in Figure 2:

Figure. 2 Schematic diagram of coal sample

3.2 Experimental Settings

The determination of about 60 groups of coal samples has been completed. Three groups of coal samples are taken as examples to simulate the desorption under the conditions of 1/3, 2/3 and unflooded coal samples under the conditions of 0.4mpa, 0.6mpa and 0.8mpa respectively. The soaking time of coal samples under the conditions of flooded is 48 hours, and the data of no. 1, 2 and 3 coal are shown in Table 1.

Table 1 Dimensional data of test coal samples

Coal sample No.	Coal sample Volume	Coal sample weight
1	$60cm^3$	83.6g
2	$72cm^3$	86.1g
3	$65cm^3$	85.1g

Before the test, the coal samples shall be dried and degassed for 12h at 70 ℃ to ensure the complete evaporation of water in the coal samples and the accuracy of the experiment. The 9 groups of 3 coal samples are shown in Table 2.

Table 2 Coal sample desorption test scheme

Test number	Coal sample number	Desorption temperature	Adsorption equilibrium pressure	Volume of soaking water
1	1	25°C	0.4MPa	Do not soak
2	1	25°C	0.4MPa	Submerge the coal body one third of the height
3	1	25°C	0.4MPa	Submerged coal two-thirds of the height
4	2	25°C	0.6MPa	Do not soak
5	2	25°C	0.6MPa	Submerge the coal body one third of the height
6	2	25°C	0.6MPa	Submerged coal two-thirds of the height
7	3	25°C	0.8MPa	Do not soak
8	3	25°C	0.8MPa	Submerge the coal body one third of the height
9	3	25°C	0.8MPa	Submerged coal two-thirds of the height

4 Analysis of experimental data

4.1 Analysis of coal sample desorption data

Taking the desorption data of three groups of coal samples as an example, the desorption rates of soaked coal samples of different volumes under three adsorption equilibrium pressures are shown in Figure 3~Figure 5. The desorption rates have been converted into the desorption rates per g of coal samples, and counting is stopped when the desorption rates are lower than 0.001L/min.

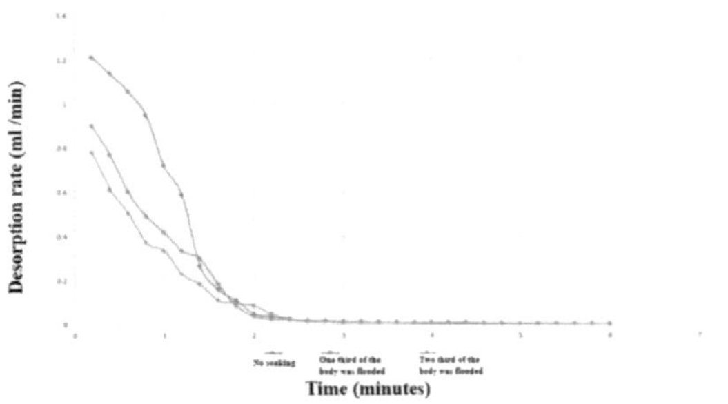

Figure. 3 Gas desorption rate curves of coal samples per unit mass at different flooded volumes under the adsorption equilibrium pressure of 0.4MPa

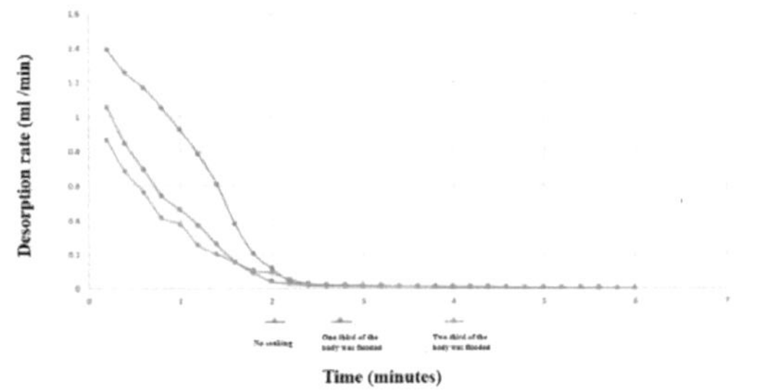

Figure. 4 Gas desorption rate curves of coal samples per unit mass at different flooded volumes under the adsorption equilibrium pressure of 0.6MPa

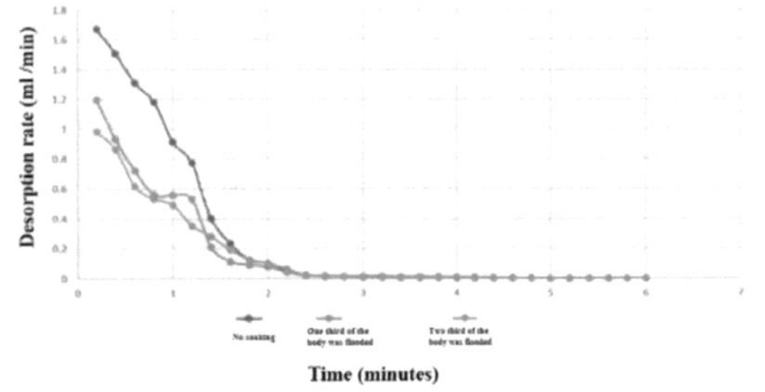

Figure. 5 Gas desorption rate curves of coal samples per unit mass at different flooded volumes under the adsorption equilibrium pressure of 0.8 MPa

It can be seen from the figure that the initial desorption rate of waterlogged coal decreases significantly. Under 0.4mpa adsorption equilibrium pressure, the initial desorption rate of waterlogged coal is 34% higher than that of soaked one-third coal, and 56% higher than that of soaked two-thirds coal. The difference in desorption rate between the two kinds of volumetric soaking coal samples is 22%. Under 0.6mpa adsorption equilibrium pressure, the initial desorption rate of waterless coal is 32% higher than that of immersed one-third coal, and 62% higher than that of immersed two-thirds coal, and the difference of desorption rate of coal samples under two volume immersion conditions is 30%. Under 0.8mpa adsorption equilibrium pressure, the initial desorption rate of waterless coal is 53% higher than that of immersed one-third coal, and 70% higher than that of immersed two-thirds coal, and the difference of desorption rate of coal samples under two volumetric soaking conditions is 17%. There is a significant difference between the gas desorption rate of the coal sample without waterlogging and that of the coal sample soaked in a certain volume, indicating that different volume waterlogging has a significant impact on the desorption rate of the coal sample with large particle size. However, the desorption rate of the coal sample decreases rapidly in all three cases. After two minutes, the desorption rate of the coal sample in the three cases tends to be flat until the counting stops. From the desorption data, we can see that under the adsorption equilibrium pressure of 0.4mpa, the cumulative gas desorption capacity of No. 1 coal sample is 51.2% higher than that under the condition of 1/3 flooding, and 78.6% higher than that under the condition of 2/3 flooding. Under the equilibrium pressure of 0.6mpa, the cumulative gas desorption capacity of No.2 coal sample is 57.3% and 82.5% higher than that of 2/3 under the condition of no flooding. Under the adsorption equilibrium pressure of 0.8mpa, the cumulative gas desorption capacity of No. 3 coal sample is 63.7% and 89.2% higher than that of no. 3 coal sample under the condition of 1/3 and 2/3 flooding.

To sum up, when the adsorption equilibrium pressure increases, the difference between the desorption capacity of unwaterlogged coal sample and that of waterlogged coal sample further increases. In the case of two volume waterlogging conditions, the difference between the cumulative desorption capacity of coal sample gas under three adsorption equilibrium pressures is 27.4%, 25.2% and 25.5%, respectively, which are relatively close. Therefore, by increasing the adsorption equilibrium pressure, and the difference of gas desorption capacity between waterlogged coal and unwaterlogged coal gradually increases, while the difference of gas desorption capacity under different waterlogged coal volumes is relatively small.

4.2 Derivation of empirical formula for desorption rateDrilling equipment

Derived from experimental data under different water volume of coal gas desorption empirical formula, at 0.6 MPa pressure consists of desorption of adsorption equilibrium time t (min), set the volume of water desorption rate influence coefficient for R, water volume factor for p (%), and coal desorption rate under different adsorption equilibrium pressure difference can be calculated by previous research, The formula of gas desorption rate V (mL /min) of coal body (g) with different waterlogged volume is:

$$V = \left[(1.38 - 0.60 \times t + 0.061 \times t^2)^2\right] \times R \quad (1)$$

By analyzing the difference of gas desorption rate of block coal under different soaking volumes and fitting the difference data, it can be deduced that the influence coefficient of water flooded volume on desorption rate is R and the calculation formula is as follows:

$$R = 1.05 - 0.06 \times p + 0.001 \times p^2 - 7.6 \times p^{2.5} + 0.048 \times p^{0.5} \times Ln(p) \quad (2)$$

4.3 Validation of empirical formulas

Three groups of new coal samples 1, 2 and 3 were selected to soak 15%, 45% and 75% of their volume respectively into water for desorption, and the empirical formula was first used to fit the desorption data and compare the calculation results and test results. The comparison results are shown in Table 3.

Table 3 Data table of empirical formula verification

Coal sample number	Cumulative desorption	Empirical formula fitting quantity	alignment
1	471.05ml	504.05ml	92.9%
2	369.23ml	421.62ml	85.8%
3	342.11ml	308.36ml	90.1%
		Average:	89.6%

5 Conclusion

(1) In order to verify the law of gas desorption of block coal under different water flooding conditions, an experimental device for gas desorption of block coal was trial-produced. The device could accurately control the adsorption equilibrium pressure, water flooding volume and time, which met the experimental requirements.

(2) on the coal sample processing multiple sets of desorption experiment was carried out at the same time, different volume water can be found through the experiment of big diameter coal desorption exists significant influence, with the increase of water volume, the desorption rate is reduced, at the same time improve the adsorption equilibrium pressure, water and water coal gas desorption amount gap will gradually increase, The difference of gas desorption capacity under different volume flooding is relatively small. The empirical formula of coal gas desorption with different watered-out volume unit and the influence coefficient of watered-out volume on desorption rate are deduced by experimental data.

(3) The empirical formula was verified, and the coincidence degree between the calculated data of the empirical formula and the measured data reached 89.6%, proving the accuracy of the empirical formula, which can well reflect the law of coal gas desorption under different water flooded volumes under experimental conditions.

Acknowledgments

This work was financially supported by the s Shanxi Province Science and Technology Major Special Project: Coal and CBM Co-production Theory and Key Technology System and Demonstration Project(No.20201102001).

References

1. Zeng Shejiao, Ma Dongmin, Wang Penggang. Journal of Xi 'an University of Science and Technology, 2009,29(4):449-453.
2. Nie Baisheng, Yang Tao, Li Xiangchun, et al. Journal of China University of Mining and Technology, 2013,42(6):975-981.
3. Wang Yibo, Li Hongtao, Qi Liming. Study on gas desorption law of coal body under low temperature [J]. China Coal,2011,37(5):103-104,107.
4. Chen Xiangjun, Cheng Yuanping, Wang Lin. Experimental study on inhibition effect of added water on gas desorption in coal [J]. Journal of Mining and Safety Engineering,2013, 30(2):296-301.
5. Yang Qiluan. Experimental study on the law of coal dust gas emission [J]. Safety in Coal Mines, 1986,18(2):9-17.
6. Wang Zhaofeng. Study on gas desorption law and application of coal in air, water and mud media [D]. Xuzhou: China University of Mining and Technology, 2001.
7. Jiang Yongdong, Xiong Ling, Yang Xingyang, et al. Journal of China Coal Society,2010,35(10):1649-1653.
8. Nie Baisheng, He Xueqiu, Wang Enyuan, et al. Study on the mechanism of the influence of electromagnetic field on methane adsorption in coal seams [J]. Natural Gas Industry, 2004,24(10):32-34.
9. Wang Ran. Experimental study on residual gas content and desorption law of coal mine goaf [D]. Chongqing: Chongqing Research Institute Co., LTD., China Coal Science & Engineering Group,2015.

Research and application of coal and gas outburst early warning system based on gas geological features

Yunlong Zou [1, 2, 3, *]

[1] State Key Laboratory of the Gas Disaster Detecting, Preventing and Emergency Controlling, Chongqing, 400037, China
[2] Chongqing Research Institute of China Coal Technology and Engineering Group Crop., Chongqing, 400037, China
[3] School of Resources and Safety Engineering, Chongqing University, Chongqing, 400044, China

Abstract. The geological structure of coal mines has always been a dangerous object of attention in coal mine outburst prevention work. In order to realize coal mine safety information management and early warning of gas disasters, comprehensive use of gas geological theory, coal mine disaster warning theory, computer information technology and other analysis methods, considering the influence of geological structure, coal seam occurrence parameters, and gas parameters, an early warning indicator system for identifying the risk of coal and gas outbursts reflecting the geological characteristics of gas has been constructed. The coal and gas outburst risk identification and early warning system is constructed using the principle of multi-index step-by-step identification and extreme value determination, and it is applied on-site in the 3303 Measure Lane in the East Shaft Area of Sihe Mine. The research results show that the constructed early warning system can provide accurate early warning for the area (belt) affected by the geological structure by 10m, and can provide accurate early warning of coal and gas outbursts based on the outburst signs of gas geology such as the thickness of soft layers and changes in coal seam thickness. This technology provides effective support for coal mines to effectively prevent gas disasters and ensure coal mine production safety.

1 Introduction

Coal and gas outburst disasters occur frequently, causing mass deaths and injuries [1-2]. There are many factors influencing the risk of coal and gas outbursts, and there are many indicators that can directly or indirectly reflect one or more aspects of the outburst risk of coal seams [3-4]. The overall highlight of the hazardous environment, and the factor that can reflect the outstanding hazardous environment is the gas geological environment. The gas geological environment (geological structural environment and gas pressure and other occurrence parameters) objectively determines the overall severity of the outburst hazard of the working face, and belongs to the grasp of the macroscopic primitive environment of the coal seam. The theory of gas geological zoning believes that: geological structure has obvious control effect on the outburst risk of coal seam, and geological structure controls the distribution of outburst risk by controlling the occurrence of gas, coal body structure type and tectonic stress. The statistical results of the outbursts show that more than 80% of outbursts occur near faults and other geological structures [5-6].

From the analysis of outburst mechanism, geological structure has three effects on gas outburst: (1) due to the influence of geological activities, the coal seam structure near the geological structure is often destroyed, most of which are accompanied by structural soft coal. The coal body has low strength and resists outburst damage. (2) there are often high residual tectonic stresses near geological structures, and the distribution is uneven, which increases the kinetic energy of outbursts. (3) some geological structural environments will create a closed environment and cause high gas pressure in the coal body. Therefore, geological structure has always been a dangerous object of great concern for outburst prevention work [7-10]. The establishment of a gas geological early warning system suitable for the specific conditions of mines can realize the information management of mine safety and the early warning of mine gas disasters. It will be one of the important technical means for coal mines to effectively prevent gas disasters and ensure coal mine production safety in the future.

2 Establishment of gas geological early warning indicator system

The coal seam in the area where the 3303 Measure Lane in the east shaft area of Sihe Mine is located is about 6.6m thick, the floor elevation is 610~630m, and the burial depth is about 440m. According to the gas geological map of the east zone, the original gas content of the coal seam in the area where the 3303 Measure Lane is located is about 14m^3/t. After the formation of the 3303 working face, bedding boreholes were constructed along the lanes 33033 and 33032 from both sides of the trench, and gas was pre-drained. Therefore, the gas content of the coal in the 3303 working face has been greatly reduced.

* Corresponding author: 2010152@cqccteg.com

Nevertheless, the gas gushing out during the 3303 working face was still large, and gas exceeding the limit occurred. It can be inferred that the gas content in this area is still high. In addition, the coal body in this area is affected by the thinning zone of the coal seam, and there is a large amount of soft coal. Therefore, the outburst risk of the coal seam in the area where the 3303 Measure Lane is located is relatively high.

The 3303 measure lane was excavated from 2010/9/25 to the end of 2010/10/30. The tunneling method was blasting, using bolt-net support. During the tunneling process, it was managed according to the outburst prevention face, and local comprehensive outburst prevention measures were taken. The 3303 measure lane is located at the 3303 working face of the Dongsanpan area. There is a coal thinning zone in front of the roadway. The purpose of this roadway is to provide a place for the construction of the 3303 working face to cross the coal thinning zone. According to the design, the 3303 Measure Alley is driven northward from Lane 33033 (54# Hengchuankou location). It first excavates 10.5m uphill at 16°, and then excavates 108m horizontally along the roof of the coal seam, as shown in Figure 1.

The gas geological early warning index system is established by mainly considering the geological structure, coal seam occurrence parameters, and gas parameters [11-12], as shown in Figure 2.

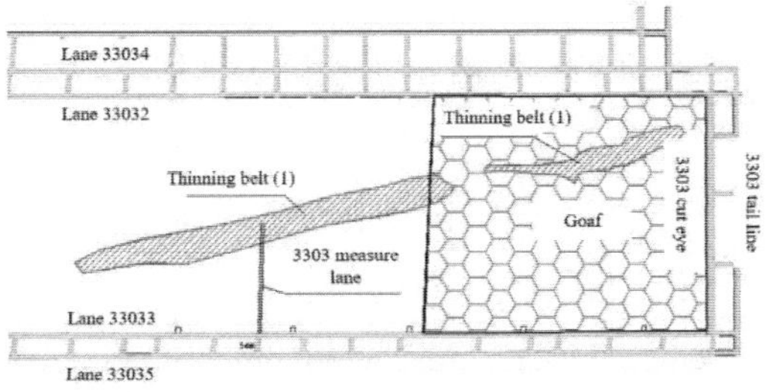

Figure 1. Layout of 3303 measure lane

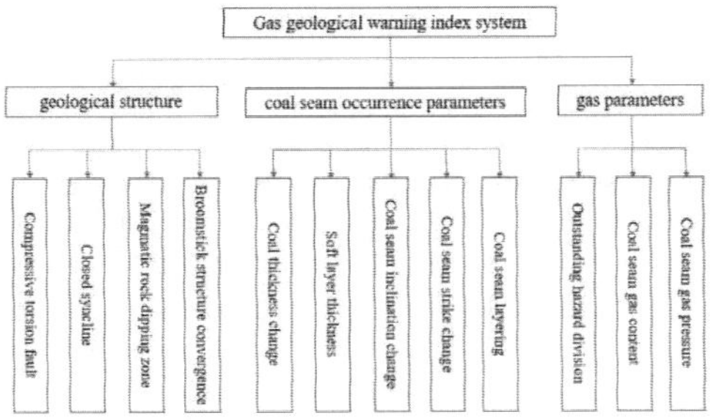

Figure 2. Gas geological early warning indicator system

3 Construction of risk identification model for coal and gas outburst

According to the implementation steps of highlight early warning, it is necessary to analyze, process and calculate the collected various hazard source information related to highlight to obtain various specific and quantitative highlight early warning indicators. On this basis, according to the police analysis model, the possibility, hazard and development trend of disasters are analyzed to determine the degree of disaster risk. Automated early warning needs to rely on computer technology to abstract the entire early warning realization process by computer, and also need to build a corresponding algorithm model. Therefore, according to the coal and gas outburst early warning process, the gas outburst early warning model based on the geological characteristics of gas mainly includes the following three parts:

(1) The comprehensive early warning model is used to analyze the warning situation and determine the warning level according to the warning indicators.

The development trend of outstanding danger is to judge whether the degree of outstanding danger on the working face tends to increase or decrease with the continuation of time or space according to the changes of monitoring indicators. For example, as the distance between the working face and the geological structure shrinks, the outburst hazard gradually becomes more serious, and as the working face enters the effective pressure relief area of the protective layer, the outburst hazard tends to be reduced or eliminated. Since the development trend generally belongs to the macroscopic identification, referring to the alertness expression methods in other fields, the alert level of the prominent trend is divided into three levels, which are represented by green, orange and red respectively. For this reason, the coal and gas outburst early warning system has been established, and its classification, grade and meaning are shown in Table 1.

Table.1 Coal and gas outburst early warning system and description

Type	Grade	Explanation
Status warning	Normal	Various indicators of the working face are normal and safe operation is possible.
	Threat	Prominent prediction of the working face is not dangerous or needs to be determined, but it needs to focus on and strengthen management.
	Danger	The working face has protruding danger, it is necessary to stop the operation and take anti-outburst measures or further confirm the protruding danger.
	Catastrophe	The working face has been prominent, and emergency response measures should be taken, power off and evacuation.
Trend warning	Green	The protruding danger ahead tends towards safety.
	Orange	There may be danger at a certain distance ahead, please call attention.
	Red	The prominent danger ahead tends to be serious, and we should focus on, strengthen management, and strengthen measures.

(2) The preprocessing model is used for the acquisition, prediction and calculation of various early warning indicators.

Early warning indicators are the basis for monitoring and early warning. In the established coal and gas outburst comprehensive early warning indicator system, some indicators can be directly obtained and used to determine the outburst hazard level, such as the amount of drill dust *S* in the daily forecast index, and the drilling cuttings gas desorption index K_1, $\triangle h_2$, etc. Some indicators need to establish a certain prediction model to obtain through calculations, such as gas pressure or gas content, and the scope of the division of prominent dangerous areas. Some indicators need to be calculated professionally, such as the range of influence of geological structure, the rate of change of coal seam thickness, and gas emission dynamic indicators. Some indicators belong to the qualitative category and need to be quantitatively processed through a certain model before they can be used, such as blowholes, top drills, etc., outburst coal seam damage types, and blank areas of outburst prevention measures. Therefore, building a series of specialized analysis models and pre-calculating and processing early warning indicators can achieve indicator acquisition and early warning.

(3) Method realization model, used for the realization of related algorithms of computer graphics processing.

The early warning system needs to dynamically monitor various hazards in the production process, and needs to quickly transmit and calculate a large amount of monitored data. It must rely on computer technology to realize it, especially the computer realization of some pre-processing models. Drawing and correction of gas pressure contours, delimitation of outlier dangerous areas, automatic identification of stress superimposition, automatic calculation of space distance, automatic correction of floor contours, drawing of coal seam depth based on floor contours and topographic maps value lines, etc., not only need to be calculated but also need to be visualized, but also need to involve the generation of isoline models and isoline conversion space surface models using the grid method in computer graphics processing.

4 Analysis of the application effect of the early warning model

4.1. Geological structure influence area (belt)

During the excavation of the 3303 Measure Lane, a total of 50 early warnings were carried out for the geological structure-influenced area (belt), and the early warning conditions are shown in Table 2. After investigation and analysis, the above-mentioned early warning results are all correct. The boundary of coal seam erosion zone in 3303 Measure Lane is located 81m away from the opening. In the process of roadway excavation, the change of soft coal layer thickness and daily prediction index with the roadway length is shown in Figure 3. It can be seen from the figure that within 40m from the coal seam erosion zone, the thickness of the soft layer and the daily predictive indicators are significantly different from other areas. In this area, soft coal stratification is common and thick (all larger than 0.5m, and the thickest part of the roadway is all soft coal). At the same time, the daily forecast index is too large, and some circulation forecast indexes exceed the standard (the maximum K1 is 0.62 mL/g.min$^{1/2}$, and the maximum S is 8.4kg/m). The above analysis fully shows: near the coal seam erosion zone, the outburst risk of coal seams is greater; and the influence range of the coal seam erosion zone in the Sihe mining area is determined to be 30m, and the outer 10m is used as a buffer area, and early warning control is very scientific, reasonable.

Table.2 Special warning statistics of gas geological and geological structure affected area (belt) in 3303 Measure Lane

Warning type	Warning result	Total number of times
Status warning	Normal	27
	Threat	23
	Danger	0
	Total	50
Trend warning	Green	26
	Orange	5
	Red	29
	Total	50

The 3303 Measure Lane reached a position 42m away from the opening after 2m of excavation on 2010/10/18/6. At this time, the working face is 39m away from the boundary of the coal erosion zone and has entered within 10m of the periphery of the erosion zone. The trend warning result of the corresponding warning system gas geology special warning is "orange". Afterwards, the working face continued to excavate and reached a position 52m away from the opening after 2m of excavation on the shift working face on 2010/10/20/6. The working face was 29m away from the boundary of the coal scour zone and entered the affected area of the scour zone. Correspondingly, the state warning result of the special warning of gas geology is upgraded to "threat", and the trend warning result is upgraded to "red". From then on to the shift of 2010/10/28/6, the 3303 measure road was dug in place (83m from the opening), and the working face has been within the erosion range of the coal seam thinning zone. The corresponding state warning result of the special warning result of gas geology is "threatening", and the trend warning result is "red".

 https://doi.org/10.1051/e3sconf/202132901086

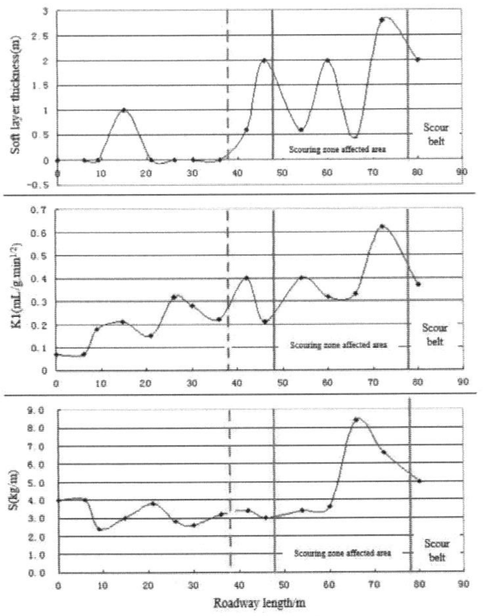

Figure 3. The daily forecast index and soft layer thickness of 3303 Measure Lane change with the length of the roadway

4.2. Prominent signs

Table 3 shows the statistics of the special warning results of gas geology under the condition of abnormal coal seam structure, coal seam occurrence, and soft layer thickness during the follow-up investigation. It can be seen from the table that the outburst signs of gas geology in Sihe Mine are mainly soft layer thickness and coal seam thickness changes. During the trial operation of the early warning system, when there are outliers such as abnormal soft layer thickness or drastic changes in coal seam thickness on the working face, the special warning of gas geology of the early warning system accurately gives the status warning results of the "threat" level. It can give early warnings to coal and gas outbursts based on the outburst signs of gas geology such as soft layer thickness and coal seam thickness changes.

Table.3 Special early warning statistics of gas geological signs and abnormalities in the follow-up investigation of the working face

No.	Working face name	Shift	Prominent signs	Warning result
1	3303 Measure Lane	2010/10/6/6 o'clock	1m soft coal	Threat
2		2010/10/18/18 o'clock	0.6m soft coal	Threat
3		2010/10/21/6 o'clock	0.6m soft coal	Threat
4		2010/10/23/0 o'clock	0.4m soft coal	Threat
5		2010/10/24/12 o'clock	2.8m soft coal	Threat
6		2010/10/28/0 o'clock	The thickness of the coal seam changes drastically	Threat

5 Conclusion

(1) Research and establish an early warning indicator system based on gas geological characteristics and three types of models required for early warning realization based on the model of highlighting the cause of the accident and the basic theory of highlighting prevention technology.

(2) The constructed gas geology special early warning system can provide accurate warnings of 10m ahead of the geological structure influence area (belt) of the 3303 Measure Alley, and at the same time, it can monitor the coal and gas outbursts according to the gas geological outbursts such as the thickness of soft layers and changes in coal seam thickness. The gas outburst made early warning reminders, which provided decision-making support for the coal mine on-site outburst prevention work.

Acknowledgments

The study was supported by the Special Fund Project for Safety Production in Chongqing (Coal Development) (2020B182 [5985] 20015501), Chongqing Research Institute Independence Project (202001120).

References

1. Zhang Chaolin, Wang Enyuan, Wang Yibo, et al. The temporal and spatial distribution of coal and gas outburst accidents in my country in the past 20 years and suggestions for prevention and control[J]. Coal Geology and Prospecting, 2021, 49(04): 134-141.
2. Tan Guowen. Highlight the construction of a diversified precision prevention and control system for mine gas disasters[J]. Mining Safety and Environmental Protection, 2020, 47(03): 49-53.
3. Zhang Qinghua, Lin Huiqin, Wang Qixiang, et al. Exploratory analysis of gas occurrence and distribution based on gas geological information platform[J]. Mining Safety and Environmental Protection, 2014, 41(05): 101-104.
4. Wei Zhonghua. Research on the influence of geological structure on the risk of coal and gas outburst[J]. Energy Technology and Management, 2020, 45(02): 31-32+78.
5. Li Junhu. Safety analysis of roadway driving through fault under deep mining conditions in coal mines[J]. Energy and Conservation, 2021(02): 43-45.
6. Lin Xiaobo, Guo Liquan. Comprehensive analysis and evaluation of the safety risk of tunneling through faults in deep mining of coal mines[J]. Coal Mine Safety, 2020, 51(03): 200-206.
7. Yao Yahu. Application of multi-parameter outburst warning technology based on gas geological characteristics in new coal mines[J]. Mining Safety and Environmental Protection, 2021, 48(03): 94-99.
8. Wen Hengcong, Li Xuechen, Liu Baobao, et al. Advanced detection system for geological structure

in high gas mining area[J]. Coal Mine Safety, 2021, 52(03): 165-169.

9. Qin Muguang. Application of gas geological dynamic analysis technology system in Yanghe Coal Industry[J]. Energy and Environmental Protection, 2020, 42(06): 69-72+76.
10. Pan Xin, Yu Baozhong. Research and application of multi-stage dynamic gas geological analysis system [J]. Energy and Environmental Protection, 2020, 42(04): 56-60.
11. Qin Muguang, Zhao Xusheng, Zhang Qinghua, et al. Research on outburst warning model based on gas geological characteristics[J]. Coal Science and Technology, 2018, 46(12): 138-144.
12. Zheng Lei, Jiang Qian. Research on the role of gas geology in the establishment of outburst warning index system[J]. Zhongzhou Coal, 2016(08): 65-68+109.

https://doi.org/10.1051/e3sconf/202132901087

Electric power emergency drill evaluation based on factor analysis

Xiaoguang Sun*, Jianyu Geng, Kai Liu, Yuanzhi Wang and Zhipeng Wang

The Changchun Power Supply Company of State Grid Jilin Electric Power Company, Changchun, Jilin, China

Abstract. This paper explores an evaluation method to evaluate the performance of participants in electric power emergency drill, expounds various important activities in electric power emergency drill, and summarizes the general mode of emergency drill. This paper introduces the principle of factor analysis and the principal component method, then uses this method to find common factors in a series of evaluation indexes, finally takes factor score as evaluation score directly. This comprehensive evaluation method can reduce the number of evaluation indexes by looking for commonness among many indexes.

1 Introduction

Electric power emergency drill is an important part of electric power emergency and has irreplaceable evaluation function. On the one hand, to evaluate the power grid's ability to deal with accidents, the drill will show the power grid's technology and methods to deal with power accidents. On the other hand, for the evaluation of power grid employees' ability to deal with accidents, the drill subject is the drill personnel, and the drill performance is the best prediction of power grid employees' ability to deal with real accidents.

Starting from the general drill mode, this paper introduces the electric power emergency drill evaluation. At present, many evaluation index systems have been established. Many evaluation indexes can make the evaluation results more comprehensive, but too many indexes may lead to too much information and complicated items.

2 Electric power emergency drill

Literature [1] introduces the main process of emergency drill in detail. With this as a reference, the key content of electric power emergency drill is summarized, and appropriate evaluation indexes are selected.

2.1 Key contents of the drill

2.1.1 News and communication.

This part of the drill is mainly aimed at the news center, the customer service center, the science and technology information department and other relevant departments, mainly covering the release of all kinds of information.

2.1.1.1 Warning information

The source of warning information for natural disasters can be meteorological departments or historical data. News department releases warning information in accordance with formal procedures, and warning information should contain at least the following two aspects [2].

(1) Notify the relevant personnel to enter the warning state, release warning information to the public, and take appropriate measures to try to eliminate the factors that produce warning.

(2) The operation personnel shall strengthen the operation duty, and the emergency repair personnel shall make preparations for various maintenance. In case of power grid failure, they shall quickly respond to reduce losses.

2.1.1.2 Troubleshooting information

The progress of handling accidents should be made public, including the blackout area and blackout time.

2.1.1.3 Accident summary information

Including accident causes, treatment methods and results as well as the reward and punishment information of relevant departments.

2.1.2 Plant and station operation.

Plant and station operation mainly includes collecting, reporting and related operations of fault information. Fault information collection covers the monitoring system background machine alarm, switch position, power flow indication, optical display plate, etc. In the whole process of accident handling, the operator of plant and station should report the accident to the dispatcher for several times, including two reports at the beginning of the occurrence of fault information (reflecting the operation status of the primary equipment and the action of the secondary equipment), and reports before and after the operation of the substation. Substation operation is the use of substation simulation system to simulate the operation of real scene.

* Corresponding author: 15601259722@163.com

2.1.3 Dispatching command.

Dispatching command and the plant and station operation are the two most important parts of electric power emergency drill. Especially in the climax of disaster accident, dispatching is very important. Dispatching is the brain of the drill subject and issues important instructions. The content to be drilled mainly includes the following two aspects.

2.1.3.1 Accident analysis

Generally, the dispatch personnel cannot visit the fault site in person. Therefore, they need to determine the fault location and severity according to the operation report.

2.1.3.2 Output control and load transfer

In the case of the outage of main transformer substation, or even the power outage of whole station, load transfer, power pulling, power limiting and other means become extremely important. Dispatching personnel need to give instructions to adjust the operation mode of power grid according to power grid structure and power flow distribution. This part of drill is based on the dispatcher training simulation system [3].

2.1.4 Maintenance.

This part of drill is mainly aimed at maintenance department and material department. The fault handling of maintenance department is different from the plant and station operation, and it is not substation operation, but the replacement and maintenance of equipment parts, so it is necessary for maintenance personnel to fully understand the structure of equipment and accurately judge the damage degree of equipment, which helps to accurately estimate the maintenance time of equipment.

As the coordination department of maintenance department, the drill personnel of material department need to master various reserve information of emergency materials, including location, type, item and call time [4].

2.2 General process of emergency drill

The emergency drill process was set according to each stage of disaster accidents to coordinate with the specific activities of emergency drill. The general process of emergency drill is shown in Table 1.

Table 1. General process of emergency drill.

Drill stage	Drill content	Corresponding disaster accident stage
Preparation	Meteorological information and historical data analysis Warning	Produce
Response	Fault information collection	Development
	Failure report Accident analysis Substation operation Material call and equipment repair Output control Load transfer	Climax
Recovery	Repair report Recovery operation Accident evaluation	Descending

3 Factor analysis

3.1 Overview of factor analysis

Principal component analysis (PCA) has been applied in many fields of electric power, and factor analysis is a kind of extension of PCA. The basic purpose of factor analysis is to describe the covariance relationship between many variables with a few random variables. These random variables are not observable and are usually called factors. In essence, the idea is to group variables according to the magnitude of correlation, so that the correlation between variables in the same group is high, and the correlation between variables in different groups is low. Due to the numerous items of evaluation indexes, although they look very detailed and comprehensive, the final conclusion should be drawn through induction and summary, which requires to explore the internal relationship between each index.

3.2 Principle of factor analysis

Factor analysis model assume that X linearly depends on a few unobservable random variables F1, F2, ···, Fm (are called common factors, where M < P), and ε1, ε2, ···, εp are called special factors or error. Its mathematical model is:

$$\begin{cases} X_1 - \mu_1 = l_{11}F_1 + l_{12}F_2 + \cdots + l_{1m}F_m + \varepsilon_1 \\ X_2 - \mu_2 = l_{21}F_1 + l_{22}F_2 + \cdots + l_{2m}F_m + \varepsilon_2 \\ \cdots \\ X_p - \mu_p = l_{p1}F_1 + l_{p2}F_2 + \cdots + l_{pm}F_m + \varepsilon_p \end{cases} \tag{1}$$

Where, l_{ij} is the load of the variable i on the factor j, which can be regarded as the effect of the common factor j on the variable i. Unobservable random variables F_1, F_2, …, F_m and ε_1, ε_2, …, ε_p are used to represent deviations X_1-μ_1, X_2-μ_2, …, X_p-μ_p.

At present, there are two most common parameter estimation methods to estimate factor load: the principal component method and the maximum likelihood method. In the principal component method, n groups of data are set, and each group of data includes 6 variables, namely 6 indexes X_1, X_2, X_3, X_4, X_5 and X_6. The original matrix can be obtained. Each element in the matrix is the evaluation result collected, and x_{ij} represents the value of the index j in the group i [5].

$$X = \begin{bmatrix} x_{11} & x_{12} & x_{13} & x_{14} & x_{15} & x_{16} \\ x_{21} & x_{22} & x_{23} & x_{24} & x_{25} & x_{26} \\ \vdots & \vdots & \vdots & \vdots & \vdots & \vdots \\ x_{n1} & x_{n2} & x_{n3} & x_{n4} & x_{n5} & x_{n6} \end{bmatrix} \tag{2}$$

The standardized data matrix is as follows:

$$Z = \left(z_{ij}\right)_n \tag{3}$$

$$z_{ij} = \left(x_{ij} - \overline{x_j}\right) / \sqrt{S_j} \tag{4}$$

Among them:

$$\overline{x_j} = \frac{1}{n}\sum_{i=1}^{n} x_{ij} \quad (5)$$

$$S_j = \frac{1}{n-1}\sum_{i=1}^{n}\left(x_{ij} - \overline{x_j}\right)^2 \quad (6)$$

Then the correlation matrix R is established, and its eigenvalues and eigenvectors are calculated:

$$R = Z^T Z / (N-1) \quad (7)$$

Find the eigenvalues of the matrix R: $\lambda_1 \geq \lambda_2 \geq ... \geq \lambda_6$, and the corresponding normalized orthogonal eigenvectors can be obtained:

$$e = \left(e_1, e_2, L, e_6\right)^T \quad (8)$$

The principal component expression is:

$$Y = Ze \quad (9)$$

Calculate the contribution rate of each common factor to sample variance:

$$\beta_i = \lambda_i / \sum_{i=1}^{6} \lambda_i \quad (10)$$

Cumulative contribution:

$$\beta(i) = \sum_{k=1}^{i} \lambda_k / \sum_{k=1}^{6} \lambda_k \quad (11)$$

According to the principal component method, except for the constant factor $\sqrt{\lambda_j}$, the load on the factor j is the coefficient e_j of the population principal component $_j$, and the factor load estimation can be obtained:

$$l_{ij} = \sqrt{\lambda_j e_{ij}} \quad (12)$$

Rotate the axes of the factor loads so that all points fall in the first quadrant. This is done to achieve an ideal situation where each variable has a large load on only one factor and smaller loads on the others.

4 Application of factor analysis in electric power emergency drill

The literature [6] list 12 indexes. Preliminary screening is carried out from these 12 indexes, and the principle of screening is: the acquisition of indexes is relatively direct and can be accompanied by the process of drill. And after the drill, the acquisition of various evaluation indexes can also be completed.

Combined with the specific assessment content, 6 indexes are selected from the 12 indexes, namely, safety production knowledge, safety operation reaction, safety operation process, safety attention, safety operation skill and safety production memory.

4.1 Extraction of drill evaluation factors

In the electric power emergency drill, several groups of drill evaluation data are found, then each index is similarity processed, and the factor analysis is carried out on the selected 6 indexes. Firstly, the correlation coefficient matrix of 6 variables was calculated, as shown in Table 2.

Table 2. Correlation coefficient matrix of 6 variables.

Variable	1	2	3	4	5	6
1	1.00	0.58	0.59	0.48	0.73	0.41
2	0.58	1.00	0.42	0.86	0.41	0.63
3	0.59	0.42	1.00	0.57	0.66	0.37
4	0.48	0.86	0.57	1.00	0.52	0.65
5	0.73	0.41	0.66	0.52	1.00	0.27
6	0.41	0.63	0.37	0.65	0.27	1.00

Note that the two data 0.86 and 0.73, two common factors can be taken according to the factor analysis model. The eigenvalue and principal component contribution rate are shown in Table 3.

Table 3. Eigenvalue and principal component contribution rate.

Eigenvalue	Value	Contribution rate/%	Cumulative contribution/%
1	3.74	62.28	62.28
2	1.06	17.61	79.89
3	0.46	7.69	87.58
4	0.42	6.96	94.54
5	0.26	4.38	98.92
6	0.06	1.08	100.00

This indicates that the contribution rate of two main components reaches 79.89%, which has important practical explanation.

After factor rotation, the rotation factor load is estimated.

4.2 Data analysis

As can be seen from Table 4, the indexes 1, 3, 5 have a large load on F_2, which is called "professional factor" by analyzing the commonness of all indexes. The indexes 2, 4, 6 have a large load on F_1, which is called "status factor". The score of professional factor reflects the level of professional accomplishment of drill personnel, which can be improved through acquired professional training. The status factor is associated with many factors, including the physical state, work attitude, adaptability, environment of the drill personnel.

Table 4. Factor load estimation after rotation.

Number	Variable	Factor load estimation after rotation	
		F1	F2
1	Safety production knowledge	0.33	0.81
2	Safety operation reaction	0.87	0.31
3	Safety operation process	0.28	0.79
4	Safety attention	0.84	0.40
5	Safety operation skill	0.16	0.91
6	Safety production memory	0.86	0.13

In the process of drill evaluation, using multiple indexes can make the evaluation mechanism more comprehensive and multi-directional. Inspired by the McClelland iceberg model, the paper tries to find a potential measurement method closer to self-concept from the surface phenomena reflected by indexes.

4.3 Factor score based on big data technology

In addition to the weighted least square method and the regression method, another factor score can be constructed by using a simpler method. The specific method is as follows: firstly, find out the variable with large load for each factor, and the factor score is equal to the algebraic sum of the standardized observation values of all variables with large load according to the load symbols.

This simple factor score has a high correlation with the factor score obtained by the least square method and the regression method. Although the evaluation method looks very simple, it has good summative and generalization.

5 Conclusion

Electric power emergency drill evaluation is the summary of emergency drill, the main content is to evaluate the emergency performance of drill personnel. By introducing the general process and key contents of the drill, the evaluation index system of the emergency drill is introduced, and some indexes in the evaluation index system are analyzed to explore their relevance and find out the common factors, so as to reduce the index dimension and keep the important information in the drill evaluation.

The evaluation methods before and after factor analysis are combined as common methods of emergency drill evaluation, so that the evaluation is comprehensive and detailed, and the key points are highlighted.

Acknowledgments

This work was financially supported by the Science and Technology Project of State Grid Jilin Electric Power Company (2020-4).

References

1. YUAN Guidong. The planning and implementation for the project of power public emergency full-scale exercise[D]. North China Electric Power University, 2009.
2. LI Yigang, YIN Guanghui, HUANG Jianfa, et al. Several key problems in emergency drill[J]. China Emergency Rescue, 2007(03): 33-35.
3. TAMURA Y, HORIE E D, et al. Development of power system simulator for research and education[J]. IEEE Trans on Power System, 1990, 5(2): 492-498.
4. HOU Hui, ZHOU Jianzhong, ZHANG Yongchuan, et al. Analysis of power emergency drills system at home and abroad and its inspiration for China[J]. Power System Protection and Control, 2010, 38(24): 236-241.
5. DAI Xinbo, CUI Yong, ZHOU Dexiang, et al. LS-SVM short-term load forecasting based on principal component analysis and improved particle swarm optimization [J]. Electrical Measurement & Instrumentation, 2012, 49(06): 5-9.
6. DONG Yu, ZHUANG Wei, TAO Hongzhu, et al. Organization and implementation of large-scale joint anti-accident training of power system[J]. Power System Technology, 2007, 31(02): 1-4.

Discussion on numerical simulation method of salt-resistant polymer flooding in three types of reservoirs

Jiqiang Chen

Geological Brigade of the Fourth Oil Production Plant of Daqing Oilfield Co., Ltd., Heilongjiang, Daqing 163511, China.

Abstract. According to the development characteristics of three types of reservoirs, this paper analyzes the main characteristics of salt-resistant polymer, and forms a highly targeted setting method for the key parameters of numerical simulation. At the same time, according to the actual situation of the salt-resistant test area, different injection parameter schemes are predicted, and the optimal design scheme is determined. Finally, the numerical simulation method of salt-resistant polymer flooding in three types of reservoirs is preliminarily formed.

Keywords: Three types of oil layers; Numerical simulation; Salt-resistant polymert

1 Introduction

Chemical flooding numerical simulation has many adjustment parameters, complex adjustment methods and difficult prediction. Salt-resistant polymer flooding in three types of reservoirs has the characteristics of poor reservoir development and special polymer, which further increases the difficulty of numerical simulation. Therefore, the numerical simulation method of salt-resistant polymer flooding in three types of reservoirs is discussed to ensure the prediction accuracy of development effect in the test area.

2 Set method for forming shear coefficient

In the numerical simulation of chemical flooding, the sensitivity of thick viscosity parameters is the strongest, which has a great influence on the water cut decline rate, the minimum water cut and the final fitting result, and the polymer will also be subjected to pore shear when migrating in the reservoir. The viscosity of polymer will lose a part, which must be considered in numerical simulation. The setting of shear coefficient is related to the rationality of thick viscosity parameters AP1, AP2 and AP3 (Formula 1), so the setting method of shear coefficient plays a vital role in the final prediction accuracy. The results of laboratory experiments show that the salt-resistant polymer has strong salt resistance, so the thick viscosity curve measured under the condition of high salinity brine should be selected when calculating the thick viscosity parameters (Figure 1); At the same time, the experimental results show that the shear resistance of salt-resistant polymer has no advantage compared with ordinary polymer (Table 1), so the setting of shear coefficient of this polymer can follow the law of other ordinary polymers.

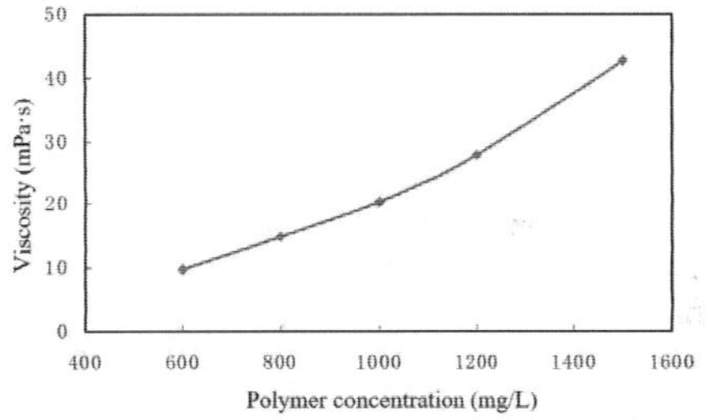

Figure 1 Thick viscosity curve of salt-resistant polymer

Table 1 Evaluation results of shear resistance of salt-resistant polymer

Polymer	Salt-resistant polymer		Ordinary 8 million	
	1000(mg/L)	1400(mg/L)	1000(mg/L)	1400(mg/L)
Before shearing(mPa·s)	17.6	32.6	11.7	19.5
After shearing,(mPa·s)	9.6	20.3	8.5	14.3
Viscosity retention rate(%)	54.5	62.3	72.9	73.3

The shear action of polymer in reservoir is mainly produced when polymer passes through reservoir pores.

* Corresponding author: chenjiqiang@nepu.edu.cn

The median particle size is the main parameter reflecting the pore structure. By studying the rule of shear coefficient and median particle size of the previous chemical flooding block tracking model, it is found that there is an approximate linear relationship between them (Figure 2). The reasonable initial value of shear coefficient can be accurately calculated by using this regression formula.

$$\mu_p^0 = \mu_w\left(1+\left(A_{P1}C_P + A_{P2}C_P^2 + A_{P3}C_P^3\right)C_{SEP}^{S_P}\right) \quad (1)$$

Among them: μ_p^0——Viscosity of polymer solution at zero shear rate;

C_P——Polymer concentration;

C_{SEP}——Effective salt content in aqueous phase;

S_P——Test constant;

μ_w——Water viscosity.

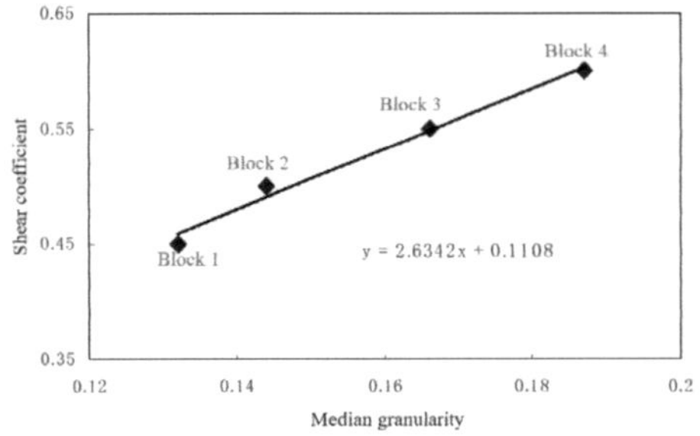

Figure 2 Relationship curve between median particle size and shear coefficient

3 Determine the fitting method of oil and water well measures

Compared with the main oil layer, the target layer in the salt-resistant test area is relatively poor in development (Table 2). With the deepening of polymer injection, the injection-production capacity of the test area will gradually decline, and the test area will inevitably improve the injection-production capacity by means of measures. It is necessary to simulate this phenomenon in the process of digital-analog prediction, so that the whole prediction process is closer to the actual production situation. Compared with the previous three types of oil layer blocks, the oil layer development status of the test area and block A is relatively close (Table 3 and Table 4); In the process of measure prediction, the oil layer development status in the test area is analyzed in depth, and combined with the actual oil layer production status of Block A (Table 5), 44 wells and 35 wells of oil-water well measures are simulated in the low water cut period and the water cut recovery period of the test area respectively.

Fracturing measures mainly improve the percolation capacity of the target layer by means of fracture-making. In numerical simulation, this phenomenon is mainly simulated by improving the conductivity of oil layer. Permeability is the main parameter reflecting the percolation capacity of oil layer. The permeability improvement multiple of the post-fracturing measures layer can be approximately defined as the conductivity improvement multiple. Different developed reservoirs have different permeability improvement before and after measures. If the improvement multiples of the measures horizon conductivity are assigned to an empirical coefficient, the measures phenomenon will not be simulated realistically, which will eventually affect the prediction accuracy. Therefore, a scientific calculation method must be applied to determine the conductivity coefficient. The regression curve of permeability after fracturing (Figure 3) is summarized by studying the results of previous measures in polymer injection block. The permeability improvement times after measures can be calculated by applying the permeability before fracturing and regression formula, and the scientific conductivity coefficient can be obtained.

Table 2 Basic situation of salt resistance test area

Projects	Development horizon	injection well	Production well	Development			Shoot open		
				Sandstone (m)	Effective (m)	Permeability (mD)	Sandstone (m)	Effective (m)	Permeability (mD)
Whole region	SaII11 ~ 16	41	41	9.8	5.2	156	8.5	5.2	179

Table 3 Sandstone thickness statistical table of reservoir permeability classification

Block	Permeability classification				
	<50	50-100	100-200	200-300	≥300
Block a	15.3	24.7	33.3	12.8	13.9
Block b	30.1	30.1	24.7	12.3	2.8
Test area	8.2	21.8	38.9	14.6	16.5

Table 4 Statistical table of effective thickness of reservoir permeability classification

Block	Permeability classification				
	<50	50-100	100-200	200-300	≥300
Block a	10.7	20.2	30.3	16.5	22.3
Block b	13	30.5	30.5	19.8	6.2
Test area	1.4	13.4	30.8	20.3	34.1

https://doi.org/10.1051/e3sconf/202132901088

Table 5 Statistical table of production status of oil layer a in block

Stage	Deposition unit								
	S211-1	S211-2	S211-3	S212	S213	S214	S215	S215-1	S216
Blank water drive	60.6	46.8	47.4	77.2	60.0	43.0	52.3	31.0	59.3
Pre-polymer flooding	77.9	50.5	75.0	59.4	45.8	55.0	51.6	18.7	59.3
Ternary stage	69.2	45.9	81.6	76.7	65.8	58.9	51.0	40.9	71.9

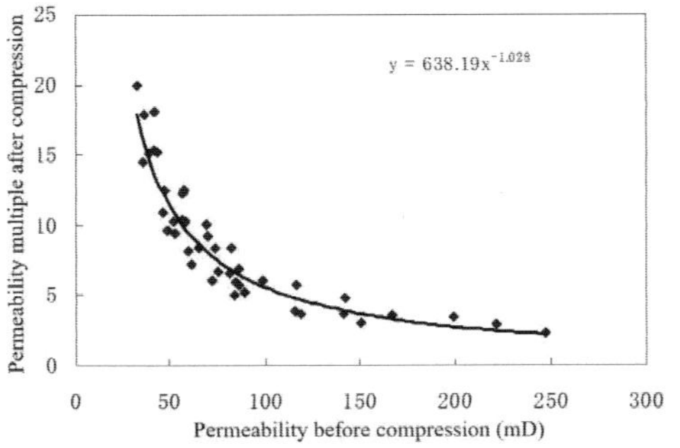

Figure 3 Regression curve of permeability after fracturing

4 Optimized injection parameter scheme

The target horizon of the salt-resistant test area is three types of oil layers. In order to maintain the injection-production capacity of the test area and the uniform production of each horizon, the rationality of injection parameter design is the key. The injection-production well spacing in the test area is 125m m. According to the calculation formula of injection rate (Formula 2), the reasonable injection rate in the test area is 0.18 ~ 0.20 PV/a; The design of injection concentration is the core work of the whole injection parameter design. In the past, the injection parameter matching template was established according to the matching of common polymer in the main oil layer, which could not be directly applied in the test area, so the parameter matching template of salt-resistant polymer was established (Figure 4). According to the parameter matching template, in order to ensure the smooth injection in the test area, the upper limit of injection concentration should be 1200mg/l; At the same time, different injection slugs, such as single concentration and stepped concentration reduction, were designed, and seven injection parameter schemes were numerically simulated. The numerical simulation results show that the fifth scheme has the best development effect (Table 6). In the process of numerical simulation, when the water cut in the test area rises to 96%, the polymerization is stopped until the water cut in the test area rises to 98%.

$$P_{max} = L^2 \phi V / (180 N_{min}) \qquad (2)$$

Among them, P_{max}——Maximum injection pressure at wellhead, MPa;

L——Injection-production well spacing, m;

ϕ——Reservoir porosity, decimal;

V——Injection speed, PV/a;

N_{min}——Lowest apparent water absorption index of oil layer, m3/d.m.MPa.

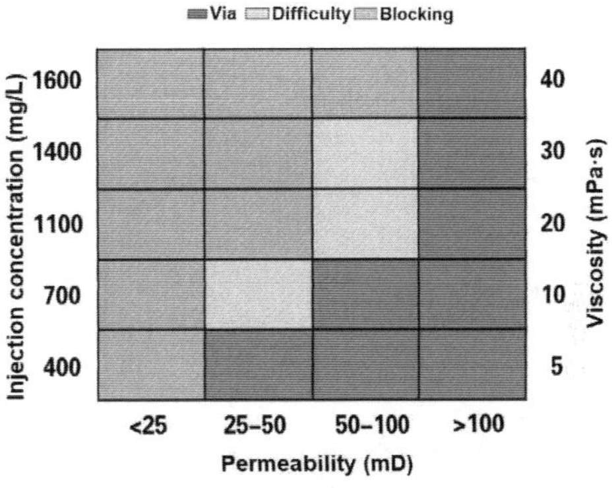

Figure 4 Parameter matching template of ds800 polymer

Table 6 Statistical table of numerical simulation results of injection parameter scheme

Programme	Injection parameters			Mathematical model prediction results	
	PV number	Injection velocity (PV/a)	Injection concentration (mg/L)	Minimum water cut (%)	Enhanced oil recovery (percentage points)
1	0.8	0.20	800	93.25	9.14
2	0.8	0.18	1000	92.18	9.50
3	0.8	0.16	1200	91.42	9.31
4	0.1/0.7	0.16/0.18	1200/1000	91.16	9.65
5	0.1/0.7	0.18/0.20	1000/800	91.07	10.37
6	0.1/0.4/0.3	0.16/0.18/0.20	1200/1000/800	91.14	10.19
7	0.05/0.6/0.15	0.16/0.18/0.20	1200/1000/800	91.21	9.72

https://doi.org/10.1051/e3sconf/202132901088

5 Actual application situation

The above method is applied to the salt-resistant test area for numerical prediction. According to the results of numerical prediction (Figure 5), the lowest water cut in the test area is 91.07%, the final recovery rate is 11.39%, the planned salt-resistant polymer consumption is 1256.89t, and the cumulative oil increase in the test area is estimated to be 9.24×10^4t.

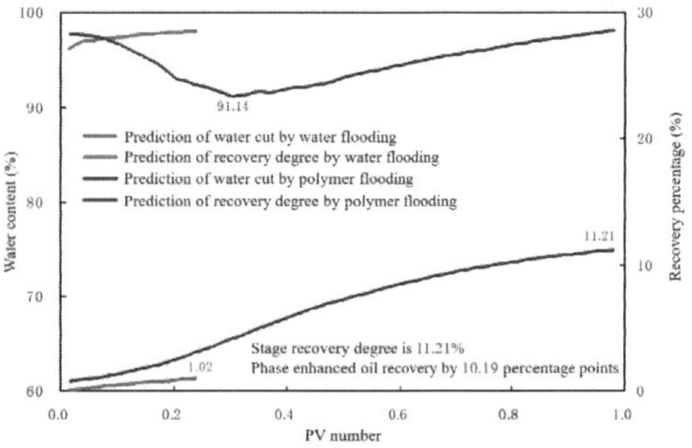

Figure 5 Figure of numerical simulation prediction results in salt resistance test area

6 Summary

(1) Numerical simulation prediction of polymer flooding must deeply understand the characteristics of polymer, and it is the key to ensure the prediction accuracy to set important numerical simulation parameters with scientific calculation methods;
(2) Combined with the development of oil layers in the block, it is necessary to make reasonable prediction of important measures in the future, which is a necessary method to further improve the prediction accuracy.
(3) Scientific injection parameter scheme prediction can effectively improve the rationality of numerical simulation prediction, and also improve the practical guiding significance of numerical simulation curve to field production and development.

References

1. Wang Xinhai. Determination of main parameters of numerical simulation of polymer flooding [J]. Petroleum Exploration and Development. 1990,17(3): 69-76.
2. Zhang Yanhui, Zeng Xuemei, Wang Yingbiao, et al. Numerical Simulation of Polymer Flooding in Three Types of Reservoirs in Daqing Oilfield [J]. Fault Block Oil and Gas Field, 2011, 18(2):232-234.
3. Lu kewei. parameter estimation technology and application of polymer flooding numerical simulation [A]. petroleum geology and engineering, 2013,27(1):114-116.
4. Wang Zhengbo, Wang Qiang, Ye Yinzhu, et al. History fitting method of polymer flooding with multi-parameter constraints [J]. Petroleum Exploration and Development, 2010, 18(2):216-219.

https://doi.org/10.1051/e3sconf/202132901089

Study on Pre Pumping and Outburst Elimination Technology of "one hole and two elimination" in Bedding Directional Long Drilling

Benqing Yuan [1, 2, 3, *] Yongjiang Zhang [1, 3]

[1] Chongqing Research Institute CO., Ltd of China Coal Technology Engineering Group, Chongqing, China
[2] Anhui University of Science and Technology, Huainan 232001, China
[3] National Key Laboratory of Gas Disaster Monitoring and Emergency Technology, Chongqing, China

Abstract. Aiming at the problem of gas control in the working face and the roadway to be excavated at the same time due to the small construction length of bedding borehole under the condition of the occurrence of soft coal, a technology of "one hole and two elimination" pre-drainage to eliminate gas outburst by using air screw motor drilling was put forward, and the field test was carried out in the 17102(3) working face of Pansu Coal Mine.The results show that this technology can effectively control the borehole trajectory, ensure the uniform and reasonable range of borehole outburst elimination, solve the problems of long gas treatment cycle caused by the construction of gas treatment roadway, improve the effective drainage time of borehole, ensure enough time and space for gas treatment, and be beneficial to the mining and replacement of mine.

1 "One hole two elimination" technology principle

"One hole two elimination " technique is to use the roadway drivage face side has the construction site, the rig directional construction some bedding boring (generally more than 200 m), bedding near borehole control with mining face mining areas and with digging coal roadway stripe area (Fig.1), then use bedding directional borehole extraction control area of coal seam gas drilling,In order to solve the problem of small control range of short drilling, many drilling holes, long moving time during construction, short extraction time, uncontrollable deviation of drilling holes resulting in drilling holes can not be uniformly distributed in the coal seam according to the design, and relieve the tension of extraction and mining replacement[1-3].

2. Engineering geological condition

The 17102(3) working face of Pan-San Mine is located in Dong-si Mining Area with elevation of -680~-724m. The mining strike length of the working face is 1820m(including 1048m solid section), the dip width is 200m, the 13-1 coal thickness is 3.2~4.7m, the average thickness is 4.0m, the average dip Angle is 8°, and the measured gas pressure is 2.8MPa.The gas content is 8.4m^3/t, and the permeability coefficient of coal seam is 0.022m^2/(MPa^2·d).

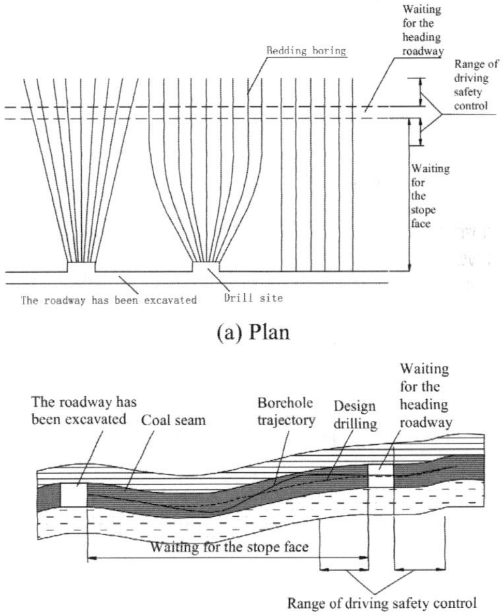

(a) Plan

(b) Profile map

Fig. 1 Schematic diagram of "one hole and two elimination" drilling layout

17102(3) track grooves are goaf tunneling roadways. In order to solve the problem of gas outburst on the coal working face and transportation grooves at the same time after tunneling is completed, 17102(3) track grooves are considered as the drilling site to drill long boreholes into

* Corresponding author: 2012156@cqccteg.com

the coal working face and transportation grooves for regional pre-drainage and outburst elimination, that is, "one hole and two elimination".(3) The design length of the transport route is 1,820m, of which 1,048m in the western section is located in the solid section of the outburst danger zone. In the solid section, 100m area is selected as the "one hole and two elimination" test area.However, directional drilling of bedding borehole is difficult, the borehole trajectory is difficult to effectively control, and the gas treatment period is long caused by the construction of gas treatment roadway, etc., which are difficult problems for the implementation of "one hole and two elimination" technology.

3. Drilling equipment technology

3.1 Drilling equipment

3.1.1 Rig equipment

ZDY6000LD (F) type crawler full hydraulic tunnel drill produced by Xi 'an Coal Research Institute is selected as the construction rig, as shown in Fig 2.

Fig.2 ZDY6000LD (F) model crawler type hydraulic tunnel drill for coal mine

3.1.2 Drilling assembly

1) Drilling BHA

Φ108mm drill bit + 73/83mm outer screw pneumatic motor + 81mm outer screw non-magnetic drill pipe + 81mm outer screw probe pipe + insulation sub + 81mm non-magnetic drill pipe + top 73mm auger pipe, as shown in Fig. 3.

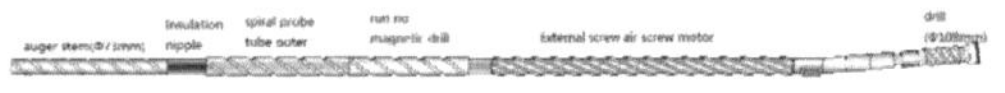

Fig.3 Drilling BHA

2) Run a casing assembly

Φ92mm steerable bit + 73-40mm integrated large hole auger. The 92mm steerable bit is shown in Fig. 4.

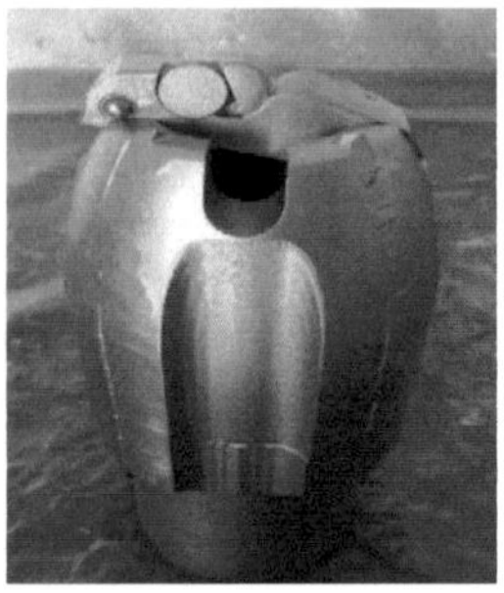

Fig.4 Φ92mm guided open and close drill bit

3.1.3 Air screw motor

Air screw motor surface smooth type outside diameter 73mm, spiral type outside diameter 85mm, length 4 ~ 4.4m, front deflection 1.25 ~ 1.5°, as shown in Fig.5.

Fig.5 Air screw motor (flat)

3.1.4 Wireless measuring system

Wireless measurement system transmission distance is not less than 500m, the transmission rate is as high as 50bps, the measurement time is about 40s, the measurement accuracy: Angle ±0.2°, azimuth Angle ±1.2°, tool facing Angle ±1.2°.This is shown in Fig. 6.

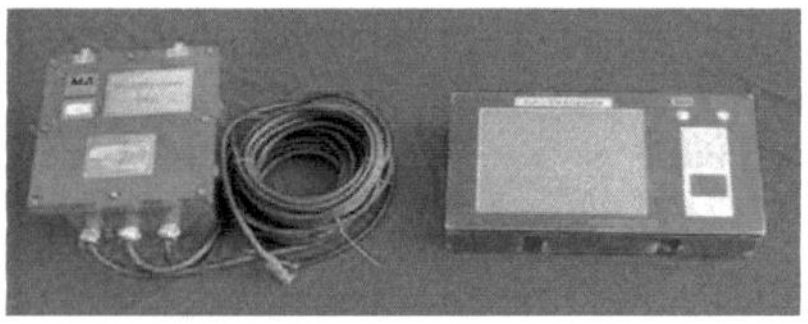

Fig.6 Wireless measuring system (outside hole part)

3.1.5 Lubricating and cooling devices

The hydraulic oil lubrication device improved by the foam generator lubricates and cools the motor to ensure the normal operation of the motor.The main structure includes storage tank, storage hydraulic oil, pressure gauge, communication pipeline, circulation pipeline, gate valve, flowmeter, flowmeter pipeline, structural support, etc., as shown in Fig. 7.

https://doi.org/10.1051/e3sconf/202132901089

Fig.7 Lubricating device

3.1.6 Air compressor

The construction of bedding drilling adopts the process of pressing air and discharging slag, and the air pressure in the tunnel system is low. In order to ensure that the air pressure of drilling meets the requirements of drilling construction, the MLGF17/12.5-132G air compressor is selected, with an exhaust capacity of $17m^3$/min and an exhaust pressure of 1.25MPa.

3.2 Technical parameters

3.2.1 Construction technology

The construction technology of bedding directional drilling for soft coal seam is shown in Fig. 8. In view of the stress concentration area in the open hole section (generally 0~50m) of the construction area, the coal seam is extremely broken. Rotary drilling and grouting treatment are adopted in this section to ensure the stability of the open hole section[4].For some coal seams are loose and broken, and collapse holes are easy to occur. The coal seams in local hole sections are wet, and it is difficult for directional drilling to discharge slag. Therefore, rotary composite drilling should be the main method to ensure normal drilling[5].

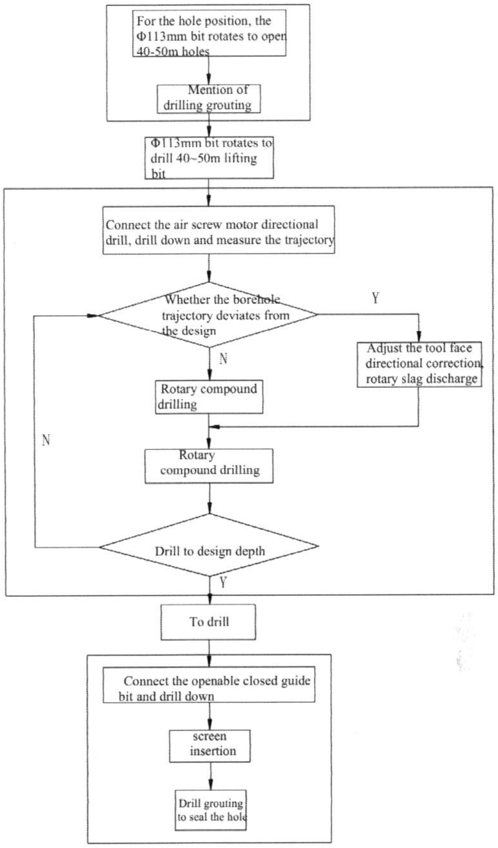

Fig.8 Construction process of directional long drilling along bedding

3.2.2 Key parameter

1) Drilling tools

Φ108mm drill bit + 73/83mm external spiral pneumatic motor + 81mm non-magnetic drill pipe under spiral + 81mm external spiral probe pipe + insulation sub + 81mm non-magnetic on spiral + 73mm auger string.

2) Drilling parameters of air screw motor

Rotary drilling speed of 80~120r/min, air volume of more than 400m3/h;The air volume of sliding directional drilling is more than 500m3/h;When the air volume is lower than 400m3/h, the rotary sweeper can fully discharge slag.

3) Directional drilling

The rotary composite drilling is the main method, and the sliding directional drilling is the auxiliary method. The directional drilling is 2~3m, and the rotary sweeping hole is used to drain the slag, and then the directional and rotary sweeping hole is continued to be repeated, so as to ensure the smooth inside the hole and the construction safety.

4) Finish the hole of screen tube

Adopt Φ89mm openable closed guide bit + A73mm integrated large hole auger drill, among which openable closed guide bit ensures the smooth running into the hole during the second trip and the reliable opening during the running of the screen.

After the completion of directional drilling, connect the openable closed guide bit and the large hole drill pipe to drill again. When the drilling reaches a predetermined depth, stop drilling and insert the hole protection screen pipe from the large hole drill pipe. When the screen pipe pushes the openable closed guide bit position at the bottom of the hole, open the head of the drill and the screen pipe can enter the hole through the drill bit.After the screen is in place, start pulling out the drill pipe, leaving the screen in the hole and the bit and drill pipe out of the hole.After the hole shielding screen tube is placed into the hole, 15~20m outside the hole is covered with PVC pipe.

5) Drilling and sealing holes

The hole sealing process of "two plugging and one injection" was used to seal the hole. The hole sealing depth was 20m, polyurethane plug was used in the range of 5m at the orifice and bottom of the hole, and grouting capsule was used for high pressure grouting at the middle 15m, with the grouting pressure not less than 2MPa.

4. "One hole two elimination" technical characteristics

Existing single unprotected recoverable coal seams in or protective layer for the coal seams, occurrence condition of coal seam bedding directional drilling crossing with mining face and with digging a tunnel gas extraction of elimination technique, solve the end of the alley pumping + layer drilling or bedding combined gas control technology of drilling is large volume, high cost, long cycle, etc.It has the following characteristics[6]:

(1) The directional drilling track across the working face can be oriented to ensure that the drilling hole is in the coal seam.Using the air screw motor directional drilling method can solve the problem that the roof or floor drilling is not in place because of the fluctuation of coal seam, which can easily be seen by the conventional rotary drilling, improve the drilling depth and coal seam drilling encounter rate of soft coal seam, realize the precise control of bedding drilling trajectory, and ensure the coverage of drilling to eliminate the blind area of gas extraction.

(2) The screen pipe under the whole hole section of directional drilling can guarantee the gas drainage channel, realize the gas pre-drainage and outburst elimination of the working face and roadway to be excavated, and ensure the safe excavation of the roadway and the safe mining of the coal working face.

(3) It can save the engineering quantity of floor roadway and through layer hole, and reduce the production cost while ensuring the safety of coal roadway excavation and working face mining.

5. Application effect

In order to solve the problem of gas outrush elimination in coal working face and transport roadway (return air roadway) at the same time, reduce the comprehensive cost input of supporting floor roadway and relieve the tension of mine production replacement, 17102 (3) working face with more complex gas and coal seam occurrence conditions was selected for field application of directional long drilling drainage technology along bed in Pansan Mine of Huainan mining area.

5.1 Drilling construction

17102 (3) A total of 19 directional long boreholes along the bedding were constructed in the return air roadway (Fig. 9). The completion parameters of the boreholes are shown in Table 1.According to the actual formation conditions of Pansan Mine, the directional long hole diameter of 17102 (3) working face is determined to be Φ108mm. The boreholes are arranged in two rows, with opening heights of 1.2m and 1.8m respectively. The spacing of each row of holes is 0.6m, and the spacing of the final hole is 5m.The directional drilling construction technology is adopted to realize the drilling construction along the coal seam, improve the coal penetration rate of the long borehole along the bed, basically realize the control of the drilling trajectory, realize the precise directional construction, and the drilling penetration rate of 8 boreholes in the coal seam reaches 100%.

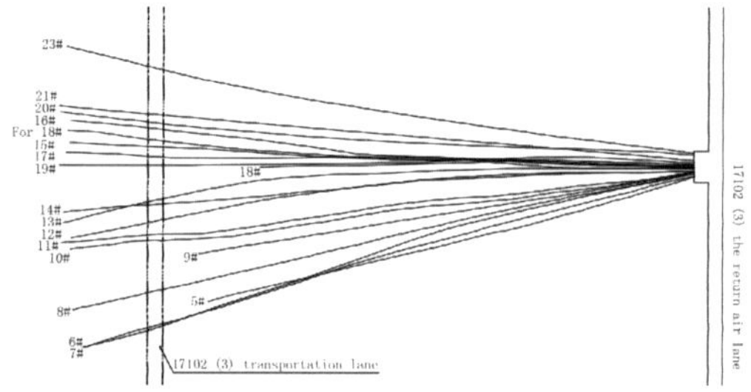

Fig. 9 Pore drawing of directional drilling construction in bedding

Table 1 17102 (3) Construction of directional drilling for bedding in return air roadway

Serial number	Boring number	Aperture /mm	Hole depth /m	Drill encounter rate/%	Sieve tube length /m	Upper and lower deviation /m	Left-right deviation /m
1	23#	108	231	85.54	231	-0.87	-
2	21#	108	224.5	93.7	220	-0.40	-0.7
3	20#	108	225	93.7	214	-1.03	1.9
4	19#	108	225	91.6	220	-0.92	-1.5
5	18#	108	160	100	-	-	-
6	For 18#	108	221	100	220	-1.88	1.4
7	17#	108	223	100	220	-0.61	-2.0
8	16#	108	221	89.6	220	-1.94	-
9	15#	108	221	100	202	-1.55	
10	14#	108	224.5	100	220	-1.81	-2.1
11	13#	108	222.5	100	220	-1.4	-1.4
12	12#	108	224	100	220	-1.7	-1.9
13	11#	108	224	91.5	220	-2.0	-0.9
14	10#	108	224	100	220	-2.04	2.0
15	8#	108	224	100	218	-1.9	1.1
16	7#	108	227	90.7	220	-1.9	-2.0
17	6#	108	227	93	220	-1.72	-
18	9#	108	170	100	170	-	-
19	5#	108	175	100	171	-	-

5.2 Application effect

Based on the failure characteristics of surrounding rock of deep mining roadway in huainan mining area, the grouting pressure of "two blocks and one injection" with pressure grouting and hole sealing method is calculated.According to the results of fluid mechanics and field grouting technology, the P_Z expression of grouting pressure is obtained as follows.

In order to test the drainage effect, six drill holes were constructed corresponding to 17102(3) transport lane. The maximum distance from the bedding directional drill was 6.4m, the measured maximum residual gas pressure was 0.35MPa, and the residual gas content was 3.69m^3/t.After inspection, the extraction within the control range of drilling is up to standard.

17102(3) in the area where bedding directional drilling extraction technology is adopted in the transport roadway, the average driving footage is 9.3m/d. Outburst risk cycle prediction is performed in the roadway driving process, and the maximum prediction indexes for outburst prevention are S_{max}=3.5kg/m and Q_{max} =2.91L/min.During tunneling, the roadway air distribution volume was 960m^3/min, and the maximum gas concentration of return air was 0.23%, which realized safe and efficient tunneling.

6 Conclusion

(1) The technology innovation of "one hole and two elimination" of bedding long drilling is carried out. One bedding long drilling is carried out, and at the same time, the coal mining face and transportation groove are pre-drained and outburst is eliminated in the region, which saves a lot of engineering work of bedding through drilling.

(2) A good drainage effect was obtained. A total of 322,000 m3 of gas was extracted from 14 boreholes, the extraction rate was 61.3%, the residual gas pressure was 0.35MPa, and the residual gas content was 3.69m3/t.The average gas concentration was still about 50% after 7 months, which was 3-4 times that of conventional drilling.

(3) The directional drilling technology of "one hole and two elimination" bedding long borehole can effectively control the borehole trajectory, ensure the uniform and reasonable range of borehole outpouring elimination, and at the same time can fully discharge slag, ensure the unblocked drainage channel in the borehole, and improve the drainage effect.

(4) "One hole and two elimination" bedding long borehole is constructed in the groove by the track constructed in advance, which solves the problem of long gas treatment cycle caused by the construction of gas treatment roadway, improves the effective drainage time of borehole, ensures enough time and space for gas treatment, and is conducive to the production replacement of mine.

Acknowledgments

This work is financially supported by National Natural Science Foundation of China (No. 52004005), Anhui Provincial Natural Science Foundation (No. 2008085QE222), China Postdoctoral Science Foundation (2021M691185), Anhui University of Science and Technology Introduction of Talents Research Fund Project, Scholastic Key Project (No. QN2019113), Patent Transformation and Cultivation Project (No. ZL201907).

References

1. Chen Jian, Luo Yong. Research on the technology of "one hole and two elimination" pre-drainage to eliminate outburst in bedding long borehole [J]. Mining Technology, 2020,20(5):47-50.
2. ZHANG Jie, WANG Yi, HUANG Hanjing, et al.Research on directional drilling technology of air srew motor in soft coal seam[J].Coal Science and Technology, 2018, 46(11):114-118.
3. WANG Xian, XU Chao, LIU Fei, JIANG Lei, et al. Key equipment and application effect analysis of ultra - long directional drilling along coal seam [J].Coal Engineering, 2019, 51(11):46-50.
4. WANG Yong, ZHOU Yan-an, TONG Xiao-zhang. Optimization and Application of Regional Anti-outburst Measures in Pre-pumping Coal Seam Gas [J].Coal Technology, 2021, 40(02):133-136.
5. LI Yanqing,TANG Yongzhi, TANG Bin, et al. Innovation and development of coal and gas co–mining technology in Huainan mining area[J].Safety in Coal Mines, 2020, 51(8):77-81.
6. FANG Jun, LI Quanxin, XU Chao, et al.Construction technology and development tendency of gas drainage borehole in soft and outburst seam[J]. Coal Science and Technology, 2018, 46(5): 130-137, 172.

Analysis of main points of channel freezing-thawing disease in cold region and influence of channel base soil on the disease

Liqing Liang *

Jilin agriculture science and technology university, Jilin, Jilin, China

Abstract. The frozen soil area in China is more than two thirds of the total territory, so the problem of frost heave is obvious. Especially in northeast, northwest, north China and other cold regions, the problem of frost heave of hydraulic structures is very common. Canal is a common hydraulic structure in agricultural water, which is affected by seasonal frozen soil and may cause problems such as lining damage, seepage and irrigation efficiency. Therefore, this paper mainly summarizes the necessity of research on channel freezing-thawing damage, the research direction of channel freezing-thawing damage, and expounds the influence of seasonal frozen soil on freezing and thawing diseases in cold regions by taking the particle size of saturated soil based on channel as an example.

Keywords: Seasonal frozen ground; Channel; Freeze-thaw disease; Canal base

1 Research necessity analysis

Permafrost covers about one-fifth of the earth's surface area. China's permafrost area is 2.15 million square kilometers, ranking the third in the world, second only to the Soviet Union and Canada. If seasonal frozen ground is included, the frozen ground area in China is more than two-thirds of the total territory. Freezing damage to hydraulic structures is very common and serious in northeast, northwest and North China [1]. The winter climate in China's cold regions is cold and dry, and water delivery channels in irrigated areas are vulnerable to the damage of seasonal frozen soil freezing-thawing [2]. Small will cause leakage problems, affect the efficiency of irrigation, large will cause water interruption or even dike break and other accidents.

The investigation shows that the freezing damage of canal hydraulic structures in seasonal frozen regions is mainly caused by the influence of soil frost heaving and thawing on the structure and durability of materials. , this article analyzes the mechanism of these diseases, and cold and irrigation channels freeze-thaw including channel under the condition of freezing and thawing process, causes, occurrence regularity, disease characteristics, such as analysis, cold and freezing and thawing channel is determined based on the above disease types, for normal operation of the service life of cold and irrigation channels, which provides the theoretical foundation and maintenance management, etc.

2 The main research direction is analysis

Based on the 2015 project "Research on Soil Hydrothermal and Vertical Change of Frozen Layer in Winter -- Taking Jiuzhan Campus of Jilin University of Agricultural Science and Technology as an example", the research on frozen soil parameters in winter was carried out by investigating disease phenomenon and analyzing disease occurrence process. The following analysis directions were mainly sorted out through experiments and literature analysis.

2.1 Analyze the characteristics of diseases in irrigation channels in cold area

Relatively high latitude in the northeast in our country, the region in the channel project is located in the seasonal frozen, channel and linear engineering [3], along the channels affected by freeze-thaw action is bigger, frost heaving soil engineering area are widely distributed, and channel engineering winter without water requirement, water freeze damage did not occur because the channel, The diseases of all channels are mainly concentrated in lining diseases. Through field investigation and literature analysis, the causes and specific characteristics of soil freezing-thawing diseases of channels in northeast China are analysed[4], laying a foundation for the study of disease mechanism.

* Corresponding author: hhs_liangliqing@126.com

2.2 The mechanism of channel freezing-thawing disease was analyzed

Channel soil frost heave soil is not only the central plains the freezing of water, but with winter canal in the inland waters not completely emptying and lead to freeze the underground water level is higher, the frozen soil depth within the scope of the capillary water, weak bonding water freezing continuously to cover the migration, frost heave effect is strong, and the frost heave deformation, destruction is more serious. Through field investigation, through the indoor experiment, such as the frost heave susceptibility analysis, systems apply soil permeability under impact experiment and so on, combined with literature analysis research channels disease occurs under the condition of cold and freezing and thawing process, the conditions, influencing factors, based on the division of channel under the condition of freezing and thawing of disease types, and preliminary study channels disease occurrence regularity of freeze-thaw conditions.

2.3 The influence of channel concrete material on prevention of freezing injury was analyzed

The frost resistance of channel lining concrete also has an influence on channel damage, mainly including the study of concrete material mix ratio and concrete performance. The combination of concrete materials, such as the amount of admixture, water-cement ratio, and the amount of new materials, such as differential foam, etc. [5]; On the basis of the above, the frost resistance and durability of concrete are studied, including the pore structure, freezing temperature, freezing rate, saturation, reliability analysis and carbonization of concrete[6-8].

Through the above three contents, how to control the disease in the design and construction process of the channel in cold areas, including new structure types, new impermeable materials, etc., to provide a basis for the service life, normal operation and maintenance management of the channel in cold areas irrigation area

3 Effect of seasonal frozen soil as channel base on canal diseases

According to the analysis of the causes and mechanism of the disease, the damage types of the freeze-thaw channel in the irrigation area in cold area can be divided into two types: frost heaving failure of the land at the bottom of the channel and freezing failure of water body in the channel. If the channel is not carried out in winter, the destruction of the channel due to the freezing and thawing action of soil base is the primary factor. If there is water in the channel in winter, the freezing damage caused by water freezing in the channel is the primary factor [9]. In northeast China, channel water is generally drained in winter due to channel maintenance and seasonal requirements, so the damage of channel caused by freezing and thawing of channel foundation soil becomes the main type.

Unsaturated soil, such as water seepage or groundwater recharge, replenish the gap between soil particles, so the frost heave is small, while saturated soil in negative temperature environment, soil particles pore water crystallization, when water recharge pore water, crystallization will continue to increase, causing soil expansion, thus affecting channel lining and causing frost heave failure. Therefore, the temperature field experiment of saturated soil was designed to analyze the temperature variation trend of each layer of saturated soil samples with different particle sizes in the freezing process under the same temperature condition. Taking this as an example, the influence of channel foundation soil on channel freezing-thawing disease was analyzed.

Seasonal frozen soil is unidirectional freezing at the top. In order to understand the internal temperature field of soil samples in the freezing process, unidirectional freezing at the top is adopted in order to simulate the conditions consistent with the real environment. First of all, the temperature in the box and the temperature at the top and bottom were set at 1℃ for constant temperature. Then, according to the external environment of the frozen soil sample ice rate measurement test, the roof temperature was adjusted to -20℃ to monitor the internal temperature change of the sample.

3.1 The experiment design

Sample preparation: a layer of vaseline was coated on the plexiglass cylinder, and sand with particle size ≤0.25mm, 0.25-0.5mm, 0.5-1mm were layered into the plexiglass cylinder. In order to ensure the consistency with the samples measured by volume ice content of frozen soil samples, 353g sand was packed into the cylinder each time, tamped to 3cm high, and 1530g sand was loaded into the cylinder four times to prepare dry soil samples with dry density of 1.5g/cm3. The filter was arranged at the bottom of the cylinder to ensure that no sand flowed out during the movement of the sand column. The prepared sand column was placed in a bucket, and the sample was saturated with water slowly added to the bottom. After water saturation, gravity water release was carried out to obtain saturated soil samples.

Sample installation: place the prepared soil sample on the bottom plate of the freeze-thaw test device, fix the fixing plate with screws, and wrap the insulation layer in the outer layer. To ensure the stable operation of the test box, connect the water refill device outside the box to the water refill channel at the bottom plate inside the box.

Temperature monitoring: In the whole freezing test process, the temperature of box body, roof and bottom plate is adjusted to 1℃ for 6 hours, so that the temperature of each part of the sample is uniform, and then the roof temperature is adjusted to -20℃, until the soil reaches the frozen stable state. Temperature changes at 0cm, 1.5cm, 3cm, 4.5cm, 6cm, 7.5cm, 9cm and 10.5cm were monitored at the bottom of the test.

3.2 Test results and analysis

A probe temperature sensor was inserted into the soil every 15mm, and the reading time interval was 10min to monitor the internal temperature field of saturated sand samples with different particle sizes during freezing process. The dry density of soil samples used in the three tests was the same, and the temperature control of the top, bottom and three ends of the box body was the same. The freezing of soil samples with different particle sizes is shown in Fig 1-3. retention.

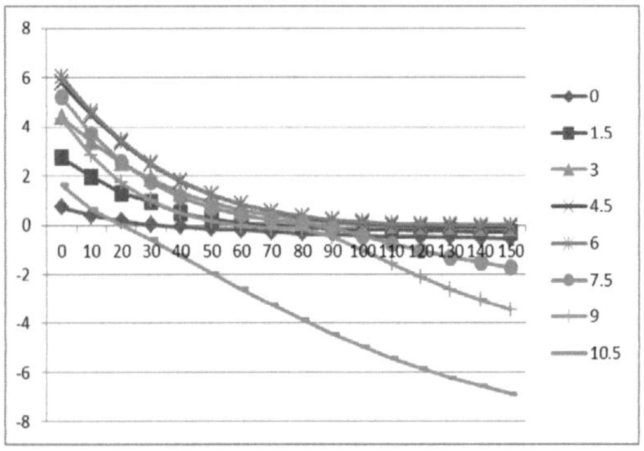

Fig.1 ≤0.25mm sand column freezing temperature field

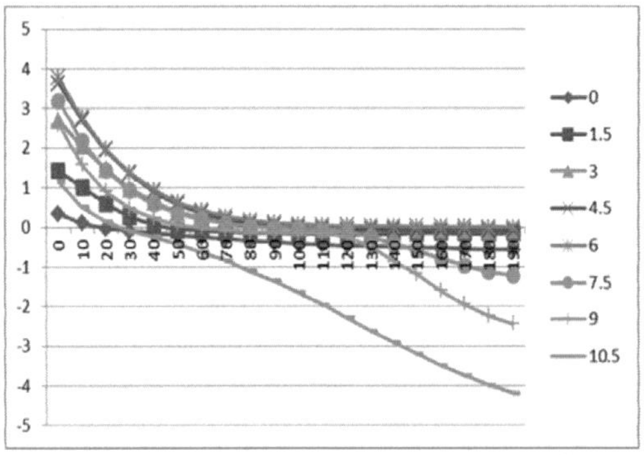

Fig.2 0.25-0.5 sand column freezing temperature field

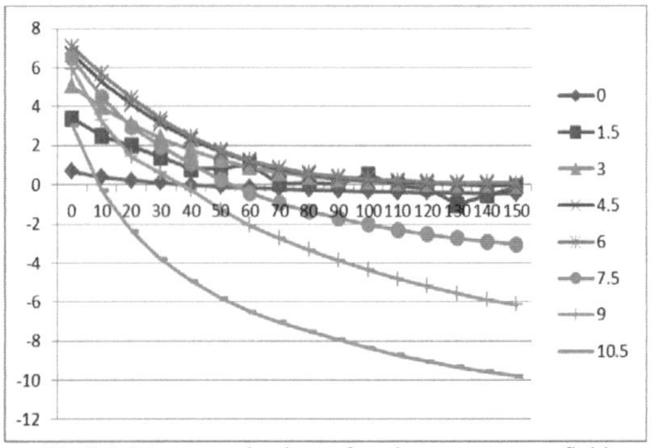

Fig.3 0.5-1mm sand column freezing temperature field

As can be seen from the figure, saturated soil samples with the same dry density and different particle sizes show basically the same temperature variation trend in each layer during the freezing process under the same temperature condition. The temperature of the soil samples at the top decreases rapidly under the influence of roof temperature, while the temperature of the soil in the middle decreases slowly, while the temperature of the soil at the bottom changes the most slowly with a very small amplitude. In the low-temperature environment of -20℃ at the top, the freezing time of soil samples with different particle sizes is different. It takes 250min to form stable low temperature inside the sand column with particle size of 0.5-1mm, 190min to reach negative temperature completely inside the sand column with particle size of 0.25-0.5mm, and only 140min to reach negative temperature inside the sand column with particle size of ≤0.25mm. It can be seen from the experiment that the freezing time of saturated coarse sand sample is significantly longer than that of fine sand sample, which is because the coarse sand particle gap is large, water crystallization is slow and frost heaving time is long in the freezing-thawing process. It can be seen that the soil frost heave time is affected by the particle size of channel foundation soil.

4 Summary

To sum up, under the action of seasonal frozen soil, diseases caused by channel freezing and thawing damage mainly include channel lining and channel bottom foundation, etc. The research direction of channel diseases can be analyzed from the generation and classification of diseases, channel foundation soil freezing and thawing effect and influencing factors, and channel lining materials.

For seasonal frozen soil, the influence of freezing and thawing damage of channel foundation soil is mainly water condition and temperature. Through the temperature field experiment, it can be seen that the smaller the soil particles are, the more likely the saturated soil is to frost heave, and the greater the influence on channel is.

As an important water conservancy project to improve irrigation water condition and adjust regional microclimate, the water conveyance channel plays an important economic, social and ecological benefits. The analysis of soil freezing-thawing mechanism is of great theoretical and practical significance to the study of soil moisture in irrigation and water conservancy and solute migration in agriculture and environment. It provides theoretical and technical basis for the construction plan, structure design, operation management and winter maintenance of the channel in the cold area irrigation area in the future, so as to improve the service life of the channel project in the cold area and reduce the cost.

4 Acknowledgements

Natural science research project of Jilin Institute of agricultural science and technology: " Mechanism analysis of freezing and thawing diseases in irrigation channels in cold regions " (no. : 20190625)

References

1. SUI Tieling. Simple Test of anti-freezing technology -- Lecture on anti-freezing technology knowledge of Canal System Hydraulic building [J]. Water Resources World,1988(06): 22-23.
2. Jia Zhiguo, Liu Guanglou, Han Yiting. Anti-frost heaving design of canal system in Inner Mongolia irrigation area downstream of Muzhuang Hydraulic Project [J]. Inner Mongolia Water Resources, 2019(03): 38-39.
3. HE SHANshan. Study on anti-frost heave design of lining channel in Liaoning Coastal irrigation area [D]. Shenyang Agricultural University,2017.
4. Li Haojun. Characteristics of freezing damage of canal seepage prevention project [J]. Private Science and Technology,2011(08):322.
5. Wang Yi, Liu Jincheng, Liu Quanhong, Wang Zhengzhong. Shape optimization of trapezoidal lining channel structure under temperature-water-soil-structure coupling effect in cold area [J]. Journal of tsinghua university (natural science edition) : 1-10 [2019-05-17]. https://doi.org/10.16511/j.cnki.qhdxxb. 2019.22.010.
6. Ma Jin-long, LI Zhao-yu, TIAN Wen, ZHANG Wan-rui, LI Jia-min, SU An-shuang, CHANG Jun-de. Characteristics and causes of freezing damage of channel in cold area [J]. Hydraulic science and cold area engineering,2018,1(11):28-33.
7. Zhang Xiaohui, Tong Yuxuan, Zhang Zhicheng. Characteristics and causes of freezing damage in lined channels and underground pipelines [J]. Forestry science and technology information,2011,43(03):73-74.
8. Nan Ruifang. Study on the law of frost resistance and durability deterioration of Hydraulic Concrete materials [D]. North China University of Water Resources and Electric Power,2015.
9. Xiu Lu. Analysis on frost heaving failure factors of channel slope protection in seasonal frozen region [D]. Heilongjiang University,2015.

AUTHOR INDEX

AUTHOR INDEX

AUTHOR INDEX